Manifolds
and Differential
Geometry

GRADUATE STUDIES
IN MATHEMATICS **107**

Manifolds and Differential Geometry

Jeffrey M. Lee

AMERICAN
MATHEMATICAL
SOCIETY
Providence, Rhode Island

EDITORIAL COMMITTEE
David Cox (Chair)
Steven G. Krantz
Rafe Mazzeo
Martin Scharlemann

2000 *Mathematics Subject Classification.* Primary 58A05, 58A10, 53C05, 22E15, 53C20, 53B30, 55R10, 53Z05.

For additional information and updates on this book, visit
www.ams.org/bookpages/gsm-107

Library of Congress Cataloging-in-Publication Data
Lee, Jeffrey M., 1956–
  Manifolds and differential geometry / Jeffrey M. Lee.
   p. cm. — (Graduate studies in mathematics ; v. 107)
  Includes bibliographical references and index.
  ISBN 978-0-8218-4815-9 (alk. paper) | Softcover ISBN 978-1-4704-6982-5
  1. Geometry, Differential. 2. Topological manifolds. 3. Riemannian manifolds. I. Title.

QA641.L38 2009
516.3′6—dc22

2009012421

Copying and reprinting. Individual readers of this publication, and nonprofit libraries acting for them, are permitted to make fair use of the material, such as to copy a chapter for use in teaching or research. Permission is granted to quote brief passages from this publication in reviews, provided the customary acknowledgment of the source is given.

Republication, systematic copying, or multiple reproduction of any material in this publication is permitted only under license from the American Mathematical Society. Requests for such permission should be addressed to the Acquisitions Department, American Mathematical Society, 201 Charles Street, Providence, Rhode Island 02904-2294 USA. Requests can also be made by e-mail to reprint-permission@ams.org.

© 2009 by the American Mathematical Society. All rights reserved.
The American Mathematical Society retains all rights
except those granted to the United States Government.
Printed in the United States of America.
Reprinted by the American Mathematical Society, 2022.
∞ The paper used in this book is acid-free and falls within the guidelines
established to ensure permanence and durability.
Visit the AMS home page at http://www.ams.org/
  11 10 9 8 7 6 5 4 3   24 23 22

Contents

Preface	xi
Chapter 1. Differentiable Manifolds	1
§1.1. Preliminaries	2
§1.2. Topological Manifolds	6
§1.3. Charts, Atlases and Smooth Structures	11
§1.4. Smooth Maps and Diffeomorphisms	22
§1.5. Cut-off Functions and Partitions of Unity	28
§1.6. Coverings and Discrete Groups	31
§1.7. Regular Submanifolds	46
§1.8. Manifolds with Boundary	48
Problems	51
Chapter 2. The Tangent Structure	55
§2.1. The Tangent Space	55
§2.2. Interpretations	65
§2.3. The Tangent Map	66
§2.4. Tangents of Products	72
§2.5. Critical Points and Values	74
§2.6. Rank and Level Set	78
§2.7. The Tangent and Cotangent Bundles	81
§2.8. Vector Fields	87
§2.9. 1-Forms	110
§2.10. Line Integrals and Conservative Fields	116

§2.11. Moving Frames	120
Problems	122
Chapter 3. Immersion and Submersion	**127**
§3.1. Immersions	127
§3.2. Immersed and Weakly Embedded Submanifolds	130
§3.3. Submersions	138
Problems	140
Chapter 4. Curves and Hypersurfaces in Euclidean Space	**143**
§4.1. Curves	145
§4.2. Hypersurfaces	152
§4.3. The Levi-Civita Covariant Derivative	165
§4.4. Area and Mean Curvature	178
§4.5. More on Gauss Curvature	180
§4.6. Gauss Curvature Heuristics	184
Problems	187
Chapter 5. Lie Groups	**189**
§5.1. Definitions and Examples	189
§5.2. Linear Lie Groups	192
§5.3. Lie Group Homomorphisms	201
§5.4. Lie Algebras and Exponential Maps	204
§5.5. The Adjoint Representation of a Lie Group	220
§5.6. The Maurer-Cartan Form	224
§5.7. Lie Group Actions	228
§5.8. Homogeneous Spaces	240
§5.9. Combining Representations	249
Problems	253
Chapter 6. Fiber Bundles	**257**
§6.1. General Fiber Bundles	257
§6.2. Vector Bundles	270
§6.3. Tensor Products of Vector Bundles	282
§6.4. Smooth Functors	283
§6.5. Hom	285
§6.6. Algebra Bundles	287
§6.7. Sheaves	288

§6.8.	Principal and Associated Bundles	291
	Problems	303

Chapter 7.	Tensors	307
§7.1.	Some Multilinear Algebra	308
§7.2.	Bottom-Up Approach to Tensor Fields	318
§7.3.	Top-Down Approach to Tensor Fields	323
§7.4.	Matching the Two Approaches to Tensor Fields	324
§7.5.	Tensor Derivations	327
§7.6.	Metric Tensors	331
	Problems	342

Chapter 8.	Differential Forms	345
§8.1.	More Multilinear Algebra	345
§8.2.	Differential Forms	358
§8.3.	Exterior Derivative	363
§8.4.	Vector-Valued and Algebra-Valued Forms	367
§8.5.	Bundle-Valued Forms	370
§8.6.	Operator Interactions	373
§8.7.	Orientation	375
§8.8.	Invariant Forms	384
	Problems	388

Chapter 9.	Integration and Stokes' Theorem	391
§9.1.	Stokes' Theorem	394
§9.2.	Differentiating Integral Expressions; Divergence	397
§9.3.	Stokes' Theorem for Chains	400
§9.4.	Differential Forms and Metrics	404
§9.5.	Integral Formulas	414
§9.6.	The Hodge Decomposition	418
§9.7.	Vector Analysis on $\mathbb{R}^3$	425
§9.8.	Electromagnetism	429
§9.9.	Surface Theory Redux	434
	Problems	437

Chapter 10.	De Rham Cohomology	441
§10.1.	The Mayer-Vietoris Sequence	447
§10.2.	Homotopy Invariance	449

§10.3.	Compactly Supported Cohomology	456
§10.4.	Poincaré Duality	460
Problems		465

Chapter 11. Distributions and Frobenius' Theorem — 467

§11.1.	Definitions	468
§11.2.	The Local Frobenius Theorem	471
§11.3.	Differential Forms and Integrability	473
§11.4.	Global Frobenius Theorem	478
§11.5.	Applications to Lie Groups	484
§11.6.	Fundamental Theorem of Surface Theory	486
§11.7.	Local Fundamental Theorem of Calculus	494
Problems		498

Chapter 12. Connections and Covariant Derivatives — 501

§12.1.	Definitions	501
§12.2.	Connection Forms	506
§12.3.	Differentiation Along a Map	507
§12.4.	Ehresmann Connections	509
§12.5.	Curvature	525
§12.6.	Connections on Tangent Bundles	530
§12.7.	Comparing the Differential Operators	532
§12.8.	Higher Covariant Derivatives	534
§12.9.	Exterior Covariant Derivative	536
§12.10.	Curvature Again	540
§12.11.	The Bianchi Identity	541
§12.12.	G-Connections	542
Problems		544

Chapter 13. Riemannian and Semi-Riemannian Geometry — 547

§13.1.	Levi-Civita Connection	550
§13.2.	Riemann Curvature Tensor	553
§13.3.	Semi-Riemannian Submanifolds	560
§13.4.	Geodesics	567
§13.5.	Riemannian Manifolds and Distance	585
§13.6.	Lorentz Geometry	588
§13.7.	Jacobi Fields	594

§13.8.	First and Second Variation of Arc Length	599
§13.9.	More Riemannian Geometry	612
§13.10.	Cut Locus	617
§13.11.	Rauch's Comparison Theorem	619
§13.12.	Weitzenböck Formulas	623
§13.13.	Structure of General Relativity	627
Problems		634
Appendix A.	The Language of Category Theory	637
Appendix B.	Topology	643
§B.1.	The Shrinking Lemma	643
§B.2.	Locally Euclidean Spaces	645
Appendix C.	Some Calculus Theorems	647
Appendix D.	Modules and Multilinearity	649
§D.1.	R-Algebras	660
Bibliography		663
Index		667

Preface

Classical differential geometry is the approach to geometry that takes full advantage of the introduction of numerical coordinates into a geometric space. This use of coordinates in geometry was the essential insight of René Descartes that allowed the invention of analytic geometry and paved the way for modern differential geometry. The basic object in differential geometry (and differential topology) is the smooth manifold. This is a topological space on which a sufficiently nice family of coordinate systems or "charts" is defined. The charts consist of locally defined n-tuples of functions. These functions should be sufficiently independent of each other so as to allow each point in their common domain to be specified by the values of these functions. One may start with a topological space and add charts which are compatible with the topology or the charts themselves can generate the topology. We take the latter approach. The charts must also be compatible with each other so that changes of coordinates are always smooth maps. Depending on what type of geometry is to be studied, extra structure is assumed such as a distinguished group of symmetries, a distinguished "tensor" such as a metric tensor or symplectic form or the very basic geometric object known as a *connection*. Often we find an interplay among many such elements of structure.

Modern differential geometers have learned to present much of the subject without constant direct reference to locally defined objects that depend on a choice of coordinates. This is called the "invariant" or "coordinate free" approach to differential geometry. The only way to really see exactly what this all means is by diving in and learning the subject.

The relationship between geometry and the physical world is fundamental on many levels. Geometry (especially differential geometry) clarifies, codifies and then generalizes ideas arising from our intuitions about certain aspects of our world. Some of these aspects are those that we think of as forming the spatiotemporal background of our activities, while other aspects derive from our experience with objects that have "smooth" surfaces. The Earth is both a surface and a "lived-in space", and so the prefix "geo" in the word geometry is doubly appropriate. Differential geometry is also an appropriate mathematical setting for the study of what we classically conceive of as continuous physical phenomena such as fluids and electromagnetic fields.

Manifolds have dimension. The surface of the Earth is two-dimensional, while the configuration space of a mechanical system is a manifold which may easily have a very high dimension. Stretching the imagination further we can conceive of each possible field configuration for some classical field as being an abstract point in an infinite-dimensional manifold.

The physicists are interested in geometry because they want to understand the way the physical world is in "actuality". But there is also a discovered "logical world" of pure geometry that is in some sense a part of reality too. This is the reality which Roger Penrose calls the Platonic world.[1] Thus the mathematicians are interested in the way worlds *could be in principle* and geometers are interested in what might be called "possible geometric worlds". Since the inspiration for what we find interesting has its roots in our experience, even the abstract geometries that we study retain a certain physicality. From this point of view, the intuition that guides the pure geometer is fruitfully enhanced by an explicit familiarity with the way geometry plays a role in physical theory.

Knowledge of differential geometry is common among physicists thanks to the success of Einstein's highly geometric theory of gravitation and also because of the discovery of the differential geometric underpinnings of modern gauge theory[2] and string theory. It is interesting to note that the gauge field concept was introduced into physics within just a few years of the time that the notion of a connection on a fiber bundle (of which a gauge field is a special case) was making its appearance in mathematics. Perhaps the most exciting, as well as challenging, piece of mathematical physics to come along in a while is string theory mentioned above.

The usefulness of differential geometric ideas for physics is also apparent in the conceptual payoff enjoyed when classical mechanics is reformulated in the language of differential geometry. Mathematically, we are led to the subjects of symplectic geometry and Poisson geometry. The applicability of differential geometry is not limited to physics. Differential geometry is

---

[1] Penrose seems to take this Platonic world rather literally giving it a great deal of ontological weight as it were.

[2] The notion of a connection on a fiber bundle and the notion of a gauge field are essentially identical concepts discovered independently by mathematicians and physicists.

also of use in engineering. For example, there is the increasingly popular differential geometric approach to control theory.

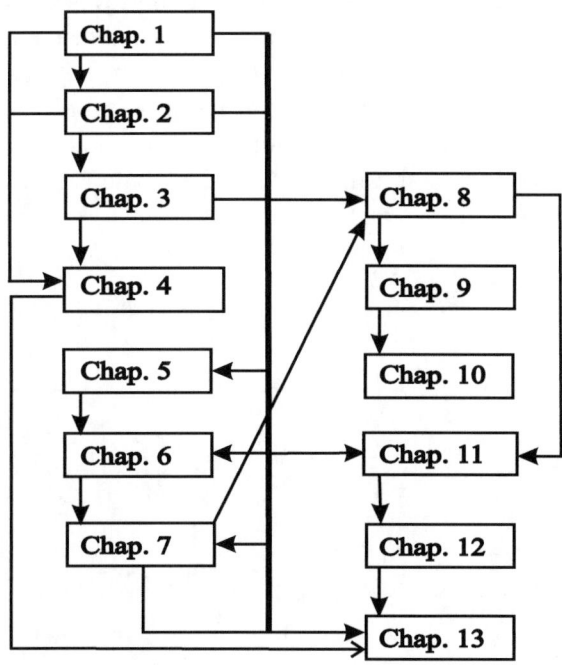

There is a bit more material in this book than can be comfortably covered in a two semester course. A course on manifold theory would include Chapters 1, 2, 3, and then a selection of material from Chapters 5, 7, 8, 9, 10, and 11. A course in Riemannian geometry would review material from the first three chapters and then cover at least Chapters 8 and 13. A more leisurely course would also include Chapter 4 before getting into Chapter 13. The book need not be read in a strictly linear manner. We included here a flow chart showing approximate chapter dependence. There are exercises throughout the text and problems at the end of each chapter. The reader should at least read and think about every exercise. Some exercises are rather easy and only serve to keep the reader alert. Other exercises take a bit more thought.

Differential geometry is a huge field, and even if we had restricted our attention to just manifold theory or Riemannian geometry, only a small fragment of what might be addressed at this level could possibly be included. In choosing what to include in this book, I was guided by personal interest and, more importantly, by the limitations of my own understanding. While preparing this book I used too many books and papers to list here but a few that stand out as having been especially useful include [**A-M-R**], [**Hicks**], [**L1**], [**Lee, John**], [ON1], [ON2], and [**Poor**].

I would like to thank Lance Drager, Greg Friedman, Chris Monico, Mara Neusel, Efton Park, Igor Prokhorenkov, Ken Richardson, Magdalena Toda, and David Weinberg for proofreading various portions of the book. I am especially grateful to Lance Drager for many detailed discussions concerning some of the more difficult topics covered in the text, and also for the many hours he spent helping me with an intensive final proofreading. It should also be noted that certain nice ideas for improvements of a few proofs are due to Lance. Finally, I would like to thank my wife Maria for her support and patience and my father for his encouragement.

Chapter 1

Differentiable Manifolds

"Besides language and music, mathematics is one of the primary manifestations of the free creative power of the human mind."

– Hermann Weyl

In this chapter we introduce differentiable manifolds and smooth maps. A differentiable manifold is a topological space on which there are defined coordinates allowing basic notions of differentiability. The theory of differentiable manifolds is a natural result of extending and clarifying notions already familiar from multivariable calculus. Consider the task of writing out clearly, in terms of sets and maps, what is going on when one does calculus in polar coordinates on the plane or in spherical coordinates on a sphere. If this were done with sufficient care about domains and codomains, a good deal of ambiguity in standard notation would be discovered and clarifying the situation would almost inevitably lead to some of the very definitions that we will see shortly. In some sense, a good part of manifold theory is just multivariable calculus done carefully. Unfortunately, this care necessitates an increased notational burden that can be intimidating at first. The reader should always keep in mind the example of surfaces and the goal of doing calculus on surfaces. Another point is that manifolds can have nontrivial topology, and this is one reason the subject becomes so rich. In this chapter we will make some connection with some basic ideas from topology such as covering spaces and the fundamental group.

Manifold theory and differential geometry play a role in an increasingly large amount of modern mathematics and have long played an important role in physics. In Einstein's general theory of relativity, spacetime is taken to be a 4-dimensional manifold. Manifolds of arbitrarily high dimension play a role in many physical theories. For example, in classical mechanics, the set of all possible locations and orientations of a rigid body is a manifold of dimension six, and the phase space of a system of N Newtonian particles moving in 3-dimensional space is a manifold of dimension $6N$.

1.1. Preliminaries

To understand the material to follow, it is necessary that the reader have a good background in the following subjects.

1) **Linear algebra**. The reader should be familiar with the idea of the dual space of a vector space and also with the notion of a quotient vector space. A bit of the language of module theory will also be used and is outlined in Appendix D.

2) **Point set topology**. We assume familiarity with the notions of subspace topology, compactness and connectedness. The reader should know the definitions of Hausdorff topological spaces, regular spaces and normal spaces. The reader should also have been exposed to quotient topologies. Some of the needed concepts are reviewed in the online supplement [**Lee, Jeff**].

Convention: A neighborhood of a point in a topological space is often defined to be a set whose interior contains the point. Our convention in the sequel is that the word "neighborhood" will always mean *open* neighborhood unless otherwise indicated. Nevertheless, we will sometimes write "open neighborhood" for emphasis.

3) **Abstract algebra**. The reader will need a familiarity with the basics of abstract algebra at least to the level of the basic isomorphism theorems for groups and rings.

4) **Multivariable calculus**. The reader should be familiar with the idea that the derivative at a point p of a map between open sets of (normed) vector spaces is a linear transformation between the vector spaces. Usually the normed spaces are assumed to be the Euclidean coordinate spaces such as $\mathbb{R}^n$ with the norm $\|x\| = \sqrt{x \cdot x}$. A reader who felt the need for a review could do no better than to study roughly the first half of the classic book "Calculus on Manifolds" by Michael Spivak. Also, the online supplement ([**Lee, Jeff**]) gives a brief treatment of differential calculus on Banach spaces. Here we simply review a few definitions and notations.

Notation 1.1. The elements of $\mathbb{R}^n$ are n-tuples of real numbers, and we shall write the indices as superscripts, so $v = (v^1, \ldots, v^n)$. Furthermore, when using matrix algebra, elements of $\mathbb{R}^n$ are most often written as column

1.1. Preliminaries

vectors so in this context, $v = [v^1, \ldots, v^n]^t$. On the other hand, elements of the dual space of $\mathbb{R}^n$ are often written as row vectors and the indices are written as subscripts. With this convention, an element of the dual space $(\mathbb{R}^n)^*$ acts on an element of $\mathbb{R}^n$ by matrix multiplication. We shall only be careful to write elements of $\mathbb{R}^n$ as column vectors if necessary (as when we write Av for some $m \times n$ matrix A and $v \in \mathbb{R}^n$). If $f : U \to \mathbb{R}^m$ then $f = (f^1, \ldots, f^m)$ for real-valued functions $f^1, \ldots, f^m$.

Definition 1.2. Let U be an open subset of $\mathbb{R}^n$. A map $f : U \to \mathbb{R}^m$ is said to be **differentiable at** $a \in U$ if and only if there is a linear map $A_a : \mathbb{R}^n \to \mathbb{R}^m$ such that

$$\lim_{\|h\| \to 0} \frac{\|f(a+h) - f(a) - A_a(h)\|}{\|h\|} = 0.$$

The map A_a is uniquely determined by f and a and is denoted by $Df(a)$ or $Df|_a$.

Notation 1.3. If $L : V \to W$ is a linear transformation and $v \in V$, we shall often denote $L(v)$ by either $L \cdot v$ or Lv depending on which is clearer in a given context. This applies to $Df(a)$, so we shall have occasion to write things like $Df(a)(v)$, $Df(a) \cdot v$, $Df(a)v$ or $Df|_a v$. The space of linear maps from V to W is denoted by $L(V, W)$ and also by $\mathrm{Hom}(V, W)$.

With respect to standard bases, $Df(a)$ is given by the $m \times n$ matrix of partial derivatives (the **Jacobian matrix**). Thus if $w = Df(a)v$, then

$$w^i = \sum_j \frac{\partial f^i}{\partial x^j}(a) v^j.$$

Note that for a differentiable real-valued function f, we will often denote $\partial f / \partial x^i$ by $\partial_i f$.

Recall that if U, V and W are vector spaces, then a map $\beta : U \times V \to W$ is called **bilinear** if for each fixed $u_0 \in U$ and fixed $v_0 \in V$, the maps $v \mapsto \beta(u_0, v)$ and $u \mapsto \beta(u, v_0)$ are linear. If $\beta : V \times V \to W$ is bilinear and $\beta(u, v) = \beta(v, u)$ for all $u, v \in V$, then β is said to be **symmetric**. Similarly, antisymmetry or skewsymmetry is defined by the condition $\beta(u, v) = -\beta(v, u)$. If V is a real (resp. complex) vector space, then a bilinear map $V \times V \to \mathbb{R}$ (resp. $\mathbb{C}$) is called a **bilinear form**.

If $Df(a)$ is defined for all $a \in U$, then we obtain a map $Df : U \to L(\mathbb{R}^n, \mathbb{R}^m)$, and since $L(\mathbb{R}^n, \mathbb{R}^m) \cong \mathbb{R}^{mn}$, we can consider the differentiability of Df. Thus if Df is differentiable at a, then we have a second derivative $D^2 f(a) : \mathbb{R}^n \to L(\mathbb{R}^n, \mathbb{R}^m)$, and so if $v, w \in \mathbb{R}^n$, then $\left(D^2 f(a)v\right) w \in \mathbb{R}^m$. This allows us to think of $D^2 f(a)$ as a bilinear map:

$$D^2 f(a)(v, w) := \left(D^2 f(a)v\right) w \text{ for } v, w \in \mathbb{R}^n.$$

If $u = D^2 f(a)(v, w)$, then with respect to standard bases, we have

$$u^i = \sum_{j,k} \frac{\partial f^i}{\partial x^j \partial x^k}(a) v^j w^k.$$

Higher derivatives $D^r f$ can be defined similarly as multilinear maps (see D.13 of Appendix D), and if $D^r f$ exists and is continuous on U, then we say that f is r-times continuously differentiable, or C^r, on U. A map f is C^r if and only if all partial derivatives of order less than or equal to r of the component functions f^i exist and are continuous on U. The vector space of all such maps is denoted by $C^r(U, \mathbb{R}^m)$, and $C^r(U, \mathbb{R})$ is abbreviated to $C^r(U)$. If f is C^r for all $r \geq 0$, then we say that f is **smooth** or C^∞. The vector space of all smooth maps from U to $\mathbb{R}^m$ is denoted by $C^\infty(U, \mathbb{R}^m)$ and we thereby include ∞ as a possible value for r. Also, f is said to be C^r at p if its restriction to some open neighborhood of p is of class C^r. If f is C^2 near a then $D^2 f(a)$ is symmetric and this is reflected in the fact that we have equality of mixed second order partial derivatives; $\partial f^i / \partial x^j \partial x^k = \partial f^i / \partial x^k \partial x^j$. Similarly, if f is C^r near a then the order of partial differentiations can be rearranged in any mixed partial derivative of order less than or equal to r.

Definition 1.4. A bijection f between open sets $U \subset \mathbb{R}^n$ and $V \subset \mathbb{R}^m$ is called a C^r **diffeomorphism** if and only if f and f^{-1} are both C^r differentiable. If $r = \infty$, then we simply call f a **diffeomorphism**.

Definition 1.5. Let U be open in $\mathbb{R}^n$. A map $f : U \to \mathbb{R}^n$ is called a **local C^r diffeomorphism** if and only if for every $p \in U$ there is an open set $U_p \subset U$ with $p \in U_p$ such that $f(U_p)$ is open and $f|_{U_p} : U_p \to f(U_p)$ is a C^r diffeomorphism.

We will sometimes think of the derivative of a curve[1] $c : I \subset \mathbb{R} \to \mathbb{R}^m$ at $t_0 \in I$, as a velocity vector and so we are identifying $Dc|_{t_0} \in L(\mathbb{R}, \mathbb{R}^m)$ with $Dc|_{t_0} \cdot 1 \in \mathbb{R}^m$. Here the number 1 is playing the role of the unit vector in $\mathbb{R}$. Especially in this context, we write the velocity vector using the notation $\dot{c}(t_0)$ or $c'(t_0)$.

Let $f : U \subset \mathbb{R}^n \to \mathbb{R}^m$ be a map and suppose that we write $\mathbb{R}^n = \mathbb{R}^k \times \mathbb{R}^l$. Let (x, y) denote a generic element of $\mathbb{R}^k \times \mathbb{R}^l$. For every $(a, b) \in U \subset \mathbb{R}^k \times \mathbb{R}^l$ the partial maps $f_a : y \mapsto f(a, y)$ and $f^b : x \mapsto f(x, b)$ are defined in some neighborhood of b (resp. a). We define the partial derivatives, when they exist, by $D_2 f(a, b) := Df_a(b)$ and $D_1 f(a, b) := Df^b(a)$. These are, of course,

---

[1] We will often use the letter I to denote a generic (usually open) interval in the real line.

linear maps.
$$D_1 f(a,b) : \mathbb{R}^k \to \mathbb{R}^m,$$
$$D_2 f(a,b) : \mathbb{R}^l \to \mathbb{R}^m.$$

Remark 1.6. Notice that if we consider the maps $\iota_a : x \mapsto (a,x)$ and $\iota^b : x \mapsto (x,b)$, then $D_2 f(a,b) = D(f \circ \iota_a)(b)$ and $D_1 f(a,b) = D(f \circ \iota^b)(a)$.

Proposition 1.7. *If f has continuous partial derivatives $D_i f(x,y)$, $i = 1,2$ near $(x,y) \in \mathbb{R}^k \times \mathbb{R}^l$, then $Df(x,y)$ exists and is continuous. In this case, we have for $\mathrm{v} = (\mathrm{v}_1, \mathrm{v}_2) \in \mathbb{R}^k \times \mathbb{R}^l$,*
$$Df(x,y) \cdot (\mathrm{v}_1, \mathrm{v}_2) = D_1 f(x,y) \cdot \mathrm{v}_1 + D_2 f(x,y) \cdot \mathrm{v}_2.$$

Clearly we can consider maps on several factors $f : \mathbb{R}^{k_1} \times \mathbb{R}^{k_2} \times \cdots \times \mathbb{R}^{k_r} \to \mathbb{R}^m$ and then we can define partial derivatives $D_i f : \mathbb{R}^{k_i} \to \mathbb{R}^m$ for $i = 1, \ldots, r$ in the obvious way. Notice that the meaning of $D_i f$ depends on how we factor the domain. For example, we have both $\mathbb{R}^3 = \mathbb{R}^2 \times \mathbb{R}$ and also $\mathbb{R}^3 = \mathbb{R} \times \mathbb{R} \times \mathbb{R}$. Let U be an open subset of $\mathbb{R}^n$ and let $f : U \to \mathbb{R}$ be a map. Note that for the factorization $\mathbb{R}^n = \mathbb{R} \times \cdots \times \mathbb{R}$, the linear map $(D_i f)(a)$ is often identified with the number $\partial_i f(a)$.

Theorem 1.8 (Chain Rule). *Let U be an open subset of $\mathbb{R}^n$ and V an open subset of $\mathbb{R}^m$. If $f : U \to \mathbb{R}^m$ and $g : V \to \mathbb{R}^d$ are maps such that f is differentiable at $a \in U$ and g is differentiable at $f(a)$, then $g \circ f$ is differentiable at a and*
$$D(g \circ f)(a) = Dg(f(a)) \circ Df(a).$$
Furthermore, if f and g are C^r at a and $f(a)$ respectively, then $g \circ f$ is C^r at a.

The chain rule may also be written $D_i(g \circ f)(a) = \sum D_j g(f(a)) D_i f^j(a)$ where $f = (f^1, \ldots, f^m)$.

Notation 1.9. Einstein Summation Convention. Summations such as
$$h^i = \sum_{j=1}^n \tau_j^i \alpha^j$$
occur often in differential geometry. It is often convenient to employ a convention whereby summation over *repeated indices* is implied. This convention is attributed to Einstein and is called the Einstein summation convention. Using this convention, the above equation would be written
$$h^i = \tau_j^i \alpha^j.$$
The range of the indices is either determined by context or must be explicitly mentioned. We shall use this convention in some later chapters.

Finally, we will use the notion of a commutative diagram. The reader unfamiliar with this notion should consult Appendix A.

1.2. Topological Manifolds

We recall a few concepts from point set topology. A **cover** of a topological space X is a family of sets $\{U_\beta\}_{\beta \in B}$ such that $X = \bigcup_\beta U_\beta$. If all the sets U_β are open, we call it an **open cover**. A **refinement** of a cover $\{U_\beta\}_{\beta \in B}$ of a topological space X is another cover $\{V_i\}_{i \in I}$ such that every set from the second cover is contained in at least one set from the original cover. This means that if $\{U_\beta\}_{\beta \in B}$ is the given cover of X, then a refinement may be described as a cover $\{V_i\}_{i \in I}$ together with a set map $i \mapsto \beta(i)$ of the indexing sets $I \to B$ such that $V_i \subset U_{\beta(i)}$ for all i. Two covers $\{U_\alpha\}_{\alpha \in A}$ and $\{U_\beta\}_{\beta \in B}$ have a common refinement. Indeed, we simply let $I = A \times B$ and then let $U_i = U_\alpha \cap U_\beta$ if $i = (\alpha, \beta)$. This common refinement will obviously be open if the two original covers were open. We say that a cover $\{V_i\}_{i \in I}$ of X (by not necessarily open sets) is a **locally finite** cover if every point of X has a neighborhood that intersects only a finite number of sets from the cover. A topological space X is called **paracompact** if every *open* cover of X has a refinement which is a locally finite open cover.

Definition 1.10. Let X be a set. A collection $\mathfrak{B}$ of subsets of X is called a **basis** of subsets of X if the following conditions are satisfied:

(i) $X = \bigcup_{B \in \mathfrak{B}} B$;
(ii) If $B_1, B_2 \in \mathfrak{B}$ and $x \in B_1 \cap B_2$, then there exists a set $B \in \mathfrak{B}$ with $x \in B \subset B_1 \cap B_2$.

It is a fact of elementary point set topology that if $\mathfrak{B}$ is a basis of subsets of a set X, then the family $\mathcal{T}$ of all possible unions of members of $\mathfrak{B}$ is a topology on X. In this case, we say that $\mathfrak{B}$ is a **basis** for the topology $\mathcal{T}$ and $\mathcal{T}$ is said to be **generated by the basis**. In thinking about bases for topologies, it is useful to introduce a certain technical notion as follows: If $\mathfrak{B}$ is any family of subsets of X, then we say that a subset $U \subset X$ satisfies the **basis criterion** with respect to $\mathfrak{B}$ if for any $x \in U$, there is a $B \in \mathfrak{B}$ with $x \in B \subset U$. Then we have the following technical lemma which is sometimes used without explicit mention.

Lemma 1.11. *If $\mathfrak{B}$ is a basis of subsets of X, then the topology generated by $\mathfrak{B}$ is exactly the family of all subsets of X which satisfy the basis criterion with respect to the family $\mathfrak{B}$.* (See Problem 2.)

A topological space is called **second countable** if its topology has a countable basis. The space $\mathbb{R}^n$ with the usual topology derived from the Euclidean distance function is second countable since we have a basis for

1.2. Topological Manifolds

the topology consisting of open balls with rational radii centered at points with rational coordinates.

Definition 1.12. An n-dimensional **topological manifold** is a paracompact Hausdorff topological space, say M, such that every point $p \in M$ is contained in some open set U_p that is homeomorphic to an open subset of the Euclidean space $\mathbb{R}^n$. Thus we say that a topological manifold is "locally Euclidean". The integer n is referred to as the **dimension** of M, and we denote it by $\dim(M)$.

Note: At first it may seem that a locally Euclidean space must be Hausdorff, but this is not the case.

Example 1.13. $\mathbb{R}^n$ is trivially a topological manifold of dimension n.

Example 1.14. The unit circle $S^1 := \{(x,y) \in \mathbb{R}^2 : x^2 + y^2 = 1\}$ is a 1-dimensional topological manifold. Indeed, the map $\mathbb{R} \to S^1$ given by $\theta \mapsto (\cos\theta, \sin\theta)$ has restrictions to small open sets which are homeomorphisms. The boundary of a square in the plane is a topological manifold homeomorphic to the circle, and so we say that it is a topological circle. More generally, the n-sphere

$$S^n := \left\{ (x^1, \ldots, x^{n+1}) \in \mathbb{R}^{n+1} : \sum (x^i)^2 = 1 \right\}$$

is a topological manifold.

If M_1 and M_2 are topological manifolds of dimensions n_1 and n_2 respectively, then $M_1 \times M_2$, with the product topology, is a topological manifold of dimension $n_1 + n_2$. Such a manifold is called a product manifold. $M_1 \times M_2$ is locally Euclidean. Indeed, the required homeomorphisms are constructed in the obvious way from those defined on M_1 and M_2: If $\phi_p : U_p \subset M_1 \to V_{\phi(p)} \subset \mathbb{R}^{n_1}$ and $\psi_q : U_q \subset M_2 \to V_{\psi(q)} \subset \mathbb{R}^{n_2}$ are homeomorphisms, then $U_p \times U_q$ is a neighborhood of (p,q) and we have a homeomorphism

$$\phi_p \times \psi_q : U_p \times U_q \to V_{\phi(p)} \times V_{\psi(q)} \subset \mathbb{R}^{n_1} \times \mathbb{R}^{n_2},$$

where $(\phi_p \times \psi_q)(x,y) := (\phi_p(x), \psi_q(y))$. That $M_1 \times M_2$ is Hausdorff is an easy exercise. Then, by Proposition B.5 of Appendix B, $M_1 \times M_2$ is paracompact since it is metrizable.

The product manifold construction can obviously be extended to products of a finite number of manifolds.

Example 1.15. The n-torus

$$T^n := S^1 \times S^1 \times \cdots \times S^1 \ (n \text{ factors})$$

is a topological manifold of dimension n.

Recall that a topological space is **connected** if it is not the disjoint union of nonempty open subsets. A subset of a topological space is said to be connected if it is connected with the subspace topology. If a space is not connected, then it can be decomposed into components:

Definition 1.16. Let X be a topological space and $x \in X$. The **component** containing x, denoted $C(x)$, is defined to be the union of all connected subsets of X that contain x. A subset of a topological space X is a (connected) component if it is $C(x)$ for some $x \in X$.

A topological space is called **locally connected** if the topology has a basis consisting of connected sets. If a topological space is locally connected, then the connected components of each open set (considered with its subspace topology) are all open. In particular, each connected component of a locally connected space is open (and also closed). This clearly applies to manifolds.

A continuous curve $\gamma : [a, b] \to M$ is said to connect a point p to a point q in M if $\gamma(a) = p$ and $\gamma(b) = q$. We define an equivalence relation on a topological space M by declaring $p \sim q$ if and only if there is a continuous curve connecting p to q. The equivalence classes are called **path components**, and if there is only one path component, then we say that M is **path connected**.

Exercise 1.17. The path components of a manifold M are exactly the connected components of M. Thus, a manifold is connected if and only if it is path connected.

We shall not give many examples of topological manifolds at this time because our main concern is with *smooth* manifolds defined below. We give plenty of examples of smooth manifolds, and every smooth manifold is also a topological manifold.

Manifolds are often defined with the requirement of second countability because, when the manifold is Hausdorff, this condition implies paracompactness. Paracompactness is important in connection with the notion of a "partition of unity" discussed later in this book. *It is known that for a locally Euclidean Hausdorff space, paracompactness is equivalent to the property that each connected component is second countable.* Thus if a locally Euclidean Hausdorff space has at most a countable number of components, then paracompactness implies second countability. Proposition B.5 of Appendix B gives a list of conditions equivalent to paracompactness for a locally Euclidean Hausdorff space. One theorem that fails if the manifold has an uncountable number of components is Sard's Theorem 2.34. Our approach will be to add in the requirement of second countability when needed.

1.2. Topological Manifolds

In defining a topological manifold, one could allow the dimension n of the Euclidean space to depend on the homeomorphism ϕ and so on the point $p \in M$. However, it is a consequence of a result of Brouwer called "invariance of domain" that n would be constant on connected components of M. This result is rather easy to prove if the manifold has a differentiable structure (defined below), but is more difficult in general. We shall simply record Brouwer's theorem:

Theorem 1.18 (Invariance of Domain). *The image of an open set $U \subset \mathbb{R}^n$ under an injective continuous map $f : U \to \mathbb{R}^n$ is open and f is a homeomorphism from U to $f(U)$. It follows that if $U \subset \mathbb{R}^n$ is homeomorphic to $V \subset \mathbb{R}^m$, then $m = n$.*

Let us define n-dimensional closed **Euclidean half-space** to be $\mathbb{H}^n := \mathbb{R}^n_{x^n \geq 0} := \{(a^1, \ldots, a^n) \in \mathbb{R}^n : a^n \geq 0\}$. The **boundary** of $\mathbb{H}^n$ is $\partial \mathbb{H}^n = \mathbb{R}^n_{x^n = 0} =: \{(a^1, \ldots, a^n) : a^n = 0\}$. The **interior** of $\mathbb{H}^n$ is $\text{int}(\mathbb{H}^n) = \mathbb{H}^n \setminus \partial \mathbb{H}^n$. There are other half-spaces homeomorphic to $\mathbb{H}^n = \mathbb{R}^n_{x^n \geq 0}$. In fact, for a fixed $c \in \mathbb{R}$ and for any nonzero linear function $\lambda \in (\mathbb{R}^n)^*$, we define

$$\mathbb{R}^n_{\lambda \geq c} = \{a \in \mathbb{R}^n : \lambda(a) \geq c\},$$

which includes $\mathbb{R}^n_{x^k \geq c}$ and $\mathbb{R}^n_{x^k \leq c}$. We will also use the notations $\mathbb{R}^n_{\lambda < c}$, $\mathbb{R}^n_{\lambda = c}$, etc., whose meanings are obvious.

The space $\mathbb{H}^n$ is not a manifold because points on the boundary do not have open neighborhoods homeomorphic to any open set in a Euclidean space. However, $\mathbb{H}^n$ will be our model for the following generalization. A **topological manifold with boundary** (of dimension n) is a paracompact Hausdorff topological space M such that each point $p \in M$ is contained in some open set U_p that is homeomorphic to an open subset[2] in $\mathbb{H}^n$. Clearly, we could also use charts with images in other half-spaces as mentioned above, but we initially stick with $\mathbb{H}^n$ for purposes of the theoretical development. A point $p \in M$ that is mapped to $\partial \mathbb{H}^n$ under some homeomorphism $\psi : U_p \to V_{\psi(p)}$ is called a boundary point and the set of all boundary points of M is called the **boundary** of M and is denoted ∂M. The manifold's **interior** consists of those points of M that are mapped to points of $\text{int}(\mathbb{H}^n)$. It is a corollary to Brouwer's theorem that these concepts are well-defined independently of the homeomorphism used.

In developing manifolds with boundary, a preference is often given to $\mathbb{H}^n = \mathbb{R}^n_{x^n \geq 0}$. However, this is just a convenience that reduces notation. In fact, from the point of view of "manifold orientation" developed in a later chapter, the spaces of the form $\mathbb{R}^n_{x^1 \leq 0}$ have distinct and often unnoticed advantages which we explain in due course.

---

[2]The reader should recall carefully the meaning of an open set in $\mathbb{H}^n$; these certainly need not be open in the containing $\mathbb{R}^n$.

Exercise 1.19. Show that $\text{int}(M) \cap \partial M = \emptyset$.

Exercise 1.20. The boundary ∂M of an n-dimensional topological manifold with boundary is an $(n-1)$-dimensional topological manifold (without boundary).

From now on, by "manifold" we shall mean a manifold without boundary unless otherwise indicated or implied by context.

Example 1.21. Let $[a,b] \subset \mathbb{R}$ be a closed interval. If N is a topological manifold of dimension $n-1$, then $N \times [a,b]$ is an n-dimensional topological manifold with boundary $\partial (N \times [a,b]) = (N \times \{a\}) \cup (N \times \{b\})$.

Example 1.22. If N is a topological manifold with boundary ∂N, then $N \times \mathbb{R}$ is a topological manifold with boundary and $\partial(N \times \mathbb{R}) = \partial N \times \mathbb{R}$.

Example 1.23. The closed cube $\overline{C^n} = \{x : \max\{|x^1|, \ldots, |x^n|\} \leq 1\}$ with its subspace topology inherited from $\mathbb{R}^n$ is a topological manifold with boundary. The boundary is (homeomorphic to) an $(n-1)$-dimensional sphere.

Recall that the (topological) boundary of a set S in a topological space X is defined as the set of all points $p \in X$ with the property that every open set containing p also contains elements of both S and $X \backslash S$. This is not quite the same idea as boundary in the sense of manifold with boundary defined above. For example, the subset of $\mathbb{R}^2$ defined by

$$S := \{(x,y) \in \mathbb{R}^2 : 1 \leq x^2 + y^2 < 2 \text{ or } 2 < x^2 + y^2 \leq 3\}$$

is a manifold with boundary

$$\partial S = \{(x,y) \in \mathbb{R}^2 : x^2 + y^2 = 1 \text{ or } x^2 + y^2 = 3\}.$$

On the other hand, the topological boundary of S also contains the circle given by $x^2 + y^2 = 2$.

Topological manifolds, both with or without boundary, are paracompact, Hausdorff and hence also **normal**. This means that given any pair of disjoint closed sets $F_1, F_2 \subset M$, there are open sets U_1 and U_2 containing F_1 and F_2 respectively such that U_1 and U_2 are also disjoint. Recall that a topological space X (or its topology) is called **metrizable** if there exists a metric on the space which induces the given topology of X. Since they are normal, manifolds are, a fortiori, regular. According to the Urysohn metrization theorem, every second countable regular Hausdorff space is metrizable. Now if every connected component of a space is metrizable, then the whole space is also metrizable by a standard trick which modifies the metric so as to make the distance between points in a given component always less than one while a pair of points from distinct components has distance one. Since every paracompact Hausdorff manifold has second countable components, we see that all manifolds, as defined here, are metrizable. Thus, the reader may as well think of manifolds as special kinds of metric spaces. For more on manifold topology, consult [**Matsu**].

1.3. Charts, Atlases and Smooth Structures

In this section we introduce the notion of charts or coordinate systems. The existence of such charts is what allows for a well-defined notion of what it means for a function on a manifold to be differentiable and also what it means for a map from one manifold to another to be differentiable.

Definition 1.24. Let M be a set. A **chart** on M is a bijection of a subset $U \subset M$ onto an open subset of some Euclidean space $\mathbb{R}^n$.

We say that the chart takes values in $\mathbb{R}^n$ or simply that the chart is $\mathbb{R}^n$-valued. A chart $\mathbf{x} : U \to \mathbf{x}(U) \subset \mathbb{R}^n$ is traditionally indicated by the pair $(U, \mathbf{x})$. If $\mathrm{pr}_i : \mathbb{R}^n \to \mathbb{R}$ is the projection onto the i-th factor given by $\mathrm{pr}_i(a^1, \ldots, a^n) = a^i$, then x^i is the function defined by $x^i = \mathrm{pr}_i \circ \mathbf{x}$ and is called the i-th **coordinate function** for the chart $(U, \mathbf{x})$. We often write $\mathbf{x} = (x^1, \ldots, x^n)$. If $p \in M$ and $(U, \mathbf{x})$ is a chart with $p \in U$ and $\mathbf{x}(p) = 0 \in \mathbb{R}^n$, then we say that the chart is **centered** at p.

Definition 1.25. Let $\mathcal{A} = \{(U_\alpha, \mathbf{x}_\alpha)\}_{\alpha \in A}$ be a collection of $\mathbb{R}^n$-valued charts on a set M. We call $\mathcal{A}$ an $\mathbb{R}^n$-valued **atlas of class** C^r if the following conditions are satisfied:

(i) $\bigcup_{\alpha \in A} U_\alpha = M$.

(ii) The sets of the form $\mathbf{x}_\alpha(U_\alpha \cap U_\beta)$ for $\alpha, \beta \in A$ are all open in $\mathbb{R}^n$.

(iii) Whenever $U_\alpha \cap U_\beta$ is not empty, the map
$$\mathbf{x}_\beta \circ \mathbf{x}_\alpha^{-1} : \mathbf{x}_\alpha(U_\alpha \cap U_\beta) \to \mathbf{x}_\beta(U_\alpha \cap U_\beta)$$
is a C^r diffeomorphism.

Remark 1.26. Here $\mathbf{x}_\beta \circ \mathbf{x}_\alpha^{-1}$ is really a shorthand for
$$\mathbf{x}_\beta|_{U_\alpha \cap U_\beta} \circ \mathbf{x}_\alpha^{-1}|_{\mathbf{x}_\alpha(U_\alpha \cap U_\beta)},$$
but this notation is far too pedantic and cluttered for most people's tastes.

The maps $\mathbf{x}_\beta \circ \mathbf{x}_\alpha^{-1}$ in the definition are called **overlap maps** or **change of coordinate maps**. It is exactly the way we have required the overlap maps to be diffeomorphisms that will allow us to have a well-defined and useful notion of what it means for a function on M to be differentiable of class C^r. An **atlas of class** C^r is also called a C^r **atlas**.

There exist various simplifying but occasionally ambiguous notational conventions regarding coordinates. Consider an arbitrary pair of charts $(U, \mathbf{x})$ and $(V, \mathbf{y})$ and the overlap map $\mathbf{y} \circ \mathbf{x}^{-1} : \mathbf{x}(U \cap V) \to \mathbf{y}(U \cap V)$. If we write $\mathbf{x} = (x^1, \ldots, x^n)$ and $\mathbf{y} = (y^1, \ldots, y^n)$, then we have

$$(1.1) \qquad y^i(p) = y^i \circ \mathbf{x}^{-1}(x^1(p), \ldots, x^n(p))$$

for any $p \in U \cap V$, which makes sense, but in the literature we also see

$$(1.2) \qquad y^i = y^i(x^1, \ldots, x^n).$$

In considering this last expression, one might wonder if the x^i are functions or numbers. But this ambiguity is sort of purposeful. For if (1.1) is true for all $p \in U \cap V$, then (1.2) is true for all $(x^1, \ldots, x^n) \in \mathbf{x}(U \cap V)$, and so we are unlikely to be led into error.

Definition 1.27. Two C^r atlases $\mathcal{A}_1$ and $\mathcal{A}_2$ on M are said to be **equivalent** provided that $\mathcal{A}_1 \cup \mathcal{A}_2$ is also a C^r atlas for M. A C^r **differentiable structure** on M is an equivalence class of C^r atlases. A C^∞ differentiable structure will also be called a **smooth structure**.

Exercise 1.28. Show that the notion of equivalence of atlases given above defines an equivalence relation.

The union of all the C^r atlases in an equivalence class, is itself an atlas that is in the equivalence class. Such an atlas is called a **maximal C^r atlas** since it is not properly contained in any larger atlas. Atlases obtained in this way are precisely those that are maximal with respect to partial ordering of atlases by inclusion. Thus every C^r atlas is contained in a unique maximal C^r atlas which is the union of all the atlases equivalent to it. We thereby obtain a 1-1 correspondence of the set of equivalence classes of C^r differentiable structures and the set of maximal C^r atlases. Thus an

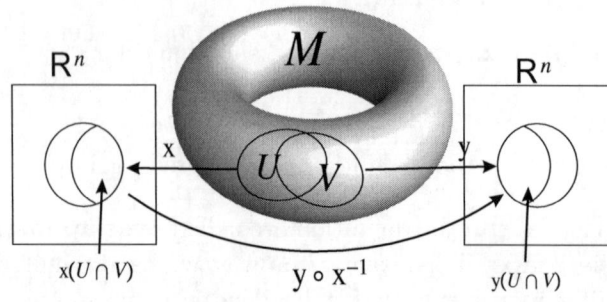

Figure 1.1. Chart overlaps

1.3. Charts, Atlases and Smooth Structures

alternative way to define a C^r differentiable structure is as a maximal C^r atlas. We shall use this alternative quite often *without* comment. As soon as we have any C^r atlas, we have a determined C^r differentiable structure. Indeed, we just take the equivalence class of this atlas. Alternatively, we take the maximal atlas that contains the given atlas.

A pair of charts, say $(U, \mathbf{x})$ and $(V, \mathbf{y})$, are said to be C^r-**related** if either $U \cap V = \emptyset$ or both $\mathbf{x}(U \cap V)$ and $\mathbf{y}(U \cap V)$ are open and $\mathbf{x} \circ \mathbf{y}^{-1}$ and $\mathbf{y} \circ \mathbf{x}^{-1}$ are C^r maps. Thus an atlas is just a family of mutually C^r-related charts whose domains form a cover of the manifold. We say that a chart $(U, \mathbf{x})$ on M is **compatible** with a C^r atlas $\mathcal{A}$ if $\mathcal{A} \cup \{(U, \mathbf{x})\}$ is also a C^r atlas. This just means that $(U, \mathbf{x})$ is C^r-related to every chart in $\mathcal{A}$. The maximal C^r atlas determined by $\mathcal{A}$ is exactly composed of all charts compatible with $\mathcal{A}$. We also say that a chart from the maximal atlas that gives the smooth structure is **admissible**. Charts will be assumed admissible unless otherwise indicated.

Example 1.29. The space $\mathbb{R}^n$ itself has an atlas consisting of the single chart $(\mathrm{id}, \mathbb{R}^n)$, where $\mathrm{id} : \mathbb{R}^n \to \mathbb{R}^n$ is just the identity map. This atlas determines a differentiable structure.

Lemma 1.30. *Let M be a set with a C^r structure given by an atlas $\mathcal{A} = \{(U_\alpha, \mathbf{x}_\alpha)\}_{\alpha \in A}$. If $(U, \mathbf{x})$ and $(V, \mathbf{y})$ are charts compatible with $\mathcal{A}$ such that $U \cap V \neq \emptyset$, then the charts $(U \cap V, \mathbf{x}|_{U \cap V})$ and $(U \cap V, \mathbf{y}|_{U \cap V})$ are also compatible with $\mathcal{A}$ and hence are in the maximal atlas generated by $\mathcal{A}$. Furthermore, if O is an open subset of $\mathbf{x}(U)$ for some compatible chart $(U, \mathbf{x})$, then taking $V = \mathbf{x}^{-1}(O)$ we have that $(V, \mathbf{x}|_V)$ is also a compatible chart.*

Proof. The assertions of the lemma are almost obvious: If $\mathbf{x} \circ \mathbf{x}_\alpha^{-1}$, $\mathbf{x}_\alpha \circ \mathbf{x}^{-1}$, $\mathbf{y} \circ \mathbf{x}_\alpha^{-1}$, $\mathbf{x}_\alpha \circ \mathbf{y}^{-1}$ are all C^r diffeomorphisms, then certainly the restrictions $\mathbf{x}|_{U \cap V} \circ \mathbf{x}_\alpha^{-1}$, $\mathbf{x}_\alpha \circ \mathbf{x}|_{U \cap V}^{-1}$, $\mathbf{y}|_{U \cap V} \circ \mathbf{x}_\alpha^{-1}$ and $\mathbf{x}_\alpha \circ \mathbf{y}|_{U \cap V}^{-1}$ are also. One might just check that the natural domains of these maps are indeed open in $\mathbb{R}^n$. For example, the domain of $\mathbf{x}|_{U \cap V} \circ \mathbf{x}_\alpha^{-1}$ is $\mathbf{x}_\alpha(U_\alpha \cap U \cap V) = \mathbf{x}_\alpha(U_\alpha \cap U) \cap \mathbf{x}_\alpha(U_\alpha \cap V)$ and both $\mathbf{x}_\alpha(U_\alpha \cap U)$ and $\mathbf{x}_\alpha(U_\alpha \cap V)$ are open because of what it means for $(\mathbf{y}, U)$ to be compatible. The last assertion of the lemma is equally easy to prove. $\square$

It follows that the family of sets which are the domains of charts from a maximal atlas provide a basis for a topology on M, which we call the **topology induced by the C^r structure** on M or simply the **manifold topology** if the C^r structure is understood. Thus the open sets are exactly the empty set plus arbitrary unions of chart domains from the maximal atlas. Since a C^r atlas determines a C^r structure, we will also call this the **topology induced by the atlas**. This topology can be characterized as follows: A subset $V \subset M$ is open if and only if $\mathbf{x}_\alpha(U_\alpha \cap V)$ is an open subset of Euclidean space for all charts $(U_\alpha, \mathbf{x}_\alpha)$ in any atlas $\{(U_\alpha, \mathbf{x}_\alpha)\}_{\alpha \in A}$ giving the C^r structure.

Exercise 1.31. Show that if $\mathcal{A}_1$ is a subatlas of $\mathcal{A}_2$, then they both induce the same topology.

Proposition 1.32. *Let M be a set with a C^r structure given by an atlas $\mathcal{A}$. We have the following:*

(i) *If for every two distinct points $p, q \in M$, we have that either p and q are respectively in disjoint chart domains U_α and U_β from the atlas, or they are both in a common chart domain, then the topology induced by the atlas is Hausdorff.*

(ii) *If $\mathcal{A}$ is countable, or has a countable subatlas, then the topology induced by the atlas is second countable.*

(iii) *If the collection of chart domains $\{U_\alpha\}_{\alpha \in A}$ from the atlas $\mathcal{A}$ is such that for every fixed $\alpha_0 \in A$ the set $\{\alpha \in A : U_\alpha \cap U_{\alpha_0} \neq \emptyset\}$ is at most countable, then the topology induced by the atlas is paracompact. Thus, if this condition holds and if M is connected, then the topology induced by the atlas is second countable.*

Proof. We leave the proofs of (i) and (ii) as a problem, or the reader may consult the online supplement [**Lee, Jeff**].

We prove (iii). Give M the topology induced by the atlas. It is enough to show that each connected component has a countable basis. Thus we may as well assume that M is connected. By (ii) it suffices to show that $\mathcal{A}$ is countable. Let U_{α_1} be a particular chart domain from the atlas. We proceed inductively to define a sequence of sets starting with $X_1 = U_{\alpha_1}$. Now given X_{n-1} let X_n be the union of those chart domains U_α which intersect X_{n-1}. It follows (inductively) that each X_n is a countable union of chart domains and hence the same is true of the union $X = \bigcup_n X_n$. By construction, if some chart domain U_α meets X, then it is actually contained in X since to meet X_{n-1} is to be contained in X_n. All that is left is to show that $M = X$. We have reduced to the case that M is connected, and since X is open, it will suffice to show that $M \backslash X$ is also open. If $M \backslash X = \emptyset$ we are done. If $p \in M \backslash X$, then it is in some U_α, and as we said, U_α cannot meet X without being contained in X. Thus it must be the case that $U_\alpha \cap X = \emptyset$ and so $U_\alpha \subset M \backslash X$. We see that $M \backslash X$ is open as is X. Since M is connected, we conclude that $M = X$ (and $M \backslash X = \emptyset$ after all). $\square$

This leads us to a principal definition:

Definition 1.33. A **differentiable manifold of class C^r** is a set M together with a specified C^r structure on M such that the topology induced

1.3. Charts, Atlases and Smooth Structures

by the C^r structure is Hausdorff and paracompact. If the charts are $\mathbb{R}^n$-valued, then we say the manifold has **dimension** n. We write $\dim(M)$ for the dimension of M.

In other words, an n-dimensional differentiable manifold of class C^r is a pair $(M, \mathcal{A})$, where $\mathcal{A}$ is a maximal ($\mathbb{R}^n$-valued) C^r atlas and such that the topology induced by the atlas makes M a topological manifold. A differentiable manifold of class C^r is also referred to as a C^r manifold. A differentiable manifold of class C^∞ is also called a **smooth** manifold and a C^∞ atlas is also called a **smooth atlas**.[3] Notice that if $0 \leq r_1 < r_2$, then any C^{r_2} atlas is also a C^{r_1} atlas and so any C^{r_2} manifold is also a C^{r_1} manifold ($0 \leq r_1 < r_2$). In fact, it is a result of Hassler Whitney that for $r > 0$, every maximal C^r atlas contains a C^∞ atlas. For this and other reasons, we will be mostly concerned with the C^∞ case. For every integer $n \geq 0$, the Euclidean space $\mathbb{R}^n$ is a smooth manifold where, as noted above, there is an atlas whose only member is the chart $(\mathbb{R}^n, \mathrm{id})$ and this atlas determines a maximal atlas providing the usual smooth structure for $\mathbb{R}^n$. If V is an n-dimensional real vector space, then it is a smooth manifold of dimension n in a natural way. Indeed, for each choice of basis $(e_1, \ldots, e_n)$ we obtain a chart whose coordinate functions x^i are defined so that $x^i(v) = a^i$ when $v = \sum a^i e_i$. The overlap maps between any two such charts are linear and hence smooth. Thus we have a smooth atlas which defines a smooth structure.

It is important to notice that if $r > 0$, then a C^r manifold is much more than merely a topological manifold. Note also that we have defined manifolds in such a way that they are necessarily paracompact and Hausdorff. For many purposes, neither assumption is necessary. We could have just defined a C^r manifold to be a set with a C^r structure and with the topology induced by the C^r structure. Proposition 1.32 tells us how to determine, from knowledge about a given atlas, whether the topology is indeed Hausdorff and/or paracompact. In Problem 3, we ask the reader to check that these topological conditions hold for the examples of smooth manifolds that we give in this chapter.

Notation 1.34. As defined, a C^r manifold is a pair $(M, \mathcal{A})$. However, we follow the tradition of using the single symbol M itself to denote the differentiable manifold if the atlas is understood.

Now we come to an important point. Suppose that M already has some natural or previously given topology. For example, perhaps it is already known that M is a topological manifold. If M is given a C^r structure, then it is important to know whether this topology is the same as the topology

---

[3]In some contexts we will just say "atlas" when we mean "smooth atlas".

induced by the C^r structure. For this consideration we have the following lemma which we ask the reader to prove in Problem 8:

Lemma 1.35. *Let $(M, \mathcal{T})$ be a topological space which also has a C^r atlas. If each chart of the atlas has open domain and is a homeomorphism with respect to this topology, then $\mathcal{T}$ will be the same as the topology induced by the C^r structure.*

A good portion of the examples of C^r manifolds that we provide will be of the type described by this lemma. In fact, one often finds differentiable manifolds defined as topological manifolds that have a C^r atlas consisting of charts that are homeomorphisms. In expositions that use this alternative definition, the fact that one can start out with a set, provide charts, and then end up with an appropriate topology is presented as a separate lemma (see for example [**Lee, John**] or [**ON1**]).

Exercise 1.36. Let M be a smooth manifold of dimension n and let $p \in M$. Show that for any $r > 0$ there is a chart $(U, \mathbf{x})$ with $p \in U$ and such that $\mathbf{x}(U) = B(0, r) := \{x \in \mathbb{R}^n : |x| < r\}$. Show that for any $p \in U$ we may further arrange that $\mathbf{x}(p) = 0$.

Remark 1.37. From now on all manifolds in this book will be assumed to be smooth manifolds unless otherwise indicated. Also, let us refer to an n-dimensional smooth manifold as an "n-**manifold**".

As mentioned above, it is certainly possible for there to be two different differentiable structures on the same topological manifold. For example, if $\phi : \mathbb{R}^1 \to \mathbb{R}^1$ is given by $\phi(x) = x^3$, then $\{(\mathbb{R}^1, \phi)\}$ is a smooth atlas for $\mathbb{R}^1$, but the resulting smooth structure is different from the usual structure provided by the atlas $\{(\mathbb{R}^1, \mathrm{id})\}$. The problem is that the inverse of $x \mapsto x^3$ is not differentiable (in the usual sense) at the origin. Now we have two differentiable structures on the line $\mathbb{R}^1$. Actually, although the two atlases do give distinct differentiable structures, they are equivalent in another sense (Definition 1.62 below).

If U is some open subset of a smooth manifold M with atlas $\mathcal{A}_M$, then U is itself a differentiable manifold with an atlas of charts being given by all the restrictions $(U_\alpha \cap U, \mathbf{x}_\alpha|_{U_\alpha \cap U})$ where $(U_\alpha, \mathbf{x}_\alpha) \in \mathcal{A}_M$. We shall refer to such an open subset $U \subset M$ with this differentiable structure as an **open submanifold** of M. Open subsets of $\mathbb{R}^n$ might seem to be very uninteresting manifolds, but in fact they can be quite complex. For example, much can be learned about a knot $K \subset \mathbb{R}^3$ by studying its complement $\mathbb{R}^3 \setminus K$ and the latter is an open subset of $\mathbb{R}^3$.

We now give several examples of smooth manifolds. All of the examples are easily seen to be Hausdorff and paracompact.

1.3. Charts, Atlases and Smooth Structures

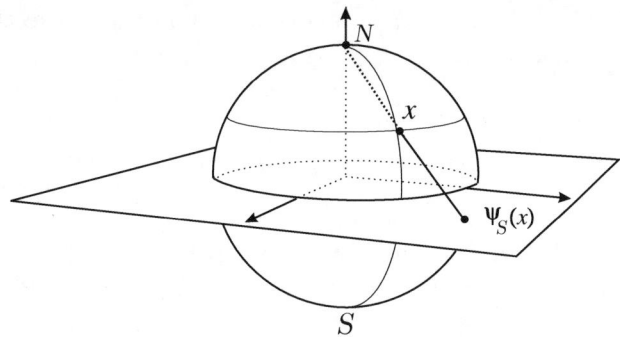

Figure 1.2. Stereographic projection

Example 1.38. Consider the sphere $S^2 \subset \mathbb{R}^3$. We have the usual spherical coordinates (ϕ, θ), where ϕ is the polar angle measured from the north pole ($z = 1$). We want the domain of this chart to be open so we restrict to the set where $0 < \phi < \pi$ and $0 < \theta < 2\pi$. We can also use projection onto the coordinate planes as charts. For instance, let U_z^+ be the set of all $(x, y, z) \in S^2$ such that $z > 0$. Then $(x, y, z) \mapsto (x, y)$ provides a chart $U_z^+ \to \mathbb{R}^2$. The various overlap maps of all of these charts are smooth (some can be computed explicitly without much trouble). It is easy to show that the topology induced by the atlas is the usual topology.

Example 1.39. We can also use **stereographic projection** to give charts on S^2. More generally, we can provide the n-sphere $S^n \subset \mathbb{R}^{n+1}$ with a smooth structure using two charts (U_S, ψ_S) and (U_N, ψ_N). Here,

$$U_S = \{x = (x_1, \ldots, x_{n+1}) \in S^n : x_{n+1} \neq 1\},$$
$$U_N = \{x = (x_1, \ldots, x_{n+1}) \in S^n : x_{n+1} \neq -1\}$$

and $\psi_S : U_S \to \mathbb{R}^n$ (resp. $\psi_N : U_N \to \mathbb{R}^n$) is stereographic projection from the north pole $p_N = (0, 0, \ldots, 0, 1)$ (resp. south pole $p_S = (0, 0, \ldots, 0, -1)$). Note that ψ_S maps from the southern open set containing p_S. Explicitly we have

$$\psi_S(x) = \frac{1}{(1 - x_{n+1})}(x_1, \ldots, x_n) \in \mathbb{R}^n,$$
$$\psi_N(x) = \frac{1}{(1 + x_{n+1})}(x_1, \ldots, x_n) \in \mathbb{R}^n.$$

Exercise 1.40. Show that $\psi_S(U_N \cap U_S) = \psi_N(U_N \cap U_S) = \mathbb{R}^n \setminus \{0\}$ and that $\psi_S \circ \psi_N^{-1}(y) = y/\|y\|^2 = \psi_N \circ \psi_S^{-1}(y)$ for all $y \in \mathbb{R}^n \setminus \{0\}$. Thus we have an atlas $\{(U_S, \psi_S), (U_N, \psi_N)\}$. Verify that the topology induced by this atlas is the same as the usual topology on S^n (as a subspace of $\mathbb{R}^{n+1}$) and that all the maps involved are smooth.

If we identify $\mathbb{R}^2$ with $\mathbb{C}$, then the overlap maps for charts on S^2 from the last example become

$$(1.3) \qquad \psi_S \circ \psi_N^{-1}(z) = \bar{z}^{-1} = \psi_N \circ \psi_S^{-1}(z)$$

for all $z \in \mathbb{C}\backslash\{0\}$. This observation will come in handy later.

Example 1.41 (Projective spaces). The set of all lines through the origin in $\mathbb{R}^3$ is denoted $\mathbb{R}P^2$ and is called the **real projective plane**. Let U_z be the set of all lines $\ell \in \mathbb{R}P^2$ not contained in the x,y plane. Every line $\ell \in U_z$ intersects the plane $z = 1$ at exactly one point of the form $(x(\ell), y(\ell), 1)$. We can define a bijection $\varphi_z : U_z \to \mathbb{R}^2$ by letting $\ell \mapsto (x(\ell), y(\ell))$. This is a chart for $\mathbb{R}P^2$, and there are obviously two other analogous charts (U_x, φ_x) and (U_y, φ_y). These charts cover $\mathbb{R}P^2$ and form a smooth atlas since they have smooth overlap maps.

More generally, the set $\mathbb{R}P^n$ of all lines through the origin in $\mathbb{R}^{n+1}$ is called **real projective n-space**. We have the surjective map $\pi : \mathbb{R}^{n+1}\backslash\{0\} \to \mathbb{R}P^n$ given by letting $\pi(x)$ be the line through x and the origin. We give $\mathbb{R}P^n$ the quotient topology where $U \subset \mathbb{R}P^n$ is open if and only if $\pi^{-1}(U)$ is open. Also $\mathbb{R}P^n$ is given an atlas consisting of charts of the form (U_i, φ_i), where

$$U_i = \{\ell \in \mathbb{R}P^n : \ell \text{ is not contained in the hyperplane } x^i = 0\},$$

and $\varphi_i(\ell)$ is the unique $(u^1, \ldots, u^n)$ such that $(u^1, \ldots u^{i-1}, 1, u^i, \ldots, u^n) \in \ell$. Once again it can be checked that the overlap maps are smooth so that we have a smooth atlas. The topology induced by the atlas is exactly the quotient topology, and we leave it as an exercise to show that it is both paracompact and Hausdorff.

It is often useful to view $\mathbb{R}P^n$ as a quotient of the sphere S^n. Consider the map $S^n \to \mathbb{R}P^n$ given by $x \mapsto \ell_x$, where ℓ_x is the unique line through the origin in $\mathbb{R}^{n+1}$ which contains x. Notice that if $\ell_x = \ell_y$ for $x, y \in S^n$, then $x = \pm y$. It is not hard to show that $\mathbb{R}P^n$ is homeomorphic to $S^n/\sim$, where $x \sim y$ if and only if $x = \pm y$. We can use this homeomorphism to transfer the differentiable structure to $S^n/\sim$ (see Exercise 1.64). We often identify $S^n/\sim$ with $\mathbb{R}P^n$.

Exercise 1.42. Show that the overlap maps for $\mathbb{R}P^n$ are indeed smooth.

Example 1.43. In this example we consider a more general way of getting charts for the projective space $\mathbb{R}P^n$. Let $\alpha : \mathbb{R}^n \to \mathbb{R}^{n+1}$ be an affine map whose image does not contain the origin. Thus α has the form $\alpha(x) = Lx + b$, where $L : \mathbb{R}^n \to \mathbb{R}^{n+1}$ is linear and $b \in \mathbb{R}^{n+1}$ is nonzero. Let $\pi : \mathbb{R}^{n+1}\backslash\{0\} \to \mathbb{R}P^n$ be the projection defined above. The composition $\pi \circ \alpha$ can be easily shown to be a homeomorphism onto its image, and we

1.3. Charts, Atlases and Smooth Structures

call this type of map an affine parametrization. The inverses of these maps form charts for a smooth atlas. The charts described in the last example are essentially special cases of these charts and give the same smooth structure.

Notation 1.44 (Homogeneous coordinates). For $(x_1, \ldots, x_{n+1}) \in \mathbb{R}^{n+1}$, let $[x_1, \ldots, x_{n+1}]$ denote the unique $l \in \mathbb{R}P^n$ such that the line l contains the point $(x_1, \ldots, x_{n+1})$. The numbers $(x_1, \ldots, x_{n+1})$ are said to provide **homogeneous coordinates** for l because $[\lambda x_1, \ldots, \lambda x_{n+1}] = [x_1, \ldots, x_{n+1}]$ for any nonzero $\lambda \in \mathbb{R}$. In terms of homogeneous coordinates, the chart map $\varphi_i : U_i \to \mathbb{R}^n$ is given by

$$\varphi_i([x_1, \ldots, x_{n+1}]) = (x_1 x_i^{-1}, \ldots, \widehat{1}, \ldots, x_{n+1} x_i^{-1}),$$

where the caret symbol ˆ means we have omitted the 1 in the i-th slot to get an element of $\mathbb{R}^n$.

Example 1.45. By analogy with the real case we can construct the complex projective n-space $\mathbb{C}P^n$. As a set, $\mathbb{C}P^n$ is the family of all 1-dimensional complex subspaces of $\mathbb{C}^{n+1}$ (each of these has real dimension 2). In tight analogy with the real case, $\mathbb{C}P^n$ can be given an atlas consisting of charts of the form (U_i, φ_i), where

$$U_i = \{\ell \in \mathbb{C}P^n : \ell \text{ is not contained in the complex hyperplane } z^i = 0\}$$

and $\varphi_i(\ell)$ is the unique $(z^1, \ldots, z^n)$ such that $(z^1, \ldots z^{i-1}, 1, z^i \ldots, z^n) \in \ell$. Here $\varphi_i : U_i \to \mathbb{C}^n \cong \mathbb{R}^{2n}$ and so $\mathbb{C}P^n$ is a manifold of (real) dimension $2n$. Homogeneous coordinate notation $[z_1, \ldots, z_{n+1}]$ is defined as in the real case, but now the multiplier λ is complex.

Exercise 1.46. In reference to the last example, compute the overlap maps $\varphi_i \circ \varphi_j^{-1} : U_i \cap U_j \to \mathbb{C}^n$. For $\mathbb{C}P^1$, show that $U_1 \cap U_2 = \mathbb{C}\setminus\{0\}$, and that $\varphi_2 \circ \varphi_1^{-1}(z) = z^{-1} = \varphi_1 \circ \varphi_2^{-1}(z)$ for $z \in \mathbb{C}\setminus\{0\}$. Show also that if we define $\bar{\varphi}_1 : U_1 \to \mathbb{C}$ by $\bar{\varphi}_1(\ell) = \overline{\varphi_1(\ell)}$, where the bar denotes complex conjugation, then $\{(U_1, \bar{\varphi}_1), (U_2, \varphi_2)\}$ is an atlas for $\mathbb{C}P^1$ giving the same smooth structure as before. Show that

$$\varphi_2 \circ \bar{\varphi}_1^{-1}(z) = \bar{z}^{-1} = \varphi_1 \circ \varphi_2^{-1}(z) \text{ for } z \in \mathbb{C}\setminus\{0\}.$$

Notice that with the atlas $\{(U_1, \bar{\varphi}_1), (U_2, \varphi_2)\}$ for $\mathbb{C}P^1$ from the last example, we have the same overlap maps as for S^2 (see (1.3)). This suggests that $\mathbb{C}P^1$ is diffeomorphic to S^2. Let us construct a diffeomorphism $S^2 \to \mathbb{C}P^1$. For each $x = (x_1, x_2, x_3) \in S^2$ with $x_3 \neq -1$, let $z(x) := \frac{x_1}{1+x_3} + \frac{x_2}{1+x_3}i$, and for each $x = (x_1, x_2, x_3) \in S^2$ with $x_3 \neq 1$ let $w(x) := \frac{x_1}{1-x_3} - \frac{x_2}{1-x_3}i$. Define f by

$$f(x) = \begin{cases} [z(x), 1] & \text{if } x_3 \neq -1, \\ [1, w(x)] & \text{if } x_3 \neq 1. \end{cases}$$

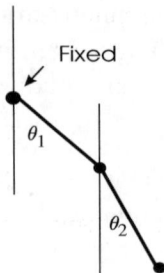

Figure 1.3. Double pendulum

Using the fact that $1 - x_3^2 = x_1^2 + x_2^2$, one finds that for $x = (x_1, x_2, x_3) \in S^2$ with $-1 < x_3 < 1$, we have $w(x) = z(x)^{-1}$. It follows that for such x, we have $[z(x), 1] = [1, w(x)]$. This means that f is well-defined.

Exercise 1.47. Show that the map $f : S^2 \to \mathbb{C}P^1$ defined above is a diffeomorphism.

Exercise 1.48. Show that $\mathbb{R}P^1$ is diffeomorphic to S^1.

Example 1.49. The set of all $m \times n$ real matrices $M_{m \times n}(\mathbb{R})$ is an mn-manifold. We only need one chart since it is clear that $M_{m \times n}(\mathbb{R})$ is in one-to-one correspondence with $\mathbb{R}^{mn}$ by the map $[a_{ij}] \mapsto (a_{11}, a_{12}, \ldots, a_{mn})$. Also, the set of all nonsingular matrices $\mathrm{GL}(n, \mathbb{R})$ is an open submanifold of $M_{n \times n} \cong \mathbb{R}^{n^2}$.

If we have two manifolds M_1 and M_2 of dimensions n_1 and n_2 respectively, we can form the topological Cartesian product $M_1 \times M_2$. We may give $M_1 \times M_2$ a differentiable structure in the following way: Let $\mathcal{A}_{M_1}$ and $\mathcal{A}_{M_2}$ be atlases for M_1 and M_2. Take as charts on $M_1 \times M_2$ the maps of the form
$$\mathbf{x} \times \mathbf{y} : U_\alpha \times V_\gamma \to \mathbb{R}^{n_1} \times \mathbb{R}^{n_2},$$
where $(U, \mathbf{x})$ is a chart from $\mathcal{A}_{M_1}$ and $(V, \mathbf{y})$ a chart from $\mathcal{A}_{M_2}$. This gives $M_1 \times M_2$ an atlas called the product atlas, which induces a maximal atlas and hence a differentiable structure. With this product differentiable structure, $M_1 \times M_2$ is called a **product manifold**. The product of several manifolds is also possible by an obvious iteration. The topology induced by the product atlas is the product topology, and so the underlying topological manifold is the product topological manifold discussed earlier.

Example 1.50. The circle S^1 is a 1-manifold, and hence so is the product $T^2 = S^1 \times S^1$, which is a **torus**. The set of all configurations of a double pendulum constrained to a plane and where the arms are free to swing past each other can be taken to be modeled by $T^2 = S^1 \times S^1$. See Figure 1.3.

1.3. Charts, Atlases and Smooth Structures

Example 1.51. For any smooth manifold M we can construct the "cylinder" $M \times I$, where $I = (a, b)$ is some open interval in $\mathbb{R}$.

We now discuss an interesting class of examples. Let $G(n, k)$ denote the set of k-dimensional subspaces of $\mathbb{R}^n$. We will exhibit a natural differentiable structure on this set. The idea is the following: An alternative way of defining the points of projective space is as equivalence classes of n-tuples $(v^1, \ldots, v^n) \in \mathbb{R}^n \backslash \{0\}$, where $(v^1, \ldots, v^n) \sim (\lambda v^1, \ldots, \lambda v^n)$ for any nonzero λ. This is clearly just a way of specifying a line through the origin. Generalizing, we shall represent a k-plane as an $n \times k$ matrix whose column vectors span the k-plane. Thus we are putting an equivalence relation on the set of $n \times k$ matrices where $A \sim Ag$ for any nonsingular $k \times k$ matrix g. Let $\mathbb{M}_{n \times k}^{\text{full}}$ be the set of $n \times k$ matrices with rank $k < n$ (full rank). Two matrices from $\mathbb{M}_{n \times k}^{\text{full}}$ are equivalent exactly if their columns span the same k-dimensional subspace. Thus the set $G(k, n) := \mathbb{M}_{n \times k}^{\text{full}} / \sim$ of equivalence classes is in one-to-one correspondence with the set of k-dimensional subspaces of $\mathbb{R}^n$.

Let U be the set of all $[A] \in G(k, n)$ such that any representative A has its first k rows linearly independent. This property is independent of the representative A of the equivalence class $[A]$, and so U is a well-defined set. Now every element $[A] \in U \subset G(k, n)$ is an equivalence class that has a unique member A_0 of the form

$$\begin{bmatrix} I_{k \times k} \\ Z \end{bmatrix}$$

which is obtained by Gaussian column reduction. Thus we have a map on U defined by $\Psi : [A] \mapsto Z \in \mathbb{M}_{(n-k) \times k} \cong \mathbb{R}^{k(n-k)}$. We wish to cover $G(k, n)$ with sets similar to U and define similar maps. Consider the set $U_{i_1 \ldots i_k}$ of all $[A] \in G(k, n)$ such that any representative A has the property that the k rows indexed by $i_1, \ldots, i_k$ are linearly independent. The permutation that puts the k rows indexed by $i_1, \ldots, i_k$ into the positions $1, \ldots, k$ without changing the relative order of the remaining rows induces an obvious bijection $\sigma_{i_1 \ldots i_k}$ from $U_{i_1 \ldots i_k}$ onto $U = U_{1 \ldots k}$. We now have maps $\Psi_{i_1 \ldots i_k} : U_{i_1 \ldots i_k} \to \mathbb{M}_{(n-k) \times k} \cong \mathbb{R}^{k(n-k)}$ given by composition $\Psi_{i_1 \ldots i_k} := \Psi \circ \sigma_{i_1 \ldots i_k}$. These maps form an atlas $\{(U_{i_1 \ldots i_k}, \Psi_{i_1 \ldots i_k})\}$ for $G(k, n)$ that gives it the structure of a smooth manifold called the **Grassmann manifold** of real k-planes in $\mathbb{R}^n$. The topology induced by the atlas is the same as the quotient topology, and one can check that this topology is Hausdorff and paracompact.

We have defined C^r manifold for $0 \le r \le \infty$ for r an integer or ∞. We can also define C^ω manifolds (analytic manifolds) by requiring that the charts are related by analytic maps. This means that the overlaps maps have component functions that may be expressed as convergent powers series in a neighborhood of any point in their domains. For convenience, we agree to take $\infty < \omega$.

1.4. Smooth Maps and Diffeomorphisms

Definition 1.52. Let M and N be smooth manifolds with corresponding maximal atlases $\mathcal{A}_M$ and $\mathcal{A}_N$. We say that a map $f : M \to N$ is of **class** C^r (or r-**times continuously differentiable**) **at** $p \in M$ if there exists a chart (V, y) from $\mathcal{A}_N$ with $f(p) \in V$, and a chart (U, x) from $\mathcal{A}_M$ with $p \in U$, such that $f(U) \subset V$ and such that $\mathsf{y} \circ f \circ \mathsf{x}^{-1}$ is of class C^r. If f is of class C^r at every point $p \in M$, then we say that f is of **class** C^r (or that f is a C^r map). Maps of class C^∞ are called **smooth maps**.

Exercise 1.53. Show that a C^r map is continuous. [Hint: Consider compositions $\mathsf{y}^{-1} \circ \left(\mathsf{y} \circ f \circ \mathsf{x}^{-1}\right) \circ \mathsf{x}$.]

Exercise 1.54. Show that a composition of C^r maps is a C^r map.

The family of C^r manifolds together with the family of smooth maps determines a category called the C^r category (see Appendix A). The C^∞ category is called the **smooth category**. Smooth structures are often tailor made so that certain maps are smooth. For example, given smooth manifolds M_1 and M_2, the smooth structure on a product manifold $M_1 \times M_2$ is designed to make the projection maps onto M_1 and M_2 smooth.

Even when dealing with smooth manifolds, we may still be interested in maps which are only of class C^r for some $r < \infty$. This is especially so when one wants to do analysis on smooth manifolds. In fact, one could define what it means for a map to be Lebesgue measurable in a similar way. It is obvious from the way we have formulated the definition that the property of being of class C^r is a local property.

Let $f : M \to N$ be a map and suppose that (U, x) and (V, y) are admissible charts for M and N respectively. If $f^{-1}(V) \cap U$ is not empty, then we have a composition

$$\mathsf{y} \circ f \circ \mathsf{x}^{-1} : \mathsf{x}\left(f^{-1}(V) \cap U\right) \to \mathsf{y}(V).$$

Maps of this form are called the local **representative maps** for f. Notice that if f is continuous, then $f^{-1}(V) \cap U$ is open. Definition 1.52 does not start out with the assumption that f is continuous, but is constructed carefully so as to *imply* that a function that is of class C^r (at a point) according to the definition is automatically continuous (at the point). But if f is known to be continuous, then we may check C^r differentiability using representative maps with respect to atlases that are not necessarily maximal:

Proposition 1.55. *Let $\{(U_\alpha, \mathsf{x}_\alpha)\}_{\alpha \in A}$ and $\{(V_\beta, \mathsf{y}_\beta)\}_{\beta \in B}$ be (not necessarily maximal) C^r atlases for M and N respectively. A continuous map*

1.4. Smooth Maps and Diffeomorphisms

$f : M \to N$ is of class C^r if for each α and β, the representative map $\mathbf{y}_\beta \circ f \circ \mathbf{x}_\alpha^{-1}$ is C^r on its domain $\mathbf{x}_\alpha \left(f^{-1}(V_\beta) \cap U_\alpha \right)$.

Proof. Suppose that a continuous f is given and that all the representative maps $\mathbf{y}_\beta \circ f \circ \mathbf{x}_\alpha^{-1}$ are C^r. Let $p \in M$ and choose $(U_\alpha, \mathbf{x}_\alpha)$ and $(V_\beta, \mathbf{y}_\beta)$ with $p \in U_\alpha$ and $f(p) \in V_\beta$. Letting $U := f^{-1}(V_\beta) \cap U_\alpha$, we have a chart $(U, \mathbf{x}_\alpha|_U)$ with $p \in U$ and $f(U) \subset V_\beta$ such that $\mathbf{y} \circ f \circ \mathbf{x}|_\alpha^{-1}$ is C^r. Thus f is C^r at p by definition. Since p was arbitrary, we see that f is of class C^r. $\square$

If f is continuous, then the condition that f be C^r at $p \in M$ for $r > 0$ can be seen to be equivalent to the condition that for some (and hence every) choice of charts $(U, \mathbf{x})$ from $\mathcal{A}_M$ and $(V, \mathbf{y})$ from $\mathcal{A}_N$ such that $p \in U$ and $f(p) \in V$, the map

$$\mathbf{y} \circ f \circ \mathbf{x}^{-1} : \mathbf{x}(f^{-1}(V) \cap U) \to \mathbf{y}(V)$$

is C^r. Note the use of the phrase "and hence every" above. The point is that if we choose another pair of charts $(\mathbf{x}', U')$ and $(\mathbf{y}', V')$ with $p \in U'$ and $f(p) \in V'$, then $\mathbf{y}' \circ f \circ \mathbf{x}'^{-1}$ must be C^r on some open neighborhood of $\mathbf{x}'(p)$ if and only if $\mathbf{y} \circ f \circ \mathbf{x}^{-1}$ is C^r on some neighborhood[4] of $\mathbf{x}(p)$. This is true because the overlap maps $\mathbf{x}' \circ \mathbf{x}^{-1}$ and $\mathbf{y}' \circ \mathbf{y}^{-1}$ are C^r diffeomorphisms (the chain rule is at work here of course). Without worrying about domains, the point is that

$$\begin{aligned}\mathbf{y}' &\circ f \circ \mathbf{x}'^{-1} \\&= \mathbf{y}' \circ \left(\mathbf{y}^{-1} \circ \mathbf{y}\right) \circ f \circ \left(\mathbf{x}^{-1} \circ \mathbf{x}\right) \circ \mathbf{x}'^{-1} \\&= \left(\mathbf{y}' \circ \mathbf{y}^{-1}\right) \circ \left(\mathbf{y} \circ f \circ \mathbf{x}^{-1}\right) \circ \left(\mathbf{x}' \circ \mathbf{x}^{-1}\right)^{-1}.\end{aligned}$$

Now the reader should be able to see quite clearly why we required overlap maps to be diffeomorphisms.

A representative map $\bar{f} = \mathbf{y} \circ f \circ \mathbf{x}^{-1}$ is defined on an open subset of $\mathbb{R}^n$ where $n = \dim(M)$. If $\dim(N) = k$, then $\bar{f} = (\bar{f}^1, \ldots, \bar{f}^k)$ and each $\bar{f}^i$ is a function of n variables. If we denote generic points in $\mathbb{R}^n$ as $(u^1, \ldots, u^n)$, and those in $\mathbb{R}^k$ as $(v^1, \ldots, v^k)$, then we may write $v^i = \bar{f}^i(u^1, \ldots, u^n)$, $1 \le i \le k$. It is also common and sometimes psychologically helpful to simply write $y^i = \bar{f}^i(x^1, \ldots, x^n)$. The bars over the f's are also sometimes dropped. Another common way to indicate $\mathbf{y} \circ f \circ \mathbf{x}^{-1}$ is with the notation f_{VU} which is very suggestive and tempting, but it has a slight logical defect since there may be many charts with domain U and many charts with domain V.

Exercise 1.56. Consider the map $\pi : S^2 \to \mathbb{R}P^2$ given by taking the point (x, y, z) to the line through this point. Using an atlas on each of these manifolds such as the atlases introduced previously, show that π is smooth.

---

[4] Recall that our convention is that a neighborhood is assumed to be open unless indicated.

(At least check one of the representative maps with respect to a chart on S^2 and a chart on $\mathbb{R}P^2$.)

As a special case of the above, we note that a function $f: M \to \mathbb{R}$ (resp. $\mathbb{C}$) is C^r differentiable at $p \in M$ if and only if it is continuous and

$$f \circ \mathbf{x}^{-1} : \mathbf{x}(U) \to \mathbb{R} \quad (\text{resp. } \mathbb{C})$$

is C^r-differentiable for some admissible chart $(U, \mathbf{x})$ with $p \in U$. And, f is of class C^r if it is of class C^r at every p. The set of all C^r maps $M \to N$ is denoted $C^r(M, N)$ and $C^r(M, \mathbb{R})$ is abbreviated to $C^r(M)$. Both $C^r(M, \mathbb{R})$ and $C^r(M, \mathbb{C})$ are rings and also algebras over the respective fields $\mathbb{R}$ and $\mathbb{C}$ (Definition in Appendix D). The addition, scaling, and multiplication are defined pointwise so that $(f+g)(p) := f(p) + g(p)$, etc.

Definition 1.57. Let $(U, \mathbf{x})$ be a chart on an n-manifold M with $p \in U$. We write $\mathbf{x} = (x^1, \ldots, x^n)$ as usual. For $f \in C^1(M)$, define a function $\frac{\partial f}{\partial x^i}$ on U by

$$\frac{\partial f}{\partial x^i}(p) := \lim_{h \to 0} \left[\frac{f \circ \mathbf{x}^{-1}(a^1, \ldots, a^i + h, \ldots, a^n) - f \circ \mathbf{x}^{-1}(a^1, \ldots, a^n)}{h} \right],$$

where $\mathbf{x}(p) = (a^1, \ldots, a^n)$. In other words,

$$\frac{\partial f}{\partial x^i}(p) := \partial_i \left(f \circ \mathbf{x}^{-1} \right) (\mathbf{x}(p)) = \frac{\partial \left(f \circ \mathbf{x}^{-1} \right)}{\partial u^i}(\mathbf{x}(p)),$$

where $(u^1, \ldots, u^n)$ denotes the standard coordinates on $\mathbb{R}^n$.

Recall that D_i is the notation for i-th partial derivative with respect to the decomposition $\mathbb{R}^n = \mathbb{R} \times \cdots \times \mathbb{R}$. Thus if $g: U \subset \mathbb{R} \to \mathbb{R}$ is differentiable at $a \in \mathbb{R}$, then $D_i g(a) : \mathbb{R} \to \mathbb{R}$ is a linear map but is often identified with the single entry $\partial_i g(a)$ of the 1×1 matrix that represents it with respect to the standard basis on $\mathbb{R}$. Thus one sometimes sees the definition above written as

$$\frac{\partial f}{\partial x^i}(p) := D_i \left(f \circ \mathbf{x}^{-1} \right) (\mathbf{x}(p)).$$

If f is a C^r function, then $\partial f / \partial x^i$ is clearly C^{r-1}. Notice also that f really only needs to be defined in a neighborhood of p and differentiable at p for the expression $\frac{\partial f}{\partial x^i}(p)$ to make sense. This definition makes precise the notation that is often encountered in calculus courses. For example, if T is the "temperature" on a sphere S^2, then T takes as arguments points p on S^2. On the other hand, using spherical coordinates, we often consider $\partial T / \partial \phi$ and $\partial T / \partial \theta$ as being defined *on* S^2 rather than on some open set in a "ϕ, θ-space".

1.4. Smooth Maps and Diffeomorphisms

Finally, notice that if f and g are C^1 and defined at least on the domain of the chart $(U, \mathbf{x})$, then we easily obtain that on U

$$\frac{\partial (af + bg)}{\partial x^i} = a\frac{\partial f}{\partial x^i} + b\frac{\partial g}{\partial x^i} \text{ for any } a, b \in \mathbb{R},$$

and

$$\frac{\partial (fg)}{\partial x^i} = f\frac{\partial g}{\partial x^i} + g\frac{\partial f}{\partial x^i} \text{ (the product rule)}.$$

Let $(U, \mathbf{x})$ and $(V, \mathbf{y})$ charts on an n-manifold with $p \in U \cap V$. Then it is easy to check using the usual chain rule and the definitions above that for any smooth function f defined at least on a neighborhood of p we have the following version of the chain rule:

$$\frac{\partial f}{\partial y^i}(p) = \sum_{j=1}^{n} \frac{\partial f}{\partial x^j}(p) \frac{\partial x^j}{\partial y^i}(p).$$

A map f which is defined only on some proper open subset of a manifold is said to be C^r if it is C^r as a map of the corresponding open submanifold, but this is again just to say that it is C^r at each point in the open set. We shall often need to consider maps that are defined on subsets $S \subset M$ that are not necessarily open.

Definition 1.58. Let S be an arbitrary subset of a smooth manifold M. Let $f : S \to N$ be a continuous map where N is a smooth manifold. The map f is said to be C^r if for every $s \in S$ there is an open set $O \subset M$ containing s and a map $\widetilde{f}$ that is C^r on O and such that $\widetilde{f}\big|_{S \cap O} = f$.

In a later exercise we ask the reader to show that a function f with domain S is smooth if and only if it has a smooth extension to some open set containing all of S. In particular, a curve defined on a closed interval $[a, b]$ is smooth if it has a smooth extension to an open interval containing $[a, b]$.

We already have the notion of a diffeomorphism between open sets of some Euclidean space $\mathbb{R}^n$. We are now in a position to extend this notion to the realm of smooth manifolds.

Definition 1.59. Let M and N be smooth (or C^r) manifolds. A homeomorphism $f : M \to N$ such that f and f^{-1} are C^r differentiable with $r \geq 1$ is called a C^r **diffeomorphism**. In the case $r = \infty$, we shorten C^∞ diffeomorphism to just **diffeomorphism**. The set of all C^r diffeomorphisms of a manifold M onto itself is a group under the operation of composition. This group is denoted $\text{Diff}^r(M)$. In the case $r = \infty$, we simply write $\text{Diff}(M)$ and refer to it as the diffeomorphism group of M.

We will use the convention that $\text{Diff}^0(M)$ denotes the group of homeomorphisms of M onto itself. Also, it should be pointed out that if we refer to a map between open subsets of manifolds as being a C^r diffeomorphism, we mean that the map is a C^r diffeomorphism of the corresponding open submanifolds.

Example 1.60. The map $r_\theta : S^2 \to S^2$ given by
$$r_\theta(x,y,z) = (x\cos\theta - y\sin\theta, x\sin\theta + y\cos\theta, z)$$
for $x^2 + y^2 + z^2 = 1$ is a diffeomorphism.

Exercise 1.61. Let $0 < \theta < 2\pi$. Consider the map $f : S^2 \to S^2$ given by $f_\theta(x,y,z) = (x\cos((1-z^2)\theta) - y\sin((1-z^2)\theta), x\sin((1-z^2)\theta) + y\cos((1-z^2)\theta), z)$. Is this map a diffeomorphism? Try to picture this map.

Definition 1.62. C^r manifolds M and N will be called (C^r) **diffeomorphic** and then said to be in the same **diffeomorphism class** if and only if there is a C^r diffeomorphism $f : M \to N$.

Exercise 1.63. Let M and N be smooth manifolds with respective maximal atlases $\mathcal{A}_M$ and $\mathcal{A}_N$. Show that a bijection $f : M \to N$ is a diffeomorphism if and only if the following condition holds:
$$(U, \mathbf{y}) \in \mathcal{A}_N \text{ if and only if } (f^{-1}(U), \mathbf{y} \circ f) \in \mathcal{A}_M.$$

Exercise 1.64. Show that if M is a C^r manifold and $\phi : M \to X$ is any bijection, then there is a unique C^r structure on X such that ϕ is a diffeomorphism. This process is called a **transfer of structure**. If X is a topological space and ϕ is a homeomorphism, then the topology induced by the transferred structure is the original topology.

For another example, consider the famous Cantor set $C \subset [0,1] \subset \mathbb{R}$. Consider $\mathbb{R}$ as a coordinate axis subspace of $\mathbb{R}^2$ set $M_C := \mathbb{R}^2 \setminus C$. It can be shown that M_C is diffeomorphic to a surface suggested by Figure 1.4. Once again we see that open sets in a Euclidean space can have interesting differential topology.

In the definition of diffeomorphism, we have suppressed explicit reference to the maximal atlases, but note that whether or not a map is differentiable (C^r or smooth) essentially involves the choice of differentiable structures on the manifolds. Recall that we can put more than one differentiable structure on $\mathbb{R}$ by using the function x^3 as a chart. This generalizes in the obvious way: The map $\varepsilon : (x^1, x^2, \ldots, x^n) \mapsto ((x^1)^3, x^2, \ldots, x^n)$ is a chart for $\mathbb{R}^n$, but is not C^∞-related with the standard (identity) chart. It is globally defined and so provides an atlas that induces the usual topology again, but the resulting maximal atlas is different! Thus we seem to have two smooth manifolds $(\mathbb{R}^n, \mathcal{A}_1)$ and $(\mathbb{R}^n, \mathcal{A}_2)$ both with the *same* underlying topological

1.4. Smooth Maps and Diffeomorphisms

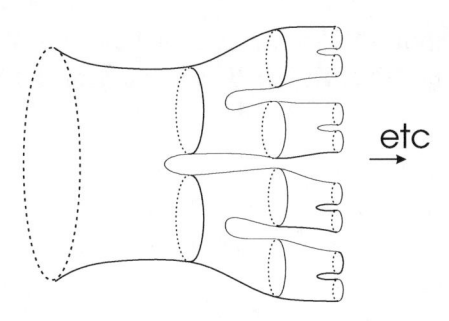

Figure 1.4. Interesting surface

space. Indeed, this is true. Technically, they *are* different. But they are equivalent and therefore the same in another sense. Namely, they are diffeomorphic via the map ε. So it may be that the same underlying topological space M carries two different differentiable structures, and so we really have two differentiable manifolds with the same underlying set. It remains to ask whether they are nevertheless diffeomorphic. It is an interesting question whether a given topological manifold can carry differentiable structures that are not diffeomorphic. It has been shown that there are 28 pairwise nondiffeomorphic smooth structures on the topological space S^7 and more than 16 million on S^{31}. Each $\mathbb{R}^k$ for $k \neq 4$ has only one diffeomorphism class compatible with the usual topology. On the other hand, it is a deep result that there exist infinitely many truly different (nondiffeomorphic) differentiable structures on $\mathbb{R}^4$. The existence of exotic differentiable structures on $\mathbb{R}^4$ follows from the results of [**Donaldson**] and [**Freedman**]. The reader ought to be wondering what is so special about dimension four. Note that when we mention $\mathbb{R}^4$, S^7, S^{31}, etc. as smooth manifolds, we shall normally assume the usual smooth structures unless otherwise indicated.

Definition 1.65. Let N and M be smooth manifolds of the same dimension. A map $f : M \to N$ is called a local diffeomorphism if and only if every point $p \in M$ is contained in an open subset $U \subset M$ such that $f|_U : U \to f(U)$ is a diffeomorphism onto an open subset of N. For C^r manifolds, a C^r local diffeomorphism is defined similarly.

Example 1.66. The map $\pi : S^2 \to \mathbb{R}P^2$ given by taking the point (x, y, z) to the line through this point and the origin is a local diffeomorphism, but is not a diffeomorphism since it is 2-1 rather than 1-1.

Example 1.67. The map $(x, y) \mapsto (x/z(x, y), y/z(x, y))$, where

$$z(x, y) = \sqrt{1 - x^2 - y^2},$$

is a diffeomorphism from the open disk $B(0,1) = \{(x,y) : x^2 + y^2 < 1\}$ onto the whole plane. Thus $B(0,1)$ and $\mathbb{R}^2$ are diffeomorphic and in this sense are the "same" differentiable manifold.

Sometimes it is only important how maps behave *near a certain point*. Let M and N be smooth manifolds and consider the set $S(p, M, N)$ of all smooth maps into N which are defined on some open neighborhood of the fixed point $p \in M$. Thus,

$$S(p, M, N) := \bigcup_{U \in \mathcal{N}_p} C^\infty(U, N),$$

where $\mathcal{N}_p$ denotes the set of all open neighborhoods of $p \in M$. On this set we define the equivalence relation where f and g are equivalent at p if and only if they agree on a neighborhood of p. The equivalence class of f is denoted $[f]$, or by $[f]_p$ if the point in question needs to be made clear. The set of equivalence classes $S(p, M, N)/\sim$ is denoted $C_p^\infty(M, N)$.

Definition 1.68. Elements of $C_p^\infty(M, N)$ are called **germs**, and if f and g are in the same equivalence class, we write $f \sim_p g$ and we say that f and g have the same **germ at** p.

The value of a germ at p is well-defined by $[f](p) = f(p)$. Taking $N = \mathbb{R}$ we see that $C_p^\infty(M, \mathbb{R})$ is a commutative $\mathbb{R}$-algebra if we make the definitions

$$a[f] + b[g] := [af + bg] \text{ for } a, b \in \mathbb{R},$$
$$[f][g] := [fg].$$

The $\mathbb{C}$-algebra of complex-valued germs $C_p^\infty(M, \mathbb{C})$ is defined similarly.

1.5. Cut-off Functions and Partitions of Unity

There is a special and extremely useful kind of function called a bump function or **cut-off function**, which we now take the opportunity to introduce. Recall that given a topological space X, the **support**, $\operatorname{supp}(f)$, of a function $f : X \to \mathbb{R}$ is the closure of the subset on which f takes nonzero values. The same definition applies for vector space-valued functions $f : X \to V$. It is a standard fact that there exist smooth functions defined on $\mathbb{R}$ that have compact support. For example, we have the smooth function $\Psi : \mathbb{R} \to \mathbb{R}$ defined by

$$\Psi(t) = \begin{cases} e^{-1/(1-x^2)} & \text{for } |t| < 1, \\ 0 & \text{otherwise.} \end{cases}$$

Lemma 1.69 (Existence of cut-off functions). *Let M be a smooth manifold. Let K be a compact subset of M and O an open set containing K. There exists a smooth function β on M that is identically equal to 1 on K, takes values in the interval $[0, 1]$, and has compact support in O.*

1.5. Cut-off Functions and Partitions of Unity

Proof. Special case 1: Assume that $M = \mathbb{R}^n$ and that $O = B(0, R)$ and $K = \overline{B}(0, r)$ for $0 < r < R$. In this case we may take

$$\phi(x) = \frac{\int_{|x|}^{R} g(t)\, dt}{\int_{r}^{R} g(t)\, dt},$$

where

$$g(t) = \begin{cases} e^{-(t-r)^{-1}} e^{(t-R)^{-1}} & \text{if } r < t < R, \\ 0 & \text{otherwise}. \end{cases}$$

It is an exercise in calculus to show that g is a smooth function and thus that ϕ is smooth. Clearly, a composition with a translation gives the result for a ball centered at an arbitrary point.

Special case 2: Assume again that $M = \mathbb{R}^n$. Let $K \subset O$ be as in the hypotheses. For each point $p \in K$ let U_p be an open ball centered at p and contained in O. Let K_p be the closed ball centered at p of half the radius of U_p. The interiors of the K_p's form an open cover for K and so by compactness we can reduce to a finite subcover. Thus we have a finite family $\{K_i\}$ of closed balls of various radii such that $K \subset \bigcup K_i$, and with corresponding concentric open balls $U_i \subset O$. For each U_i, let ϕ_i be the corresponding function provided in the proof of Special case 1 so that ϕ_i has support in U_i and is identically 1 on K_i. Examination of the following function will convince the reader that it is well-defined and provides the needed cut-off function:

$$\beta(x) = 1 - \prod_i (1 - \phi_i(x)).$$

General case: From the second special case above it is clear that we have the result if K is contained in the domain U of a chart $(U, \mathbf{x})$. If K is not contained in such a chart, then we may take a finite number of charts $(U_1, \mathbf{x}_1), \ldots, (U_k, \mathbf{x}_k)$ and compact sets $K_1, \ldots, K_k$ with $K \subset \bigcup_{i=1}^{k} K_i$, $K_i \subset U_i$, and $\bigcup U_i \subset O$. Now let ϕ_i be identically 1 on K_i and identically 0 on $U_i^c = M \setminus U_i$. Then the function β we are looking for is given by

$$\beta = 1 - \prod_{i=1}^{k} (1 - \phi_i). \qquad \square$$

Let $[f] \in C_p^\infty(M, \mathbb{R})$ (or $\in C_p^\infty(M, \mathbb{C})$) and let f be a representative of the equivalence class $[f]$. We can find an open set U containing p such that $\overline{U}$ is compact and contained in the domain of f. If β is a cut-off function that is identically equal to 1 on $\overline{U}$, and has support inside the domain of f, then βf is smooth and it can be extended to a globally defined smooth function that is zero outside of the domain of f. Denote this extended function by $(\beta f)_{\text{ext}}$. Then $(\beta f)_{\text{ext}} \in [f]$ (usually, the extended function is just written as βf). Thus every element of $C_p^\infty(M,\mathbb{R})$ has a representative in $C^\infty(M,\mathbb{R})$. In short, each germ has a global representative. (The word "global" means defined on, or referring to, the whole manifold.)

A partition of unity is a technical tool that can help one piece together locally defined smooth objects with some desirable properties to obtain a globally defined object that also has the desired properties. For example, we will use this tool to show that on any (paracompact) smooth manifold there exists a Riemannian metric tensor. As we shall see, the metric tensor is the basic object whose existence allows the introduction of notions such as length and volume.

Definition 1.70. A **partition of unity** on a smooth manifold M is a collection $\{\varphi_\alpha\}_{\alpha \in A}$ of smooth functions on M such that

(i) $0 \leq \varphi_\alpha \leq 1$ for all α.

(ii) The collection of supports $\{\mathrm{supp}(\varphi_\alpha)\}_{\alpha \in A}$ is locally finite; that is, each point p of M has a neighborhood W_p such that $W_p \cap \mathrm{supp}(\varphi_\alpha) = \emptyset$ for all but a finite number of $\alpha \in A$.

(iii) $\sum_{\alpha \in A} \varphi_\alpha(p) = 1$ for all $p \in M$ (this sum has only finitely many nonzero terms by (ii)).

If $\mathcal{O} = \{O_\alpha\}_{\alpha \in A}$ is an open cover of M and $\mathrm{supp}(\varphi_\alpha) \subset O_\alpha$ for each $\alpha \in A$, then we say that $\{\varphi_\alpha\}_{\alpha \in A}$ is a partition of unity **subordinate to** $\mathcal{O} = \{O_\alpha\}_{\alpha \in A}$.

Remark 1.71. Let $\mathcal{U} = \{U_\alpha\}_{\alpha \in A}$ be a cover of M and suppose that $\mathcal{W} = \{W_\beta\}_{\beta \in B}$ is a refinement of $\mathcal{U}$. If $\{\psi_\beta\}_{\beta \in B}$ is a partition of unity subordinate to $\mathcal{W}$, then we may obtain a partition of unity $\{\varphi_\alpha\}$ subordinate to $\mathcal{U}$. Indeed, if $f : B \to A$ is such that $W_\beta \subset U_{f(\beta)}$ for every $\beta \in B$, then we may let $\varphi_\alpha := \sum_{\beta \in f^{-1}(\alpha)} \psi_\beta$.

Our definition of a smooth manifold M includes the requirement that M be paracompact (and Hausdorff). Paracompact Hausdorff spaces are normal spaces, but the following theorem would be true for a normal locally Euclidean space with smooth structure even without the assumption of paracompactness. The reason is that we *explicitly assume* the local finiteness of the cover. For this reason we put the word "normal" in parentheses as a pedagogical device.

Theorem 1.72. *Let M be a (normal) smooth manifold and $\{U_\alpha\}_{\alpha \in A}$ be a locally finite cover of M. If each U_α has compact closure, then there is a partition of unity $\{\varphi_\alpha\}_{\alpha \in A}$ subordinate to $\{U_\alpha\}_{\alpha \in A}$.*

Proof. We shall use a well-known result about normal spaces sometimes called the "shrinking lemma". Namely, if $\{U_\alpha\}_{\alpha \in A}$ is a locally finite (and hence "point finite") cover of a normal space M, then there exists another cover $\{V_\alpha\}_{\alpha \in A}$ of M such that $\overline{V}_\alpha \subset U_\alpha$. This is Theorem B.4 proved in Appendix B.

We do this to our cover and then notice that since each U_α has compact closure, each $\overline{V}_\alpha$ is compact. We apply Lemma 1.69 to obtain nonnegative smooth functions ψ_α such that $\operatorname{supp} \psi_\alpha \subset U_\alpha$ and $\psi_\alpha|_{\overline{V}_\alpha} \equiv 1$. Let $\psi := \sum_{\alpha \in A} \psi_\alpha$ and notice that for each $p \in M$, the sum $\sum_{\alpha \in A} \psi_\alpha(p)$ is a finite sum and $\psi(p) > 0$. Let $\varphi_\alpha := \psi_\alpha/\psi$. It is now easy to check that $\{\varphi_\alpha\}_{\alpha \in A}$ is the desired partition of unity. $\square$

If we use the paracompactness assumption, then we can show that there exists a partition of unity that is subordinate to *any* given cover.

Theorem 1.73. *Let M be a (paracompact) smooth manifold and $\{U_\alpha\}_{\alpha \in A}$ a cover of M. Then there is a partition of unity $\{\varphi_\alpha\}_{\alpha \in A}$ subordinate to $\{U_\alpha\}_{\alpha \in A}$.*

Proof. By Remark 1.71 and the fact that M is locally compact we may assume without loss of generality that each U_α has compact closure. Then since M is paracompact, we may find a locally finite refinement of $\{U_\alpha\}_{\alpha \in A}$ which we denote by $\{V_i\}_{i \in I}$. Now use the previous theorem to get a partition of unity subordinate to $\{V_i\}_{i \in I}$. Finally use remark 1.71 one more time to get a partition of unity subordinate to $\{U_\alpha\}_{\alpha \in A}$. $\square$

Exercise 1.74. Show that if a function is smooth on an arbitrary set $S \subset M$ as defined earlier, then it has a smooth extension to an open set that contains S.

Now that we have established the existence of partitions of unity we may show that the analogue of Lemma 1.69 works with K closed but not necessarily compact:

Exercise 1.75. Let M be a smooth manifold. Let K be a closed subset of M and O an open set containing K. Show that there exists a smooth function β on M that is identically equal to 1 on K, takes values in the interval $[0, 1]$, and has compact support in O.

1.6. Coverings and Discrete Groups

1.6.1. Covering spaces and the fundamental group.
In this section, and later when we study fiber bundles, many of the results are interesting and true in either the purely topological category or in the smooth category. Let us agree that a C^0 manifold is simply a topological manifold. Thus all

relevant maps in this section are to be C^r, where if $r = 0$ we just mean continuous and then only require that the spaces be sufficiently nice topological spaces. Also, "C^0 diffeomorphism" just means homeomorphism.

In the definition of path connectedness and path component given before, we used *continuous* paths, but it is not hard to show that if two points on a smooth manifold can be connected by a continuous path, then they can be connected by a smooth path. Thus the notion of path component remains unchanged by the use of smooth paths.

Definition 1.76. Let $f_0 : X \to Y$ and $f_1 : X \to Y$ be C^r maps. A C^r **homotopy** from f_0 to f_1 is a C^r map $H : X \times [0,1] \to Y$ such that

$$H(x,0) = f_0(x) \quad \text{and}$$
$$H(x,1) = f_1(x)$$

for all x. If there exists such a C^r homotopy, we then say that f_0 is C^r **homotopic** to f_1 and write $f_0 \stackrel{C^r}{\simeq} f_1$. If $A \subset X$ is a closed subset and if $H(a,s) = f_0(a) = f_1(a)$ for all $a \in A$ and all $s \in [0,1]$, then we say that f_0 is C^r homotopic to f_1 **relative** to A and we write $f_0 \stackrel{C^r}{\simeq} f_1$ (rel A). The map H is called a C^r **homotopy**.

In the above definition, the condition that $H : X \times [0,1] \to Y$ be C^r for $r > 0$ can be understood by considering $X \times [0,1]$ as a subset of $X \times \mathbb{R}$. Homotopy is obviously an equivalence relation.

Exercise 1.77. Show that $f_0 : X \to Y$ and $f_1 : X \to Y$ are C^r homotopic (rel A) if and only if there exists a C^r map $H : X \times \mathbb{R} \to Y$ such that $H(x,s) = f_0(x)$ for all x and $s \leq 0$, $H(x,s) = f_1(x)$ for all x and $s \geq 1$ and $H(a,s) = f_0(a) = f_1(a)$ for all $a \in A$ and all s.

At first it may seem that there could be a big difference between C^∞ and C^0 homotopies, but if all the spaces involved are smooth manifolds, then the difference is not big at all. In fact, we have the following theorems which we merely state. Proofs may be found in [**Lee, John**].

Theorem 1.78. *If $f : M \to N$ is a continuous map on smooth manifolds, then f is homotopic to a smooth map $f_0 : M \to N$. If the continuous map $f : M \to N$ is smooth on a closed subset A, then it can be arranged that $f \stackrel{C^\infty}{\simeq} f_0$ (rel A).*

Theorem 1.79. *If $f_0 : M \to N$ and $f_1 : M \to N$ are homotopic smooth maps, then they are smoothly homotopic. If f_0 is homotopic to f_1 relative to a closed subset A, then f_0 is smoothly homotopic to f_1 relative to A.*

Because of these last two theorems, we will usually simply write $f \simeq f_0$ instead of $f \stackrel{C^r}{\simeq} f_0$, the value of r being of little significance in this setting.

1.6. Coverings and Discrete Groups

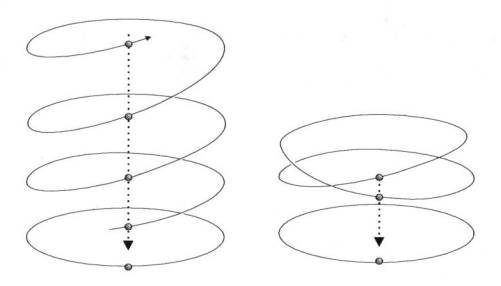

Figure 1.5. Coverings of a circle

Definition 1.80. Let $\widetilde{M}$ and M be C^r manifolds. A surjective C^r map $\wp : \widetilde{M} \to M$ is called a C^r **covering map** if every point $p \in M$ has an open connected neighborhood U such that each connected component $\widetilde{U}_i$ of $\wp^{-1}(U)$ is C^r diffeomorphic to U via the restrictions $\wp|_{\widetilde{U}_i} : \widetilde{U}_i \to U$. In this case, we say that U is **evenly covered** by $\wp$ (or by the sets $\widetilde{U}_i$). The triple $(\widetilde{M}, \wp, M)$ is called a covering space. We also refer to the space $\widetilde{M}$ (somewhat informally) as a **covering space** for M.

Example 1.81. The map $\mathbb{R} \to S^1$ given by $t \mapsto e^{it}$ is a covering. The set of points $\{e^{it} : \theta - \pi < t < \theta + \pi\}$ is an open set evenly covered by the intervals I_n in the real line given by $I_n := (\theta - \pi + 2\pi n, \theta + \pi + 2\pi n)$ for $n \in \mathbb{Z}$.

Exercise 1.82. Explain why the map $(-2\pi, 2\pi) \to S^1$ given by $t \mapsto e^{it}$ is *not* a covering map.

Definition 1.83. A continuous map f is said to be **proper** if $f^{-1}(K)$ is compact whenever K is compact.

Exercise 1.84. Show that a C^r *proper* map between connected smooth manifolds is a smooth covering map if and only if it is a local C^r diffeomorphism.

The set of all C^r covering spaces is the set of objects of a category. A morphism between C^r covering spaces, say $(\widetilde{M}_1, \wp_1, M_1)$ and $(\widetilde{M}_2, \wp_2, M_2)$, is a pair of C^r maps $(\widetilde{f}, f)$ such that the following diagram commutes:

$$\begin{array}{ccc} \widetilde{M}_1 & \xrightarrow{\widetilde{f}} & \widetilde{M}_2 \\ \wp_1 \downarrow & & \downarrow \wp_2 \\ M_1 & \xrightarrow{f} & M_2 \end{array}$$

This means that $f \circ \wp_1 = \wp_2 \circ \widetilde{f}$. Similarly, the coverings of a fixed space M are the objects of a category where the morphisms are maps $\Phi : \widetilde{M}_1 \to \widetilde{M}_2$

that make the following diagram commute:

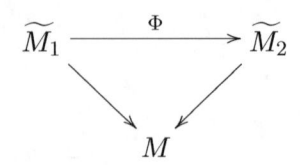

$$\widetilde{M}_1 \xrightarrow{\Phi} \widetilde{M}_2$$
$$\searrow \swarrow$$
$$M$$

meaning that $\wp_1 = \wp_2 \circ \Phi$. Let $(\widetilde{M}, \wp, M)$ be a C^r covering space. The C^r diffeomorphisms Φ that are automorphisms in the above category, that is, diffeomorphisms for which $\wp = \wp \circ \Phi$, are called **deck transformations** or **covering transformations**. The set of deck transformations is a group of C^r diffeomorphisms of $\widetilde{M}$ called the deck transformation group, which we denote by $\mathrm{Deck}(\wp)$ or sometimes by $\mathrm{Deck}(\widetilde{M})$. A deck transformation permutes the elements of each fiber $\wp^{-1}(p)$. In fact, it is not hard to see that if $U \subset M$ is evenly covered, then Φ permutes the connected components of $\wp^{-1}(U)$.

Proposition 1.85. *If $\wp : \widetilde{M} \to M$ is a C^r covering map with M path connected, then the cardinality of $\wp^{-1}(p)$ is independent of p. In the latter case, the cardinality of $\wp^{-1}(p)$ is called the **multiplicity** of the covering.*

Proof. Fix a cardinal number k. Let B_k be the set of all points such that $\wp^{-1}(p)$ has cardinality k. For any fixed $p \in M$, there is a connected neighborhood U_p that is evenly covered, and it is easy to see that the cardinality of $\wp^{-1}(q)$ is the same for all points $q \in U_p$. Using this, it is easy to show that B_k is both open and closed and so, since M is connected, B_k is either empty or all of M. $\square$

Since we are mainly interested in the smooth case, the following theorem is quite useful:

Theorem 1.86. *Let M be a C^r manifold with $r > 0$ and suppose that $\wp : \widetilde{M} \to M$ is a C^0 covering map with $\widetilde{M}$ paracompact. Then there exists a (unique) C^r structure on $\widetilde{M}$ making $\wp$ a C^r covering map.*

Proof. Choose an atlas $\{(U_\alpha, \mathbf{x}_\alpha)\}_{\alpha \in A}$ such that each domain U_α is small enough to be evenly covered by $\wp$. Thus we have that $\wp^{-1}(U_\alpha)$ is a disjoint union of open sets U_α^i with each restriction $\wp|_{U_\alpha^i}$ a homeomorphism. We now construct charts on $\widetilde{M}$ using the maps $\mathbf{x}_\alpha \circ \wp|_{U_\alpha^i}$ defined on the sets U_α^i

1.6. Coverings and Discrete Groups

(which cover $\widetilde{M}$). The overlap maps are smooth since if $U^i_\alpha \cap U^j_\beta \neq \emptyset$ then

$$\left(\mathbf{x}_\alpha \circ \wp|_{U^i_\alpha}\right) \circ \left(\mathbf{x}_\beta \circ \wp|_{U^j_\beta}\right)^{-1}$$
$$= \mathbf{x}_\alpha \circ \wp|_{U^i_\alpha} \circ \left(\wp|_{U^j_\beta}\right)^{-1} \circ \mathbf{x}_\beta^{-1}$$
$$= \mathbf{x}_\alpha \circ \mathbf{x}_\beta^{-1}.$$

We leave it to the reader to show that $\widetilde{M}$ is Hausdorff if M is Hausdorff. □

The following is a special case of Definition 1.76.

Definition 1.87. Let $\alpha : [0,1] \to M$ and $\beta : [0,1] \to M$ be two C^r maps (paths) both starting at $p \in M$ and ending at q. A C^r **fixed endpoint homotopy** from α to β is a family of C^r maps $H_s : [0,1] \to M$ parameterized by $s \in [0,1]$ such that

1) $H : [0,1] \times [0,1] \to M$ defined by $H(t,s) := H_s(t)$ is C^r;
2) $H_0 = \alpha$ and $H_1 = \beta$;
3) $H_s(0) = p$ and $H_s(1) = q$ for all $s \in [0,1]$.

Definition 1.88. If there is a C^r homotopy from α to β, then we say that α is C^r **homotopic** to β and write $\alpha \simeq \beta$ (C^r). If $r = 0$, we speak of paths being continuously homotopic.

Remark 1.89. By Theorems 1.79 and 1.78 above we know that in the case of smooth manifolds, if α and β are smooth paths, then we have that $\alpha \simeq \beta$ (C^0) if and only if $\alpha \simeq \beta$ (C^r) for $r > 0$. Thus we can just say that α is homotopic to β and write $\alpha \simeq \beta$. In case α and β are only continuous, they may be replaced by smooth paths α' and β' with $\alpha' \simeq \alpha$ and $\beta' \simeq \beta$.

It is easily checked that homotopy is an equivalence relation. Let $P(p,q)$ denote the set of all continuous (or smooth) paths from p to q defined on $[0,1]$. Every $\alpha \in P(p,q)$ has a unique inverse (or reverse) path $\alpha^{\leftarrow}$ defined by

$$\alpha^{\leftarrow}(t) := \alpha(1-t).$$

If p_1, p_2 and p_3 are three points in M, then for $\alpha \in P(p_1, p_2)$ and $\beta \in P(p_2, p_3)$ we can "multiply" the paths to get a path $\alpha * \beta \in P(p_1, p_3)$ defined by

$$\alpha * \beta(t) := \begin{cases} \alpha(2t) & \text{for } 0 \leq t < 1/2, \\ \beta(2t-1) & \text{for } 1/2 \leq t \leq 1. \end{cases}$$

Notice that $\alpha * \beta$ is a path that follows along α and then β in that order.[5] An important observation is that if $\alpha_1 \simeq \alpha_2$ and $\beta_1 \simeq \beta_2$, then $\alpha_1 * \beta_1 \simeq \alpha_2 * \beta_2$.

---
[5] In some settings, it is convenient to reverse this convention.

The homotopy between $\alpha_1 * \beta_1$ and $\alpha_2 * \beta_2$ is given in terms of the homotopies $H_\alpha : \alpha_1 \simeq \alpha_2$ and $H_\beta : \beta_1 \simeq \beta_2$ by

$$H(t,s) := \begin{cases} H_\alpha(2t, s) & \text{for } 0 \le t < 1/2, \\ H_\beta(2t-1, s) & \text{for } 1/2 \le t < 1, \end{cases} \text{ and } 0 \le s < 1.$$

Similarly, if $\alpha_1 \simeq \alpha_2$, then $\alpha_1^{\leftarrow} \simeq \alpha_2^{\leftarrow}$. Using this information, we can define a group structure on the set of homotopy equivalence classes of loops, that is, of paths in $P(p,p)$ for some fixed $p \in M$. First of all, we can always form $\alpha * \beta$ for any $\alpha, \beta \in P(p,p)$ since we always start and stop at the same point p. Secondly, we have the following result.

Proposition 1.90. *Let $\pi_1(M,p)$ denote the set of fixed endpoint homotopy classes of paths from $P(p,p)$. For $[\alpha], [\beta] \in \pi_1(M,p)$, define $[\alpha] \cdot [\beta] := [\alpha * \beta]$. This is a well-defined multiplication, and with this multiplication $\pi_1(M,p)$ is a group. The identity element of the group is the homotopy class 1 of the constant map $1_p : t \mapsto p$, the inverse of a class $[\alpha]$ is $[\alpha^{\leftarrow}]$.*

Proof. We have already shown that $[\alpha] \cdot [\beta] := [\alpha * \beta]$ is well-defined. One must also show that

1) For any α, the paths $\alpha \circ \alpha^{\leftarrow}$ and $\alpha^{\leftarrow} \circ \alpha$ are both homotopic to the constant map 1_p.

2) For any $\alpha \in P(p,p)$, we have $1_p * \alpha \simeq \alpha$ and $\alpha * 1_p \simeq \alpha$.

3) For any $\alpha, \beta, \gamma \in P(p,p)$, we have $(\alpha * \beta) * \gamma \simeq \alpha * (\beta * \gamma)$.

Proof of 1): 1_p is homotopic to $\alpha \circ \alpha^{\leftarrow}$ via

$$H(t,s) = \begin{cases} \alpha(2t) & \text{for } 0 \le 2t \le s, \\ \alpha(s) & \text{for } s \le 2t \le 2-s, \\ \alpha^{\leftarrow}(2t-1) & \text{for } 2-s \le 2t \le 2, \end{cases}$$

where $0 \le s \le 1$. Interchanging the roles of α and $\alpha^{\leftarrow}$ we also get that 1_p is homotopic to $\alpha^{\leftarrow} \circ \alpha$.

Proof of 2): Use the homotopy

$$H(t,s) = \begin{cases} \alpha(\frac{2}{1+s}t) & \text{for } 0 \le t \le 1/2 + s/2, \\ p & \text{for } 1/2 + s/2 \le t \le 1. \end{cases}$$

Proof of 3): Use the homotopy

$$H(t,s) = \begin{cases} \alpha(\frac{4}{1+s}t) & \text{for } 0 \le t \le \frac{1+s}{4}, \\ \beta(4(t - \frac{1+s}{4})) & \text{for } \frac{1+s}{4} \le t \le \frac{2+s}{4}, \\ \gamma(\frac{4}{2-s}(t - \frac{2+s}{4})) & \text{for } \frac{2+s}{4} \le t \le 1. \end{cases} \qquad \square$$

The group $\pi_1(M,p)$ is called the **fundamental group** of M at p. If desired, one can take the equivalence classes in $\pi_1(M,p)$ to be represented

1.6. Coverings and Discrete Groups

by smooth maps. If $\gamma : [0,1] \to M$ is a path from p to q, then we have a group isomorphism $\pi_1(M,q) \to \pi_1(M,p)$ given by

$$[\alpha] \mapsto [\gamma * \alpha * \gamma^{\leftarrow}].$$

It is easy to show that this prescription is a well-defined group isomorphism. Thus, for any two points p,q in the same path component of M, the groups $\pi_1(M,p)$ and $\pi_1(M,q)$ are isomorphic. In particular, if M is connected, then the fundamental groups based at different points are all isomorphic. Because of this, if M is connected, we may simply refer to *the* fundamental group of M, which we write as $\pi_1(M)$.

Definition 1.91. A path connected topological space is called **simply connected** if $\pi_1(M) = \{1\}$.

The fundamental group is actually the result of applying a functor (see Appendix A). Consider the category whose objects are pairs (M,p), where M is a C^r manifold and p is a distinguished point (base point), and whose morphisms $f : (M,p) \to (N,q)$ are C^r maps $f : M \to N$ such that $f(p) = q$. The pairs are called pointed C^r spaces and the morphisms are called pointed C^r maps (or base point preserving maps). To every pointed space (M,p), we assign the fundamental group $\pi_1(M,p)$, and to every pointed C^r map $f : (M,p) \to (N, f(p))$ we may assign a group homomorphism $\pi_1(f) : \pi_1(M,p) \to \pi_1(N, f(p))$ by

$$\pi_1(f)([\alpha]) = [f \circ \alpha].$$

It is easy to check that this is a covariant functor, and so for pointed maps f and g that can be composed, $(M,x) \xrightarrow{f} (N,y) \xrightarrow{g} (P,z)$, we have $\pi_1(g \circ f) = \pi_1(g) \circ \pi_1(f)$.

Notation 1.92. To avoid notational clutter, we will often denote $\pi_1(f)$ by $f_\#$.

Definition 1.93. Let $\wp : \widetilde{M} \to M$ be a C^r covering and let $f : P \to M$ be a C^r map. A map $\tilde{f} : P \to \widetilde{M}$ is said to be a **lift** of the map f if $\wp \circ \tilde{f} = f$.

Theorem 1.94. *Let $\wp : \widetilde{M} \to M$ be a C^r covering, let $\gamma : [a,b] \to M$ be a C^r curve and pick a point y in $\wp^{-1}(\gamma(a))$. Then there exists a unique C^r lift $\tilde{\gamma} : [a,b] \to \widetilde{M}$ of γ such that $\tilde{\gamma}(a) = y$. Thus the following diagram commutes:*

$$\begin{array}{ccc} & & \widetilde{M} \\ & \tilde{\gamma} \nearrow & \downarrow \wp \\ [a,b] & \xrightarrow{\gamma} & M \end{array}$$

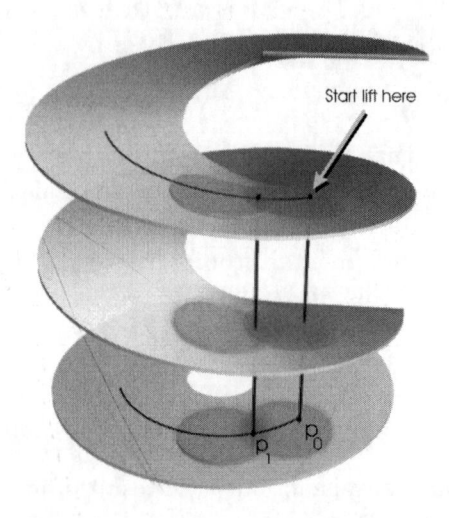

Figure 1.6. Lifting a path to a cover

If two paths α and β with $\alpha(a) = \beta(a)$ are fixed endpoint homotopic via a homotopy h, then for a given point y in $\wp^{-1}(\gamma(a))$, we have the corresponding lifts $\tilde{\alpha}$ and $\tilde{\beta}$ starting at y. In this case, the homotopy h lifts to a fixed endpoint homotopy $\tilde{h}$ between $\tilde{\alpha}$ and $\tilde{\beta}$. In short, homotopic paths lift to homotopic paths.

Proof. We just give the basic idea and refer the reader to the extensive literature for details (see [**Gre-Hrp**]). Figure 1.6 shows the way. Decompose the curve γ into segments that lie in evenly covered open sets by using the Lebesgue number lemma. Lift inductively starting by using the inverse of $\wp$ in the first evenly covered open set. It is clear that in order to connect up continuously, each step is forced and so the lifted curve is unique. A similar argument shows how to lift the homotopy h. A little thought reveals that if $\wp$ is a C^r covering, then the lifts of C^r maps are C^r. □

There are several important corollaries to this result. One is simply that if $\alpha : [0,1] \to M$ is a path starting at a base point $p \in M$, then since there is one and only one lift $\tilde{\alpha}$ starting at a given p' in the fiber $\wp^{-1}(p)$, the endpoint $\tilde{\alpha}(1)$ is completely determined by the path α and by the point p' from which we want the lifted path to start. In fact, the endpoint only depends on the homotopy class of α (and the choice of starting point p'). To see this, note that if $\alpha, \beta : [0,1] \to M$ are fixed endpoint homotopic paths in M beginning at p, and if $\tilde{\alpha}$ and $\tilde{\beta}$ are the corresponding lifts with $\tilde{\alpha}(0) = \tilde{\beta}(0) = p'$, then by the second part of the theorem, any homotopy $h_t : \alpha \simeq \beta$ lifts to a unique fixed endpoint homotopy $\tilde{h}_t : \tilde{\alpha} \simeq \tilde{\beta}$. This then

1.6. Coverings and Discrete Groups

implies that $\widetilde{\alpha}(1) = \widetilde{\beta}(1)$. Applying these ideas to loops based at $p \in M$, we will next see that the fundamental group $\pi_1(M, p)$ acts on the fiber $\wp^{-1}(p)$ as a group of permutations. (This is a right action as we will see.) In case the covering space $\widetilde{M}$ is simply connected, we will also obtain an isomorphism of the group $\pi_1(M, p)$ with the deck transformation group (which acts from the *left* on $\widetilde{M}$). Before we delve into these matters, we state, without proof, two more standard results (see [**Gre-Hrp**]):

Theorem 1.95. *Let* $\wp : \widetilde{M} \to M$ *be a* C^r *covering. Fix a point* $q \in Q$ *and a point* $\widetilde{p} \in \widetilde{M}$. *Let* $\phi : Q \to M$ *be a* C^r *map with* $\phi(q) = \wp(\widetilde{p})$. *If* Q *is connected, then there is at most one lift* $\widetilde{\phi} : Q \to \widetilde{M}$ *of* ϕ *such that* $\widetilde{\phi}(p) = \widetilde{p}$. *If* $\phi_\#(\pi_1(Q, q)) \subset \wp_\#(\pi_1(\widetilde{M}, \widetilde{p}))$, *then* ϕ *has such a lift. In particular, if* Q *is simply connected, then the lift exists.*

Theorem 1.96. *Every connected topological manifold* M *has a* C^0 *simply connected covering space which is unique up to isomorphism of coverings. This is called the* **universal cover**. *Furthermore, if* H *is any subgroup of* $\pi_1(M, p)$, *then there is a connected covering* $\wp : \widetilde{M} \to M$ *and a point* $\widetilde{p} \in \widetilde{M}$ *such that* $\wp_\#(\pi_1(\widetilde{M}, \widetilde{p})) = H$.

If follows from this and Theorem 1.86 that if M is a C^r manifold, then there is a unique C^r structure on the universal covering space $\widetilde{M}$ so that $\wp : \widetilde{M} \to M$ is a C^r covering.

Since a deck transformation is a lift, we have the following corollary.

Corollary 1.97. *Let* $\wp : \widetilde{M} \to M$ *be a* C^r *covering map and choose a base point* $p \in M$. *If* $\widetilde{M}$ *is connected, there is at most one deck transformation* ϕ *that maps a given* $p_1 \in \wp^{-1}(p)$ *to a given* $p_2 \in \wp^{-1}(p)$. *If* $\widetilde{M}$ *is simply connected, then such a deck transformation exists and is unique.*

Theorem 1.98. *If* $\widetilde{M}$ *is the universal cover of* M *and* $\wp : \widetilde{M} \to M$ *is the corresponding universal covering map, then for any base point* $p_0 \in M$, *there is an isomorphism* $\pi_1(M, p_0) \cong \text{Deck}(\wp)$.

Proof. Fix a point $\widetilde{p} \in \wp^{-1}(p_0)$. Let $a \in \pi_1(M, p_0)$ and let α be a loop representing a. Lift to a path $\widetilde{\alpha}$ starting at $\widetilde{p}$. As we have seen, the point $\widetilde{\alpha}(1)$ depends only on the choice of $\widetilde{p}$ and $a = [\alpha]$. Let ϕ_a be the unique deck transformation such that $\phi_a(\widetilde{p}) = \widetilde{\alpha}(1)$. The assignment $a \mapsto \phi_a$ gives a map $\pi_1(M, p_0) \to \text{Deck}(\wp)$. For $a = [\alpha]$ and $b = [\beta]$ chosen from $\pi_1(M, p_0)$, we have the lifts $\widetilde{\alpha}$ and $\widetilde{\beta}$, and we see that $\phi_a \circ \widetilde{\beta}$ is a path from $\phi_a(\widetilde{p})$ to $\phi_a(\widetilde{\beta}(1)) = \phi_a(\phi_b(\widetilde{p}))$. Thus the path $\widetilde{\gamma} := \widetilde{\alpha} * \left(\phi_a \circ \widetilde{\beta} \right)$ is defined. Since

$\wp \circ \phi_a = \wp$, we have

$$\wp \circ \widetilde{\gamma} = \wp \circ \left[\widetilde{\alpha} * \left(\phi_a \circ \widetilde{\beta}\right)\right]$$
$$= (\wp \circ \widetilde{\alpha}) * \left(\wp \circ \left(\phi_a \circ \widetilde{\beta}\right)\right)$$
$$= (\wp \circ \widetilde{\alpha}) * \left((\wp \circ \phi_a) \circ \widetilde{\beta}\right)$$
$$= (\wp \circ \widetilde{\alpha}) * \left(\wp \circ \widetilde{\beta}\right) = \alpha * \beta.$$

Since $\alpha * \beta$ represents the element $ab \in \pi_1(M, p_0)$, we have $\phi_{ab}(\widetilde{p}) = \widetilde{\gamma}(1) = \phi_a(\phi_b(\widetilde{p}))$. Since $\widetilde{M}$ is connected, Corollary 1.97 gives $\phi_{ab} = \phi_a \circ \phi_b$. Thus the map is a group homomorphism.

It is easy to see that the map $a \mapsto \phi_a$ is onto. Indeed, given $f \in \text{Deck}(\wp)$, we simply take a curve $\widetilde{\gamma}$ from $\widetilde{p}$ to $f(\widetilde{p})$, and then we have $f = \phi_g$, where $g = [\wp \circ \widetilde{\gamma}] \in \pi_1(M, p_0)$.

Finally, if $\phi_a = \text{id}$, then we conclude that any loop $\alpha \in [\alpha] = a$ lifts to a loop $\widetilde{\alpha}$ based at $\widetilde{p}$. But $\widetilde{M}$ is simply connected, and so $\widetilde{\alpha}$ is homotopic to a constant map to $\widetilde{p}$, and its projection α is therefore homotopic to a constant map to p. Thus $a = [\alpha] = 0$ and so the homomorphism is 1-1. $\square$

1.6.2. Discrete group actions. Groups actions are ubiquitous in differential geometry and in mathematics generally. Felix Klein emphasized the role of group actions in the classical geometries (see [**Klein**]). More on classical geometries can be found in the online supplement [**Lee, Jeff**]. In this section, we discuss *discrete* group actions on smooth manifolds and show how they give rise to covering spaces.

Definition 1.99. Let G be a group and M a set. A **left group action** on M is a map $l : G \times M \to M$ such that

1) $l(g_2, l(g_1, x)) = l(g_2 g_1, x)$ for all $g_1, g_2 \in G$ and all $x \in M$;

2) $l(e, x) = x$ for all $x \in M$, where e is the identity element of G.

We often write $g \cdot x$ or just gx in place of the more pedantic notation $l(g, x)$. Using this notation, we have $g_2(g_1 x) = (g_2 g_1)x$ and $ex = x$. Similarly, we define a **right group action** as a map $r : M \times G \to M$ with $r(r(x, g_1), g_2) = r(x, g_1 g_2)$ for all $g_1, g_2 \in G$ and all $x \in M$ and $r(x, e) = x$ for all $x \in M$. In the case of right actions, we write $r(x, g)$ as $x \cdot g$ or xg.

If $l : G \times M \to M$ is a left action, then for every $g \in G$ we have a map $l_g : M \to M$ defined by $l_g(x) = l(g, x)$, and similarly a right action gives for every g, a map $r_g : M \to M$. For every result about left actions, there is an analogous result for right actions. However, mathematical conventions are such that while $g \mapsto l_g$ is a group homomorphism from G to the group of

1.6. Coverings and Discrete Groups

permutations of M, the map $g \mapsto r_g$ is a group *anti-homomorphism* which means that $r_{g_1} \circ r_{g_2} = r_{g_2 g_1}$ for all $g_1, g_2 \in G$ (notice the order reversal).

Given a left action, the sets of the form $Gx = \{gx : g \in G\}$ are called **orbits**. The set Gx is called the orbit of x. Two points x and y are in the same orbit if and only if there is a group element g such that $gx = y$. The orbits are equivalence classes and so they partition M. Let $G \backslash M$ be the set of orbits and let $\wp : M \to G \backslash M$ be the projection taking each x to Gx. We give $G \backslash M$ the quotient topology. By definition, U is open in $G \backslash M$ if and only if $\wp^{-1}(U)$ is open. This makes $\wp$ continuous, but in this case, it is also an open map. To see this, let U be open. Then $\wp^{-1}(\wp(U))$ is the union $\bigcup_{g \in G} gU$ which is open and so $\wp(U)$ is open by the definition of quotient topology.

Definition 1.100. Suppose G acts on a set M by $l : G \times M \to M$. We say that G acts **transitively** if for any $x, y \in M$ there is a g such that $gx = y$. Equivalently, the action is **transitive** if the action has only one orbit. We say that the action is **effective** provided that $l_g = \text{id}_M$ implies that $g = e$. If the action has the property that $gx = x$ for some $x \in M$ only when $g = e$, then we say that G acts **freely** (or that the action is **free**). In other words, an action is free provided that the only element of G that fixes any element of M is the identity element.

Similar statements and definitions apply for right actions except that the orbits have the form xG. The quotient space (space of orbits) will then be denoted by M/G.

Warning: The notational distinction between $G \backslash M$ and M/G is not universal.

Example 1.101. Let $\wp : \widetilde{M} \to M$ be a covering map. Fix a base point $p_0 \in M$ and a base point $\widetilde{p}_0 \in \wp^{-1}(p_0)$. If $a \in \pi_1(M, p_0)$, then for each $x \in \wp^{-1}(p_0)$, we define $r_a(x) := xa := \widetilde{\alpha}(1)$, where $\widetilde{\alpha}$ is the lift of any loop α representing a. The reader may check that r_a is a right action on the set $\wp^{-1}(p_0)$.

Example 1.102. Recall that if $\wp : \widetilde{M} \to M$ is a universal C^r covering map (so that $\widetilde{M}$ is simply connected), we have an isomorphism $\pi_1(M, p_0) \to \text{Deck}(\wp)$, which we denote by $a \mapsto \phi_a$. This means that $l(a, x) = \phi_a(x)$ defines a left action of $\pi_1(M, p_0)$ on $\widetilde{M}$.

Let G be a group and endow G with the discrete topology so that, in particular, every point is an open set. In this case, we call G a discrete group. If M is a topological space, then we endow $G \times M$ with the product topology. What does it mean for a map $\alpha : G \times M \to M$ to be continuous? The topology of $G \times M$ is clearly generated by sets of the form $S \times U$, where

S is an arbitrary subset of G and U is open in M. The map $\alpha : G \times M \to M$ will be continuous if for any point $(g_0, x_0) \in G \times M$ and any open set $U \subset M$ containing $\alpha(g_0, x_0)$ we can find an open set $S \times V$ containing (g_0, x_0) such that $\alpha(S \times V) \subset U$. Since the topology of G is discrete, it is necessary and sufficient that there is an open V such that $\alpha(g_0 \times V) \subset U$. It is easy to see that a necessary and sufficient condition for α to be continuous on all of $G \times M$ is that the partial maps $\alpha_g := \alpha(g, \cdot)$ are continuous for every $g \in G$.

Definition 1.103. Let G be a discrete group and M a manifold. A left **discrete group action** is a group action $l : G \times M \to M$ such that for every $g \in G$ the partial map $l_g(\cdot) := l(g, \cdot)$ is continuous. A right discrete group action is defined similarly.

It follows that if $l : G \times M \to M$ is a discrete action, then each partial map l_g is a homeomorphism with $l_g^{-1} = l_{g^{-1}}$.

Definition 1.104. A discrete group action is C^r if M is a C^r manifold and each l_g (resp. r_g) is a C^r map.

Example 1.105. Let $\phi : M \to M$ be a diffeomorphism and let $\mathbb{Z}$ act on M by $n \cdot x := \phi^n(x)$ where

$$\phi^0 := \mathrm{id}_M,$$
$$\phi^n := \phi \circ \cdots \circ \phi \text{ for } n > 0,$$
$$\phi^{-n} := (\phi^{-1})^n \text{ for } n > 0.$$

This gives a discrete action of $\mathbb{Z}$ on M.

Definition 1.106. A discrete group action $l : G \times M \to M$ is said to be **proper** if for every two points $x, y \in M$ there are open neighborhoods U_x and U_y respectively such that the set $\{g \in G : gU_x \cap U_y \neq \emptyset\}$ is finite.

There is a more general notion of proper action which we shall meet later. For free *and* proper discrete actions, we have the following useful characterization.

Proposition 1.107. *A discrete group action $l : G \times M \to M$ is **proper and free** if and only if the following two conditions hold:*

(i) *Each $x \in M$ has an open neighborhood U such that $gU \cap U = \emptyset$ for all g except the identity e. We shall call such open sets **self-avoiding**.*

(ii) *If $x, y \in M$ are not in the same orbit, then they have self-avoiding neighborhoods U_x and U_y such that $gU_x \cap U_y = \emptyset$ for all $g \in G$.*

Proof. Suppose that the action l is proper and free. Let x be given. We then know that there is an open V containing x such that $gV \cap V = \emptyset$ except

1.6. Coverings and Discrete Groups

for a finite number of g, say, $g_1, \ldots, g_k$, which are distinct. One of these, say g_1, must be e. Since the action is free, we know that for each fixed $i > 1$ we have $g_i x \in M \setminus \{x\}$. By using continuity and then the fact that M is a regular topological space, we can replace V by a smaller open set (called V again) such that $g_i \overline{V} \subset M \setminus \{x\}$ for all $i = 2, \ldots, k$ or, in other words, that $x \notin g_2 \overline{V} \cup \cdots \cup g_k \overline{V}$. Let $U = V \setminus (g_2 \overline{V} \cup \cdots \cup g_k \overline{V})$. Notice that U is open and we have arranged that U contains x. We show that $U \cap gU$ is empty unless $g = e$. So suppose $g \neq e = g_1$. Since $U \cap gU \subset V \cap gV$, we know that this is empty for sure in all cases except maybe where $g = g_i$ for $i = 2, \ldots, k$. If $x \in U \cap g_i U$ for such an i, then $x \in U$ and so $x \notin g_i V$ by the definition of U. But we also have $x \in g_i U \subset g_i V$, which is a contradiction. We conclude that (i) holds.

Now suppose that $x, y \in M$ are not in the same orbit. We know that there exist open sets U_x and U_y with $x \in U_x$, $y \in U_y$ and such that $gU_x \cap U_y$ is empty except possibly for some finite set of elements which we denote by $g_1, \ldots, g_k$. Since the action is free, $g_1 x, \ldots, g_k x$ are distinct. We also know that y is not equal to any of $g_1 x, \ldots, g_k x$, and so since M is a Hausdorff space, there exist pairwise disjoint open sets $O_1, \ldots, O_k$, O_y with $g_i x \in O_i$ and $y \in O_y$. By continuity, we may shrink U_x so that $g_i U_x \subset O_i$ for all $i = 1, \ldots, k$, and then we also replace U_y with $O_y \cap U_y$ (renaming this U_y again). As a result we now see that $gU_x \cap U_y = \emptyset$ for $g = g_1, \ldots, g_k$ and hence for all g. By shrinking the sets U_x and U_y further we may make them self-avoiding.

Next we suppose that (i) and (ii) always hold for a given discrete action l. First we show that l is free. Suppose that $x = gx$. Then for every open neighborhood U of x the set $gU \cap U$ is nonempty, which by (i) means that $g = e$. Thus the action is free. Next pick $x, y \in M$. If x, y are not in the same orbit, then by (ii) we may pick U_x and U_y so that $\{g \in G : gU_x \cap U_y \neq \emptyset\}$ is empty and so certainly a finite set. If x, y are in the same orbit, then $y = g_0 x$ for a unique g_0 since we now know that the action is free. Choose an open neighborhood U of x so that $gU \cap U = \emptyset$ for $g \neq e$. Let $U_x = U$ and $U_y = g_0 U$. Then, $gU_x \cap U_y = gU \cap g_0 U$. If $gU \cap g_0 U \neq \emptyset$, then $g_0^{-1} gU \cap U \neq \emptyset$ and so $g_0^{-1} g = e$ and $g = g_0$. Thus the only way that $gU_x \cap U_y$ is nonempty is if $g = g_0$ and so the set $\{g \in G : gU_x \cap U_y \neq \emptyset\}$ has cardinality one. In either case, we may choose U_x and U_y so that the set is finite, which is what we wanted to show. $\square$

It is easy to see that if $U \subset M$ is self-avoiding, then any open subset $V \subset U$ is also self-avoiding. Thus, if the discrete group G acts freely and properly on M, then the open sets of (i) and (ii) in the above proposition can be taken to be connected chart domains.

Proposition 1.108. *Let M be an n-manifold and let $l : G \times M \to M$ be a smooth discrete action which is free and proper. Then the quotient space $G\backslash M$ has a natural smooth structure such that the quotient map is a smooth covering map.*

Proof. Giving $G\backslash M$ the quotient topology makes $\wp : M \to G\backslash M$ continuous. Using (ii) of Proposition 1.107, it is easy to show that the quotient topology on $G\backslash M$ is Hausdorff. By Proposition 1.107, we may cover M by charts whose domains are self-avoiding and connected. Let $(U, \mathbf{x})$ be such a chart and consider the restriction $\wp|_U$. This restricted map is open since, as remarked above, $\wp$ is an open map. If $x, y \in U$ and $\wp(x) = \wp(y)$, then x and y are in the same orbit and so $y = gx$ for some g. Therefore $y \in gU \cap U$, which means that $gU \cap U$ is not empty and so $g = e$ since U is self-avoiding. Thus $x = y$ and we conclude that $\wp|_U$ is injective. Since $\wp|_U$ is also surjective, we see that it is a bijection and hence a homeomorphism. I.e., since $\wp|_U$ is also open, it has a continuous inverse and so it is a homeomorphism. Since U is connected, the connected components of $\wp^{-1}(\wp(U))$ are exactly the sets gU for $g \in G$. Since $\wp \circ l_g = \wp$ for all G, it is easy to see that $\wp$ restricts to a homeomorphism on each connected component gU of $\wp^{-1}(\wp(U))$. Thus, $\wp(U)$ is evenly covered by $\wp$ and so $\wp$ is a covering map. For every such chart $(U, \mathbf{x})$, we have a map

$$\mathbf{x} \circ (\wp|_U)^{-1} : \wp(U) \to \mathbf{x}(U),$$

which is a chart on $G\backslash M$. This map is clearly a homeomorphism. Given any other map constructed in this way, say $\mathbf{y} \circ (\wp|_V)^{-1}$, the domains $\wp(U)$ and $\wp(V)$ only meet if there is a $g \in G$ such that gU meets V and l_g maps an open subset of U diffeomorphically onto a subset of V. In fact, by Exercise 1.109 below, the map $(\wp|_V)^{-1} \circ \wp|U$ is defined on an open set each point of which has a neighborhood on which this map is a restriction of l_g for some g. Thus $(\wp|_V)^{-1} \circ \wp|_U$ is smooth and for the overlap map we have

$$\mathbf{y} \circ (\wp|_V)^{-1} \circ \left(\mathbf{x} \circ (\wp|_U)^{-1} \right)^{-1} = \mathbf{y} \circ (\wp|_V)^{-1} \circ \wp|_U \circ \mathbf{x}^{-1},$$

which is smooth. Thus we have an atlas on $G\backslash M$ and the topology induced by the atlas is the same as the quotient topology since we have already established that the charts are homeomorphisms. $\square$

Exercise 1.109. In the context of the proof above, show that $(\wp|_V)^{-1} \circ \wp|_U$ is defined on an open set $O = (\wp|_U)^{-1}(\wp(U) \cap \wp(V))$. Show that each $x \in O$ has a neighborhood on which the map $(\wp|_V)^{-1} \circ \wp|_U$ coincides with a restriction of a map $x \mapsto gx$ for some fixed g. Conclude that $(\wp|_V)^{-1} \circ \wp|_U$ is a C^r map. [Outline of solution: For $x \in O$, we must have $(\wp|_V)^{-1} \circ \wp|_U (x) = gx$ for some g. Now $O' = U \cap g^{-1}V$ is an open set that contains x. But also,

1.6. Coverings and Discrete Groups

$\wp(g^{-1}V) = \wp(V)$ so $O' = U \cap g^{-1}V \subset (\wp|U)^{-1}(\wp(U) \cap \wp(V))$. Let $x' \in O'$. Then since $\wp(x') \in \wp(U) \cap \wp(V)$, it follows that

$$(\wp|_V)^{-1} \circ \wp|_U (x') = x'',$$

where x'' is the unique point in V such that $\wp(x'') = \wp(x')$. But $gx' \in V$ and $\wp(gx') = x''$ so $gx' = x''$.]

Example 1.110. We have seen the torus previously presented as $T^2 = S^1 \times S^1$. Another presentation that uses a group action is given as follows: Let the group $\mathbb{Z} \times \mathbb{Z} = \mathbb{Z}^2$ act on $\mathbb{R}^2$ by

$$(m,n) \cdot (x,y) := (x+m, y+n).$$

It is easy to check that Proposition 1.108 applies to give a manifold $\mathbb{R}^2/\mathbb{Z}^2$. This is actually the torus in another guise, and we have the diffeomorphism $\phi : \mathbb{R}^2/\mathbb{Z}^2 \to S^1 \times S^1 = T^2$ given by $[(x,y)] \mapsto (e^{i2\pi x}, e^{i2\pi y})$. The following diagram commutes:

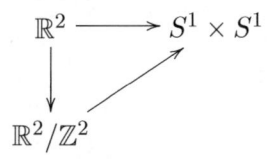

$$\begin{array}{ccc} \mathbb{R}^2 & \longrightarrow & S^1 \times S^1 \\ \downarrow & \nearrow & \\ \mathbb{R}^2/\mathbb{Z}^2 & & \end{array}$$

Exercise 1.111. Let $\wp : M \to G\backslash M$ be the covering arising from a free and proper discrete action of G on M and suppose that M is *connected*. Let $\Gamma_G := \{l_g \in \text{Diff}(M) : g \in G\}$; then G is isomorphic to Γ_G by the obvious map $g \mapsto l_g$ and furthermore $\Gamma_G = \text{Deck}(\wp)$.

Covering spaces $\wp : \widetilde{M} \to M$ that arise from a proper and free discrete group action are special in that if M is connected, then the covering is a normal covering, which means that the group $\text{Deck}(\wp)$ acts transitively on each fiber $\wp^{-1}(p)$ (why?).

Example 1.112. Recall the multiplicative abelian group $\mathbb{Z}_2 = \{1, -1\}$ of two elements. Let $\mathbb{Z}_2$ act on the sphere $S^n \subset \mathbb{R}^{n+1}$ by $(\pm 1) \cdot x := \pm x$. Thus the action is generated by letting -1 send a point on the sphere to its antipode. This action is also easily seen to be free and proper. The quotient space is the real projective space $\mathbb{R}P^n$ (See Example 1.41),

$$\mathbb{R}P^n = S^n/\mathbb{Z}_2.$$

If M is simply connected and we have a free and proper action by G as above, then we can define a map $\phi : G \to \pi_1(G\backslash M, b_0)$ as follows: Fix a base point $x_0 \in M$ with $\wp(x_0) = b_0$. Given $g \in G$, let $\gamma : [0,1] \to M$ be a path with $\gamma(0) = x_0$ and $\gamma(1) = gx_0$. Then

$$\phi(g) := [\wp \circ \gamma] \in \pi_1(G\backslash M, b_0).$$

This is well-defined because M is simply connected. In fact, we already know from Theorem 1.98 that there is an isomorphism $\pi_1(G\backslash M, b_0) \cong \mathrm{Deck}(\wp)$. But by Exercise 1.111, we know that $G \cong \mathrm{Deck}(\wp)$ by the map $g \mapsto l_g$. Composing, we obtain an isomorphism $\psi : \pi_1(G\backslash M, b_0) \to G$. Recalling the definition of the isomorphism constructed in the proof of Theorem 1.98, we see that the map $\phi : G \to \pi_1(G\backslash M, b_0)$ defined above is just the inverse of the isomorphism $\psi : \pi_1(G\backslash M, b_0) \to G$. Thus we obtain the following theorem, which is essentially a variation of Theorem 1.98.

Theorem 1.113. *If M is simply connected, then the map*
$$\phi : G \to \pi_1(G\backslash M, b_0)$$
defined above is a group isomorphism.

The reader may wish to try to prove directly that ϕ is a group isomorphism.

Corollary 1.114. $\pi_1(\mathbb{R}P^n) \cong \mathbb{Z}_2$.

1.7. Regular Submanifolds

A subset S of a smooth n-manifold M is called a **regular submanifold** of dimension k if every point $p \in S$ is in the domain of a chart $(U, \mathbf{x})$ that has the following **regular submanifold property** with respect to S:
$$\mathbf{x}(U \cap S) = \mathbf{x}(U) \cap (\mathbb{R}^k \times \{c\}) \text{ for some } c \in \mathbb{R}^{n-k}.$$

Usually c is chosen to be 0, which can always be accomplished by composition with a translation of $\mathbb{R}^n$. The terminology here does not seem to be quite standardized. If a subset $S \subset M$ is covered by charts of M of the above type, then S itself is said to have the **(regular) submanifold property**. We will refer to such charts as being **single-slice charts** (adapted to S). For every such single-slice chart $(U, \mathbf{x})$, we obtain a chart $(U \cap S, \mathbf{x}_S)$ on S, where $\mathbf{x}_S := \mathrm{pr} \circ \mathbf{x}|_{U \cap S}$ and $\mathrm{pr} : \mathbb{R}^k \times \mathbb{R}^{n-k} \to \mathbb{R}^k$ is projection onto the first k coordinates. In other words, if $x^1, \ldots, x^n$ are the coordinate functions of a single-slice chart, then the restrictions of $x^1, \ldots, x^k$ to $U \cap S$ are coordinate functions on S. These charts provide an atlas for S (called a submanifold atlas) making it a smooth manifold in its own right. Indeed, one checks that the overlap maps for such charts are smooth.

Exercise 1.115. Prove this last statement.

We will see more general types of submanifolds in the sequel. An important aspect of regular submanifolds is that the topology induced by the smooth structure is the same as the relative topology. The integer $n - k$ is called the **codimension** of S (in M), and we say that S is a regular submanifold of codimension $n - k$.

1.7. Regular Submanifolds

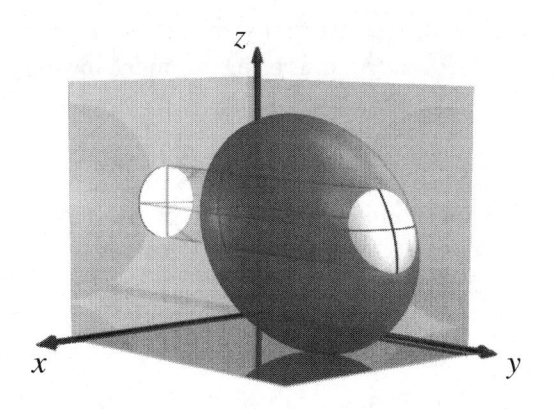

Figure 1.7. Projection chart

Example 1.116. The unit sphere $S^n \subset \mathbb{R}^{n+1}$ is a regular submanifold of $\mathbb{R}^{n+1}$. To see this, let $W_i^{\pm} := \{(a^1, \ldots, a^{n+1}) \in \mathbb{R}^{n+1} : \pm a^i > 0\}$. Then define $\psi_i^{\pm} : W_i^{\pm} \to \psi_i^{\pm}(\mathbb{R}^{n+1})$ by

$$\psi_i^{\pm}(a^1, \ldots, a^{n+1}) = \left(a^1, \ldots, a^{i-1}, a^{i+1}, \ldots, a^{n+1}, \|a\| - 1\right).$$

These can easily be checked to give charts on $\mathbb{R}^{n+1}$ smoothly related to the standard chart. If $p \in S^2$ then p is in the domain of one of the charts $\psi_i^{\pm}$. On the other hand, identifying $\mathbb{R}^{n+1}$ with $\mathbb{R}^n \times \mathbb{R}$, we have $\psi_i^{\pm}(W_i^{\pm} \cap S^n) = \psi_i^{\pm}(U) \cap (\mathbb{R}^n \times \{0\})$ and so these charts have the submanifold property with respect to S^n. Let $\mathrm{pr} : \mathbb{R}^{n+1} = \mathbb{R}^{n-1} \times \mathbb{R} \to \mathbb{R}^{n-1}$. The resulting charts on S^n given by $\mathrm{pr} \circ \psi_i^{\pm}|_{W_i^{\pm} \cap S^n}$ have the form $(a^1, \ldots, a^{n+1}) \mapsto (a^1, \ldots, a^{i-1}, a^{i+1}, \ldots, a^{n+1})$. These are projections from S^n onto coordinate hyperplanes in $\mathbb{R}^{n+1}$ and are easily checked to give the same smooth structure as the stereographic charts given earlier.

Exercise 1.117. Show that a continuous map $f : N \to M$ that has its image contained in a regular submanifold S is differentiable with respect to the submanifold atlas if and only if it is differentiable as a map into M.

Exercise 1.118. Show that the graph of any smooth map $\mathbb{R}^n \to \mathbb{R}^m$ is a regular submanifold of $\mathbb{R}^n \times \mathbb{R}^m$.

Exercise 1.119. Show that if M is a k-dimensional regular submanifold of $\mathbb{R}^n$, then for every $p \in M$, there exists at least one k-dimensional coordinate plane P such that the orthogonal projection $\mathbb{R}^n \to P \cong \mathbb{R}^k$ restricts to a coordinate chart for M defined on some neighborhood of p. [Hint: If $(U, \mathbf{x})$ is a single-slice chart for M so that $x^1, \ldots, x^k$ restrict to coordinates on M, then $x^{k+1}, \ldots, x^n$ together give a map $f : U \subset \mathbb{R}^n = \mathbb{R}^k \times \mathbb{R}^{n-k} \to \mathbb{R}^{n-k}$

such that $M \cap U = f^{-1}(0)$. Argue that if $u^1, \ldots, u^n$ are standard coordinates on $\mathbb{R}^n$, then after a suitable renumbering we must have

$$\frac{\partial(x^{k+1}, \ldots, x^n)}{\partial(u^{k+1}, \ldots, u^n)} \neq 0.$$

Now use the implicit mapping theorem to show that M is locally the graph of a smooth function.]

1.8. Manifolds with Boundary

For the general Stokes theorem, where the notion of flux has its natural setting, we will need to have the concept of a *smooth* manifold with boundary. We have already introduced the notion of a topological manifold with boundary, but now we want to see how to handle the issue of the smooth structures. Some basic two-dimensional examples to keep in mind are the upper half-plane $\mathbb{R}^2_{y \geq 0} := \{(x,y) \in \mathbb{R}^2 : y \geq 0\}$, the closed unit disk $D = \{(x,y) \in \mathbb{R}^2 : x^2 + y^2 \leq 1\}$, and the closed hemisphere which is the set of all $(x,y,z) \in S^2$ with $z \geq 0$. Recall that in Section 1.2 we defined the closed n-dimensional Euclidean half-spaces $\mathbb{R}^n_{\lambda \geq c} := \{a \in \mathbb{R}^n : \lambda(a) \geq c\}$. Of course, $\mathbb{R}^n_{\lambda \leq c} = \mathbb{R}^n_{-\lambda \geq -c}$ so we are including both $\mathbb{R}^n_{x^k \leq 0}$ and $\mathbb{R}^n_{x^k \geq 0}$. We also write $\mathbb{R}^n_{\lambda = c} = \{a \in \mathbb{R}^n : \lambda(a) = c\}$. Give $\mathbb{R}^n_{\lambda \geq c}$ the relative topology as a subset of $\mathbb{R}^n$. Since $\mathbb{R}^n_{\lambda \geq c} \subset \mathbb{R}^n$, we already have a notion of differentiability for a map $U \to \mathbb{R}^m$, where U is a relatively open subset of $\mathbb{R}^n_{\lambda \geq c}$. We just invoke Definition 1.58. We can extend definitions a bit more:

Definition 1.120. Let $U \subset \mathbb{R}^n_{\lambda_1 \geq c_1}$ and $f : U \to \mathbb{R}^m_{\lambda_2 \geq c_2}$. We say that f is C^r if it is C^r as a map into $\mathbb{R}^m$. If both $f : U \to f(U)$ and $f^{-1} : f(U) \to U$ are homeomorphisms of relatively open sets and C^r in this sense, then f is called a C^r diffeomorphism.

For convenience, let us introduce for an open set $U \subset \mathbb{R}^n_{\lambda \geq c}$ (relatively open) the following notations: Let ∂U denote $\mathbb{R}^n_{\lambda = c} \cap U$ and $\text{int}(U)$ denote $U \setminus \partial U$. In particular, $\partial \mathbb{R}^n_{\lambda \geq c} = \mathbb{R}^n_{\lambda = c}$. Notice that ∂U is clearly an $(n-1)$-manifold.

We have the following three facts:

(1) First, let $f : U \subset \mathbb{R}^n \to \mathbb{R}^k$ be C^r differentiable (with $r \geq 1$) and g another such map with the same domain. If $f = g$ on $\mathbb{R}^n_{\lambda \geq c} \cap U$, then $Df(x) = Dg(x)$ for all $x \in \mathbb{R}^n_{\lambda \geq c} \cap U$.

(2) If $f : U \subset \mathbb{R}^n \to \mathbb{R}^n_{\lambda \geq c}$ is C^r differentiable (with $r \geq 1$) and $f(x) \in \mathbb{R}^n_{\lambda = c} = \partial \mathbb{R}^n_{\lambda \geq c}$ for all $x \in U$, then $Df(x)$ must have its image in $\mathbb{R}^n_{\lambda = 0}$.

1.8. Manifolds with Boundary

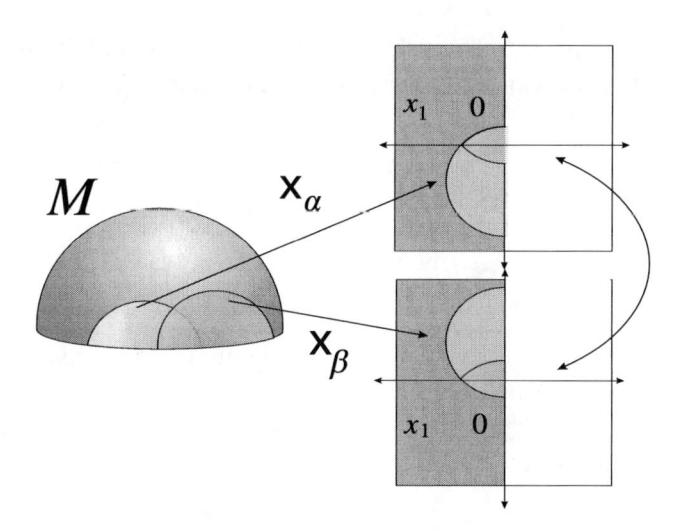

Figure 1.8. Manifold with boundary

(3) Let $f: U_1 \subset \mathbb{R}^n_{\lambda_1 \geq c_1} \to U_2 \subset \mathbb{R}^m_{\lambda_2 \geq c_2}$ be a diffeomorphism (in our new extended sense). Assume that ∂U_1 and ∂U_2 are not empty. Then f induces diffeomorphisms $\partial U_1 \to \partial U_2$ and $\text{int}(U_1) \to \text{int}(U_2)$.

These three claims are not exactly obvious, but they are very intuitive. On the other hand, none of them is difficult to prove (see Problem 19).

We can now form a definition of smooth manifold with boundary in a fashion completely analogous to the definition of a smooth manifold without boundary. A **half-space chart** x for a set M is a bijection of some subset U of M onto an open subset of some half-space $\mathbb{R}^n_{\lambda \geq c}$. A C^r **half-space atlas** is a collection $(U_\alpha, \mathbf{x}_\alpha)$ of half-space charts such that for any two, say $(U_\alpha, \mathbf{x}_\alpha)$ and $(U_\beta, \mathbf{x}_\beta)$, the map $\mathbf{x}_\alpha \circ \mathbf{x}_\beta^{-1}$ is a C^r diffeomorphism on its natural domain. Notice carefully that we allow the half-space to vary from chart to chart, but we will keep n fixed for a given M and refer to the charts as n-**dimensional half-space charts**.

Definition 1.121. An n-**dimensional C^r manifold with boundary** is a pair $(M, \mathcal{A})$ consisting of a set M together with a maximal atlas of n-dimensional half-space charts $\mathcal{A}$. The manifold topology is that generated by the domains of the charts in the maximal atlas. The **boundary** of M is denoted by ∂M and is the set of points whose image under any chart is contained in the boundary of the associated half-space.

The three facts listed above show that the notion of a boundary is a well-defined concept and is a natural notion in the context of smooth manifolds; it is a "differentiable invariant".

Colloquially, one usually just refers to M as a manifold with boundary and forgoes the explicit reference to the atlas. Also we refer to an n-dimensional C^∞ manifold with boundary as an n-**manifold with boundary**. The **interior** of a manifold with boundary is $M \setminus \partial M$. It is a manifold without boundary and is denoted $\text{int}(M)$ or $\overset{\circ}{M}$.

Exercise 1.122. Show that ∂M is a closed set in M.

If no component of a manifold without boundary is compact, it is called an **open manifold**. For example, the interior $\text{int}(M)$ of a connected manifold M with nonempty boundary is never compact and is an open manifold in the above sense if every component of M contains part of the boundary.

Remark 1.123. We avoid the phrase "closed manifold", which is sometimes taken to refer to a compact manifold without boundary.

Let M be an n-dimensional C^r manifold with boundary and $p \in \partial M$. Then by definition there is a chart $(U, \mathbf{x})$ with $\mathbf{x}(p) \in \partial \mathbb{R}^n_{\lambda \geq c}$. The image of the restriction $\mathbf{x}|_{U \cap \partial M}$ is contained in $\partial \mathbb{R}^n_{\lambda \geq c}$ for some λ and c depending on the chart. By composing this restriction with any fixed linear isomorphism $\partial \mathbb{R}^n_{\lambda \geq c} \to \mathbb{R}^{n-1}$, we obtain a bijection, say $\mathbf{x}_{\partial M}$, of $U \cap \partial M$ onto an open subset of $\mathbb{R}^{n-1}$ which provides a chart $(U_\alpha \cap \partial M, \mathbf{x}_{\partial M})$ for ∂M. The family of charts obtained in this way is an atlas for ∂M. The overlaps are smooth and so we have the following:

Proposition 1.124. *If M is an n-manifold with boundary, then ∂M is an $(n-1)$-manifold.*

Exercise 1.125. Show that the overlap maps for the atlas just constructed for ∂M are smooth.

Exercise 1.126. The closed unit ball $\overline{B}(p, 1)$ in $\mathbb{R}^n$ is a smooth manifold with boundary $\partial \overline{B}(p, 1) = S^{n-1}$. Also, the closed hemisphere $S^n_+ = \{x \in S^n : x^{n+1} \geq 0\}$ is a smooth manifold with boundary.

Exercise 1.127. Is the Cartesian product of two smooth manifolds with boundary necessarily a smooth manifold with boundary?

Exercise 1.128. Show that the concept of smooth partition of unity makes sense for manifolds with boundary. Show that such exist.

Problems

(1) Prove Proposition 1.32. The online supplement [**Lee, Jeff**] outlines the proof.

(2) Prove Lemma 1.11.

(3) Check that the manifolds given as examples are indeed paracompact and Hausdorff.

(4) Let M_1, M_2 and M_3 be smooth manifolds.
   (a) Show that $(M_1 \times M_2) \times M_3$ is diffeomorphic to $M_1 \times (M_2 \times M_3)$ in a natural way.
   (b) Show that $f : M \to M_1 \times M_2$ is C^∞ if and only if the composite maps $\text{pr}_1 \circ f : M \to M_1$ and $\text{pr}_2 \circ f : M \to M_2$ are both C^∞.

(5) Show that a C^r manifold M is connected as a topological space if and only it is C^r path connected in the sense that for any two points $p_1, p_2 \in M$ there is a C^r map $c : [0,1] \to M$ such that $c(0) = p_1$ and $c(1) = p_2$.

(6) A k-frame in $\mathbb{R}^n$ is a linearly independent ordered set of vectors $(v_1, \ldots, v_k)$. Show that the set of all k-frames in $\mathbb{R}^n$ can be given the structure of a smooth manifold. This kind of manifold is called a **Stiefel manifold**.

(7) For a product manifold $M \times N$, we have the two projection maps $\text{pr}_1 : M \times N \to M$ and $\text{pr}_2 : M \times N \to N$ defined by $(x,y) \mapsto x$ and $(x,y) \mapsto y$ respectively. Show that if we have smooth maps $f_1 : P \to M$ and $f_2 : P \to N$, then the map $(f,g) : P \to M \times N$ given by $(f,g)(p) = (f(p), g(p))$ is the unique smooth map such that $\text{pr}_1 \circ (f,g) = f$ and $\text{pr}_2 \circ (f,g) = g$.

(8) Prove (i) and (ii) of Lemma 1.35.

(9) Show that the atlas obtained for a regular submanifold induces the relative topology inherited from the ambient manifold.

(10) The topology induced by a smooth structure is not necessarily Hausdorff: Let S be the subset of $\mathbb{R}^2$ given by the union $(\mathbb{R} \times 0) \cup \{(0,1)\}$. Let U be $\mathbb{R} \times 0$ and let V be the set obtained from U by replacing the point $(0,0)$ by $(0,1)$. Define a chart map $\mathbf{x}$ on U by $\mathbf{x}(x,0) = x$ and a chart $\mathbf{y}$ on V by

$$\mathbf{y}(x,0) = \begin{cases} x & \text{if } x \neq 0, \\ 0 & \text{if } x = 0. \end{cases}$$

Show that these two charts provide a C^∞ atlas on S, but that the topology induced by the atlas is *not* Hausdorff.

(11) As we have defined them, manifolds are not required to be second countable and so may have an uncountable number of connected components. Consider the set $\mathbb{R}^2$ without its usual topology. For each $a \in \mathbb{R}$, define a bijection $\phi_a : \mathbb{R} \times \{a\} \to \mathbb{R}$ by $\phi_a(x, a) = x$. Show that the family of sets of the form $U \times \{a\}$ for U open in $\mathbb{R}$ and $a \in \mathbb{R}$ provide a basis for a paracompact topology on $\mathbb{R}^2$. Show that the maps ϕ_a are charts and together provide an atlas for $\mathbb{R}^2$ with this unusual topology. Show that the resulting smooth manifold has an uncountable number of connected components (and so is not second countable).

(12) Show that every connected manifold has a countable atlas consisting of charts whose domains have compact closure and are simply connected. Hint: We are assuming that our manifolds are paracompact, so each connected component is second countable.

(13) Show that every second countable manifold has a countable fundamental group (a solution can be found in [**Lee, John**] on page 10).

(14) If $\mathbb{C} \times \mathbb{C}$ is identified with $\mathbb{R}^4$ in the obvious way, then S^3 is exactly the subset of $\mathbb{C} \times \mathbb{C}$ given by $\{(z_1, z_2) : |z_1|^2 + |z_2|^2 = 1\}$. Let p, q be coprime integers and $p > q \geq 0$. Let ω be a primitive p-th root of unity so that $\mathbb{Z}_p = \{1, \omega, \ldots, \omega^{p-1}\}$. For $(z_1, z_2) \in S^3$, let $\omega \cdot (z_1, z_2) := (\omega z_1, \omega^q z_2)$ and extend this to an action of $\mathbb{Z}_p$ on S^3 so that $\omega^k \cdot (z_1, z_2) = (\omega^k z_1, \omega^{qk} z_2)$. Show that this action is free and proper. The quotient space $\mathbb{Z}_p \backslash S^3$ is called a **lens space** and is denoted by $L(p;q)$.

(15) Let S^1 be realized as the set of complex numbers of modulus one. Define a map $\theta : S^1 \times S^1 \to S^1 \times S^1$ by $\theta(z, w) = (-z, \overline{w})$ and note that $\theta \circ \theta = \mathrm{id}$. Let G be the group $\{\mathrm{id}, \theta\}$. Show that $M := \left(S^1 \times S^1\right)/G$ is a smooth 2-manifold.

(16) Show that if S is a regular k-dimensional submanifold of an n-manifold M, then we may cover S by *special* single-slice charts from the atlas of M which are of the form $\mathbf{x} : U \to V_1 \times V_2 \subset \mathbb{R}^k \times \mathbb{R}^{n-k} = \mathbb{R}^n$ with

$$\mathbf{x}(U \cap S) = V_1 \times \{0\}$$

for some open sets $V_1 \subset \mathbb{R}^k, V_2 \subset \mathbb{R}^{n-k}$. Show that we may arrange for V_1 and V_2 to both be Euclidean balls or cubes. (This problem should be easy. Experienced readers will likely see it as merely an observation.)

(17) Show that $\pi_1(M \times N, (p, q))$ is isomorphic to $\pi_1(M, p) \times \pi_1(N, q)$.

(18) Suppose that $M = U \cup V$, where U and V are simply connected and open. Show that if $U \cap V$ is path connected, then M is simply connected.

(19) Prove the three properties about maps involving the model half-spaces $\partial \mathbb{R}^n_{\lambda \geq 0}$ listed in Section 1.8.

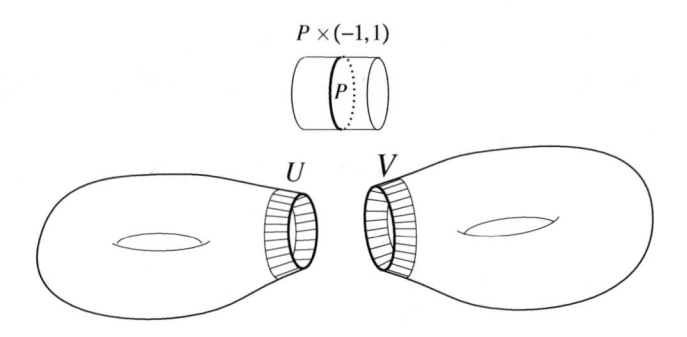

Figure 1.9. Smoothly connecting manifolds

(20) Let M and N be smooth n-manifolds with boundaries ∂M and ∂N. Let P be a smooth manifold diffeomorphic to both ∂M and ∂N via maps α and β. Suppose that there are open neighborhoods U and V of ∂M and ∂N respectively and diffeomorphisms

$$\phi_1 : U \to P \times [0, 1),$$
$$\phi_2 : V \to P \times (-1, 0]$$

such that $\phi_1(p) = (\alpha(p), 0)$ whenever $p \in \partial M \cap U$, and similarly $\phi_2(p) = (\beta(p), 0)$ whenever $p \in \partial N \cap U$. Then $\phi_2^{-1} \circ \phi_1 : U \to V$ is a diffeomorphism. Let $f := \phi_2^{-1} \circ \phi_1 \big| \partial M$ so that $f = \beta^{-1} \circ \alpha : \partial M \to \partial N$. Let $M \cup_f N$ be the topological space defined by identifying $x \in \partial M$ with $f(x) \in \partial N$ (see the online supplement [**Lee, Jeff**] for a discussion of identification spaces). Show that there is a unique smooth structure on $M \cup_f N$ such that the inclusion maps $M \hookrightarrow M \cup_f N$ and $N \hookrightarrow M \cup_f N$ are smooth and such that the induced map $U \cup_f V \to P \times (-1, 1)$ is a diffeomorphism. Here $U \cup_f V$ is the image of $U \cup V$ under the projection map $M \cup N \to M \cup_f N$.

(21) Define

$$\mathbb{R}^n_+ = \{u \in \mathbb{R}^n : u^i > 0 \text{ for } i = 1, 2, \ldots, n\},$$
$$\overline{\mathbb{R}^n_+} = \{u \in \mathbb{R}^n : u^i \geq 0 \text{ for } i = 1, 2, \ldots, n\}.$$

A boundary point of $\overline{\mathbb{R}^n_+}$ is a point such that at least one of its coordinates u^i is 0. A corner point of $\overline{\mathbb{R}^n_+}$ is a point such that at least two of its coordinates u^i, u^j are 0. We consider a set M. An $\overline{\mathbb{R}^n_+}$-valued chart on M is a pair $(U, \mathbf{x})$, where $U \subset M$ and $\mathbf{x} : U \to \mathbf{x}(U)$ is a bijection onto an open subset $\mathbf{x}(U)$ of $\overline{\mathbb{R}^n_+}$, where the latter has the relative topology as a subset of $\mathbb{R}^n$. A smooth atlas for M is a family $\{(U_\alpha, \mathbf{x}_\alpha)\}_{\alpha \in A}$ of $\overline{\mathbb{R}^n_+}$-valued charts whose domains cover M and such that whenever

$U_\alpha \cap U_\beta$ is nonempty, the composite map

$$\mathbf{x}_\beta \circ \mathbf{x}_\alpha^{-1} : \mathbf{x}_\alpha(U_\alpha \cap U_\beta) \to \mathbf{x}_\beta(U_\alpha \cap U_\beta)$$

is smooth. A maximal atlas of $\overline{\mathbb{R}^n_+}$-valued charts of this type gives M the structure of a **smooth manifold with corners** (of dimension n). We also use the terminology n-manifold with corners.

  (a) Suppose that $p \in M$ and $\mathbf{x}_\alpha(p)$ is a corner point in $\overline{\mathbb{R}^n_+}$. Show that if p is in the domain of another chart $(U_\beta, \mathbf{x}_\beta)$ in the atlas (as above), then $\mathbf{x}_\beta(p)$ is also a corner point. Use this to define "corner points" on M. Do the same for "boundary points". Thus the boundary contains the set of corner points. Explain why a manifold with corners whose set of corner points is empty is a manifold with (possibly empty) boundary.

  (b) Define the notion of smooth functions on a manifold with corners and the notion of smooth maps between manifolds with corners.

  (c) Show that the boundary of a manifold with corners is not necessarily a manifold with corners.

(22) Prove Theorem 1.113 and its corollary.

Chapter 2

The Tangent Structure

In this chapter we introduce the notions of tangent space and cotangent space of a smooth manifold. The union of the tangent spaces of a given manifold will be given a smooth structure making this union a manifold in its own right, called the tangent bundle. Similarly we introduce the cotangent bundle of a smooth manifold. We then discuss vector fields and their integral curves together with the associated dynamic notions of Lie derivative and Lie bracket. Finally, we define and discuss the notion of a 1-form (or covector field), which is the notion dual to the notion of a vector field. One can integrate 1-forms along curves. Such an integration is called a line integral. We explore the concept of exact 1-forms and nonexact 1-forms and their relation to the question of path independence of line integrals.

2.1. The Tangent Space

If $c : (-\epsilon, \epsilon) \to \mathbb{R}^N$ is a smooth curve, then it is common to visualize the "velocity vector" $\dot{c}(0)$ as being based at the point $p = c(0)$. It is often desirable to explicitly form a separate N-dimensional vector space for each point p, whose elements are to be thought of as being based at p. One way to do this is to use $\{p\} \times \mathbb{R}^N$ so that a tangent vector based at p is taken to be a pair (p, v) where $v \in \mathbb{R}^N$. The set $\{p\} \times \mathbb{R}^N$ inherits a vector space structure from $\mathbb{R}^N$ in the obvious way. In this context, we *provisionally* denote $\{p\} \times \mathbb{R}^N$ by $T_p\mathbb{R}^N$ and refer to it as the tangent space at p. If we write $c(t) = (x^1(t), \ldots, x^N(t))$, then the velocity vector of a curve c at time[1] $t = 0$ is $(p, \frac{dx^1}{dt}(0), \ldots, \frac{dx^N}{dt}(0))$, which is based at $p = c(0)$. Ambiguously, both

---

[1] It is common to refer to the parameter t for a curve as "time", although it may have nothing to do with physical time in a given situation.

$(p, \frac{dx^1}{dt}(0), \ldots, \frac{dx^N}{dt}(0))$ and $(\frac{dx^1}{dt}(0), \ldots, \frac{dx^N}{dt}(0))$ are often denoted by $\dot{c}(0)$ or $c'(0)$. A bit more generally, if V is a finite-dimensional vector space, then V is a smooth manifold and the tangent space at $p \in$ V can be provisionally taken to be the set $\{p\} \times$ V. We use the notation $v_p := (p, v)$. If $v_p := (p, v)$ is a tangent vector at p, then v is called the **principal part** of v_p.

We have a natural isomorphism between $\mathbb{R}^N$ and $T_p\mathbb{R}^N$ given by $v \mapsto (p, v)$, for any p. Of course we also have a natural isomorphism $T_p\mathbb{R}^N \cong T_q\mathbb{R}^N$ for any pair of points given by $(p, v) \mapsto (q, v)$. This is sometimes referred to as **distant parallelism**. Here we see the reason that in the context of calculus on $\mathbb{R}^N$, the explicit construction of vectors based at a point is often deemed unnecessary. However, from the point of view of manifold theory, the tangent space at a point is a fundamental construction. We will define the notion of a tangent space at a point of a differentiable manifold, and it will be seen that there is, in general, no canonical way to identify tangent spaces at different points.

Actually, we shall give several (ultimately equivalent) definitions of the tangent space. Let us start with the special case of a submanifold of $\mathbb{R}^N$. A tangent vector at p can be variously thought of as the velocity of a curve, as a direction for a directional derivative, and also as a geometric object which has components that depend in a special way on the coordinates used. Let us explore these aspects in the case of a submanifold of $\mathbb{R}^N$. If M is an n-dimensional regular submanifold of $\mathbb{R}^N$, then a smooth curve $c: (-\epsilon, \epsilon) \to M$ is also a smooth curve into $\mathbb{R}^N$ and $\dot{c}(0)$ is normally thought of as a vector based at the point $p = c(0)$. This vector is tangent to M. The set of all vectors obtained in this way from curves into M is an n-dimensional *subspace* of the tangent space of $\mathbb{R}^N$ at p (described above). In this special case, this subspace could play the role of the tangent space of M at p. Let us tentatively accept this definition of the tangent space at p and denote it by T_pM.

Let $v_p := (p, v) \in T_pM$. There are three things we should notice about v_p. First, there are many different curves $c: (-\epsilon, \epsilon) \to M$ with $c(0) = p$ which all give the same tangent vector v_p, and there is an obvious equivalence relation among these curves: two curves passing through p at $t = 0$ are equivalent if they have the same velocity vector. Already one can see that perhaps this could be turned around so that we can think of a tangent vector as an equivalence class of curves. Curves would be equivalent if they agree infinitesimally in some appropriate sense.

The second thing that we wish to bring out is that a tangent vector can be used to construct a directional derivative operator. If $v_p = (p, v)$ is a tangent vector in $T_p\mathbb{R}^N$, then we have a directional derivative operator at p which is a map $C^\infty(\mathbb{R}^N) \to \mathbb{R}$ given by $f \mapsto Df(p)v$. Now if v_p is tangent

2.1. The Tangent Space

to M, we would like a similar map $C^\infty(M) \to \mathbb{R}$. If f is only defined on M, then we do not have Df to work with but we can just take our directional derivative to be the map given by

$$D_{v_p} : f \mapsto (f \circ c)'(0),$$

where $c : I \to M$ is any curve whose velocity at $t = 0$ is v_p. Later we use the abstract properties of such a directional derivative to actually *define* the notion of a tangent vector.

Finally, notice how v_p relates to charts for the submanifold. If $(U, \mathbf{y})$ is a chart on M with $p \in U$, then by inverting we obtain a map $\mathbf{y}^{-1} : V \to M$, which we may then think of as a map into the ambient space $\mathbb{R}^N$. The map $\mathbf{y}^{-1}$ parameterizes a portion of M. For convenience, let us suppose that $\mathbf{y}^{-1}(0) = p$. Then we have the "coordinate curves" $y^i \mapsto \mathbf{y}^{-1}(0, \ldots, y^i, \ldots, 0)$ for $i = 1, \ldots, n$. The resulting tangent vectors E_i at p have principal parts given by the partial derivatives so that

$$E_i := \left(p, \frac{\partial \mathbf{y}^{-1}}{\partial y^i}(0) \right).$$

It can be shown that $(E_1, \ldots, E_n)$ is a basis for $T_p M$. For another coordinate system $\bar{\mathbf{y}}$ with $\bar{\mathbf{y}}^{-1}(0) = p$, we similarly define a basis $(\bar{E}_1, \ldots, \bar{E}_n)$. If $v_p = \sum_{i=1}^n a^i E_i = \sum_{i=1}^n \bar{a}^i \bar{E}_i$, then letting $a = (a^1, \ldots, a^n)$ and $\bar{a} = (\bar{a}^1, \ldots, \bar{a}^n)$, the chain rule can be used to show that

$$\bar{a} = D(\bar{\mathbf{y}} \circ \mathbf{y}^{-1})\big|_{\mathbf{y}(p)} a,$$

which is classically written as

$$\bar{a}^i = \sum_{j=1}^n \frac{\partial \bar{y}^i}{\partial y^j} a^j.$$

Both $(a^1, \ldots, a^n)$ and $(\bar{a}^1, \ldots, \bar{a}^n)$ represent the tangent vector v_p, but with respect to different charts. This is a simple example of a transformation law.

The various definitions for the notion of a tangent vector given below in the general setting will be based in turn on the following three ideas: (1) Equivalence classes of curves through a point. (2) Transformation laws for the components of a tangent vector with respect to various charts. (3) The idea of a "derivation" which is a kind of abstract directional derivative. Of course we will also have to show how to relate these various definitions to see that they are really equivalent.

2.1.1. Tangent space via curves. Let p be a point in a smooth n-manifold M. Suppose that we have smooth curves c_1 and c_2 mapping into M, each with open interval domains containing $0 \in \mathbb{R}$ and with $c_1(0) = c_2(0) = p$. We say that c_1 is tangent to c_2 at p if for all smooth real-valued functions

f defined on an open neighborhood of p, we have $(f \circ c_1)'(0) = (f \circ c_2)'(0)$. This is an equivalence relation on the set of all such curves. The reader should check that this really is an equivalence relation and also do so when we introduce other simple equivalence relations later. Define a **tangent vector at** p to be an equivalence class under this relation.

Notation 2.1. The equivalence class of c will be denoted by $[c]$, but we also denote tangent vectors by notation such as v_p or X_p, etc. Eventually we will often denote tangent vectors simply as v, w, etc., but for the discussion to follow we reserve these letters without the subscript for elements of $\mathbb{R}^n$ for some n.

If $v_p = [c]$ then we will also write $\dot{c}(0) = v_p$. The **tangent space** T_pM is defined to be the set of all tangent vectors at $p \in M$. A simple cut-off function argument shows that c_1 is equivalent to c_2 if and only if $(f \circ c_1)'(0) = (f \circ c_2)'(0)$ for all globally defined smooth functions $f : M \to \mathbb{R}$.

Lemma 2.2. *c_1 is tangent to c_2 at p if and only if $(f \circ c_1)'(0) = (f \circ c_2)'(0)$ for all $\mathbb{R}^k$-valued functions f defined on an open neighborhood of p.*

Proof. If $f = (f^1, \ldots, f^n)$, then $(f \circ c_1)'(0) = (f \circ c_2)'(0)$ if and only if $(f^i \circ c_1)'(0) = (f^i \circ c_2)'(0)$ for $i = 1, \ldots, k$. Thus $(f \circ c_1)'(0) = (f \circ c_2)'(0)$ if c_1 is tangent to c_2 at p. Conversely, let g be a smooth real-valued function defined on an open neighborhood of p and consider the map $f = (g, 0, \ldots, 0)$. Then the equality $(f \circ c_1)'(0) = (f \circ c_2)'(0)$ implies that $(g \circ c_1)'(0) = (g \circ c_2)'(0)$. $\square$

The definition of tangent space just given is very geometric, but it has one disadvantage. Namely, it is not immediately obvious that T_pM is a vector space in a natural way. The following principle is used to obtain a vector space structure:

Proposition 2.3 (Consistent transfer of linear structure). *Suppose that S is a set and $\{V_\alpha\}_{\alpha \in A}$ is a family of n-dimensional vector spaces. Suppose that for each α we have a bijection $b_\alpha : V_\alpha \to S$. If for every $\alpha, \beta \in A$ the map $b_\beta^{-1} \circ b_\alpha : V_\alpha \to V_\beta$ is a linear isomorphism, then there is a unique vector space structure on the set S such that each b_α is a linear isomorphism.*

Proof. Define addition in S by $s_1 + s_2 := b_\alpha(b_\alpha^{-1}(s_1) + b_\alpha^{-1}(s_2))$. This definition is independent of the choice of α. Indeed,

$$b_\alpha(b_\alpha^{-1}(s_1) + b_\alpha^{-1}(s_2)) = b_\alpha[b_\alpha^{-1} \circ b_\beta \circ b_\beta^{-1}(s_1) + b_\alpha^{-1} \circ b_\beta \circ b_\beta^{-1}(s_2)]$$

$$= b_\alpha \circ b_\alpha^{-1} \circ b_\beta \left[b_\beta^{-1}(s_1) + b_\beta^{-1}(s_2)\right]$$

$$= b_\beta \left(b_\beta^{-1}(s_1) + b_\beta^{-1}(s_2)\right).$$

2.1. The Tangent Space

The definition of scalar multiplication is $a \cdot s := b_\alpha(a\, b_\alpha^{-1}(s))$, and this is shown to be independent of α in a similar way. The axioms of a vector space are satisfied precisely because they are satisfied by each V_α. □

We will use the above proposition to show that there is a natural vector space structure on T_pM. For every chart $(\mathbf{x}_\alpha, U_\alpha)$ with $p \in U$, we have a map $b_\alpha : \mathbb{R}^n \to T_pM$ given by $v \mapsto [\gamma_v]$, where $\gamma_v : t \mapsto \mathbf{x}_\alpha^{-1}(\mathbf{x}_\alpha(p) + tv)$ for t in a sufficiently small but otherwise irrelevant interval containing 0.

Lemma 2.4. *For each chart $(U_\alpha, \mathbf{x}_\alpha)$, the map $b_\alpha : \mathbb{R}^n \to T_pM$ is a bijection and $b_\beta^{-1} \circ b_\alpha = D\left(\mathbf{x}_\beta \circ \mathbf{x}_\alpha^{-1}\right)(\mathbf{x}_\alpha(p))$.*

Proof. We have
$$(\mathbf{x}_\alpha \circ \gamma_v)'(0) = \left.\frac{d}{dt}\right|_{t=0} \mathbf{x}_\alpha \circ \mathbf{x}_\alpha^{-1}(\mathbf{x}_\alpha(p) + tv)$$
$$= \left.\frac{d}{dt}\right|_{t=0} (\mathbf{x}_\alpha(p) + tv) = v.$$

Suppose that $[\gamma_v] = [\gamma_w]$ for $v, w \in \mathbb{R}^n$. Then by Lemma 2.2 we have
$$v = (\mathbf{x}_\alpha \circ \gamma_v)'(0) = (\mathbf{x}_\alpha \circ \gamma_w)'(0) = w.$$

This means that b_α is injective.

Next we show that b_α is surjective. Let $[c] \in T_pM$ be represented by $c : (-\epsilon, \epsilon) \to M$. Let $v := (\mathbf{x}_\alpha \circ c)'(0) \in \mathbb{R}^n$. Then we have $b_\alpha(v) = [\gamma_v]$, where $\gamma_v : t \mapsto \mathbf{x}_\alpha^{-1}(\mathbf{x}_\alpha(p)+tv)$. But $[\gamma_v] = [c]$ since for any smooth f defined near p we have

$$(f \circ \gamma_v)'(0) = \left.\frac{d}{dt}\right|_{t=0} f \circ \mathbf{x}_\alpha^{-1}(\mathbf{x}_\alpha(p) + tv) = D\left(f \circ \mathbf{x}_\alpha^{-1}\right)(\mathbf{x}_\alpha(p)) \cdot v$$
$$= D\left(f \circ \mathbf{x}_\alpha^{-1}\right)(\mathbf{x}_\alpha(p)) \cdot (\mathbf{x}_\alpha \circ c)'(0) = (f \circ c)'(0).$$

Thus b_α is surjective. From Lemma 2.2 we see that the map $[c] \mapsto (\mathbf{x}_\alpha \circ c)'(0)$ is well-defined, and from the above we see that this map is exactly b_α^{-1}. Thus
$$b_\beta^{-1} \circ b_\alpha(v) = \left.\frac{d}{dt}\right|_{t=0} \mathbf{x}_\beta \circ \mathbf{x}_\alpha^{-1}(\mathbf{x}_\alpha(p) + tv)) = D\left(\mathbf{x}_\beta \circ \mathbf{x}_\alpha^{-1}\right)(\mathbf{x}_\alpha(p))v. \quad □$$

The above lemma and proposition combine to provide a vector space structure on the set of tangent vectors. Let us temporarily call the tangent space defined above, the **kinematic tangent space** and denote it by $(T_pM)_{\text{kin}}$. Thus, if $\mathcal{C}_p$ is the set of smooth curves c defined on some open interval containing 0 such that $c(0) = p$, then
$$(T_pM)_{\text{kin}} = \mathcal{C}_p/\sim,$$
where the equivalence is as described above.

Exercise 2.5. Let c_1 and c_2 be smooth curves mapping into a smooth manifold M, each with open interval domains containing $0 \in \mathbb{R}$ and with $c_1(0) = c_2(0) = p$. Show that

$$(f \circ c_1)'(0) = (f \circ c_2)'(0)$$

for all smooth f if and only if the curves $\mathbf{x} \circ c_1$ and $\mathbf{x} \circ c_2$ have the same velocity vector in $\mathbb{R}^n$ for some and hence any chart $(U, \mathbf{x})$.

2.1.2. Tangent space via charts. Let $\mathcal{A}$ be the maximal atlas for an n-manifold M. For fixed $p \in M$, consider the set Γ_p of all triples $(p, v, (U, \mathbf{x})) \in \{p\} \times \mathbb{R}^n \times \mathcal{A}$ such that $p \in U$. Define an equivalence relation on Γ_p by requiring that $(p, v, (U, \mathbf{x})) \sim (p, w, (V, \mathbf{y}))$ if and only if

(2.1) $$w = D(\mathbf{y} \circ \mathbf{x}^{-1})\big|_{\mathbf{x}(p)} \cdot v.$$

In other words, the derivative at $\mathbf{x}(p)$ of the coordinate change $\mathbf{y} \circ \mathbf{x}^{-1}$ "identifies" v with w. The set $\Gamma_p/\sim$ of equivalence classes can be given a vector space structure as follows: For each chart $(U, \mathbf{x})$ containing p, we have a map $b_{(U,\mathbf{x})} : \mathbb{R}^n \to \Gamma_p/\sim$ given by $v \mapsto [p, v, (U, \mathbf{x})]$, where $[p, v, (U, \mathbf{x})]$ denotes the equivalence class of $(p, v, (U, \mathbf{x}))$. To see that this map is a bijection, notice that if $[p, v, (U, \mathbf{x})] = [p, w, (U, \mathbf{x})]$, then

$$v = D(\mathbf{x} \circ \mathbf{x}^{-1})\big|_{\mathbf{x}(p)} \cdot v = w$$

by definition. By Proposition 2.3 we obtain a vector space structure on $\Gamma_p/\sim$ whose elements are **tangent vectors**. This is another version of the tangent space at p, and we shall (temporarily) denote this by $(T_pM)_{\text{phys}}$. The subscript "phys" refers to the fact that this version of the tangent space is based on a "transformation law" and corresponds to a way of looking at things that has traditionally been popular among physicists. If $v_p = [p, v, (U, \mathbf{x})] \in (T_pM)_{\text{phys}}$, then we say that $v \in \mathbb{R}^n$ represents v_p with respect to the chart $(U, \mathbf{x})$.

This viewpoint takes on a more familiar appearance if we use a more classical notation. Let $(U, \mathbf{x})$ and $(V, \mathbf{y})$ be two charts containing p in their domains. If an n-tuple $(v^1, \ldots, v^n)$ represents a tangent vector at p from the point of view of $(U, \mathbf{x})$, and if the n-tuple $(w^1, \ldots, w^n)$ represents the same vector from the point of view of $(V, \mathbf{y})$, then (2.1) is expressed in the form

(2.2) $$w^i = \sum_{j=1}^n \frac{\partial y^i}{\partial x^j}\bigg|_{\mathbf{x}(p)} v^j,$$

where we write the change of coordinates as $y^i = y^i(x^1, \ldots, x^n)$ with $1 \leq i \leq n$.

Notation 2.6. It is sometimes convenient to index the maximal atlas: $\mathcal{A} = \{(U_\alpha, \mathbf{x}_\alpha)\}_{\alpha \in A}$. Then we would consider triples of the form (p, v, α) and let

2.1. The Tangent Space

the defining equivalence relation for $(T_pM)_{\text{phys}}$ be $(p,v,\alpha) \sim (p,w,\beta)$ if and only if
$$D(\mathbf{x}_\beta \circ \mathbf{x}_\alpha^{-1})\big|_{\mathbf{x}_\alpha(p)} \cdot v = w.$$

2.1.3. Tangent space via derivations. We abstract the notion of directional derivative for our next approach to the tangent space. There are actually at least two common versions of this approach, and we explain both. Let M be a smooth manifold of dimension n. A **tangent vector** v_p at p is a linear map $v_p : C^\infty(M) \to \mathbb{R}$ with the property that for $f, g \in C^\infty(M)$,
$$v_p(fg) = g(p)v_p(f) + f(p)v_p(g).$$
This is the **Leibniz law**. We may say that a tangent vector at p is a **derivation** of the algebra $C^\infty(M)$ with respect to the evaluation map ev_p at p defined by $\text{ev}_p(f) := f(p)$. Alternatively, we say that v_p is a **derivation at** p. The set of such derivations at p is easily seen to be a vector space which is called the **tangent space** at p and is denoted by T_pM. We temporarily distinguish this version of the tangent space from $(T_pM)_{\text{kin}}$ and $(T_pM)_{\text{phys}}$ defined previously by denoting it $(T_pM)_{\text{alg}}$ and referring to it as the **algebraic tangent space**. We could also consider the vector space of derivations of $C^r(M)$ at a point for $r < \infty$, but this would not give a finite-dimensional vector space and so is not a good candidate for the definition of the tangent space (see Problem 18). Recall that if $(U, \mathbf{x})$ is a chart on an n-manifold M, we have defined $\partial f/\partial x^i$ by
$$\frac{\partial f}{\partial x^i}(p) := D_i\left(f \circ \mathbf{x}^{-1}\right)(\mathbf{x}(p))$$
(see Definition 1.57).

Definition 2.7. Given $(U, \mathbf{x})$ and p as above, define the operator $\frac{\partial}{\partial x^i}\big|_p : C^\infty(M) \to \mathbb{R}$ by
$$\frac{\partial}{\partial x^i}\bigg|_p f := \frac{\partial f}{\partial x^i}(p).$$

It is often helpful to use the easily verified fact that if $c^i : (-\epsilon, \epsilon) \to M$ is the curve defined for sufficiently small ϵ by
$$c_i(t) := \mathbf{x}^{-1}(\mathbf{x}(p) + \mathbf{e}_i),$$
where $\mathbf{e}_i$ is the i-th member of the standard basis of $\mathbb{R}^n$, then
$$\frac{\partial}{\partial x^i}\bigg|_p f = \lim_{h \to 0} \frac{f(c_i(h)) - f(p)}{h}.$$
From the usual product rule it follows that $\frac{\partial}{\partial x^i}\big|_p$ is a derivation at p and so is an element of $(T_pM)_{\text{alg}}$. We will show that $\left(\frac{\partial}{\partial x^1}\big|_p, \ldots, \frac{\partial}{\partial x^n}\big|_p\right)$ is a basis for the vector space $(T_pM)_{\text{alg}}$.

Lemma 2.8. *Let $v_p \in (T_pM)_{\text{alg}}$. Then*

  (i) *if $f, g \in C^\infty(M)$ are equal on some neighborhood of p, then $v_p(f) = v_p(g)$;*

  (ii) *if $h \in C^\infty(M)$ is constant on some neighborhood of p, then $v_p(h) = 0$.*

Proof. (i) Since v_p is a linear map, it suffices to show that if $f = 0$ on a neighborhood U of p, then $v_p(f) = 0$. Of course $v_p(0) = 0$. Let β be a cut-off function with support in U and $\beta(p) = 1$. Then we have that βf is identically zero and so

$$0 = v_p(\beta f) = f(p)v_p(\beta) + \beta(p)v_p(f)$$
$$= v_p(f) \quad (\text{since } \beta(p) = 1 \text{ and } f(p) = 0).$$

(ii) From what we have just shown, it suffices to assume that h is equal to a constant c globally on M. In the special case $c = 1$, we have

$$v_p(1) = v_p(1 \cdot 1) = 1 \cdot v_p(1) + 1 \cdot v_p(1) = 2v_p(1),$$

so that $v_p(1) = 0$. Finally we have $v_p(c) = v_p(1c) = c(v_p(1)) = 0$. □

Notation 2.9. We shall often write $v_p f$ or $v_p \cdot f$ in place of $v_p(f)$.

We must now deal with a technical issue. We anticipate that the action of a derivation is really a differentiation and so it seems that a derivation at p should be able to act on a function defined only in some neighborhood U of p. It is pretty easy to see how this would work for $\frac{\partial}{\partial x^i}\big|_p$. But the domain of a derivation as defined is the ring $C^\infty(M)$ and not $C^\infty(U)$. There is nothing in the definition that immediately allows an element of $(T_pM)_{\text{alg}}$ to act on $C^\infty(U)$ unless $U = M$. It turns out that we can in fact identify $(T_pU)_{\text{alg}}$ with $(T_pM)_{\text{alg}}$, and the following discussion shows how this is done. Once we reach a fuller understanding of the tangent space, this identification will be natural and automatic. So, let $p \in U \subset M$ with U open. We construct a rather obvious map $\Phi : (T_pU)_{\text{alg}} \to (T_pM)_{\text{alg}}$ by using the restriction map $C^\infty(M) \to C^\infty(U)$. For each $w_p \in T_pU$, we define $\widetilde{w_p} : C^\infty(M) \to \mathbb{R}$ by $\widetilde{w_p}(f) := w_p(f|_U)$. It is easy to show that $\widetilde{w_p}$ is a derivation of the appropriate type and so $\widetilde{w_p} \in (T_pM)_{\text{alg}}$. Thus we get a linear map $\Phi : (T_pU)_{\text{alg}} \to (T_pM)_{\text{alg}}$. We want to show that this map is an isomorphism, but notice that we have not yet established the finite-dimensionality of either $(T_pU)_{\text{alg}}$ or $(T_pM)_{\text{alg}}$. First we show that $\Phi : w_p \mapsto \widetilde{w_p}$ has trivial kernel. So suppose that $\widetilde{w_p} = 0$, i.e. $\widetilde{w_p}(f) = 0$ for all $f \in C^\infty(M)$. Let $h \in C^\infty(U)$. Pick a cut-off function β with support in U so that βh extends by zero to a smooth function f on all of M that agrees with h on a neighborhood of

2.1. The Tangent Space

p. Then by the above lemma, $w_p(h) = w_p(f|_U) = \widetilde{w_p}(f) = 0$. Thus, since h was arbitrary, we see that $w_p = 0$ and so Φ has trivial kernel.

Next we show that Φ is *onto*. Let $v_p \in (T_pM)_{\text{alg}}$. We wish to define $w_p \in (T_pU)_{\text{alg}}$ by $w_p(h) := v_p(\beta h)$, where β is as above and βh is extended by zero to a function in $C^\infty(M)$. If β_1 is another similar choice of cut-off function, then βh and $\beta_1 h$ (both extended to all of M) agree on a neighborhood of p, and so by Lemma 2.8, $v_p(\beta h) = v_p(\beta_1 h)$. Thus w_p is well-defined. Thinking of $\beta(f|_U)$ as defined on M, we have $\widetilde{w_p}(f) := w_p(f|_U) = v_p(\beta f|_U) = v_p(f)$ since $\beta f|_U$ and f agree on a neighborhood of p. Thus $\Phi : (T_pU)_{\text{alg}} \to (T_pM)_{\text{alg}}$ is an isomorphism.

Because of this isomorphism, we tend to identify $(T_pU)_{\text{alg}}$ with $(T_pM)_{\text{alg}}$ and in particular, if $(U, \mathbf{x})$ is a chart, we think of the derivations $\frac{\partial}{\partial x^i}\big|_p$, $1 \leq i \leq n$ as being simultaneously elements of both $(T_pU)_{\text{alg}}$ and $(T_pM)_{\text{alg}}$. In either case the formula is the same: $\frac{\partial}{\partial x^i}\big|_p f = \frac{\partial(f \circ \mathbf{x}^{-1})}{\partial u^i}(\mathbf{x}(p))$.

Notice that agreeing on a neighborhood of a point is an important relation here and this provides motivation for employing the notion of a germ of a function (Definition 1.68). First we establish the basis theorem:

Theorem 2.10. *Let M be an n-manifold and $(U, \mathbf{x})$ a chart with $p \in U$. Then the n-tuple of vectors (derivations) $\left(\frac{\partial}{\partial x^1}\big|_p, \ldots, \frac{\partial}{\partial x^n}\big|_p\right)$ is a basis for $(T_pM)_{\text{alg}}$. Furthermore, for each $v_p \in (T_pM)_{\text{alg}}$ we have*

$$v_p = \sum_{i=1}^n v_p(x^i) \frac{\partial}{\partial x^i}\bigg|_p.$$

Proof. From our discussion above we may assume that $\mathbf{x}(U)$ is a convex set such as a ball of radius ε in $\mathbb{R}^n$. By composing with a translation we assume that $\mathbf{x}(p) = 0$. This makes no difference for what we wish to prove since v_p applied to a constant is 0. For any smooth function g defined on the convex set $\mathbf{x}(U)$ let

$$g_i(u) := \int_0^1 \frac{\partial g}{\partial u^i}(tu)\, dt \text{ for all } u \in \mathbf{x}(U).$$

The fundamental theorem of calculus can be used to show that $g = g(0) + \sum g_i u^i$. We see that $g_i(0) = \frac{\partial g}{\partial u^i}\big|_0$. For a function $f \in C^\infty(U)$, we let $g := f \circ \mathbf{x}^{-1}$. Using the above, we arrive at the expression $f = f(p) + \sum f_i x^i$, and applying $\frac{\partial}{\partial x^i}\big|_p$ we get $f_i(p) = \frac{\partial f}{\partial x^i}\big|_p$. Now apply the derivation v_p to

$f = f(p) + \sum f_i x^i$ to obtain
$$v_p f = 0 + \sum v_p(f_i x^i)$$
$$= \sum v_p(x^i) f_i(p) + \sum 0 \, v_p f_i$$
$$= \sum v_p(x^i) \left.\frac{\partial f}{\partial x^i}\right|_p.$$

This shows that $v_p = \sum v_p(x^i) \left.\frac{\partial}{\partial x^i}\right|_p$ and thus we have a spanning set. To see that $(\left.\frac{\partial}{\partial x^i}\right|_p, \ldots, \left.\frac{\partial}{\partial x^i}\right|_p)$ is a linearly independent set, let us assume that $\sum a^i \left.\frac{\partial}{\partial x^i}\right|_p = 0$ (the zero derivation). Applying this to x^j gives $0 = \sum a^i \left.\frac{\partial x^j}{\partial x^i}\right|_p = \sum a^i \delta_i^j = a^j$, and since j was arbitrary, we get the result. □

Remark 2.11. On the manifold $\mathbb{R}^n$, we have the identity map $\mathrm{id} : \mathbb{R}^n \to \mathbb{R}^n$ which gives the standard chart. As is often the case, the simplest situations have the most confusing notation because of the various identifications that may exist. On $\mathbb{R}$, there is one coordinate function, which we often denote by either u or t. This single function is just $\mathrm{id}_\mathbb{R}$. The basis vector at $t_0 \in \mathbb{R}$ associated to this coordinate is $\left.\frac{\partial}{\partial u}\right|_{t_0}$ (or $\left.\frac{\partial}{\partial t}\right|_{t_0}$). If we think of the tangent space at $t_0 \in \mathbb{R}$ as being $\{t_0\} \times \mathbb{R}$, then $\left.\frac{\partial}{\partial u}\right|_{t_0}$ is just $(t_0, 1)$. It is also common to denote $\left.\frac{\partial}{\partial u}\right|_{t_0}$ by "1" regardless of the point t_0.

Above, we used the notion of a derivation as one way to define a tangent vector. There is a slight variation of this approach that allows us to worry a bit less about the relation between $(T_p U)_{\mathrm{alg}}$ and $(T_p M)_{\mathrm{alg}}$. Let $\mathcal{F}_p = C_p^\infty(M, \mathbb{R})$ be the algebra of germs of functions defined near p. Recall that if f is a representative for the equivalence class $[f] \in \mathcal{F}_p$, then we can unambiguously define the value of $[f]$ at p by $[f](p) = f(p)$. Thus we have an evaluation map $\mathrm{ev}_p : \mathcal{F}_p \to \mathbb{R}$.

Definition 2.12. A **derivation** (with respect to the evaluation map ev_p) of the algebra $\mathcal{F}_p$ is a map $\mathcal{D}_p : \mathcal{F}_p \to \mathbb{R}$ such that $\mathcal{D}_p([f][g]) = f(p)\mathcal{D}_p[g] + g(p)\mathcal{D}_p[f]$ for all $[f], [g] \in \mathcal{F}_p$.

The set of all these derivations on $\mathcal{F}_p$ is easily seen to be a real vector space and is sometimes denoted by $\mathrm{Der}(\mathcal{F}_p)$.

Remark 2.13. The notational distinction between a function and its germ at a point is not always maintained; $\mathcal{D}_p f$ is taken to mean $\mathcal{D}_p[f]$.

Let M be a smooth manifold of dimension n. Consider the set of all germs of C^∞ functions $\mathcal{F}_p$ at $p \in M$. The vector space $\mathrm{Der}(\mathcal{F}_p)$ of derivations of $\mathcal{F}_p$ with respect to the evaluation map ev_p could also be taken as the definition of the tangent space at p. This would be a slight variation of what we have called the algebraic tangent space.

2.2. Interpretations

We will now show how to move from one definition of tangent vector to the next. Let M be a (smooth) n-manifold. Consider a tangent vector v_p as an equivalence class of curves represented by $c : I \to M$ with $c(0) = p$. We obtain a derivation by defining

$$v_p f := \left.\frac{d}{dt}\right|_{t=0} f \circ c.$$

This gives a map $(T_p M)_{\text{kin}} \to (T_p M)_{\text{alg}}$ which can be shown to be an isomorphism. We also have a natural isomorphism $(T_p M)_{\text{kin}} \to (T_p M)_{\text{phys}}$. Given $[c] \in (T_p M)_{\text{kin}}$, we obtain an element $v_p \in (T_p M)_{\text{phys}}$ by letting v_p be the equivalence class of the triple $(p, v, (U, \mathbf{x}))$, where $v^i := \left.\frac{d}{dt}\right|_{t=0} x^i \circ c$ for a chart $(U, \mathbf{x})$ with $p \in U$.

If v_p is a derivation at p and $(U, \mathbf{x})$ an admissible chart with domain containing p, then v_p, as a tangent vector in the sense of Definition 2.1.2, is represented by the triple $(p, v, (U, \mathbf{x}))$, where $v = (v^1, \ldots, v^n)$ is given by

$$v^i = v_p x^i \quad (v_p \text{ is acting as a derivation}).$$

This gives us an isomorphism $(T_p M)_{\text{alg}} \to (T_p M)_{\text{phys}}$.

Next we exhibit the inverse isomorphism $(T_p M)_{\text{phys}} \to (T_p M)_{\text{alg}}$. Suppose that $[(p, v, (U, \mathbf{x}))] \in (T_p M)_{\text{phys}}$ where $v \in \mathbb{R}^n$. We obtain a derivation by defining

$$v_p f = D(f \circ \mathbf{x}^{-1})\big|_{\mathbf{x}(p)} \cdot v.$$

In other words,

$$v_p f = \sum_{i=1}^n v^i \left.\frac{\partial}{\partial x^i}\right|_p f$$

for $v = (v^1, \ldots, v^n)$. It is an easy exercise that v_p defined in this way is independent of the representative triple $(p, v, (U, \mathbf{x}))$.

We now adopt the explicitly flexible attitude of interpreting a tangent vector in any of the ways we have described above depending on the situation. Thus we effectively identify the spaces $(T_p M)_{\text{kin}}, (T_p M)_{\text{phys}}$ and $(T_p M)_{\text{alg}}$. Henceforth we use the notation $T_p M$ for the **tangent space** of a manifold M at a point p.

Definition 2.14. The dual space to a tangent space $T_p M$ is called the **cotangent space** and is denoted by $T_p^* M$. An element of $T_p^* M$ is referred to as a **covector**.

The basis for $T_p^* M$ that is dual to the coordinate basis ($\left.\frac{\partial}{\partial x^1}\right|_p, \ldots, \left.\frac{\partial}{\partial x^n}\right|_p$) described above is denoted ($\left. dx^1\right|_p, \ldots, \left. dx^n\right|_p$). By definition $\left.dx^i\right|_p \left(\left.\frac{\partial}{\partial x^j}\right|_p\right) = \delta^i_j$. (Note that $\delta^i_j = 1$ if $i = j$ and $\delta^i_j = 0$ if $i \neq j$. The symbols δ_{ij} and

δ^{ij} are defined similarly.) The reason for the differential notation dx^i will be explained below. Sometimes one abbreviates $\frac{\partial}{\partial x^j}\big|_p$ and $dx^i\big|_p$ to $\frac{\partial}{\partial x^j}$ and dx^i respectively, but there is some risk of confusion since later $\frac{\partial}{\partial x^j}$ and dx^i will more properly denote not elements of the vector spaces $T_p M$ and $T_p^* M$, but rather *fields* defined over a chart domain. More on this shortly.

2.2.1. Tangent space of a vector space. Our provisional definition of the tangent space at point p in a vector space V was the set $\{p\} \times V$, but this set does not immediately fit any of the definitions of tangent space just given. This is remedied by finding a natural isomorphism $\{p\} \times V \cong T_p V$. One may pick a version of the tangent space and then exhibit a natural isomorphism directly, but we take a slightly different approach. Namely, we first define a natural map $\jmath_p : V \to T_p V$. We think in terms of equivalence classes of curves. For each $v \in V$, let $c_{p,v} : \mathbb{R} \to V$ be the curve $c_{p,v}(t) := p + tv$. Then

$$\jmath_p(v) := [c_{p,v}] \in T_p V.$$

As a derivation, $\jmath_p(v)$ acts according to

$$f \longmapsto \frac{d}{dt}\bigg|_0 f(p + tv).$$

On the other hand, we have the obvious projection $\mathrm{pr}_2 : \{p\} \times V \to V$. Then our natural isomorphism $\{p\} \times V \cong T_p V$ is just $\jmath_p \circ \mathrm{pr}_2$. The isomorphism between the vector spaces $\{p\} \times V$ and $T_p V$ is so natural that they are often identified. Of course, since $T_p V$ itself has various manifestations $((T_p V)_{\mathrm{alg}}, (T_p V)_{\mathrm{phys}}$, and $(T_p V)_{\mathrm{kin}}$, we now have a multitude of spaces which are potentially being identified in the case of a vector space.

Note: Because of the identification of $\{p\} \times V$ with $T_p V$, we shall often denote by pr_2 the map $T_p V \to V$. Furthermore, in certain contexts, $T_p V$ is identified with V itself. The potential identifications introduced here are often referred to as "canonical" or "natural" and $\jmath_p : V \to T_p V$ is often called the **canonical** or **natural** isomorphism. The inverse map $T_p V \to V$ is also referred to as the **canonical** or **natural** isomorphism. Context will keep things straight.

2.3. The Tangent Map

The first definition given below of the tangent map at $p \in M$ of a smooth map $f : M \to N$ will be considered our main definition, but the others are actually equivalent. Given f and p as above, we wish to define a linear map $T_p f : T_p M \to T_{f(p)} N$. Since we have several definitions of tangent space, we expect to see several equivalent definitions of the tangent map. For the first definition we think of $T_p M$ as $(T_p M)_{\mathrm{kin}}$.

2.3. The Tangent Map

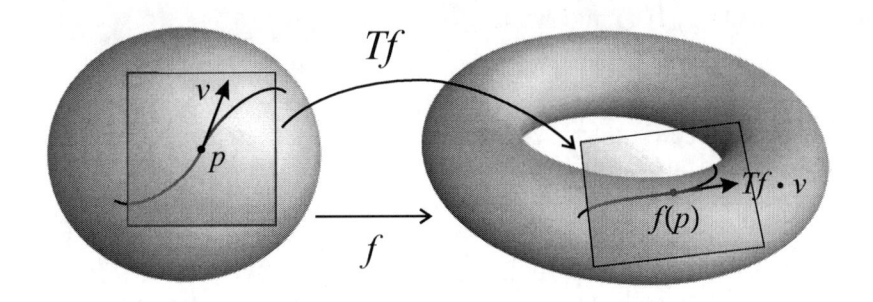

Figure 2.1. Tangent map

Definition 2.15 (Tangent map I). If we have a smooth function between manifolds
$$f : M \to N,$$
and we consider a point $p \in M$ and its image $q = f(p) \in N$, then we define the **tangent map** at p,
$$T_p f : T_p M \to T_q N,$$
in the following way: Suppose that $v_p \in T_p M$ and we pick a curve c with $c(0) = p$ so that $v_p = [c]$; then by definition
$$T_p f \cdot v_p = [f \circ c] \in T_q N,$$
where $[f \circ c] \in T_q N$ is the vector represented by the curve $f \circ c$. (Recall Notation 1.3.)

Another popular way to denote the tangent map $T_p f$ is f_{p*}, but a further abbreviation to f_* is dangerous since it conflicts with a related meaning for f_* introduced later.

Exercise 2.16. Let $f : M \to N$ be a smooth map. Show that $T_p f : T_p M \to T_q N$ is a linear map and that if f is a diffeomorphism, then $T_p f$ is a linear isomorphism. [Hint: Think about how the linear structure on $(T_p M)_{\text{kin}}$ was defined.]

We have the following version of the **chain rule for tangent maps**:

Theorem 2.17. *Let $f : M \to N$ and $g : N \to P$ be smooth maps. For each $p \in M$ we have $T_p(g \circ f) = (T_{f(p)} g) \circ T_p f$.*

Proof. Let $v \in T_pM$ be represented by the curve c so that $v = [c]$. Then $f \circ c$ represents $T_pf(v)$ and we have

$$T_p(g \circ f)(v) = [(g \circ f) \circ c] = [g \circ (f \circ c)]$$
$$= \left(T_{f(p)}g\right)(T_pf(v)) = \left(\left(T_{f(p)}g\right) \circ T_pf\right)(v). \qquad \square$$

For the next alternative definition of tangent map, we consider T_pM as $(T_pM)_{\text{phys}}$.

Definition 2.18 (Tangent map II). Let $f : M \to N$ be a smooth map and consider a point $p \in M$ with image $q = f(p) \in N$. Choose any chart $(U, \mathbf{x})$ containing p and a chart $(V, \mathbf{y})$ containing $q = f(p)$ so that for $v_p \in T_pM$ we have the representative $(p, v, (U, \mathbf{x}))$. Then the **tangent map** $T_pf : T_pM \to T_{f(p)}N$ is defined by letting the representative of $T_pf \cdot v_p$ in the chart $(V, \mathbf{y})$ be given by $(q, w, (V, \mathbf{y}))$, where

$$w = D(\mathbf{y} \circ f \circ \mathbf{x}^{-1}) \cdot v.$$

This uniquely determines $T_pf \cdot v$, and the chain rule guarantees that this is well-defined (independent of the choice of charts).

Another alternative definition of tangent map is given in terms of derivations:

Definition 2.19 (Tangent map III). Let M be a smooth n-manifold. Continuing our set up above, we define $T_pf \cdot v_p$ as a derivation by

$$(T_pf \cdot v_p)g = v_p(g \circ f)$$

for each smooth function g. It is easy to check that this defines a derivation and so a tangent vector in T_qM. This map is yet another version of the **tangent map** T_pf.

In the above definition, one could take g to be the germ of a smooth function defined on a neighborhood of $f(p)$ and then $T_pf \cdot v_p$ would act as a derivation of such germs.

One can check the chain rule for tangent maps using the above definition in terms of derivations as follows: If $f : M \to N$ and $g : N \to P$ are smooth maps and $v_p \in T_pM$, then for $h \in C^\infty(M)$ we have

$$(T_p(g \circ f) \cdot v_p)h = v_p(h \circ (g \circ f)) = v_p((h \circ g) \circ f)$$
$$= (T_pf \cdot v_p)(h \circ g) = T_{f(p)}g \cdot (T_pf \cdot v_p)h,$$

so since h and v_p were arbitrary, we conclude again that $T_p(g \circ f) = \left(T_{f(p)}g\right) \circ T_pf$. We have just proved the same thing using different interpretations of tangent space and tangent map, but the isomorphisms between the versions are so natural that we may use any version convenient for a given purpose and then draw conclusions about all versions. This could be formalized

2.3. The Tangent Map

using category theory arguments but we shall forgo the endeavor. Just as with the idea of the tangent space, we will think of the above versions of the tangent map as a single abstract thing with more than one interpretation depending on the interpretation of tangent space in play.

Now we introduce the differential of a function.

Definition 2.20. Let M be a smooth manifold and let $p \in M$. For $f \in C^\infty(M)$, we define the **differential** of f at p as the linear map $df(p) : T_pM \to \mathbb{R}$ given by

$$df(p) \cdot v_p = v_p f$$

for all $v_p \in T_pM$. Thus $df(p) \in T^*M$.

The notation df_p or $df|_p$ is also used in place of $df(p)$. One may view $df_p(v_p)$ as an "infinitesimal" aspect of the composition $f \mapsto f \circ \gamma$, where $\gamma'(0) = v_p$. It is easy to see that $df(p)$ is just $T_p f$ followed by the natural map $T_{f(p)}\mathbb{R} \to \mathbb{R}$. In a way, $df(p)$ is just a version of the tangent map that takes advantage of the identification of $T_{f(p)}\mathbb{R}$ with $\mathbb{R}$ (recall Remark 2.11).

Let $(U, \mathbf{x})$ be a chart with $\mathbf{x} = (x^1, \ldots, x^n)$ and let $p \in U$. We previously denoted the basis dual to $\left(\frac{\partial}{\partial x^1}(p), \ldots, \frac{\partial}{\partial x^n}(p)\right)$ by $(dx^1|_p, \ldots, dx^n|_p)$, and now this notation is justified since we can check directly that we really do have $dx^i|_p \left(\frac{\partial}{\partial x^j}|_p\right) = \delta^i_j$.

Definition 2.21. Let $I = (a, b)$ be an interval in $\mathbb{R}$. If $c : I \to M$ is a smooth map (a curve), then the **velocity** at $t_0 \in I$ is the vector $\dot{c}(t_0) \in T_{c(t_0)}M$ defined by

$$\dot{c}(t_0) := T_{t_0} c \cdot \left.\frac{\partial}{\partial u}\right|_{t_0},$$

where $\left.\frac{\partial}{\partial u}\right|_{t_0}$ is the coordinate basis vector at $t_0 \in T_{t_0}I = T_{t_0}\mathbb{R}$ associated to the standard coordinate function on $\mathbb{R}$ (denoted here by u).

Note: We may also occasionally write c' for $\dot{c}$.

Thus if f is a smooth function defined in a neighborhood of $c(t_0)$, then $\dot{c}(t_0)$ acts as a derivation as follows:

$$\dot{c}(t_0) \cdot f = \left.\frac{d}{dt}\right|_{t=t_0} f \circ c;$$

$\dot{c}(t_0)$ is also denoted by $\left.\frac{d}{dt}\right|_{t_0} c$. In this notation, $\dot{c} = \frac{d}{dt}c$ is a map that assigns to every $t \in I$ the tangent vector $\dot{c}(t) := T_t c \cdot \left.\frac{\partial}{\partial u}\right|_t$, and this is referred to as the velocity field *along* the curve c. Also note that we may view $\dot{c}(t_0)$ as the equivalence class of the curve $t \mapsto c(t + t_0)$.

The differential can be generalized:

Definition 2.22. Let V be a vector space. For a smooth $f : M \to V$ with $p \in M$ as above, the differential $df(p) : T_pM \to V$ is the composition of the tangent map T_pf and the canonical map $T_yV \to V$ where $y = f(p)$,

$$df(p) : T_pM \xrightarrow{Tf} T_yV \to V.$$

The notational distinction between T_pf and df_p is not universal, and df_p is itself often used to denote T_pf.

Exercise 2.23. Let $f : M \to N$ be a smooth map. Show that if $T_pf = 0$ for all $p \in M$, then f is locally constant (constant on connected components of M).

We now consider the inclusion map $\iota : U \hookrightarrow M$ where U is open. For $p \in U$, we get the tangent map $T_p\iota : T_pU \to T_pM$. Let us look at this map from several points of view corresponding to the various ways one can define the tangent space. First, consider tangent spaces from the derivation point of view. From this point of view the map $T_p\iota$ is defined for $v_p \in T_pU$ as acting on $C^\infty(M)$ as follows: $T_p\iota(v_p)f = v_p(f \circ \iota) = v_p(f|U)$. We have seen this map before where we called it $\Phi : (T_pU)_{\text{alg}} \to (T_pM)_{\text{alg}}$, and it was observed to be an isomorphism and we decided to identify $(T_pU)_{\text{alg}}$ with $(T_pM)_{\text{alg}}$. From the point of view of equivalence classes of curves, the map $T_p\iota$ sends $[\gamma]$ to $[\iota \circ \gamma]$. But while γ is a curve into U, the map $\iota \circ \gamma$ is simply the same curve, but thought of as mapping into M. We leave it to the reader to verify the expected fact that $T_p\iota$ is a linear isomorphism. Thus it makes sense to identify $[\gamma]$ with $[\iota \circ \gamma]$ and so again to identify T_pU with T_pM via this isomorphism. Next consider $v_p \in T_pU$ to be represented by a triple $(p, v, (U_\alpha, \mathbf{x}_\alpha))$ where $(U_\alpha, \mathbf{x}_\alpha)$ is a chart on the open manifold U. Since $(U_\alpha, \mathbf{x}_\alpha)$ is also a chart on M, the triple also represents an element of T_pM which is none other than $T_p\iota \cdot v_p$. The map $T_p\iota$ looks more natural and trivial than ever, and we once again see the motivation for identifying T_pU and T_pM.

More generally, when S is a regular submanifold of M, then the tangent space T_pS at $p \in S \subset M$ is intuitively a subspace of T_pM. Again, this is true as long as one is not bent on distinguishing a curve in S through p from the "same" curve thought of as a map into M. If one wants to be pedantic, then we have the inclusion map $\iota : S \hookrightarrow M$, and if $c : I \to S$ is a curve into S, then $\iota \circ c : I \to M$ is a map into M. At the tangent level this means that $\dot{c}(0) \in T_pS$ while $(\iota \circ c)'(0) \in T_pM$. Thus $T_p\iota : T_pS \to T_p\iota(T_pS) \subset T_pM$. When convenient, we just identify T_pS with $T_p\iota(T_pS)$ and so think of T_pS as a subspace of T_pM and take $T_p\iota$ to be an inclusion.

2.3. The Tangent Map

Recall that a smooth pointed map $f : (M, p) \to (N, q)$ is a smooth map $f : M \to N$ such that $f(p) = q$. Taking the set of all pairs (M, p) as objects and pointed maps as morphisms we have an obvious category.

Definition 2.24. The "pointed" version of the **tangent functor** T takes (M, p) to $T_p M$ and a map $f : (M, p) \to (N, q)$ to the linear map $T_p f : T_p M \to T_q M$.

The following theorem is the **inverse mapping theorem for manifolds** and will be used repeatedly.

Theorem 2.25. *If $f : M \to N$ is a smooth map such that $T_p f : T_p M \to T_q N$ is an isomorphism, then there exists an open neighborhood O of p such that $f(O)$ is open and $f|_O : O \to f(O)$ is a diffeomorphism. If $T_p f$ is an isomorphism for all $p \in M$, then $f : M \to N$ is a local diffeomorphism.*

Proof. The proof is a simple application of the inverse mapping theorem, Theorem C.1 found in Appendix C. Let $(U, \mathbf{x})$ be a chart centered at p and let $(V, \mathbf{y})$ be a chart centered at $q = f(p)$ with $f(U) \subset V$. From the fact that $T_p f$ is an isomorphism we easily deduce that M and N have the same dimension, say n, and then $D(\mathbf{y} \circ f \circ \mathbf{x}^{-1})(\mathbf{x}(p)) : \mathbb{R}^n \to \mathbb{R}^n$ is an isomorphism. From Theorem C.1 if follows that $\mathbf{y} \circ f \circ \mathbf{x}^{-1}$ restricts to a diffeomorphism on some neighborhood O' of $0 \in \mathbb{R}^n$. From this we obtain that $f|_O : O \to f(O)$ is a diffeomorphism where $O = \mathbf{x}^{-1}(O')$. The second part follows from the first. $\square$

2.3.1. Tangent spaces on manifolds with boundary. Recall that a manifold with boundary is modeled on the half-spaces $\mathbb{R}^n_{\lambda \geq c} := \{a \in \mathbb{R}^n : \lambda(a) \geq c\}$. If M is a manifold with boundary, then the tangent space $T_p M$ is defined as before. For instance, even if $p \in \partial M$, the fiber $T_p M$ may still be thought of as consisting of equivalence classes where $(p, v, \alpha) \sim (p, w, \beta)$ if and only if $D(\mathbf{x}_\beta \circ \mathbf{x}_\alpha^{-1})|_{\mathbf{x}_\alpha(p)} \cdot v = w$. Notice that for a given chart $(U_\alpha, \mathbf{x}_\alpha)$, the vectors v in (p, v, α) still run through all of $\mathbb{R}^n$ and so $T_p M$ still has dimension n even if $p \in \partial M$. On the other hand, if $p \in \partial M$, then for any half-space chart $\mathbf{x} : U \to \mathbb{R}^n_{\lambda \geq c}$ with p in its domain, $T\mathbf{x}^{-1}(T_{\mathbf{x}(p)} \mathbb{R}^n_{\lambda = c})$ is a subspace of $T_p M$. This is the subspace of vectors tangent to the boundary and is identified with $T_p \partial M$, the tangent space to ∂M (also a manifold).

Exercise 2.26. Show that this subspace does not depend on the choice of chart.

If one traces back through the definitions, it becomes clear that because of the way charts and differentiability are defined for manifolds with boundary, any smooth function defined on a neighborhood of a boundary point can be thought of as being the restriction of a smooth function defined

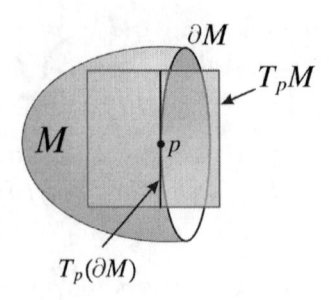

Figure 2.2. Tangents at a boundary point

slightly "outside" M. More precisely, the representative function always has a smooth extension from a (relatively open) neighborhood in $\mathbb{R}^n_{\lambda \geq c}$ to a neighborhood in $\mathbb{R}^n$. The derivatives of the extended function at points of $\partial \mathbb{R}^n_{\lambda \geq c} = \mathbb{R}^n_{\lambda = c}$ are independent of the extension. These considerations can be used to show that tangent vectors at boundary points of a smooth n-manifold with boundary can still be considered as derivations of germs of smooth functions. A closed interval $[a, b]$ is a one-dimensional manifold with boundary, and with only minor modifications in our definitions we can also consider equivalence classes of curves to define the full tangent space at a boundary point. It also follows that if c is a smooth curve with domain $[a, b]$, then we can make sense of the velocities $\dot{c}(a)$ and $\dot{c}(b)$, and this is true even if $c(a)$ or $c(b)$ is a boundary point. The major portion of the theory of manifolds extends in a natural way to manifolds with boundary.

2.4. Tangents of Products

Suppose that $f : M_1 \times M_2 \to N$ is a smooth map. For fixed $p \in M_1$ and fixed $q \in M_2$, consider the "insertion" maps $\iota_p : y \mapsto (p, y)$ and $\iota^q : x \mapsto (x, q)$. Then $f \circ \iota^q$ and $f \circ \iota_p$ are the maps sometimes denoted by $f(\cdot, q)$ and $f(p, \cdot)$.

Definition 2.27. Let $f : M_1 \times M_2 \to N$ be as above. Define the **partial tangent maps** $\partial_1 f$ and $\partial_2 f$ by

$$(\partial_1 f)(p, q) := T_p (f \circ \iota^q) : T_p M_1 \to T_{f(p,q)} N,$$
$$(\partial_2 f)(p, q) := T_q (f \circ \iota_p) : T_q M_2 \to T_{f(p,q)} N.$$

Next we introduce another natural identification. It is obvious that a curve $c : I \to M_1 \times M_2$ is equivalent to a pair of curves

$$c_1 : I \to M_1,$$
$$c_2 : I \to M_2.$$

2.4. Tangents of Products

The infinitesimal version of this fact gives rise to a natural identification on the tangent level. If $c(t) = (c_1(t), c_2(t))$ and $c(0) = (p, q)$, then the map

$$T_{(p,q)}\mathrm{pr}_1 \times T_{(p,q)}\mathrm{pr}_2 : T_{(p,q)}(M_1 \times M_2) \to T_pM_1 \times T_qM_2$$

is given by $[c] \mapsto ([c_1], [c_2])$, which is quite natural. This map is an isomorphism. Indeed, consider the insertion maps $\iota_p : q \mapsto (p, q)$ and $\iota^q : p \mapsto (p, q)$,

$$(M_1, p) \underset{\iota^q}{\overset{\mathrm{pr}_1}{\rightleftarrows}} (M_1 \times M_2, (p,q)) \underset{\iota_p}{\overset{\mathrm{pr}_2}{\rightleftarrows}} (M_2, p) \ .$$

We have linear monomorphisms $T\iota^q(p) : T_pM_1 \to T_{(p,q)}(M_1 \times M_2)$ and $T\iota_p(q) : T_qM_2 \to T_{(p,q)}(M_1 \times M_2)$,

$$T_pM_1 \underset{T_p\iota^q}{\overset{T_{(p,q)}\mathrm{pr}_1}{\rightleftarrows}} T_{(p,q)}M_1 \times M_2 \underset{T_q\iota_p}{\overset{T_{(p,q)}\mathrm{pr}_2}{\rightleftarrows}} T_pM_2 \ .$$

Then, we have the map

$$T\iota^q + T\iota_p : T_pM_1 \times T_qM_2 \to T_{(p,q)}(M_1 \times M_2),$$

which sends $(v, w) \in T_pM_1 \times T_qM_2$ to $T\iota^q(v) + T\iota_p(w)$. It can be checked that this map is the inverse of $[c] \mapsto ([c_1], [c_2])$. Thus we may identify $T_{(p,q)}(M_1 \times M_2)$ with $T_pM_1 \times T_qM_2$. Let us say a bit about the naturalness of this identification. In the smooth category, there is a direct product operation. The essential point is that for any two manifolds M_1 and M_2, the manifold $M_1 \times M_2$ together with the two projection maps serves as the direct product in the technical sense that for any smooth maps $f : N \longrightarrow M_1$ and $g : N \longrightarrow M_2$ we always have the unique map $f \times g : N \longrightarrow M_1 \times M_2$ which makes the following diagram commute:

$$\begin{array}{c} N \\ {}^f\swarrow \ {\downarrow}{\scriptstyle f\times g}\ \searrow^g \\ M_1 \xleftarrow{\mathrm{pr}_1} M_1 \times M_2 \xrightarrow{\mathrm{pr}_2} M_2 \end{array}$$

For a point $x \in N$, write $p = f(x)$ and $q = g(x)$. On the tangent level we have

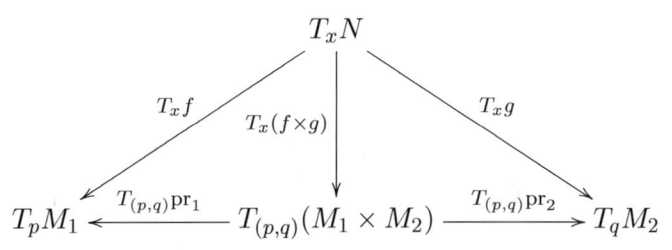

$$\begin{array}{c} T_xN \\ {}^{T_xf}\swarrow \ {\downarrow}{\scriptstyle T_x(f\times g)}\ \searrow^{T_xg} \\ T_pM_1 \xleftarrow{T_{(p,q)}\mathrm{pr}_1} T_{(p,q)}(M_1 \times M_2) \xrightarrow{T_{(p,q)}\mathrm{pr}_2} T_qM_2 \end{array}$$

which is a diagram in the vector space category. In the category of vector spaces, the product of T_pM_1 and T_pM_2 is $T_pM_1 \times T_pM_2$ together with the projections onto the two factors. Corresponding to the maps $T_p f$ and $T_q g$ we have the map $T_p f \times T_q g$. But

$$\begin{aligned}
\left(T_{(p,q)} \mathrm{pr}_1 \times T_{(p,q)} \mathrm{pr}_2\right) &\circ T_x \left(f \times g\right) \\
&= \left(T_{(p,q)} \mathrm{pr}_1 \circ T_x \left(f \times g\right)\right) \times \left(T_{(p,q)} \mathrm{pr}_2 \circ T_x \left(f \times g\right)\right) \\
&= T_x \left(\mathrm{pr}_1 \circ \left(f \times g\right)\right) \times T_x \left(\mathrm{pr}_2 \circ \left(f \times g\right)\right) \\
&= T_x f \times T_x g.
\end{aligned}$$

Thus under the identification introduced above the map $T_x (f \times g)$ corresponds to $T_x f \times T_x g$. Now if $v \in T_p M_1$ and $w \in T_q M_2$, then (v, w) represents an element of $T_{(p,q)}(M_1 \times M_2)$, and so it should act as a derivation. In fact, we can discover how this works by writing $(v, w) = (v, 0) + (0, w)$. If c_1 is a curve that represents v and c_2 is a curve that represents w, then $(v, 0)$ and $(0, w)$ are represented by $t \mapsto (c_1(t), q)$ and $t \mapsto (p, c_2(t))$ respectively. Then for any smooth function f on $M_1 \times M_2$ we have

$$(v, w) f = (v, 0) f + (0, w) f = \left.\frac{d}{dt}\right|_0 f(c_1(t), q) + \left.\frac{d}{dt}\right|_0 f(p, c_2(t))$$
$$= v[f \circ \iota^q] + w[f \circ \iota_p].$$

Lemma 2.28 (Partials lemma). *For a map $f : M_1 \times M_2 \to N$, we have*

$$T_{(p,q)} f \cdot (v, w) = (\partial_1 f)_{(p,q)} \cdot v + (\partial_2 f)_{(p,q)} \cdot w,$$

where we have used the aforementioned identification $T_{(p,q)}(M_1 \times M_2) = T_p M_1 \times T_q M_2$.

Proving this last lemma is much easier and more instructive than reading the proof so we leave it to the reader in good conscience.

2.5. Critical Points and Values

Definition 2.29. Let $f : M \to N$ be a C^r-map and $p \in M$. We say that p is a **regular point** for the map f if $T_p f$ is a surjection. Otherwise, p is called a **critical point** or **singular point**. A point q in N is called a **regular value** of f if every point in the inverse image $f^{-1}\{q\}$ is a regular point for f. This includes the case where $f^{-1}\{q\}$ is empty. A point of N that is not a regular value is called a **critical value**.

Most values of a smooth map are regular values. In order to make this precise, we will introduce the notion of measure zero on a second countable smooth manifold. It is actually no problem to define a Lebesgue measure on such a manifold, but for now the notion of measure zero is all we need.

Definition 2.30. A subset A of $\mathbb{R}^n$ is said to be of **measure zero** if for any $\epsilon > 0$ there is a sequence of cubes $\{W_i\}$ such that $A \subset \bigcup W_i$ and $\sum \mathrm{vol}(W_i) < \epsilon$. Here, $\mathrm{vol}(W_i)$ denotes the volume of the cube.

2.5. Critical Points and Values

In the definition above, if the W_i are taken to be balls, then we arrive at the very same notion of measure zero. It is easy to show that a countable union of sets of measure zero is still of measure zero. Our definition is consistent with the usual definition of Lebesgue measure zero as defined in standard measure theory courses.

Lemma 2.31. *Let $U \subset \mathbb{R}^n$ be open and $f : U \to \mathbb{R}^n$ a C^1 map. If $A \subset U$ has measure zero, then $f(A)$ has measure zero.*

Proof. Since A is certainly contained in the countable union of compact balls (all of which are translates of a ball at the origin), we may as well assume that $U = B(0,r)$ and that A is contained in a slightly smaller ball $B(0, r-\delta) \subset B(0,r)$. By the mean value theorem (see Appendix C), there is a constant c depending only on f and its domain such that for $x, y \in B(0,r)$ we have $\|f(y) - f(x)\| \leq c \|x - y\|$. Let $\epsilon > 0$ be given. Since A has measure zero, there is a sequence of balls $B(x_i, \epsilon_i)$ such that $A \subset \bigcup B(x_i, \epsilon_i)$ and
$$\sum \operatorname{vol}(B(x_i, \epsilon_i)) < \frac{\epsilon}{2^n c^n}.$$
Thus $f(B(x_i, \epsilon_i)) \subset B(f(x_i), 2c\epsilon_i)$, and while $f(A) \subset \bigcup B(f(x_i), 2c\epsilon_i)$, we also have
$$\operatorname{vol}\left(\bigcup B(f(x_i), 2c\epsilon_i)\right) \leq \sum \operatorname{vol}(B(f(x_i), 2c\epsilon_i))$$
$$\leq \sum \operatorname{vol}(B_1)(2c\epsilon_i)^n \leq 2^n c^n \sum \operatorname{vol}(B(x_i, \epsilon_i)) \leq \epsilon,$$
where $B_1 = B(0,1)$ is the ball of radius one centered at the origin. Since ϵ was arbitrary, it follows that A has measure zero. $\square$

The previous lemma allows us to make the following definition:

Definition 2.32. *Let M be an n-manifold that is second countable. A subset $A \subset M$ is said to be of **measure zero** if for every admissible chart $(U, \mathbf{x})$ the set $\mathbf{x}(A \cap U)$ has measure zero in $\mathbb{R}^n$.*

In order for this to be a reasonable definition, the manifold must be second countable so that every atlas has a countable subatlas. This way we may be assured that every set that we have defined to be measure zero is the countable union of sets that are measure zero as viewed in some chart. It is not hard to see that in this more general setting it is still true that a countable union of sets of measure zero has measure zero. Also, we still have that the image of a set of measure zero under a smooth map, has measure zero.

Proposition 2.33. *Let M be second countable as above and $\mathcal{A} = \{(U_\alpha, \mathbf{x}_\alpha)\}$ a fixed atlas for M. If $\mathbf{x}_\alpha(A \cap U_\alpha)$ has measure zero for all α, then A has measure zero.*

Proof. The atlas $\mathcal{A}$ has a countable subatlas, so we may as well assume from the start that $\mathcal{A}$ is countable. We need to show that given any admissible chart $(U, \mathbf{x})$ the set $\mathbf{x}(A \cap U)$ has measure zero. We have

$$\mathbf{x}(A \cap U) = \bigcup_\alpha \mathbf{x}(A \cap U \cap U_\alpha) \text{ (a countable union)}.$$

Since $\mathbf{x}_\alpha(A \cap U \cap U_\alpha) \subset \mathbf{x}_\alpha(A \cap U_\alpha)$, we see that $\mathbf{x}_\alpha(A \cap U \cap U_\alpha)$ has measure zero for all α. But $\mathbf{x}(A \cap U \cap U_\alpha) = \mathbf{x} \circ \mathbf{x}_\alpha^{-1} \circ \mathbf{x}_\alpha(A \cap U \cap U_\alpha)$, and so by the lemma above, $\mathbf{x}(A \cap U \cap U_\alpha)$ also has measure zero. Thus $\mathbf{x}(A \cap U)$ has measure zero since it is a countable union of sets of measure zero. $\square$

We now state the famous and useful theorem of Arthur Sard.

Theorem 2.34 (Sard). *Let N be an n-manifold and M an m-manifold, both assumed second countable. For a smooth map $f : N \to M$, the set of critical values has measure zero.*

The somewhat technical proof may be found in the online supplement **[Lee, Jeff]** or in **[Bro-Jan]**.

Corollary 2.35. *If M and N are second countable manifolds, then the set of regular values of a smooth map $f : M \to N$ is dense in N.*

2.5.1. Morse lemma. If we consider a smooth function $f : M \to \mathbb{R}$, and assume that M is a compact manifold (without boundary), then f must achieve both a maximum at one or more points of M and a minimum at one or more points of M. Let p_e be one of these points. The usual argument shows that $df|_{p_e} = 0$. (Recall that under the usual identification of $\mathbb{R}$ with any of its tangent spaces we have $df|_{p_e} = T_{p_e} f$.) Now let p be some point for which $df|_p = 0$, i.e. p is a critical point for f. Does f achieve either a maximum or a minimum at p? How does the function behave in a neighborhood of p? As the reader may well be aware, these questions are easier to answer in case the second derivative of f at p is nondegenerate. But what is the second derivative in this case?

Definition 2.36. The **Hessian** matrix of f at one of its critical points p and with respect to coordinates $\mathbf{x} = (x^1, \ldots, x^n)$, is the matrix of second partials:

$$[H_{f,p}]_\mathbf{x} = \begin{bmatrix} \frac{\partial^2 f \circ \mathbf{x}^{-1}}{\partial x^1 \partial x^1}(x_0) & \cdots & \frac{\partial^2 f \circ \mathbf{x}^{-1}}{\partial x^1 \partial x^n}(x_0) \\ \vdots & & \vdots \\ \frac{\partial^2 f \circ \mathbf{x}^{-1}}{\partial x^n \partial x^1}(x_0) & \cdots & \frac{\partial^2 f \circ \mathbf{x}^{-1}}{\partial x^n \partial x^n}(x_0) \end{bmatrix},$$

where $x_0 = \mathbf{x}(p)$. The critical point p is called **nondegenerate** if H is nonsingular.

2.5. Critical Points and Values

Any such matrix H is symmetric, and by Sylvester's law of inertia, it is congruent to a diagonal matrix whose diagonal entries are either 0 or 1 or -1. The number of -1's occurring in this diagonal matrix is called the **index** of the critical point. According to Problem 13 we may define the Hessian $H_{f,p} : T_pM \times T_pM \to \mathbb{R}$, which is a symmetric bilinear form at each critical point p of f, by letting $H_{f,p}(v,w) = X_p(Yf) = Y_p(Xf)$ for any vector fields X and Y which respectively take the values v and w at p. Thus, we may give a coordinate free definition of a nondegenerate point for f. Namely, p is a nondegenerate point for f if and only if $H_{f,p}$ is a nondegenerate bilinear form. The form $H_{f,p}$ is nondegenerate if for each fixed nonzero $v \in T_pM$ the map $H_{f,p}(v,\cdot) : T_pM \to \mathbb{R}$ is a nonzero element of the dual space T_p^*M.

Exercise 2.37. Show that the nondegeneracy is well-defined by either of the two definitions given above and that the definitions agree.

Exercise 2.38. Show that nondegenerate critical points are isolated. Show by example that this need not be true for general critical points.

The structure of a function near one of its nondegenerate critical points is given by the following famous theorem of M. Morse:

Theorem 2.39 (Morse lemma). *Let $f : M \to \mathbb{R}$ be a smooth function and let x_0 be a nondegenerate critical point for f of index ν. Then there is a local coordinate system $(U, \mathbf{x})$ containing x_0 such that the local representative $f_U := f \circ \mathbf{x}^{-1}$ for f has the form*

$$f_U(x^1, \ldots, x^n) = f(x_0) + \sum_{i,j} h_{ij} x^i x^j$$

and it may be arranged that the matrix $h = (h_{ij})$ is a diagonal matrix of the form $\mathrm{diag}(-1,\ldots,-1,1,\ldots,1)$ for some number (perhaps zero) of ones and minus ones. The number of minus ones is exactly the index ν.

Proof. This is clearly a local problem and so it suffices to assume that $f : U \to \mathbb{R}$ for some open $U \subset \mathbb{R}^n$ and also that $f(0) = 0$. Our task is to show that there exists a diffeomorphism $\phi : \mathbb{R}^n \to \mathbb{R}^n$ such that $f \circ \phi(x) = x^t h x$ for a matrix of the form described. The first step is to observe that if $g : U \subset \mathbb{R}^n \to \mathbb{R}$ is any function defined on a convex open set U and $g(0) = 0$, then

$$g(u_1, \ldots, u_n) = \int_0^1 \frac{d}{dt} g(tu_1, \ldots, tu_n)\, dt$$
$$= \int_0^1 \sum_{i=1}^n u_i \partial_i g(tu_1, \ldots, tu_n)\, dt.$$

Thus g is of the form $g = \sum_{i=1}^{n} u_i g_i$ for certain smooth functions $g_i, 1 \leq i \leq n$ with the property that $\partial_i g(0) = g_i(0)$. Now we apply this procedure first to f to get $f = \sum_{i=1}^{n} u_i f_i$ where $\partial_i f(0) = f_i(0) = 0$ and then apply the procedure to each f_i and substitute back. The result is that

$$(2.3) \qquad f(u_1, \ldots, u_n) = \sum_{i,j=1}^{n} u_i u_j h^{ij}(u_1, \ldots, u_n)$$

for some functions h^{ij} with the property that h^{ij} is nonsingular at, and therefore near 0. Next we symmetrize the matrix $h = (h^{ij})$ by replacing h^{ij} with $\frac{1}{2}(h^{ij} + h^{ji})$ if necessary. This leaves (2.3) untouched. The index of the matrix $(h^{ij}(0))$ is ν, and this remains true in a neighborhood of 0. The trick is to find a matrix $C(x)$ for each x in the neighborhood that effects the diagonalization guaranteed by Sylvester's theorem: $D = C(x)h(x)C(x)^{-1}$. The remaining details, including the fact that the matrix $C(x)$ may be chosen to depend smoothly on x, are left to the reader. $\square$

2.6. Rank and Level Set

Definition 2.40. The **rank** of a smooth map f at p is defined to be the rank of $T_p f$.

If $f : M \to N$ is a smooth map that has the same rank at each point, then we say it has **constant rank**. Similarly, if f has the same rank for each p in a open subset U, then we say that f has constant rank on U.

Theorem 2.41 (Level submanifold theorem). *Let $f : M \to N$ be a smooth map and consider the level set $f^{-1}(q_0)$ for $q_0 \in N$. If f has constant rank k on an open neighborhood of each $p \in f^{-1}(q_0)$, then $f^{-1}(q_0)$ is a closed regular submanifold of codimension k.*

Proof. Clearly $f^{-1}(q_0)$ is a closed subset of M. Let $p_0 \in f^{-1}(q_0)$ and consider a chart (U, φ) centered at p_0 and a chart (V, ψ) centered at q_0 with $f(U) \subset V$. We may choose U small enough that f has rank k on U. By Theorem C.5, we may compose with diffeomorphisms to replace (U, φ) by a new chart $(U', \mathbf{x})$ also centered at p_0 and replace (V, ψ) by a chart $(V', \mathbf{y})$ centered at q_0 such that $\bar{f} := \mathbf{y} \circ f \circ \mathbf{x}^{-1}$ is given by $(a^1, \ldots, a^n) \mapsto (a^1, \ldots, a^k, 0, \ldots, 0)$, where $n = \dim(M)$. We show that

$$U' \cap f^{-1}(q_0) = \{p \in U' : x^1(p) = \cdots = x^k(p) = 0\}.$$

If $p \in U' \cap f^{-1}(q_0)$, then $\mathbf{y} \circ f(p) = 0$ and $\mathbf{y} \circ f \circ \mathbf{x}^{-1}(x^1(p), \ldots, x^n(p)) = 0$ or

$$x^1(p) = \cdots = x^k(p) = 0.$$

On the other hand, suppose that $p \in U'$ and $x^1(p) = \cdots = x^k(p) = 0$. Then we can reverse the logic to obtain that $\mathbf{y} \circ f(p) = 0$ and hence $f(p) = q_0$.

Since p_0 was arbitrary, we have verified the existence of a cover of $f^{-1}(q_0)$ by single-slice charts (see Section 1.7). □

Proposition 2.42. *Let M and N be smooth manifolds of dimension m and n respectively with $n > m$. Consider any smooth map $f : M \to N$. Then if $q \in N$ is a regular value, the inverse image set $f^{-1}(q)$ is a regular submanifold.*

Proof. It is clear that since f must have maximal rank in a neighborhood of $f^{-1}(q)$, it also has constant rank there. We may now apply Theorem 2.41. □

Example 2.43 (The unit sphere). The set $S^{n-1} = \{x \in \mathbb{R}^n : \sum (x^i)^2 = 1\}$ is a codimension 1 submanifold of $\mathbb{R}^n$. For this we apply the above proposition with the map $f : \mathbb{R}^n \to \mathbb{R}$ given by $x \mapsto \sum (x^i)^2$ and with the choice $q = 1 \in \mathbb{R}$.

Example 2.44. The set of all square matrices $M_{n \times n}$ is a manifold by virtue of the obvious isomorphism $M_{n \times n} \cong \mathbb{R}^{n^2}$. The set $\text{sym}(n, \mathbb{R})$ of all symmetric matrices is a smooth $n(n+1)/2$-dimensional manifold by virtue of the obvious 1-1 correspondence $\text{sym}(n, \mathbb{R}) \cong \mathbb{R}^{n(n+1)/2}$ given by using $n(n+1)/2$ entries in the upper triangle of the matrix as coordinates. It can be shown that the map $f : M_{n \times n} \to \text{sym}(n, \mathbb{R})$ given by $A \mapsto A^t A$ has full rank on $O(n, \mathbb{R}) = f^{-1}(I)$ and so we can apply Proposition 2.42. Thus the set $O(n, \mathbb{R})$ of all $n \times n$ orthogonal matrices is a submanifold of $M_{n \times n}$. We leave the details to the reader, but note that we shall prove a more general theorem later (Theorem 5.107).

The following proposition shows an example of the simultaneous use of Sard's theorem and Proposition 2.42.

Proposition 2.45. *Let S be a connected submanifold of $\mathbb{R}^n$ and let L be a codimension one linear subspace of $\mathbb{R}^n$. Then there exist $x \in \mathbb{R}^n$ such that $(x + L) \cap S$ is a submanifold of S.*

Proof. Start with a line l through the origin that is normal to L. Let $\text{pr} : \mathbb{R}^n \to S$ be orthogonal projection onto l. The restriction $\pi := \text{pr}|_S \to l$ is easily seen to be smooth. If $\pi(S)$ were just a single point x, then $\pi^{-1}(x) = (x+L) \cap S$ would be all of S, so let us assume that $\pi(S)$ contains more than one point. Now, $\pi(S)$ is a connected subset of $l \cong \mathbb{R}$, so it must contain an open interval. This implies that $\pi(S)$ has positive measure. Thus by Sard's theorem there must be a point $x \in \pi(S) \subset l$ that is a regular value of π. Then Theorem 2.42 implies that $\pi^{-1}(x)$ is a submanifold of S. But this is the conclusion since $\pi^{-1}(x) = (x+L) \cap S$. □

We can generalize Theorem 2.42 using the concept of transversality.

Definition 2.46. Let $f : M \to N$ be a smooth map and $S \subset N$ a submanifold of N. We say that f is **transverse to** S if for every $p \in f^{-1}(S)$ we have
$$T_{f(p)}N = T_{f(p)}S + T_p f(T_p M).$$
If f is **transverse to** S, we write $f \pitchfork S$.

Theorem 2.47. *Let $f : M \to N$ be a smooth map and $S \subset N$ a submanifold of N of codimension k and suppose that $f \pitchfork S$ and $f^{-1}(S) \neq \emptyset$. Then $f^{-1}(S)$ is a submanifold of M with codimension k. Furthermore we have $T_p(f^{-1}(S)) = Tf^{-1}(T_{f(p)}S)$ for all $p \in f^{-1}(S)$.*

Proof. Let $q = f(p) \in S$ and choose a single-slice chart $(V, \mathbf{x})$ centered at $q \in V$ so that $\mathbf{x}(S \cap V) = \mathbf{x}(V) \cap (\mathbb{R}^{n-k} \times 0)$. Let $U := f^{-1}(V)$ so that $p \in U$. If $\pi : \mathbb{R}^{n-k} \times \mathbb{R}^k \to \mathbb{R}^k$ is the second factor projection, then the transversality condition on U implies that 0 is a regular value of $\pi \circ \mathbf{x} \circ f|_U$. Thus $(\pi \circ \mathbf{x} \circ f|_U)^{-1}(0) = f^{-1}(S) \cap U$ is a submanifold of U of codimension k. Since this is true for all $p \in f^{-1}(S)$, the result follows. □

We can also define when a pair of maps are transverse to each other:

Definition 2.48. If $f_1 : M_1 \to N$ and $f_2 : M_2 \to N$ are smooth maps, we say that f_1 and f_2 are **transverse** at $q \in N$ if
$$T_{f(p)}N = T_{p_1}f_1(T_{p_1}M) + T_{p_2}f_2(T_{p_2}M) \text{ whenever } f_1(p_1) = f_2(p_2) = q.$$
(Note that f_1 is transverse to f_2 at any point not in the image of one of the maps f_1 and f_2.) If f_1 and f_2 are transverse for all $q \in N$, then we say that f_1 and f_2 are transverse and we write $f_1 \pitchfork f_2$.

One can check that if $f : M \to N$ is a smooth map and S is a submanifold of N, then f and the inclusion $\iota : S \hookrightarrow N$ are transverse if and only if $f \pitchfork S$ according to Definition 2.47.

If $f_1 : M_1 \to N$ and $f_2 : M_2 \to N$ are smooth maps, then we can consider the set
$$(f_1 \times f_2)^{-1}(\Delta) := \{(p_1, p_2) \in M_1 \times M_2 : f_1(p_1) = f_2(p_2)\},$$
which is the inverse image of the diagonal $\Delta := \{(q_1, q_2) \in N \times N : q_1 = q_2\}$.

Corollary 2.49 (Transverse pullbacks). *If $f_1 : M_1 \to N$ and $f_2 : M_2 \to N$ are transverse smooth maps, then $(f_1 \times f_2)^{-1}(\Delta)$ is a submanifold of $M_1 \times M_2$. If $g_1 : P \to M_1$ and $g_2 : P \to M_2$ are any smooth maps with the property $f_1 \circ g_1 = f_2 \circ g_2$, then the map $(g_1, g_2) : P \to (f_1 \times f_2)^{-1}(\Delta)$ given by $(g_1, g_2)(x) = (g_1(x), g_2(x))$ is smooth and is the unique smooth map such that $\mathrm{pr}_1 \circ (g_1, g_2) = g_1$ and $\mathrm{pr}_2 \circ (g_1, g_2) = g_2$.*

Proof. We leave the proof as an exercise. Hint: $f_1 \times f_2$ is transverse to Δ if and only if $f_1 \pitchfork f_2$. □

2.7. The Tangent and Cotangent Bundles

We define the **tangent bundle** of a manifold M as the (disjoint) union of the tangent spaces; $TM = \bigcup_{p \in M} T_p M$. We show in Proposition 2.55 below that TM is a smooth manifold, but first we introduce a couple of definitions.

Definition 2.50. Given a smooth map $f : M \to N$ as above, the tangent maps $T_p f$ on the individual tangent spaces combine to give a map

$$Tf : TM \to TN$$

on the tangent bundle which is linear on each fiber. This map is called the **tangent map** or sometimes the **tangent lift** of f.

For smooth maps $f : M \to N$ and $g : N \to M$ we have the following simple looking version of the chain rule:

$$T(g \circ f) = Tg \circ Tf.$$

If U is an open set in a finite-dimensional vector space V, then the tangent space at $x \in U$ can be viewed as $\{x\} \times $ V. For example, recall that an element $v_p = (p, v)$ corresponds to the derivation $f \mapsto v_p f := \frac{d}{dt}\big|_{t=0} f(p+tv)$. Thus the tangent bundle of U can be viewed as the product $U \times$ V. Let U_1 and U_2 be open subsets of vector spaces V and W respectively and let $f : U_1 \to U_2$ be smooth (or at least C^1). Then we have the tangent map $Tf : TU_1 \to TU_2$. Viewing TU_1 as $U_1 \times$ V and similarly for TU_2, the tangent map Tf is given by $(p, v) \mapsto (f(p), Df(p) \cdot v)$.

Definition 2.51. If $f : M \to$ V, where V is a finite-dimensional vector space, then we have the differential $df(p) : T_p M \to$ V for each p. These maps can be combined to give a single map $df : TM \to$ V (also called the **differential**) which is defined by $df(v) = df(p)(v)$ when $v \in T_p M$.

If we identify TV with the product V $\times$ V, then $df = \text{pr}_2 \circ Tf$, where $\text{pr}_2 : T\text{V} = \text{V} \times \text{V} \to \text{V}$ is the projection onto the second factor.

Remark 2.52 (Warning). The notation "df" is subject to interpretation. Besides the map $df : TM \to$ V described above it could also refer to the map $df : p \mapsto df(p)$ or to another map on vector fields which we describe later in this chapter.

Definition 2.53. The map $\pi_{TM} : TM \to M$ defined by $\pi_{TM}(v) = p$ if $v \in T_p M$ is called the **tangent bundle projection map**. (The set TM together with the map $\pi_{TM} : TM \to M$ is an example of a vector bundle which is defined later.)

Whenever possible, we abbreviate π_{TM} to π. For every chart $(U, \mathbf{x})$ on M, we obtain a chart $(\widetilde{U}, \widetilde{\mathbf{x}})$ on TM by letting

$$\widetilde{U} := TU = \pi^{-1}(U) \subset TM$$

and by defining $\widetilde{\mathbf{x}}$ on $\widetilde{U}$ by the prescription

$$\widetilde{\mathbf{x}}(v_p) = (x^1(p), \ldots, x^n(p), v^1, \ldots, v^n), \text{ where } v_p \in T_p M,$$

and where $v^1, \ldots, v^n$ are the (unique) coefficients in the coordinate expression $v = \sum v^i \frac{\partial}{\partial x^i}\big|_p$. Thus $\widetilde{\mathbf{x}}^{-1}(u^1, \ldots, u^n, v^1, \ldots, v^n) = \sum v^i \frac{\partial}{\partial x^i}\big|_{\mathbf{x}^{-1}(u)}$. Recall that if $v_p = \sum v^i \frac{\partial}{\partial x^i}\big|_p$, then $v^i = dx^i(v_p)$. From this we see that $\widetilde{\mathbf{x}} = (x^1 \circ \pi, \ldots, x^n \circ \pi, dx^1, \ldots, dx^n)$.

For any $(U, \mathbf{x})$, we have the tangent lift $T\mathbf{x} : TU \to TV$ where $V = \mathbf{x}(U)$. Since $V \subset \mathbb{R}^n$, we can identify $T_{\mathbf{x}(p)} V$ with $\{\mathbf{x}(p)\} \times \mathbb{R}^n$. Let us invoke this identification. Now let $v_p \in T_p U$ and let γ be a curve that represents v_p so that $\gamma'(0) = v_p$.

Exercise 2.54. Under the identification of $T_{\mathbf{x}(p)} V$ with $\{\mathbf{x}(p)\} \times \mathbb{R}^n$ we have $T_p \mathbf{x} \cdot v_p = (\mathbf{x}(p), \frac{d}{dt}\big|_{t=0} (\mathbf{x} \circ \gamma))$. [Hint: Interpret both sides as derivations.]

If $v_p = \frac{\partial}{\partial x^i}\big|_p$, then we can take $\gamma(t) := \mathbf{x}^{-1}(\mathbf{x}(p) + t\mathbf{e}_i)$, where $\mathbf{e}_i$ is the i-th member of the standard basis of $\mathbb{R}^n$. Thus

$$T_p \mathbf{x} \cdot \frac{\partial}{\partial x^i}\bigg|_p = \left(\mathbf{x}(p), \frac{d}{dt}\bigg|_{t=0} (\mathbf{x}(p) + t\mathbf{e}_i) \right) = (\mathbf{x}(p), \mathbf{e}_i).$$

Now suppose that $v_p = \sum v^i \frac{\partial}{\partial x^i}\big|_p$. Then

$$T_p \mathbf{x} \cdot v_p = T_p \mathbf{x} \cdot \left(\sum v^i \frac{\partial}{\partial x^i}\bigg|_p \right)$$

$$= \left(\mathbf{x}(p), \sum v^i \mathbf{e}_i \right) = (x^1(p), \ldots, x^n(p), v^1, \ldots, v^n).$$

From this we see that $T\mathbf{x}$ is none other than $\widetilde{\mathbf{x}}$ defined above, and since $\widetilde{U} = TU$, we see that an alternative and suggestive notation for $(\widetilde{U}, \widetilde{\mathbf{x}})$ is $(TU, T\mathbf{x})$, and we adopt this notation below. This notation reminds one that the charts we have constructed are not just any charts on TM, but are each associated naturally with a chart on M and are essentially the tangent lifts of charts on M. They are called **natural charts**.

Proposition 2.55. *For any smooth n-manifold M, the set TM is a smooth $2n$-manifold in a natural way and $\pi_{TM} : TM \to M$ is a smooth map. Furthermore, for a smooth map $f : M \to N$, the tangent map Tf is smooth and*

2.7. The Tangent and Cotangent Bundles

the following diagram commutes:

$$\begin{array}{ccc} TM & \xrightarrow{Tf} & TN \\ \downarrow & & \downarrow \\ M & \xrightarrow{f} & N \end{array}$$

Proof. For every chart $(U, \mathbf{x})$, let $TU = \pi^{-1}(U)$ and let $T\mathbf{x}$ be the map $T\mathbf{x} : TU \to \mathbf{x}(U) \times \mathbb{R}^n$. The pair $(TU, T\mathbf{x})$ is a chart on TM. Suppose that $(TU, T\mathbf{x})$ and $(TV, T\mathbf{y})$ are two such charts constructed as above from two charts $(U, \mathbf{x})$ and $(V, \mathbf{y})$ and that $U \cap V \neq \emptyset$. Then $TU \cap TV \neq \emptyset$ and on the overlap we have the coordinate transitions $T\mathbf{y} \circ T\mathbf{x}^{-1} : (x, v) \mapsto (y, w)$ where

$$y = \mathbf{y} \circ \mathbf{x}^{-1}(x),$$
$$w = D(\mathbf{y} \circ \mathbf{x}^{-1})\big|_x v.$$

Thus the overlap maps are smooth. It is easy to see that $T\mathbf{x}(TU \cap TV)$ and $T\mathbf{y}(TU \cap TV)$ are open. Thus we obtain a smooth atlas on TM from an atlas on M and this generates a topology. It follows from Proposition 1.32 that TM is Hausdorff and paracompact.

To test for the smoothness of π, we look at maps of the form $\mathbf{x} \circ \pi \circ (T\mathbf{x})^{-1}$. We have

$$\mathbf{x} \circ \pi \circ (T\mathbf{x})^{-1}(x, v) = \mathbf{x} \circ \pi \left(v^i \frac{\partial}{\partial x^i} \bigg|_{\mathbf{x}^{-1}(x)} \right) = \mathbf{x} \circ \mathbf{x}^{-1}(x) = x,$$

which is just a projection and so clearly smooth. The remainder is left for the exercise below. $\square$

In the above proof we observed that $T\mathbf{x}(TU \cap TV)$ and $T\mathbf{y}(TU \cap TV)$ are open. This must be checked because of (ii) in Definition 1.25 and is the kind of detail we may leave to the reader as we move forward.

Exercise 2.56. For a smooth map $f : M \to N$, the map

$$Tf : TM \to TN$$

is itself a smooth map.

If $p \in U \cap V$ and $\mathbf{x}(p) = (x^1(p), \ldots, x^n(p))$, then, as in the proof above, $T\mathbf{y} \circ T\mathbf{x}^{-1}$ sends $(x^1(p), \ldots, x^n(p), v^1, \ldots, v^n)$ to $(y^1(p), \ldots, y^n(p), w^1, \ldots, w^n)$, where

$$w^i = \sum \frac{\partial (\mathbf{y} \circ \mathbf{x}^{-1})^i}{\partial x^k} v^k.$$

If we abbreviate the i-th component of $\mathbf{y} \circ \mathbf{x}^{-1}(x^1(p), \ldots, x^n(p))$ to $y^i = y^i(x^1(p), \ldots, x^n(p))$, then we could express the tangent bundle overlap map by the relations

$$y^i = y^i(x^1(p), \ldots, x^n(p)) \text{ and } w^i = \sum \frac{\partial y^i}{\partial x^k} v^k.$$

Since this is true for all $p \in \mathbf{x}(U \cap V)$, we can write the very classical looking expressions

$$y^i = y^i(x^1, \ldots, x^n) \text{ and } w^i = \sum \frac{\partial y^i}{\partial x^k} v^k,$$

where we now can interpret $(x^1, \ldots, x^n)$ as an n-tuple of numbers. Once again we note that local expression could either be interpreted as living on the manifold in the chart domain or equally, in Euclidean space on the image of the chart domain. This should not be upsetting since, after all, one could argue that the charts are there to *identify* chart domains in the manifold with open sets in Euclidean space.

Definition 2.57. The **tangent functor** is defined by assigning to a manifold M its tangent bundle TM and to any map $f : M \to N$ the tangent map $Tf : TM \to TN$. The chain rule shows that this is a covariant functor (see Appendix A).

Recall that we also defined a "pointed" tangent functor.

We have seen that if U is an open set in a vector space V, then the tangent bundle is often taken to be $U \times$ V. Suppose that for some smooth n-manifold M, there is a diffeomorphism $F : TM \to M \times$ V such that the restriction of F to each tangent space is a linear isomorphism $T_pM \to \{p\} \times$ V and such that the following diagram commutes:

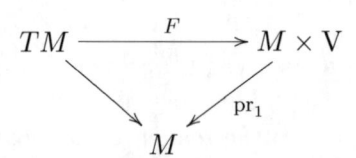

$$\begin{array}{ccc} TM & \xrightarrow{F} & M \times V \\ & \searrow & \downarrow \text{pr}_1 \\ & & M \end{array}$$

Then for some purposes, we can identify TM with $M \times$ V.

Definition 2.58. A diffeomorphism $F : TM \to M \times$ V such that the map $F|_{T_pM} : T_pM \to \{p\} \times$ V is linear for each p and such that the above diagram commutes is called a (global) **trivialization** of TM. If a (global) trivialization exists, then we say that TM is trivial. For an open set $U \subset M$, a trivialization of TU is called a **local trivialization** of TM over U.

For most manifolds, there does not exist a global trivialization of the tangent bundle. On the other hand, every point p in a manifold M is contained in an open set U so that TM has a local trivialization over U. The

2.7. The Tangent and Cotangent Bundles

existence of these local trivializations is quickly deduced from the existence of the special charts which we constructed above for a tangent bundle.

Next we introduce the cotangent bundle. Recall that for each $p \in M$, the tangent space T_pM has a dual space T_p^*M called the cotangent space at p.

Definition 2.59. Define the **cotangent bundle** of a manifold M to be the set
$$T^*M := \bigcup_{p \in M} T_p^*M$$
and define the map $\pi_{T^*M} : T^*M \to M$ to be the obvious projection taking elements in each space T_p^*M to the corresponding point p.

Remark 2.60. We will denote both the tangent bundle projection and the cotangent bundle projection simply by π whenever no confusion is likely.

Remark 2.61. Suppose that $f : M \to N$ is a smooth map. It is important to notice that even though for each $p \in M$, the map $T_pf : T_pM \to T_{f(p)}N$ has a dual map $(T_pf)^* : (T_{f(p)}N)^* \to (T_pM)^*$, these maps *do not* generally combine to give a map from T^*N to T^*M. In general, there is nothing like a "cotangent lift". To see this, just consider the case where f is a constant map.

We now show that T^*M is also a smooth manifold. Let $\mathcal{A}$ be an atlas on M. For each chart $(U, \mathbf{x}) \in \mathcal{A}$, we obtain a chart $(T^*U, T^*\mathbf{x})$ for T^*M which we now describe. First, $T^*U = \pi_{T^*M}^{-1}(U) = \bigcup_{p \in U} T_p^*M$. Secondly, $T^*\mathbf{x}$ is a map which we now define directly and then show that, in some sense, it is dual to the map $T\mathbf{x}$. For convenience, consider the map $p_i : \theta_p \mapsto \xi_i$ which just peals off the coefficients in the expansion of any $\theta_p \in T_p^*M$ in the basis $(dx^1|_p, \ldots, dx^n|_p)$:
$$p_i(\theta_p) = p_i\left(\sum \xi_j \, dx^i\big|_p\right) := \xi_i.$$
Notice that we have
$$\theta_p\left(\frac{\partial}{\partial x^i}\bigg|_p\right) = \sum \xi_j \, dx^i\big|_p\left(\frac{\partial}{\partial x^i}\bigg|_p\right) = \sum \xi_j \delta_i^j = \xi_j = p_i(\theta_p),$$
and so
$$p_i(\theta_p) = \theta_p\left(\frac{\partial}{\partial x^i}\bigg|_p\right).$$
With this definition of the p_i in hand, we can define
$$T^*\mathbf{x} = (x^1 \circ \pi, \ldots, x^n \circ \pi, p_1, \ldots, p_n)$$
on T^*U. We call $(T^*U, T^*\mathbf{x})$ a **natural chart**. If $\mathbf{x} = (x^1, \ldots, x^n)$, then for the natural chart $(T^*U, T^*\mathbf{x})$, we could use the abbreviation $T^*\mathbf{x} =$

$(x^1, \ldots, x^n, p_1, \ldots, p_n)$. Another common notation is $q^i := x^i \circ \pi$. This notation is very popular in applications to mechanics.

We claim that if we take advantage of the identifications of $T_x \mathbb{R}^n = \mathbb{R}^n = (\mathbb{R}^n)^* = T^* \mathbb{R}^n$ where $(\mathbb{R}^n)^*$ is the dual space of $\mathbb{R}^n$, then $T^* \mathbf{x}$ acts on each fiber $T_p^* M$ as the dual of the inverse of the map $T_p \mathbf{x}$, i.e. the *contragredient* of $T_p \mathbf{x}$:

$$\left((T_p \mathbf{x})^{-1}\right)^* (\theta_p) \cdot (v) = \theta_p \left((T_p \mathbf{x})^{-1} \cdot v\right).$$

Let us unravel this. If $\theta_p \in T_p^* M$ for some $p \in U$, then we can write

$$\theta_p = \sum \xi_i \, dx^i \big|_p$$

for some numbers ξ_i depending on θ_p which are what we have called $p_i(\theta_p)$. We have

$$\left((T_p \mathbf{x})^{-1}\right)^* (\theta_p) \cdot (v) = \theta_p \left((T_p \mathbf{x})^{-1} \cdot v\right)$$
$$= \sum \xi_i \, dx^i \big|_p \cdot \left((T_p \mathbf{x})^{-1} \cdot v\right)$$
$$= \sum \xi_i \, dx^i \big|_p \left(\sum v^k \frac{\partial}{\partial x^k}\bigg|_p\right) = \sum \xi_i v^i.$$

Thus, under the identification of $\mathbb{R}^n$ with its dual we see that $\left((T_p \mathbf{x})^{-1}\right)^* (\theta_p)$ is just $(\xi_1, \ldots, \xi_n)$. But recall that $T^* \mathbf{x}(\theta_p) = (x^1(p), \ldots, x^n(p), \xi_1, \ldots, \xi_n)$. Thus for $\theta_p \in T_p^* M$ we have

$$T^* \mathbf{x}(\theta_p) = \left(\mathbf{x}(p), \left((T_p \mathbf{x})^{-1}\right)^* (\theta_p)\right).$$

Suppose that $(T^* U, T^* \mathbf{x})$ and $(T^* V, T^* \mathbf{y})$ are the coordinates constructed as above from two charts $(U, \mathbf{x})$ and $(V, \mathbf{y})$ respectively with $U \cap V \neq \emptyset$. Then on the overlap $T^* U \cap T^* V$ we have

$$T^* \mathbf{y} \circ (T^* \mathbf{x})^{-1} : \mathbf{x}(U \cap V) \times \mathbb{R}^{n*} \to \mathbf{y}(U \cap V) \times \mathbb{R}^{n*}.$$

This last map will send something of the form $(x, \xi) \in U \times \mathbb{R}^{n*}$ to $(\bar{x}, \bar{\xi}) = (\mathbf{y} \circ \mathbf{x}^{-1}(x), D(\mathbf{x} \circ \mathbf{y}^{-1})^* \cdot \xi)$, where $D(\mathbf{x} \circ \mathbf{y}^{-1})^*$ is the dual map to $D(\mathbf{x} \circ \mathbf{y}^{-1})$, which is the contragredient of the map $D(\mathbf{y} \circ \mathbf{x}^{-1})$. If we identify $\mathbb{R}^{n*}$ with $\mathbb{R}^n$ and write $\xi = (\xi_1, \ldots, \xi_n)$ and $\bar{\xi} = (\bar{\xi}_1, \ldots, \bar{\xi}_n)$, then in the classical style we have:

$$\bar{y}^i = y^i(x^1, \ldots, x^n) \text{ and } \bar{\xi}_i = \sum_k \xi_k \frac{\partial x^k}{\partial y^i}.$$

This should be compared to the expression (2.2). It is now clear that we have an atlas on $T^* M$ constructed from an atlas on M. The topology of $T^* M$ (induced by the above atlas) is easily seen to be paracompact and Hausdorff.

In summary, both TM and T^*M are smooth manifolds whose smooth structure is derived from the smooth structure on M in a natural way. In both cases, the charts are derived from charts on the base M and are given by the n coordinates of the base point together with the n components of the element of TM (or T^*M) in the corresponding coordinate frame.

2.8. Vector Fields

In this section we introduce vector fields. Roughly, a vector field is a smooth assignment of a tangent vector to each point of a manifold.

Definition 2.62. If $\pi : M \to N$ is a smooth map, then a (global) **section** of π is a map $\sigma : N \to M$ such that $\pi \circ \sigma = \text{id}$. If σ is defined only on an open subset U of N and $\pi \circ \sigma = \text{id}_U$, then we call σ a **local section**. In case the section σ is a smooth (or C^r) map, we call σ a **smooth** (or C^r) **section**.

Clearly, if $\pi : M \to N$ has a (global) section, then it must be surjective.

Definition 2.63. A smooth **vector field** on M is a smooth map $X : M \to TM$ such that $X(p) \in T_pM$ for all $p \in M$. In other words, a vector field on M is a **smooth section** of the tangent bundle $\pi : TM \to M$. We often write $X_p = X(p)$.

Convention: Obviously the notion of a section or field that is not smooth makes sense. Sometimes one is interested in merely continuous sections or measurable sections. In this book, by "vector field" or "section", we will always mean "*smooth* vector field" or "*smooth* section" unless otherwise indicated explicitly or by context.

A local section of TM defined on an open set U is just the same thing as a vector field on the open manifold U. If $(U, \mathbf{x})$ is a chart on a smooth n-manifold, then writing $\mathbf{x} = (x^1, \ldots, x^n)$, we have vector fields defined on U by

$$\frac{\partial}{\partial x^i} : p \mapsto \left.\frac{\partial}{\partial x^i}\right|_p.$$

The ordered set of fields $(\frac{\partial}{\partial x^1}, \ldots, \frac{\partial}{\partial x^n})$ is called a **coordinate frame field** (or also "holonomic frame field"). If X is a smooth vector field defined on some set including this chart domain U, then for some smooth functions X^i defined on U we have

$$X(p) = \sum X^i(p) \left.\frac{\partial}{\partial x^i}\right|_p,$$

or in other words

$$X|_U = \sum X^i \frac{\partial}{\partial x^i}.$$

Notation 2.64. In this context, we will not usually bother to distinguish X from its restrictions to chart domains and so we just write $X = \sum X^i \frac{\partial}{\partial x^i}$.

Lemma 2.65. *If $v \in T_p M$ then there exists a vector field X such that $X(p) = v$.*

Proof. Write $v = \sum v^i \frac{\partial}{\partial x^i}\big|_p$. Define a field X_U by the formula $\sum v^i \frac{\partial}{\partial x^i}$ where the v^i are taken as constant functions on U. Let β be a cut-off function with support in U and such that $\beta(p) = 1$. Then let $X := \beta X_U$ on U and extended to zero outside of U. $\square$

Let us unravel what the smoothness condition means for a vector field. Let $(TU, T\mathbf{x})$ be one of the natural charts that we constructed for TM from a corresponding chart $(U, \mathbf{x})$ on M. To test the smoothness of X, we look at the composition $T\mathbf{x} \circ X \circ \mathbf{x}^{-1}$. For $x \in \mathbf{x}(U)$, we have

$$T\mathbf{x} \circ X \circ \mathbf{x}^{-1}(x)$$
$$= T\mathbf{x} \circ \left(\sum X^i \frac{\partial}{\partial x^i}\right) \circ \mathbf{x}^{-1}(x)$$
$$= T\mathbf{x}\left(\sum X^i(\mathbf{x}^{-1}(x)) \frac{\partial}{\partial x^i}\bigg|_{\mathbf{x}^{-1}(x)}\right)$$
$$= \left(x,\, T_{\mathbf{x}^{-1}(x)}\mathbf{x}\left(\sum X^i(\mathbf{x}^{-1}(x)) \frac{\partial}{\partial x^i}\bigg|_{\mathbf{x}^{-1}(x)}\right)\right)$$
$$= \left(x,\, X^1 \circ \mathbf{x}^{-1}(x), \ldots, X^n \circ \mathbf{x}^{-1}(x)\right).$$

Our chart was arbitrary, and so we see that the smoothness of X is equivalent to the smoothness of the component functions X^i in every chart of an atlas for the smooth structure.

Exercise 2.66. Show that if $X : M \to TM$ is continuous and $\pi \circ X = \mathrm{id}$, then X is smooth if and only if $Xf : p \mapsto X_p f$ is a smooth function for every locally defined smooth function f on M. Show that it is enough to consider globally defined smooth functions.

Notation 2.67. The set of all smooth vector fields on M is denoted by $\mathfrak{X}(M)$. Smooth vector fields may at times be defined only on some open set $U \subset M$ so we also have the notation $\mathfrak{X}(U) = \mathfrak{X}_M(U)$ for these fields.

We define the addition of vector fields, say X and Y, by
$$(X + Y)(p) := X(p) + Y(p),$$
and scaling by real numbers, by
$$(cX)(p) := cX(p).$$

2.8. Vector Fields

Then the set $\mathfrak{X}(M)$ is a real vector space. If we define multiplication of a smooth vector field X by a smooth function f by

$$(fX)(p) := f(p)X(p),$$

then the expected algebraic properties hold making $\mathfrak{X}(M)$ a module over the ring $C^\infty(M)$ (see Appendix D). It should be clear how to define vector fields of class C^r on M and the set of these is denoted $\mathfrak{X}^r(M)$ (a module over $C^r(M)$).

The notion of a vector field *along a map* is often useful.

Definition 2.68. Let $f : N \to M$ be a smooth map. A vector field **along** f is a smooth map $X : N \to TM$ such that $\pi_{TM} \circ X = f$. A vector field along a regular submanifold $S \subset M$ is a vector field along the inclusion map $S \hookrightarrow M$. (Note that we include the case where S is an open submanifold.) We let $\mathfrak{X}_f$ denote the space of vector fields along f.

It is easy to check that for a smooth map $f : N \to M$, the set $\mathfrak{X}_f$ is a $C^\infty(N)$-module in a natural way.

We have seen how individual tangent vectors in T_pM can be identified as derivations at p. The derivation idea can be globalized. We explain how we may view vector fields as derivations.

Definition 2.69. Let M be a smooth manifold. A (global) **derivation** on $C^\infty(M)$ is a linear map $\mathcal{D} : C^\infty(M) \to C^\infty(M)$ such that

$$\mathcal{D}(fg) = \mathcal{D}(f)g + f\mathcal{D}(g).$$

We denote the set of all such derivations of $C^\infty(M)$ by $\mathrm{Der}(C^\infty(M))$.

Notice the difference between a derivation in this sense and a derivation at a point.

Definition 2.70. To a vector field X on M, we associate the map $\mathcal{L}_X : C^\infty(M) \to C^\infty(M)$ defined by

$$(\mathcal{L}_X f)(p) := X_p f.$$

$\mathcal{L}_X$ is called the **Lie derivative** on functions.

It is important to notice that $(\mathcal{L}_X f)(p) = X_p \cdot f = df(X_p)$ for any p and so $\mathcal{L}_X f = df \circ X$. If X is a vector field on an open set U, and if f is a function on a domain $V \subset U$, then we take $\mathcal{L}_X f$ to be the function defined on V by $p \mapsto X_p f$ for all $p \in V$. It is easy to see that we have $\mathcal{L}_{aX+bY} = a\mathcal{L}_X + b\mathcal{L}_Y$ for $a, b \in \mathbb{R}$ and $X, Y \in \mathfrak{X}(M)$.

Lemma 2.71. *Let $U \subset M$ be an open set and $X \in \mathfrak{X}(M)$. If $\mathcal{L}_X f = 0$ for all $f \in C^\infty(U)$, then $X|_U = 0$.*

Proof. Let $p \in U$ be given. Working locally in a chart (V, x), let $X = \sum X^i \partial / \partial x^i$. We may assume $p \in V \subset U$. Using a cut-off function we may find functions f^i defined on U such that f^i coincides with x^i on a neighborhood of p. Then we have $X^i(p) = X_p x^i = X_p f^i = \left(\mathcal{L}_X f^i \right)(p) = 0$. Thus $X(p) = 0$ for an arbitrary $p \in U$. $\square$

The next result is a very important characterization of smooth vector fields. In particular, it paves the way for the definition of the bracket of vector fields which plays a central role in differential geometry.

Theorem 2.72. *For $X \in \mathfrak{X}(M)$, we have $\mathcal{L}_X \in \mathrm{Der}(C^\infty(M))$, and if $\mathcal{D} \in \mathrm{Der}(C^\infty(M))$, then $\mathcal{D} = \mathcal{L}_X$ for a uniquely determined $X \in \mathfrak{X}(M)$.*

Proof. That $\mathcal{L}_X$ is in $\mathrm{Der}(C^\infty(M))$ follows from the Leibniz law, in other words, from the fact that X_p is a derivation at p for each p. If we are given a derivation $\mathcal{D}$, we define a derivation X_p at p (i.e. a tangent vector) by the rule $X_p f := (\mathcal{D} f)(p)$. We need to show that the assignment $p \mapsto X_p$ is smooth. Recall that any locally defined function can be extended to a global one by using a cut-off function. Because of this, it suffices to show that $p \mapsto X_p f$ is smooth for any $f \in C^\infty(M)$. But this is clear since $X_p f := (\mathcal{D} f)(p)$ and $\mathcal{D} f \in C^\infty(M)$. Suppose now that $\mathcal{D} = \mathcal{L}_{X_1} = \mathcal{L}_{X_2}$. Notice that $\mathcal{L}_{X_1} - \mathcal{L}_{X_2} = \mathcal{L}_{X_1 - X_2}$ and so $\mathcal{L}_{X_1 - X_2}$ is the zero derivation on $C^\infty(M)$. By Lemma 2.71, we have $X_1 - X_2 = 0$. $\square$

Because of this theorem, we can identify $\mathrm{Der}(C^\infty(M))$ with $\mathfrak{X}(M)$ and we can and often will write Xf in place of $\mathcal{L}_X f$:

$$Xf := \mathcal{L}_X f.$$

The derivation law (also called the Leibniz law) $\mathcal{L}_X(fg) = g\mathcal{L}_X f + f\mathcal{L}_X g$ becomes simply $X(fg) = gXf + fXg$. Another thing worth noting is that if we have a derivation of $C^\infty(M)$, then from our discussion above we know that it corresponds to a vector field. As such, it can be restricted to any open set $U \subset M$, and thus we get a derivation of $C^\infty(U)$. If $f \in C^\infty(U)$ we write Xf instead of the more pedantic $X|_U f$.

While it makes sense to talk of vector fields on M of differentiability r where $0 < r < \infty$ and these do act as derivations on $C^r(M)$, it is only in the smooth case ($r = \infty$) that we can say that vector fields account for *all* derivations of $C^r(M)$.

Theorem 2.73. *If $\mathcal{D}_1, \mathcal{D}_2 \in \mathrm{Der}(C^\infty(M))$, then $[\mathcal{D}_1, \mathcal{D}_2] \in \mathrm{Der}(C^\infty(M))$ where*

$$[\mathcal{D}_1, \mathcal{D}_2] := \mathcal{D}_1 \circ \mathcal{D}_2 - \mathcal{D}_2 \circ \mathcal{D}_1.$$

2.8. Vector Fields

Proof. We compute

$$\mathcal{D}_1\left(\mathcal{D}_2\left(fg\right)\right) = \mathcal{D}_1\left(\mathcal{D}_2(f)g + f\mathcal{D}_2(g)\right)$$
$$= \left(\mathcal{D}_1\mathcal{D}_2 f\right)g + \mathcal{D}_2 f \mathcal{D}_1 g + \mathcal{D}_1 f \mathcal{D}_2 g + f\mathcal{D}_1\mathcal{D}_2 g.$$

Writing out the similar expression for $\mathcal{D}_2\left(\mathcal{D}_1\left(fg\right)\right)$ and then subtracting we obtain, after a cancellation,

$$[\mathcal{D}_1, \mathcal{D}_2]\left(fg\right) = \left(\mathcal{D}_1\mathcal{D}_2 f\right)g + f\mathcal{D}_1\mathcal{D}_2 g - \left(\left(\mathcal{D}_2\mathcal{D}_1 f\right)g + f\mathcal{D}_2\mathcal{D}_1 g\right)$$
$$= \left([\mathcal{D}_1, \mathcal{D}_2]f\right)g + f[\mathcal{D}_1, \mathcal{D}_2]g. \qquad \square$$

Corollary 2.74. *If $X, Y \in \mathfrak{X}(M)$, then there is a unique vector field $[X, Y]$ such that $\mathcal{L}_{[X,Y]} = \mathcal{L}_X \circ \mathcal{L}_Y - \mathcal{L}_Y \circ \mathcal{L}_X$.*

Since $\mathcal{L}_X f$ is also written Xf, we have $[X, Y]f = X\left(Yf\right) - Y\left(Xf\right)$ or

$$[X, Y] = XY - YX.$$

Definition 2.75. *The vector field $[X, Y]$ from the previous corollary is called the **Lie bracket** of X and Y.*

Proposition 2.76. *The map $(X, Y) \mapsto [X, Y]$ is bilinear over $\mathbb{R}$, and for $X, Y, Z \in \mathfrak{X}(M)$ we have*

  (i) $[X, Y] = -[Y, X]$;
  (ii) $[X, [Y, Z]] + [Y, [Z, X]] + [Z, [X, Y]] = 0$ (*Jacobi Identity*);
  (iii) $[fX, gY] = fg[X, Y] + f\left(Xg\right)Y - g\left(Yf\right)X$ for all $f, g \in C^\infty(M)$.

Proof. These results follow from direct calculation and the previously mentioned fact that $\mathcal{L}_{aX+bY} = a\mathcal{L}_X + b\mathcal{L}_Y$ for $a, b \in \mathbb{R}$ and $X, Y \in \mathfrak{X}(M)$. $\square$

The map $(X, Y) \mapsto [X, Y]$ is bilinear over $\mathbb{R}$, but by (iii) above, it is not bilinear over $C^\infty(M)$. Also notice that in (ii) above, X, Y, Z are permuted cyclically.

We ought to see what the local formula for the Lie derivative looks like in conventional "index" notation. Suppose we have $X = \sum X^i \frac{\partial}{\partial x^i}$ and $Y = \sum Y^i \frac{\partial}{\partial x^i}$. Then we have the local formula

$$[X, Y] = \sum_{ij}\left(X^j \frac{\partial Y^i}{\partial x^j} - Y^j \frac{\partial X^i}{\partial x^j}\right)\frac{\partial}{\partial x^i}.$$

Exercise 2.77. Verify this last formula.

The $\mathbb{R}$-vector space $\mathfrak{X}(M)$ together with the $\mathbb{R}$-bilinear map $(X, Y) \mapsto [X, Y]$ is an example of an extremely important abstract algebraic structure:

Definition 2.78 (Lie algebra). A vector space $\mathfrak{a}$ (over a field $\mathbb{F}$) is called a **Lie algebra** if it is equipped with a bilinear map $\mathfrak{a} \times \mathfrak{a} \to \mathfrak{a}$ (a multiplication) denoted $(v, w) \mapsto [v, w]$ such that
$$[v, w] = -[w, v]$$
and such that we have the **Jacobi identity**
$$[x, [y, z]] + [y, [z, x]] + [z, [x, y]] = 0$$
for all $x, y, z \in \mathfrak{a}$.

Definition 2.79. A Lie algebra $\mathfrak{a}$ is called **abelian** (or commutative) if $[v, w] = 0$ for all $v, w \in \mathfrak{a}$. A subspace $\mathfrak{h}$ of $\mathfrak{a}$ is called a **Lie subalgebra** if it is closed under the bracket operation, and it is called an **ideal** if $[v, w] \in \mathfrak{h}$ for any $v \in \mathfrak{a}$ and $w \in \mathfrak{h}$. (We indicate this by writing $[\mathfrak{a}, \mathfrak{h}] \subset \mathfrak{h}$.)

Notice that the Jacobi identity may be restated as $[x, [y, z]] = [[x, y], z] + [y, [x, z]]$, which just says that for fixed x the map $y \mapsto [x, y]$ is a derivation of the Lie algebra $\mathfrak{a}$. This is significant mathematically and also an easy way to remember the Jacobi identity. The Lie algebra $\mathfrak{X}(M)$ is infinite-dimensional (unless M is zero-dimensional), but later we will be very interested in certain finite-dimensional Lie algebras which are subalgebras of $\mathfrak{X}(M)$.

Given a diffeomorphism $\phi : M \to N$, we define the **pull-back** $\phi^* Y \in \mathfrak{X}(M)$ for $Y \in \mathfrak{X}(N)$ and the **push-forward** $\phi_* X \in \mathfrak{X}(N)$ of $X \in \mathfrak{X}(M)$ by ϕ by
$$\phi^* Y = T\phi^{-1} \circ Y \circ \phi \quad \text{and}$$
$$\phi_* X = T\phi \circ X \circ \phi^{-1}.$$
In other words, $(\phi^* Y)(p) = T\phi^{-1} \cdot Y_{\phi(p)}$ and $(\phi_* X)(p) = T\phi \cdot X_{\phi^{-1}(p)}$. Notice that $\phi^* Y$ and $\phi_* X$ are both smooth vector fields. **Warning**: Since many authors use the notation f_* for the tangent map Tf, the notation $f_* X$ might be interpreted to mean $Tf \circ X$, which is actually a vector field *along the map f* rather than an element of $\mathfrak{X}(N)$. We shall not use f_* as a notation for Tf.

To summarize a bit, if $f : M \to N$ is a smooth map, then for each p we have the tangent map $T_p f : T_p M \to T_{f(p)} N$, the tangent lift $Tf : TM \to TN$ (a "bundle map"), and if f is a diffeomorphism, we have the induced maps on the level of fields $f_* : \mathfrak{X}(M) \to \mathfrak{X}(N)$ and $f^* : \mathfrak{X}(N) \to \mathfrak{X}(M)$. Notice that if $\phi : M \to N$ and $\psi : N \to P$ are diffeomorphisms, then we have
$$(\psi \circ \phi)_* = \psi_* \circ \phi_* : \mathfrak{X}(M) \to \mathfrak{X}(P),$$
$$(\psi \circ \phi)^* = \phi^* \circ \psi^* : \mathfrak{X}(P) \to \mathfrak{X}(M).$$

We have right and left actions of the diffeomorphism group $\mathrm{Diff}(M)$ on the space of vector fields. The left action $\mathrm{Diff}(M) \times \mathfrak{X}(M) \to \mathfrak{X}(M)$ is given by

2.8. Vector Fields

$(\phi, X) \mapsto \phi_* X$, and the right action $\mathfrak{X}(M) \times \text{Diff}(M) \to \mathfrak{X}(M)$ is given by $(X, \phi) \mapsto \phi^* X$.

On functions, the **pull-back** is defined by $\phi^* g := g \circ \phi$ for any smooth map, but if ϕ is a diffeomorphism, then we can also define a **push-forward** $\phi_* := (\phi^{-1})^*$. With this notation we have the following proposition.

Proposition 2.80. *The Lie derivative on functions is natural with respect to pull-back and push-forward by diffeomorphisms. In other words, if $\phi : M \to N$ is a diffeomorphism and $f \in C^\infty(M)$, $g \in C^\infty(N)$, $X \in \mathfrak{X}(M)$ and $Y \in \mathfrak{X}(N)$, then*

$$\mathcal{L}_{\phi^* Y} \phi^* g = \phi^* \mathcal{L}_Y g$$

and

$$\mathcal{L}_{\phi_* X} \phi_* f = \phi_* \mathcal{L}_X f.$$

Proof. We use Definition 2.19. For any p we have

$$\begin{aligned}(\mathcal{L}_{\phi^* Y} \phi^* g)(p) &= (\phi^* Y)_p \phi^* g = \left(T\phi^{-1} \circ Y \circ \phi\right)_p [g \circ \phi] \\ &= \left(T\phi^{-1} \cdot Y_{\phi(p)}\right)[g \circ \phi] = T\phi\left(T\phi^{-1} Y_{\phi(p)}\right) g \\ &= Y_{\phi(p)} g = (\mathcal{L}_Y g)(\phi(p)) = (\phi^* \mathcal{L}_Y g)(p).\end{aligned}$$

The second statement follows from the first since $\phi_* = (\phi^{-1})^*$. □

Even if $f : M \to N$ is *not* a diffeomorphism, it *may* still be that there is a vector field $Y \in \mathfrak{X}(N)$ such that

$$Tf \circ X = Y \circ f.$$

In other words, it may happen that $Tf \cdot X_p = Y_{f(p)}$ for all p in M. In this case, we say that Y is f-**related** to X and write $X \sim_f Y$. It is not hard to check that if X_i is f-related to Y_i for $i = 1, 2$, then $aX_1 + bX_1$ is f-related to $aY_1 + bY_1$.

Example 2.81. Let M and N be smooth manifolds and consider the projections $\text{pr}_1 : M \times N \to M$ and $\text{pr}_2 : M \times N \to N$. Since $T_{(p,q)}(M \times N)$ can be identified with $T_p M \times T_q N$, we see that for $X \in \mathfrak{X}(M)$, $Y \in \mathfrak{X}(N)$ we obtain a vector field $X \times Y \in \mathfrak{X}(M \times N)$ defined by $(X \times Y)(p, q) = (X(p), Y(p))$. Then one can check that

$$X \times Y \text{ and } X \text{ are } \text{pr}_1\text{-related}$$

and

$$X \times Y \text{ and } Y \text{ are } \text{pr}_2\text{-related}.$$

Exercise 2.82. Let M, N, X, Y and $X \times Y$ be as in the example above. Show that if $\iota_q : M \to M \times N$ is the insertion map $p \mapsto (p, q)$, then X and $X \times Y$ are ι_q-related if and only if $Y(q) = 0$.

Lemma 2.83. *Suppose that $f : M \to N$ is a smooth map, $X \in \mathfrak{X}(M)$ and $Y \in \mathfrak{X}(N)$. Then X and Y are f-related if and only if $X(g \circ f) = (Yg) \circ f$ for all $g \in C^\infty(N)$.*

Proof. Let $p \in M$ and let $g \in C^\infty(N)$. Then
$$X(g \circ f)(p) = X_p(g \circ f) = (T_p f \cdot X_p) g$$
and
$$(Yg \circ f)(p) = Y_{f(p)} g$$
so that $X(g \circ f) = (Yg) \circ f$ for all such g if and only if $T_p f \cdot X_p = Y_{f(p)}$. $\square$

Proposition 2.84. *If $f : M \to N$ is a smooth map and X_i is f-related to Y_i for $i = 1, 2$, then $[X_1, X_2]$ is f-related to $[Y_1, Y_2]$. In particular, if ϕ is a diffeomorphism, then $[\phi_* X_1, \phi_* X_2] = \phi_* [X_1, X_2]$ for all $X_1, X_2 \in \mathfrak{X}(M)$.*

Proof. We use the previous lemma: Let $g \in C^\infty(N)$. Then $X_1 X_2(g \circ f) = X_1((Y_2 g) \circ f) = (Y_1 Y_2 g) \circ f$. In the same way, $X_2 X_1(g \circ f) = (Y_2 Y_1 g) \circ f$ and subtracting we obtain
$$\begin{aligned}[X_1, X_2](g \circ f) &= X_1 X_2(g \circ f) - X_2 X_1(g \circ f) \\ &= (Y_1 Y_2 g) \circ f - (Y_2 Y_1 g) \circ f \\ &= ([Y_1, Y_2] g) \circ f.\end{aligned}$$
Using the lemma one more time, we have the result. $\square$

If S is a submanifold of M and $X \in \mathfrak{X}(M)$, then the restriction $X|_S \in \mathfrak{X}(S)$ defined by $X|_S(p) = X(p)$ for all $p \in S$ is ι-related to X where $\iota : S \hookrightarrow M$ is the inclusion map. Thus for $X, Y \in \mathfrak{X}(M)$ we always have that $[X|_S, Y|_S]$ is ι-related to $[X, Y]$. This just means that $[X, Y](p) = [X|_S, Y|_S](p)$ for all p.

We also have

Proposition 2.85. *Let $f : M \to N$ be a smooth map and suppose that $X \sim_f Y$. Then we have $\mathcal{L}_X(f^* g) = f^* \mathcal{L}_Y g$ for any $g \in C^\infty(N)$.*

The proof is similar to what we did above and is left to the reader.

2.8.1. Integral curves and flows. Recall that if $c : I \to M$ is a smooth curve, then the velocity at "time" t is
$$\frac{d}{dt} c(t) = \dot{c}(t) = T_t c \cdot \left.\frac{\partial}{\partial u}\right|_t,$$
where $\frac{\partial}{\partial u}$ is the standard field on $\mathbb{R}$ given at $a \in \mathbb{R}$ as the equivalence class of the curve $t \mapsto a + t$ or by the derivation $\left.\frac{\partial}{\partial u}\right|_a f = f'(a)$.

2.8. Vector Fields

Definition 2.86. Let X be a smooth vector field on M. A curve $c : I \to M$ is called an **integral curve** for X if for all $t \in I$, the velocity of c at time t is equal to $X(c(t))$, that is, if

$$\dot{c} = X \circ c.$$

Thus if c is an integral curve for X and f is a smooth function, then

$$X_{c(t)} f = (f \circ c)'(t)$$

for all t in the domain of c. If the image of an integral curve c lies in U for a chart $(U, \mathbf{x})$, and if $X = \sum X^i \frac{\partial}{\partial x^i}$, then $\dot{c} = X \circ c$ gives the local expressions

$$\frac{d}{dt} x^i \circ c = X^i \circ c \text{ for } i = 1, \ldots, n,$$

which constitute a system of ordinary differential equations for the functions $x^i \circ c$. These equations are classically written as $\frac{dx^i}{dt} = X^i$.

A **(complete) flow** is a map $\Phi : \mathbb{R} \times M \to M$ such that if $\Phi_t(x) := \Phi(t, x)$ for each $x \in M$, then $t \mapsto \Phi_t$ is a group homomorphism from the additive group $\mathbb{R}$ to the diffeomorphism group of M. More generally, a **flow** is defined similarly except that $\Phi(t, x)$ may not be defined on all of $\mathbb{R} \times M$, but rather on some open neighborhood of $\{0\} \times M \subset \mathbb{R} \times M$, and so we explicitly require that $\Phi_t \circ \Phi_s = \Phi_{t+s}$ and $\Phi_t^{-1} = \Phi_{-t}$ for all t and s such that both sides of these equations are defined. **Convention:** We shall also loosely refer to the map $t \mapsto \Phi_t$ as the flow.

Using a smooth flow, we can define a vector field X^Φ by

$$X^\Phi(p) = \left. \frac{d}{dt} \right|_0 \Phi(t, p) \in T_p M \text{ for } p \in M.$$

If one computes the velocity vector $\dot{c}(0)$ of the curve $c : t \mapsto \Phi(t, p)$, one gets $X^\Phi(p)$. In fact, because of the two properties assumed above, we get $\dot{c}(t) = X^\Phi(c(t))$ for any t for which $\Phi(t, p)$ is defined.

We would like to start with a vector field and produce a flow at least locally. Our study of the flows of vector fields begins with a quick recounting of a basic **existence and uniqueness** theorem for differential equations stated here in the setting of real Banach spaces. If desired, the reader may take the Banach space to be a finite-dimensional normed space such as $\mathbb{R}^n$.

Theorem 2.87. *Let E be a Banach space and let $F : U \subset \mathsf{E} \to \mathsf{E}$ be a smooth map with open domain U. Given any $x_0 \in U$, there is a smooth curve $c : (-\epsilon, \epsilon) \to U$ with $c(0) = x_0$ such that $c'(t) = F(c(t))$ for all $t \in (-\epsilon, \epsilon)$. If $c_1 : (-\epsilon_1, \epsilon_1) \to U$ is another such curve with $c_1(0) = x_0$ and $c_1'(t) = F(c(t))$ for all $t \in (-\epsilon_1, \epsilon_1)$, then $c = c_1$ on the intersection $(-\epsilon_1, \epsilon_1) \cap (-\epsilon, \epsilon)$. Furthermore, given any fixed $x_0 \in U$, there is an $a > 0$, an open set V with $x_0 \in V \subset U$, and a smooth map $\Phi : (-a, a) \times V \to U$*

such that $t \mapsto c_x(t) := \Phi(t, x)$ is a curve satisfying $c'_x(t) = F(c_x(t))$ for all $t \in (-a, a)$ and $c_x(0) = x$.

Example 2.88. Consider the differential equation on the line given by
$$c'(t) = (c(t))^{2/3}.$$
There are two distinct solutions with initial condition $c(0) = 0$. Namely,
$$c(t) = 0 \text{ for all } t$$
and
$$c(t) = \frac{1}{27}t^3 \text{ for all } t.$$
The reason uniqueness fails is the fact that the function $F(x) = x^{2/3}$ is not differentiable at $x = 0$.

Now let $X \in \mathfrak{X}(M)$ and consider a point p in the domain of a chart $(U, \mathbf{x})$. The local expression for the integral curve equation $\dot{c}(t) = X(c(t))$ is of the form treated in the last theorem, and so we see that there certainly exists an integral curve for X through p defined on at least some small interval $(-\epsilon, \epsilon)$. We will now use this theorem to obtain similar but more global results on smooth manifolds. First of all, we can get a more global version of uniqueness:

Lemma 2.89. *If $c_1 : (-\epsilon_1, \epsilon_1) \to M$ and $c_2 : (-\epsilon_2, \epsilon_2) \to M$ are integral curves of a vector field X with $c_1(0) = c_2(0)$, then $c_1 = c_2$ on the intersection of their domains.*

Proof. Let $K = \{t \in (-\epsilon_1, \epsilon_1) \cap (-\epsilon_2, \epsilon_2) : c_1(t) = c_2(t)\}$. The set K is closed since M is Hausdorff. It follows from Theorem 2.87 that K contains a (small) open interval $(-\epsilon, \epsilon)$. Let t_0 be any point in K and consider the translated curves $c_1^{t_0}(t) = c_1(t_0 + t)$ and $c_2^{t_0}(t) = c_2(t_0 + t)$. These are also integral curves of X and they agree at $t = 0$, and by Theorem 2.87 again we see that $c_1^{t_0} = c_2^{t_0}$ on some open neighborhood of 0. But this means that c_1 and c_2 agree in this neighborhood, so in fact this neighborhood is contained in K implying that K is also open since t_0 was an arbitrary point in K. Thus, since $I = (-\epsilon_1, \epsilon_1) \cap (-\epsilon_2, \epsilon_2)$ is connected, it must be that $I = K$ and so c_1 and c_2 agree on $I = (-\epsilon_1, \epsilon_1) \cap (-\epsilon_2, \epsilon_2)$. $\square$

Let X be a C^∞ vector field on M. A **flow box** for X at a point $p \in M$ is a triple (U, a, φ^X), where

(1) U is an open set in M containing p.

(2) $\varphi^X : (-a, a) \times U \to M$ is a C^∞ map and $0 < a \le \infty$.

(3) For each $p \in U$, the curve $t \mapsto c_p(t) = \varphi^X(t, p)$ is an integral curve of X with $c_p(0) = p$.

2.8. Vector Fields

(4) The map $\varphi_t^X : U \to M$ given by $\varphi_t^X(p) = \varphi^X(t,p)$ is a diffeomorphism onto its image for all $t \in (-a, a)$.

We sometimes refer to φ^X as a **local flow** for X. Before we prove that flow boxes actually exist, we make the following observation: If we have a triple that satisfies (1)–(3) above, then both $c_1 : t \mapsto \varphi_{t+s}^X(p)$ and $c_2 : t \mapsto \varphi_t^X(\varphi_s^X(p))$ are integral curves of X with $c_1(0) = c_2(0) = \varphi_s^X(p)$, so by uniqueness (Lemma 2.89) we conclude that $\varphi_t^X(\varphi_s^X(p)) = \varphi_{t+s}^X(p)$ as long as both sides are defined. This also shows that

$$\varphi_s^X \circ \varphi_t^X = \varphi_{t+s}^X = \varphi_t^X \circ \varphi_s^X$$

whenever defined. This is the local group property, so called because if φ_t^X were defined for all $t \in \mathbb{R}$ (and X a global vector field), then $t \mapsto \varphi_t^X$ would be a group homomorphism from $\mathbb{R}$ into the diffeomorphism group $\text{Diff}(M)$. Whenever this happens, that is, whenever φ^X is defined for all (t, p), we say that X is a **complete vector field**. In other words, a vector field is complete if all its integral curves are defined on all of $\mathbb{R}$. The local group property also implies that $\varphi_t^X \circ \varphi_{-t}^X = \text{id}$, and so in general φ_t^X must at least be a locally defined diffeomorphism with inverse φ_{-t}^X.

Notice that whereas $\frac{\partial}{\partial x}$ is a complete vector field on $\mathbb{R}^2$ (using standard coordinates x, y), this vector field restricted to $\mathbb{R}^2 \setminus \{0\}$ is *not* complete on the manifold $\mathbb{R}^2 \setminus \{0\}$. The reason is that integral curves starting at points on the x-axis will run up against the missing origin in finite time. This "running up to missing points" is not the only way a vector field can fail to be complete. Consider the vector field $(1 + x^2) \frac{\partial}{\partial x}$ on $\mathbb{R}$. The integral curve that is at the origin at time zero is obtained by solving the initial value problem $x' = 1 + x^2$, $x(0) = 0$. The unique solution is $x(t) = \tan t$, and since $\lim_{t \to \pm \pi/2} \tan t = \pm \infty$, the solution cannot be extended beyond $\pm \pi/2$.

Exercise 2.90. Show that on $\mathbb{R}^2$ the vector fields $y^2 \frac{\partial}{\partial x}$ and $x^2 \frac{\partial}{\partial y}$ are complete, but $y^2 \frac{\partial}{\partial x} + x^2 \frac{\partial}{\partial y}$ is not complete. In particular, the set of complete vector fields is not generally a vector space (but this is true if M is compact).

Theorem 2.91 (Flow box). *Let X be a C^∞ vector field on an n-manifold M with $r \geq 1$. Then for every point $p_0 \in M$ there exists a flow box for X at p_0. If (U_1, a_1, φ_1^X) and (U_2, a_2, φ_2^X) are two flow boxes for X at p_0, then $\varphi_1^X = \varphi_2^X$ on $(-a_1, a_1) \cap (-a_2, a_2) \times U_1 \cap U_2$.*

Proof. First of all, notice that the U in the triple (U, a, φ^X) does not have to be contained in a chart or even be homeomorphic to an open set in $\mathbb{R}^n$. However, to prove that there *are* flow boxes at any point we can work in the domain of a chart $(U, \mathbf{x})$ and so we might as well assume that the vector field is defined on an open set in $\mathbb{R}^n$. Of course, we may have to choose a to be smaller so that the flow stays within the range of the chart map $\mathbf{x}$. In this setting, a vector field can be taken to be a map $U \to \mathbb{R}^n$, so Theorem 2.87 provides us with the flow box data (V, a, Φ), where we have taken $a > 0$ small enough that $V_t = \Phi(t, V) \subset U$ for all $t \in (-a, a)$. Now the flow box is transferred back to the manifold via $\mathbf{x}$,

$$U = \mathbf{x}^{-1}(V),$$
$$\varphi^X(t, p) = \mathbf{x}^{-1}\left[\Phi(t, \mathbf{x}(p))\right].$$

If we have two such flow boxes (U_1, a_1, φ_1^X) and (U_2, a_2, φ_2^X), then by Lemma 2.89, we see that for any $x \in U_1 \cap U_2$ we must have $\varphi_1^X(t, x) = \varphi_2^X(t, x)$ for all $t \in (-a_1, a_1) \cap (-a_2, a_2)$.

Finally, since $\varphi_t^X = \varphi^X(t, \cdot)$ and $\varphi_{-t}^X = \varphi^X(-t, \cdot)$ are both smooth and inverses of each other, we see that φ_t^X is a diffeomorphism onto its image $U_t = \mathbf{x}^{-1}(V_t)$. □

Lemma 2.92. *Suppose that $X_1, \ldots, X_k$ are smooth vector fields on M and let $p_0 \in M$ be given and O be an open set containing p_0. If $\varphi^{X_1}, \ldots, \varphi^{X_k}$ are the local flows corresponding to flow boxes whose domains $U_1, \ldots, U_k$ all contain p_0, then there is an open set $U \subset U_1 \cap \cdots \cap U_k$ and an $\epsilon > 0$ such that the composition*

$$\varphi_{t_k}^{X_k} \circ \cdots \circ \varphi_{t_1}^{X_1}$$

is defined on U and maps U into O whenever $t_1, \ldots, t_k \in (-\epsilon, \epsilon)$.

Proof. If the flow box corresponding to φ^{X_i} is $(U_i, \epsilon_i, \varphi_i^X)$, then by shrinking U_1 further we may arrange things so that $\varphi_t^{X_1}$ maps U_1 into O for all $t \in (-\epsilon_1, \epsilon_1)$, and then inductively we arrange for φ^{X_i} to map U_i into U_{i-1} for all $t \in (-\epsilon_i, \epsilon_i)$. Now let $\epsilon = \min\{\epsilon_1, \ldots, \epsilon_k\}$. □

Remark 2.93. When making compositions of local flows, we will not always make careful statements about domains, but the previous lemma will be invoked implicitly.

If $c_p(t)$ is an integral curve of X defined on some interval (a, b) containing 0 and $c_p(0) = p$, then we may consider the limit

$$\lim_{t \to b-} c_p(t).$$

If this limit exists as a point $p_1 \in M$, then we may consider the integral curve c_{p_1} beginning at p_1. One may now use Lemma 2.89 to combine $t \mapsto c_p(t)$ with $t \mapsto c_{p_1}(t-b)$ to produce an extended integral curve beginning at p. We may repeat this process as long as the limit exists. We may do a similar thing in the negative direction. This suggests that there is a **maximal integral curve** defined on a maximal interval $J_p^X := (T_{p,X}^-, T_{p,X}^+)$, where $T_{p,X}^-$ might be $-\infty$ and $T_{p,X}^+$ might be $+\infty$. We produce this maximal integral curve

2.8. Vector Fields

as follows: Consider the collection $\mathcal{J}_p$ of all pairs (J, α), where J is an open interval containing 0 and $\alpha : J \to M$ is an integral curve of X with $\alpha(0) = p$. Then let $J_p^X = \bigcup_{(J,\alpha) \in \mathcal{J}_p} J$ and define $c_{\max}(t) := \alpha(t)$ whenever $t \in J$ for $(J, \alpha) \in \mathcal{J}_p$. By existence and uniqueness, this definition is unambiguous. The curve $c_{\max} : J_p^X \to M$ is the desired maximal integral curve and is easily seen to be unique.

Definition 2.94. Let X be a C^∞ vector field on M. For any given $p \in M$, let $J_p^X := (T_{p,X}^-, T_{p,X}^+) \subset \mathbb{R}$ be the domain of the maximal integral curve $c : J_p^X \to M$ of X with $c(0) = p$. The **maximal flow** φ^X is defined on the set (called the maximal flow domain)

$$\mathcal{D}_X = \bigcup_{p \in M} J_p^X \times \{p\}$$

by the prescription that $t \mapsto \varphi^X(t, p)$ is the maximal integral curve of X such that $\varphi^X(0, p) = p$.

Thus by definition, X is a complete vector field if and only if $\mathcal{D}_X = \mathbb{R} \times M$. We will abbreviate $\left(T_{p,X}^-, T_{p,X}^+\right)$ to $\left(T_p^-, T_p^+\right)$.

Theorem 2.95. *For $X \in \mathfrak{X}(M)$, the set $\mathcal{D}_X$ is an open neighborhood of $\{0\} \times M$ in $\mathbb{R} \times M$ and the map $\varphi^X : \mathcal{D}_X \to M$ is smooth. Furthermore,*

$$\varphi^X(t + s, p) = \varphi^X(t, \varphi^X(s, p)) \tag{2.4}$$

whenever both sides are defined. If the right hand side is defined, then the left hand side is defined. Suppose that $t, s \geq 0$ or $t, s \leq 0$. Then if the left hand side is defined so is the right hand side.

Proof. Let $q = \varphi^X(s, p)$. If the right hand side is defined, then $s \in \left(T_p^-, T_p^+\right)$ and $t \in \left(T_q^-, T_q^+\right)$. The curve $\psi : \tau \mapsto \varphi^X(s + \tau, p)$ is defined for $\tau \in \left(T_p^- - s, T_p^+ - s\right)$, and this is the maximal domain for ψ. We have

$$\left.\frac{d}{d\tau}\right|_\tau \psi = \frac{d}{d\tau}\varphi^X(s + \tau, p) = \left.\frac{d}{du}\right|_{u=s+\tau} \varphi^X(u, p)$$
$$= X(\varphi^X(s + \tau, p)) = X(\psi(\tau)).$$

We also have $\psi(0) = \varphi^X(\tau + s, p)\big|_{\tau=0} = \varphi^X(s, p)$ so ψ is an integral curve starting at $q = \varphi^X(s, p)$. Thus $\left(T_p^- - s, T_p^+ - s\right) \subset \left(T_q^-, T_q^+\right)$ and $\psi = \varphi^X(\cdot, q)$ on $\left(T_p^- - s, T_p^+ - s\right)$. But the *maximal* domain for ψ is $\left(T_p^- - s, T_p^+ - s\right)$ and so in fact $\left(T_p^- - s, T_p^+ - s\right) = \left(T_q^-, T_q^+\right)$ for otherwise $\varphi^X(\cdot, q)$ would be a proper extension. But then, since $t \in \left(T_q^-, T_q^+\right)$, we have that $t \in \left(T_p^- - s, T_p^+ - s\right)$ and so $\psi(t) = \varphi^X(t + s, p)$ is defined and

$$\varphi^X(t + s, p) = \varphi^X(t, \varphi^X(s, p)).$$

Now let us assume that $t, s \geq 0$ and that $\varphi^X(s+t, p)$ is defined (the case of $t, s \leq 0$ is similar). Then since $s, t \geq 0$, we have

$$s, \ t, \ t+s \in \left(T_p^-, T_p^+\right).$$

Let $q = \varphi^X(s, p)$ as before and let $\theta(u) = \varphi^X(s+u, p)$ be defined for u with $0 \leq u \leq t$. But $\theta(u)$ is an integral curve with $\theta(0) = q$. Thus we have that $\varphi^X(u, q)$ must also be defined for $u = t$ and $\theta(t) = \varphi^X(t, q)$. But $\varphi^X(t, q) = \varphi^X(t, \varphi^X(s, p))$, which is thereby defined, and we have

$$\varphi^X(s+t, p) := \theta(t) = \varphi^X(t, \varphi^X(s, p)).$$

Now we will show that $\mathcal{D}_X$ is open in $\mathbb{R} \times M$ and that $\varphi^X : \mathcal{D}_X \to M$ is smooth. We carefully define a subset $\mathcal{S} \subset \mathcal{D}_X$ by the condition that $(t, p) \in \mathcal{S}$ exactly if there exists an interval J containing 0 and t and also an open set $U \subset M$ such that the restriction of φ^X to $J \times U$ is smooth. Notice that $\mathcal{S}$ is open by construction. We intend to show that $\mathcal{S} = \mathcal{D}_X$. Suppose not. Then let $(t_0, p_0) \in \mathcal{D}_X \cap \mathcal{S}^c$. We will assume that $t_0 > 0$ since the case $t_0 < 0$ is proved in a similar way. Now let $\tau := \sup\{t : (t, p_0) \in \mathcal{S}\}$. We know that $(0, p_0)$ is contained in some flow box and so $\tau > 0$. We also have $\tau \leq t_0$ by the definition of t_0. Thus $\tau \in J_{p_0}^X$ and we define $q_0 := \varphi^X(\tau, p_0)$. Now applying the local theory we know that q_0 is contained in an open set U_0 such that φ^X is defined and smooth on $(-\epsilon, \epsilon) \times U_0$ for some $\epsilon > 0$. We will now show that φ^X is actually defined and smooth on a set of the form $(-\delta, r) \times O$ where O is open, $\delta, r > 0$, and $(\tau, p_0) \in (-\delta, r) \times O$. Since this contradicts the definition of τ, we will be done.

We may choose $t_1 > 0$ so that $\tau \in (t_1, t_1 + \epsilon)$ and so that $\varphi^X(t_1, p_0) \in U_0$. Note that $(t_1, p_0) \in \mathcal{S}$ since $t_1 < \tau$. So on some neighborhood $(-\delta, t_1 + \delta) \times U_1$ of (t_1, p_0) the flow φ^X is smooth. By choosing U_1 smaller if necessary we can arrange that $\varphi^X(\{t_1\} \times U_1) \subset U_0$. Now consider the equation

$$\varphi^X(t, p) = \varphi^X(t - t_1, \varphi^X(t_1, p)).$$

If $|t - t_1| < \epsilon$ and $p \in U_1$, then both sides are defined and the right hand side is smooth near such (t, p). But the right hand side is already known to be smooth on $(-\delta, t_1 + \delta) \times U_1$. We now see that φ^X is smooth on $(-\delta, t_1 + \epsilon) \times U_1$, which contains (τ, p_0) contradicting the definition of τ. Thus $\mathcal{S} = \mathcal{D}_X$. □

Remark 2.96. In this text, φ^X will either refer to the *unique* maximal flow defined on $\mathcal{D}_X$ or to its restriction to the domain of a flow box. In the latter case we call φ^X a local flow. We could have introduced notation such as $\varphi_{\max}^X$, but prefer not to clutter up the notation to that extent unless necessary. We hope that the reader will be able to tell from context what we are referring to when we write φ^X.

2.8. Vector Fields

If φ^X is a flow of X, then we write φ^X_t for the map $p \mapsto \varphi^X_t(p)$. The (maximal) domain of this map is $\mathcal{D}^t_X = \{p : t \in (T^-_{p,X}, T^+_{p,X})\}$. Note that, in general, the domain of φ^X_t depends on t. Also, we have the tangent map $T_0 \varphi^X_p : T_0 \mathbb{R} \to T_p M$ and

$$\left.\frac{d}{dt}\right|_{t=0} \varphi^X_p(t) = T_0 \varphi^X_p \left.\frac{\partial}{\partial u}\right|_0 = X_p,$$

where $\left.\frac{\partial}{\partial u}\right|_0$ is the vector at 0 associated to the standard coordinate function on $\mathbb{R}$ (denoted by u here).

Exercise 2.97. Let s and t be real numbers. Show that the domain of $\varphi^X_s \circ \varphi^X_t$ is contained in $\mathcal{D}^{s+t}_X$ and show that for each t, $\mathcal{D}^t_X$ is open. Show that $\varphi^X_t(\mathcal{D}^t_X) = \mathcal{D}^{-t}_X$.

Definition 2.98. The **support** of a vector field X is the *closure* of the set $\{p : X(p) \neq 0\}$ and is denoted $\mathrm{supp}(X)$.

Lemma 2.99. *Every vector field that has compact support is a complete vector field. In particular, if M is compact, then every vector field is complete.*

Proof. Let c^X_p be the maximal integral curve through p and $J^X_p = (T^-, T^+)$ its domain. If $X(p) = 0$, then the constant curve, $c(t) = p$ for all p, is the unique integral curve through p and is defined for all t so $J^X_p = \mathbb{R}$. Now suppose $X(p) \neq 0$. If $t \in (T^-, T^+)$, then the image point $c^X_p(t)$ must always lie in the support of X. Indeed, since $X(p)$ is not zero, c^X_p is not constant. If $c^X_p(t)$ were in $M \setminus \mathrm{supp}(X)$, then $c^X_p(t)$ would be contained in some ball on which X vanishes and then by uniqueness this implies that c^X_p is constantly equal to $c^X_p(t)$ for all time—a contradiction. But we show that if $T^+ < \infty$, then given any compact set $K \subset M$, for example the support of X, there is an $\epsilon > 0$ such that for all $t \in (T^+ - \epsilon, T^+)$, the image $c^X_p(t)$ is outside K. If not, then we may take a sequence t_i converging to T^+ such that $c^X_p(t_i) \in K$. But then going to a subsequence if necessary, we have $x_i := c^X_p(t_i) \to x \in K$. Now there must be a flow box (U, a, x), so that for large enough k, we have that t_k is within distance a of T^+ and $x_k = c^X_p(t_k)$ is inside U. We are then guaranteed to have an integral curve $c^X_{x_k}(t)$ of X that continues beyond T^+ and thus can be used to extend c^X_p which is a contradiction of the maximality of T^+. Hence we must have $T^+ = \infty$. A similar argument gives the result that $T^- = -\infty$. $\square$

Exercise 2.100. Let $a > 0$ be any fixed positive real number. Show that if for a given vector field X, the flow φ^X is defined on $(-a, a) \times M$, then in fact the (maximal) flow is defined on $\mathbb{R} \times M$ and so X is a complete vector field.

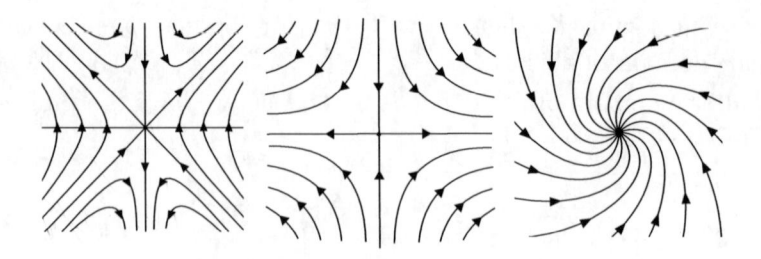

Figure 2.3. Isolated vanishing points

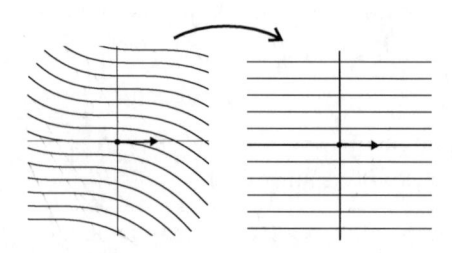

Figure 2.4. Straightening

If a vector field is zero at some point but nonzero elsewhere in a neighborhood of that point, then we say that the field has an isolated zero. The structure of a vector field near such an isolated zero can be quite complex and interesting. The qualitative structure of three such possibilities are show in Figure 2.3 for dimension two. Near zeros that are not isolated, the situation is also potentially complex. On the other hand, at non-vanishing points, all vector fields are the same up to a local diffeomorphism. This is the content of the following theorem which is sometimes called the **straightening theorem**.

Theorem 2.101. *Let X be a smooth vector field on M with $X(p) \neq 0$ for some $p \in M$. Then there is a chart $(U, \mathbf{x})$ with $p \in U$ such that*

$$X = \frac{\partial}{\partial x^1} \text{ on } U.$$

Proof. Since this is clearly a local problem, it will suffice to assume that $M = \mathbb{R}^n$ and $p = 0$. Let $(u^1, \ldots, u^n)$ be standard coordinates on $\mathbb{R}^n$. By a rotation and translation if necessary, we may assume that $X(0) = \frac{\partial}{\partial u^1}\big|_0$. The idea is that there must be a unique integral curve through each point of the hyperplane $\{u^1 = 0\}$. We wish to arrange for the new coordinates of $q \in \mathbb{R}^n$ to be such that if an integral curve of X passes through $(0, a^2, \ldots, a^n)$ at time zero and hits q at time t, then $x^i(p) = a^i$ for $i = 2, \ldots, n$ while $x^1(p) = t$. Let φ be a local flow for X near 0, and define χ in some sufficiently

2.8. Vector Fields

small neighborhood of 0 by

$$\chi(a^1, \ldots, a^n) := \varphi_{a^1}(0, a^2, \ldots, a^n).$$

For $a = (a^1, a^2, \ldots, a^n)$ in the domain of χ, and $f \in C^\infty(M)$, we have

$$T\chi \cdot \left.\frac{\partial}{\partial u^1}\right|_a f = \left.\frac{\partial}{\partial u^1}\right|_a (f \circ \chi)$$

$$= \lim_{h \to 0} \frac{1}{h} \left[f(\chi(a^1 + h, a^2, \ldots, a^n)) - f(\chi(a^1, a^2, \ldots, a^n)) \right]$$

$$= \lim_{h \to 0} \frac{1}{h} \left[f\left(\varphi_{a^1+h}(0, a^2, \ldots, a^n)\right) - f(\chi(a^1, a^2, \ldots, a^n)) \right]$$

$$= \lim_{h \to 0} \frac{1}{h} \left[f(\varphi_h(a)) - f(\chi(a))) \right] = (Xf)(\chi(a)).$$

In particular, $T\chi \cdot \left.\frac{\partial}{\partial u^1}\right|_0 = \left.\frac{\partial}{\partial u^1}\right|_0$. If $i > 0$, then at 0 we have

$$T\chi \cdot \left.\frac{\partial}{\partial u^i}\right|_0 f = \left.\frac{\partial}{\partial u^i}\right|_0 f \circ \chi$$

$$= \lim_{h \to 0} \frac{1}{h} \left[f(\chi(0, .., h, \ldots) - f(0) \right]$$

$$= \lim_{h \to 0} \frac{1}{h} \left[f(0, \ldots, h, \ldots) - f(0) \right] = \left.\frac{\partial}{\partial u^i}\right|_0 f.$$

Thus $T_0 \chi = \text{id}$ and so by the inverse mapping theorem (Theorem 2.25) we see that after restricting χ to a smaller neighborhood of zero, the map $\mathbf{x} := \chi^{-1}$ is a chart map. We have already seen that $T\chi \left.\frac{\partial}{\partial u^1}\right| = X \circ \chi$. But then for $f \in C^\infty(M)$ we have

$$\left.\frac{\partial}{\partial x^1}\right|_p f = \left.\frac{\partial}{\partial u^1}\right|_{\mathbf{x}(p)} f \circ \mathbf{x}^{-1} = \left.\frac{\partial}{\partial u^1}\right|_{\mathbf{x}(p)} f \circ \chi$$

$$= T\chi \left.\frac{\partial}{\partial u^1}\right|_{\mathbf{x}(p)} f = (X \circ \chi)(\mathbf{x}(p))f = X_p f,$$

so that $\frac{\partial}{\partial x^1} = X$. $\square$

2.8.2. Lie derivative. We now introduce the important concept of the Lie derivative of a vector field extending the previous definition. The Lie derivative will be extended further to tensor fields.

Definition 2.102. Given a vector field X, we define a map $\mathcal{L}_X : \mathfrak{X}(M) \to \mathfrak{X}(M)$ by

$$\mathcal{L}_X Y := [X, Y].$$

This map is called the **Lie derivative** (with respect to X).

The Jacobi identity for the Lie bracket easily implies the following two identities for any $X, Y, Z \in \mathfrak{X}(M) \to \mathfrak{X}(M)$:

$$\mathcal{L}_X[Y, Z] = [\mathcal{L}_X Y, Z] + [Y, \mathcal{L}_X Z],$$
$$\mathcal{L}_{[X,Y]} = [\mathcal{L}_X, \mathcal{L}_Y] \quad \text{(i.e. } \mathcal{L}_{[X,Y]} = \mathcal{L}_X \circ \mathcal{L}_Y - \mathcal{L}_Y \circ \mathcal{L}_X\text{)}.$$

We will see below that $\mathcal{L}_X Y$ measures the rate of change of Y in the direction X. To be a bit more specific, $(\mathcal{L}_X Y)(p)$ measures how Y changes in comparison with the field obtained by "dragging" Y_p along the flow of X. Recall our definition of the Lie derivative of a function (Definition 2.70). The following is an alternative characterization in terms of flows: For a smooth function $f : M \to \mathbb{R}$ and a smooth vector field $X \in \mathfrak{X}(M)$, the **Lie derivative** $\mathcal{L}_X$ of f with respect to X is given by

$$\mathcal{L}_X f(p) = \left.\frac{d}{dt}\right|_0 f \circ \varphi^X(t, p).$$

Exercise 2.103. Explain why the above formula is compatible with Definition 2.70.

We will also characterize the Lie derivative on vector fields in terms of flows. First, we need a technical lemma:

Lemma 2.104. Let $X \in \mathfrak{X}(M)$ and $f \in C^\infty(U)$ with U open and $p \in U$. There is an interval $I_\delta := [-\delta, \delta]$ and an open set V containing p such that $\varphi^X(I_\delta \times V) \subset U$ and a function $g \in C^\infty(I_\delta \times V)$ such that

$$f(\varphi^X(t, q)) = f(q) + t g(t, q)$$

for all $(t, q) \in I_\delta \times V$ and such that $g(0, q) = X_q f$ for all $q \in V$.

Proof. The existence of the set $I_\delta \times V$ with $\varphi^X(I_\delta \times V) \subset U$ follows from our study of flows. The function $r(\tau, q) := f(\varphi^X(\tau, q)) - f(q)$ is smooth on $I_\delta \times V$ and $r(0, q) = 0$. Let

$$g(t, q) := \int_0^1 \frac{\partial r}{\partial \tau}(st, q)\, ds,$$

so that

$$t g(t, q) = \int_0^1 \frac{\partial r}{\partial \tau}(st, q) t\, ds = \int_0^1 \frac{\partial}{\partial s} r(st, q)\, ds = r(t, q).$$

Then $f(\varphi^X(t, q)) = f(q) + t g(t, q)$. Also

$$g(0, q) = \lim_{t \to 0} \frac{1}{t} r(t, q) = \lim_{t \to 0} \frac{f(\varphi^X(t, q)) - f(q)}{t} = X_q f. \qquad \square$$

2.8. Vector Fields

Proposition 2.105. *Let X and Y be smooth vector fields on M. Let $\varphi = \varphi^X$ be the flow. The function $t \mapsto T\varphi_{-t} \cdot Y_{\varphi_t(p)}$ is differentiable at $t = 0$ and*

$$(2.5) \qquad \left.\frac{d}{dt}\right|_{t=0} T\varphi_{-t} \cdot Y_{\varphi_t(p)} = [X,Y]_p = (\mathcal{L}_X Y)(p).$$

Proof. Let $f \in C^\infty(U)$ with $p \in U$ as in the previous lemma. We have

$$\left.\frac{d}{dt}\right|_{t=0} T\varphi_{-t} \cdot Y_{\varphi_t(p)} f = \lim_{t \to 0} \frac{T\varphi_{-t} \cdot Y_{\varphi_t(p)} - Y_p}{t} f = \lim_{t \to 0} \frac{Y_p f - \left(T\varphi_t \cdot Y_{\varphi_{-t}(p)}\right) f}{t}$$

by replacing t by $-t$. Thus we study the difference quotient in the second limit above,

$$\frac{Y_p f - \left(T\varphi_t \cdot Y_{\varphi_{-t}(p)}\right) f}{t} = \frac{Y_p f - Y_{\varphi_{-t}(p)} (f \circ \varphi_t)}{t}$$

$$= \frac{Y_p f - Y_{\varphi_{-t}(p)} (f + t g_t)}{t},$$

where g is as in the lemma and $g_t(q) = g(t,q)$. Continuing, we have

$$\frac{Y_p f - Y_{\varphi_{-t}(p)} (f + t g_t)}{t} = \frac{(Yf)(p) - (Yf)(\varphi_{-t}(p))}{t} - Y_{\varphi_{-t}(p)} g_t.$$

Taking the limit as $t \to 0$ and recalling that $g_0 = Xf$ on V, the right hand side above becomes

$$\lim_{t \to 0} \frac{(Yf)(p) - (Yf)(\varphi_{-t}(p))}{t} - \lim_{t \to 0} Y_{\varphi_{-t}(p)} g_t$$

$$= \lim_{t \to 0} \frac{(Yf)(\varphi_t(p)) - (Yf)(p)}{t} - Y_p X f$$

$$= X_p Y f - Y_p X f = [X,Y]_p f.$$

All told we have

$$\left.\frac{d}{dt}\right|_{t=0} \left(T\varphi_{-t} \cdot Y_{\varphi_t(p)}\right) f = [X,Y]_p f$$

for all $f \in C^\infty(U)$. If we let U be the domain of a chart $(U, \mathbf{x})$, then letting f be each of the coordinate functions we see that each component of the T_pM-valued function $t \mapsto T\varphi_{-t} \cdot Y_{\varphi_t(p)}$ is differentiable at $t = 0$ and so the function is also. Then we can conclude that $\left.\frac{d}{dt}\right|_{t=0} \left(T\varphi_{-t} \cdot Y_{\varphi_t(p)}\right) = [X,Y]_p$. $\square$

We see from this characterization that in order for $(\mathcal{L}_X Y)(p)$ to make sense, Y only needs to be defined along the integral curve of X that passes through p.

Discussion: Notice that if X is a complete vector field, then for each $t \in \mathbb{R}$ the map φ_t^X is a diffeomorphism $M \to M$ and

$$(2.6) \qquad \left(\varphi_t^X\right)^* Y = (T\varphi_t^X)^{-1} \circ Y \circ \varphi_t^X.$$

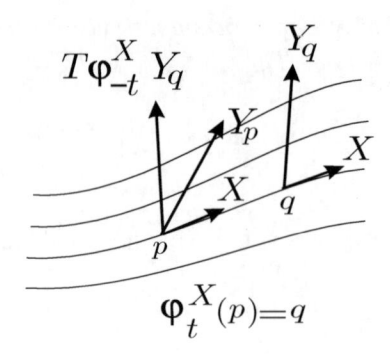

Figure 2.5. Lie derivative by flow

One may write
$$\mathcal{L}_X Y = \left.\frac{d}{dt}\right|_0 (\varphi_t^X)^* Y.$$

On the other hand, if X is not complete, then there exists no t such that φ_t^X is a diffeomorphism of M since for any specific t there might be points of M for which φ_t^X is not even defined! For an X which is not necessarily complete, it is best to consider the map $\varphi^X : (t,x) \mapsto \varphi^X(t,x)$ which is defined on some open neighborhood of $\{0\} \times M$ in $\mathbb{R} \times M$ which may not contain any set of the form $[-\epsilon, \epsilon] \times M$ unless $\epsilon = 0$. In fact, suppose that the domain of φ^X contains such a set with $\epsilon > 0$. It follows that for all $0 \le t \le \epsilon$ the map φ_t^X is defined on all of M and $\varphi_t^X(p)$ exists for $0 \le t \le \epsilon$ independent of p. But a standard argument (recall Exercise 2.100) shows that $t \mapsto \varphi_t^X(p)$ is defined for *all* t, which means that X is a complete vector field. If X is not complete, then φ_t^X is defined only on $\mathcal{D}_X^t$. Thus $(\varphi_t^X)^* : \mathfrak{X}(\mathcal{D}_X^{-t}) \to \mathfrak{X}(\mathcal{D}_X^t)$. (We will abbreviate $(\varphi_t^X)^*$ to φ_t^{X*}.) If $Y \in \mathfrak{X}(M)$, then we may interpret $\varphi_t^{X*} Y$ to mean $\varphi_t^{X*}(Y|_{\mathcal{D}_X^{-t}}) \in \mathfrak{X}(\mathcal{D}_X^t)$. Now $M \subset \bigcup_{t \ne 0} \mathcal{D}_X^t$, so both $\varphi_t^X(p)$ and $(\varphi_t^{X*} Y)(p)$ make sense for any p as long as t is sufficiently small and
$$(\varphi_t^{X*} Y)(p) = (T\varphi_t^X)^{-1} Y(\varphi_t^X(p)).$$

In fact, one should take the last equation for the definition of $(\varphi_t^{X*} Y)(p)$ for the case of an incomplete X. At any rate, the function $t \mapsto (\varphi_t^{X*} Y)(p)$ has a well-defined germ at $t = 0$. With this in mind we might still write $\mathcal{L}_X Y = \left.\frac{d}{dt}\right|_0 (\varphi_t^{X*} Y)$ even for vector fields that are not complete as long as we take the right interpretation of $\varphi_t^{X*} Y$.

Theorem 2.106. *Let X, Y be vector fields on a smooth manifold M. Then*
$$\frac{d}{dt}(\varphi_t^{X*} Y)(p) = (\varphi_t^{X*} \mathcal{L}_X Y)(p)$$
for any $p \in M$.

2.8. Vector Fields

Proof. Let $\varphi := \varphi^X$. Suppressing the point p we have

$$\left.\frac{d}{dt}\right|_t \varphi_t^* Y = \left.\frac{d}{ds}\right|_0 \varphi_{t+s}^* Y = \left.\frac{d}{ds}\right|_0 \varphi_t^*(\varphi_s^{X*} Y)$$

$$= \varphi_t^* \left.\frac{d}{ds}\right|_0 (\varphi_s^* Y) = \varphi_t^* \mathcal{L}_X Y. \qquad \square$$

Exercise 2.107. Show that $\left.\frac{d}{dt}\right|_0 (\varphi_{t*}^X Y)(p) = -(\mathcal{L}_X Y) = -[X, Y]$.

Proposition 2.108. *Let $X \in \mathfrak{X}(M)$ and $Y \in \mathfrak{X}(N)$ be f-related vector fields for a smooth map $f : M \to N$. Then*

$$f \circ \varphi_t^X = \varphi_t^Y \circ f$$

whenever both sides are defined. Suppose that $f : M \to N$ is a diffeomorphism and $X \in \mathfrak{X}(M)$. Then the flow of $f_* X = (f^{-1})^* X$ is $f \circ \varphi_t^X \circ f^{-1}$ and the flow of $f^* X$ is $f^{-1} \circ \varphi_t^X \circ f$.

Proof. For any $p \in M$, we have $\frac{d}{dt}(f \circ \varphi_t^X)(p) = Tf \cdot \frac{d}{dt} \varphi_t^X(p) = Tf \circ X \circ \varphi_t^X(p) = Y \circ f \circ \varphi_t^X(p)$. But $f \circ \varphi_0^X(p) = f(p)$ and so $t \mapsto f \circ \varphi_t^X(p)$ is an integral curve of Y starting at $f(p)$. By uniqueness we have $f \circ \varphi_t^X(p) = \varphi_t^Y(f(p))$. The second part follows from the first. $\qquad \square$

Theorem 2.109. *For $X, Y \in \mathfrak{X}(M)$, the following are equivalent:*

  (i) $\mathcal{L}_X Y = [X, Y] = 0$.

  (ii) $(\varphi_t^X)^* Y = Y$ whenever defined.

  (iii) *The flows of X and Y commute:*

$$\varphi_t^X \circ \varphi_s^Y = \varphi_s^Y \circ \varphi_t^X \text{ whenever defined.}$$

Proof. (Sketch) The equivalence of (i) and (ii) follows easily from Proposition 2.105 and Theorem 2.106. Using Proposition 2.108, the equivalence of (ii) and (iii) can be seen by noticing that $\varphi_t^X \circ \varphi_s^Y = \varphi_s^Y \circ \varphi_t^X$ is defined and true exactly when $\varphi_s^Y = \varphi_{-t}^X \circ \varphi_s^Y \circ \varphi_t^X$ is defined and true, which happens exactly when

$$\varphi_s^Y = \varphi_s^{(\varphi_t^X)^* Y}$$

is defined and true. This happens, in turn, exactly when $Y = (\varphi_t^X)^* Y$. $\qquad \square$

Example 2.110. On $\mathbb{R}^2$ we have the flows given by $\phi(t, (x,y)) = \phi_t(x, y) := (x + ty, y)$ and $\psi(t, (x,y)) = \psi_t(x, y) := (x, y + t)$. We have $\psi_1 \circ \phi_1(0, 0) = (1, 1)$ while $\phi_1 \circ \psi_1(0, 0) = (0, 1)$. These noncommuting flows correspond to the noncommuting vector fields $X = y\partial/\partial x$ and $Y = \partial/\partial y$.

Exercise 2.111. Find global noncommuting flows on S^2.

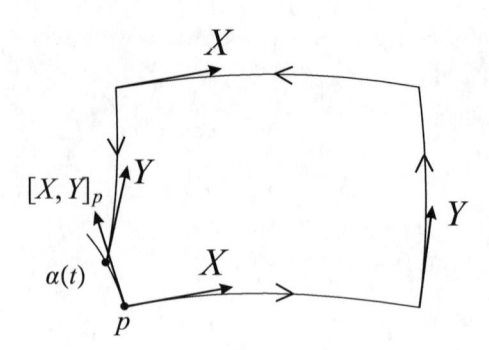

Figure 2.6. Bracket measures lack of commutativity of flows

The Lie derivative and the Lie bracket are essentially the same object and are defined for local sections $X \in \mathfrak{X}_M(U)$ as well as global sections. This is obvious anyway since open subsets are themselves manifolds. As is so often the case for operators in differential geometry, the Lie derivative is natural with respect to restriction, so we have the commutative diagram

$$\begin{array}{ccc} \mathfrak{X}(U) & \stackrel{\mathcal{L}_{X|_U}}{\longrightarrow} & \mathfrak{X}(U) \\ r_V^U \downarrow & & \downarrow r_V^U \\ \mathfrak{X}(V) & \stackrel{\mathcal{L}_{X|_U}}{\longrightarrow} & \mathfrak{X}(V) \end{array}$$

where $X|_U$ denotes the restriction of $X \in \mathfrak{X}(M)$ to the open set U and r_V^U is the map that restricts from U to $V \subset U$.

If X and Y are smooth vector fields with flows φ^X and φ^Y, then starting at some $p \in M$, if we flow with X for time $\sqrt{t}$, then flow with Y for time $\sqrt{t}$, and then flow backwards along X and then Y for time $\sqrt{t}$, we arrive at a point $\alpha(t)$ given by

$$\alpha(t) := \varphi^Y_{-\sqrt{t}} \circ \varphi^X_{-\sqrt{t}} \circ \varphi^Y_{\sqrt{t}} \circ \varphi^X_{\sqrt{t}}.$$

It turns out that $\alpha(t)$ is usually not p (see Figure 2.6). In fact, we have the following theorem:

Theorem 2.112. *With $\alpha(t)$ as above we have*

$$\left.\frac{d}{dt}\right|_{t=0} \alpha(t) = [X, Y](p).$$

Proof. See Problem 8. □

We know by direct calculation that if $(x^1, \ldots, x^n)$ are the coordinate functions of a chart, then

$$\left[\frac{\partial}{\partial x^i}, \frac{\partial}{\partial x^j}\right] = 0$$

2.8. Vector Fields

for all $i,j = 1,\ldots,n$. The converse is also *locally* true. This follows from the next theorem, which can be thought of as saying that we can simultaneously "straighten" commuting local fields.

Theorem 2.113. *Let M be an n-manifold. Suppose that on a neighborhood V of a point $p \in M$ we have vector fields $X_1, \ldots, X_k$ such that $X_1(x), \ldots, X_k(x)$ are linearly independent for all $x \in V$. If $[X_i, X_j] = 0$ on V for all i,j, then there exist a possibly smaller neighborhood U of p and a chart $\mathbf{x} : U \to \mathbb{R}^n$ such that*

$$\frac{\partial}{\partial x^i} = X_i \text{ on } U \text{ for } i = 1, \ldots, k$$

and such that for the corresponding flows we have

$$\varphi_t^{X_i} \circ \mathbf{x}^{-1}(u^1, \ldots, u^i, \ldots, u^n) = (u^1, \ldots, u^i + t, \ldots, u^n)$$

for $i = 1, \ldots, k$.

Proof. Let $p \in V$ and choose a chart $(O, \mathbf{y})$ centered at p with $O \subset V$. By rearranging the coordinate functions if necessary and using a simple linear independence argument, we may assume that the vectors

$$X_1(p), \ldots, X_k(p), \left.\frac{\partial}{\partial x^{k+1}}\right|_p, \ldots, \left.\frac{\partial}{\partial x^n}\right|_p$$

form a basis for $T_p M$. Let $\varphi_{t_1}^{X_1}, \ldots, \varphi_{t_k}^{X_k}$ be local flows for $X_1, \ldots, X_k$ and let W be an open neighborhood of p contained in O such that the composition

$$\varphi_{t_k}^{X_k} \circ \cdots \circ \varphi_{t_1}^{X_1}$$

is defined on W and maps W into O for all $t_1, \ldots, t_k \in (-\epsilon, \epsilon)$ as in Lemma 2.92. Define

$$S := \{(a^{k+1}, \ldots, a^n) : \mathbf{y}^{-1}(0, \ldots, 0, a^{k+1}, \ldots, a^n)\} \in W,$$

and define the map $\psi : (-\epsilon, \epsilon)^k \times S \to U$ by

$$\psi(u^1, \ldots, u^k, u^{k+1}, \ldots, u^n) = \varphi_{u_k}^{X_k} \circ \cdots \circ \varphi_{u_1}^{X_1} \circ \mathbf{y}^{-1}(0, \ldots, 0, u^{k+1}, \ldots, u^n).$$

Since $[X_i, X_j] = 0$ for $0 \leq i, j \leq k$, we know that the flows in the composition above commute. Hence for any $a \in \mathbf{y}(W)$ and smooth function f, we have for $1 \leq i \leq k$,

$$T_a \psi \cdot \left.\frac{\partial}{\partial u^i}\right|_a = \left.\frac{\partial}{\partial u^i}\right|_a f \circ \psi(u^1, \ldots, u^n)$$

$$= \left.\frac{\partial}{\partial u^i}\right|_a f(\varphi_{u_k}^{X_k} \circ \cdots \circ \varphi_{u_1}^{X_1} \circ \mathbf{y}^{-1}(0, \ldots, 0, u^{k+1}, \ldots, u^n))$$

$$= \left.\frac{\partial}{\partial u^i}\right|_a f(\varphi_{u_i}^{X_i} \circ \varphi_{u_k}^{X_k} \circ \cdots \circ \widehat{\varphi_{u_i}^{X_i}} \circ \cdots \circ \varphi_{u_1}^{X_1} \circ \mathbf{y}^{-1}(0, \ldots, 0, u^{k+1}, \ldots, u^n))$$

$$= X_i(\psi(a))f,$$

where the caret indicates omission. Thus we have
$$T_a\psi \cdot \left.\frac{\partial}{\partial u^i}\right|_a = X_i(\psi(a)) \text{ for } 1 \le i \le k,$$
and in particular,
$$T_0\psi \cdot \left.\frac{\partial}{\partial u^i}\right|_0 = X_i(p) \text{ for } 1 \le i \le k.$$
For $k+1 \le i \le n$, we have
$$\psi(0,\ldots,0,u^{k+1},\ldots,u^n) = \mathbf{y}^{-1}(0,\ldots,0,u^{k+1},\ldots,u^n),$$
so
$$T_0\psi \cdot \left.\frac{\partial}{\partial u^i}\right|_0 = \left.\frac{\partial}{\partial y^i}\right|_p.$$
Thus $T_0\psi$ maps a basis of $T_0\mathbb{R}^n$ to a basis of T_pM. We can use the inverse mapping theorem (Theorem 2.25) to conclude that ψ is a diffeomorphism from some open neighborhood of $0 \in \mathbb{R}^n$ onto an open neighborhood U of p in M. It is now straightforward to check that $\mathbf{x} = \psi^{-1}$ is a chart map of the desired type. □

For later reference, we note that if $(U,\mathbf{x})$ is a chart of the type produced in the theorem above, then we can easily arrange that $\mathbf{x}(U)$ is of the form $V \times W \subset \mathbb{R}^k \times \mathbb{R}^{n-k}$, so that the fields $\frac{\partial}{\partial x^1}, \ldots, \frac{\partial}{\partial x^k}$ are tangent to the submanifolds of the form
$$S := \{p \in U : x^{k+1}(p) = a^{k+1}, \ldots, x^{k+1}(p) = a^n\}.$$

The reader has probably surmised that the mathematics of vector fields and flows can be applied to fluid mechanics. This is quite true, but one needs to deal with time dependent vector fields. There is a trick that allows time dependent vector fields to be treated as ordinary vector fields on a manifold of one higher dimension, but it is not always best to think in those terms. The author has included a bit about time dependent vector fields in the online supplement [**Lee, Jeff**].

2.9. 1-Forms

Definition 2.114. A smooth (resp. C^r) section of the cotangent bundle is called a smooth (resp. C^r) **1-form** or also a smooth (resp. C^r) **covector field**. The set of all C^r 1-forms is denoted by $\mathfrak{X}^{r*}(M)$ and the set of smooth 1-forms is denoted by $\mathfrak{X}^*(M)$.

The set $\mathfrak{X}^{r*}(M)$ is a module over $C^r(M)$. Later we will have reason to denote $\mathfrak{X}^*(M)$ also by $\Omega^1(M)$. The analogue of Lemma 2.65 is true. That is, if $\alpha_p \in T_p^*M$, then there is a 1-form α such that $\alpha(p) = p$. The proof is again a simple cut-off function argument.

2.9. 1-Forms

Definition 2.115. Let $f : M \to \mathbb{R}$ be a C^r function with $r \geq 1$. The map $df : M \to T^*M$ is defined by $p \mapsto df(p)$ where $df(p)$ is the differential at p as defined in Definitions 2.20 and 2.22; df is a 1-form called the **differential** of f.

As a consequence of the definitions of we have

$$d(fg) = g\,df + f\,dg.$$

Indeed, for $v_p \in T_pM$ we have

$$d(fg)(p) \cdot v_p = v_p(fg) = g(p)v_p(f) + f(p)v_p(g)$$
$$= g(p)(df)(p) \cdot v_p + f(p)dg(p) \cdot v_p$$
$$= (g\,df + f\,dg)(p) \cdot v_p.$$

Three views on a 1-form: The novice may easily become confused about what should be the argument of a 1-form. The reason for this is that one can view a 1-form in at least three different ways. If α is a smooth 1-form, then we have the following interpretations of α:

(1) We may view α as a map $\alpha : M \to T^*M$ (as in the definition) so that $\alpha(.)$ takes points as arguments; $\alpha(p) \in T_p^*M$. We also sometimes need to write $\alpha_p = \alpha(p)$ just as for a vector field X we sometimes write $X(p)$ as X_p.

(2) We may view α as a smooth map $\alpha : TM \to \mathbb{R}$ so that for $v_p \in T_pM$ we make sense of $\alpha(v_p)$ by $\alpha(v_p) = \alpha_p(v_p)$.

(3) We may view α as a map $\alpha : \mathfrak{X}(M) \to C^\infty(M)$, where for $X \in \mathfrak{X}(M)$ we interpret $\alpha(X)$ as the smooth function $p \mapsto \alpha_p(X_p)$.

The second and third interpretations are dependent on the first. The third view is very important, and in that view, $\alpha : \mathfrak{X}(M) \to C^\infty(M)$ is a $C^\infty(M)$-linear map. We have already mentioned that given a chart $(U, \mathbf{x})$ with $\mathbf{x} = (x^1, \ldots, x^n)$, the differentials $dx^i : p \mapsto dx^i\big|_p$ define 1-forms on U such that $dx^1\big|_p, \ldots, dx^n\big|_p$ form a basis of T_p^*M for each $p \in U$. If α is any smooth 1-form (covector field) defined at least on U, then

$$\alpha|_U = \sum a_i dx^i$$

for uniquely determined functions a_i. In fact, $\alpha(\frac{\partial}{\partial x^i}) = a_i$ (using the third view from above). As with vector fields we will usually just write $\alpha = \sum a_i dx^i$. In particular, $df(\frac{\partial}{\partial x^i}) = \frac{\partial f}{\partial x^i}$, and so we have the following familiar looking formula:

$$df = \sum \frac{\partial f}{\partial x^i} dx^i,$$

which is interpreted to mean that at each $p \in U_\alpha$ we have
$$df(p) = \sum \left.\frac{\partial f}{\partial x^i}\right|_p \left.dx^i\right|_p.$$
The covector fields dx^i form what is called a **coordinate coframe field** or **holonomic coframe field** [2] over U. Note that the component functions X^i of a vector field with respect to the chart above are given by $X^i = dx^i(X)$, where by definition $dx^i(X)$ is the function $p \mapsto \left.dx^i\right|_p (X_p)$. Thus
$$\left.X\right|_U = \sum dx^i(X) \frac{\partial}{\partial x^i}.$$

Note. If α is a 1-form on M and $p \in M$, then one can always find many functions f such that $df(p) = \alpha(p)$, but there may not be a single function f such that this is true for all points in a neighborhood, let alone all points on M. If in fact $df = \alpha$ for some $f \in C^\infty(M)$, then we say that α is **exact**. More on this later.

Let us try to picture 1-forms. As a warm up, let us recall how we might picture a tangent vector v_p at a point $p \in \mathbb{R}^n$. Let γ be a curve with $\gamma'(0) = v_p$. If we zoom in on the curve near p, then it appears to straighten out and so begins to look like the curve $t \mapsto p + tv_p$. So one might say that a tangent vector is the "infinitesimal representation" of a (parameterized) curve.

At each point a 1-form gives a linear functional in that tangent space, and as we know, the level sets of a linear functional are parallel affine subspaces or hyperplanes. We should imagine level sets as being labeled by the values of the function. A covector puts a ruling in the tangent space that measures tangent vectors stretching across this ruling. For example, the 1-form df in $\mathbb{R}^3$ gives a ruling in each tangent space as suggested by Figure 2.7a. For a given p, the level sets of df_p are what we see if we zoom in on the level sets of f near p. The fact that the individual df_p's living in the tangent spaces somehow coalesce into the level sets of the global function f as shown in 2.7b, is due to the fact that the 1-form df is the differential of the function f.

A more general 1-form α is still pictured as straight parallel hyperplanes in each tangent space. Because these level sets live in the tangent space, we might call them infinitesimal level sets. These (value labeled) level sets may not coalesce into the level sets of any global smooth function on the manifold. There are various, increasingly severe ways coalescing may fail to happen. The least severe situation is when α is not the differential of a global function but is still locally a differential near each point. For example,

---

[2] The word holonomic comes from mechanics and just means that the frame field derives from a chart. A related fact is that $[\frac{\partial}{\partial x^i}, \frac{\partial}{\partial x^j}] = 0$.

2.9. 1-Forms

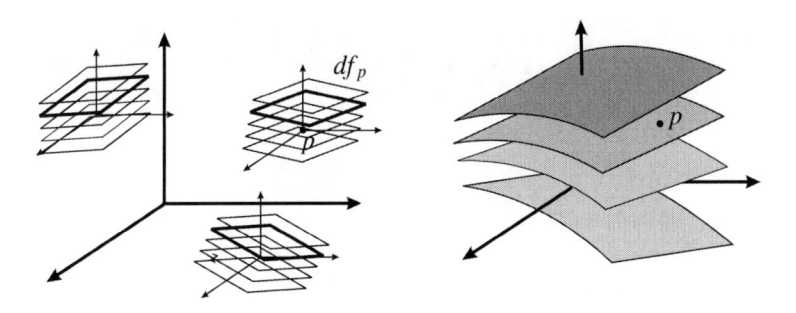

Figure 2.7. The form *df* follows level sets

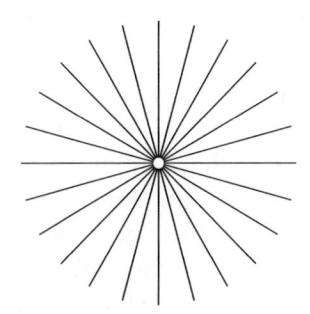

Figure 2.8. Level sets of overlapping angle functions

if $M = \mathbb{R}^2\backslash\{0\}$, then the familiar 1-form $\alpha = (x^2 + y^2)^{-1}(-y\,dx + x\,dy)$ is locally equal to $d\theta$ for some "angle function" θ which measures the angle from some fixed ray such as the positive x axis. But there is no such single smooth angle function defined on *all* of $\mathbb{R}^2\backslash\{0\}$. Thus, globally speaking, α is not the differential of any function. In Figure 2.8, we see the coalesced result of "integrating" the infinitesimal level sets which live in the tangent spaces. While these suggest an angular function, we see that if we try to picture rising as we travel around the origin, we find that we do not return to the same level in one full circulation, but rather keep rising. Locally, however, we really *do* have level sets of smooth functions.

Now the second more severe way that a 1-form may fail to be the differential of a function is where there is not even a *local* function whose differential agrees with the 1-form near a point. The infinitesimal level sets do not coalesce to the level sets of a smooth function, even in small neighborhoods. This is much harder to represent, but Figure 2.9 is meant to at least be suggestive. Nearby curves cross inconsistent numbers of level sets. As an example consider the 1-form

$$\beta = -y\,dx + x\,dy.$$

The astute reader may object that surely radial rays *do* match up with the

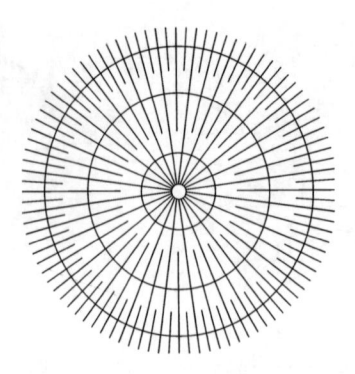

Figure 2.9. Suggestive representation of a form which is not closed

directions described by this 1-form. However, the point is that a covector in a tangent space is not completely described by the level sets as such, but rather the level sets are to be thought of as labeled according to the values they represent in the individual tangent spaces. Here we have a case where the infinitesimal level sets coalesce, but the values assigned to them do not; they are 1-dimensional submanifolds that fit the 1-form but they are not level sets of even a local smooth function whose differential agrees with the 1-form. This brings us to the most severe case which only happens in dimension 3 or greater. It can be the case that there is no nice family of $(n-1)$-dimensional submanifolds that line up in *any* reasonable sense with the 1-form (either globally or locally). This is the topic of the Frobenius integrability theory for tangent distributions that we study in Chapter 11, and we shall forgo any further discussion until then except to say that the reader should be ready to understand much of that chapter, including the Frobenius theorem, after finishing the next chapter.

Definition 2.116. If $\phi : M \to N$ is a C^∞ map, the **pull-back** of a 1-form $\alpha \in \mathfrak{X}^*(N)$ by ϕ is defined by

$$(\phi^* \alpha)_p \cdot v = \alpha_{\phi(p)}(T_p \phi \cdot v)$$

for $v \in T_p M$.

This extends the notion of the pull-back of a function defined earlier. If we view a 1-form on M as a map $TM \to \mathbb{R}$, then the pull-back is given by $\phi^* \alpha = \alpha \circ T\phi$.

Exercise 2.117. The pull-back is contravariant in the sense that if $\phi_1 : M_1 \to M_2$ and $\phi_2 : M_2 \to N$, then for $\alpha \in \mathfrak{X}^*(N)$ we have $(\phi_2 \circ \phi_1)^* = \phi_1^* \circ \phi_2^*$.

Next we describe the local expression for the pull-back of a 1-form. Let $(U, \mathbf{x})$ be a chart on M and $(V, \mathbf{y})$ a coordinate chart on N with $\phi(U) \subset V$. A

2.9. 1-Forms

typical 1-form has a local expression on V of the form $\alpha = \sum a_i dy^i$ for $a_i \in C^\infty(V)$. The local expression for $\phi^* \alpha$ on U is $\phi^* \alpha = \sum (a_i \circ \phi) \, d(y^i \circ \phi) = \sum (a_i \circ \phi) \frac{\partial (y^i \circ \phi)}{\partial x^j} dx^j$. Thus we get a local pull-back formula[3] convenient for computations:

$$(2.7) \qquad \phi^* \left(\sum a_i \, dy^i \right) = \sum (a_i \circ \phi) \frac{\partial (y^i \circ \phi)}{\partial x^j} dx^j.$$

The pull-back of a function or 1-form is defined whether $\phi : M \to N$ happens to be a diffeomorphism or not. On the other hand, the pull-back of a vector field only works in special circumstances such as where ϕ is a diffeomorphism. Let $\phi : M \to N$ be a C^∞ diffeomorphism with $r \geq 1$. Recall that the **push-forward** of a function $f \in C^\infty(M)$ is denoted $\phi_* f$ and defined by $\phi_* f(p) := f(\phi^{-1}(p))$. We can also define the **push-forward** of a 1-form as $\phi_* \alpha = \alpha \circ T\phi^{-1}$.

Exercise 2.118. Find the local expression for $\phi_* f$ and $\phi_* \alpha$. Explain why we need ϕ to be a diffeomorphism.

Lemma 2.119. *The differential is natural with respect to pull-back. In other words, if $\phi : M \to N$ is a C^∞ map and $f : N \to \mathbb{R}$ a C^∞ function, then $d(\phi^* f) = \phi^* df$. Consequently, the differential is also natural with respect to restrictions.*

Proof. We wish to show that

$$(\phi^* df)_p = d(\phi^* f)_p$$

for all $p \in M$. Let $v \in T_p M$ and write $q = f(p)$. Then

$$\begin{aligned} (\phi^* df)|_p v &= df|_q (T_p \phi \cdot v) \\ &= ((T_p \phi \cdot v) f)(q) \\ &= v (f \circ \phi)(p) = d(\phi^* f)|_p v. \end{aligned}$$

The second statement is obvious from local coordinate expressions, but also notice that if U is open in M and $\iota : U \hookrightarrow M$ is the inclusion map (i.e. the identity map id_M restricted to U), then $f|_U = \iota^* f$ and $df|_U = \iota^* df$. So the statement about restrictions is just a special case of the first part. $\square$

The tangent and cotangent bundles TM and T^*M are themselves manifolds and so have their own tangent and cotangent bundles. Among other things, this means that there exist 1-forms and vector fields on these manifolds. Here we introduce the canonical 1-form on T^*M. We denote this form by θ_{can} and note that it is a section of $T^*(T^*M)$. Let $a \in T^*M$ and suppose that a is based at p so that $a \in T_p^* M$. Consider a vector $u_a \in T_a(T^*M)$.

---

[3] To ensure clarity we have not used the Einstein summation convention here.

Notice that since $\pi : T^*M \to M$, we have $T_a\pi : T_a(T^*M) \to T_pM$. Thus $T_a\pi \cdot u_a \in T_pM$. We define
$$\theta_{\text{can}}(u_a) = a(T_a\pi \cdot u_a).$$
The definition makes sense because $a \in T_p^*M$ and $T_a\pi \cdot u_a \in T_pM$. Let $(U,\mathbf{x})$ be a chart containing p and let $(x^1 \circ \pi, \ldots, x^n \circ \pi, p_1, \ldots, p_n) = (q^1, \ldots, q^n, p_1, \ldots, p_n)$ be the associated natural coordinates for T^*M.

Exercise 2.120. It is geometrically clear that $T_a\pi \cdot \left.\frac{\partial}{\partial p^i}\right|_a = 0$ since $\left.\frac{\partial}{\partial p^i}\right|_a$ is tangent to the fiber $T^*_{\pi(a)}M$ along which π is constant. Deduce this directly from the definitions. Hint: $\left.\frac{\partial}{\partial p^i}\right|_a$ can be represented by a curve in $T^*_{\pi(a)}M$.

We wish to show that locally $\theta_{\text{can}} = \sum p_i\, dq^i$. It will suffice to show that $\theta_{\text{can}}(\left.\frac{\partial}{\partial q^i}\right|_a) = p_i(a)$ and $\theta_{\text{can}}(\left.\frac{\partial}{\partial p^i}\right|_a) = 0$ for all i. We have
$$\theta_{\text{can}}\left(\left.\frac{\partial}{\partial p^i}\right|_a\right) = a\left(T_a\pi \cdot \left.\frac{\partial}{\partial p^i}\right|_a\right) = 0$$
since in fact $T_a\pi \cdot \left.\frac{\partial}{\partial p^i}\right|_a = 0$ by Exercise 2.120. Also, we have
$$\theta_{\text{can}}\left(\left.\frac{\partial}{\partial q^i}\right|_a\right) = a\left(T_a\pi \cdot \left.\frac{\partial}{\partial q^i}\right|_a\right)$$
$$= a\left(\left.\frac{\partial}{\partial x^i}\right|_p\right) = a^i = p_i(a),$$
where we have used the fact that $T_a\pi \cdot \left.\frac{\partial}{\partial q^i}\right|_a = \left.\frac{\partial}{\partial x^i}\right|_p$, which follows from the definition $q^i = x^i \circ \pi$. Indeed, we know that $T_a\pi \cdot \left.\frac{\partial}{\partial q^i}\right|_a = c_i^k \left.\frac{\partial}{\partial x^k}\right|_p$ for some constants c_i^k, but we have
$$c_i^k = dx^k\left(T_a\pi \cdot \left.\frac{\partial}{\partial q^i}\right|_a\right) = \pi^*dx^k\left(\left.\frac{\partial}{\partial q^i}\right|_a\right)$$
$$= d\left(x^k \circ \pi\right)\left(\left.\frac{\partial}{\partial q^i}\right|_a\right) = dq^k\left(\left.\frac{\partial}{\partial q^i}\right|_a\right) = \delta_i^k.$$

This 1-form plays a basic role in symplectic geometry and classical mechanics. For more about symplectic geometry see the online supplement [**Lee, Jeff**].

2.10. Line Integrals and Conservative Fields

Just as in calculus on Euclidean space, we can consider line integrals on manifolds, and it is exactly the 1-forms that are the appropriate objects to integrate. First notice that all 1-forms on open sets in $\mathbb{R}^1$ must be of the

2.10. Line Integrals and Conservative Fields

form $f\,dt$ for some smooth function f, where t is the coordinate function on $\mathbb{R}^1$. We begin by defining the line integral of a 1-form defined and smooth on an interval $[a,b] \subset \mathbb{R}^1$. If $\beta = f\,dt$ is such a 1-form, then

$$\int_{[a,b]} \beta := \int_a^b f(t)\,dt.$$

Any smooth map $\gamma : [a,b] \to M$ is the restriction of a smooth map on some larger open interval $(a - \varepsilon, b + \varepsilon)$, and so there is no problem defining the pull-back $\gamma^*\alpha$. If $\gamma : [a,b] \to M$ is a smooth curve, then we define the **line integral** of a 1-form α along γ to be

$$\int_\gamma \alpha := \int_{[a,b]} \gamma^*\alpha = \int_a^b f(t)\,dt.$$

where $\gamma^*\alpha = f\,dt$. Now if $t = \phi(s)$ is a smooth increasing function, then we obtain a positive reparametrization $\widetilde{\gamma} = \gamma \circ \phi : [c,d] \to M$, where $\phi(c) = a$ and $\phi(d) = b$. With such a reparametrization we have

$$\int_{[c,d]} \widetilde{\gamma}^*\alpha = \int_{[c,d]} \phi^*\gamma^*\alpha$$
$$= \int_{[c,d]} \phi^*(f\,dt) = \int_{[c,d]} f\circ\phi \frac{d\phi}{ds} ds$$
$$= \int_c^d f(\phi(s))\phi'(s)\,ds = \int_a^b f(t)\,dt,$$

where the last line is the standard change of variable formula and where we have used $\phi^*(f\,dt) = \frac{d(f\circ\phi)}{ds} ds$, which is a special case of the pull-back formula mentioned above. We see now that we get the same result as before. This is just as in ordinary multivariable calculus. We have just transferred the usual calculus ideas to the manifold setting.

Definition 2.121. A continuous curve $\gamma : [a,b] \to M$ into a smooth manifold is called **piecewise smooth** if there exists a partition $a = t_0 < t_1 < \cdots < t_k = b$ such that γ restricted to $[t_i, t_{i+1}]$ is smooth for $0 \le i \le k-1$ (in the sense of Definition 1.58).

It is convenient to extend the definitions a bit to include integration along piecewise smooth curves. Thus if $\gamma : [a,b] \to M$ is such a curve, then we define for a 1-form α,

$$\int_\gamma \alpha = \sum_{i=0}^{k-1} \int_{[t_i, t_{i+1}]} \gamma_i^*\alpha,$$

where γ_i is the restriction of γ to the interval $[t_i, t_{i+1}]$.

Just as in ordinary multivariable calculus we have the following:

Proposition 2.122. *Let $\gamma : [a,b] \to M$ be a piecewise smooth curve with $\gamma(a) = p_1$ and $\gamma(b) = p_2$. If $\alpha = df$, then*
$$\int_\gamma \alpha = \int_\gamma df = f(p_2) - f(p_1).$$
In particular, $\int_\gamma \alpha$ is path independent in the sense that it is equal to $\int_c \alpha$ for any other piecewise smooth path c that also begins at p_1 and ends at p_2.

Definition 2.123. *If α is a 1-form on a smooth manifold M such that $\int_c \alpha = 0$ for all closed piecewise smooth curves c, then we say that α is* **conservative**.

We will need a lemma on differentiability.

Lemma 2.124. *Suppose f is a function defined on a smooth manifold M, and let α be a smooth 1-form on M. Suppose that for any $p \in M$, $v_p \in TM$ and smooth curve c with $\dot{c}(0) = v_p$, the derivative $\frac{d}{dt}\big|_0 f(c(t))$ exists and*
$$\frac{d}{dt}\bigg|_0 f(c(t)) = \alpha(v_p).$$
Then f is smooth and $df = \alpha$.

Proof. We work in a chart $(U, \mathbf{x})$. If we take $c(t) := \mathbf{x}^{-1}(\mathbf{x}(p) + t\mathbf{e}_i)$, then the hypotheses lead to the conclusion that all the first order partial derivatives of $f \circ \mathbf{x}^{-1}$ exist and are continuous. Thus f is C^1. But then also $df_p \cdot v_p = \frac{d}{dt}\big|_0 f(c(t)) = \alpha_p(v_p)$ for all v_p, and it follows that $df = \alpha$, and this also implies that f is actually smooth. $\square$

Proposition 2.125. *If α is a 1-form on a smooth manifold M, then α is conservative if and only if it is exact.*

Proof. We know already that if $\alpha = df$, then α is conservative. Now suppose α is conservative. Fix $p_0 \in M$. Then we can define $f(p) = \int_\gamma \alpha$, where γ is any curve beginning at p_0 and ending at p. Given any $v_p \in T_pM$, we pick a curve $c : [-1, \varepsilon)$ with $\varepsilon > 0$ such that $c(-1) = p_0$, $c(0) = p$ and $c'(0) = v_p$. Then
$$\frac{d}{d\tau}\bigg|_0 f(c(\tau)) = \frac{d}{d\tau}\bigg|_0 \int_{c|[-1,\tau]} \alpha$$
$$= \frac{d}{d\tau}\bigg|_0 \int_{c|[-1,0]} \alpha + \frac{d}{d\tau}\bigg|_0 \int_{c|[0,\tau]} \alpha$$
$$= 0 + \frac{d}{d\tau}\bigg|_0 \int_0^\tau c^*\alpha = \frac{d}{d\tau}\bigg|_0 \int_0^\tau g(t)\,dt$$
$$= g(0),$$

2.10. Line Integrals and Conservative Fields

where $c^*\alpha = g\,dt$. On the other hand,

$$\alpha(v_p) = \alpha(c'(0)) = \alpha\left(T_0 c \cdot \left.\frac{d}{dt}\right|_0\right)$$
$$= c^*\alpha\left(\left.\frac{d}{dt}\right|_0\right) = g(0)\,dt|_0\left(\left.\frac{d}{dt}\right|_0\right) = g(0),$$

where t is the standard coordinate on $\mathbb{R}$. Thus $\left.\frac{d}{d\tau}\right|_0 f(c(\tau)) = \alpha(v_p)$ for any $v_p \in T_p M$ and any $p \in M$. Now the result follows from the previous lemma. □

It is important to realize that when we say that a form is conservative in this context, we mean that it is *globally* conservative. It may also be the case that a form is *locally* conservative. This means that if we restrict the 1-form to an open set which is diffeomorphic to a Euclidean ball, then the result is conservative on that ball. The following examples explore in simple terms these issues.

Example 2.126. Let $\alpha = (x^2 + y^2)^{-1}(-y\,dx + x\,dy)$. Consider the small circular path c given by $(x, y) = (x_0 + \varepsilon \cos t, y_0 + \varepsilon \sin t)$ with $0 \leq t \leq 2\pi$ and $\varepsilon > 0$. If $(x_0, y_0) = (0, 0)$, we obtain

$$\int_c \alpha = \int_0^{2\pi} \frac{1}{\varepsilon^2}\left(-y(t)\frac{dx}{dt} + x(t)\frac{dy}{dt}\right)dt$$
$$= \int_0^{2\pi} \frac{1}{\varepsilon^2}\left(-(\varepsilon \sin t)(-\varepsilon \sin t) + (\varepsilon \cos t)(\varepsilon \cos t)\right)dt = 2\pi.$$

Thus α is not conservative and hence not exact. On the other hand, if $(x_0, y_0) \neq (0, 0)$, then we pick a ray R_0 that does *not* pass through (x_0, y_0) and a function $\theta(x, y)$ which gives the angle of the ray R passing through (x, y) measured counterclockwise from R_0. This angle function is smooth and defined on $U = \mathbb{R}^2 \setminus R_0$. If $\varepsilon < \frac{1}{2}\sqrt{x_0^2 + y_0^2}$, then c has image inside the domain of θ and we have that $\alpha|U = d\theta$. Thus $\int_c \alpha = \theta(c(0)) - \theta(c(2\pi)) = 0$. We see that α is locally conservative.

Example 2.127. Consider $\beta = y\,dx - x\,dy$ on $\mathbb{R}^2$. If it were the case that for some small open set $U \subset \mathbb{R}^2$ we had $\beta|_U = df$, then for a closed path c with image in that set, we would expect that $\int_c \beta = f(c(2\pi)) - f(c(0)) = 0$. However, if c is the curve going around a circle of radius ε centered at

(x_0, y_0), then we have

$$\int_c \beta = \int_c \left(y(t) \frac{dx}{dt} - x(t) \frac{dy}{dt} \right) dt$$
$$= \int_0^{2\pi} \left((x_0 + \varepsilon \sin t)(-\varepsilon \sin t) - (y_0 + \varepsilon \cos t)(\varepsilon \cos t) \right) dt$$
$$= -2\varepsilon^2 \pi,$$

so we do not get zero no matter what the point (x_0, y_0) and no matter how small ε. We conclude that β is not even locally conservative.

The distinction between (globally) conservative and locally conservative is often not made sufficiently clear in the physics and engineering literature.

Example 2.128. In classical physics, the static electric field set up by a fixed point charge of magnitude q can be described, with an appropriate choice of units, by the 1-form

$$\frac{q}{\rho^3} x \, dx + \frac{q}{\rho^3} y \, dy + \frac{q}{\rho^3} z \, dz,$$

where we have imposed Cartesian coordinates centered at the point charge and where $\rho = \sqrt{x^2 + y^2 + z^2}$. Notice that the domain of the form is the punctured space $\mathbb{R}^3 \setminus \{0\}$. In spherical coordinates (ρ, θ, ϕ), this same form is

$$\frac{q}{\rho^2} d\rho = d\left(\frac{-q}{\rho} \right),$$

so we see that the form is exact and the field is conservative.

2.11. Moving Frames

It is important to realize that it is possible to get a family of locally defined vector (resp. covector) fields that are linearly independent at each point in their mutual domain and yet are not necessarily of the form $\frac{\partial}{\partial x^i}$ (resp. dx^i) for any coordinate chart. In fact, this may be achieved by carefully choosing n^2 smooth functions f_k^i (resp. a_i^k) and then letting $E_k := \sum_i f_k^i \frac{\partial}{\partial x^i}$ (resp. $\theta^k := \sum_i a_i^k dx^i$).

Definition 2.129. Let $E_1, E_2, \ldots, E_n$ be smooth vector fields defined on some open subset U of a smooth n-manifold M. If $E_1(p), E_2(p), \ldots, E_n(p)$ form a basis for T_pM for each $p \in U$, then we say that $(E_1, E_2, \ldots, E_n)$ is a (non-holonomic) **moving frame** or a **frame field** over U.

If $E_1, E_2, \ldots, E_n$ is a moving frame over $U \subset M$ and X is a vector field defined on U, then we may write

$$X = \sum X^i E_i \text{ on } U,$$

2.11. Moving Frames

for some functions X^i defined on U. If the moving frame $(E_1, \ldots, E_n)$ is not identical to some frame field $(\frac{\partial}{\partial x^1}, \ldots, \frac{\partial}{\partial x^n})$ arising from a coordinate chart on U, then we say that the moving frame is **non-holonomic**. It is often possible to find such moving frame fields with domains that could never be the domain of any chart (consider a torus).

Definition 2.130. If $E_1, E_2, \ldots, E_n$ is a frame field with domain equal to the whole manifold M, then we call it a **global frame field**.

Most manifolds do not have global frame fields.

Taking the basis dual to $(E_1(p), \ldots, E_n(p))$ in $T_p^* M$ for each $p \in U$ we get a moving coframe field $(\theta^1, \ldots, \theta^n)$. The θ^i are 1-forms defined on U. Any 1-form α can be expanded in terms of these basic 1-forms as $\alpha = \sum a_i \theta^i$. Actually it is the restriction of α to U that is being expressed in terms of the θ^i, but we shall not be so pedantic as to indicate this in the notation. In a manner similar to the case of a coordinate frame, we have that for a vector field X defined at least on U, the components with respect to $(E_1, \ldots, E_n)$ are given by $\theta^i(X)$:

$$X = \sum \theta^i(X) E_i \text{ on } U.$$

Let us consider an important special situation. If $M \times N$ is a product manifold and $(U, \mathbf{x})$ is a chart on M and $(V, \mathbf{y})$ is a chart on N, then we have a chart $(U \times V, \mathbf{x} \times \mathbf{y})$ on $M \times N$ where the individual coordinate functions are $x^1 \circ \mathrm{pr}_1, \ldots, x^m \circ \mathrm{pr}_1, y^1 \circ \mathrm{pr}_2, \ldots, y^n \circ \mathrm{pr}_2$, which we temporarily denote by $\widetilde{x}^1, \ldots, \widetilde{x}^m, \widetilde{y}^1, \ldots, \widetilde{y}^n$. Now we consider what is the relation between the coordinate frame fields $(\frac{\partial}{\partial x^1}, \ldots \frac{\partial}{\partial x^m}), (\frac{\partial}{\partial y^1}, \ldots \frac{\partial}{\partial y^n})$ and the frame field $(\frac{\partial}{\partial \widetilde{x}^1}, \ldots, \frac{\partial}{\partial \widetilde{y}^n})$. The latter set of $n+m$ vector fields is certainly a linearly independent set at each point $(p,q) \in U \times V$. The crucial relations are $\frac{\partial}{\partial \widetilde{x}^i} f = \frac{\partial}{\partial x^i}(f \circ \mathrm{pr}_1)$ and $\frac{\partial}{\partial \widetilde{y}^i} = \frac{\partial}{\partial y^i}(f \circ \mathrm{pr}_2)$.

Exercise 2.131. Show that $\frac{\partial}{\partial x^i}\big|_p = T\mathrm{pr}_1 \frac{\partial}{\partial \widetilde{x}^i}\big|_{(p,q)}$ and $\frac{\partial}{\partial y^i}\big|_q = T\mathrm{pr}_2 \frac{\partial}{\partial \widetilde{y}^i}\big|_{(p,q)}$.

Remark 2.132. In some circumstances, it is safe to abuse notation and denote $x^i \circ \mathrm{pr}_1$ by x^i and $y^i \circ \mathrm{pr}_2$ by y^i. Of course we are denoting $\frac{\partial}{\partial \widetilde{x}^i}$ by $\frac{\partial}{\partial x^i}$ and so on.

A warning (*The second fundamental confusion of calculus*[4]): For a chart $(U, \mathbf{x})$ with $\mathbf{x} = (x^1, \ldots, x^n)$, we have defined $\frac{\partial f}{\partial x^i}$ for any appropriately defined smooth (or C^1) function f. However, this notation can be

---

[4] In [**Pen**], Penrose attributes this cute terminology to Nick Woodhouse.

ambiguous. For example, the meaning of $\frac{\partial f}{\partial x^1}$ is not determined by the coordinate function x^1 alone, but implicitly depends on the rest of the coordinate functions. For example, in thermodynamics we see the following situation. We have three functions P, V and T which are not independent but may be interpreted as functions on some 2-dimensional manifold. Then it may be the case that any two of the three functions can serve as a coordinate system. The meaning of $\frac{\partial f}{\partial P}$ depends on whether we are using the coordinate functions (P, V) or alternatively (P, T). We must know not only which function we are allowing to vary, but also which other functions are held fixed. To get rid of the ambiguity, one can use the notations $\left(\frac{\partial f}{\partial P}\right)_V$ and $\left(\frac{\partial f}{\partial P}\right)_T$. In the first case, the coordinates are (P, V), and V is held fixed, while in the second case, we use coordinates (P, T), and T is held fixed. Another way to avoid ambiguity would be to use different names for the same functions depending on the chart of which they are considered coordinate functions. For example, consider the following change of coordinates:

$$y^1 = x^1 + x^2,$$
$$y^2 = x^1 - x^2 + x^3,$$
$$y^3 = x^3.$$

Here $y^3 = x^3$ as functions on the underlying manifold, but we use different symbols. Thus $\frac{\partial}{\partial x^3}$ may not be the same as $\frac{\partial}{\partial y^3}$. The chain rule shows that in fact $\frac{\partial}{\partial x^3} = \frac{\partial}{\partial y^2} + \frac{\partial}{\partial y^3}$. This latter method of destroying ambiguity is not very helpful in our thermodynamic example since the letters P, V and T are chosen to stand for the physical quantities of pressure, volume and temperature. Giving these functions more than one name would only be confusing.

Problems

(1) Show that if $f : M \to N$ is a diffeomorphism, then for each $p \in M$ the tangent map $T_p f : T_p M \to T_{f(p)} N$ is a vector space isomorphism.

(2) Let M and N be smooth manifolds, and $f : M \to N$ a C^∞ map. Suppose that M is compact and that N is connected. If f is injective and $T_p f$ is an isomorphism for each $p \in M$, then show that f is a diffeomorphism. (Use the inverse mapping theorem.)

(3) Find the integral curves in $\mathbb{R}^2$ of the vector field $X = e^{-x}\frac{\partial}{\partial x} + \frac{\partial}{\partial y}$ and determine if X is complete or not.

(4) Which integral curves of the field $X = x^2\frac{\partial}{\partial x} + y\frac{\partial}{\partial y}$ are defined for all times t?

Problems

(5) Find a concrete description of the tangent bundle for each of the following manifolds:
   (a) Projective space $\mathbb{R}P^n$.
   (b) The Grassmann manifold $G(k, n)$.

(6) Recall that we have charts on $\mathbb{R}P^2$ given by
$$[x, y, z] \mapsto (u_1, u_2) = (x/z, y/z) \text{ on } U_3 = \{z \neq 0\},$$
$$[x, y, z] \mapsto (v_1, v_2) = (x/y, z/y) \text{ on } U_2 = \{y \neq 0\},$$
$$[x, y, z] \mapsto (w_1, w_2) = (y/x, z/x) \text{ on } U_1 = \{x \neq 0\}.$$

Show that there is a vector field on $\mathbb{R}P^2$ which in the last coordinate chart above has the following coordinate expression:
$$w_1 \frac{\partial}{\partial w_1} - w_2 \frac{\partial}{\partial w_2}.$$
What are the expressions for this vector field in the other two charts? (Caution: Your guess may be wrong!).

(7) Show that the graph $\Gamma(f) = \{(p, f(p)) \in M \times N : p \in M\}$ of a smooth map $f : M \to N$ is a smooth manifold and that we have an isomorphism $T_{(p,f(p))}(M \times N) \cong T_{(p,f(p))}\Gamma(f) \oplus T_{f(p)}N$.

(8) Prove Theorem 2.112.

(9) Show that a manifold supports a frame field defined on the whole of M exactly when there is a trivialization of TM (see Definitions 2.130 and 2.58).

(10) Prove Proposition 2.85.

(11) Find natural coordinates for the double tangent bundle TTM. Show that there is a nice map $s : TTM \to TTM$ such that $s \circ s = \mathrm{id}_{TTM}$ and such that $T\pi \circ s = T\pi_{TM}$ and $T\pi_{TM} \circ s = T\pi$. Here $\pi : TM \to M$ and $\pi_{TM} : TTM \to TM$ are the appropriate tangent bundle projection maps.

(12) Let N be the subset of $\mathbb{R}^{n+1} \times \mathbb{R}^{n+1}$ defined by $N = \{(x, y) : \|x\| = 1 \text{ and } x \cdot y = 0\}$ is a smooth manifold that is diffeomorphic to TS^n.

(13) (Hessian) Suppose that $f \in C^\infty(M)$ and that $df_p = 0$ for some $p \in M$. Show that for any smooth vector fields X and Y on M we have that $Y_p(Xf) = X_p(Yf)$. Let $H_{f,p}(v, w) := X_p(Yf)$, where X and Y are such that $X_p = v$ and $Y_p = w$. Show that $H_{f,p}(v, w)$ is independent of the extension vector fields X and Y and that the resulting map $H_{f,p} : T_pM \times T_pM \to \mathbb{R}$ is bilinear. $H_{f,p}$ is called the Hessian of f at p. Show that the assumption $df_p = 0$ is needed.

(14) Show that for a smooth map $F : M \to N$, the (bundle) tangent map $TF : TM \to TN$ is smooth. Sometimes it is supposed that one can

obtain a well-defined map $F_* : \mathfrak{X}(M) \to \mathfrak{X}(N)$ by thinking of vector fields as derivations on functions and then letting $(F_*X)f = X(f \circ F)$ for $f \in C^\infty(N)$. Show why this is misguided. Recall that the proper definition of $F_* : \mathfrak{X}(M) \to \mathfrak{X}(N)$ would be $F_*X := TF \circ X \circ F^{-1}$ and is defined in case F is a diffeomorphism. What if F is merely surjective?

(15) Show that if $\psi : M' \to M$ is a smooth covering map, then so also is $T\psi : TM' \to TM$.

(16) Define the map $f : M_{n \times n}(\mathbb{R}) \to \text{sym}(M_{n \times n}(\mathbb{R}))$ by $f(A) := A^T A$, where $M_{n \times n}(\mathbb{R})$ and $\text{sym}(M_{n \times n}(\mathbb{R}))$ are the manifolds of $n \times n$ matrices and $n \times n$ symmetric matrices respectively. Identify $T_A(M_{n \times n}(\mathbb{R}))$ with $M_{n \times n}(\mathbb{R})$ and $T_{f(A)}\text{sym}(M_{n \times n}(\mathbb{R}))$ with $\text{sym}(M_{n \times n}(\mathbb{R}))$ in the natural way for each A. Calculate $T_I f : M_{n \times n}(\mathbb{R}) \to \text{sym}(M_{n \times n}(\mathbb{R}))$ using these identifications.

(17) Let $f_1, \ldots, f_N$ be a set of smooth functions defined on an open subset of a smooth manifold. Show that if $df_1(p), \ldots, df_N(p)$ spans $T_p^* M$ for some $p \in U$, then some ordered subset of $\{f_1, \ldots, f_N\}$ provides a coordinate system on some open subset V of U containing p.

(18) Let Δ_r be the vector space of derivations on $C^r(M)$ at $p \in M$, where $0 < r \leq \infty$ is a positive integer or ∞. Fill in the details in the following outline which studies Δ_r. It will be shown that Δ_r is not finite-dimensional unless $r = \infty$.

(a) We may assume that $M = \mathbb{R}^n$ and $p = 0$ is the origin. Let $\mathfrak{m}_r := \{f \in C^r(\mathbb{R}^n) : f(0) = 0\}$ and let $\mathfrak{m}_r^2$ be the subspace spanned by the functions of the form fg for $f, g \in \mathfrak{m}_r$. We form the quotient space $\mathfrak{m}_r/\mathfrak{m}_r^2$ and consider its vector space dual $(\mathfrak{m}_r/\mathfrak{m}_r^2)^*$. Show that if $\delta \in \Delta_r$, then δ restricts to a linear functional on $\mathfrak{m}_r$ and is zero on all elements of $\mathfrak{m}_r^2$. Conclude that δ gives a linear functional on $\mathfrak{m}_r/\mathfrak{m}_r^2$. Thus we have a linear map $\Delta_r \to (\mathfrak{m}_r/\mathfrak{m}_r^2)^*$.

(b) Show that the map $\Delta_r \to (\mathfrak{m}_r/\mathfrak{m}_r^2)^*$ given above has an inverse. Hint: For a $\lambda \in (\mathfrak{m}_r/\mathfrak{m}_r^2)^*$, consider $\delta_\lambda(f) := \lambda([f - f(0)])$, where $f \in C^r(\mathbb{R}^n)$ and hence $[f - f(0)] \in \mathfrak{m}_r/\mathfrak{m}_r^2$. Conclude that by taking $r = \infty$ we have $T_0\mathbb{R}^n = \Delta_\infty \cong (\mathfrak{m}_r/\mathfrak{m}_r^2)^*$. The case $r < \infty$ is different as we see next.

(c) Let $r < \infty$. The goal from here on is to show that $\mathfrak{m}_r/\mathfrak{m}_r^2$ and hence $(\mathfrak{m}_r/\mathfrak{m}_r^2)^*$ are infinite-dimensional. We start out with the case $\mathbb{R}^n = \mathbb{R}$. First show that if $f \in \mathfrak{m}_r$, then $f(x) = xg(x)$ for $g \in C^{r-1}(\mathbb{R})$. Also if $f \in \mathfrak{m}_r^2$, then $f(x) = x^2g(x)$ for $g \in C^{r-1}(\mathbb{R})$.

(d) For each $r \in \{1, 2, 3, \ldots\}$ and each $\varepsilon \in (0, 1)$, define

$$g_\varepsilon^r(x) := \begin{cases} x^{r+\varepsilon} & \text{for } x > 0, \\ 0 & \text{for } x \leq 0. \end{cases}$$

Then $g_\varepsilon^r \in \mathfrak{m}_r$, but $g_\varepsilon^r \notin C^{r+1}(\mathbb{R})$. Show that for any fixed $r \in \{1, 2, 3, \ldots\}$, the set of elements of the form $[g_\varepsilon^r] := g_\varepsilon^r + \mathfrak{m}_r^2$ for $\varepsilon \in (0, 1)$ is linearly independent in the quotient. Hint: Use induction on r. In the case of $r = 1$, it would suffice to show that if we are given $0 < \varepsilon_1 < \cdots < \varepsilon_l < 1$ and if $\sum_{i=1}^{l} a_j g_{\varepsilon_j}^1 \in \mathfrak{m}_r^2$, then $a_j = 0$ for all j.

(Thanks to Lance Drager for donating this problem and its solution.)

(19) Find the integral curves of the vector field on $\mathbb{R}^2$ given by $X(x, y) := x^2 \frac{\partial}{\partial x} + xy \frac{\partial}{\partial y}$.

(20) Show that it is possible that a vector field defined on an open subset of a smooth manifold M may have no smooth extension to all of M.

(21) Find the integral curves (and hence the flow) for the vector field on $\mathbb{R}^2$ given by $X(x, y) := -y \frac{\partial}{\partial x} + x \frac{\partial}{\partial y}$.

(22) Let N be a point in the unit sphere S^2. Find a vector field on $S^2 \setminus \{N\}$ that is not complete and one that is complete.

(23) Using the usual spherical coordinates (ϕ, θ) on S^2, calculate the bracket $[\phi \frac{\partial}{\partial \theta}, \theta \frac{\partial}{\partial \phi}]$.

(24) Show that if X and Y are (time independent) vector fields that have flows φ_t^X and φ_t^Y, then if $[X, Y] = 0$, the flow of $X + Y$ is $\varphi_t^X \circ \varphi_t^Y$.

(25) Recall that the tangent bundle of the open set $\mathrm{GL}(n, \mathbb{R})$ in $M_{n \times n}(\mathbb{R})$ is identified with $\mathrm{GL}(n, \mathbb{R}) \times M_{n \times n}(\mathbb{R})$. Consider the vector field on $\mathrm{GL}(n, \mathbb{R})$ given by $X : g \mapsto (g, g^2)$. Find the flow of X.

(26) Let
$$t \mapsto Q_t = \begin{pmatrix} \cos t & -\sin t & 0 \\ \sin t & \cos t & 0 \\ 0 & 0 & 1 \end{pmatrix}$$
for $t \in \mathbb{R}$. Let $\phi(t, P) := Q_t P$, where P is a plane in $\mathbb{R}^3$. Show that this defines a flow on the Grassmann manifold $G(3, 2)$. Find the local expression in some coordinate system of the vector field X^Q that gives this flow. Do the same thing for the flow
$$t \mapsto R_t = \begin{pmatrix} \cos t & 0 & -\sin t \\ 0 & 1 & 0 \\ \sin t & 0 & \cos t \end{pmatrix}$$
and find the vector field X^R. Find the bracket $[X^R, X^Q]$.

(27) Develop definitions for tangent bundle and cotangent bundle for manifolds with corners. (See Problem 21.) [Hint: A curve into an n-manifold with corners should be considered smooth only if, when viewed in a chart, it has an extension to a map into $\mathbb{R}^n$. Similarly, a functions is smooth at a corner (or boundary) point only if its local representative

in some chart containing the point can be extended to an open set in $\mathbb{R}^n$.]

(28) Show that if $p(x) = p(x_1, \ldots, x_n)$ is a homogeneous polynomial, so that for some $m \in \mathbb{N}$,
$$p(tx_1, \ldots, tx_n) = t^m p(x_1, \ldots, x_n),$$
then as long as $c \neq 0$, the set $p^{-1}(c)$ is an $(n-1)$-dimensional submanifold of $\mathbb{R}^n$.

(29) Suppose that $g : M \to N$ is transverse to a submanifold $W \subset N$. For another smooth map $f : Y \to M$, show that $f \pitchfork g^{-1}(N)$ if and only if $(g \circ f) \pitchfork W$.

(30) Let $M \times N$ be a product manifold. Show that for each $X \in \mathfrak{X}(M)$ there is a vector field $\widetilde{X} \in \mathfrak{X}(M \times N)$ that is pr_1-related to X and pr_2-related to the zero field on N. We call $\widetilde{X}$ the **lift** of X. Similarly, we may lift a field on N to $M \times N$.

Chapter 3

Immersion and Submersion

Suppose we are given a smooth map $f : M \to N$. Near a point $p \in M$, the tangent map $T_p f : T_p M \to T_p N$ is a linear approximation of f. A very important invariant of a linear map is its rank, which is the dimension of its image. Recall that the **rank** of a smooth map f at p is defined to be the rank of $T_p f$. It turns out that under certain conditions on the rank of f at p, or near p, we can draw conclusions about the behavior of f near p. The basic idea is that f behaves very much like $T_p f$. If $L : V \to W$ is a linear map of finite-dimensional vector spaces, then $\operatorname{Ker} L$ and $L(V)$ are subspaces (and hence submanifolds). We study the extent to which something similar happens for smooth maps between manifolds. In this chapter we make heavy use of some basic theorems of multivariable calculus such as the implicit and inverse mapping theorems as well as the constant rank theorem. These can be found in Appendix C (see Theorems C.1, C.2 and C.5). More on calculus, including a proof of the constant rank theorem, can be found in the online supplement to this text [**Lee, Jeff**].

3.1. Immersions

Definition 3.1. A map $f : M \to N$ is called an **immersion at** $p \in M$ if $T_p f : T_p M \to T_{f(p)} N$ is an injection. A map $f : M \to N$ is called an **immersion** if it is an immersion at every $p \in M$.

Note that $T_p f : T_p M \to T_{f(p)} N$ is an injection if and only if its rank is equal to $\dim(M)$. Thus an immersion has constant rank equal to the dimension of its domain.

127

Immersions of open subsets of $\mathbb{R}^2$ into $\mathbb{R}^3$ appear as surfaces that may self-intersect, or periodically retrace themselves, or approach themselves in various limiting ways. The map $\mathbb{R}^2 \to \mathbb{R}^3$ given by $(u,v) \mapsto (\cos u, \sin u, v)$ is an immersion as is the map $(u,v) \mapsto (\cos u \sin v, \sin u \sin v, (1 - 2\cos^2 v)\cos v)$. The map $S^2 \to \mathbb{R}^3$ given by $(x,y,z) \mapsto (x,y,z-2z^3)$ is also an immersion. By contrast, the map $f : S^2 \to \mathbb{R}^3$ given by $(x,y,z) \mapsto (x,y,0)$ is not an immersion at any point on the equator $S^2 \cap \{z=0\}$.

Example 3.2. We describe an immersion of the torus $T^2 := S^1 \times S^1$ into $\mathbb{R}^3$. We can represent points in T^2 as pairs $(e^{i\theta_1}, e^{i\theta_2})$. It is easy to see that, for fixed $a, b > 0$, the following map is well-defined:

$$(e^{i\theta_1}, e^{i\theta_2}) \mapsto (x(e^{i\theta_1}, e^{i\theta_2}), y(e^{i\theta_1}, e^{i\theta_2}), z(e^{i\theta_1}, e^{i\theta_2})),$$

where

$$x(e^{i\theta_1}, e^{i\theta_2}) = (a + b\cos\theta_1)\cos\theta_2,$$
$$y(e^{i\theta_1}, e^{i\theta_2}) = (a + b\cos\theta_1)\sin\theta_2,$$
$$z(e^{i\theta_1}, e^{i\theta_2}) = b\sin\theta_1.$$

Exercise 3.3. Show that the map of the above example is an immersion. Give conditions on a and b that guarantee that the map is a 1-1 immersion.

Theorem 3.4. *Let $f : M \to N$ be a smooth map that is an immersion at p. Then for any chart $(\mathbf{x}, U)$ centered at p, there is a chart $(\mathbf{y}, V)$ centered at $f(p)$ such that $f(U) \subset V$ and such that the corresponding coordinate expression for f is $(x^1, \ldots, x^k) \mapsto (x^1, \ldots, x^k, 0, \ldots, 0) \in \mathbb{R}^n$. Here, n is the dimension of N and $k = \dim(M)$ is the rank of $T_p f$.*

Proof. Follows easily from Corollary C.3. □

Theorem 3.5. *If $f : M \to N$ is an immersion (so an immersion at every point), and if f is a homeomorphism onto its image $f(M)$ (using the relative topology on $f(M)$), then $f(M)$ is a regular submanifold of N.*

Proof. Let k be the dimension of M and let n be the dimension of N. Clearly f is injective since it is a homeomorphism. Let $f(p) \in f(M)$ for a unique p. By the previous theorem, there are charts $(U, \mathbf{x})$ with $p \in U$ and $(V, \mathbf{y})$ with $f(p) \in V$ such that the corresponding coordinate expression for f is $(x^1, \ldots, x^k) \mapsto (x^1, \ldots, x^k, 0, \ldots, 0) \in \mathbb{R}^n$. We arrange to have $f(U) \subset V$. But $f(U)$ is open in the relative topology on $f(M)$, so there is an open set $O \subset V$ in M such that $f(U) = f(M) \cap O$. Now it is clear that $(O, \mathbf{y}|_O)$ is a chart with the regular submanifold property, and so since p was arbitrary, we conclude that $f(M)$ is a regular submanifold. □

If $f : M \to N$ is an immersion that is a homeomorphism onto its image (as in the theorem above), then we say that f is an **embedding**.

3.1. Immersions

Exercise 3.6. Show that every injective immersion of a compact manifold is an embedding.

Exercise 3.7. Show that if $f : M \to N$ is an immersion and $p \in M$, then there is an open U containing p such that $f|_U$ is an embedding.

Exercise 3.8. Recall the definition of a vector field along a map (Definition 2.68). Let X be a vector field along $f : N \to M$. Show that if f is an embedding, then there is an open neighborhood U of $f(N)$ and a vector field $\overline{X} \in \mathfrak{X}(U)$ such that $X = \overline{X} \circ f$.

Recall that a continuous map f is said to be **proper** if $f^{-1}(K)$ is compact whenever K is compact.

Exercise 3.9. Show that a proper 1-1 immersion is an embedding. [Hint: This is mainly a topological argument. You may assume (without loss of generality) that the spaces involved are Hausdorff and second countable. The slightly more general case of paracompact Hausdorff spaces follows.]

Definition 3.10. Let S and M be smooth manifolds. A smooth map $f : S \to M$ will be called **smoothly universal** if for any smooth manifold N, a mapping $g : N \to S$ is smooth if and only if $f \circ g$ is smooth.

$$\begin{array}{ccc} S & \xrightarrow{f} & M \\ {\scriptstyle g}\uparrow & \nearrow {\scriptstyle f \circ g} & \\ N & & \end{array}$$

Definition 3.11. A **weak embedding** is a 1-1 immersion which is smoothly universal.

Let $f : S \to M$ be a weak embedding and let $\mathcal{A}$ be the maximal atlas that gives the differentiable structure on S. Suppose we consider a different differentiable structure on S given by a maximal atlas $\mathcal{A}_2$. Now suppose that $f : S \to M$ is also a weak embedding with respect to $\mathcal{A}_2$. Resorting to seldom used pedantic notation, we are supposing that both $f : (S, \mathcal{A}) \to M$ and $f : (S, \mathcal{A}_2) \to M$ are weak embeddings. From this it is easy to show that the identity map gives smooth maps $(S, \mathcal{A}) \to (S, \mathcal{A}_2)$ and $(S, \mathcal{A}_2) \to (S, \mathcal{A})$. This means that in fact $\mathcal{A} = \mathcal{A}_2$, so that the smooth structure of S is uniquely determined by the fact that f is a weak embedding.

Exercise 3.12. Show that every embedding is a weak embedding.

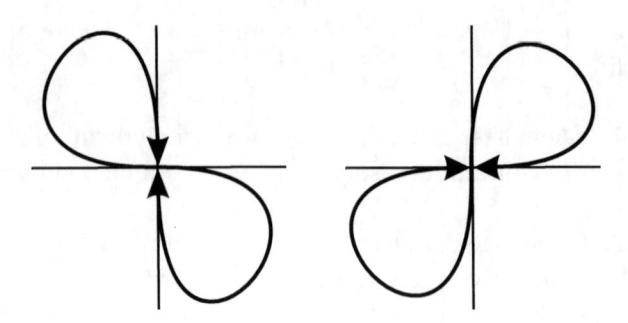

Figure 3.1. Figure eight immersions

In terms of 1-1 immersions, we have the following inclusions:

{proper embeddings} ⊂ {embeddings}
$$\subset \{\text{weak embeddings}\} \subset \{\text{1-1 immersions}\}.$$

3.2. Immersed and Weakly Embedded Submanifolds

We have already seen the definition of a regular submanifold. The more general notion of a submanifold is supposed to realize the "subobject" in the category of smooth manifolds and smooth maps. Submanifolds are to manifolds what subsets are to sets in general. However, what exactly should be the definition of a submanifold? The fact is that there is some disagreement on this point. From the category-theoretic point of view it seems natural that a submanifold of M should be some kind of smooth map $I : S \to M$. This is not quite in line with our definition of regular submanifold, which is, after all, a type of *subset* of M. There is considerable motivation to define submanifolds in general as certain subsets; perhaps the images of certain nice smooth maps. We shall follow this route.

Definition 3.13. Let S be a subset of a smooth manifold M. If S is a smooth manifold such that the inclusion map $\iota : S \to M$ is an injective immersion, then we call S an **immersed submanifold**.

Notice that in the above definition, S certainly need not have the subspace topology! Its topology is that induced by its own smooth structure. The reader may rightfully wonder just how S could acquire such a smooth structure in the first place. If $f : N \to M$ is an injective immersion, then $S := f(N)$ can be given a smooth structure so that it is an immersed submanifold. Indeed, we can simply transfer the structure from N via the bijection $f : N \to f(N)$. However, this may not be the only possible smooth structure on $f(N)$ which makes it an immersed submanifold. Thus it is imperative to specify what smooth structure is being used. Simply looking at

3.2. Immersed and Weakly Embedded Submanifolds

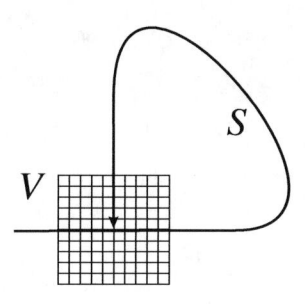

Figure 3.2. Immersions can approach themselves

the set is not enough. For example, in Figure 3.1 we see the same figure eight shaped subset drawn twice, but with arrows suggesting that it is the image of two quite different immersions which provide two quite different smooth structures.

Suppose that S is a k-dimensional immersed submanifold of a smooth n-manifold M, and let $p \in S$. Then using Theorem 3.4, we see that there is a chart $(O, \mathbf{x})$ on S, and a chart $(V, \mathbf{y})$ on M, with $p \in O \subset V$, such that

$$\mathbf{y} \circ \iota \circ \mathbf{x}^{-1} = \mathbf{y} \circ \mathbf{x}^{-1} : \mathbf{x}(O) \to \mathbf{y}(V)$$

has the form $(a^1, \ldots, a^k) \mapsto (a^1, \ldots, a^k, 0, \ldots, 0)$. This means that $\mathbf{y}(O) = \mathbf{y} \circ \mathbf{x}^{-1}(\mathbf{x}(O))$ is a relatively open subset of $\mathbb{R}^k \times \{0\}$. Thus there is an open subset W of $\mathbf{y}(V) \subset \mathbb{R}^n$ such that $\mathbf{y}(O) = W \cap (\mathbb{R}^k \times \{0\})$. Letting $U_1 := \mathbf{y}^{-1}(W)$, we see that

$$\mathbf{y}(U_1 \cap O) = \mathbf{y}(U_1) \cap (\mathbb{R}^k \times \{0\}).$$

Thus the chart $(\mathbf{y}|_{U_1}, U_1)$ has the submanifold property with respect to O (but not necessarily with respect to S). The set O has a smooth structure as an open submanifold of S. But this is the same smooth structure O has as a regular submanifold. To see this note that the restrictions $y^1|_O, \ldots, y^k|_O$ combine to give an admissible chart on S. Indeed, using the functions we obtain a bijection of O with an open subset of $\mathbb{R}^k$. We only need to show that this bijection is smoothly related to the chart $(O, \mathbf{x})$, and this amounts to showing that $y^1|_O, \ldots, y^k|_O$ are smooth. But this follows immediately from the fact that $\mathbf{y} \circ \mathbf{x}^{-1}$ is smooth. Notice that unlike the case of a regular submanifold, it may be that no matter how small V,

$$\mathbf{y}(V \cap O) \neq \mathbf{y}(V) \cap (\mathbb{R}^k \times \{0\}),$$

as indicated in Figure 3.2. So in summary, each point of an immersed submanifold has a neighborhood that is a regular submanifold.

Proposition 3.14. *Let $S \subset M$ be an immersed submanifold of dimension k and let $f : N \to S$ be a map. Suppose that $\iota \circ f : N \to M$ is smooth,* where $\iota : S \hookrightarrow M$ is the inclusion. Then if $f : N \to S$ is continuous, it is also smooth.

Proof. We wish to show that $f : N \to S$ is smooth if $f^{-1}(O)$ is open for every set $O \subset S$ that is open in the manifold topology on S. Let $p \in N$ and choose a chart $(V, \mathbf{y})$ for M centered at $\iota \circ f(p)$, so that
$$U = \{q \in V : y^{k+1}(q) = \cdots = y^n(q) = 0\}$$
is an open neighborhood of p in S and such that $y^1\big|_U, \ldots, y^k\big|_U$ are coordinates for S on U. By assumption $f^{-1}(U)$ is open. Thus $(\iota \circ f)\left(f^{-1}(U)\right) \subset U$. In other words, $\iota \circ f$ maps an open neighborhood of p into U. To test for the smoothness of f, we consider the functions $\left(y^i\big|_U\right) \circ f$ on the set $f^{-1}(U)$. But
$$\left(y^i\big|_U\right) \circ f = y^i \circ \iota \circ f,$$
and these are clearly smooth by the assumption that $\iota \circ f$ is smooth. $\square$

Sometimes the previous result is stated differently (and somewhat imprecisely): Suppose that $f : N \to M$ is a smooth map with image inside an immersed submanifold S; then f is smooth as a map into S if it is continuous as a map into S. The lack of a notational distinction between f as a map into S and f as a map into M is what makes this way of stating things less desirable.

Let $S \subset M$ and suppose that S has a smooth structure. To say that an inclusion $S \hookrightarrow M$ is an embedding is easily seen to be the same as saying that S is a regular submanifold, and so we also say that S is **embedded** in M.

Corollary 3.15. *Suppose that $S \subset M$ is a regular submanifold. Let $f : N \to S$ be a map such that $\iota \circ f : N \to M$ is smooth. Then $f : N \to S$ is smooth.*

Proof. The map $\iota : S \hookrightarrow M$ is certainly an immersion, and so by the previous theorem we need only check that $f : N \to S$ is continuous. Let O be open in S. Then since S has the relative topology, $O = U \cap S$ for some open set U in M. Then $f^{-1}(O) = f^{-1}(U \cap S) = f^{-1}(\iota^{-1}(U)) = (\iota \circ f)^{-1}(U)$, which is open since $\iota \circ f$ is continuous. Thus f is continuous. $\square$

Definition 3.16. Let S be a subset of a smooth manifold M. If S is a smooth manifold such that the inclusion map $\iota : S \to M$ is a weak embedding, then we say that S is a **weakly embedded submanifold**.

From the properties of weak embeddings we know that for any given subset $S \subset M$ there is at most one smooth structure on S that makes it a weakly embedded submanifold.

3.2. Immersed and Weakly Embedded Submanifolds

Corresponding to each type of injective immersion considered so far we have in their images different notions of submanifold:

$$\{\text{proper submanifolds}\} \subset \{\text{regular submanifolds}\}$$
$$\subset \{\text{weakly embedded submanifolds}\}$$
$$\subset \{\text{immersed submanifolds}\}.$$

We wish to further characterize the weakly embedded submanifolds.

Definition 3.17. Let S be any subset of a smooth manifold M. For any $x \in S$, denote by $C_x(S)$ the set of all points of S that can be connected to x by a smooth curve with image entirely inside S.

It is important to be clear that $C_x(S)$ is not necessarily the connected component of S with its relative topology since, for example, S could be the image of an injective nowhere differentiable curve. In the latter case, $C_x(S) = \{x\}$ for all $x \in S$!

Definition 3.18. We say that a subset S of an n-manifold M has **property W**(k) if for each $s_0 \in S$ there exists a chart $(U, \mathbf{x})$ centered at s_0 such that $\mathbf{x}(C_{s_0}(U \cap S)) = \mathbf{x}(U) \cap (\mathbb{R}^k \times \{0\})$. Here $\mathbb{R}^n = \mathbb{R}^k \times \mathbb{R}^{n-k}$.

Together, the next two propositions show that weakly embedded submanifolds are exactly those subsets that have property W(k) for some k. Our proof follows that of Michor [**Mich**], who refers to subsets with property W(k), for some k, as **initial submanifolds**. With Michor's terminology, the result will be that the initial submanifolds are the same as the weakly embedded submanifolds.

Proposition 3.19. *If an injective immersion $I : S \to M$ is smoothly universal, then the image $I(S)$ has property W(k) where $k = \dim(S)$. In particular, if $S \subset M$ is a weakly embedded submanifold of M, then it has property W(k) where $k = \dim(S)$.*

Proof. Let $\dim(S) = k$ and $\dim(M) = n$. Choose $s_0 \in S$. Since I is an immersion, we may pick a coordinate chart $(W, \mathbf{w})$ for S centered at s_0 and a chart $(V, \mathbf{v})$ for M centered at $I(s_0)$ such that

$$\mathbf{v} \circ I \circ \mathbf{w}^{-1}(y) = (y, 0) = (y^1, \ldots, y^k, 0, \ldots, 0).$$

Choose an $r > 0$ small enough that $B^k(0, 2r) \subset \mathbf{w}(W)$ and $B^n(0, 2r) \subset \mathbf{v}(V)$. Let $U = \mathbf{v}^{-1}(B^n(0, r))$ and $W_1 = \mathbf{w}^{-1}(B^k(0, r))$. Let $\mathbf{x} := \mathbf{v}|_U$. We show that the coordinate chart $(U, \mathbf{x})$ satisfies the conditions of Definition 3.18:

$$\mathbf{x}^{-1}(\mathbf{x}(U) \cap (\mathbb{R}^k \times \{0\})) = \mathbf{x}^{-1}\{(y, 0) : \|y\| < r\}$$
$$= I \circ \mathbf{w}^{-1} \circ (\mathbf{x} \circ I \circ \mathbf{w}^{-1})^{-1}(\{(y, 0) : \|y\| < r\})$$
$$= I \circ \mathbf{w}^{-1}(\{y : \|y\| < r\}) = I(W_1).$$

Clearly $I(W_1) \subset I(S)$, but we also have
$$\mathbf{v} \circ I(W_1) \subset \mathbf{v} \circ I \circ \mathbf{w}^{-1}(B^k(0,r))$$
$$= B^n(0,r) \cap \{\mathbb{R}^k \times \{0\}\} \subset B^n(0,r),$$
so that $I(W_1) \subset \mathbf{v}^{-1}(B^n(0,r)) = U$. Thus $I(W_1) \subset U \cap I(S)$. Since $I(W_1)$ is smoothly contractible to $I(s_0)$, every point of $I(W_1)$ is connected to $I(s_0)$ by a smooth curve completely contained in $I(W_1) \subset U \cap I(S)$. This implies that $I(W_1) \subset C_{I(s_0)}(U \cap I(S))$. Thus $\mathbf{x}^{-1}(\mathbf{x}(U) \cap (\mathbb{R}^k \times \{0\})) \subset C_{I(s_0)}(U \cap I(S))$ or
$$\mathbf{x}(U) \cap (\mathbb{R}^k \times \{0\}) \subset \mathbf{x}(C_{I(s_0)}(U \cap I(S))).$$
Conversely, let $z \in C_{I(s_0)}(U \cap I(S))$. By definition there must be a smooth curve $c : [0,1] \to M$ starting at $I(s_0)$ and ending at z with $c([0,1]) \subset U \cap I(S)$. Since $I : S \to M$ is injective and smoothly universal, there is a unique smooth curve $c_1 : [0,1] \to S$ with $I \circ c_1 = c$.

Claim: $c_1([0,1]) \subset W_1$. Assume not. Then there is a number $t \in [0,1]$ with $c_1(t) \in \mathbf{w}^{-1}(\{r \leq \|y\| < 2r\})$. Therefore,
$$(\mathbf{v} \circ I)(c_1(t)) \in (\mathbf{v} \circ I \circ \mathbf{w}^{-1})(\{r \leq \|y\| < 2r\})$$
$$= \{(y,0) : r \leq \|y\| < 2r\} \subset \{z \in \mathbb{R}^n : r \leq \|z\| < 2r\}.$$
This implies that $(\mathbf{v} \circ I \circ c_1)(t) = (\mathbf{v} \circ c)(t) \in \{z \in \mathbb{R}^n : r \leq \|z\| < 2r\}$, which in turn implies the contradiction $c(t) \notin U$. The claim is proven.

The fact that $c_1([0,1]) \subset W_1$ implies $c_1(1) = I^{-1}(z) \in W_1$, and so $z \in I(W_1)$. As a result we have $C_{I(s_0)}(U \cap I(S)) = I(W_1)$ which together with the first half of the proof gives the result:
$$I(W_1) = \mathbf{x}^{-1}(\mathbf{x}(U) \cap (\mathbb{R}^k \times \{0\})) \subset C_{I(s_0)}(U \cap I(S)) = I(W_1)$$
$$\implies \mathbf{x}^{-1}(\mathbf{x}(U) \cap (\mathbb{R}^k \times \{0\})) = C_{I(s_0)}(U \cap I(S))$$
$$\implies \mathbf{x}(U) \cap (\mathbb{R}^k \times \{0\}) = \mathbf{x}(C_{I(s_0)}(U \cap I(S))). \qquad \square$$

Proposition 3.20. *If $S \subset M$ has property $W(k)$, then there is a unique smooth structure on S which makes it a k-dimensional weakly embedded submanifold of M.*

Proof. (Sketch) We are given that for every $s \in S$, there exists a chart $(U_s, \mathbf{x}_s)$ with $\mathbf{x}_s(s) = 0$ and with $\mathbf{x}(C_s(U_s \cap S)) = \mathbf{x}(U_s) \cap (\mathbb{R}^k \times \{0\})$. The charts on S will be the restrictions of the charts $(U_s, \mathbf{x}_s)$ to the sets $C_s(U_s \cap S)$. The overlap maps are smooth because they are restrictions of overlap maps on M to subsets of the form $V \cap (\mathbb{R}^k \times \{0\})$ for V open in $\mathbb{R}^n$. If $(U_{s_1}, \mathbf{x}_{s_1})$ and $(U_{s_2}, \mathbf{x}_{s_2})$ are two such charts with corresponding sets $S_1 := C_{s_1}(U_{s_1} \cap S)$ and $S_2 := C_{s_2}(U_{s_2} \cap S)$, then we need to check that $\mathbf{x}_{s_1}(S_1 \cap S_1)$ is open in $\mathbb{R}^k \times \{0\}$ (recall the definition of smooth atlas). For each $p \in U_{s_1} \cap U_{s_2} \cap S$ consider the set $C(p) := C_p(U_{s_1} \cap U_{s_2} \cap S)$. We

3.2. Immersed and Weakly Embedded Submanifolds

leave it to the reader to show that $C(p) \subset S_1 \cap S_2$ and that if $p \ne q$ then $C(p) \cap C(q) = \emptyset$. Thus the sets $C(p)$ form a partition of $U_{s_1} \cap U_{s_2} \cap S$. It is not hard to see that each $C(p)$ maps onto a connected path component of $\mathbf{x}_{s_1}(U_{s_1} \cap U_{s_2} \cap S)$ and that every path component of this set is the image of some $C(p)$. But this implies that $\mathbf{x}_{s_1}(U_{s_1} \cap U_{s_2} \cap S)$ open. Notice however that the topology induced by the smooth structure thus obtained on S is finer than the relative topology that S inherits from M. This is because the sets of the form $C_s(U \cap S)$ are not necessarily open in the relative topology. Since it is a finer topology, it is also Hausdorff.

It is clear that with this smooth structure on S, the inclusion $\iota : S \hookrightarrow M$ is an injective immersion. We now show that the inclusion map $\iota : S \hookrightarrow M$ is smoothly universal and hence a weak embedding. By the comments following Definition 3.11 the smooth structure on S is unique. Let $g : N \to S$ be a map and suppose that $\iota \circ g$ is smooth. Given $x \in M$, choose a chart $(U_s, \mathbf{x}_s)$ where $s = g(x)$. The set $g^{-1}(U_s)$ is open since $\iota \circ g$ is continuous and $(\iota \circ g)^{-1}(U_s) = g^{-1} \circ \iota^{-1}(U_s) = g^{-1}(U_s)$. We may choose a chart $(V, \mathbf{y})$ centered at x with $V \subset g^{-1}(U_s)$ and we may arrange that $\mathbf{y}(V)$ is a ball centered at the origin. This means that $\iota \circ g(V)$ is smoothly contractible in $U_{g(x)} \cap S$ and hence $g(V) \subset C_{g(x)}(U_{g(x)} \cap S)$. But then

$$\mathbf{x}_s|_{C_s(U_s \cap S)} \circ g \circ \mathbf{y}^{-1} = \mathbf{x}_s \circ (\iota \circ g) \circ \mathbf{y}^{-1},$$

and so g is smooth because $\iota \circ g$ is smooth.

To be completely finished, we need to show that with the topology induced by the atlas, each connected component of S is second countable. We can give a quick proof, but it depends on Riemannian metrics which we have yet to discuss. The idea is that on any paracompact smooth manifold, there are plenty of Riemannian metrics. A choice of Riemannian metric gives a notion of distance making every connected component a second countable metric space. If we put such a Riemannian metric on M, then it induces one on S (by restriction). This means that each component of S is also a separable metric space and hence a second countable Hausdorff topological space. □

We say that two immersions $I_1 : N_1 \to M$ and $I_2 : N_2 \to M$ are **equivalent** if there exists a diffeomorphism $\Phi : N_1 \to N_2$ such that $I_2 \circ \Phi = I_1$; i.e., such that the following diagram commutes:

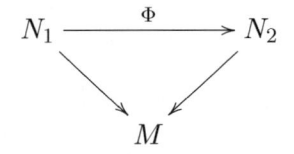

$$\begin{array}{ccc} N_1 & \xrightarrow{\Phi} & N_2 \\ & \searrow \swarrow & \\ & M & \end{array}$$

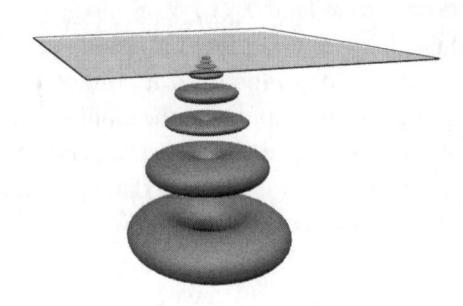

Figure 3.3. Tori converge to a point

If $I : N \to M$ is a weak embedding (resp. embedding), then there is a unique smooth structure on $S = I(N)$ such that S is a weakly embedded (resp. embedded) submanifold and $I : N \to M$ is equivalent to the inclusion $\iota : S \hookrightarrow M$ in the above sense.

We shall follow the convention that the word "submanifold", when used without a qualifier such as "immersed" or "weakly embedded", is to mean a *regular* submanifold unless otherwise indicated. What R. Sharpe [**Shrp**] calls a "submanifold", refers to something more restrictive than the weakly embedded submanifolds, but still less restrictive that the regular submanifolds. Sharpe's definition of "submanifold" seems designed to exclude examples like that shown in Figure 3.3. Here the tori converge to a point on the plane (which is taken to be part of the manifold). Every neighborhood of that point will contain an infinite number of tori. Such a behavior is excluded by Sharpe's definition, but this is still a weakly embedded submanifold. One can also imagine the tori flattening while only the holes converge to a point. This example can easily be modified to be path connected and yet, it could never be the maximal integral manifold of a tangent distribution (see Chapter 11 for definitions).

The celebrated Whitney embedding theorem states that any second-countable n-manifold can be embedded in a Euclidean space of dimension $2n$. We do not prove the full theorem, but we will settle for the following easier result.

Theorem 3.21. *Suppose that M is an n-manifold that has a finite atlas. Then there exists an injective immersion of M into $\mathbb{R}^{2n+1}$. Consequently, every compact n-manifold can be embedded into $\mathbb{R}^{2n+1}$.*

Proof. Let M be a smooth manifold with a finite atlas. In particular, M is second countable. Initially, we will settle for an immersion into $\mathbb{R}^D$ for

3.2. Immersed and Weakly Embedded Submanifolds

some possibly very large dimension D. Let $\{O_i, \varphi_i\}_{i \in I}$ be an atlas with cardinality $N < \infty$. By applying Lemma B.4 twice, the cover $\{O_i\}$ may be refined to two other covers $\{U_i\}_{i \in I}$ and $\{V_i\}_{i \in I}$ such that $\overline{U_i} \subset V_i \subset \overline{V_i} \subset O_i$. Also, we may find smooth functions $f_i : M \to [0, 1]$ with $\mathrm{supp}(f_i) \subset O_i$ and such that $f_i(x) = 1$ for all $x \in U_i$ and $f_i(x) < 1$ for $x \notin V_i$. Next we write $\varphi_i = (x_i^1, \ldots, x_i^n)$ so that $x_i^j : O_i \to \mathbb{R}$ is the j-th coordinate function of the i-th chart, and then form the product

$$f_{ij} := f_i x_i^j,$$

which is defined and smooth on all of M after extension by zero.

Now we put the functions f_i together with the functions f_{ij} to get a map $f : M \to \mathbb{R}^{n+Nn}$:

$$f = (f_1, \ldots, f_n, f_{11}, f_{12}, \ldots, f_{21}, \ldots, f_{Nn}).$$

Now we show that f is injective. Suppose that $f(x) = f(y)$. Note that $f_k(x)$ must be 1 for some k since $x \in U_k$ for some k. But then $f_k(y) = 1$ also, and this means that $y \in V_k$ (why?). Since $f_k(x) = f_k(y) = 1$, it follows that $f_{kj}(x) = f_{kj}(y)$ for all j. Remembering how things were defined, we see that x and y have the same image under $\varphi_k : O_k \to \mathbb{R}^n$ and thus $x = y$.

To show that $T_x f$ is injective for all $x \in M$, we fix an arbitrary such x; then $x \in U_k$ for some k. But then near this x, the functions $f_{k1}, f_{k2}, \ldots, f_{kn}$, are equal to $x_k^1, \ldots, x_k^n$ and so the rank of f must be *at least* n and in fact equal to n since $\dim T_x M = n$.

So far we have an injective immersion of M into $\mathbb{R}^D$ where $D = n + Nn$. We show that there is a projection $\pi : \mathbb{R}^D \to L \subset \mathbb{R}^D$, where $L \cong \mathbb{R}^{2n+1}$ is a $(2n+1)$-dimensional subspace of $\mathbb{R}^D$ such that $\pi \circ f$ is an injective immersion. The proof of this will be inductive. So suppose that there is an injective immersion f of M into $\mathbb{R}^d$ for some d with $D \geq d > 2n + 1$. We show that there is a projection $\pi_d : \mathbb{R}^d \to L^{d-1} \cong \mathbb{R}^{d-1}$ such that $\pi_d \circ f$ is still an injective immersion. To this end, define a map $h : M \times M \times \mathbb{R} \to \mathbb{R}^d$ by $h(x, y, t) := t(f(x) - f(y))$. Since $d > 2n + 1$, Sard's theorem (Theorem 2.34) implies that there is a vector $z \in \mathbb{R}^d$ which is neither in the image of the map h nor in the image of the map $df : TM \to \mathbb{R}^d$. This z cannot be 0 since 0 is certainly in the image of both of these maps. If $\mathrm{pr}_{\perp z}$ is projection onto the orthogonal complement of z, then $\mathrm{pr}_{\perp z} \circ f$ is injective; for if $\mathrm{pr}_{\perp z} \circ f(x) = \mathrm{pr}_{\perp z} \circ f(y)$, then $f(x) - f(y) = az$ for some $a \in \mathbb{R}$. But suppose $x \neq y$. Since f is injective, we must have $a \neq 0$. This state of affairs is impossible since it results in the equation $h(x, y, 1/a) = z$, which contradicts our choice of z. Thus $\mathrm{pr}_{\perp z} \circ f$ is injective.

Next we examine $T_x(\mathrm{pr}_{\perp z} \circ f)$ for an arbitrary $x \in M$. Suppose that $T_x(\mathrm{pr}_{\perp z} \circ f)v = 0$. Then $d(\mathrm{pr}_{\perp z} \circ f)|_x v = 0$, and since $\mathrm{pr}_{\perp z}$ is linear, this amounts to $\mathrm{pr}_{\perp z} \circ df|_x v = 0$, which gives $df|_x v = az$ for some number $a \in \mathbb{R}$, and which cannot be 0 since f is assumed to be an immersion. But then $df|_x \frac{1}{a} v = z$, which also contradicts our choice of z.

We conclude that $\mathrm{pr}_{\perp z} \circ f$ is an injective immersion. Repeating this process inductively we finally get a composition of projections $\mathrm{pr} : \mathbb{R}^D \to \mathbb{R}^{2n+1}$ such that $\mathrm{pr} \circ f : M \to \mathbb{R}^{2n+1}$ is an injective immersion. The final statement for compact manifolds follows from Exercise 3.6. □

3.3. Submersions

Definition 3.22. A map $f : M \to N$ is called a **submersion at** $p \in M$ if $T_p f : T_p M \to T_{f(p)} N$ is a surjection. $f : M \to N$ is called a **submersion** if f is a submersion at every $p \in M$.

Example 3.23. The map of the punctured space $\mathbb{R}^3 \backslash \{0\}$ onto the sphere S^2 given by $x \mapsto x/|x|$ is a submersion. To see this, use any spherical coordinates (ρ, ϕ, θ) on $\mathbb{R}^3 \backslash \{0\}$ and the induced submanifold coordinates (ϕ, θ) on S^2. Expressed with respect to these coordinates, the map becomes $(\rho, \phi, \theta) \mapsto (\phi, \theta)$ on the domain of the spherical coordinate chart. Here we ended up locally with a projection onto a second factor $\mathbb{R} \times \mathbb{R}^2 \to \mathbb{R}^2$, but this is clearly good enough to prove the point.

As in the last example, to show that a map is a submersion at some p it is enough to find charts containing p and $f(p)$ so that the coordinate representative of the map is just a projection. Conversely, we have

Theorem 3.24. *Let M be an m-manifold and N a k-manifold and let $f : M \to N$ be a smooth map that is a submersion at p. Then for any chart $(V, \mathbf{y})$ centered at $f(p)$ there is a chart $(U, \mathbf{x})$ centered at p with $f(U) \subset V$ such that $\mathbf{y} \circ f \circ \mathbf{x}^{-1}$ is given by $(x^1, \ldots, x^k, \ldots, x^m) \mapsto (x^1, \ldots, x^k) \in \mathbb{R}^k$. Here k is both the dimension of N and the rank of $T_p f$.*

Proof. Follows directly from Theorem C.4 of Appendix C. □

In certain contexts, submersions, especially surjective submersions, are referred to as projections. We often denote such a map by the letter π. Recall that if $\pi : M \to N$ is a smooth map, then a smooth **local section** of π is a smooth map $\sigma : V \to M$ defined on an open set V and such that $\pi \circ \sigma = \mathrm{id}_V$. Also, we adopt the terminology that subsets of M of the form $\pi^{-1}(q)$ are called **fibers** of the submersion.

Proposition 3.25. *If $\pi : M \to N$ is a submersion, then it is an open map and every point $p \in M$ is in the image of a smooth local section.*

3.3. Submersions

Proof. Let $p \in M$ be arbitrary. We choose a chart $(U, \mathbf{x})$ centered at p and a chart $(V, \mathbf{y})$ centered at $\pi(p)$ such that $\mathbf{y} \circ \pi \circ \mathbf{x}^{-1}$ is of the form $(x^1, \ldots, x^k, x^{k+1}, \ldots, x^m) \mapsto (x^1, \ldots, x^k)$, where $\dim M = m$ and $\dim N = k$. By shrinking the domains if necessary, we can arrange that $\mathbf{x}(U)$ has the form $A \times B \subset \mathbb{R}^k \times \mathbb{R}^{m-k}$ and $\mathbf{y}(V) = B \subset \mathbb{R}^l$. Then we may transfer the section $i_b : a \to (a, b)$ where $b = \mathbf{x}(p)$. More precisely, the desired local section is $\sigma := \mathbf{x}^{-1} \circ i_b \circ \mathbf{y}$ on A.

We now use the existence of local sections to show that π is an open map. Let O be any open set in M. To show that $\pi(O)$ is open, we pick any $q \in \pi(O)$ and choose $p \in O$ with $p \in \pi^{-1}(q)$. Now we choose a chart $(U, \mathbf{x})$ as above with $p \in U \subset O$. Then q is in the domain of a section which is open and contained in $\pi(O)$. □

Proposition 3.26. *Let $\pi : M \to N$ be a surjective submersion. If $f : N \to P$ is any map, then f is smooth if and only if $f \circ \pi$ is smooth:*

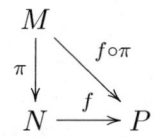

$$\begin{array}{ccc} M & & \\ \pi \downarrow & \searrow f \circ \pi & \\ N & \xrightarrow{f} & P \end{array}$$

Proof. One direction is trivial. For the other direction, assume that $f \circ \pi$ is smooth. We check for smoothness of f about an arbitrary point $q \in N$. Pick $p \in \pi^{-1}(q)$. By the previous proposition p is in the image of a smooth section $\sigma : V \to M$. This means that f and $(f \circ \pi) \circ \sigma$ agree on a neighborhood of q, and since the latter is smooth, we are done. □

Next suppose that we have a surjective submersion $\pi : M \to N$ and consider a smooth map $g : M \to P$ which is constant on fibers. That is, we assume that if $p_1, p_2 \in \pi^{-1}(q)$ for some $q \in N$, then $f(p_1) = f(p_2)$. Clearly there is a unique induced map $f : N \to P$ so that $g = f \circ \pi$. By the above proposition f must be smooth. This we record as a corollary:

Corollary 3.27. *If $g : M \to P$ is a smooth map which is constant on the fibers of a surjective submersion $\pi : M \to N$, then there is a unique smooth map $f : N \to P$ such that $g = f \circ \pi$.*

The following technical lemma is needed later and represents one more situation where second countability is needed.

Lemma 3.28. *Suppose that M is a second countable smooth manifold. If $f : M \to N$ is a smooth map with constant rank that is also surjective, then it is a submersion.*

Proof. Let $\dim M = m$, $\dim N = n$ and $\text{rank}(f) = k$ and choose $p \in M$. Suppose that f is not a submersion so that $k < n$. We can cover M

by a countable collection of charts $(U_\alpha, \mathbf{x}_\alpha)$ and cover N by charts $(V_i, \mathbf{y}_i)$ such that for every α, there is an $i = i(\alpha)$ with $f(U_\alpha) \subset V_i$ and $\mathbf{y}_i \circ f \circ \mathbf{x}_\alpha^{-1}(x^1, \ldots, x^n) = (x^1, \ldots, x^k, 0, \ldots, 0)$. But this means that $f(U_\alpha)$ has measure zero. However, $f(M) = \bigcup_\alpha f(U_\alpha)$ and so $f(M)$ is also of measure zero which contradicts the surjectivity of f. This contradiction means that f must be a submersion after all. $\square$

Problems

(1) Let $0 < a < b$. Show that the subset of $\mathbb{R}^3$ described by the equation
$$\left(\sqrt{x^2 + y^2} - b\right)^2 + z^2 = a^2$$
is a submanifold. Show that the resulting manifold is diffeomorphic to $S^1 \times S^1$.

(2) Show that the map $S^2 \to \mathbb{R}^3$ given by $(x, y, z) \mapsto (x, y, z - 2z^3)$ is an immersion. Try to determine what the image of this map looks like.

(3) Show that if M is compact and N is connected, then a submersion $f : M \to N$ must be surjective.

(4) Let $f : M \to N$ be an immersion.
   (a) Let $(U, \mathbf{x})$ be a chart for M with $p \in U$, and let $(V, \mathbf{y})$ be a chart for N with $f(p) \in V$ such that
   $$\mathbf{y} \circ f \circ \mathbf{x}^{-1}(a^1, a^2, \ldots, a^n) = (a^1, a^2, \ldots, a^n, 0, \ldots, 0).$$
   Show that
   $$T_p f \cdot \left.\frac{\partial}{\partial x^i}\right|_p = \left.\frac{\partial}{\partial y^i}\right|_{f(p)} \quad \text{for } i = 1, \ldots, n.$$
   (b) Show that if f is as in part (a) and $Y \in \mathfrak{X}(N)$ is such that $Y(p) \in T_p f(T_p M)$ for all p, then there is a unique $X \in \mathfrak{X}(M)$ such that X is f-related to Y.

(5) Define a function $s : \mathbb{R}^{n+1} \setminus \{0\} \to \mathbb{R}P^n$ by the rule that $s(x)$ is the line through x and the origin. Show that s is a submersion.

(6) Show that there is a continuous map $f : \mathbb{R}^2 \to \mathbb{R}^2$ such that $f(B(0, 1)) \subset B(0, 1)$, $f(\mathbb{R}^2 \setminus B(0, 1)) \subset f(\mathbb{R}^2 \setminus B(0, 1))$ and $f_{\partial B(0,1)} = \mathrm{id}_{\partial B(0,1)}$ and with the properties that f is C^∞ on $\overline{B}(0, 1)$ and on $\mathbb{R}^2 \setminus B(0, 1)$, while f is not C^∞ on $\mathbb{R}^2$.

(7) Construct an embedding of $\mathbb{R} \times S^n$ into $\mathbb{R}^{n+1}$.

(8) Embed the Stiefel manifold of k-frames in $\mathbb{R}^n$ into a Euclidean space $\mathbb{R}^N$ for some large N.

(9) Construct an embedding of $G(n,k)$ into $G(n,k+l)$ for each $l \geq 1$.

(10) Show that the map $f : \mathbb{R}P^2 \to \mathbb{R}^3$ defined by $f([x,y,z]) = (yz, xz, xy)$ is an immersion at all but six points $p \in \mathbb{R}P^2$. The image is called the Roman surface, and nice images can be found on the web. Show that it is a topological immersion (locally a topological embedding). Show that the map $g : \mathbb{R}P^2 \to \mathbb{R}^4$ given by $g([x,y,z]) = (yz, xz, xy, x^2 + 2y^2 + 3z^2)$ is a smooth embedding.

(11) Let $h : M \to \mathbb{R}^n$ be smooth and let $N \subset \mathbb{R}^n$ be a regular submanifold. Prove that for each $\varepsilon > 0$ there exists a $v \in \mathbb{R}^n$, with $|v| < \varepsilon$, such that the map $p \mapsto h(p) + v$ is transverse to N. (Think about the map $M \times N \to \mathbb{R}^n$ given by $(p, y) \mapsto y - f(p)$.)

(12) Define $\phi : S^1 \to \mathbb{R}$ by $e^{i\theta} \mapsto \theta$ for $0 \leq \theta < 2\pi$. Define $\lambda : \mathbb{R} \to S^1$ by $\theta \mapsto e^{i\theta}$. Show that λ is an immersion, that $\lambda \circ \phi$ is smooth, but that ϕ is not differentiable (it is not even continuous).

Chapter 4

Curves and Hypersurfaces in Euclidean Space

So far we have been studying manifold theory, which is foundational for modern differential geometry. In this chapter we change direction a bit to introduce some ideas from classical differential geometry. We do this for pedagogical reasons. The reader will see several ideas introduced here that will only be treated in generality later in the book. These ideas include parallelism, covariant derivative, metric and curvature. We concentrate on the geometry of one-dimensional submanifolds of $\mathbb{R}^n$ (geometric curves) and submanifolds of *co*dimension one in $\mathbb{R}^n$, which we refer to as **hypersurfaces**.[1] Recall that a vector in $T_p\mathbb{R}^n$ can be viewed as a pair $(p,v) \in \{p\} \times \mathbb{R}^n$, and we sometimes write v_p or $(v^1, \ldots, v^n)_p$ for (p,v). The element v is called the principal part of v_p. Recall also that in $\mathbb{R}^n$ we have a natural notion of what it means for vectors in different tangent spaces to be parallel. By definition, $v_p \in T_p\mathbb{R}^n$ is **parallel** to $w_q \in T_q\mathbb{R}^n$ if $v = w \in \mathbb{R}^n$. For any $w_p \in T_p\mathbb{R}^n$, there is a unique vector $w_q \in T_q\mathbb{R}^n$ which is parallel to w_p. We call w_q the parallel translate of w_p to the tangent space $T_q\mathbb{R}^n$. We often identify $T_p\mathbb{R}^n$ with $\mathbb{R}^n$. If $\mathsf{e}_1, \ldots, \mathsf{e}_n$ is the standard orthonormal basis for $\mathbb{R}^n$, then $\widehat{\mathsf{e}}_1, \ldots, \widehat{\mathsf{e}}_n$ will denote the **standard global frame field** on $\mathbb{R}^n$ defined by $\widehat{\mathsf{e}}_i(p) := \frac{\partial}{\partial x^i}\big|_p$ for $i = 1, \ldots, n$, where $x^1, \ldots, x^n$ are the standard coordinate functions on $\mathbb{R}^n$. The reason for this separate notation is to emphasize the fiducial role of this frame field. If $v_p \in T_p\mathbb{R}^n$ and $w_q \in T_q\mathbb{R}^n$, then v_p is

---

[1] Unless otherwise stated, we will assume that a hypersurface is a manifold without boundary.

143

parallel to w_q if $v^i = w^i$ for all i, where
$$v_p = \sum v^i \widehat{e}_i(p) \text{ and } w_q = \sum w^i \widehat{e}_i(q).$$

Recall that if $c : I \to M$ is a smooth curve into an n-manifold M, then a smooth vector field *along* c is a smooth map $Y : I \to TM$ such that $\pi \circ Y = c$. The velocity of the map Y is an element of $T(TM)$ rather than TM. On the other hand, if $c : I \to \mathbb{R}^n$, then the special structure of $\mathbb{R}^n$ allows us to take a derivative of a vector field along c and end up with another vector field along c. If Y is such a vector field along c, then it can be written $Y = \sum Y^i \widehat{e}_i \circ c$ for some smooth functions $Y^i : I \to \mathbb{R}$. In other words, $Y(t) = \sum Y^i(t) \widehat{e}_i(c(t))$ for $t \in I$. For such a field, we define
$$\frac{dY}{dt}(t) = Y'(t) = \sum \frac{dY^i}{dt}(t) \widehat{e}_i(c(t)).$$

In notation that keeps track of base points, we can write $Y = (Y^1, \ldots, Y^n)_c$ so that $Y(t) = (Y^1(t), \ldots, Y^n(t))_{c(t)}$, and then
$$\frac{dY}{dt}(t) = \left(\frac{d}{dt} Y^1(t), \ldots, \frac{d}{dt} Y^n(t) \right)_{c(t)}.$$

Since $\frac{dY}{dt}$ is obviously also a vector field along c, we can repeat the process to obtain higher derivatives $\frac{d^k Y}{dt^k} = Y^{(k)}$. In particular, we have the **velocity** c', **acceleration** c'', and higher derivatives $c^{(k)}$. Except for the current emphasis on the base point of vectors, these definitions are the usual definitions from multivariable calculus.

Another thing that is special about $\mathbb{R}^n$ is the availability of the natural inner product (the dot product) in every tangent space $T_p \mathbb{R}^n$. This inner product on $T_p \mathbb{R}^n$ is denoted $\langle \cdot, \cdot \rangle_p$ and is given by
$$(v_p, w_p) \mapsto \langle v_p, w_p \rangle = \sum v^i w^i,$$
where $v_p = (p, v)$ and $w_p = (p, w)$ as explained above. This gives what is called a **Riemannian metric** on $\mathbb{R}^n$, and it is obtained from the canonical identification of the tangent spaces with $\mathbb{R}^n$ itself. We study Riemannian metrics on general manifolds later in the book. For smooth vector fields on $\mathbb{R}^n$, say X and Y, the function $\langle X, Y \rangle$ defined by $p \mapsto \langle X_p, Y_p \rangle$ is smooth. If X and Y are fields along a curve $c : I \to \mathbb{R}^n$, then $\langle X, Y \rangle$ is a function on I and we clearly have
$$\frac{d}{dt} \langle X, Y \rangle = \left\langle \frac{dX}{dt}, Y \right\rangle + \left\langle X, \frac{dY}{dt} \right\rangle.$$

Remark 4.1. We shall sometimes leave out the subscript p in the notation v_p if the base point is understood from the context. In fact, we often just identify v_p with its principal part $v \in \mathbb{R}^n$.

An ordered basis $v_1, \ldots, v_n$ is positively oriented if $\det(v_1, \ldots, v_n)$ is positive. For $v_p, w_p \in T_p\mathbb{R}^3$, we can obviously define a cross product $v_p \times w_p := (v \times w)_p$, which results in a vector based at p which is then orthogonal to both v_p and w_p. If v_p and w_p are orthonormal, then $(v_p, w_p, (v \times w)_p)$ is a positively oriented (right-handed) orthonormal basis for $T_p\mathbb{R}^3$. The following lemma allows us to do something similar in higher dimensions.

Lemma 4.2. *If $v_1, \ldots, v_{n-1} \in \mathbb{R}^n$, then there is a unique vector $N(v_1, \ldots, v_{n-1})$ such that*

(i) $N(v_1, \ldots, v_{n-1})$ *is orthogonal to each of $v_1, \ldots, v_{n-1}$.*

(ii) *If $\{v_1, \ldots, v_{n-1}\}$ is an orthonormal set of vectors, then the list $(v_1, \ldots, v_{n-1}, N)$ is a positively oriented orthonormal basis.*

(iii) $N(v_1, \ldots, v_{n-1})$ *depends smoothly on $v_1, \ldots, v_{n-1}$.*

Proof. Let L be the linear functional defined by $L(v) := \det(v_1, \ldots, v_{n-1}, v)$. There is a unique $N \in \mathbb{R}^n$ such that $L(v) = \langle N, v \rangle$. Now (i), (ii) and (iii) follow from the properties of the determinant. $\square$

4.1. Curves

If C is a one-dimensional submanifold of $\mathbb{R}^n$ and $p \in C$, then there is a chart $(V, \mathbf{y})$ of C containing p such that $\mathbf{y}(V)$ is a connected open interval $I \subset \mathbb{R}$. The inverse map $\mathbf{y}^{-1} : I \to V \subset M$ is a **local parametrization**. Thus for local properties, we are reduced to studying curves into $\mathbb{R}^n$ which are embeddings of intervals. We can be even more general and study immersions. The idea is to extract information that is appropriately independent of the parametrization. If $\gamma : I \to \mathbb{R}^n$ and $c : J \to \mathbb{R}^n$ are curves with the same image, then we say that c is a positive reparametrization of γ if there is a smooth function $h : J \to I$ with $h' > 0$ such that $c = \gamma \circ h$. In this case, we say that γ and c have the same **sense** and provide the same **orientation** on the image. We assume that $\gamma : I \to \mathbb{R}^n$ has $\|\gamma'\| > 0$, which is the case of interest. Such a curve is called **regular**, which just means that the curve is an immersion.

Definition 4.3. If $\gamma : I \to \mathbb{R}^n$ is a regular curve, then $\mathbf{T}(t) := \gamma'(t)/\|\gamma'(t)\|$ defines the **unit tangent field** along γ. (Of course, we then have $\|\mathbf{T}\| = 1$.)

We have the familiar notion of the length of a curve defined on a closed interval $\gamma : [t_1, t_2] \to \mathbb{R}^n$:

$$L = \int_{t_1}^{t_2} \|\gamma'(t)\| \, dt.$$

One can define an **arc length function** for a curve $\gamma : I \to \mathbb{R}^n$ by choosing $t_0 \in I$ and, then defining

$$s = h(t) := \int_{t_0}^{t} \|\gamma'(\tau)\| \, d\tau.$$

Notice that s takes on negative values if $t < t_0$, so it does not always represent the length in the ordinary sense. If the curve is smooth and regular, then $h' = \|\gamma'(\tau)\| > 0$, and so by the inverse function theorem, h has a smooth inverse. We then have the familiar fact that if $c(s) = \gamma \circ h^{-1}(s)$, then $\|c'\|(s) := \|c'(s)\| = 1$ for all s. Curves which are parametrized in terms of arc length are referred to as **unit speed curves**. For a unit speed curve, $\frac{dc}{ds}(s) = \mathbf{T}(s)$. Since parametrization by arc length eliminates any component of acceleration in the direction of the curve, the acceleration must be due only to the shape of the curve.

Definition 4.4. Let $c : I \to \mathbb{R}^n$ be a unit speed curve. The vector-valued function

$$\boldsymbol{\kappa}(s) := \frac{d\mathbf{T}}{ds}(s)$$

is called the **curvature vector**. The function κ defined by

$$\kappa(s) := \|\boldsymbol{\kappa}(s)\| = \left\| \frac{d\mathbf{T}}{ds}(s) \right\|$$

is called the **curvature function**. If $\kappa(s) > 0$, then we also define the **principal normal**

$$\mathbf{N}(s) := \left\| \frac{d\mathbf{T}}{ds}(s) \right\|^{-1} \frac{d\mathbf{T}}{ds}(s),$$

so that $\frac{d\mathbf{T}}{ds} = \kappa \mathbf{N}$.

Let $\gamma : I \to \mathbb{R}^n$ be a regular curve. An **adapted orthonormal moving frame along** γ is a list $(\mathbf{E}_1, \ldots, \mathbf{E}_n)$ of smooth vector fields along γ such that $\mathbf{E}_1(t) = \gamma'(t)/\|\gamma'(t)\|$ and such that $(\mathbf{E}_1(t), \ldots, \mathbf{E}_n(t))$ is a basis of $T_{\gamma(t)}\mathbb{R}^n$ for each $t \in I$. Identifying $\mathbf{E}_1(t), \ldots, \mathbf{E}_n(t)$ with elements of $\mathbb{R}^n$ written as column vectors, we say that the orthonormal moving frame is positively oriented if

$$Q(t) = [\mathbf{E}_1(t), \ldots, \mathbf{E}_n(t)]$$

is an orthogonal matrix of determinant one for each t.

Definition 4.5. A moving frame $\mathbf{E}_1(t), \ldots, \mathbf{E}_n(t)$ along a curve $\gamma : I \to \mathbb{R}^n$ is a **Frenet frame** for γ if $\gamma^{(k)}(t)$ is in the span of $\mathbf{E}_1(t), \ldots, \mathbf{E}_k(t)$ for all t and $1 \leq k \leq n$.

As we have defined them, Frenet frames are *not* unique. However, under certain circumstances we may single out special Frenet frames. For example,

4.1. Curves

if $c : I \to \mathbb{R}^3$ is a unit speed curve with $\kappa > 0$, then the principal normal $\mathbf{N}$ is defined. By letting

$$\mathbf{B} = \mathbf{T} \times \mathbf{N}$$

we obtain a Frenet frame $\mathbf{T}, \mathbf{N}, \mathbf{B}$. It is an easy exercise to show that we obtain

$$\begin{aligned} \frac{d\mathbf{T}}{ds} &= & \kappa \mathbf{N} & \\ \frac{d\mathbf{N}}{ds} &= -\kappa \mathbf{T} & + & \tau \mathbf{B} \\ \frac{d\mathbf{B}}{ds} &= & -\tau \mathbf{N} & \end{aligned}$$

for some function τ called the **torsion**. This is the familiar form presented in many calculus texts. In matrix notation,

$$\frac{d}{ds}[\mathbf{T}, \mathbf{N}, \mathbf{B}] = [\mathbf{T}, \mathbf{N}, \mathbf{B}] \begin{bmatrix} 0 & -\kappa & 0 \\ \kappa & 0 & -\tau \\ 0 & \tau & 0 \end{bmatrix}.$$

Notice that for a regular curve, $\kappa \geq 0$, while τ may assume any real value. Another special feature of the frame $\mathbf{T}, \mathbf{N}, \mathbf{B}$ is that it is positively oriented. If γ is injective, then we can think of κ and τ as defined on the geometric image $\gamma(I)$. Thus, if $p = \gamma(s_0)$, then $\kappa(p)$ is defined to be equal to $\kappa(s_0)$.

Exercise 4.6. Let $c : I \to \mathbb{R}^3$ be a unit speed regular curve with $\kappa > 0$. Show that

$$\tau(s) = \frac{\langle (c'(s) \times c''(s)), c'''(s) \rangle}{\kappa(s)^2} \text{ for all } s \in I.$$

Exercise 4.7. Let $c : I \to \mathbb{R}^3$ be a unit speed curve and $s_0 \in I$. Show that we have a Taylor expansion of the form

$$\begin{aligned} \gamma(s) - \gamma(s_0) &= \left((s - s_0) - \frac{1}{6}(s - s_0)^3 \kappa^2(s_0) \right) \mathbf{T}(s_0) \\ &+ \left(\frac{1}{2}(s - s_0)^2 \kappa(s_0) + \frac{1}{6}(s - s_0)^3 \frac{d\kappa}{ds}(s_0) \right) \mathbf{N}(s_0) \\ &+ \left(\frac{1}{6}(s - s_0)^3 \kappa(s_0) \tau(s_0) \right) \mathbf{B}(s) + o((s - s_0)^3). \end{aligned}$$

We wish to generalize the special properties of the Frenet frame $\mathbf{T}, \mathbf{N}, \mathbf{B}$ to higher dimensions thereby obtaining a notion of a *distinguished* Frenet frame. For maximum generality, we do not assume that the curve is unit speed. A curve γ in $\mathbb{R}^n$ is called k-**regular** if $\{\gamma'(t), \gamma''(t), \ldots, \gamma^{(k)}(t)\}$ is a linearly independent set for each t in the domain of the curve. For an $(n-1)$-regular curve, the existence of a special orthonormal moving frame can be easily proved. One applies the Gram-Schmidt process: If $\mathbf{E}_1(t), \ldots, \mathbf{E}_k(t)$

are already defined for some $k < n-1$, then

$$\mathbf{E}_{k+1}(t) := c_k \left[\gamma^{(k+1)}(t) - \sum_{j=1}^{k} \left\langle \gamma^{(k+1)}(t), \mathbf{E}_j(t) \right\rangle \mathbf{E}_j(t) \right],$$

where c_k is a positive constant chosen so that $\|\mathbf{E}_{k+1}(t)\| = 1$. Inductively, this gives us $\mathbf{E}_1(t), \ldots, \mathbf{E}_{n-1}(t)$, and it is clear that the $\mathbf{E}_k(t)$ are all smooth. Now we choose $\mathbf{E}_n(t)$ to complete our frame by letting it be of unit length and orthogonal to $\mathbf{E}_1(t), \ldots, \mathbf{E}_{n-1}(t)$. By making one possible adjustment of sign on $\mathbf{E}_n(t)$ we obtain a moving frame that is positively oriented. In fact, $\mathbf{E}_n(t)$ is given by the construction of Lemma 4.2, from which it follows that $\mathbf{E}_n(t)$ is smooth in t. By construction we have a nice list of properties:

(1) For $1 \leq k \leq n$, the vectors $\mathbf{E}_1(t), \ldots, \mathbf{E}_k(t)$ have the same linear span as $\gamma'(t), \ldots, \gamma^{(k)}(t)$ so that there is an $n \times n$ upper triangular matrix function $U(t)$ such that

$$[\gamma'(t), \ldots, \gamma^{(n)}(t)] U(t) = [\mathbf{E}_1(t), \ldots, \mathbf{E}_n(t)].$$

(2) For $1 \leq k \leq n-1$, the vectors $\mathbf{E}_1(t), \ldots, \mathbf{E}_k(t)$ have the same orientation as $\gamma'(t), \ldots, \gamma^{(k)}(t)$. Thus $U(t)$ has diagonal elements which are all positive except possibly the last one.

(3) $(\mathbf{E}_1(t), \ldots, \mathbf{E}_n(t))$ is positively oriented as a basis of $T_{\gamma(t)} \mathbb{R}^n \cong \mathbb{R}^n$.

Exercise 4.8. Show that the moving frame we have constructed is the unique one with these properties.

We call a moving frame satisfying the above properties a **distinguished Frenet frame** along γ. For any orthonormal moving frame, the derivative of each $\mathbf{E}_i(t)$ is certainly expressible as a linear combination of the basis $\mathbf{E}_1(t), \ldots, \mathbf{E}_n(t)$, and so we may write

$$\frac{d}{dt} \mathbf{E}_j(t) = \sum_{i=1}^{n} \omega_{ij}(t) \mathbf{E}_i(t).$$

Of course, $\omega_{ij}(t) = \langle \mathbf{E}_i(t), \frac{d}{dt} \mathbf{E}_j(t) \rangle$, but since $\langle \mathbf{E}_i(t), \mathbf{E}_j(t) \rangle = \delta_{ij}$, we conclude that $\omega_{ij}(t) = -\omega_{ji}(t)$, i.e., the matrix $\omega(t) = [\omega_{ij}(t)]$ is antisymmetric. However, for a distinguished Frenet frame, more is true. Indeed, if $(\mathbf{E}_1(t), \ldots, \mathbf{E}_n(t))$ is such a distinguished Frenet frame, then for $1 \leq j < n$ we have $\mathbf{E}_j(t) = \sum_{k=1}^{j} U_{kj} \gamma^{(k)}(t)$, where $U(t) = [U_{kj}(t)]$ is the upper triangular matrix mentioned above. Using the fact that U, $\frac{d}{dt} U$, and U^{-1} are all upper triangular, we have

$$\frac{d}{dt} \mathbf{E}_j(t) = \sum_{k=1}^{j} \left(\frac{d}{dt} U_{kj} \right) \gamma^{(k)}(t) + \sum_{k=1}^{j} U_{kj} \gamma^{(k+1)}(t).$$

4.1. Curves

But $\gamma^{(k+1)}(t) = \sum_{r=1}^{k+1} \left(U^{-1}\right)_{r,k+1} \mathbf{E}_r(t)$, and $\gamma^{(k)}(t) = \sum_{r=1}^{k} \left(U^{-1}\right)_{rk} \mathbf{E}_r(t)$ so that

$$\frac{d}{dt}\mathbf{E}_j(t) = \sum_{k=1}^{j} \left(\frac{d}{dt}U_{kj}\right) \gamma^{(k)}(t) + \sum_{k=1}^{j} U_{kj} \gamma^{(k+1)}(t)$$

$$= \sum_{k=1}^{j} \left(\frac{d}{dt}U_{kj}\right) \sum_{r=1}^{k} \left(U^{-1}\right)_{rk} \mathbf{E}_r(t)$$

$$+ \sum_{k=1}^{j} U_{kj} \sum_{r=1}^{k+1} \left(U^{-1}\right)_{r,k+1} \mathbf{E}_r(t).$$

From this we see that $\frac{d}{dt}\mathbf{E}_j(t)$ is in the span of $(\mathbf{E}_r(t))_{1 \le r \le j+1}$. Thus $\omega(t) = (\omega_{ij}(t))$ can have no nonzero entries below the subdiagonal. But ω is antisymmetric, so we conclude that ω has the form

$$\omega(t) = \begin{bmatrix} 0 & -\omega_{21}(t) & 0 & \cdots & 0 \\ \omega_{21}(t) & 0 & -\omega_{32}(t) & & \\ 0 & \omega_{32}(t) & 0 & & \\ & & & \ddots & \\ & & & 0 & -\omega_{n,n-1}(t) \\ & & & \omega_{n,n-1}(t) & 0 \end{bmatrix}.$$

We define the i-th **generalized curvature function** by

$$\kappa_i(t) := \frac{\omega_{i+1,i}(t)}{\|\gamma'(t)\|}.$$

Thus if $\gamma : I \to \mathbb{R}^n$ is a unit speed curve, we have

$$\omega(s) = \begin{bmatrix} 0 & -\kappa_1(s) & 0 & \cdots & 0 \\ \kappa_1(s) & 0 & -\kappa_2(s) & & \\ & \kappa_2(s) & 0 & & \\ & & & \ddots & \\ & & & 0 & -\kappa_{n-1}(s) \\ & & & \kappa_{n-1}(s) & 0 \end{bmatrix}.$$

Note: Our matrix ω is the transpose of the ω presented in some other expositions. The source of the difference is that we write a basis as a formal *row* matrix of vectors.

Lemma 4.9. *If $\gamma : I \to \mathbb{R}^n$ ($n \ge 3$) is $(n-1)$-regular, then for $1 \le i \le n-2$, the generalized curvatures κ_i are positive.*

Proof. By construction, for $1 \leq i \leq n-1$

$$\mathbf{E}_i(t) = \sum_{j=1}^{i} U_{ji}(t)\gamma^{(j)}(t),$$

$$\gamma^{(i)}(t) = \sum_{j=1}^{i} (U)_{ji}^{-1}(t)\mathbf{E}_j(t),$$

with $U_{ii} > 0$, and hence $(U^{-1})_{ii} > 0$. Thus if $1 \leq i \leq n-2$, we have

$$\omega_{i+1,i}(t) = \left\langle \mathbf{E}_{i+1}, \frac{d}{dt}\mathbf{E}_i \right\rangle = \left\langle \mathbf{E}_{i+1}, \frac{d}{dt} \sum_{j=1}^{i} U_{ji}(t)\gamma^{(j)}(t) \right\rangle$$

$$= \left\langle \mathbf{E}_{i+1}, \sum_{j=1}^{i} \gamma^{(j)}(t)\frac{d}{dt}U_{ji}(t) \right\rangle + \left\langle \mathbf{E}_{i+1}, \sum_{j=1}^{i} U_{ji}(t)\gamma^{(j+1)}(t) \right\rangle$$

$$= U_{ii} \left\langle \mathbf{E}_{i+1}(t), \gamma^{(i+1)}(t) \right\rangle = U_{ii}\left(U^{-1}\right)_{i+1,i+1} > 0.$$

In passing from the second to the third line above, we have used the fact that $\mathbf{E}_{i+1}$ is orthogonal to all $\gamma^{(j)}$ for $j \leq i$ since these are in the span of $\{\mathbf{E}_j\}_{j=1,\ldots,i}$. $\square$

The last generalized curvature function κ_{n-1} is sometimes called the **torsion**. It may take on negative values.

Exercise 4.10. If $\gamma : I \to \mathbb{R}^n$ is $(n-1)$-regular, show that $\mathbf{E}_1 = \mathbf{T}$ and $\mathbf{E}_2 = \mathbf{N}$. If γ is parametrized by arc length, then $\frac{d}{ds}\gamma = \mathbf{E}_1$ and $\frac{d^2}{ds^2}\gamma = \kappa_1\mathbf{E}_2$. Conclude that $\kappa_1 = \kappa$ (the curvature defined earlier).

The orthogonal group $\mathrm{O}(\mathbb{R}^n)$ is the group of linear transformations $A : \mathbb{R}^n \to \mathbb{R}^n$ such that $\langle Av, Aw \rangle = \langle v, w \rangle$ for all $v, w \in \mathbb{R}^n$. The group $\mathrm{O}(\mathbb{R}^n)$ is identified with the group of orthogonal $n \times n$ matrices denoted $\mathrm{O}(n)$. The Euclidean group $\mathrm{Euc}(\mathbb{R}^n)$ is generated by translations and elements of $\mathrm{O}(\mathbb{R}^n)$. Every element $\phi \in \mathrm{Euc}(\mathbb{R}^n)$ can be represented by a pair (A, b), where $A \in \mathrm{O}(\mathbb{R}^n)$ and $b \in \mathbb{R}^n$ and where $\phi(v) = Av + b$. Note that in this case, $D\phi = A$ (the derivative of ϕ is A). The elements of the Euclidean group are called Euclidean motions or isometries of $\mathbb{R}^n$. If $\phi \in \mathrm{Euc}(\mathbb{R}^n)$, then for each $p \in \mathbb{R}^n$, the tangent map $T_p\phi : T_p\mathbb{R}^n \to T_{\phi(p)}\mathbb{R}^n$ is a linear isometry. In other words, $\langle T_p\phi \cdot v_p, T_p\phi \cdot w_p \rangle_{f(p)} = \langle v_p, w_p \rangle_p$ for all $v_p, w_p \in T_p\mathbb{R}^n$. The group $\mathrm{SO}(\mathbb{R}^n)$ is the special linear group on $\mathbb{R}^n$ and consists of the elements of $\mathrm{O}(\mathbb{R}^n)$ which preserve orientation. The corresponding matrix group is $\mathrm{SO}(n)$ and is the subgroup of $\mathrm{O}(n)$ consisting of elements of determinant 1. The subgroup $S\,\mathrm{Euc}(\mathbb{R}^n) \subset \mathrm{Euc}(\mathbb{R}^n)$ is the group generated by translations and elements of $\mathrm{SO}(\mathbb{R}^n)$. It is called the **special Euclidean group**.

4.1. Curves

Theorem 4.11. *Let $\gamma : I \to \mathbb{R}^n$ and $\widetilde{\gamma} : I \to \mathbb{R}^n$ be two $(n-1)$-regular curves with corresponding curvature functions κ_i and $\widetilde{\kappa}_i$ $(1 \leq i \leq n-1)$. If $\|\gamma'(t)\| = \|\widetilde{\gamma}'(t)\|$ and $\kappa_i(t) = \widetilde{\kappa}_i(t)$ for all $t \in I$ and $1 \leq i \leq n-1$, then there exists a unique isometry $\phi \in S\operatorname{Euc}(\mathbb{R}^n)$ such that*

$$\widetilde{\gamma} = \phi \circ \gamma.$$

Proof. Let $(\mathbf{E}_1(t), \ldots, \mathbf{E}_n(t))$ and $(\widetilde{\mathbf{E}}_1(t), \ldots, \widetilde{\mathbf{E}}_n(t))$ be the distinguished Frenet frames for γ and $\widetilde{\gamma}$ respectively and let ω_{ij} and $\widetilde{\omega}_{ij}$ be the corresponding matrix elements as above. Fix $t_0 \in I$ and consider the unique isometry ϕ represented by (A, b) such that $\phi(\gamma(t_0)) = \widetilde{\gamma}(t_0)$ and such that

$$A(\mathbf{E}_i(t_0)) = \widetilde{\mathbf{E}}_i(t_0) \text{ for } 1 \leq i \leq n.$$

Since $\|\gamma'(t)\| = \|\widetilde{\gamma}'(t)\|$ and $\kappa_i(t) = \widetilde{\kappa}_i(t)$, we have that $\omega_{ij}(t) = \widetilde{\omega}_{ij}(t)$ for all i, j and t. Thus we have both

$$\frac{d}{dt}\widetilde{\mathbf{E}}_i(t) = \sum_{j=1}^n \omega_{ji}(t)\widetilde{\mathbf{E}}_j(t),$$

and

$$\frac{d}{dt}A\mathbf{E}_i(t) = \sum_{j=1}^n \omega_{ji}(t) A\mathbf{E}_j(t).$$

Hence $\widetilde{\mathbf{E}}_i$ and $A\mathbf{E}_i$ satisfy the same linear differential equation, and since $A(\mathbf{E}_i(t_0)) = \widetilde{\mathbf{E}}_i(t_0)$, we conclude that $A(\mathbf{E}_i(t)) = \widetilde{\mathbf{E}}_i(t)$ for all t and $1 \leq i \leq n$. In particular, $A\gamma'(t) = \|\gamma'(t)\| A\mathbf{E}_1(t) = \|\widetilde{\gamma}'(t)\| \widetilde{\mathbf{E}}_1(t) = \widetilde{\gamma}'(t)$. Thus

$$\phi(\gamma(t)) - \phi(\gamma(t_0)) = \int_{t_0}^t (\phi \circ \gamma)'(\tau)\, d\tau = \int_{t_0}^t D\phi \cdot \gamma'(\tau)\, d\tau$$
$$= \int_{t_0}^t A\gamma'(\tau)\, d\tau = \int_{t_0}^t \widetilde{\gamma}'(\tau)\, d\tau = \widetilde{\gamma}(t) - \widetilde{\gamma}(t_0),$$

from which we conclude that $\phi(\gamma(t)) = \widetilde{\gamma}(t)$.

For uniqueness, we argue as follows. Suppose that $\psi \circ \gamma = \widetilde{\gamma}$ for $\psi \in \operatorname{Euc}(\mathbb{R}^n)$ and suppose that ψ is represented by (B, c). The fact that $D\psi$ must take the Frenet frame of γ to that of $\widetilde{\gamma}$ means that $A = D\psi = D\phi = B$. The fact that $\psi(\gamma(t_0)) = \widetilde{\gamma}(t_0)$ implies that $b = c$ and so $\psi = \phi$. $\square$

Conversely, we have

Theorem 4.12. *If $\kappa_1, \ldots, \kappa_{n-1}$ are smooth functions on a neighborhood of $s_0 \in \mathbb{R}$ such that $\kappa_i > 0$ for $i < n-1$, then there exists an $(n-1)$-regular unit speed curve γ defined on some interval containing s_0 such that $\kappa_1, \ldots, \kappa_{n-1}$ are the curvature functions of γ.*

Proof. We merely sketch the proof: Let

$$A(s) := \begin{bmatrix} 0 & -\kappa_1 & 0 & \cdots & 0 \\ \kappa_1 & 0 & & & \vdots \\ 0 & & \ddots & & 0 \\ \vdots & & & 0 & -\kappa_{n-1} \\ 0 & \cdots & 0 & \kappa_{n-1} & 0 \end{bmatrix}$$

and consider the matrix initial value problem

$$X' = XA,$$
$$X(s_0) = I.$$

This has a unique smooth solution X on some interval I which contains s_0. The skew-symmetry of A implies that $X(s)$ is orthogonal for all $s \in I$. If we let $\mathbf{x}_1$ be the first column of X, then

$$\gamma(s) := \int_{s_0}^{s} \mathbf{x}_1(t) \, dt$$

defines a unit speed $(n-1)$-regular curve with the required curvature functions. □

Exercise 4.13. Fill in the details of the previous proof.

If $n > 3$, then a regular curve γ need not have a Frenet frame. However, a regular curve still has a curvature function, and if the curve is 2-regular, then we have a **principal normal N** which is defined so that **T** and **N** are the Gram-Schmidt orthogonalization of γ' and γ''. We will sometimes denote this principal normal by $\mathbf{E}_2$ in order to avoid confusion with the normal to a hypersurface, which is denoted below by N.

4.2. Hypersurfaces

Suppose that Y is vector field on an open set in $\mathbb{R}^n$. For p in the domain of Y, and $v_p \in T_p \mathbb{R}^n$, let c be a curve with $c(0) = p$ and $\dot{c}(0) = v_p$. For any t near 0, we can look at the value of Y at $c(t)$. We let $\overline{\nabla}_{v_p} Y := (Y \circ c)'(0)$, which is defined since $Y \circ c$ is a vector field along c. Note that in this context $(Y \circ c)'(0)$ is taken to be based at $p = c(0)$. If X is a vector field, then a vector field $\overline{\nabla}_X Y$ is given by $\overline{\nabla}_X Y : p \mapsto \overline{\nabla}_{X_p} Y$. In fact, it is easy to see that if $X = \sum X^i \widehat{\mathbf{e}}_i$ and $Y = \sum Y^i \widehat{\mathbf{e}}_i$, then

$$\overline{\nabla}_X Y = \sum (XY^i) \widehat{\mathbf{e}}_i$$

since $(XY^i)(p) = X_p Y^i = \frac{d}{dt}\big|_0 Y^i \circ c$. We have presented things as we have because we wish to prime the reader for the general concept of a covariant derivative that we will meet in later chapters. However, it must be confessed

4.2. Hypersurfaces

that under the canonical identification of $\mathbb{R}^n$ with each tangent space, $\overline{\nabla}_{X_p} Y$ is just the directional derivative of Y in the direction X_p.

The map $(X, Y) \mapsto \overline{\nabla}_X Y$ is $C^\infty(\mathbb{R}^n)$-linear in X but not in Y. Rather, it is $\mathbb{R}$-linear in Y and we have a product rule:

$$\overline{\nabla}_X fY = (Xf) Y + f \overline{\nabla}_X Y.$$

Because of these properties, the operator $\overline{\nabla}_X : Y \mapsto \overline{\nabla}_X Y$, which is given for any X, is called a **covariant derivative** (or **Koszul connection**). In Chapter 12 we study covariant derivatives in a more general context. Since we shall soon consider covariant derivatives on submanifolds, let us refer to $\overline{\nabla}$ as the **ambient** covariant derivative. Notice that we have $\overline{\nabla}_X Y - \overline{\nabla}_Y X = [X, Y]$.

There is another property that our ambient covariant derivative $\overline{\nabla}$ satisfies. Namely, it respects the metric:

$$X \langle Y, Z \rangle = \langle \overline{\nabla}_X Y, Z \rangle + \langle Y, \overline{\nabla}_X Z \rangle.$$

Similarly, if $v_p \in T_p \mathbb{R}^n$, then $v_p \langle Y, Z \rangle = \langle \overline{\nabla}_{v_p} Y, Z_p \rangle + \langle Y_p, \overline{\nabla}_{v_p} Z \rangle$.

Consider a hypersurface M in $\mathbb{R}^n$. By definition, M is a regular $(n-1)$-dimensional submanifold of $\mathbb{R}^n$. (If $n = 3$, then such a submanifold has dimension two and we also just refer to it as a *surface in* $\mathbb{R}^3$.) A vector field **along** an open set $O \subset M$ is a map $X : O \to T\mathbb{R}^n$ such that the following diagram commutes:

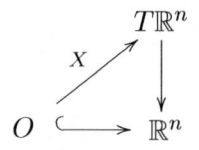

$$\begin{array}{ccc} & & T\mathbb{R}^n \\ & \overset{X}{\nearrow} & \downarrow \\ O & \hookrightarrow & \mathbb{R}^n \end{array}$$

Here the horizontal map is inclusion of O into $\mathbb{R}^n$. If $X(p) \in T_p M$ for all $p \in O$, then X is nothing more than a tangent vector field on O. If N is a field along O such that $\langle N(p), v_p \rangle = 0$ for all $v_p \in T_p M$ and all $p \in O$, then we call N a (smooth) **normal field**. If N is a normal field such that $\langle N(p), N(p) \rangle = 1$ for all $p \in O$, then N is called a **unit normal field**. Note that because of examples such as embedded Möbius bands in $\mathbb{R}^3$, it is not always the case that there exists a globally defined smooth unit normal field.

Definition 4.14. A hypersurface M in $\mathbb{R}^n$ is called **orientable** if there exists a smooth global unit normal vector field N defined along M. We say that M is **oriented** by N.

We will come to a more general and sophisticated notion of orientable manifold later. That definition will be consistent with the one above.

Exercise 4.15. Show that for a connected orientable hypersurface, there are exactly two choices of unit normal vector field.

In this chapter, we study mainly *local* geometry. We focus attention near a point $p \in M$. We consider a chart $(\widetilde{O}, \widetilde{\mathbf{u}})$ for $\mathbb{R}^n$ that is a single-slice chart centered at p and adapted to M. Thus if $\widetilde{\mathbf{u}} = (\widetilde{u}^1, \ldots, \widetilde{u}^n)$, then the restrictions of the functions $\widetilde{u}^1, \ldots, \widetilde{u}^{n-1}$ give coordinates for M on the set $O = \widetilde{O} \cap M = \{\widetilde{u}^n = 0\}$. We denote these restrictions by $u^1, \ldots, u^{n-1}$. Thus if we write $\mathbf{u} := (u^1, \ldots, u^{n-1})$, then $(O, \mathbf{u})$ is a chart on M. We may further arrange that $\widetilde{\mathbf{u}}(\widetilde{O})$ is a cube centered at the origin in $\mathbb{R}^n$ and so, in particular, O is connected and orientable. Let us temporarily call such charts **special**. The coordinate vector fields $\frac{\partial}{\partial \widetilde{u}^i}$ for $i = 1, \ldots, n$ are defined on $\widetilde{O}$, while the vector fields $\frac{\partial}{\partial u^i}$ are defined on O. We have

$$\frac{\partial}{\partial \widetilde{u}^i}(p) = \frac{\partial}{\partial u^i}(p) \text{ for all } p \in O \text{ and } i = 1, \ldots, n-1.$$

If $\widetilde{X}$ is a vector field on an open set $\widetilde{O}$, then its restriction to $O = \widetilde{O} \cap M$ is a vector field *along* O, which certainly need not be tangent to M. If X is a vector field along O, then there must be smooth functions X^i on O such that

$$X(p) = \sum_{i=1}^{n} X^i(p) \frac{\partial}{\partial \widetilde{u}^i}(p)$$

for all $p \in O$. Then X is a *tangent* vector field on O precisely when the last component X^n is identically zero on O so that

$$X = \sum_{i=1}^{n-1} X^i \frac{\partial}{\partial u^i}.$$

Now if X is a field *along* M, then it can be extended to a field $\widetilde{X}$ on $\widetilde{O} \subset \mathbb{R}^n$ by considering the component functions X^i as functions on $\widetilde{O}$ which happen to be constant with respect to the last coordinate variable $\widetilde{u}^n$. In other words, if $\pi : \widetilde{O} \to O$ is the map $(a^1, \ldots, a^n) \mapsto (a^1, \ldots, a^{n-1}, 0)$, then

$$\widetilde{X} = \sum_{i=1}^{n} \widetilde{X}^i \frac{\partial}{\partial \widetilde{u}^i},$$

where $\widetilde{X}^i = X^i \circ \widetilde{\mathbf{u}}^{-1} \circ \pi \circ \widetilde{\mathbf{u}}$. This last composition makes good sense because $(\widetilde{O}, \widetilde{\mathbf{u}})$ is a special single-slice chart as described above. We will refer to $\widetilde{X}$ as an extension of X, but note that the extension is based on a particular special choice of single-slice chart. The constructions on the hypersurface that we consider below do not depend on the extension.

Given a choice of unit normal N along a neighborhood of $p \in M$, $\overline{\nabla}_{v_p} N$ is defined for any $v_p \in T_p M$ by virtue of the fact that N only needs to be defined along a curve with tangent v_p. Alternatively, we can define $\overline{\nabla}_{v_p} N$ to be equal to $\overline{\nabla}_{v_p} \widetilde{N}$ for an extension $\widetilde{N}$ of N in a special single-slice chart

4.2. Hypersurfaces

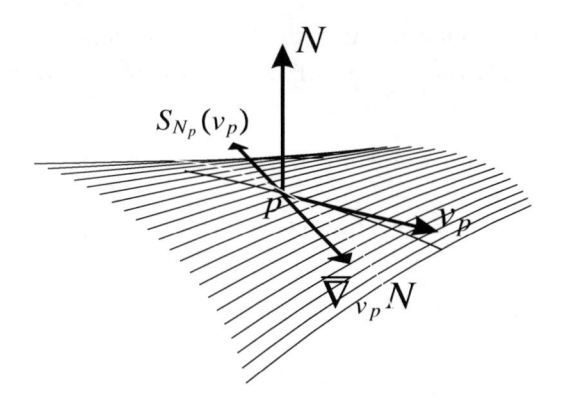

Figure 4.1. Shape operator

adapted to M and containing p. We note that $\langle \overline{\nabla}_{v_p} N, N(p) \rangle = 0$. Indeed, since $\langle N, N \rangle \equiv 1$, we have

$$0 = v_p \langle N, N \rangle = 2 \langle \overline{\nabla}_{v_p} N, N \rangle.$$

Thus $\overline{\nabla}_{v_p} N \in T_p M$ since $T_p M$ is exactly the set of vectors in $T_p \mathbb{R}^n$ perpendicular to $N_p = N(p)$.

Exercise 4.16. Suppose we are merely given a unit normal vector at p. Show that we can extend it to a smooth normal field near p. For dimensional reasons, any two such extensions must agree on some neighborhood of p.

Definition 4.17. Given a choice of unit normal N_p at $p \in M$, the map $S_{N_p} : T_p M \to T_p M$ defined by

$$S_{N_p}(v_p) := -\overline{\nabla}_{v_p} N$$

for any local unit normal field N with $N_p = N(p)$, is called the **shape operator** or **Weingarten map** at p.

From the definitions it follows that if c is a curve with $c(t_0) = p$ and $\dot{c}(t_0) = v_p$, then $S_{N_p}(v_p) = -(N \circ c)'(t_0)$. If $O \subset M$ is an open set oriented by a choice of unit normal field N along O, then we obtain a map $S_N : TO \to TO$ by $S_N|_{T_p M} := S_{N_p}$. This map is also called the shape operator (on O).

We have the inner product $\langle \cdot, \cdot \rangle_p$ on each tangent space $T_p \mathbb{R}^n$ and $T_p M \subset T_p \mathbb{R}^n$. For each $p \in M$, the restriction of this inner product to each tangent space $T_p M$ is also denoted by $\langle \cdot, \cdot \rangle_p$ or just $\langle \cdot, \cdot \rangle$. We also denote this inner product on $T_p M$ by g_p so that $g_p(\cdot, \cdot) = \langle \cdot, \cdot \rangle_p$. The map $p \mapsto g_p$ is smooth in the sense that $p \mapsto g_p(X(p), Y(p))$ is smooth whenever X and Y are smooth vector fields on M. Such a smooth assignment of inner product to the tangent spaces of M provides a **Riemannian metric** on M, a concept studied in more generality in later chapters. We denote the function $p \mapsto g_p(X(p), Y(p))$ simply by $g(X, Y)$ or $\langle X, Y \rangle$. In short, a Riemannian metric is a smooth assignment of an inner product to each tangent space.

Definition 4.18. A diffeomorphism $f : M_1 \to M_2$ between hypersurfaces in $\mathbb{R}^n$ is called an **isometry** if $T_p f : T_p M_1 \to T_{f(p)} M_2$ is an isometry of inner product spaces for all $p \in M_1$. In this case, we say that M_1 is isometric to M_2.

Proposition 4.19. $S_{N_p} : T_p M \to T_p M$ is self-adjoint with respect to $\langle \cdot, \cdot \rangle_p$.

Proof. Let $(\widetilde{O}, \widetilde{y})$ be a special chart centered at p and let $O = \widetilde{O} \cap M$ as above. Let X_p and Y_p be elements of $T_p M$ and extend them to vector fields X and Y on O. Then extend X and Y to fields $\widetilde{X}$ and $\widetilde{Y}$ on $\widetilde{O}$. Similarly, extend N_p to N and then $\widetilde{N}$. Note that $X_p \langle \widetilde{N}, \widetilde{X} \rangle = X_p \langle N, X \rangle = 0$ and $Y_p \langle \widetilde{N}, \widetilde{Y} \rangle = 0$. Using this, we have

$$-\langle S_{N_p} X_p, Y_p \rangle + \langle X_p, S_{N_p} Y_p \rangle$$
$$= \langle \overline{\nabla}_{X_p} N, Y_p \rangle - \langle X_p, \overline{\nabla}_{Y_p} N \rangle = \langle \overline{\nabla}_{\widetilde{X}} \widetilde{N}, \widetilde{Y} \rangle_p - \langle \widetilde{X}, \overline{\nabla}_{\widetilde{Y}} \widetilde{N} \rangle_p$$
$$= X_p \langle \widetilde{N}, \widetilde{Y} \rangle - \langle \widetilde{N}, \overline{\nabla}_{\widetilde{X}} \widetilde{Y} \rangle_p - Y_p \langle \widetilde{N}, \widetilde{X} \rangle + \langle \widetilde{N}, \overline{\nabla}_{\widetilde{Y}} \widetilde{X} \rangle_p$$
$$= \langle \overline{\nabla}_{\widetilde{Y}} \widetilde{X} - \overline{\nabla}_{\widetilde{X}} \widetilde{Y}, \widetilde{N} \rangle_p = \langle [\widetilde{Y}, \widetilde{X}], \widetilde{N} \rangle_p$$
$$= \langle [\widetilde{Y}, \widetilde{X}](p), \widetilde{N}(p) \rangle = \langle [Y, X](p), N(p) \rangle = 0$$

since $[\widetilde{Y}, \widetilde{X}](p) = [Y, X](p) \in T_p M$ by Proposition 2.84 applied to the inclusion map. $\square$

Definition 4.20. The symmetric bilinear form II_p on $T_p M$ defined by

$$II_p(v_p, w_p) := \langle v_p, S_{N_p} w_p \rangle = \langle S_{N_p} v_p, w_p \rangle$$

is called the **second fundamental form** at p. If N is a unit normal on an open subset of M, then for smooth tangent vector fields X, Y, the function $II(X, Y)$ defined by $p \mapsto II_p(X_p, Y_p)$ is smooth. The assignment $(X, Y) \mapsto II(X, Y)$ defined on pairs of tangent vector fields is also called the second fundamental form.

The form II is bilinear over $C^\infty(O)$, where O is the domain of N. As we shall see, the shape operator can be recovered from the second fundamental form II together with the metric $g = \langle \cdot, \cdot \rangle$ (first fundamental form). If the reader keeps the definition in mind, he or she will recognize that the second fundamental form appears implicitly in much that follows. We return to the second fundamental form again explicitly later.

Exercise 4.21. Let $X_p \in T_p \mathbb{R}^n$ and let Y be any smooth vector field on $\mathbb{R}^n$. Show that if $f : \mathbb{R}^n \to \mathbb{R}^n$ is a Euclidean motion, then we have

4.2. Hypersurfaces

$Tf \cdot \overline{\nabla}_{X_p} Y = \overline{\nabla}_{Tf \cdot X_p} f_* Y$. More generally, show that this is true if f is affine (i.e. if f is of the form $f(x) = Ax + b$ for some linear map A and $b \in \mathbb{R}^n$).

Theorem 4.22. *If M_1 is a connected hypersurface in $\mathbb{R}^n$ and if $f : \mathbb{R}^n \to \mathbb{R}^n$ is a Euclidean motion, then $M_2 = f(M_1)$ is also a hypersurface and*

(i) *the induced map $f|_{M_1} : M_1 \to M_2$ is an isometry;*

(ii) *if M_1 and M_2 are oriented by unit normals N_1 and N_2 respectively, then, after replacing N_1 or N_2 by its negative if necessary, we have*

$$Tf \circ S_{N_1} = S_{N_2} \circ Tf.$$

Proof. That $f(M_1)$ is a hypersurface is an easy exercise, which we leave to the reader. After noticing that $f|_{M_1} : M_1 \to M_2$ is smooth (why?), we argue as follows: Let $\phi := f|_{M_1}$ and notice that $T\phi \cdot v = Tf \cdot v$ for all v tangent to M_1. Since Tf preserves the inner products on $T\mathbb{R}^n$, we see that $T_p\phi$ is an isometry for each $p \in M_1$. Since ϕ is clearly a bijection we conclude that (i) holds.

Note that we must have $T_p f \cdot N_1(p) = \pm N_2(p)$ and, since M_1 is connected, one possible change of sign on unit normals gives

$$Tf \circ N_1 \circ f^{-1} = N_2.$$

Then by Exercise 4.21

$$Tf \cdot S_{N_1} v = -Tf \cdot \overline{\nabla}_v N_1 = -\overline{\nabla}_{Tf \cdot v} f_* N_1 = -\overline{\nabla}_{Tf \cdot v} N_2 = S_{N_2}(Tf \cdot v),$$

or $Tf \circ S_{N_1} = S_{N_2} \circ Tf$. □

Notice that an arbitrary isometry $M_1 \to M_2$ need *not* be the restriction of an isometry of the ambient $\mathbb{R}^n$ and need not preserve shape operators.

Definition 4.23. Let M be a hypersurface in $\mathbb{R}^n$, take a point $p \in M$, and let N_p be a unit normal at p. The **mean curvature** $H(p)$ at p in the direction N_p is defined by

$$H(p) := \frac{1}{n-1} \operatorname{trace}(S_{N_p}).$$

Notice that changing N_p to $-N_p$ changes $H(p)$ to $-H(p)$. If N is a unit normal field along an open set $U \subset M$, then the function $p \mapsto H(p)$ is smooth. It is called a **mean curvature function**. If M is an orientable hypersurface, then it has a global mean curvature function for each unit normal field.

Definition 4.24. Let M be a hypersurface in $\mathbb{R}^n$. Let a unit normal N_p be given at p. The **Gauss curvature** $K(p)$ at p is defined by

$$K(p) := \det(S_{N_p}).$$

If N is a unit normal field along an open set $U \subset M$, then the function $p \mapsto K(p) = \det(S_{N(p)})$ is smooth. It is called a Gauss curvature function associated to the normal field, and always exists locally. If M is orientable, then it has a global Gauss curvature function for every choice of unit normal field. Notice that changing N_p to $-N_p$ changes $K(p)$ to $(-1)^{n-1}K(p)$. It follows that if $n-1$ is even, then there is a unique global Gauss curvature function regardless of whether M is orientable or not. In particular, this is the case for surfaces in $\mathbb{R}^3$.

The shape operator S_{N_p} encodes the local geometry of the submanifold at p and measures the way M bends and twists through the ambient Euclidean space. Since S_{N_p} is self-adjoint, there is a basis for $T_p M$ consisting of eigenvectors of S_{N_p}. An eigenvector for S_{N_p} is called a **principal vector**, and a unit principal vector is called a **principal direction** or a **direction of curvature**. The eigenvalues are called **principal curvatures** at p. If $k_1, \ldots, k_{n-1}$ are the principal curvatures at p, then

$$H(p) = \frac{1}{n-1} \sum_{i=1}^{n-1} k_i \quad \text{and} \quad K(p) = \prod_{i=1}^{n-1} k_i.$$

If $u \in T_p M$ is a tangent vector with $\langle u, u \rangle = 1$, then $k(u) = \langle S_{N_p} u, u \rangle$ is the **normal curvature** in the direction u. Of course, if u is a unit length eigenvector (principal direction), then the corresponding principal curvature k is just the normal curvature in that direction. Notice that $k(-u) = k(u)$. A vector $v \in T_p M$ is called **asymptotic** if $\langle S_{N_p} v, v \rangle = 0$.

Proposition 4.25. *Let M be a hypersurface and N a unit normal field. Let $\gamma : I \to M \subset \mathbb{R}^n$ be a regular curve with image in the domain of N. Then $\langle S_{N_p} \gamma'(t), \gamma'(t) \rangle = \langle N(\gamma(t)), \gamma''(t) \rangle$.*

Proof. Since $\langle N(\gamma(t)), \gamma'(t) \rangle = 0$, differentiation gives

$$\left\langle N(\gamma(t)), \gamma''(t) \right\rangle + \left\langle \frac{d}{dt} N(\gamma(t)), \frac{d}{dt} \gamma(t) \right\rangle = 0.$$

Thus

$$\left\langle S_{N_p} \gamma'(t), \gamma'(t) \right\rangle = \left\langle -\overline{\nabla}_{\gamma'(t)} N, \gamma'(t) \right\rangle$$
$$= \left\langle -\frac{d}{dt} N(\gamma(t)), \frac{d}{dt} \gamma(t) \right\rangle = \left\langle N(\gamma(t)), \gamma''(t) \right\rangle. \qquad \square$$

In particular, if u is a unit vector at p and $u = \dot{c}(0)$ for some unit speed curve c, then $k(u) = \langle N_p, c''(0) \rangle$. This shows that all unit speed curves with a given velocity u have the same normal component given by the normal curvature in that direction. This curvature is forced by the shape of M and we see how normal curvatures measure the shape of M.

4.2. Hypersurfaces

Corollary 4.26. *Let $c : I \to M \subset \mathbb{R}^n$ be a unit speed curve. If $\kappa(s_0) = 0$ for some $s_0 \in I$, then $k(\dot{c}(s_0)) = 0$. If $\kappa(s) > 0$ and $\mathbf{E}_2$ is the principal normal defined near s, then $k(\dot{c}(s)) = \kappa(s)\cos\theta(s)$, where $\theta(s)$ is the angle between $N(c(s))$ and $\mathbf{E}_2(s)$.*

Proof. Suppose $\kappa(0) = 0$. Then $k(c'(0)) = \langle N(c(0)), c''(0)\rangle = 0$. If $\kappa(s) > 0$, then $\mathbf{E}_2$ is defined for an interval containing s. We have

$$k(c'(s)) = \langle N(c(s)), c''(s)\rangle = \langle N(c(s)), \kappa \mathbf{E}_2(c(s))\rangle$$
$$= \kappa(s)\langle N(c(s)), \mathbf{E}_2(s)\rangle = \kappa(s)\cos\theta(s). \qquad \square$$

If P is a 2-plane containing $p \in M$ and such that N_p is tangent to P, then for a small enough open neighborhood $\widetilde{O}$ of p, the set $C = \widetilde{O} \cap P \cap M$ is a regular one-dimensional submanifold of $\mathbb{R}^n$. We parameterize C by a unit speed curve c with $c(0) = p$. This curve c is called a **normal section** at p. Notice that in this case $E_2(0) = \pm N_p$. Then $k(\dot{c}(0)) = \pm\kappa_1(0)$ where the "$-$" sign is chosen in case $N_p = -E_2(0)$. Thus we see that $k(\dot{c}(0))$ is positive if the normal section c bends away from N_p.

Definition 4.27 (Curve types). Let M be a hypersurface in $\mathbb{R}^n$ and let $\gamma : I \to M$ be a regular curve.

(i) γ is called a **geodesic** if the acceleration $\gamma''(t)$ is normal to M for all $t \in I$.

(ii) γ is called a **principal curve** (or line of curvature) if $\dot{\gamma}(t)$ is a principal vector for all $t \in I$.

(iii) γ is called an **asymptotic curve** if $\dot{\gamma}(t)$ is an asymptotic vector for all $t \in I$.

Proposition 4.28. *Let M be a surface in $\mathbb{R}^3$. Suppose that a regular curve $\gamma : I \to M$ is contained in the intersection of M and a plane P. If the angle between M and P is constant along γ, then γ is a principal curve.*

Proof. The result is local, and so we assume that M is oriented by a unit normal N. Let ν be a unit normal along P. Since P is a plane, ν is constant. By assumption, $\langle N, \nu \rangle$ is constant along γ. Thus

$$0 = \frac{d}{dt}\langle N \circ \gamma, \nu \circ \gamma\rangle = \langle \overline{\nabla}_{\dot{\gamma}} N, \nu \circ \gamma\rangle,$$

so $\overline{\nabla}_{\dot{\gamma}} N$ is orthogonal to ν along γ. By the same token, $\overline{\nabla}_{\dot{\gamma}} N$ is orthogonal to N since $\langle N, N\rangle = 1$. Thus $\overline{\nabla}_{\dot{\gamma}} N$ must be collinear with $\dot{\gamma}$. In other words, $S_N \dot{\gamma} = -\overline{\nabla}_{\dot{\gamma}} N = \lambda \dot{\gamma}$ for some scalar function λ. $\qquad \square$

Example 4.29 (Surface of revolution). Let $t \mapsto (g(t), h(t))$ be a regular curve in $\mathbb{R}^2$ defined on an open interval I. Assume that $h > 0$. Call this curve the **profile curve**. Define $\mathbf{x} : I \times \mathbb{R} \to \mathbb{R}^3$ by

$$\mathbf{x}(u,v) = (g(u), h(u)\cos v, h(u)\sin v).$$

This is periodic in v and its image is a surface. The curves of constant u and curves of constant v are contained in planes of the form $\{x = c\}$ and $\{z = my\}$ (or $\{y = mz\}$) and so they are principal curves.

For a surface of revolution, the circles generated by rotating a fixed point of the profile are called **parallels**. These are the constant u curves in the example above. The curves that are copies of the profile curve are called the **meridians**. In the example above, the meridians are the constant v curves.

We know from standard linear algebra that principal directions at a point in a hypersurface corresponding to distinct principal curvatures are orthogonal. Generically, each eigenspace will be one-dimensional, but in general they may be of higher dimension. In fact, if S_{N_p} is a multiple of the identity operator, then there is only one eigenvalue and the eigenspace is all of $T_p M$. In this case, every direction is a principal direction. It may even be the case that $S_{N_p} = 0$.

If $c : I \to M$ is a unit speed curve into a hypersurface, then rather than use the distinguished Frenet frame of c, we can use a frame which incorporates the unit normal to the hypersurface:

Definition 4.30. Let $c : I \to M$ be a unit speed curve into a surface in $\mathbb{R}^3$ and let N be a unit normal defined at least on an open set containing the image of c. A frame field $(\mathbf{D}_1, \mathbf{D}_2, \mathbf{D}_3)$ (along c) such that $\mathbf{D}_1 = \mathbf{T} \ (= \dot{c})$, $\mathbf{D}_3 = N \circ c$ and $\mathbf{D}_2 = \mathbf{D}_3 \times \mathbf{D}_1$ is called a **Darboux frame**.

Exercise 4.31. Let $c : I \to M$ be a unit speed curve into a surface M in $\mathbb{R}^3$ and let $\mathbf{D}_1, \mathbf{D}_2, \mathbf{D}_3$ be an associated Darboux frame. Show that there exist smooth functions g_1, g_2 and g_3 such that

$$\begin{aligned} d\mathbf{D}_1/ds &= & g_1 \mathbf{D}_2 &+ g_2 \mathbf{D}_3, \\ d\mathbf{D}_2/ds &= -g_1 \mathbf{D}_1 & &+ g_3 \mathbf{D}_3, \\ d\mathbf{D}_3/ds &= -g_2 \mathbf{D}_1 &- g_3 \mathbf{D}_2. & \end{aligned}$$

Show that $g_1 = 0$ along c if and only if c is a geodesic. Show that $g_2 = 0$ along c if and only if c is asymptotic, and $g_3 = 0$ along c if and only if c is principal.

The function g_1 from the previous exercise is called the **geodesic curvature** function and is often denoted κ_g. Let us obtain a formula for κ_g. Define $J : T_p M \to T_p M$ by

$$Jv_p := N_p \times v_p.$$

4.2. Hypersurfaces

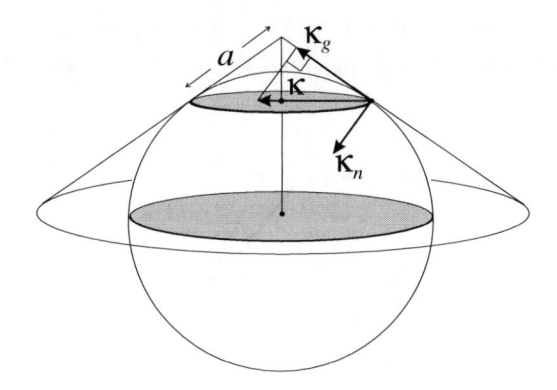

Figure 4.2. Curvature vectors

Notice that $\|Jv_p\| = \|v_p\|$ and so J is an isometry of the 2-dimensional inner product space $(T_pM, \langle \cdot, \cdot \rangle_p)$. But J also satisfies $J^2 = -\text{id}$, and in dimension 2 this determines J up to sign since it must be a rotation by $\pm \pi/2$. We now assume that N is globally defined so that the map J extends to a smooth map $TM \to TM$. For a unit speed curve $c : I \to M$, the geodesic curvature is given by

$$\kappa_g = \langle c'', Jc' \rangle = \left\langle \frac{d\mathbf{T}}{ds}, J\mathbf{T} \right\rangle. \tag{4.1}$$

Indeed, abbreviating $N \circ c$ to N we have from the first equation in Exercise 4.31 the following:

$$\frac{d\mathbf{T}}{ds} = c'' = \kappa_g N \times c' + g_2 N = \kappa_g J c' + g_2 N.$$

Taking inner products with Jc' gives formula (4.1).

Let $c : I \to M$ be a curve in a surface that is parametrized by arc length. The curvature vector $\boldsymbol{\kappa}(s)$ can be decomposed into a component $\boldsymbol{\kappa}_g(s)$ tangent to M and a component $\boldsymbol{\kappa}_n(s)$ normal to M:

$$\boldsymbol{\kappa}(s) = \boldsymbol{\kappa}_g(s) + \boldsymbol{\kappa}_n(s).$$

The curve will be a geodesic if and only if $\boldsymbol{\kappa}_g(s) = 0$ for all s. The vector $\boldsymbol{\kappa}_g(s)$ is the **geodesic curvature vector** at $c(s)$. It is easy to show that $\kappa_g = \pm \|\boldsymbol{\kappa}_g(s)\|$. Figure 4.2 depicts a sphere of radius R with a conical "hat". The cone intersects the sphere in a curve of latitude. Since the cone is tangent to the sphere, the vectors $\boldsymbol{\kappa}$, $\boldsymbol{\kappa}_g$ and $\boldsymbol{\kappa}_n$ apply equally well to both surfaces at least along the curve. Using the Pythagorean theorem and similar triangles, it is possible to show that $\|\boldsymbol{\kappa}_g\| = 1/a$, where a is the distance from the curve to the vertex of the cone. Now imagine cutting the cone along a generating line and unrolling it as shown in Figure 4.3. The result is a planar region with a circular arc of radius a and curvature whose magnitude is $1/a$. The reader might want to try and give a reason why this is to be expected. We will answer this later in this chapter.

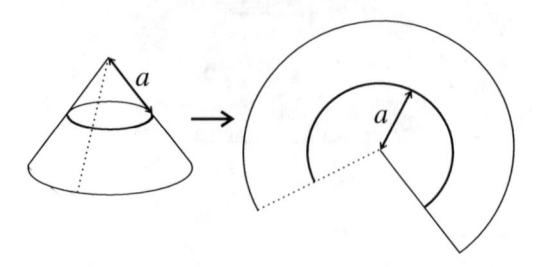

Figure 4.3. Unrolling a cone

Definition 4.32. If S_{N_p} is a multiple of the identity operator, then p is called an **umbilic point**. If $S_{N_p} = 0$, then p is called a **flat point**. If every point of a hypersurface is umbilic, then we say that the hypersurface is **totally umbilic**.

Example 4.33. If $a_1 x^1 + a_2 x^2 + \cdots + a_n x^n = 0$ is the equation of a hyperplane P in $\mathbb{R}^n$, then
$$N = \sum_{i=1}^n a_i \widehat{\mathbf{e}}_i$$
is a normal field when restricted to P. Since clearly $S_{N_p} = 0$ for all $p \in P$, we see that every point of P is flat and that the Gauss and mean curvatures are identically zero.

Example 4.34. Let S^{n-1} be the unit sphere in $\mathbb{R}^n$. Then the map $p = (a^1, \ldots, a^n) \mapsto N(p) := \sum_{i=1}^n a_i \widehat{\mathbf{e}}_i$ is a unit normal field along S^{n-1}. We can calculate $S_N(v) = -\overline{\nabla}_v N$. Let c be a curve in S^{n-1} with $\dot{c}(0) = v$. Then
$$-\overline{\nabla}_v N = -\left.\frac{d}{dt}\right|_{t=0} N(c(t))$$
$$= -\sum_{i=1}^n \left.\frac{dc^i}{dt}\right|_{t=0} \widehat{\mathbf{e}}_i = -v.$$

We are really just using the fact that up to a change in base point we have $N(c(t)) = c(t)$. Thus $S_N = -\operatorname{id}$ and we see that every point is umbilic and every principal curvature is unity. Also, $K = 1$ and the mean curvature $H = -1$ everywhere.

Exercise 4.35. What is the shape operator on a sphere of radius r?

Composition with a Euclidean motion preserves local geometry so if we want to study a hypersurface near a point p, then we may as well assume

4.2. Hypersurfaces

that p is the origin and that $N_p = \widehat{\mathsf{e}}_n(0) = \frac{\partial}{\partial x^n}\big|_0$. Locally, M is then the graph of a function $f : \mathbb{R}^{n-1} \to \mathbb{R}$ with $\frac{\partial f}{\partial x^i}(0) = 0$ for $i = 1, \ldots, n-1$. A normal field N which extends $N_0 = \widehat{\mathsf{e}}_n(0)$ is given by

$$\left(1 + \sum_{i=1}^{n-1} (\partial f/\partial x^i)^2\right)^{-1/2} \left(\widehat{\mathsf{e}}_n - \sum_{i=1}^{n-1} \frac{\partial f}{\partial x^i} \widehat{\mathsf{e}}_i\right).$$

Let $Z = \widehat{\mathsf{e}}_n - \sum_{i=1}^{n-1} \frac{\partial f}{\partial x^i} \widehat{\mathsf{e}}_i$ so that $N = gZ$, where g is the first factor in the formula above. For $v = \sum_{i=1}^{n-1} v^i \widehat{\mathsf{e}}_i$ tangent to M at the origin we have

$$\overline{\nabla}_v (gZ) = (vg) Z + g \overline{\nabla}_v Z,$$

where vg means that v acts on g as a derivation. But vg vanishes at the origin since g takes on a maximum there. Since $g(0) = 1$, we have $\overline{\nabla}_v (gZ) = \overline{\nabla}_v Z$. Thus we may compute S_{N_0} using Z:

$$S_{N_0} v = -\overline{\nabla}_v Z.$$

We have

$$S_{N_0} v = \sum_{i=1}^{n-1} v\left(\partial f/\partial x^i\right) \widehat{\mathsf{e}}_i(0)$$

$$= \sum_{i=1}^{n-1} \sum_{j=1}^{n-1} v^j \left.\frac{\partial f}{\partial x^i \partial x^j}\right|_0 \widehat{\mathsf{e}}_i(0).$$

We conclude that the shape operator at the origin is represented by the $(n-1) \times (n-1)$ matrix

$$[D^2 f](0) = \left[\frac{\partial f}{\partial x^i \partial x^j}(0)\right]_{1 \le i,j \le n-1}.$$

This is only valid at the origin. In other words, $[D^2 f]$ does not give us a representation of the shape operator except at the origin where M is tangent to $\mathbb{R}^{n-1}$. We can arrange, by rotating further if necessary, that $\mathsf{e}_1, \ldots, \mathsf{e}_{n-1}$ are directions of curvature. In this case, the above matrix is diagonal with the principal curvatures $k_1, \ldots, k_{n-1}$ down the diagonal.

Example 4.36. Let M be the graph of the function $f(x, y) = xy$. We look at $0 \in M$. We have

$$[D^2 f] = \begin{bmatrix} 0 & 1 \\ 1 & 0 \end{bmatrix},$$

which diagonalizes to $\begin{bmatrix} 1 & 0 \\ 0 & -1 \end{bmatrix}$ with corresponding eigenvectors $\frac{1}{\sqrt{2}}(\mathsf{e}_1 + \mathsf{e}_2)$ and $\frac{1}{\sqrt{2}}(\mathsf{e}_1 - \mathsf{e}_2)$. Thus the Gauss curvature at the origin is -1 and the mean curvature is 0. We can understand this graph geometrically. The normal sections created by intersecting the graph with the planes $y = 0$ and $x = 0$ are straight lines, and so the normal curvatures in those directions are zero. However, the normal section given by intersecting with the plane $y = x$ is concave up, while that created by the plane $y = -x$ is concave down.

Proposition 4.37. *Let $M \subset \mathbb{R}^3$ be a surface. If $v_1, v_2 \in T_pM$ are linearly independent, then for a given unit normal N_p we have*

$$S_{N_p}v_1 \times S_{N_p}v_2 = K(p)(v_1 \times v_2),$$
$$S_{N_p}v_1 \times v_2 + v_1 \times S_{N_p}v_2 = 2H(p)(v_1 \times v_2).$$

Proof. We prove the first equation and leave the second as an easy exercise. Let (s^i_j) be the matrix of S_{N_p} with respect to the basis v_1, v_2. Then

$$S_{N_p}v_1 \times S_{N_p}v_2$$
$$= (v^1 s^1_1 + v^2 s^1_2) \times (v^1 s^2_1 + v^2 s^2_2)$$
$$= s^1_1 s^2_2 - s^1_2 s^2_1 = \det(S_{N_p}). \qquad \square$$

We remind the reader of the easily checked **Lagrange identity**:

$$\langle v \times w, a \times b \rangle = \begin{vmatrix} \langle v, a \rangle & \langle v, b \rangle \\ \langle w, a \rangle & \langle w, b \rangle \end{vmatrix}$$

for any $a, b, v, w \in \mathbb{R}^3$ (or in any $T_p\mathbb{R}^3$). If N is a normal field defined over an open set in the surface M, and if X and Y are linearly independent vector fields over the same domain, then we can apply the Lagrange identity and the above proposition at each point to obtain

$$K = \frac{\begin{vmatrix} \langle S_N X, X \rangle & \langle S_N X, Y \rangle \\ \langle S_N Y, X \rangle & \langle S_N Y, Y \rangle \end{vmatrix}}{\begin{vmatrix} \langle X, X \rangle & \langle X, Y \rangle \\ \langle Y, X \rangle & \langle Y, Y \rangle \end{vmatrix}}$$

and also

$$H = \frac{1}{2} \frac{\begin{vmatrix} \langle S_N X, X \rangle & \langle S_N X, Y \rangle \\ \langle Y, X \rangle & \langle Y, Y \rangle \end{vmatrix} + \begin{vmatrix} \langle X, X \rangle & \langle X, Y \rangle \\ \langle S_N Y, X \rangle & \langle S_N Y, Y \rangle \end{vmatrix}}{\begin{vmatrix} \langle X, X \rangle & \langle X, Y \rangle \\ \langle Y, X \rangle & \langle Y, Y \rangle \end{vmatrix}}.$$

These formulas show clearly the smoothness of K and H over any region where a smooth unit normal is defined. Also, the formula

$$k^{\pm} = H \pm \sqrt{H^2 - K}$$

gives the two functions k^+ and k^- such that $k^+(p)$ and $k^-(p)$ are principal curvatures at p. These functions are clearly smooth on any region where $k^+ > k^-$ and are continuous on all of the surface. Furthermore, if every point of an open set in M is an umbilic point, then $k(p) := k^+(p) = k^-(p)$ defines a smooth function on this open set. This continuity is important for obtaining some global results on compact surfaces. Notice that for a surface in $\mathbb{R}^3$ the set of nonumbilic points is exactly the set where $k^+ > k^-$, and so this is an open set.

Definition 4.38. A frame field $E_1, \ldots, E_{n-1}$ on an open region in a hypersurface M is called an **orthonormal frame field** if $E_1(p), \ldots, E_{n-1}(p)$ is an orthonormal basis for T_pM for each p in the region. An orthonormal frame field is called **principal frame field** if each $E_i(p)$ is a principal vector at each point in the region.

Theorem 4.39. *Let M be a surface in $\mathbb{R}^3$. If $p \in M$ is a point that is not umbilic, then there is a principal frame field defined on a neighborhood of p.*

Proof. The set of nonumbilic points is open, and so we can start with any frame field (say a coordinate frame field) on a neighborhood of p. We may take this open set to be oriented by a unit normal N. We then apply the Gram-Schmidt orthogonalization process simultaneously over the open set to obtain a frame field F_1, F_2. Since p is not umbilic, we can multiply by an orthogonal matrix to assure that F_1, F_2 are not principal at p and hence not principal in a neighborhood of p. On this smaller neighborhood we have

$$S_N F_1 = aF_1 + bF_2,$$
$$S_N F_2 = bF_1 + cF_2$$

for functions a, b, c with $b \neq 0$. Now define G_1, G_2 by

$$G_1 = bF_1 + (k^+ - a)F_2,$$
$$G_2 = (k^- - c)F_1 + bF_2$$

and check by direct computation that $S_N G_1 = k^+ G_1$ and $S_N G_2 = k^+ G_2$. We have used a standard linear algebra technique for changing to an eigenbasis. Since $b \neq 0$, we see that $\|G_1\|$ and $\|G_2\|$ are not zero. Finally, let

$$E_1 = G_1/\|G_1\| \text{ and } E_2 = G_2/\|G_2\|. \qquad \square$$

4.3. The Levi-Civita Covariant Derivative

The Levi-Civita covariant derivative is studied here only in the special case of a hypersurface in $\mathbb{R}^n$. The more general case is studied in Chapters 12 and 13. We derive some central equations, which include the Gauss formula, the Gauss curvature equation and the Codazzi-Mainardi equation.

Let M be a hypersurface. For $X_p \in T_pM$, and Y a tangent vector field on M (or an open subset of M), define $\nabla_{X_p} Y$ by

$$\nabla_{X_p} Y := \operatorname{proj}_{T_pM} \overline{\nabla}_{X_p} Y,$$

where $\operatorname{proj}_{T_pM} : T_p\mathbb{R}^n \to T_pM$ is orthogonal projection onto T_pM. For convenience, let us agree to denote the orthogonal projection of a vector

$v \in T_p\mathbb{R}^n$ onto T_pM by $v^\top$ and the orthogonal projection onto the normal direction by $v^\perp$. We call $v^\top$ the **tangent part** of v and $v^\perp$ the **normal part**. Then $\nabla_{X_p} Y = \left(\overline{\nabla}_{X_p} Y\right)^\top$.

Exercise 4.40. Show that for f a smooth function and X_p and Y as above, we have $\nabla_{X_p} fY = (X_p f) Y(p) + f(p) \nabla_{X_p} Y$.

If N is any unit normal field defined near p, then
$$\overline{\nabla}_{X_p} Y = \nabla_{X_p} Y + \langle N_p, \overline{\nabla}_{X_p} Y_p \rangle N_p.$$
Since
$$0 = X_p \langle N, Y \rangle = \langle -S_N X_p, Y \rangle + \langle N, \overline{\nabla}_{X_p} Y \rangle,$$
we obtain the **Gauss formula**:
$$\nabla_{X_p} Y = \overline{\nabla}_{X_p} Y - \langle S_{N_p} X_p, Y_p \rangle N_p. \tag{4.2}$$
Notice that the right hand side of the above equation is unchanged if N is replaced by $-N$. It follows that if X and Y are smooth tangent vector fields, then $p \mapsto \nabla_{X_p} Y$ is smooth and we may then define the field $\nabla_X Y$ by $(\nabla_X Y)(p) := \nabla_{X_p} Y$. By construction $(\nabla_X Y)(p) = (\nabla_Z Y)(p)$ if $X(p) = Z(p)$. It is also straightforward to check that the map $(X, Y) \mapsto \nabla_X Y$ is $C^\infty(M)$-linear in X, but not in Y. Rather, like $\overline{\nabla}$, it is $\mathbb{R}$-linear in Y and we have the product rule
$$\nabla_X fY = (Xf) Y + f \nabla_X Y,$$
which follows directly from Exercise 4.40. Thus ∇ is a **covariant derivative on** M and $\nabla_X Y$ is defined for $X, Y \in \mathfrak{X}(M)$. We remind the reader that $\mathfrak{X}(M)$ is the space of tangent vector fields and not to be confused with vector fields *along* M which may not be tangent to M.

Let Y and Z be smooth tangent vector fields on M and take $X_p \in T_pM$. We study the situation locally near p. Let $(\widetilde{O}, \widetilde{\mathbf{u}})$ be a special chart centered at p and let $(O, \mathbf{u})$ be the chart obtained by restriction where $O = \widetilde{O} \cap M$ as before. If $\widetilde{Y}$ and $\widetilde{Z}$ are extensions of Y and Z to $\widetilde{O}$, then since S_N is self-adjoint, we have
$$(\nabla_Y Z - \nabla_Z Y)(p) = \left(\overline{\nabla}_Y Z - \overline{\nabla}_Z Y\right)(p)$$
$$= \left(\overline{\nabla}_{\widetilde{Y}} \widetilde{Z} - \overline{\nabla}_{\widetilde{Z}} \widetilde{Y}\right)(p)$$
$$= [\widetilde{Y}, \widetilde{Z}]_p = [Y, Z]_p.$$
Thus $\nabla_Y Z - \nabla_Z Y = [Y, Z]$ for all $Y, Z \in \mathfrak{X}(M)$. This fact is expressed by saying that ∇ is **torsion free**. Also,
$$X_p \langle Y, Z \rangle = X_p \langle \widetilde{Y}, \widetilde{Z} \rangle = \langle \overline{\nabla}_{X_p} \widetilde{Y}, \widetilde{Z}_p \rangle + \langle \widetilde{Y}_p, \overline{\nabla}_{X_p} \widetilde{Z} \rangle$$
$$= \langle \nabla_{X_p} Y, Z_p \rangle + \langle Y_p, \nabla_{X_p} Z \rangle,$$

4.3. The Levi-Civita Covariant Derivative

so that if $X \in \mathfrak{X}(M)$, then $X \langle Y, Z \rangle = \langle \nabla_X Y, Z \rangle + \langle Y, \nabla_X Z \rangle$. We express this latter fact by saying that ∇ is a **metric covariant derivative** on M. There is only one covariant derivative on M that satisfies these last two properties, and it is called the **Levi-Civita covariant derivative** (or Levi-Civita connection). We prove the uniqueness later in this chapter.

With coordinates $u^1, \ldots, u^{n-1}$ as above we have functions

$$g_{ij} := \left\langle \frac{\partial}{\partial u^i}, \frac{\partial}{\partial u^j} \right\rangle.$$

If

$$X = \sum_{i=1}^{n-1} X^i \frac{\partial}{\partial u^i} \text{ and } Y = \sum_{i=1}^{n-1} Y^j \frac{\partial}{\partial u^j},$$

then

$$\langle X, Y \rangle = \sum_{i,j} g_{ij} X^i Y^j.$$

The length of a curve in M is just the same as its length as a curve in the ambient Euclidean space. One may define a distance function on M by

$$\mathrm{dist}(p, q) := \inf\{L(c)\},$$

where the infimum is over all curves connecting p and q. This gives M a metric space structure whose topology is the same as the underlying topology. This will be proved in more generality in Chapter 13. For now, the point is that metric aspects of M are determined by $\langle \cdot, \cdot \rangle$ and locally by the g_{ij} associated to each chart of an atlas for M. For example, the length of a curve $c : [a, b] \to O \subset M$ is given by

$$(4.3) \qquad L(c) = \int_a^b \sum_{i,j=1}^{n-1} g_{ij}(c(t)) \frac{dc^i}{dt} \frac{dc^j}{dt} \, dt,$$

where $c^i(t) := u^i \circ c$.

Exercise 4.41. Deduce the above local formula for length from the formula for the length of the curve in the ambient Euclidean space.

Returning to our covariant derivative ∇, we have

$$\nabla_{\frac{\partial}{\partial u^i}} \frac{\partial}{\partial u^j} = \sum_{k=1}^{n-1} \Gamma_{ij}^k \frac{\partial}{\partial u^k}$$

for smooth functions Γ_{ij}^k known as the **Christoffel symbols** of ∇. For X and Y expressed as above, we have

$$\nabla_X Y = \sum_{i=1}^{n-1} X^i \nabla_{\frac{\partial}{\partial u^i}} Y = \sum_{i=1}^{n-1} X^i \nabla_{\frac{\partial}{\partial u^i}} \left(\sum_{j=1}^{n-1} Y^j \frac{\partial}{\partial u^j} \right)$$

$$= \sum_{i=1}^{n-1} X^i \left(\sum_{j=1}^{n-1} \frac{\partial Y^j}{\partial u^i} \frac{\partial}{\partial u^j} + \sum_{j=1}^{n-1} Y^j \nabla_{\frac{\partial}{\partial u^i}} \frac{\partial}{\partial u^j} \right)$$

$$= \sum_{k=1}^{n-1}\sum_{i=1}^{n-1} X^i \frac{\partial Y^k}{\partial u^i} \frac{\partial}{\partial u^k} + \sum_{k=1}^{n-1}\sum_{i=1}^{n-1}\sum_{j=1}^{n-1} X^i Y^j \Gamma_{ij}^k \frac{\partial}{\partial u^k},$$

and so we have

$$(4.4) \qquad \nabla_X Y = \sum_{k=1}^{n-1} \left(\sum_{i=1}^{n-1} \frac{\partial Y^k}{\partial u^i} X^i + \sum_{i,j=1}^{n-1} \Gamma_{ij}^k X^i Y^j \right) \frac{\partial}{\partial u^k}.$$

Thus the functions Γ_{ij}^k determine ∇ in the coordinate chart. Also note that

$$0 = \left[\frac{\partial}{\partial u^i}, \frac{\partial}{\partial u^j} \right] = \nabla_{\frac{\partial}{\partial u^i}} \frac{\partial}{\partial u^j} - \nabla_{\frac{\partial}{\partial u^j}} \frac{\partial}{\partial u^i} = \sum_k \left(\Gamma_{ij}^k - \Gamma_{ji}^k \right) \frac{\partial}{\partial u^k}.$$

It follows that

$$(4.5) \qquad \Gamma_{ij}^k = \Gamma_{ji}^k \text{ for all } i,j,k = 1, \ldots, n-1.$$

We also have

$$\frac{\partial g_{ij}}{\partial u^k} = \frac{\partial}{\partial u^k} \left\langle \frac{\partial}{\partial u^i}, \frac{\partial}{\partial u^j} \right\rangle = \left\langle \nabla_{\frac{\partial}{\partial u^k}} \frac{\partial}{\partial u^i}, \frac{\partial}{\partial u^j} \right\rangle + \left\langle \frac{\partial}{\partial u^i}, \nabla_{\frac{\partial}{\partial u^k}} \frac{\partial}{\partial u^j} \right\rangle$$

$$= \sum_{s=1}^{n-1} \left(\Gamma_{ki}^s g_{sj} + \Gamma_{kj}^s g_{si} \right).$$

The matrix (g_{ij}) is invertible, and it is traditional to denote the components of the inverse by g^{ij}, so that $\sum_r g_{jr} g^{ri} = \delta_j^i$. One may solve to obtain

$$(4.6) \qquad \Gamma_{ij}^k = \frac{1}{2} \sum_s g^{ks} \left(\frac{\partial g_{si}}{\partial u^j} - \frac{\partial g_{ij}}{\partial u^s} + \frac{\partial g_{js}}{\partial u^k} \right).$$

Exercise 4.42. Prove formula (4.6) above. [Hint: First write the formula $\frac{\partial g_{ij}}{\partial u^k} = \sum_s \Gamma_{ki}^s g_{sj} + \Gamma_{kj}^s g_{si}$ two more times, but cyclically permuting i,j,k to obtain three expressions. Subtract the second expression from the sum of the first and third. Use equality (4.5).]

Proposition 4.43. *The Levi-Civita connection on a hypersurface is determined uniquely by the properties of being a torsion free metric connection.*

4.3. The Levi-Civita Covariant Derivative

Proof. In deriving the local formula for the Christoffel symbols, we only used the fact that ∇ is a torsion free metric connection. We do this again in Chapter 13 in a more satisfying way. $\square$

If an object is determined completely by the metric on M, then we say that the object is **intrinsic**. Equivalently, if all local coordinate expressions for an object can be written in terms of the metric coefficients g_{ij} (and their derivatives, etc.), then that object is intrinsic. We have just seen that the connection ∇ is intrinsic. It follows that if $f : M_1 \to M_2$ is an isometry of hypersurfaces and $\overset{1}{\nabla}$ and $\overset{2}{\nabla}$ are the respective Levi-Civita covariant derivatives, then

$$f_* \overset{1}{\nabla}_X Y = \overset{2}{\nabla}_{f_* X} f_* Y$$

for all vector fields $X, Y \in \mathfrak{X}(M)$. One way to see this is to examine the situation using a chart $(V, \mathbf{u})$ on M_1 and the chart $(f(V), \mathbf{u} \circ f^{-1})$ on M_2. A better way is to show that $(X, Y) \mapsto f^* \overset{2}{\nabla}_{f_* X} f_* Y$ defines a torsion free metric connection on M_1 and then use Proposition 4.43. (Exercise!)

If $Y : I \to TM$ is a vector field *along* a curve $c : I \to M \subset \mathbb{R}^n$, then we define

$$\nabla_{\frac{\partial}{\partial t}}\big|_{t_0} Y := \operatorname{proj}_{T_{c(t_0)}M} Y'(t_0) = Y'(t_0)^\top$$

for any $t_0 \in I$ and then define $\nabla_{\frac{\partial}{\partial t}} Y$ by

$$\left(\nabla_{\frac{\partial}{\partial t}} Y\right)(t) := \nabla_{\frac{\partial}{\partial t}}\big|_t Y \text{ for } t \in I.$$

Suppose that $c : I \to M$ is such that $c(J) \subset V$ for some subinterval $J \subset I$ and some chart $(V, \mathbf{u})$. We can then write

$$Y(t) = \sum_{i=1}^{n-1} Y^i(t) \frac{\partial}{\partial u^i}\bigg|_{c(t)} \text{ for all } t \in J.$$

Let us focus attention on a t_0 such that $\dot{c}(t_0) \neq 0$ and let J be an open interval with $t_0 \in J$. By restricting J if necessary we may also assume that c is an embedding. Then there is a smooth field $\widetilde{Y}$ on a neighborhood of $c(J)$ such that $\widetilde{Y} \circ c = Y$. In fact, a simple partition of unity argument shows that we may arrange that $\widetilde{Y}$ be defined on all of V (in fact, on all of

M). For $t \in J$ we have

$$\left(\nabla_{\frac{\partial}{\partial t}} Y\right)(t) = Y'(t)^\top = \left(\left(\widetilde{Y} \circ c\right)'(t)\right)^\top$$

$$= \left(\overline{\nabla}_{\dot{c}(t)} \widetilde{Y}\right)^\top = \nabla_{\dot{c}(t)} \widetilde{Y}$$

$$= \sum_{k=1}^{n-1} \left(\sum_{i=1}^{n-1} \frac{\partial \widetilde{Y}^k}{\partial u^i}(c(t)) \frac{dc^i}{dt} + \sum_{i,j=1}^{n-1} \Gamma_{ij}^k(c(t)) \frac{dc^i}{dt}(t) \widetilde{Y}^j(c(t))\right) \left.\frac{\partial}{\partial u^k}\right|_{c(t)}$$

$$= \sum_{k=1}^{n-1} \left(\frac{\partial Y^k}{\partial t} + \sum_{i,j=1}^{n-1} \left(\Gamma_{ij}^k \circ c\right)(t) \frac{dc^i}{dt}(t) Y^j(t)\right) \left.\frac{\partial}{\partial u^k}\right|_{c(t)}.$$

We arrive at

$$(4.7) \qquad \nabla_{\frac{\partial}{\partial t}} Y = \sum_{k=1}^{n-1} \left(\frac{\partial Y^k}{\partial t} + \sum_{i,j=1}^{n-1} \left(\Gamma_{ij}^k \circ c\right) \frac{dc^i}{dt} Y^j\right) \frac{\partial}{\partial u^k}.$$

We would like to argue that the above formula holds for general curves. If $\dot{c}(t_0) \neq 0$, then there is an interval around t_0 so that the formula holds as we have just seen. If $\dot{c}(t_0) = 0$, we consider two cases. If there exists a sequence t_i converging to t_0 such that $\dot{c}(t_i) = 0$ for all i, then the formula holds for each t_i and hence by continuity at t_0. Otherwise there must be an interval J containing t_0 such that c is constant on J, say $c(t) = p$ for all $t \in J$, and then in this interval Y is just a map into the vector space $T_p M$. In this case we have

$$\left(\nabla_{\frac{\partial}{\partial t}} Y\right)(t) = Y'(t)^\top = \left(\frac{d}{dt} \sum Y^k(t) \left.\frac{\partial}{\partial u^k}\right|_p\right)^\top$$

$$= \sum \frac{\partial Y^k}{\partial t}(t) \left.\frac{\partial}{\partial u^k}\right|_{c(t)}.$$

But this agrees with the formula since $\frac{dc^i}{dt} = 0$.

Exercise 4.44. Show that for a vector field X along $c : I \to M$ and a smooth function $h \in C^\infty(I)$ we have

$$\nabla_{\partial/\partial t} hX = h\nabla_{\partial/\partial t} X + h'X.$$

Exercise 4.45. Show that for vector fields X, Y along c we have

$$\frac{d}{dt}\langle X, Y\rangle = \langle \nabla_{\partial/\partial t} X, Y\rangle + \langle X, \nabla_{\partial/\partial t} Y\rangle = 0.$$

The operator $\nabla_{\partial/\partial t}$ involves the curve c despite the fact that the latter is not indicated in the notation.

4.3. The Levi-Civita Covariant Derivative

We pause to consider again the question posed earlier about the circular arc on the unrolled cone in Figure 4.3. The key lies in the fact that the **absolute geodesic curvature** $|\kappa_g|$ is intrinsic. We remarked earlier that the operator J is intrinsic up to sign (the latter being determined by orientation). On the other hand, in Problem 11 the reader is asked to derive the formula
$$\kappa_g = \frac{\langle \gamma'', J\gamma' \rangle}{\|\gamma'\|^3}.$$
But
$$|\kappa_g| = \frac{|\langle \gamma'', \pm J\gamma' \rangle|}{\|\gamma'\|^3} = \frac{|\langle \nabla_{\partial/\partial t}\gamma', \pm J\gamma' \rangle|}{\|\gamma'\|^3},$$
and since both ∇ and the pair $\pm J$ are intrinsic, we see that $|\kappa_g|$ is intrinsic. A little thought should convince the reader that if a curve in one surface is carried to a curve in another surface by an isometry, then the curves will have equal absolute geodesic curvatures at corresponding points. The unrolling of the cone in Figure 4.3 can be thought of as inducing an isometry between the cone (minus a line segment) and a region in a planar surface in $\mathbb{R}^3$. Thus we expect $|\kappa_g|$ to be the same for both curves. But $|\kappa_g|$ for a circular arc in a plane is just the reciprocal of the radius.

Definition 4.46. A tangent vector field Y along a curve $c : I \to M$ is said to be **parallel** (in M) along c if $\nabla_{\frac{\partial}{\partial t}} Y = 0$ for all $t \in I$.

If c is self-parallel, i.e. $\nabla_{\partial/\partial t}\dot{c}(t) = 0$ for all $t \in I$, then it is easy to see that c is a geodesic (in M) and in fact, this could serve as an alternative definition of geodesic curve, which will be the basis of later generalizations. Notice that if c is a curve in $\mathbb{R}^n$, then Y can be considered as taking values in $T\mathbb{R}^n$. However, Y being parallel in M is not the same as Y being parallel as a $T\mathbb{R}^n$-valued vector field along c. In particular, a geodesic in M certainly need not be a straight line in $\mathbb{R}^n$. For example, constant speed parametrizations of great circles on $S^2 \subset \mathbb{R}^3$ are geodesics.

Exercise 4.47. Given a smooth curve $c : I \to M$, show that $c'' = 0$ if and only if c is a geodesic in M such that $\langle S_N \dot{c}(t), \dot{c}(t) \rangle = 0$ for all t and choice of unit normal at $c(t)$. Suppose that c'' is never zero. Show that c is a geodesic in M if and only if c'' is normal to M (i.e. $c''(t) \perp T_{c(t)}M$ for all $t \in I$).

The following simple result follows from the preceding exercise:

Proposition 4.48. *Let M_1 and M_2 be hypersurfaces in $\mathbb{R}^n$. Suppose that $c : I \to \mathbb{R}^n$ is such that $c(t) \in M_1 \cap M_2$ for all t. If c'' is not zero on any subinterval of I, then $T_{c(t)}M_1 = T_{c(t)}M_2$ for all $t \in I$.*

Proof. By Exercise 4.47, $T_{c(t)}M_1 = c''(t)^\perp = T_{c(t)}M_2$ for all t. $\square$

Proposition 4.49. *If vector fields X, Y along a curve $c : I \to M$ are parallel, then $\langle X, Y \rangle (t) := \langle X(t), Y(t) \rangle$ is constant in t. In particular, a parallel vector field has constant length.*

Proof. $\frac{d}{dt} \langle X, Y \rangle = \langle \nabla_{\partial/\partial t} X, Y \rangle + \langle X, \nabla_{\partial/\partial t} Y \rangle = 0.$ □

Corollary 4.50. *A geodesic has a velocity vector of constant length.*

Definition 4.51. Suppose $Y \in \mathfrak{X}(M)$ is a smooth vector field on M; then Y is called a **parallel vector field** on M if $\nabla_X Y = 0$ at all points of M and for all smooth vector fields $X \in \mathfrak{X}(M)$.

Obviously, Y is a parallel field if and only if $Y \circ c$ is parallel along c for all curves c.

We now move on to prove two basic identities and introduce the curvature tensor. It is easy to check by direct computation that for vector fields $\widetilde{X}, \widetilde{Y}, \widetilde{Z}$ on an open subset of $\mathbb{R}^n$, we have

$$(4.8) \qquad \overline{\nabla}_{\widetilde{X}} \overline{\nabla}_{\widetilde{Y}} \widetilde{Z} - \overline{\nabla}_{\widetilde{Y}} \overline{\nabla}_{\widetilde{X}} \widetilde{Z} - \overline{\nabla}_{[\widetilde{X}, \widetilde{Y}]} \widetilde{Z} \equiv 0.$$

If M is a hypersurface and X, Y, Z are tangent vector fields on a neighborhood of an arbitrary $p \in M$, then we may extend these to fields $\widetilde{X}, \widetilde{Y}, \widetilde{Z}$ on a neighborhood $\widetilde{O}$ in $\mathbb{R}^n$. Then we have

$$\left(\overline{\nabla}_X \overline{\nabla}_Y Z - \overline{\nabla}_Y \overline{\nabla}_X Z - \overline{\nabla}_{[X,Y]} Z \right)(p)$$
$$= \left(\overline{\nabla}_{\widetilde{X}} \overline{\nabla}_{\widetilde{Y}} \widetilde{Z} - \overline{\nabla}_{\widetilde{Y}} \overline{\nabla}_{\widetilde{X}} \widetilde{Z} - \overline{\nabla}_{[\widetilde{X}, \widetilde{Y}]} \widetilde{Z} \right)(p) = 0,$$

and so

$$(4.9) \qquad \overline{\nabla}_X \overline{\nabla}_Y Z - \overline{\nabla}_Y \overline{\nabla}_X Z - \overline{\nabla}_{[X,Y]} Z = 0$$

wherever the fields are all defined on M. Suppose that N is a unit normal field defined on the same domain in M. We apply the Gauss formula (4.2) to equation (4.9) above and then decompose it into tangent and normal parts:

$$0 = \overline{\nabla}_X \left(\nabla_Y Z + \langle S_N Y, Z \rangle N \right) - \overline{\nabla}_Y \left(\nabla_X Z + \langle S_N X, Z \rangle N \right) - \overline{\nabla}_{[X,Y]} Z$$
$$= \nabla_X \nabla_Y Z + \langle S_N X, \nabla_Y Z \rangle N + X \langle S_N Y, Z \rangle N - \langle S_N Y, Z \rangle S_N X$$
$$\quad - \nabla_Y \nabla_X Z - \langle S_N Y, \nabla_X Z \rangle N - Y \langle S_N X, Z \rangle N + \langle S_N X, Z \rangle S_N Y$$
$$\quad - \nabla_{[X,Y]} Z - \langle S_N [X,Y], Z \rangle N.$$

Equating the tangential parts of the above gives the **Gauss curvature equation**:

$$(4.10) \quad \nabla_X \nabla_Y Z - \nabla_Y \nabla_X Z - \nabla_{[X,Y]} Z = \langle S_N Y, Z \rangle S_N X - \langle S_N X, Z \rangle S_N Y.$$

The normal parts give

$$0 = \langle S_N X, \nabla_Y Z \rangle + X \langle S_N Y, Z \rangle - \langle S_N Y, \nabla_X Z \rangle$$
$$\quad - Y \langle S_N X, Z \rangle - \langle S_N [X,Y], Z \rangle,$$

4.3. The Levi-Civita Covariant Derivative

or $\langle \nabla_X S_N Y, Z \rangle - \langle \nabla_Y S_N X, Z \rangle - \langle S_N[X,Y], Z \rangle = 0$ for all Z. From this we obtain the **Codazzi-Mainardi equation**:

(4.11) $$\nabla_X S_N Y - \nabla_Y S_N X - S_N[X,Y] = 0.$$

Let us give an application of the Codazzi-Mainardi equation and then return to the Gauss curvature equation.

Proposition 4.52. *Let M be a connected hypersurface in $\mathbb{R}^n$ oriented by a unit normal field N. If every point of M is umbilic (i.e. M is totally umbilic), then the normal curvatures are all equal and constant on M. Furthermore, M is an open subset of a hyperplane or a sphere according to whether the normal curvatures are zero or nonzero. In particular, if M is a closed subset of $\mathbb{R}^n$, then it is a sphere or a hyperplane according to whether it is compact or not.*

Proof. There is a function k such that $S_N = kI$, where I is the identity on each tangent space. This function is continuous since $k = \frac{1}{n-1} \operatorname{trace} S_N$. Let $X_p \in T_p M$ and pick $Y_p \in T_p M$ so that X_p and Y_p are linearly independent. Extend these to tangent fields X and Y on a neighborhood of p. By the Codazzi-Mainardi equation we have

$$\begin{aligned} 0 &= \nabla_X kY - \nabla_Y kX - k[X,Y] \\ &= (Xk)Y + k\nabla_X Y - ((Yk)X + k\nabla_Y X) - k[X,Y] \\ &= (Xk)Y - (Yk)X, \end{aligned}$$

where we have used $\nabla_X Y - \nabla_Y X = [X,Y]$. In particular, at p we have $(X_p k) Y_p - (Y_p k) X_p = 0$. Since X_p and Y_p are linearly independent, $X_p k = 0$. Since p and X_p were arbitrary and M is connected, we see that k is in fact constant.

If the constant is $k = 0$, then $S_N = 0$ on M, and this means that $N = N_0$ is constant along M. This implies that M is in a hyperplane normal to N_0.

If $k \neq 0$, then (changing N to $-N$ if necessary) we may assume that $k > 0$. Define a function $f : M \to \mathbb{R}^n$ by $f(p) = p + \frac{1}{k} N(p)$. We identify tangent spaces with subspaces of $\mathbb{R}^n$ and calculate $Df(p)$. Let $v \in T_p M$ and choose a curve $c : (-a, a) \to M$ with $\dot{c}(0) = v$. Then we have

$$\begin{aligned} Df(p) \cdot v &= \left. \frac{d}{dt} \right|_{t=0} f \circ c = \dot{c}(0) + \frac{1}{k} \left. \frac{d}{dt} \right|_{t=0} N \circ c \\ &= v - \frac{1}{k} S_N v = v - \frac{1}{k} kv = 0. \end{aligned}$$

Thus $Df(p) = 0$ for all $p \in M$, and since M is connected, f is constant (Exercise 2.23). Thus $p + \frac{1}{k} N(p) = q$ for some fixed q and all p. In other words, all $p \in M$ are at a distance $1/k$ from $q \in \mathbb{R}^n$, and so M is contained in that sphere of radius $1/k$. □

The left hand side of the Gauss curvature equation (4.10) is given its own notation
$$R(X,Y)Z := \nabla_X \nabla_Y Z - \nabla_Y \nabla_X Z - \nabla_{[X,Y]} Z,$$
and looking at the right side of (4.10) we see that $(R(X,Y)Z)(p)$ depends only on the values of X, Y, Z at the point p, which means that we obtain a map $R_p : T_pM \times T_pM \times T_pM \to T_pM$ defined by the formula
$$R_p(X_p, Y_p)Z_p := (R(X,Y)Z)(p).$$
We say that R is a tensor since it is linear in each variable separately. We study tensors systematically in Chapter 7. The tensor R is called the **Riemannian curvature tensor** and the map $p \mapsto R_p$ is smooth in the sense that if X, Y, and Z are smooth tangent vector fields, then $p \mapsto R_p(X_p, Y_p)Z_p$ is a smooth vector field. We often omit the subscript p and just write $R(X_p, Y_p)Z_p$. (Notice that equation (4.8) just says that the curvature of $\mathbb{R}^n$ associated to the ambient covariant derivative $\overline{\nabla}$ is identically zero.) Using (4.10) again, it is also easy to check that R_p is linear in each slot separately on T_pM. In particular, for fixed $X_p, Y_p \in T_pM$ we have a linear map $R(X_p, Y_p) : T_pM \to T_pM$.

Theorem 4.53. *If M is a surface in $\mathbb{R}^3$ and (X_p, Y_p) is an orthonormal basis for T_pM, then*
$$\langle R(X_p, Y_p)Y_p, X_p \rangle = K(p).$$

Proof. Using 4.10, and abbreviating S_{N_p} to S, we have
$$\langle R(X_p, Y_p)Y_p, X_p \rangle = \langle SY_p, Y_p \rangle \langle SX_p, X_p \rangle - \langle SX_p, Y_p \rangle \langle SY_p, X_p \rangle$$
$$= \det S = K(p). \quad \square$$

We have shown that ∇ is intrinsic and thus R is also intrinsic. Thus the previous theorem implies that the Gauss curvature K for a surface in $\mathbb{R}^3$ is intrinsic. This is the content of Gauss's *Theorema Egregium*, which we prove (again) below using local parametric notation.

If M is a k-dimensional submanifold in $\mathbb{R}^n$, and $(V, \mathbf{u})$ is a chart on M with $U = \mathbf{u}(V)$, then $\mathbf{u}^{-1} : U \to M$ is a parametrization of a portion of M, and we denote this map by
$$\mathbf{x} : U \to M.$$
This notation is traditional in surface theory. Composing with the inclusion $\iota : M \hookrightarrow \mathbb{R}^n$, we obtain an immersion $\iota \circ \mathbf{x} : U \to \mathbb{R}^n$, but we normally identify $\iota \circ \mathbf{x}$ and $\mathbf{x}$ when possible. One may study immersions that are not

4.3. The Levi-Civita Covariant Derivative

necessarily one-to-one. A reason for this extension is that it might be the case that while $\mathbf{x}: U \to \mathbb{R}^n$ is not one-to-one, its image is a submanifold M and so we can still study M via such a map. For example, consider the map $\mathbf{x}: \mathbb{R}^2 \to \mathbb{R}^3$ given by

(4.12) $\qquad \mathbf{x}(u,v) = ((a + b\cos u)\cos v, (a + b\cos u)\sin v, b\sin u)$

for $0 < b < a$. This map is periodic and its image is an embedded torus. Notice that the restriction of $\mathbf{x}$ to sets of the form $(u_0 - \pi/2, u_0 + \pi/2) \times (v_0 - \pi/2, v_0 + \pi/2)$ are parametrizations whose inverses are charts on the torus. Another example is the map $\mathbf{x}: \mathbb{R}^2 \to S^2 \subset \mathbb{R}^3$ given by

$$\mathbf{x}(\varphi, \theta) = (\cos\theta \sin\varphi, \sin\theta \sin\varphi, \cos\varphi).$$

The restriction of this map to $(0, \pi) \times (0, 2\pi)$ parametrizes all of S^2 but a set of measure zero.

If $\mathbf{x}: U \to M$ is a parametrization of a hypersurface in $\mathbb{R}^n$, then for $u \in U$, the vectors $\frac{\partial \mathbf{x}}{\partial u^i}(u)$ are tangent to M at $p = \mathbf{x}(u)$ and in fact form the coordinate basis at p in somewhat different notation. In fact, if we abuse notation and write u^i for $u^i \circ \mathbf{x}^{-1}$, we obtain a chart $(V, \mathbf{u})$ with $\mathbf{x}(U) = V$. Then $\frac{\partial \mathbf{x}}{\partial u^i}(u)$ is essentially just $\frac{\partial}{\partial u^i}\big|_{\mathbf{x}(u)}$. We have $\left\langle \frac{\partial \mathbf{x}}{\partial u^i}, \frac{\partial \mathbf{x}}{\partial u^j} \right\rangle = g_{ij} \circ \mathbf{x}$, but in the current context of viewing things in terms of the parametrization, we just change our notations slightly so that $\left\langle \frac{\partial \mathbf{x}}{\partial u^i}, \frac{\partial \mathbf{x}}{\partial u^j} \right\rangle = g_{ij}$. These g_{ij} are the components of the metric with respect to the parametrization. In these terms, some of the calculations look a bit different. For example, if $\gamma: [a,b] \to M$ is a curve whose image is in the range of $\mathbf{x}$, then the length of the curve can be computed in terms of the g_{ij}, which are now *functions of the parameters* u^i. First, γ must be of the form $t \mapsto \mathbf{x}(u^1(t), \ldots, u^{n-1}(t))$, where $u: t \mapsto (u^1(t), \ldots, u^{n-1}(t))$ is a smooth curve in U. Then we have

$$L(\gamma) = \int_a^b \|\dot\gamma(t)\|\, dt = \int_a^b \left\langle \sum \frac{du^i}{dt} \frac{\partial \mathbf{x}}{\partial u^i}, \sum \frac{du^j}{dt} \frac{\partial \mathbf{x}}{\partial u^j} \right\rangle dt$$
$$= \int_a^b \sum_{i,j}^{n-1} g_{ij}(u(t)) \frac{du^i}{dt} \frac{du^j}{dt}\, dt,$$

which is only notationally different from formula (4.3) due to our current parametric viewpoint.

Exercise 4.54. Show that there always exists a local parametrization of a hypersurface $M \subset \mathbb{R}^n$ around each of its points such that $g_{ij}(0) = \delta_{ij}$ and $\frac{\partial g_{ij}}{\partial u^k}(0) = 0$ for all i, j, k. [Hint: Argue that we may assume that M is written as a graph of a function $f: \mathbb{R}^{n-1} \to \mathbb{R}$ with $f(0) = 0$ and $Df(0) = 0$. Then let $\mathbf{x}: \mathbb{R}^{n-1} \to \mathbb{R}^n$ be defined by $(u^1, \ldots, u^{n-1}) \mapsto (u^1, \ldots, u^{n-1}, f(u^1, \ldots, u^{n-1}))$.]

Theorem 4.55 (Gauss's Theorema Egregium). *Let M be a surface in $\mathbb{R}^3$ and let $p \in M$. There exists a parametrization $\mathbf{x} : U \to M$ with $\mathbf{x}(0,0) = p$ such that $g_{ij} = \delta_{ij}$ to first order at 0 and for which we have*

$$K(p) = \frac{\partial^2 g_{12}}{\partial u \partial v}(0) - \frac{1}{2}\frac{\partial^2 g_{22}}{\partial u^2}(0) - \frac{1}{2}\frac{\partial^2 g_{11}}{\partial v^2}(0).$$

Proof. In the coordinates of the exercise above, which give the parametrization $(u,v) \mapsto (u,v,f(u,v))$, where p is the origin of $\mathbb{R}^3$, we have

$$\begin{bmatrix} g_{11}(u,v) & g_{12}(u,v) \\ g_{21}(u,v) & g_{22}(u,v) \end{bmatrix} = \begin{bmatrix} 1+(\frac{\partial f}{\partial u})^2 & \frac{\partial f}{\partial u}\frac{\partial f}{\partial v} \\ \frac{\partial f}{\partial u}\frac{\partial f}{\partial v} & 1+(\frac{\partial f}{\partial v})^2 \end{bmatrix},$$

from which we find, after a bit of straightforward calculation, that

$$\frac{\partial^2 g_{12}}{\partial u \partial v}(0) - \frac{1}{2}\frac{\partial^2 g_{22}}{\partial u^2}(0) - \frac{1}{2}\frac{\partial^2 g_{11}}{\partial v^2}(0)$$
$$= \frac{\partial^2 f}{\partial u^2}\frac{\partial^2 f}{\partial v^2} - \frac{\partial^2 f}{\partial u \partial v} = \det D^2 f(0)$$
$$= \det S(p) = K(p). \qquad \square$$

Let us introduce some traditional notation. For $\mathbf{x} : U \to M \subset \mathbb{R}^3$, and denoting coordinates in U again by (u,v), and in $\mathbb{R}^3$ by (x,y,z), we have

$$\mathbf{x}_u = \left(\frac{\partial x}{\partial u}, \frac{\partial y}{\partial u}, \frac{\partial x}{\partial u}\right)_{\mathbf{x}},$$

$$\mathbf{x}_v = \left(\frac{\partial x}{\partial v}, \frac{\partial y}{\partial v}, \frac{\partial x}{\partial v}\right)_{\mathbf{x}},$$

$$\mathbf{x}_{uv} = \left(\frac{\partial^2 x}{\partial u \partial v}, \frac{\partial^2 y}{\partial u \partial v}, \frac{\partial^2 x}{\partial u \partial v}\right)_{\mathbf{x}},$$

and so on. In this context, we always take the unit normal to be given, as a function of u and v, by

$$N(u,v) = \frac{\mathbf{x}_u \times \mathbf{x}_v}{\|\mathbf{x}_u \times \mathbf{x}_v\|} \qquad \text{(based at } \mathbf{x}(u,v)\text{)}.$$

A careful look at the definitions gives

$$\frac{\partial N}{\partial u} = S_N(\mathbf{x}_u) \quad \text{and} \quad \frac{\partial N}{\partial v} = S_N(\mathbf{x}_v).$$

Recall that the second fundamental form is defined by $II(v,w) = \langle S_N v, w \rangle$, and this makes sense if the tangent vectors v, w are replaced by fields along $\mathbf{x}$ defined on the domain of N. The traditional notation we wish to introduce is

$$E = \langle \mathbf{x}_u, \mathbf{x}_u \rangle, \quad F = \langle \mathbf{x}_u, \mathbf{x}_v \rangle, \quad G = \langle \mathbf{x}_v, \mathbf{x}_v \rangle,$$
$$\mathtt{l} = \langle S_N \mathbf{x}_u, \mathbf{x}_u \rangle, \quad \mathtt{m} = \langle S_N \mathbf{x}_u, \mathbf{x}_v \rangle, \quad \mathtt{n} = \langle S_N \mathbf{x}_v, \mathbf{x}_v \rangle.$$

4.3. The Levi-Civita Covariant Derivative

Thus the matrix of the metric $\langle \cdot, \cdot \rangle$ (sometimes called the first fundamental form) with respect to $\mathbf{x}_u, \mathbf{x}_v$ is

$$\begin{bmatrix} g_{11} & g_{12} \\ g_{21} & g_{22} \end{bmatrix} = \begin{bmatrix} \mathrm{E} & \mathrm{F} \\ \mathrm{F} & \mathrm{G} \end{bmatrix},$$

while that of the second fundamental form II is

$$\begin{bmatrix} \mathrm{l} & \mathrm{m} \\ \mathrm{m} & \mathrm{n} \end{bmatrix}.$$

The reader can check that $\|\mathbf{x}_u \times \mathbf{x}_v\|^2 = \mathrm{EG} - \mathrm{F}^2$. The formula for the length of a curve written as $t \mapsto \mathbf{x}(u(t), v(t))$ on the interval $[a, b]$ is

$$\int_a^b \sqrt{\mathrm{E}\left(\frac{du}{dt}\right)^2 + 2\mathrm{F}\frac{du}{dt}\frac{dv}{dt} + \mathrm{F}\left(\frac{dv}{dt}\right)^2}\, dt.$$

For this reason the classical notation for the metric or first fundamental form is

$$ds^2 = \mathrm{E}\, du^2 + 2\mathrm{F}\, du\, dv + \mathrm{F}\, dv^2,$$

where ds is taken to be an "infinitesimal element of arc length".

Consider the map $g : \mathbb{R}^3 \to (\mathbb{R}^3)^*$ given by $v \mapsto \langle v, \cdot \rangle$. With respect to the standard basis and its dual basis, the matrix for this map is $\begin{bmatrix} \mathrm{E} & \mathrm{F} \\ \mathrm{F} & \mathrm{G} \end{bmatrix}$. Similarly, we can consider the second fundamental form as a map $II : \mathbb{R}^3 \to (\mathbb{R}^3)^*$ given by $v \mapsto II(v, \cdot)$, and the matrix for this transformation is $\begin{bmatrix} \mathrm{l} & \mathrm{m} \\ \mathrm{m} & \mathrm{n} \end{bmatrix}$. Then since $II(v, w) = \langle S_N v, w \rangle$, we have $II = g \circ S_N$ and so

$$S_N = g^{-1} \circ II.$$

We conclude that S_N is represented by the matrix

$$\begin{bmatrix} \mathrm{E} & \mathrm{F} \\ \mathrm{F} & \mathrm{G} \end{bmatrix}^{-1} \begin{bmatrix} \mathrm{l} & \mathrm{m} \\ \mathrm{m} & \mathrm{n} \end{bmatrix}.$$

This matrix may not be symmetric even though the shape operator is symmetric with respect to the inner products on the tangent spaces. Taking the determinant and half the trace of this matrix we arrive at the formulas

$$K = \frac{\mathrm{nl} - \mathrm{m}^2}{\mathrm{EG} - \mathrm{F}^2},$$

$$H = \frac{\mathrm{Gl} + \mathrm{En} - 2\mathrm{Fm}}{2(\mathrm{EG} - \mathrm{F}^2)}.$$

Exercise 4.56. Show that $\mathrm{l} = \langle N, \mathbf{x}_{uu} \rangle$, $\mathrm{m} = \langle N, \mathbf{x}_{uv} \rangle$ and $\mathrm{n} = \langle N, \mathbf{x}_{vv} \rangle$.

Exercise 4.57. Consider the surface of revolution given parametrically by

$$\mathbf{x}(u, v) = (g(u), h(u)\cos v, h(u)\sin v)$$

with $h > 0$. Denote the principal curvature for the meridians through a point with parameters (u,v) by k_μ and that of the parallels by k_π. Show that these are functions of u only given by

$$k_\mu = \frac{-\begin{vmatrix} g' & h' \\ g'' & h'' \end{vmatrix}}{((g')^2 + (h')^2)^{3/2}},$$

$$k_\pi = \frac{g'}{h((g')^2 + (h')^2)^{1/2}}.$$

4.4. Area and Mean Curvature

In this section we give a result that provides more geometric insight into the nature of mean curvature. The basic idea is that we wish to deform a surface and keep track of how the area of the surface changes. Let $M \subset \mathbb{R}^3$ be a surface and let $\mathbf{x} : U \to V \subset M$ be a parametrization of a portion V of M. We suppose that V has compact closure. The area of V is defined by

$$A(V) := \int_U \|\mathbf{x}_u \times \mathbf{x}_v\| \, du \, dv.$$

The total area of M (if it is finite) can be obtained by breaking M up into pieces of this sort whose closures only overlap in sets of measure zero. However, our current study is local and it suffices to consider the areas of small pieces of M as above. Suppose $\mathbf{x} : U \times (-\varepsilon, \varepsilon) \to \mathbb{R}^3$ is a smooth map such that for each fixed $t \in (-\varepsilon, \varepsilon)$, the partial map $\mathbf{x}(\cdot, \cdot, t) : (u, v) \mapsto \mathbf{x}(u, v, t)$ is an embedding and such that $\frac{\partial \mathbf{x}}{\partial t}$ is normal to $\mathbf{x}_u$ and $\mathbf{x}_v$ for all t. For each t, the image $V_t = \mathbf{x}(U, t)$ is a surface. We have in mind the case where $\mathbf{x}(\cdot, \cdot, 0)$ is a parametrization of a portion of a given surface M so that $V_0 = V \subset M$ (see Figure 4.4). The normal $N = \mathbf{x}_u \times \mathbf{x}_v / \|\mathbf{x}_u \times \mathbf{x}_v\|$ depends on t and at time t provides a unit normal to the surface V_t. Thus V_t is a one parameter family of surfaces.

Theorem 4.58. *Let $\mathbf{x} : U \times (-\varepsilon, \varepsilon) \to \mathbb{R}^3$ and let $V_t = \mathbf{x}(U, t)$ be as above so that $\frac{\partial \mathbf{x}}{\partial t}$ is normal to the surface V_t. Let $H(t)$ denote the mean curvature of the surface V_t. Then*

$$\frac{d}{dt} A(V_t) = -2 \int_{V_t} \left\| \frac{\partial \mathbf{x}}{\partial t} \right\| H(t) \, dA.$$

Proof. Since $\frac{\partial \mathbf{x}}{\partial t}$ is parallel to N, we have $\frac{\partial \mathbf{x}}{\partial t} = \left\| \frac{\partial \mathbf{x}}{\partial t} \right\| N$. Thus

$$\frac{\partial \mathbf{x}_u}{\partial t} = \frac{\partial}{\partial u} \frac{\partial \mathbf{x}}{\partial t} = \frac{\partial}{\partial u} \left(\left\| \frac{\partial \mathbf{x}}{\partial t} \right\| N \right) = -\left\| \frac{\partial \mathbf{x}}{\partial t} \right\| S_N(\mathbf{x}_u) + \frac{\partial}{\partial u} \left\| \frac{\partial \mathbf{x}}{\partial t} \right\| N$$

4.4. Area and Mean Curvature

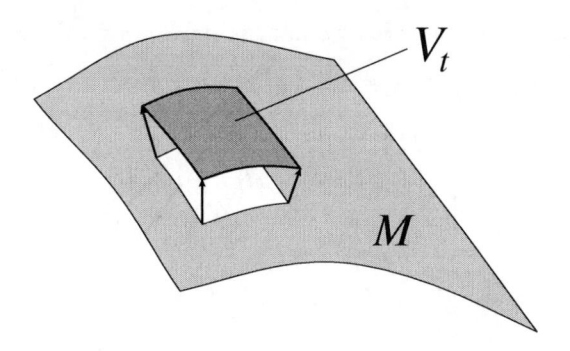

Figure 4.4. Deformation of a patch

and so
$$\left\langle N, \frac{\partial \mathbf{x}_u}{\partial t} \times \mathbf{x}_v \right\rangle = - \left\| \frac{\partial \mathbf{x}}{\partial t} \right\| \langle N, S_N(\mathbf{x}_u) \times \mathbf{x}_v \rangle.$$

Similarly,
$$\left\langle N, \mathbf{x}_u \times \frac{\partial \mathbf{x}_v}{\partial t} \right\rangle = - \left\| \frac{\partial \mathbf{x}}{\partial t} \right\| \langle N, \mathbf{x}_u \times S_N(\mathbf{x}_v) \rangle.$$

Now we calculate using the second formula of Proposition 4.37:

$$\begin{aligned}
\frac{d}{dt} A &= \frac{d}{dt} \int \|\mathbf{x}_u \times \mathbf{x}_v\| \, du \, dv = \frac{d}{dt} \int \langle N, \mathbf{x}_u \times \mathbf{x}_v \rangle \, du \, dv \\
&= \int \left\langle \frac{d}{dt} N, \mathbf{x}_u \times \mathbf{x}_v \right\rangle du \, dv + \int \left\langle N, \frac{\partial \mathbf{x}_u}{\partial t} \times \mathbf{x}_v + \mathbf{x}_u \times \frac{\partial \mathbf{x}_v}{\partial t} \right\rangle du \, dv \\
&= \int \left\langle N, \frac{\partial \mathbf{x}_u}{\partial t} \times \mathbf{x}_v + \mathbf{x}_u \times \frac{\partial \mathbf{x}_v}{\partial t} \right\rangle du \, dv \\
&= -\int \left\| \frac{\partial \mathbf{x}}{\partial t} \right\| \{ \| S_N(\mathbf{x}_u) \times \mathbf{x}_v + \mathbf{x}_u \times S_N(\mathbf{x}_v) \| \} \, du \, dv \\
&= -2 \int H(p) \left\| \frac{\partial \mathbf{x}}{\partial t} \right\| \|\mathbf{x}_u \times \mathbf{x}_v\| \, du \, dv = -2 \int \left\| \frac{\partial \mathbf{x}}{\partial t} \right\| H \, dA. \quad \square
\end{aligned}$$

In particular, we can arrange that $\left\| \frac{\partial \mathbf{x}}{\partial t} \right\| = 1$ at time $t = 0$, and then we have
$$\left. \frac{d}{dt} \right|_{t=0} A = -2 \int H \, dA.$$

Thus, H is a measure of the rate of change in area under perturbations of the surface.

Definition 4.59. A hypersurface in $\mathbb{R}^n$ for which H is identically zero is called a **minimal hypersurface** (or minimal surface if $n = 3$ so that $\dim M = 2$).

Example 4.60. For $c > 0$, the catenoid is parametrized by
$$\mathbf{x}(u, v) = (u, \, c \cosh(u/c) \cos v, \, c \cosh u \sin v)$$
and is a minimal surface. Indeed, a straightforward calculation gives
$$\begin{bmatrix} E & F \\ F & G \end{bmatrix} = \begin{bmatrix} c^2 \cosh^2(u/c) & 0 \\ 0 & \cosh^2(u/c) \end{bmatrix},$$
$$\begin{bmatrix} l & m \\ m & n \end{bmatrix} = \begin{bmatrix} -c & 0 \\ 0 & 1/c \end{bmatrix}.$$
It follows that $H = 0$.

Example 4.61. For each $b \neq 0$, the map $\mathbf{x}(u, v) = (bv, u \cos v, u \sin v)$ is a parametrization of a surface called a helicoid and this is also a minimal surface. The reader may enjoy plotting this and other surfaces using a computer algebra system such as Maple or Mathematica.

4.5. More on Gauss Curvature

In this section we construct surfaces of revolution with prescribed Gauss curvature and also prove that an oriented compact surface of constant curvature must be a sphere. Let
$$\mathbf{x}(u, v) = (g(u), h(u) \cos v, h(u) \sin v)$$
be a parametrization of a surface of revolution. By a reparametrization of the profile curve $(g(u), h(u))$ we may assume that it is a unit speed curve so that $(g')^2 + (h')^2 = 1$. When this is done, we say that we have a **canonical parametrization** of the surface of revolution. A straightforward calculation using the results of Exercise 4.57 shows that for any surface of revolution given as above we have
$$K = -\frac{g'}{h} \frac{\begin{vmatrix} g' & h' \\ g'' & h'' \end{vmatrix}}{\left((g')^2 + (h')^2\right)^2}.$$
If we assume that it is a canonical parametrization, then we have
$$K = \frac{-(g')^2 h'' + g' g'' h'}{h}.$$
On the other hand, differentiation of $(g')^2 + (h')^2 = 1$ leads to $g' g'' = -h' h''$, and so we arrive at
$$K = \frac{-h''}{h},$$
which shows the expected result that K is constant on parallels (curves along which u is constant). Now suppose we are given a smooth function K defined on some interval I, which we may as well assume to contain 0.

4.5. More on Gauss Curvature

We would like K to be our Gauss curvature and so we wish to solve the differential equation $h'' + Kh = 0$ subject to $h(0) > 0$ and $|h'(0)| < 1$. The first condition simplifies the analysis, while the second condition allows us to obtain the canonical situation $(g')^2 + (h')^2 = 1$. In fact, we let

$$g(u) = \int_0^u \sqrt{1 - (h'(t))^2}\, dt,$$

and this will give the desired solution defined on the largest open subinterval J of I such that $h > 0$ and $|h'| < 1$. With this solution, our surface of revolution will be defined, but only for $u \in J$.

Suppose we try to obtain a surface of revolution with constant positive Gauss curvature $K = 1/c^2$ for some constant c. Then a solution of the equation for h will be $h(u) = a\cos(u/c)$ for an appropriate $a > 0$. Then

$$g(u) = \int_0^u \sqrt{1 - \frac{a^2}{c^2}\sin^2(t/c)}\, dt,$$

and with the resulting profile curve, we obtain a surface of revolution with Gauss curvature $1/c^2$ at all points. There are three cases to consider. First, if $a = c$, then the interval J on which $h > 0$ and $|h'| < 1$ is easily seen to be $(-\pi c/2, \pi c/2)$ and we have

$$h(u) = c\cos(u/c) \text{ and } g(u) = c\sin(u/c).$$

This gives a semicircle which revolves to make a sphere minus the two points on the axis of revolution. We already know that these two points can be added to give the sphere of radius $c = 1/\sqrt{K}$, which is a compact surface of constant positive Gauss curvature K. It turns out that the spheres are the only compact surfaces (without boundary) of constant positive Gauss curvature. Now consider the case $0 < a < c$. The interval J is the same as before, but the surface extends in the x direction between $x_1 = \lim_{u \to -\pi c/2} g(u)$ and $x_2 = \lim_{u \to \pi c/2} g(u)$. It is easy to show that $x_1 < -a$ and $x_2 = -x_1 > a$. Since the maximum value of h is now smaller than c, the profile curve is shallower and wider than the semicircle of the $a = c$ case above. Although $\lim_{u \to \pm \pi c/2} h(u) = 0$ as before, we now have the profile curve tangents at the endpoints given by

$$\lim_{u \to \pm\pi c/2}\left(g'(u), h'(u)\right) = \lim_{u \to \pm\pi c/2}\left(\sqrt{1 - \frac{a^2}{c^2}\sin^2(u/c)}, -\frac{a}{c}\sin(u/c)\right)$$

$$= \left(1 - \frac{a^2}{c^2}, \mp\frac{a}{c}\right).$$

This shows that the revolved surface is pointed at its extremes and forms an American football shape. In this case, there is no way to add in the missing points on the axis of revolution to obtain a smooth surface. The two principal curvatures are no longer equal, but their product is still $1/c^2$.

The third case $a > 0$ gives a surface that has an extension to a surface with boundary.

Exercise 4.62. Analyze the case $a > 0$.

From above we see that there is an infinite family of surfaces with constant curvature (only one extending to a compact surface). The situation for constant *negative* curvature is similar, but we obtain no compact surfaces. One particular constant negative curvature surface is of special interest:

Example 4.63 (Bugle surface). This surface has Gauss curvature $-1/c^2$ and is given by
$$\mathbf{x}(u,v) = (u, h(u)\cos v, h(u)\sin v),$$
where h is the solution of the differential equation
$$h' = \frac{-h}{\sqrt{c^2 - h^2}},$$
subject to initial condition $\lim_{u \to 0} h(u) = c$. The function h is defined on $(0, \infty)$.

Lemma 4.64. *Let M be a surface in $\mathbb{R}^3$. Let $p \in M$ be a nonumbilic point and E_1, E_2 a principal frame on a neighborhood of p (oriented by N) so that $S_N E_1 = k^+ E_1$ and $S_N E_2 = k^- E_2$. If we define functions*
$$h_1 = \frac{-E_2 k^+}{k^+ - k^-} \quad \text{and} \quad h_2 = \frac{E_1 k^-}{k^+ - k^-},$$
then
$$K = -E_1 h_2 - E_2 h_1 - h_1^2 - h_2^2.$$

Proof. Since $\langle E_2, E_2 \rangle = 1$, we have $\langle \nabla_{E_1} E_2, E_2 \rangle = 0$ and so there is some function h_1 such that $\nabla_{E_1} E_2 = h_1 E_1$. Similarly, $\nabla_{E_2} E_1 = h_2 E_2$ for some function h_2. We find formulas for each of these functions. We have
$$0 = E_1 \langle E_1, E_1 \rangle = 2 \langle \nabla_{E_1} E_1, E_1 \rangle$$
and
$$0 = E_1 \langle E_1, E_2 \rangle = \langle \nabla_{E_1} E_1, E_2 \rangle + \langle E_1, \nabla_{E_1} E_2 \rangle,$$
from which it follows that $\nabla_{E_1} E_1 = -h_1 E_2$. Similarly, $\nabla_{E_2} E_2 = -h_2 E_1$. We have
$$[E_1, E_2] = \nabla_{E_1} E_2 - \nabla_{E_2} E_1 = h_1 E_1 - h_2 E_2.$$

4.5. More on Gauss Curvature

We now apply the Codazzi-Mainardi equations (4.11):

$$\begin{aligned}
0 &= \nabla_{E_1} S_N E_2 - \nabla_{E_2} S_N E_1 - S_N[E_1, E_2] \\
&= \nabla_{E_1} k^- E_2 - \nabla_{E_2} k^+ E_1 - S_N(h_1 E_1 - h_2 E_2) \\
&= \left(E_1 k^- \, E_2 + k^- \nabla_{E_1} E_2\right) - \left(E_2 k^+ \, E_1 + k^+ \nabla_{E_2} E_1\right) \\
&\qquad - \left(h_1 k^+ E_1 - h_2 k^- E_2\right) \\
&= \left(E_1 k^- \, E_2 + k^- h_1 E_1\right) - \left(E_2 k^+ \, E_1 + k^+ h_2 E_2\right) \\
&\qquad - \left(h_1 k^+ E_1 - h_2 k^- E_2\right) \\
&= \left(k^- h_1 - E_2 k^+ - h_1 k^+\right) E_1 + \left(E_1 k^- + h_2 k^- - k^+ h_2\right) E_2.
\end{aligned}$$

Setting the coefficients of E_1 and E_2 equal to zero we obtain

$$h_1 = \frac{-E_2 k^+}{k^+ - k^-} \text{ and } h_2 = \frac{E_1 k^-}{k^+ - k^-}.$$

We compute as follows:

$$\begin{aligned}
R(E_1, E_2) E_2 &= \nabla_{E_1} \nabla_{E_2} E_2 - \nabla_{E_2} \nabla_{E_1} E_2 - \nabla_{[E_1, E_2]} E_2 \\
&= \nabla_{E_1}(-h_2 E_1) - \nabla_{E_2}(h_1 E_1) - \nabla_{(h_1 E_1 - h_2 E_2)} E_2 \\
&= -(E_1 h_2) E_1 - h_2 \nabla_{E_1} E_1 - (E_2 h_1) E_1 - h_1 \nabla_{E_2} E_1 \\
&\qquad - h_1 \nabla_{E_1} E_2 + h_2 \nabla_{E_2} E_2 \\
&= -(E_1 h_2) E_1 - (E_2 h_1) E_1 - h_2(-h_1 E_2) - h_1(h_2 E_2) \\
&\qquad - h_1(h_1 E_1) + h_2(-h_2 E_1) \\
&= -(E_1 h_2) E_1 - (E_2 h_1) E_1 - h_1^2 E_1 - h_2^2 E_1.
\end{aligned}$$

Using the Gauss curvature equation (4.10) we arrive at

(4.13) $\qquad K = \langle R(E_1, E_2) E_2, E_1 \rangle = -E_1 h_2 - E_2 h_1 - h_1^2 - h_2^2.$ $\quad\square$

Corollary 4.65. *If p is a nonumbilic point for which both k^+ and k^- are critical, then*

$$K(p) = \frac{E_2^2 k^+ - E_1^2 k^-}{k^+ - k^-}(p).$$

Proof. Using equation (4.13), we have

$$\begin{aligned}
K &= -E_1 h_2 - E_2 h_1 - h_1^2 - h_2^2 \\
&= -E_1\left(\frac{E_1 k^-}{k^+ - k^-}\right) - E_2\left(\frac{-E_2 k^+}{k^+ - k^-}\right) - \left(\frac{-E_2 k^+}{k^+ - k^-}\right)^2 - \left(\frac{E_1 k^-}{k^+ - k^-}\right)^2.
\end{aligned}$$

At p we have $E_2 k^+ = E_1 k^- = 0$, and so (suppressing evaluations at p) we obtain

$$K(p) = \frac{-(k^+ - k^-) E_1^2 k^-}{(k^+ - k^-)^2} - \frac{-(k^+ - k^-) E_2^2 k^+}{(k^+ - k^-)^2} = \frac{E_2^2 k^+ - E_1^2 k^-}{k^+ - k^-}. \quad\square$$

Corollary 4.66 (Hilbert). *Let $M \subset \mathbb{R}^3$ be a surface. Suppose that K is a positive constant on M. Then k^+ cannot have a relative maximum at a nonumbilic point. Similarly, k^- cannot have a relative minimum at a nonumbilic point.*

Proof. Since $K = k^+ k^- > 0$ is constant, k^+ has a relative maximum exactly when k^- has a relative minimum, so the second statement follows from the first. For the first statement, suppose that k^+ has a relative maximum at p. Notice that $X^2 k^+ \leq 0$ and $X^2 k^- \geq 0$ for any tangent vector field defined near p. Near p, there is a principal frame as above, and if we use Corollary 4.65, we have

$$K(p) = \frac{E_2^2 k^+ - E_1^2 k^-}{k^+ - k^-}(p) < 0,$$

contradicting our assumption about K. $\square$

We are now able to use the above technical results to obtain a nice theorem.

Theorem 4.67. *If $M \subset \mathbb{R}^3$ is an oriented connected compact surface with constant positive Gauss curvature K, then M is a sphere of radius $1/\sqrt{K}$.*

Proof. The hypotheses give us the existence of a global unit normal field N. Note that $k^+ \geq \sqrt{K}$ at every point, and since M is compact, k^+ must have an absolute maximum at some point p. By Corollary 4.66, p is an umbilic point and so $k^+(p) = k^-(p)$. But then $(k^+(p))^2 = k^+(p) k^-(p) = K$, and thus the maximum value of k^+ is $\sqrt{K}$. We have both $k^+ \geq \sqrt{K}$ and $k^+ \leq \sqrt{K}$ on all of M, and so $k^+ = k^- = \sqrt{K}$ everywhere (M is totally umbilic). By Proposition 4.52 we see that M is a sphere of radius $1/\sqrt{K}$. $\square$

We already know from our surface of revolution examples that there are many noncompact surfaces of constant positive Gauss curvature, but now we see that the sphere is the only compact example. Actually, more is true: If a surface of constant positive Gauss curvature is a closed subset of $\mathbb{R}^3$, then it is a sphere. This follows from a theorem of Myers (see Theorem 13.143 in Chapter 13).

4.6. Gauss Curvature Heuristics

The reader may be left wondering about the geometric meaning of the Gauss curvature. We will learn more about the Gauss curvature in Section 9.9 and much more about curvature in general in Chapter 13. For now, we will simply pursue an informal understanding of the Gauss curvature.

4.6. Gauss Curvature Heuristics

Negative Curvature   Zero Curvature   Positive Curvature

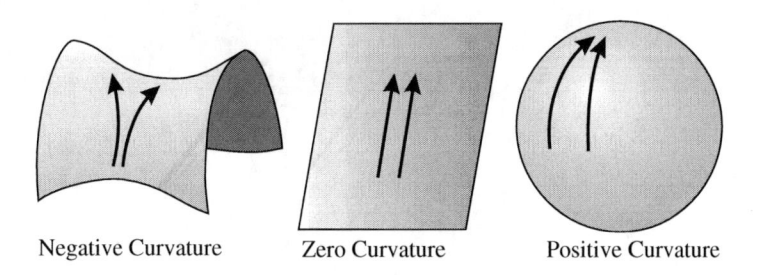

Figure 4.5. Curvature bends geodesics

On a plane, geodesics are straight lines parametrized with constant speed. If two insects start off in parallel directions and maintain a policy of not turning either left or right, then they will travel on straight lines, and if their speeds are the same, then the distance between them remains constant. On a sphere or on any surface with positive curvature, the situation is different. In this case, geodesics tend to curve toward each other. Particles (or insects) moving along geodesics that start out near each other and roughly parallel will bend toward each other if they travel at the same speed. The second derivative of the distance between them will be negative. For example, two airplanes traveling due north from the equator at constant speed and altitude will be drawn closer and, if they continue, will eventually meet at the north pole. It is the curvature of the earth that "pulls" them together. On a surface of negative curvature, initially parallel motions along geodesics will bend away from each other. The second derivative of the distance between them will be positive. These three situations are depicted in Figure 4.5. For a more precise formulation of these ideas it is best to consider a parametrized family of curves, and we will do this in Chapter 13.

It should be mentioned that according to Einstein's theory, it is the curvature of spacetime that accounts for those aspects of gravity (such as tidal forces) that cannot be nullified by a choice of frame (accelerating frames cause gravity-like effects even in a flat spacetime). For example, an initially spherical cluster of particles in free fall near the earth will be deformed into an egg shape. For a wonderful popular account of gravity as curvature, see [**Wh**].

Another way to see the effects of curvature is by considering triangles on surfaces whose sides are geodesic segments. These geodesic triangles are affected by the Gauss curvature. For instance, consider the geodesic triangle on the sphere in Figure 4.6. The angles shown are actually measured in the tangent spaces at each point based on the tangents of the curves at the endpoints of each segment. We have the following Gauss-Bonnet formula

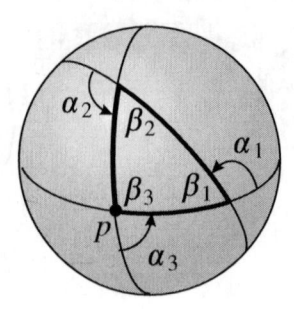

Figure 4.6. Curvature and triangle

involving the interior angles:

$$\beta_1 + \beta_2 + \beta_3 = \pi + \int_D K\,dS,$$

where D is the region interior to the geodesic triangle. This formula is true for geodesic triangles on any surface M. For a sphere of radius a, this becomes $\beta_1 + \beta_2 + \beta_3 = \pi + A/a^2$, where A is the area of the region interior to the geodesic triangle. The fact that the sum of the interior angles for a triangle in a plane is equal to π regardless of the area inside the triangle is exactly due to the fact that a plane has zero Gauss curvature. If one starts at p and moves counterclockwise around the triangle, then at the corners one must turn through angles $\alpha_1, \alpha_2, \alpha_3$. In terms of these turning angles, the statement is

$$\sum_{i=1}^{3} \alpha_i = 2\pi - \int_D K\,dS.$$

By an argument that involves triangulating a surface, it can be shown that if one integrates the Gauss curvature over a whole surface (without boundary) $M \subset \mathbb{R}^3$, then something amazing occurs. We obtain

$$\int_M K\,dS = 2\pi\chi(M).$$

This result is called the **Gauss-Bonnet theorem**. Here $\chi(M)$ is a topological invariant called the Euler characteristic of the surface and is equal to $2 - 2g$, where g is the genus of the surface (see [**Arm**]). Thus while the left hand side of the above equation involves the shape dependent curvature K, the right hand side is a purely topological invariant! A presentation of a very general version of the Gauss-Bonnet theorem may be found in [**Poor**] (also see [**Lee, Jeff**]).

Problems

(1) Let $\gamma : I \to \mathbb{R}^2$ be a regular plane curve. Let $J : \mathbb{R}^2 \to \mathbb{R}^2$ be the rotation $(x, y) \mapsto (-y, x)$. Show that the **signed curvature function**
$$\kappa_2(t) := \frac{\gamma''(t) \cdot J\gamma'(t)}{\|\gamma''(t)\|^3}$$
determines γ up to reparametrization and Euclidean motion.

(2) Let $f, g : (a, b) \to \mathbb{R}$ be differentiable functions with $f^2 + g^2 = 1$. Let $t_0 \in (a, b)$ and suppose that $f(t_0) = \cos\theta_0$ and $g(t_0) = \sin\theta_0$ for some $\theta_0 \in \mathbb{R}$. Show that there is a unique continuous function $\theta : (a, b) \to \mathbb{R}$ such that $\theta(t_0) = \theta_0$ and
$$f(t) = \cos\theta(t), \quad g(t) = \sin\theta(t) \text{ for } t \in (a, b).$$

(3) Let $\gamma : (a, b) \to \mathbb{R}^2$ be a regular plane curve.
   (a) Given $t_0 \in (a, b)$ and θ_0 with
   $$\frac{\gamma'(t_0)}{\|\gamma'(t_0)\|} = (\cos\theta_0, \sin\theta_0),$$
   show that there is a unique continuous **turning angle function** $\theta_\gamma : (a, b) \to \mathbb{R}$ such that $\theta(t_0) = \theta_0$ and
   $$\frac{\gamma'(t)}{\|\gamma'(t)\|} = (\cos\theta_\gamma(t), \sin\theta_\gamma(t)) \text{ for } t \in (a, b).$$
   (b) Show that $\theta'_\gamma(t) = \|\gamma'(t)\| \kappa_2(t)$, where κ_2 is as in Problem 1.

(4) Show that if κ_2 for a curve $\gamma : (a, b) \to \mathbb{R}^2$ is $1/r$, then γ parametrizes a portion of a circle of radius r.

(5) Let $\gamma : \mathbb{R} \to \mathbb{R}^3$ be the elliptical helix given by $t \mapsto (a \cos t, b \sin t, ct)$.
   (a) Calculate the torsion and curvature of γ.
   (b) Define a map $F : \mathbb{R} \to SO(3)$ by letting F have columns given by the Frenet frame of γ. Show that F is a periodic parameterization of a closed curve in $SO(3)$.

(6) Calculate the curvature and torsion for the twisted cubic $\gamma(t) := (t, t^2, t^3)$. Examine the behavior of the curvature, torsion, and Frenet frame as $t \to \pm\infty$.

(7) Find a unit speed parametrization of the catenary curve given by $c(t) := (a \cosh(t/a), t)$. Revolve the resulting profile curve to obtain a canonical parametrization of a catenoid and find the Gauss curvature in these terms.

(8) (Four vertex theorem) Show that the signed curvature function κ_2 for a simple, closed plane curve is either constant or has at least two local maxima and at least two local minima.

(9) If $\gamma : I \to \mathbb{R}^3$ is a regular space curve (not necessarily unit speed), then show that $\mathbf{B}(t) = \frac{\gamma' \times \gamma''}{\|\gamma' \times \gamma''\|}$, $\tau = \frac{(\gamma' \times \gamma'') \cdot \gamma'''}{\|\gamma' \times \gamma''\|^2}$ and $\kappa = \frac{\|\gamma' \times \gamma''\|}{\|\gamma''\|^3}$.

(10) With γ as in Problem 1, show that $\gamma'' = \left(\frac{d}{dt}\|\gamma'\|\right)\mathbf{T} + \|\gamma'\|^2 \kappa \mathbf{N}$.

(11) Let $\gamma : I \to M \subset \mathbb{R}^3$ be a curve which is not necessarily unit speed and suppose M is oriented by a unit normal field N. Show that $\kappa_g = \frac{\langle \gamma'', J\gamma' \rangle}{\|\gamma'\|^3}$.

(12) Show that if a surface $M \subset \mathbb{R}^3$, either the image by an immersion of an open domain in $\mathbb{R}^2$ or is the zero set of a real-valued function f (such that $df \neq 0$ on M) then it is orientable.

(13) Calculate the shape operator at a generic point on the cylinder $\{(x, y, z) : x^2 + y^2 = r^2\}$.

(14) (Euler's formula) Let p be a point on a surface M in $\mathbb{R}^3$ and let u_1, u_2 be principal directions with corresponding principal curvatures k_1, k_2 at p. If $u = (\cos\theta)\, u_1 + (\sin\theta)\, u_2$, show that $k(u) = k_1 \cos^2\theta + k_2 \sin^2\theta$.

(15) Show that a point on a hypersurface in $\mathbb{R}^n$ is umbilic if and only if there is a constant k_0 such that $k(u) = k_0$ for all $u \in T_pM$ with $\|u\| = 1$.

(16) Let Z be a nonvanishing (not necessarily unit) normal field on a surface $M \subset \mathbb{R}^3$. If X and Y are tangent fields such that $X \times Y = Z$, then show that
$$K = \frac{\langle Z, \overline{\nabla}_X Z \times \overline{\nabla}_Y Z \rangle}{\|Z\|^4} \quad \text{and} \quad H = -\frac{\langle Z, \overline{\nabla}_X Z \times Y + X \times \overline{\nabla}_Y Z \rangle}{2\|Z\|^3}.$$

(17) Find the Gauss curvature K at (x, y, z) on the ellipsoid $x^2/a^2 + y^2/b^2 + y^2/c^2 = 1$.

(18) Show that a surface in $\mathbb{R}^3$ is minimal if and only if there are orthogonal asymptotic vectors at each point.

(19) Compute $\mathbf{E}, \mathbf{F}, \mathbf{G}, \mathbf{l}, \mathbf{m}, \mathbf{n}$ as well as H and K for the following surfaces:
   (a) Paraboloid: $(u, v) \longmapsto (u, v, au^2 + bv^2)$ with $a, b > 0$.
   (b) Monkey Saddle: $(u, v) \longmapsto (u, v, u^3 - 3uv^2)$.
   (c) Torus: $(u, v) \longmapsto ((a + b\cos v)\cos u, (a + b\cos v)\sin u, b\sin v)$ with $a > b > 0$.

(20) (Enneper's surface) Show that the following is a minimal but not 1-1 immersion:
$$\mathbf{x}(u, v) := \left(u - \frac{u^3}{3} + uv^2, -v + \frac{v^3}{3} - vu^2, u^2 - v^2\right).$$

Chapter 5

Lie Groups

One approach to geometry is to view it as the study of invariance and symmetry. In our case, we are interested in studying symmetries of smooth manifolds, Riemannian manifolds, symplectic manifolds, etc. The usual way to deal with symmetry in mathematics is by the use of the notion of a transformation group. The wonderful thing for us is that the groups that arise in the study of geometric symmetries are often themselves smooth manifolds. Such "group manifolds" are called Lie groups.

In physics, Lie groups play a big role in connection with physical symmetries and conservation laws (Noether's theorem). Within physics, perhaps the most celebrated role played by Lie groups is in particle physics and gauge theory. In mathematics, Lie groups play a prominent role in harmonic analysis (generalized Fourier theory), group representations, differential equations, and in virtually every branch of geometry including Riemannian geometry, Cartan geometry, algebraic geometry, Kähler geometry, and symplectic geometry.

5.1. Definitions and Examples

Definition 5.1. A smooth manifold G is called a **Lie group** if it is a group (abstract group) such that the multiplication map $\mu : G \times G \to G$ and the inverse map $\text{inv} : G \to G$, given respectively by $\mu(g, h) = gh$ and $\text{inv}(g) = g^{-1}$, are C^∞ maps. If the group is abelian, we sometimes opt to use the additive notation $g + h$ for the group operation.

We will usually denote the identity element of any Lie group by the same letter e. Exceptions include the case of matrix or linear groups where we use the letter I or id. The map inv : $G \to G$ given by $g \mapsto g^{-1}$ is called **inversion** and is easily seen to be a diffeomorphism.

Example 5.2. $\mathbb{R}$ is a one-dimensional (abelian) Lie group, where the group multiplication is the usual addition $+$. Similarly, any real or complex vector space is a Lie group under vector addition.

Example 5.3. The circle $S^1 = \{z \in \mathbb{C} : |z|^2 = 1\}$ is a 1-dimensional (abelian) Lie group under complex multiplication. It is also traditional to denote this group by $U(1)$.

Example 5.4. Let $\mathbb{R}^* = \mathbb{R}\setminus\{0\}$, $\mathbb{C}^* = \mathbb{C}\setminus\{0\}$ and $\mathbb{H}^* = \mathbb{H}\setminus\{0\}$ (here $\mathbb{H}$ is the quaternion division ring discussed in detail later). Then, using multiplication, $\mathbb{R}^*$, $\mathbb{C}^*$, and $\mathbb{H}^*$ are Lie groups. The Lie group $\mathbb{H}^*$ is not abelian.

The group of all invertible real $n \times n$ matrices is a Lie group denoted $\mathrm{GL}(n, \mathbb{R})$. A global chart on $\mathrm{GL}(n, \mathbb{R})$ is given by the n^2 functions x^i_j, where if $A \in \mathrm{GL}(n, \mathbb{R})$ then $x^i_j(A)$ is the ij-th entry of A. We study this group and some of its subgroups below. Showing that $\mathrm{GL}(n, \mathbb{R})$ is a Lie group is straightforward. Multiplication is clearly smooth. For the inversion map one appeals to the usual formula for the inverse of a matrix, $A^{-1} = \mathrm{adj}(A)/\det(A)$. Here $\mathrm{adj}(A)$ is the adjoint matrix (whose entries are the cofactors). This shows that A^{-1} depends smoothly on the entries of A. Similarly, the group $\mathrm{GL}(n, \mathbb{C})$ of invertible $n \times n$ complex matrices is a Lie group.

Exercise 5.5. Let H be a subgroup of G and consider the cosets gH, $g \in G$. Recall that G is the disjoint union of the cosets of H. Show that if H is open, then so are all the cosets. Conclude that the complement H^c is also open and hence H is closed.

Theorem 5.6. *If G is a connected Lie group and U is a neighborhood of the identity element e, then U generates the group. In other words, every element of g is a product of elements of U.*

Proof. First note that $V = \mathrm{inv}(U) \cap U$ is an open neighborhood of the identity with the property that $\mathrm{inv}(V) = V$. We say that V is symmetric. We show that V generates G. For any open W_1 and W_2 in G, the set $W_1 W_2 = \{w_1 w_2 : w_1 \in W_1 \text{ and } w_2 \in W_2\}$ is an open set being a union of the open sets $\bigcup_{g \in W_1} g W_2$. Thus, in particular, the inductively defined sets

$$V^n = V V^{n-1}, \ n = 1, 2, 3, \ldots,$$

are open. We have

$$e \in V \subset V^2 \subset \cdots V^n \subset \cdots.$$

5.1. Definitions and Examples

It is easy to check that each V^n is symmetric and so also is the union

$$V^\infty := \bigcup_{n=1}^{\infty} V^n.$$

Moreover, V^∞ is not only closed under inversion, but also obviously closed under multiplication. Thus V^∞ is an open subgroup. From Exercise 5.5, V^∞ is also closed, and since G is connected, we obtain $V^\infty = G$. $\square$

In general, the connected component of a Lie group G that contains the identity is a Lie group denoted G_0, and it is generated by any open neighborhood of the identity. We call G_0 the **identity component** of G.

Definition 5.7. For a Lie group G and a fixed element $g \in G$, the maps $L_g : G \to G$ and $R_g : G \to G$ are defined by

$$L_g x = gx \text{ for } x \in G,$$
$$R_g x = xg \text{ for } x \in G,$$

and are called **left translation** and **right translation** (by g) respectively.

It is easy to see that L_g and R_g are diffeomorphisms with $L_g^{-1} = L_{g^{-1}}$ and $R_g^{-1} = R_{g^{-1}}$.

If G and H are Lie groups, then so is the product manifold $G \times H$, where multiplication is $(g_1, h_1) \cdot (g_2, h_2) = (g_1 g_2, h_1 h_2)$. The Lie group $G \times H$ is called the **product Lie group**. For example, the product group $S^1 \times S^1$ is called the 2-**torus group**. More generally, the higher torus groups are defined by $T^n = S^1 \times \cdots \times S^1$ (n factors).

Definition 5.8. Let H be an abstract subgroup of a Lie group G. If H is a Lie group such that the inclusion map $H \hookrightarrow G$ is an immersion, then we say that H is a **Lie subgroup** of G.

Proposition 5.9. *If H is an abstract subgroup of a Lie group G that is also a regular submanifold, then H is a closed Lie subgroup.*

Proof. The multiplication and inversion maps, $H \times H \to H$ and $H \to H$, are the restrictions of the multiplication and inversion maps on G, and since H is a regular submanifold, we obtain the needed smoothness of these maps. The harder part is to show that H is closed. So let $x_0 \in \overline{H}$ be arbitrary. Let $(U, \mathbf{x})$ be a single-slice chart adapted to H whose domain contains e. Let $\delta : G \times G \to G$ be the map $\delta(g_1, g_2) = g_1^{-1} g_2$, and choose an open set V such that $e \in V \subset \overline{V} \subset U$. By continuity of the map δ we can find an open neighborhood O of the identity element such that $O \times O \subset \delta^{-1}(V)$. Now if $\{h_i\}$ is a sequence in H converging to $x_0 \in \overline{H}$, then $x_0^{-1} h_i \to e$ and $x_0^{-1} h_i \in O$ for all sufficiently large i. Since $h_j^{-1} h_i = \left(x_0^{-1} h_j\right)^{-1} x_0^{-1} h_i$, we

have that $h_j^{-1}h_i \in V$ for sufficiently large i, j. For any sufficiently large fixed j, we have
$$\lim_{i\to\infty} h_j^{-1}h_i = h_j^{-1}x_0 \in \overline{V} \subset U.$$
Since U is the domain of a single-slice chart, $U \cap H$ is closed in U. Thus since each $h_j^{-1}h_i$ is in $U \cap H$, we see that $h_j^{-1}x_0 \in U \cap H \subset H$ for all sufficiently large j. This shows that $x_0 \in H$, and since x_0 was arbitrary, we are done. □

By a **closed Lie subgroup** we shall always mean one that is a regular submanifold as in the previous theorem. It is a nontrivial fact that an abstract subgroup of a Lie group that is also a closed subset is automatically a closed Lie subgroup in this sense (see Theorem 5.81).

Example 5.10. S^1 embedded as $S^1 \times \{1\}$ in the torus $S^1 \times S^1$ is a closed subgroup.

Example 5.11. Let S^1 be considered as the set of unit modulus complex numbers. The image in the torus $T^2 = S^1 \times S^1$ of the map $\mathbb{R}^1 \to S^1 \times S^1$ given by $t \mapsto \left(e^{i2\pi t}, e^{i2\pi at}\right)$ is a Lie subgroup. This map is a homomorphism. If a is a rational number, then the image is an embedded copy of S^1 wrapped around the torus several times depending on a. If a is irrational, then the image is still a Lie subgroup *but is now dense in T^2*.

The last example is important since it shows that a Lie subgroup might actually be a dense subset of the containing Lie group.

5.2. Linear Lie Groups

Let V be an n-dimensional vector space over $\mathbb{F}$, where $\mathbb{F} = \mathbb{R}$ or $\mathbb{C}$. The space $L(V, V)$ of linear maps from V to V is a vector space and therefore a smooth manifold. A global chart for $L(V, V)$ may be obtained by first choosing a basis for V and then defining n^2 functions $\{x_j^i\}_{1 \le i,j \le n}$ by the rule that if $A \in L(V, V)$, then $x_j^i(A)$ is the ij-th entry of the matrix that represents A with respect to the chosen basis. If the field is $\mathbb{R}$, then these are the coordinate functions of a global chart. If the field is $\mathbb{C}$, then we simply take the real and imaginary parts of the x_j^i and thereby obtain $2n^2$ coordinate functions. The various choices of basis give compatible charts, and the reader may check that these are charts from the smooth structure that $L(V, V)$ has by virtue of being a finite-dimensional vector space.

The determinant of an element $A \in L(V, V)$ is given as the determinant of any matrix which represents A with respect to some basis. The group $GL(V)$ of all linear automorphisms of V is an open submanifold of $L(V, V)$ given by the condition of nonvanishing determinant, and the restrictions of

5.2. Linear Lie Groups

the coordinate functions just introduced provide a global chart for GL(V). Let GL($n, \mathbb{F}$) denote the group of invertible matrices with entries from $\mathbb{F}$. We obtain an isomorphism of GL(V) with the matrix group GL($n, \mathbb{F}$) by choosing a basis and then simply sending each element of GL(V) to its matrix representative with respect to that basis. If we write mat(A) for the matrix that represents $A \in$ GL(V) with respect to a fixed basis, then $A \to$ mat(A) is a group isomorphism and is clearly smooth. It follows that GL(V) is a Lie group, and for each choice of basis we have an isomorphism of Lie groups GL(V) $\cong$ GL($n, \mathbb{F}$). In practice it is common to work with the matrix group GL($n, \mathbb{F}$) and its subgroups. The Lie group GL(V) is called the **general linear group of** V and is also denoted GL(V, $\mathbb{F}$) when we want to make the field apparent. The matrix group GL($n, \mathbb{F}$) is also referred to as a general linear (matrix) group and is often identified with GL($\mathbb{F}^n$). More specifically, GL($n, \mathbb{R}$) is called the *real* general linear group, and GL($n, \mathbb{C}$) is called the *complex* general linear group. Lie groups that are subgroups of GL(V) for some vector space V are referred to as **linear Lie groups** and are often realized as matrix subgroups of GL($n, \mathbb{F}$) for some n.

Definition 5.12. Let V be an n-dimensional vector space over the field $\mathbb{F}$ which we take to be either $\mathbb{R}$ or $\mathbb{C}$. Then the group SL(V) defined by

$$\text{SL(V)} = \text{SL(V}, \mathbb{F}) := \{A \in \text{GL(V)} : \det(A) = 1\}$$

is called the **special linear group** for V.

A bilinear form $\beta : \text{V} \times \text{V} \to \mathbb{F}$ on an $\mathbb{F}$-vector space V is called **nondegenerate** if the maps V $\to$ V* given by $\beta_R : v \mapsto \beta(v, \cdot)$ and $\beta_L : v \mapsto \beta(\cdot, v)$, are both linear isomorphisms. If V is finite-dimensional, then β_R is an isomorphism if and only if β_L is an isomorphism and then β is nondegenerate provided it has the property that if $\beta(v, w) = 0$ for all $w \in V$, then $v = 0$.

Definition 5.13. A (real) **scalar product** on a (real) finite-dimensional vector space V is a nondegenerate symmetric bilinear form $\beta : \text{V} \times \text{V} \to \mathbb{R}$. A (real) **scalar product space** is a pair (V, β) where V is a real vector space and β is a scalar product. (As usual we refer to V itself as the scalar product space when β is given)

Definition 5.14. Let V be a complex vector space. An $\mathbb{R}$-bilinear map $\beta : \text{V} \times \text{V} \to \mathbb{C}$ that satisfies $\beta(av, w) = \bar{a}\beta(v, w)$ and $\beta(v, aw) = a\beta(v, w)$ for all $a \in \mathbb{C}$ and $v, w \in \text{V}$ is called a **sesquilinear form**. If also $\beta(v, w) = \overline{\beta(w, v)}$, we call β a **Hermitian form**. If a Hermitian form is nondegenerate, we call it a **Hermitian scalar product**, and then (V, β) a **Hermitian scalar product space**.

The sesquilinear conditions imposed are described by saying that β is to be **conjugate linear** in the first slot and linear in the second slot. Nondegeneracy for a sesquilinear form is defined as for bilinear forms. Many authors define sesquilinear and Hermitian forms to be conjugate linear in the second slot and linear in the first. Obviously, $\beta(v,v)$ is always real for a Hermitian form.

Definition 5.15. Let β be a (real) scalar product or Hermitian scalar product on a vector space V. Then

    (i) β is **positive** (resp. negative) **definite** if $\beta(v,v) \geq 0$ (resp. $\beta(v,v) \leq 0$) for all $v \in V$ and $\beta(v,v) = 0 \Longrightarrow v = 0$;

    (ii) β is **positive** (resp. negative) **semidefinite** if $\beta(v,v) \geq 0$ (resp. $\beta(v,v) \leq 0$) for all $v \in V$.

What is called an **inner product** (a term we have already used) is a *positive definite* scalar product (or positive definite Hermitian scalar product in the complex case). In this book the term "inner product" always implies positive definiteness. Let us generalize the notion of orthonormal basis for an inner product space to include indefinite scalar product spaces. A basis $(e_1, \ldots, e_n)$ for a scalar product space (V, β) is called an **orthonormal basis** if $\beta(e_i, e_j) = 0$ when $i \neq j$, and $\beta(e_i, e_i) = \pm 1$ for all i. An orthonormal basis always exists for a finite-dimensional scalar product space.

Definition 5.16. Let V be an n-dimensional vector space over the field $\mathbb{F}$ which we take to be either $\mathbb{R}$ or $\mathbb{C}$. If β is a bilinear form or sesquilinear form on V, then $\mathrm{Aut}(V, \beta)$ is the subgroup of $\mathrm{GL}(V)$ defined by

$$\mathrm{Aut}(V, \beta) := \{A \in \mathrm{GL}(V) : \beta(Av, Aw) = \beta(v, w) \text{ for all } v, w \in V\}.$$

If β is a scalar product, then the elements of $\mathrm{Aut}(V, \beta)$ are called isometries of V.

Theorem 5.17. $\mathrm{SL}(V)$ *is a closed Lie subgroup of* $\mathrm{GL}(V)$. *If* β *is a bilinear or sesquilinear form as above, then* $\mathrm{Aut}(V, \beta)$ *and* $\mathrm{SAut}(V, \beta) := \mathrm{Aut}(V, \beta) \cap \mathrm{SL}(V)$ *are closed Lie subgroups of* $\mathrm{GL}(V)$. *(The form β is most often taken to be nondegenerate.)*

Proof. It is easy to check that the sets in question are subgroups. They are clearly *closed*. For example, $\mathrm{Aut}(V, \beta) = \bigcap_{v,w} F_{v,w}$, where

$$F_{v,w} := \{A \in \mathrm{GL}(V) : \beta(Av, Aw) = \beta(v, w)\}.$$

The fact that they are *Lie* subgroups follows from Theorem 5.81 below. However, as we shall see, most of the specific cases arising from various choices of β can be proved to be Lie groups by other means. That they are Lie subgroups follows from Proposition 5.9 once we show that they are

5.2. Linear Lie Groups

regular submanifolds of the appropriate group $GL(V, \mathbb{F})$. We will return to this later, once we have introduced another powerful theorem that will allow us to verify this without the use of Theorem 5.81. □

Let $\dim V = n$. After choosing a basis, $SL(V)$ gives the matrix version $SL(n, \mathbb{F}) := \{A \in M_{n \times n}(\mathbb{F}) : \det A = 1\}$. Notice that even when $\mathbb{F} = \mathbb{C}$, it may be that β is only required to be $\mathbb{R}$-linear. Depending on whether $\mathbb{F} = \mathbb{C}$ or $\mathbb{R}$ and on the nature of β, the notation for the linear groups takes on special conventional forms introduced below. When choosing a basis in order to represent one of the groups associated to a form β in a matrix version, it is usually the case that one uses a basis under which the matrix that represents β takes on a canonical form.

Example 5.18 (The (semi) orthogonal groups). Let (V, β) be a real scalar product space. In this case we write $\mathrm{Aut}(V, \beta)$ as $O(V, \beta)$ and refer to it as the **semiorthogonal group** associated to β. With respect to an appropriately ordered orthonormal basis, β is represented by a diagonal matrix of the form

$$\eta_{p,q} = \begin{bmatrix} 1 & 0 & \cdots & & \cdots & 0 \\ 0 & \ddots & 0 & & & \vdots \\ \vdots & 0 & 1 & \ddots & & \\ & & \ddots & -1 & 0 & \vdots \\ \vdots & & & 0 & \ddots & 0 \\ 0 & \cdots & & \cdots & 0 & -1 \end{bmatrix},$$

where there are p ones and q minus ones down the diagonal. The group of matrices arising from $O(V, \beta)$ with such a choice of basis is denoted $O(p, q)$ and consists exactly of the real matrices Q satisfying $Q\eta_{p,q}Q^t = \eta_{p,q}$. These groups are called the semiorthogonal matrix groups. With such an orthonormal choice of basis as above, the bilinear form (scalar product) is given as a canonical form on $\mathbb{R}^n$ where $(p + q = n)$:

$$\langle x, y \rangle := \sum_{i=1}^{p} x^i y^i - \sum_{i=p+1}^{n} x^i y^i,$$

and we have the alternative description

$$O(p, q) = \{Q \in GL(n) : \langle Qx, Qy \rangle = \langle x, y \rangle \text{ for all } x, y \in \mathbb{R}^n\}.$$

If β is positive definite, we then have $q = 0$, and $O(V, \beta)$ is referred to as a **real orthogonal group**. We write $O(n, 0)$ as $O(n)$ and refer to it as the **real orthogonal (matrix) group**; $Q \in O(n) \iff Q^t Q = I$.

Example 5.19. There are also **complex orthogonal groups** (not to be confused with unitary groups). In matrix representation, we have $O(n, \mathbb{C}) := \{Q \in \mathrm{GL}(n, \mathbb{C}) : Q^t Q = I\}$.

Example 5.20. Let (V, β) be a **Hermitian scalar product space**. In this case, we write $\mathrm{Aut}(V, \beta, \mathbb{C})$ as $U(V, \beta)$ and refer to it as the **semiunitary group** associated to β. If β is positive definite, then we call it a **unitary group**. Again we may choose a basis for V such that β is represented by the Hermitian form on $\mathbb{C}^n$ given by

$$\langle x, y \rangle := \sum_{i=1}^{p} \bar{x}^i y^i - \sum_{i=p+1}^{p+q=n} \bar{x}^i y^i.$$

We then obtain the semiunitary matrix group

$$\mathrm{U}(p, q) = \{A \in \mathrm{GL}(n, \mathbb{C}) : \langle Ax, Ay \rangle = \langle x, y \rangle \text{ for all } x, y \in \mathbb{R}^n\}.$$

We write $\mathrm{U}(n, 0)$ as $\mathrm{U}(n)$ and refer to it as the **unitary** (matrix) **group**. In particular, $U(1) = S^1 = \{z \in \mathbb{C} : |z| = 1\}$.

Definition 5.21. For (V, β) a real scalar product space, we have the **special orthogonal group** of (V, β) given by

$$\mathrm{SO}(V, \beta) = \mathrm{O}(V, \beta) \cap \mathrm{SL}(V, \mathbb{R}).$$

For (V, β) a Hermitian scalar product space, we have the special unitary group of (V, β) given by

$$\mathrm{SU}(V, \beta) = \mathrm{U}(V, \beta) \cap \mathrm{SL}(V, \mathbb{C}).$$

Definition 5.22. The group of $n \times n$ complex matrices of determinant one is the complex **special linear matrix group** $\mathrm{SL}(n, \mathbb{C})$. We also have the similarly defined real special linear group $\mathrm{SL}(n, \mathbb{R})$. The **special orthogonal** and **special semiorthogonal** matrix groups, $\mathrm{SO}(n)$ and $\mathrm{SO}(p, q)$, are the matrix groups defined by $\mathrm{SO}(n) = \mathrm{O}(n) \cap \mathrm{SL}(n, \mathbb{R})$ and $\mathrm{SO}(p, q) = \mathrm{O}(p, q) \cap \mathrm{SL}(n, \mathbb{R})$. The **special unitary** and **special semiunitary** matrix groups $\mathrm{SU}(n)$ and $\mathrm{SU}(p, q)$ are defined similarly.

The group $\mathrm{SO}(3)$ is the familiar matrix representation of the proper rotation group of Euclidean space and plays a prominent role in classical physics. Here "proper" refers to the fact that $\mathrm{SO}(3)$ does not contain any reflections. In the problems we ask the reader to show that $\mathrm{SO}(3)$ is the connected component of the identity in $\mathrm{O}(3)$.

Exercise 5.23. Show that $\mathrm{SU}(2)$ is simply connected while $\mathrm{SO}(3)$ is not.

Example 5.24 (Symplectic groups). We will describe both the real and the complex symplectic groups. Suppose that β is a nondegenerate skew-symmetric $\mathbb{C}$-bilinear (resp. $\mathbb{R}$-bilinear) form on a $2n$-dimensional complex

5.2. Linear Lie Groups

(resp. real) vector space V. The group $\text{Aut}(V, \beta)$ is called the **complex (resp. real) symplectic group** and is denoted by $\text{Sp}(V, \mathbb{C})$ (resp. $\text{Sp}(V, \mathbb{R})$). There exists a basis $\{f_i\}$ for V such that β is represented in the canonical form by

$$(v, w) = \sum_{i=1}^{n} v^i w^{n+i} - \sum_{j=1}^{n} v^{n+j} w^j.$$

The **symplectic matrix groups** are given by

$$\text{Sp}(2n, \mathbb{C}) := \{A \in M_{2n \times 2n}(\mathbb{C}) : (Av, Aw) = (v, w)\},$$
$$\text{Sp}(2n, \mathbb{R}) := \{A \in M_{2n \times 2n}(\mathbb{R}) : (Av, Aw) = (v, w)\},$$

where (v, w) is given as above.

Exercise 5.25. For $\mathbb{F} = \mathbb{C}$ or $\mathbb{R}$, show that $A \in \text{Sp}(2n, \mathbb{F})$ if and only if $A^t J A = J$, where

$$J = \begin{pmatrix} 0 & I \\ -I & 0 \end{pmatrix}.$$

Much of the above can be generalized somewhat more. Recall that the algebra of quaternions $\mathbb{H}$ is a copy of $\mathbb{R}^4$ endowed with a multiplication described as follows: First let a generic elements of $\mathbb{R}^4$ be denoted by $x = (x^0, x^1, x^2, x^3)$, $y = (y^0, y^1, y^2, y^3)$, etc. Thus we are using $\{0, 1, 2, 3\}$ as our index set. Let the standard basis be denoted by $\mathbf{1}, \mathbf{i}, \mathbf{j}, \mathbf{k}$. We define a multiplication by taking these basis elements as generators and insisting on the following relations:

$$\mathbf{i}^2 = \mathbf{j}^2 = \mathbf{k}^2 = -\mathbf{1},$$
$$\mathbf{ij} = -\mathbf{ji} = \mathbf{k},$$
$$\mathbf{jk} = -\mathbf{kj} = \mathbf{i},$$
$$\mathbf{ki} = -\mathbf{ik} = \mathbf{j}.$$

Of course, $\mathbb{H}$ is a vector space over $\mathbb{R}$ since it is just $\mathbb{R}^4$ with some extra structure. As a ring, $\mathbb{H}$ is a division algebra which is very much like a field, lacking only the property of commutativity. In particular, we shall see that every nonzero element of $\mathbb{H}$ has a multiplicative inverse. Elements of the form $a\mathbf{1}$ for $a \in \mathbb{R}$ are identified with the corresponding real numbers, and such quaternions are called **real** quaternions. By analogy with complex numbers, quaternions of the form $x^1\mathbf{i} + x^2\mathbf{j} + x^3\mathbf{k}$ are called **imaginary** quaternions. For a given quaternion $x = x^0\mathbf{1} + x^1\mathbf{i} + x^2\mathbf{j} + x^3\mathbf{k}$, the quaternion $x^1\mathbf{i} + x^2\mathbf{j} + x^3\mathbf{k}$ is called the imaginary part of x, and $x^0\mathbf{1} = x^0$ is called the real part of x. We also have a conjugation defined by

$$x \mapsto \bar{x} := x^0\mathbf{1} - x^1\mathbf{i} - x^2\mathbf{j} - x^3\mathbf{k}.$$

Notice that $x\bar{x} = \bar{x}x$ is real and equal to $\left(x^0\right)^2 + \left(x^1\right)^2 + \left(x^2\right)^2 + \left(x^3\right)^2$. We denote the positive square root of this by $|x|$ so that $\bar{x}x = |x|^2$.

Exercise 5.26. Verify the following for $x, y \in \mathbb{H}$ and $a, b \in \mathbb{R}$:

$$\overline{ax + by} = a\bar{x} + b\bar{y}, \qquad \overline{(\bar{x})} = x,$$
$$|xy| = |x|\,|y|, \qquad |\bar{x}| = |x|,$$
$$\overline{xy} = \bar{y}\bar{x}.$$

Now we can write down the inverse of a nonzero $x \in \mathbb{H}$:

$$x^{-1} = \frac{1}{|x|^2}\bar{x}.$$

Notice the strong analogy with complex number arithmetic.

Example 5.27. The set of unit quaternions is $U(1, \mathbb{H}) := \{|x| = 1\}$. This set is closed under multiplication. As a manifold it is (diffeomorphic to) S^3. With quaternionic multiplication, $S^3 = U(1, \mathbb{H})$ is a compact Lie group. Compare this to Example 5.3 where we saw that $U(1, \mathbb{C}) = S^1$. For the future, we unify things by letting $U(1, \mathbb{R}) := \mathbb{Z}_2 = S^0 \subset \mathbb{R}$. In other words, we take the 0-sphere to be the subset $\{-1, 1\}$ with its natural structure as a multiplicative group.

$$U(1, \mathbb{H}) = S^3,$$
$$U(1, \mathbb{C}) = S^1,$$
$$U(1, \mathbb{R}) := \mathbb{Z}_2 = S^0.$$

Exercise 5.28. Prove the assertions in the last example.

We now consider the n-fold product $\mathbb{H}^n$, which, as a real vector space (and a smooth manifold), is $\mathbb{R}^{4n}$. However, let us think of elements of $\mathbb{H}^n$ as column vectors with quaternion entries. We want to treat $\mathbb{H}^n$ as a vector space over $\mathbb{H}$ with addition defined just as for $\mathbb{R}^n$ and $\mathbb{C}^n$, but since $\mathbb{H}$ is not commutative, we are not properly dealing with a vector space. In particular, we should decide whether scalars should multiply column vectors on the right or on the left. We choose to multiply on the right, and this could take some getting used to, but there is a good reason for our choice. This puts us into the category of right $\mathbb{H}$-modules were elements of $\mathbb{H}$ are the "scalars". The reader should have no trouble catching on, and so we do not make formal definitions at this time (but see Appendix D). For $v, w \in \mathbb{H}^n$ and $a, b \in \mathbb{H}$, we have

$$v(a + b) = va + vb$$
$$(v + w)a = va + wa$$
$$(va)\,b = v\,(ab).$$

5.2. Linear Lie Groups

A map $A : \mathbb{H}^n \to \mathbb{H}^n$ is said to be $\mathbb{H}$-linear if $A(va) = A(v)a$ for all $v \in \mathbb{H}^n$ and $a \in \mathbb{H}$. There is no problem with doing matrix algebra with matrices with quaternion entries, as long as one respects the noncommutativity of $\mathbb{H}$. For example, if $A = (a^i_j)$ and $B = (b^i_j)$ are matrices with quaternion entries, then writing $C = AB$ we have

$$c^i_j = \sum a^i_k b^k_j,$$

but we *cannot* expect that $\sum a^i_k b^k_j = \sum b^k_j a^i_k$. For any $A = (a^i_j)$, the map $\mathbb{H}^n \to \mathbb{H}^n$ defined by $v \mapsto Av$ is $\mathbb{H}$-linear since $A(va) = (Av)a$.

Definition 5.29. The set of all $m \times n$ matrices with quaternion entries is denoted $M_{m \times n}(\mathbb{H})$. The subset $\mathrm{GL}(n, \mathbb{H})$ is defined as the set of all $Q \in M_{m \times n}(\mathbb{H})$ such that the map $v \mapsto Qv$ is a bijection.

We will now see that $\mathrm{GL}(n, \mathbb{H})$ is a Lie group isomorphic to a subgroup of $\mathrm{GL}(2n, \mathbb{C})$. First we define a map $\iota : \mathbb{C}^2 \to \mathbb{H}$ as follows: For $(z_1, z_2) \in \mathbb{C}$ with $z_1 = x^0 + x^1 \mathbf{i}$ and $z_2 = x^2 + x^3 \mathbf{i}$, we let $\iota(z^1, z^2) = (x^0 + x^1 \mathbf{i}) + (x^2 + x^3 \mathbf{i})\mathbf{j}$ where on the right hand side we interpret $\mathbf{i}$ as a quaternion. Note that $(x^0 + x^1 \mathbf{i}) + (x^2 + x^3 \mathbf{i})\mathbf{j} = x^0 + x^1 \mathbf{i} + x^2 \mathbf{j} + x^3 \mathbf{k}$. It is easily shown that this map is an $\mathbb{R}$-linear bijection, and we use this map to identify $\mathbb{C}^2$ with $\mathbb{H}$. Another way of looking at this is that we identify $\mathbb{C}$ with the span of 1 and $\mathbf{i}$ in $\mathbb{H}$ and then every quaternion has a unique representation as $z^1 + z^2 \mathbf{j}$ for $z^1, z^2 \in \mathbb{C} \subset \mathbb{H}$. We extend this idea to square quaternionic matrices; we can write every $Q \in M_{m \times n}(\mathbb{H})$ in the form $A + B\mathbf{j}$ for $A, B \in M_{m \times n}(\mathbb{C})$ in a unique way. This representation makes it clear that $M_{m \times n}(\mathbb{H})$ has a natural complex vector space structure, where the scalar multiplication is $z(A + B\mathbf{j}) = zA + zB\mathbf{j}$. Direct computation shows that

$$(A + B\mathbf{j})(C + D\mathbf{j}) = (AC - B\bar{D}) + (AD + B\bar{C})\mathbf{j}$$

for $A + B\mathbf{j} \in M_{m \times n}(\mathbb{H})$ and $C + D\mathbf{j} \in M_{n \times k}(\mathbb{H})$, where we have used the fact that for $Q \in M_{m \times n}(\mathbb{C})$ we have $Q\mathbf{j} = \mathbf{j}\bar{Q}$. From this it is not hard to show that the map $\vartheta_{m \times n} : M_{m \times n}(\mathbb{H}) \to M_{2m \times 2n}(\mathbb{C})$ given by

$$\vartheta_{m \times n} : A + B\mathbf{j} \longmapsto \begin{pmatrix} A & B \\ -\bar{B} & \bar{A} \end{pmatrix}$$

is an injective $\mathbb{R}$-linear map which respects matrix multiplication and thus is an $\mathbb{R}$-algebra isomorphism onto its image. We may identify $M_{m \times n}(\mathbb{H})$ with the subspace of $M_{2m \times 2n}(\mathbb{C})$ consisting of all matrices of the form $\begin{pmatrix} A & B \\ -\bar{B} & \bar{A} \end{pmatrix}$, where $A, B \in \mathbb{C}^{m \times n}$. In particular, if $m = n$, then we obtain an injective $\mathbb{R}$-linear algebra homomorphism $\vartheta_{n \times n} : M_{n \times n}(\mathbb{H}) \to M_{2n \times 2n}(\mathbb{C})$, and thus the image of this map in $M_{2n \times 2n}(\mathbb{C})$ is another realization of the matrix algebra $M_{n \times n}(\mathbb{H})$. If we specialize to the case of $n = 1$, we get a realization of $\mathbb{H}$ as the set of all 2×2 complex matrices of the form $\begin{pmatrix} z & w \\ -\bar{w} & \bar{z} \end{pmatrix}$. This set of matrices is closed under multiplication and forms an algebra over the field $\mathbb{R}$. Let us denote this algebra of matrices by the symbol $\mathcal{R}^4$ since it is diffeomorphic to $\mathbb{H} \cong \mathbb{R}^4$. We now have an algebra isomorphism $\vartheta : \mathbb{H} \to \mathcal{R}^4$ under which the quaternions $\mathbf{1}, \mathbf{i}, \mathbf{j}$ and $\mathbf{k}$ correspond to the matrices

$$\begin{pmatrix} 1 & 0 \\ 0 & 1 \end{pmatrix}, \begin{pmatrix} i & 0 \\ 0 & -i \end{pmatrix}, \begin{pmatrix} 0 & 1 \\ -1 & 0 \end{pmatrix} \text{ and } \begin{pmatrix} 0 & i \\ i & 0 \end{pmatrix}$$

respectively. Since $\mathbb{H}$ is a division algebra, each of its nonzero elements has a multiplicative inverse. Thus $\mathcal{R}^4$ must contain the matrix inverse of each of its nonzero elements. This can be seen directly:

$$\begin{pmatrix} z & w \\ -\bar{w} & \bar{z} \end{pmatrix}^{-1} = \frac{1}{|z|^2 + |w|^2} \begin{pmatrix} \bar{z} & -w \\ \bar{w} & z \end{pmatrix}.$$

Consider again the group of unit quaternions $U(1, \mathbb{H})$. We have already seen that as a smooth manifold, $U(1, \mathbb{H})$ is S^3. However, under the isomorphism $\mathbb{H} \to \mathcal{R}^4 \subset M_{2\times 2}(\mathbb{C})$ just mentioned, $U(1, \mathbb{H})$ manifests itself as SU(2). Thus we obtain a smooth map $U(1, \mathbb{H}) \to \mathrm{SU}(2)$ that is a group isomorphism. We record this as a proposition:

Proposition 5.30. *The map $U(1, \mathbb{H}) \to \mathrm{SU}(2)$ given by*

$$x = z + w\mathbf{j} \mapsto \begin{pmatrix} z & w \\ -\bar{w} & \bar{z} \end{pmatrix},$$

where $x = x^0 + x^1\mathbf{i} + x^2\mathbf{j} + x^3\mathbf{k}$, $z = x^0 + x^1\mathbf{i}$ and $w = x^2 + x^3\mathbf{i}$, is a group isomorphism. Thus $S^3 = U(1, \mathbb{H}) \cong \mathrm{SU}(2)$.

Proof. The first equality has already been established. Notice that we then have

$$|x|^2 = |z|^2 + |w|^2 = \det\begin{pmatrix} z & w \\ -\bar{w} & \bar{z} \end{pmatrix},$$

and so $x \in U(1, \mathbb{H})$ if and only if $\begin{pmatrix} z & w \\ -\bar{w} & \bar{z} \end{pmatrix}$ has determinant one. But such matrices account for all elements of SU(2) (verify this). We leave it to the reader to check that the map $U(1, \mathbb{H}) \to \mathrm{SU}(2)$ is a group isomorphism. $\square$

Exercise 5.31. Show that $Q \in \mathrm{GL}(n, \mathbb{H})$ if and only if $\det(\vartheta_{n\times n}(Q)) \neq 0$.

The set of all elements of $\mathrm{GL}(2n, \mathbb{C})$ which are of the form $\begin{pmatrix} A & B \\ -\bar{B} & \bar{A} \end{pmatrix}$ is a subgroup of $\mathrm{GL}(2n, \mathbb{C})$ and in fact a Lie group. Using the last exercise, we see that we may identify $\mathrm{GL}(n, \mathbb{H})$ as a Lie group with this subgroup of $\mathrm{GL}(2n, \mathbb{C})$. We want to find a quaternionic analogue of $\mathrm{U}(n, \mathbb{C})$, and so we define $b : \mathbb{H}^n \times \mathbb{H}^n \to \mathbb{H}$ by

$$b(v, w) = \bar{v}^t w.$$

Explicitly, if
$$v = \begin{bmatrix} v^1 \\ \vdots \\ v^n \end{bmatrix} \text{ and } w = \begin{bmatrix} w^1 \\ \vdots \\ w^n \end{bmatrix},$$
then
$$b(v,w) = \begin{bmatrix} v^1 & \cdots & v^n \end{bmatrix} \begin{bmatrix} w^1 \\ \vdots \\ w^n \end{bmatrix} = \sum \bar{v}^i w^i.$$

Note that b is obviously $\mathbb{R}$-bilinear. But if $a \in \mathbb{H}$, then we have $b(va, w) = b(v,w)\bar{a}$ and $b(v, wa) = b(v,w)a$. Notice that we consistently use *right* multiplication by quaternionic scalars. Thus b is the quaternionic analogue of an Hermitian scalar product.

Definition 5.32. We define $\mathrm{U}(n, \mathbb{H})$:
$$\mathrm{U}(n, \mathbb{H}) := \{Q \in \mathrm{GL}(n, \mathbb{H}) : b(Qv, Qw) = b(v,w) \text{ for all } v, w \in \mathbb{H}^n\}$$
$\mathrm{U}(n, \mathbb{H})$ is called the **quaternionic unitary group**.

The group $\mathrm{U}(n, \mathbb{H})$ is sometimes called the symplectic group and is denoted $\mathrm{Sp}(n)$, but we will avoid this since we want no confusion with the symplectic groups we have already defined. The group $\mathrm{U}(n, \mathbb{H})$ is in fact a Lie group (Theorem 5.17 generalizes to the quaternionic setting). The image of $\mathrm{U}(n, \mathbb{H})$ in $M_{2n \times 2n}(\mathbb{C})$ under the map $\vartheta_{n \times n}$ is denoted $\mathrm{USp}(2n, \mathbb{C})$. Since it is easily established that $\vartheta_{n \times n}|_{\mathrm{U}(n, \mathbb{H})}$ is a group homomorphism, the image $\mathrm{USp}(2n, \mathbb{C})$ is a subgroup of $\mathrm{GL}(2n, \mathbb{C})$.

Exercise 5.33. Show that $\vartheta_{n \times n}(\bar{A}^t) = \left(\overline{\vartheta_{n \times n}(A)}\right)^t$. Show that $\mathrm{USp}(2n, \mathbb{C})$ is a Lie subgroup of $\mathrm{GL}(2n, \mathbb{C})$.

Exercise 5.34. Show that $\mathrm{USp}(2n, \mathbb{C}) = \mathrm{U}(2n) \cap \mathrm{Sp}(2n, \mathbb{C})$. Hint: Show that
$$\vartheta_{n \times n}(M_{n \times n}(\mathbb{H})) = \{A \in \mathrm{GL}(2n, \mathbb{C}) : JAJ^{-1} = \bar{A}\},$$
where $J = \begin{pmatrix} 0 & \mathrm{id} \\ -\mathrm{id} & 0 \end{pmatrix}$. Next show that if $A \in \mathrm{U}(2n)$, then $JAJ^{-1} = \bar{A}$ if and only if $A^t J A = J$.

5.3. Lie Group Homomorphisms

Definition 5.35. Let G and H be Lie groups. A smooth map $f : G \to H$ that is a group homomorphism is called a **Lie group homomorphism**. A Lie group homomorphism is called a **Lie group isomorphism** in case it has an inverse that is also a Lie group homomorphism. A Lie group isomorphism $G \to G$ is called a **Lie group automorphism** of G.

If $f: G \to H$ is a Lie group homomorphism, then by definition $f(g_1 g_2) = f(g_1)f(g_2)$ for all $g_1, g_2 \in G$, and it follows that $f(e) = e$ and also that $f(g^{-1}) = f(g)^{-1}$ for all $g \in G$.

Example 5.36. The inclusion $SO(n, \mathbb{R}) \hookrightarrow GL(n, \mathbb{R})$ is a Lie group homomorphism.

Example 5.37. The circle $S^1 \subset \mathbb{C}$ is a Lie group under complex multiplication and the map
$$z = e^{i\theta} \mapsto \begin{bmatrix} \cos(\theta) & \sin(\theta) & 0 \\ -\sin(\theta) & \cos(\theta) & 0 \\ 0 & 0 & I_{n-2} \end{bmatrix}$$
is a Lie group homomorphism of S^1 into $SO(n)$.

Example 5.38. The map $U(1, \mathbb{H}) \to SU(2)$ of Proposition 5.30 is a Lie group isomorphism.

Example 5.39. The **conjugation map** $C_g : G \to G$ given by $x \mapsto gxg^{-1}$ is a Lie group automorphism. Note that $C_g = L_g \circ R_{g^{-1}}$.

Proposition 5.40. *Let $f_1 : G \to H$ and $f_2 : G \to H$ be Lie group homomorphisms that agree in a neighborhood of the identity. If G is connected, then $f_1 = f_2$.*

Proof. By Theorem 5.6, any $g \in G$ is a product of elements in the set on which f_1 and f_2 agree, so the homomorphism property forces $f_1 = f_2$. $\square$

Exercise 5.41. Show that the multiplication map $\mu : G \times G \to G$ has tangent map at $(e, e) \in G \times G$ given as $T_{(e,e)} \mu(v, w) = v + w$. Recall that we identify $T_{(e,e)}(G \times G)$ with $T_e G \times T_e G$.

Exercise 5.42. $GL(n, \mathbb{R})$ is an open subset of the vector space of all $n \times n$ matrices $M_{n \times n}(\mathbb{R})$. Using the natural identification of $T_e GL(n, \mathbb{R})$ with $M_{n \times n}(\mathbb{R})$, show that
$$T_e C_g(x) = gxg^{-1},$$
where $g \in GL(n, \mathbb{R})$ and $x \in M_{n \times n}(\mathbb{R})$.

Example 5.43. The map $t \mapsto e^{it}$ is a Lie group homomorphism from $\mathbb{R}$ to $S^1 \subset \mathbb{C}$.

Definition 5.44. A Lie group homomorphism from the additive group $\mathbb{R}$ into a Lie group is called a **one-parameter subgroup**. (Note that despite the use of the word "subgroup", a one-parameter subgroup is actually a map.)

5.3. Lie Group Homomorphisms

Example 5.45. We have seen that the torus $S^1 \times S^1$ is a Lie group under multiplication given by $(e^{i\tau_1}, e^{i\theta_1})(e^{i\tau_2}, e^{i\theta_2}) = (e^{i(\tau_1+\tau_2)}, e^{i(\theta_1+\theta_2)})$. Every homomorphism of $\mathbb{R}$ into $S^1 \times S^1$, that is, every one-parameter subgroup of $S^1 \times S^1$, is of the form $t \mapsto (e^{tai}, e^{tbi})$ for some pair of real numbers $a, b \in \mathbb{R}$.

Example 5.46. The map $R : \mathbb{R} \to \mathrm{SO}(3)$ given by

$$t \mapsto \begin{pmatrix} \cos t & -\sin t & 0 \\ \sin t & \cos t & 0 \\ 0 & 0 & 1 \end{pmatrix}$$

is a one-parameter subgroup. Also, the map

$$t \mapsto \begin{pmatrix} \cos t & -\sin t & 0 \\ \sin t & \cos t & 0 \\ 0 & 0 & e^t \end{pmatrix}$$

is a one-parameter subgroup of $\mathrm{GL}(3)$.

Recall that an $n \times n$ complex matrix A is called **Hermitian** (resp. **skew-Hermitian**) if $\bar{A}^t = A$ (resp. $\bar{A}^t = -A$). Let $\mathfrak{su}(2)$ denote the vector space of skew-Hermitian matrices with zero trace. We will later identify $\mathfrak{su}(2)$ as the "Lie algebra" of $\mathrm{SU}(2)$.

Example 5.47. Given $g \in \mathrm{SU}(2)$, we define the map $\mathrm{Ad}_g : \mathfrak{su}(2) \to \mathfrak{su}(2)$ by $\mathrm{Ad}_g : x \mapsto gxg^{-1}$. The skew-Hermitian matrices of zero trace can be identified with $\mathbb{R}^3$ by using the following matrices as a basis:

$$\begin{pmatrix} 0 & -i \\ -i & 0 \end{pmatrix}, \begin{pmatrix} 0 & -1 \\ 1 & 0 \end{pmatrix}, \begin{pmatrix} -i & 0 \\ 0 & i \end{pmatrix}.$$

These are just $-i$ times the **Pauli matrices** $\sigma_1, \sigma_2, \sigma_3$, and so the correspondence $\mathfrak{su}(2) \to \mathbb{R}^3$ is given by $-xi\sigma_1 - yi\sigma_2 - iz\sigma_3 \mapsto (x, y, z)$. Under this correspondence, the inner product on $\mathbb{R}^3$ becomes the inner product $(A, B) = \frac{1}{2}\mathrm{trace}(A\bar{B}^t) = -\frac{1}{2}\mathrm{trace}(AB)$. But then

$$(\mathrm{Ad}_g A, \mathrm{Ad}_g B) = -\frac{1}{2}\mathrm{trace}(gAg^{-1}gBg^{-1})$$
$$= -\frac{1}{2}\mathrm{trace}(AB) = (A, B).$$

So, Ad_g can be thought of as an element of $\mathrm{O}(3)$. More is true; Ad_g acts as an element of $\mathrm{SO}(3)$, and the map $g \mapsto \mathrm{Ad}_g$ is then a homomorphism from $\mathrm{SU}(2)$ to $\mathrm{SO}(\mathfrak{su}(2)) \cong \mathrm{SO}(3)$. This is a special case of the adjoint map studied later. (This example is related to the notion of "spin". For more, see the online supplement.)

Definition 5.48. If a Lie group homomorphism $\wp : \widetilde{G} \to G$ is also a covering map then we say that $\widetilde{G}$ is a covering group and $\wp$ is a covering homomorphism. If $\widetilde{G}$ is simply connected, then $\widetilde{G}$ (resp. $\wp$) is called the universal covering group (resp. universal covering homomorphism) of G.

Exercise 5.49. Show that if $\wp : \widetilde{M} \to G$ is a smooth covering map and G is a Lie group, then $\widetilde{M}$ can be given a unique Lie group structure such that $\wp$ becomes a covering homomorphism. (You may assume that $\widetilde{M}$ is paracompact.)

Example 5.50. The group Mob of Möbius transformations of the complex plane given by $T_A : z \mapsto \frac{az+b}{cz+d}$ for $A = \begin{pmatrix} a & b \\ c & d \end{pmatrix} \in \mathrm{SL}(2,\mathbb{C})$ can be given the structure of a Lie group. The map $\wp : \mathrm{SL}(2,\mathbb{C}) \to$ Mob given by $\wp : A \mapsto T_A$ is onto but not injective. In fact, it is a (two fold) covering homomorphism. When do two elements of $\mathrm{SL}(2,\mathbb{C})$ map to the same element of Mob?

5.4. Lie Algebras and Exponential Maps

Definition 5.51. A vector field $X \in \mathfrak{X}(G)$ is called **left invariant** if and only if $(L_g)_* X = X$ for all $g \in G$. A vector field $X \in \mathfrak{X}(G)$ is called **right invariant** if and only if $(R_g)_* X = X$ for all $g \in G$. The set of left invariant (resp. right invariant) vector fields is denoted $\mathfrak{X}^L(G)$ (resp. $\mathfrak{X}^R(G)$).

Recall that by definition $(L_g)_* X = TL_g \circ X \circ L_g^{-1}$, and so left invariance means that $TL_g \circ X \circ L_g^{-1} = X$ or that given any $x \in G$ we have $T_x L_g \cdot X(x) = X(gx)$ for all $g \in G$. Thus $X \in \mathfrak{X}(G)$ is left invariant if and only if the following diagram commutes for every $g \in G$:

$$\begin{array}{ccc} TG & \xrightarrow{TL_g} & TG \\ X \uparrow & & \uparrow X \\ G & \xrightarrow{L_g} & G \end{array}$$

There is a similar diagram for right invariance.

Lemma 5.52. $\mathfrak{X}^L(G)$ *is closed under the Lie bracket operation.*

Proof. Suppose that $X, Y \in \mathfrak{X}^L(G)$. Then by Proposition 2.84 we have

$$(L_g)_*[X,Y] = [L_{g*}X, L_{g*}Y] = [X,Y]. \qquad \square$$

Given a vector $v \in T_e G$, we can define a smooth left (resp. right) invariant vector field L^v (resp. R^v) such that $L^v(e) = v$ (resp. $R^v(e) = v$) by the simple prescription

$$L^v(g) = TL_g \cdot v \qquad (\text{resp. } R^v(g) = TR_g \cdot v).$$

5.4. Lie Algebras and Exponential Maps

A bit more precisely, $L^v(g) = T_e(L_g) \cdot v$. The proof that this prescription gives *smooth* invariant vector fields is left to the reader (see Problem 7). Given a vector in $T_e G$ there are various notations for denoting the corresponding left (or right) invariant vector field, and we shall have occasion to use some different notation later on. We will also write $L(v)$ for L^v and $R(v)$ for R^v. The map $v \mapsto L(v)$ (resp. $v \mapsto R(v)$) is a linear isomorphism from $T_e G$ onto $\mathfrak{X}^L(G)$ (resp. $\mathfrak{X}^R(G)$):

Exercise 5.53. Show that $v \mapsto L^v$ gives a linear isomorphism $T_e G \cong \mathfrak{X}^L(G)$. Similarly, $T_e G \cong \mathfrak{X}^R(G)$ by $v \mapsto R(v)$.

We now restrict attention to the left invariant fields but keep in mind that essentially all of what we say for this case has analogies in the right invariant case. We will discover a conduit (the adjoint map) between the two cases. The linear isomorphism $T_e G \cong \mathfrak{X}^L(G)$ just discovered shows that $\mathfrak{X}^L(G)$ is, in fact, a vector space of finite dimension equal to the dimension of G. From this and Lemma 5.52 we immediately obtain the following:

Proposition 5.54. *If G is a Lie group of dimension n, then $\mathfrak{X}^L(G)$ is an n-dimensional Lie algebra under the bracket of vector fields (see Definition 2.75).*

Using the isomorphism $T_e G \cong \mathfrak{X}^L(G)$, we can transfer the Lie algebra structure to $T_e G$. This is the content of the following:

Definition 5.55. For a Lie group G, define the bracket of any two elements $v, w \in T_e G$ by

$$[v, w] := [L^v, L^w](e).$$

With this bracket, the vector space $T_e G$ becomes a Lie algebra (see Definition 2.78), and so we now have two Lie algebras, $\mathfrak{X}^L(G)$ and $T_e G$, which are isomorphic by construction. The abstract Lie algebra isomorphic to either/both of them is often referred to as *the* Lie algebra of the Lie group G and denoted variously by $\mathfrak{L}(G)$ or $\mathfrak{g}$. Of course, we are implying that $\mathfrak{L}(H)$ is denoted $\mathfrak{h}$ and $\mathfrak{L}(K)$ by $\mathfrak{k}$, etc. In some computations we will have to use a specific realization of $\mathfrak{g}$. Our default convention will be that $\mathfrak{g} = \mathfrak{L}(G) := T_e G$ with the bracket defined above.

Definition 5.56. Given a Lie algebra $\mathfrak{g}$, we can associate to every basis $v_1, \ldots, v_n$ for $\mathfrak{g}$, the **structure constants** c_{ij}^k which are defined by

$$[v_i, v_j] = \sum_k c_{ij}^k v_k \text{ for } 1 \le i, j, k \le n.$$

It follows from the skew symmetry of the Lie bracket and the Jacobi identity that the structure constants satisfy

(5.1)
i) $\quad c_{ij}^k = -c_{ji}^k,$

ii) $\quad \sum_k c_{rs}^k c_{kt}^i + c_{st}^k c_{kr}^i + c_{tr}^k c_{ks}^i = 0.$

The structure constants characterize the Lie algebra, and structure constants are sometimes used to actually define a Lie algebra once a basis is chosen. We will meet the structure constants again later.

Let $\mathfrak{a}$ and $\mathfrak{b}$ be Lie algebras. For (a_1, b_1) and (a_2, b_2) elements of the vector space $\mathfrak{a} \times \mathfrak{b}$, define

$$[(a_1, b_1), (a_2, b_2)] := ([a_1, a_2], [b_1, b_2]).$$

With this bracket, $\mathfrak{a} \times \mathfrak{b}$ is a Lie algebra called the **Lie algebra product** of $\mathfrak{a}$ and $\mathfrak{b}$. Recall the definition of an ideal in a Lie algebra (Definition 2.79). The subspaces $\mathfrak{a} \times \{0\}$ and $\{0\} \times \mathfrak{b}$ are ideals in $\mathfrak{a} \times \mathfrak{b}$ that are clearly isomorphic to $\mathfrak{a}$ and $\mathfrak{b}$ respectively. We often identify $\mathfrak{a}$ with $\mathfrak{a} \times \{0\}$ and $\mathfrak{b}$ with $\{0\} \times \mathfrak{b}$.

Exercise 5.57. Show that if G and H are Lie groups, then the Lie algebra $\mathfrak{g} \times \mathfrak{h}$ is (up to identifications) the Lie algebra of $G \times H$.

Definition 5.58. Given two Lie algebras over a field $\mathbb{F}$, say $(\mathfrak{a}, [,]_\mathfrak{a})$ and $(\mathfrak{b}, [,]_\mathfrak{b})$, an $\mathbb{F}$-linear map σ is called a **Lie algebra homomorphism** if and only if

$$\sigma([v, w]_\mathfrak{a}) = [\sigma v, \sigma w]_\mathfrak{b}$$

for all $v, w \in \mathfrak{a}$. A Lie algebra isomorphism is defined in the obvious way. A Lie algebra **isomorphism** $\mathfrak{g} \to \mathfrak{g}$ is called an **automorphism** of $\mathfrak{g}$.

It is not hard to show that the set of all automorphisms of $\mathfrak{g}$, denoted $\text{Aut}(\mathfrak{g})$, forms a Lie group (actually a Lie subgroup of $\text{GL}(\mathfrak{g})$).

Let V be a finite-dimensional real vector space. Then $GL(V)$ is an open subset of the linear space $L(V, V)$, and we identify the tangent bundle of $GL(V)$ with $GL(V) \times L(V, V)$ (recall Definition 2.58). The tangent space at $A \in GL(V)$ is then $\{A\} \times L(V, V)$. Now the Lie algebra is $T_I GL(V)$, and it has a Lie algebra structure derived from the Lie algebra structure on $\mathfrak{X}^L(GL(V))$. The natural isomorphism of $T_I GL(V)$ with $L(V, V)$ puts a Lie algebra structure on $L(V, V)$. We now show that the resulting bracket on $L(V, V)$ is just the commutator bracket given by $[A, B] := A \circ B - B \circ A$. In the following discussion we let $\widetilde{X}$ denote the left invariant vector field corresponding to $X \in L(V, V) \cong T_I(GL(V))$. By definition we have $\widetilde{[A, B]} = [\widetilde{A}, \widetilde{B}]$. We will need some simple results from the following easy exercises.

5.4. Lie Algebras and Exponential Maps

Exercise 5.59. For a fixed $A \in GL(V)$, the map $L_A : GL(V) \to GL(V)$ given by $A \mapsto A \circ B$ has tangent map given by $(A, X) \mapsto (A \circ B, A \circ X)$ where $(A, X) \in GL(V) \times L(V, V) \cong T(GL(V))$. Show that if $\widetilde{X}$ denotes the left invariant vector field corresponding to $X \in L(V, V)$, then $\widetilde{X}(A) = (A, AX)$.

We are going to consider functions on $GL(V)$ that are restrictions of linear functionals on the vector space $L(V, V)$. We will not notationally distinguish the functional from its restriction. If f is such a linear function and $X \in L(V, V)$, let $f_{,X}$ be given by $f_{,X}(A) := f(A \circ X)$. It is easy to check that $f_{,X}$ is also a linear functional, and so for $Y \in L(V, V)$ we also have $(f_{,X})_{,Y}$. It is clear that $(f_{,X})_{,Y} = f_{,Y \circ X}$ (notice the reversal of order).

Exercise 5.60. Show that the map $L(V, V) \to (L(V, V))^*$ given by $X \mapsto f_{,X}$ is linear over $\mathbb{R}$.

Exercise 5.61. Let $\widetilde{X}$ be the left invariant field corresponding to $X \in L(V, V)$ as above. If f is the restriction to $GL(V)$ of a linear functional on $L(V, V)$ as above, then $\widetilde{X} f = f_{,X}$. Solution/Hint: $(\widetilde{X} f)(A) = df|_A (\widetilde{X}_A) = f(A \circ X)$.

Recall that if one picks a basis for V, then we obtain a global chart on $GL(V)$ which is given by coordinate functions x^i_j defined by letting $x^i_j(A)$ be the ij-th entry of the matrix of A with respect to the chosen basis. These coordinate functions are restrictions of linear functions on $L(V, V)$. Using this we see that if $f_{,X} = f_{,Y}$ for all linear f, then $f(X) = f_{,X}(I) = f_{,Y}(I) = f(Y)$ for all linear f and this in turn implies $X = Y$.

Proposition 5.62. *The Lie algebra bracket on $L(V, V)$ induced by the isomorphisms $\mathfrak{X}_L(GL(V)) \cong T_I GL(V) \cong L(V, V)$ is the commutator bracket*

$$[X, Y] := X \circ Y - Y \circ X.$$

Proof. Let $X, Y \in L(V, V)$ and let $[X, Y]$ denote the bracket induced on $L(V, V)$. Then for any linear f we have

$$f_{,[X,Y]} = \widetilde{[X,Y]} f = [\widetilde{X}, \widetilde{Y}] f = \widetilde{X}\widetilde{Y} f - \widetilde{Y}\widetilde{X} f = \widetilde{X}(f_{,Y}) - \widetilde{Y}(f_{,X})$$
$$= (f_{,Y})_{,X} - (f_{,X})_{,Y} = f_{,X \circ Y} - f_{,Y \circ X} = f_{,(X \circ Y - Y \circ X)}.$$

Since f was an arbitrary linear functional, we conclude that $[X, Y] = X \circ Y - Y \circ X$. $\square$

A choice of basis gives the algebra isomorphism $L(V, V) \cong M_{n \times n}(\mathbb{R})$ where $n = \dim(V)$. This isomorphism preserves brackets and restricts to a group isomorphism $GL(V) \cong GL(n, \mathbb{R})$. It is now easy to arrive at the following corollary.

Corollary 5.63. *The linear isomorphism of* $\mathfrak{gl}(n,\mathbb{R}) = T_I GL(n,\mathbb{R})$ *with* $M_{n\times n}(\mathbb{R})$ *induces a Lie algebra structure on* $M_{n\times n}(\mathbb{R})$ *such that the bracket is given by* $[A, B] := AB - BA$.

From now on we follow the practice of identifying the Lie algebra $\mathfrak{gl}(V)$ of $GL(V)$ with the commutator Lie algebra $L(V, V)$. Similarly for the matrix group, the Lie algebra of $GL(n,\mathbb{R})$ is taken to be $M_{n\times n}(\mathbb{R})$ with commutator bracket.

If $G \subset GL(n)$ is some matrix group, then $T_I G$ may be identified with a linear subspace of $M_{n\times n}$. This linear subspace will be closed under the commutator bracket, and so we actually have an identification of Lie algebras: $\mathfrak{g}$ is identified with a subspace of $M_{n\times n}$. It is often the case that G is defined by some matrix equation or equations. By differentiating these equations we find the defining equations for $\mathfrak{g}$ (as a subspace of $M_{n\times n}$). We first prove a general result for Lie algebras of closed subgroups, and then we apply this to some matrix groups. Recall that if N is a submanifold of M and $\iota : N \hookrightarrow M$ is the inclusion map, then we identify $T_p N$ with $T_p \iota(T_p N)$ and $T_p \iota$ is an inclusion. If H is a closed Lie subgroup of a Lie group G, and $v \in \mathfrak{h} = T_e H$, then v corresponds to a left invariant vector field on G which is obtained by using the left translation in G. But v also corresponds to a left invariant vector field on H obtained from left translations *in H*. The notation we have been using so far is not sensitive to this distinction, so let us introduce an alternative notation.

Notation 5.64. For a Lie group G, we have the **alternative notation** v^G for the left invariant vector field whose value at e is v. If H is a closed Lie subgroup of G and $v \in T_e H$, then $v^G \in \mathfrak{X}^L(G)$ while $v^H \in \mathfrak{X}^L(H)$.

Proposition 5.65. *Let H be a closed Lie subgroup of a Lie group G. Let*

$$\widetilde{\mathfrak{X}^L}(H) := \{X \in \mathfrak{X}^L(G) : X(e) \in T_e H\}.$$

Then the restrictions of elements of $\widetilde{\mathfrak{X}^L}(H)$ to the submanifold H are the elements of $\mathfrak{X}^L(H)$. This induces an isomorphism of Lie algebras of vector fields $\widetilde{\mathfrak{X}^L}(H) \cong \mathfrak{X}^L(H)$. For $v, w \in \mathfrak{h}$, we have

$$[v,w]_{\mathfrak{h}} = [v^H, w^H]_e = [v^G, w^G]_e = [v,w]_{\mathfrak{g}},$$

and so $\mathfrak{h}$ is a Lie subalgebra of $\mathfrak{g}$; the bracket on $\mathfrak{h}$ is the same as that inherited from $\mathfrak{g}$.

Proof. The Lie bracket on $\mathfrak{h} = T_e H$ is given by $[v,w] := [v^H, w^H](e)$. Notice that if H is a closed Lie subgroup of a Lie group G, then for $h \in H$ we have left translation by h as a map $G \to G$ and also as a map $H \to H$. The latter is the restriction of the former. To avoid notational clutter, let us denote

5.4. Lie Algebras and Exponential Maps

$L_h|_H$ by l_h. If $\iota : H \hookrightarrow G$ is the inclusion, then we have $\iota \circ l_h = L_h \circ \iota$, and so $T\iota \circ Tl_h = TL_h \circ T\iota$. If $v^H \in \mathfrak{X}^L(H)$, then we have

$$T_h\iota\left(v^H(h)\right) = T_h\iota(T_e l_h(v)) = T\iota \circ Tl_h(v)$$
$$= TL_h \circ T\iota(v) = T_e L_h \left(T_e\iota(v)\right)$$
$$= T_e L_h (v) = v^G(h) = \left(v^G \circ \iota\right)(h),$$

so that v^H and v^G are ι-related for any $v \in T_e H \subset T_e G$. Thus for $v, w \in T_e H$ we have

$$[v, w]_{\mathfrak{h}} = [v^H, w^H]_e = [v^G, w^G]_e = [v, w]_{\mathfrak{g}},$$

the formula we wanted. Next, notice that if we take $T_h\iota$ as an inclusion so that $T_h\iota\left(v^H(h)\right) = v^H(h)$ for all h, then we have really shown that if $v \in T_e H$, then v^H is the restriction of v^G to H. Also it is easy to see that

$$\widetilde{\mathfrak{X}^L}(H) = \{v^G : v \in T_e H\},$$

and so the restrictions of elements of $\widetilde{\mathfrak{X}^L}(H)$ are none other than the elements of $\mathfrak{X}^L(H)$. From what we have shown, the restriction map $\widetilde{\mathfrak{X}^L}(H) \to \mathfrak{X}^L(H)$ is given by $v^G \mapsto v^H$ and is a surjective Lie algebra homomorphism. It also has kernel zero since if v^H is the zero vector field, then $v = 0$, which implies that v^G is the zero vector field. $\square$

Because $[v, w]_{\mathfrak{h}} = [v, w]_{\mathfrak{g}}$, the inclusion $\mathfrak{h} \hookrightarrow \mathfrak{g}$ is a Lie algebra homomorphism. Examining the details of the previous proof we see that we have a commutative diagram

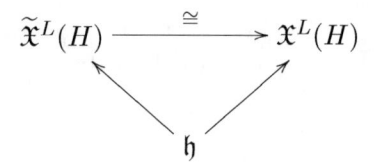

$$\widetilde{\mathfrak{X}^L}(H) \xrightarrow{\cong} \mathfrak{X}^L(H)$$
$$\nwarrow \quad \nearrow$$
$$\mathfrak{h}$$

of Lie algebra homomorphisms where the top horizontal map is a restriction to H, the left diagonal map is $v \mapsto v^G$ and the right diagonal map is $v \mapsto v^H$. In practice, what this last proposition shows is that in order to find the Lie algebra of a closed subgroup $H \subset G$, we only need to find the subspace $\mathfrak{h} = T_e H$, since the bracket on $\mathfrak{h}$ is just the restriction of the bracket on $\mathfrak{g}$. The following is also easily seen to be a commutative diagram of Lie algebra homomorphisms:

$$\begin{array}{ccc} \widetilde{\mathfrak{X}^L}(H) & \hookrightarrow & \mathfrak{X}^L(G) \\ \uparrow & & \uparrow \\ \mathfrak{h} & \hookrightarrow & \mathfrak{g} \end{array}$$

where both vertical maps are $v \mapsto v^G$.

The Lie algebra–Lie group correspondence works in the other direction too:

Theorem 5.66. *Let G be a Lie group with Lie algebra $\mathfrak{g}$. If $\mathfrak{h}$ is a Lie subalgebra of $\mathfrak{g}$, then there is a unique connected Lie subgroup H of G whose Lie algebra is $\mathfrak{h}$.*

The above theorem uses the Frobenius integrability theorem, so we defer the proof until we have that theorem in hand.

Since the Lie algebra of $\mathrm{GL}(n)$ is the set of all $n \times n$ matrices with the commutator bracket, and since we have just shown that the bracket for the Lie algebra of a subgroup is just the restriction of the bracket on the containing group, we see that the bracket on the Lie algebra of matrix subgroups can also be taken to be the commutator bracket if that Lie algebra is represented by the appropriate space of matrices. We record this as a proposition:

Proposition 5.67. *If $G \subset \mathrm{GL}(V)$ is a linear Lie group, then the Lie algebra of G may be identified with a subalgebra of $\mathrm{End}(V)$ with the commutator bracket. A similar statement holds for matrix Lie algebras.*

Example 5.68. Consider the orthogonal group $\mathrm{O}(n) \subset \mathrm{GL}(n)$. Given a curve of orthogonal matrices $Q(t)$ with $Q(0) = I$ and $\frac{d}{dt}\big|_{t=0} Q(0) = A$, we compute by differentiating the defining equation $I = Q^t Q$:

$$0 = \frac{d}{dt}\bigg|_{t=0} Q^t Q$$
$$= \left(\frac{d}{dt}\bigg|_{t=0} Q\right)^t Q(0) + Q^t(0)\left(\frac{d}{dt}\bigg|_{t=0} Q\right)$$
$$= A^t + A,$$

so that the space of skew-symmetric matrices is contained in the tangent space $T_I \mathrm{O}(n)$. But both $T_I \mathrm{O}(n)$ and the space of skew-symmetric matrices have dimension $n(n-1)/2$, so they are equal. This means that we can identify the Lie algebra $\mathfrak{o}(n) = \mathfrak{L}(\mathrm{O}(n))$ with the space of skew-symmetric matrices with the commutator bracket. One can easily check that the commutator bracket of two such matrices is skew-symmetric, as expected.

We have considered matrix groups as subgroups of $\mathrm{GL}(n)$, but it is often more convenient to consider subgroups of $\mathrm{GL}(n, \mathbb{C})$. Since $\mathrm{GL}(n, \mathbb{C})$ can be identified with a subgroup of $\mathrm{GL}(2n)$, this is only a slight change in viewpoint. The essential parts of our discussion go through for $\mathrm{GL}(n, \mathbb{C})$ without any significant change.

Example 5.69. Consider the unitary group $\mathrm{U}(n) \subset \mathrm{GL}(n, \mathbb{C})$. Given a curve of unitary matrices $Q(t)$ with $Q(0) = I$ and $\frac{d}{dt}\big|_{t=0} Q(0) = A$, we

5.4. Lie Algebras and Exponential Maps

compute by differentiating the defining equation $I = \bar{Q}^t Q$. We have

$$\begin{aligned}
0 &= \left.\frac{d}{dt}\right|_{t=0} \bar{Q}^t Q \\
&= \left(\left.\frac{d}{dt}\right|_{t=0} \bar{Q}\right)^t Q(0) + \bar{Q}^t(0) \left(\left.\frac{d}{dt}\right|_{t=0} Q\right) \\
&= \bar{A}^t + A.
\end{aligned}$$

Examining dimensions as before, we see that we can identify $\mathfrak{u}(n)$ with the space of skew-Hermitian matrices ($\bar{A}^t = -A$) under the commutator bracket.

Along the lines of the above examples, each of the familiar matrix Lie groups has a Lie algebra presented as a matrix Lie algebra with commutator bracket. This subalgebra is defined in terms of simple conditions as in the examples above, and the chart below lists some of the more common examples.

Group	Lie algebra	Conditions defining the Lie algebra
$SL(n, \mathbb{R})$	$\mathfrak{sl}(n, \mathbb{R})$	$\text{Trace}(A) = 0$
$O(n)$	$\mathfrak{o}(n)$	$A^t = -A$
$SO(n)$	$\mathfrak{so}(n)$	$A^t = -A$
$U(n)$	$\mathfrak{u}(n)$	$\bar{A}^t = -A$
$SU(n)$	$\mathfrak{su}(n)$	$\bar{A}^t = -A$, $\text{Trace}(A) = 0$
$Sp(2n, \mathbb{F})$	$\mathfrak{sp}(2n, \mathbb{F})$	$JA^t J = A$

In the chart above the matrix J is given by

$$J = \begin{pmatrix} 0 & I \\ -I & 0 \end{pmatrix}.$$

Exercise 5.70. Find the defining conditions for the Lie algebra of the semiorthogonal group $O(p, q)$.

We would like to relate Lie group homomorphisms to Lie algebra homomorphisms.

Proposition 5.71. *Let $f : G_1 \to G_2$ be a Lie group homomorphism. The map $T_e f : \mathfrak{g}_1 \to \mathfrak{g}_2$ is a Lie algebra homomorphism called the **Lie differential**, which is denoted in this context by $df : \mathfrak{g}_1 \to \mathfrak{g}_2$.*

Proof. For $v \in \mathfrak{g}_1$ and $x \in G$, we have

$$\begin{aligned}
T_x f \cdot L^v(x) &= T_x f \cdot (T_e L_x \cdot v) \\
&= T_e(f \circ L_x) \cdot v = T_e(L_{f(x)} \circ f) \cdot v \\
&= T_e L_{f(x)} (T_e f \cdot v) = T_e L_{f(x)} (T_e f \cdot v) \\
&= L^{df(v)}(f(x)),
\end{aligned}$$

so $L^v \sim_f L^{df(v)}$. Thus by Proposition 2.84 we have that for any $v, w \in \mathfrak{g}_1$,
$$L^{[v,w]} \sim_f [L^{df(v)}, L^{df(w)}].$$
In other words, $[L^{df(v)}, L^{df(w)}] \circ f = Tf \circ L^{[v,w]}$, which at e gives
$$[df(v), df(w)] = [v, w]. \qquad \square$$

Theorem 5.72. *Invariant vector fields are complete. The integral curves through the identity element are the one-parameter subgroups.*

Proof. We prove the left invariant case since the right invariant case is similar. Let X be a left invariant vector field and $c : (a, b) \to G$ be an integral curve of X with $\dot{c}(0) = X(p)$. Let $a < t_1 < t_2 < b$ and choose an element $g \in G$ such that $gc(t_1) = c(t_2)$. Let $\Delta t = t_2 - t_1$, and define $\bar{c} : (a + \Delta t, b + \Delta t) \to G$ by $\bar{c}(t) = gc(t - \Delta t)$. Then we have
$$\left.\frac{d}{dt}\right|_{t=0} \bar{c}(t) = TL_g \cdot \dot{c}(t - \Delta t) = TL_g \cdot X(c(t - \Delta t))$$
$$= X(gc(t - \Delta t)) = X(\bar{c}(t)),$$
and so $\bar{c}$ is also an integral curve of X. On the intersection $(a + \Delta t, b)$ of their domains, c and $\bar{c}$ are equal since they are both integral curves of the same field and since $\bar{c}(t_2) = gc(t_1) = c(t_2)$. Thus we can concatenate the curves to get a new integral curve defined on the larger domain $(a, b + \Delta t)$. Since this extension can be done again for the same fixed Δt, we see that c can be extended to (a, ∞). A similar argument shows that we can extend in the negative direction to get the needed extension of c to $(-\infty, \infty)$.

Next assume that c is the integral curve with $c(0) = e$. The proof that $c(s + t) = c(s)c(t)$ proceeds by considering $\gamma(t) = c(s)^{-1}c(s + t)$. Then $\gamma(0) = e$ and also
$$\dot{\gamma}(t) = TL_{c(s)^{-1}} \cdot \dot{c}(s + t) = TL_{c(s)^{-1}} \cdot X(c(s + t))$$
$$= X(c(s)^{-1}c(s + t)) = X(\gamma(t)).$$
By the uniqueness of integral curves we must have $c(s)^{-1}c(s + t) = c(t)$, which implies the result. Conversely, suppose $c : \mathbb{R} \to G$ is a one-parameter subgroup, and let $X_e = \dot{c}(0)$. There is a left invariant vector field X such that $X(e) = X_e$, namely $X = L^{X_e}$. We must show that the integral curve through e of the field X is exactly c. But for this we only need that $\dot{c}(t) = X(c(t))$ for all t. We have $c(t + s) = c(t)c(s)$ or $c(t + s) = L_{c(t)}c(s)$. Thus
$$\dot{c}(t) = \left.\frac{d}{ds}\right|_0 c(t + s) = (T_{c(t)}L) \cdot \dot{c}(0) = X(c(t)). \qquad \square$$

5.4. Lie Algebras and Exponential Maps

Proposition 5.73. *Let $v \in \mathfrak{g} = T_e G$. We have the corresponding left invariant field L^v and flow $\varphi_t^{L^v}$. Then, with $\varphi^v(t) := \varphi_t^{L^v}(e)$, we have that*

$$\varphi^v(st) = \varphi^{sv}(t). \tag{5.2}$$

A similar statement holds with R^v replacing L^v.

Proof. Let $u = st$. We have that $\frac{d}{dt}\big|_{t=0} \varphi^v(st) = \frac{d}{du}\big|_{u=0} \varphi^v(u) \frac{du}{dt}(0) = sv$ and so by uniqueness $\varphi^v(st) = \varphi^{sv}(t)$. □

Theorem 5.74. *Let G be a Lie group. For a smooth curve $c : \mathbb{R} \to G$ with $c(0) = e$ and $\dot{c}(0) = v$, the following are all equivalent:*

(i) *c is a one-parameter subgroup with $\dot{c}(0) = v$.*

(ii) *$c(t) = \varphi_t^{L^v}(e)$ for all t.*

(iii) *$c(t) = \varphi_t^{R^v}(e)$ for all t.*

(iv) *$\varphi_t^{L^v} = R_{c(t)}$ for all t.*

(v) *$\varphi_t^{R^v} = L_{c(t)}$ for all t.*

Proof. From Theorem 5.72 we already know that (ii) is equivalent to (i). The proof that (i) is equivalent to (iii) is analogous. Also, (iv) implies (i) since then $\varphi_t^{L^v}(e) = R_{c(t)}(e) = c(t)$. Now assuming (ii) we show (iv). We have $\dot{c}(t) = L^v(e) = v$ and

$$\frac{d}{dt}\bigg|_{t=0} gc(t) = \frac{d}{dt}\bigg|_{t=0} L_g(c(t))$$
$$= TL_g v = L^v(g) \text{ for any } g.$$

In other words,

$$\frac{d}{dt}\bigg|_{t=0} R_{c(t)} g = L^v(g)$$

for any g. But also, $R_{c(0)} g = eg = g$ and $\varphi_0^{L^v}(g) = g$ so by uniqueness we have $R_{c(t)} g = \varphi_t^{L^v}(g)$ for all g. We leave the remainder to the reader. □

It follows from (iv) and (v) of the previous theorem that the bracket of any left invariant vector field with any right invariant vector field is zero since their flows commute (see Theorem 2.109).

Definition 5.75 (Exponential map). For any $v \in \mathfrak{g} = T_e G$, we have the corresponding left invariant field L^v which has an integral curve $t \mapsto \varphi^v(t) := \varphi_t^{L^v}(e)$ through e. The map $\exp : \mathfrak{g} \to G$ defined by $\exp : v \mapsto \varphi^v(1)$ is referred to as the **exponential map**.

Thus for $s, t \in \mathbb{R}$ and $v \in \mathfrak{g}$ we have
$$\exp((s+t)v) = \exp(sv)\exp(tv),$$
$$\exp(-tv) = (\exp(tv))^{-1}.$$

Note that we usually use the same symbol "exp" for the exponential map of any Lie group, but we may also write $\exp^G$ to indicate that the group is G.

By Proposition 5.73 we have
$$\exp(tv) = \varphi^{tv}(1) = \varphi^v(t) = \varphi_t^{L^v}(e).$$
Thus by Theorem 5.74 we obtain the following:

Proposition 5.76. *For $v \in \mathfrak{g}$, the map $\mathbb{R} \to G$ given by $t \mapsto \exp(tv)$ is the one-parameter subgroup that is the integral curve of L^v.*

Lemma 5.77. *The map $\exp : \mathfrak{g} \to G$ is smooth.*

Proof. Consider the map $\mathbb{R} \times G \times \mathfrak{g} \to G \times \mathfrak{g}$ given by
$$(t, g, v) \mapsto (g \cdot \exp(tv), v).$$
This map is easily seen to be the flow on $G \times \mathfrak{g}$ of the vector field $\widetilde{X} : (g, v) \mapsto (L^v(g), 0)$ and so is smooth. The restriction of this smooth flow to the submanifold $\{1\} \times \{e\} \times \mathfrak{g}$ is $(1, e, v) \mapsto (\exp(v), v)$ and is also smooth. This clearly implies that $\exp$ is smooth also. $\square$

Note that $\exp(0) = e$. In the following theorem, we use the canonical identification of the tangent space of T_eG at the zero element (that is, $T_0(T_eG)$) with T_eG itself.

Theorem 5.78. *The tangent map of the exponential map $\exp : \mathfrak{g} \to G$ is the identity at $0 \in T_eG = \mathfrak{g}$, and $\exp$ is a diffeomorphism of some neighborhood of the origin onto its image in G,*
$$T_e \exp = \mathrm{id} : T_eG \to T_eG.$$

Proof. By Lemma 5.77, we know that $\exp : \mathfrak{g} \to G$ is a smooth map. Also, $\frac{d}{dt}\big|_0 \exp(tv) = v$, which means that the tangent map is $v \mapsto v$. If the reader thinks through the definitions carefully, he or she will discover that we have here used the identification of $\mathfrak{g}$ with $T_0\mathfrak{g}$. $\square$

From the definitions and Theorem 5.74 we have
$$\varphi_t^{L^v}(p) = p \exp tv,$$
$$\varphi_t^{L^v}(p) = (\exp tv) p$$
for all $v \in \mathfrak{g}$, all $t \in \mathbb{R}$ and all $p \in G$.

5.4. Lie Algebras and Exponential Maps

Proposition 5.79. *For a (Lie group) homomorphism $f : G_1 \to G_2$, the following diagram commutes:*

$$\begin{array}{ccc} \mathfrak{g}_1 & \xrightarrow{df} & \mathfrak{g}_2 \\ {\scriptstyle \exp^{G_1}}\downarrow & & \downarrow{\scriptstyle \exp^{G_2}} \\ G_1 & \xrightarrow{f} & G_2 \end{array}$$

Proof. For v in the Lie algebra of G_1, the curve $t \mapsto f(\exp^{G_1}(tv))$ is clearly a one-parameter subgroup. Also,

$$\left.\frac{d}{dt}\right|_0 f(\exp^{G_1}(tv)) = df(v),$$

and so by uniqueness of integral curves $f(\exp^{G_1}(tv)) = \exp^{G_2}(t\,df(v))$. $\square$

The Lie algebra of a Lie group and the group itself are closely related in many ways, and the exponential map is often the key to understanding the connection. One simple observation is that by Theorem 5.6, if G is a connected Lie group, then for any open neighborhood $V \subset \mathfrak{g}$ of 0 the group generated by $\exp(V)$ is all of G.

If H is a Lie subgroup of G, then the inclusion $\iota : H \hookrightarrow G$ is an injective homomorphism and Proposition 5.79 tells us that the exponential map on $\mathfrak{h} \subset \mathfrak{g}$ is the restriction of the exponential map on $\mathfrak{g}$. Thus, to understand the exponential map for linear Lie groups, we must understand the exponential map for the general linear group. Let V be a finite-dimensional vector space. It will be convenient to pick an inner product $\langle \cdot, \cdot \rangle$ on V and define the norm of $v \in$ V by $\|v\| := \sqrt{\langle v, v \rangle}$. In case V is a complex vector space, we use a Hermitian inner product. We put a norm on the set of linear transformations $L(V, V)$ by

$$\|A\| = \sup_{\|v\| \neq 0} \frac{\|Av\|}{\|v\|}.$$

We have $\|A \circ B\| \leq \|A\| \|B\|$, which implies that $\|A^k\| \leq \|A\|^k$. If we use the identification of $\mathfrak{gl}(V)$ with $L(V, V)$ (or equivalently the identification of $\mathfrak{gl}(n, \mathbb{R})$ with the vector space of $n \times n$ matrices $M_{n \times n}$), then the exponential map is given by a power series

$$A \mapsto \exp(A) = \sum_{k=0}^{\infty} \frac{1}{k!} A^k.$$

which can be seen from the following argument: The sequence of partial sums $s_N := \sum_{k=0}^{N} \frac{1}{k!} A^k$ is a Cauchy sequence in the normed space $\mathfrak{gl}(V)$,

$$\left\| \sum_{k=0}^{N} \frac{1}{k!} A^k - \sum_{k=0}^{M} \frac{1}{k!} A^k \right\| = \left\| \sum_{k=M+1}^{N} \frac{1}{k!} A^k \right\|$$

$$\leq \sum_{k=M+1}^{N} \frac{1}{k!} \|A\|^k.$$

From this we see that

$$\lim_{M,N \to \infty} \left\| \sum_{k=0}^{N} \frac{1}{k!} A^k - \sum_{k=0}^{M} \frac{1}{k!} A^k \right\| = 0,$$

and so $\{s_N\}$ is a Cauchy sequence. Since $\mathfrak{gl}(V)$ together with the given norm is known to be complete, we see that $\sum_{k=0}^{\infty} \frac{1}{k!} A^k$ converges. For a fixed $A \in \mathfrak{gl}(V)$, the function $\alpha : t \mapsto \alpha(t) = \exp(tA)$ is the unique solution of the initial value problem

$$\alpha'(t) = A\alpha(t), \qquad \alpha(0) = A.$$

This can be seen by differentiating term by term:

$$\frac{d}{dt} \exp(tA) = \sum_{k=0}^{\infty} \frac{1}{k!} t^k A^k = \sum_{k=1}^{\infty} \frac{1}{(k-1)!} t^{k-1} A^k$$

$$= A \sum_{k=1}^{\infty} \frac{1}{(k-1)!} t^{k-1} A^{k-1} = A \exp(tA).$$

Under our identifications, this says that α is the integral curve corresponding to the left invariant vector field determined by A. Thus we have a concrete realization of the exponential map for $\mathfrak{gl}(V)$ and, by restriction, each Lie subgroup of $\mathfrak{gl}(V)$. Applying what we know about exponential maps in the abstract setting of a general Lie group we have in this concrete case $\exp((s+t)A) = \exp(sA)\exp(tA)$ and $\exp(-tA) = (\exp(tA))^{-1}$. Let $A, B \in \mathfrak{gl}(V)$. Then

$$\exp(A)\exp(B) = \left(\sum_{j=0}^{\infty} \frac{1}{j!} t^j A^j \right) \left(\sum_{k=0}^{\infty} \frac{1}{k!} t^k B^k \right)$$

$$= \sum_{j=0}^{\infty} \sum_{k=0}^{\infty} \frac{1}{j!k!} t^{j+k} A^j B^k.$$

5.4. Lie Algebras and Exponential Maps

On the other hand, suppose that $A \circ B = B \circ A$. Then we have

$$\exp(A+B) = \sum_{m=0}^{\infty} \frac{1}{m!} t^m (A+B)^m = \sum_{m=0}^{\infty} \frac{1}{m!} \left(\sum_{j+k=m} \frac{m!}{j!k!} A^j B^k \right)$$

$$= \sum_{j=0}^{\infty} \sum_{k=0}^{\infty} \frac{1}{j!k!} t^{j+k} A^j B^k.$$

Thus in the case where A commutes with B, we have

$$\exp(A+B) = \exp(A)\exp(B).$$

Since most Lie groups of interest in practice are linear Lie groups, it will pay to understand the exponential map a bit better in this case. Let V be a finite-dimensional vector space equipped with an inner product as before, and take the induced norm on $\mathfrak{gl}(V)$. By Problem 18 we can define a map $\log : U \to \mathfrak{gl}(V)$, where

$$U = \{B \in \mathrm{GL}(V) : \|B\| < 1\},$$

by using the power series:

$$\log B := \sum_{k=1}^{\infty} \frac{(-1)^k}{k} (B-I)^k.$$

If we compute formally, then for $A \in \mathfrak{gl}(V)$,

$$\log(\exp A) = \left(A + \frac{1}{2!}A^2\right) - \frac{1}{2}\left(A + \frac{1}{2!}A^2\right)^2$$
$$+ \frac{1}{3}\left(A + \frac{1}{2!}A^2\right)^3 + \cdots$$
$$= A + \left(\frac{1}{2!}A^2 - \frac{1}{2}A^2\right) + \left(\frac{1}{3!}A^3 - \frac{1}{2}A^3 + \frac{1}{3}A^3\right) + \cdots.$$

We will argue that the above makes sense if $\|A\| < \log 2$ and that there must be cancellations in the last line so that $\log(\exp A) = A$. In fact, $\|\exp A - I\| \leq e^{\|A\|} - 1$, and so the double series on the first line for $\log(\exp A)$ must converge absolutely if $e^{\|A\|} - 1 < 1$ or if $\|A\| < \log 2$. This means that we may freely rearrange terms and expect the same cancellations as we find for the analogous calculation of $\log(\exp z)$ for complex z with $|z| < \log 2$. But since $\log(\exp z) = z$ for such z, we have the desired conclusion. Similarly one may argue that

$$\exp(\log B) = B \text{ if } \|B - I\| < 1.$$

Next, we prove a remarkable theorem that shows how an algebraic assumption can have implications in the differentiable category. First we need some notation.

Notation 5.80. If S is any subset of a Lie group G, then we define
$$S^{-1} = \{s^{-1} : s \in S\},$$
and for any $x \in G$ we define
$$xS = \{xs : s \in S\}.$$

Theorem 5.81. *An abstract subgroup H of a Lie group G is a (regular) submanifold and hence a closed Lie subgroup if and only if H is a closed set in G.*

Proof. First suppose that H is a (regular) submanifold. Then H is locally closed. That is, every point $x \in H$ has an open neighborhood U such that $U \cap H$ is a relatively closed set in H. Let U be such a neighborhood of the identity element e. We seek to show that H is closed in G. Let $y \in \overline{H}$ and $x \in yU^{-1} \cap H$. Thus $x \in H$ and $y \in xU$. This means that $y \in \overline{H} \cap xU$, and hence $x^{-1}y \in \overline{H} \cap U = H \cap U$. So $y \in H$, and we have shown that H is closed.

Conversely, suppose that H is a closed abstract subgroup of G. Since we can always use left/right translation to translate any point to the identity, it suffices to find a single-slice chart at e. This will show that H is a regular submanifold. The strategy is to first find $\mathfrak{L}(H) = \mathfrak{h}$ and then to exponentiate a neighborhood of $0 \in \mathfrak{h}$.

First choose any inner product on $T_e G$ so we may take norms of vectors in $T_e G$. Choose a small neighborhood $\widetilde{U}$ of $0 \in T_e G = \mathfrak{g}$ on which exp is a diffeomorphism, say $\exp : \widetilde{U} \to U$, and denote the inverse by $\log_U : U \to \widetilde{U}$. Define the set $\widetilde{H}$ in $\widetilde{U}$ by $\widetilde{H} = \log_U(H \cap U)$.

Claim. If h_n is a sequence in $\widetilde{H}$ converging to zero and such that $u_n = h_n / \|h_n\|$ converges to $v \in \mathfrak{g}$, then $\exp(tv) \in H$ for all $t \in \mathbb{R}$.

Proof of the claim: Note that $th_n / \|h_n\| \to tv$ while $\|h_n\|$ converges to zero. But since $\|h_n\| \to 0$, we must be able to find a sequence $k(n) \in \mathbb{Z}$ such that $k(n) \|h_n\| \to t$. From this we have $\exp(k(n)h_n) = \exp(k(n) \|h_n\| \frac{h_n}{\|h_n\|})$ $\to \exp(tv)$. But by the properties of exp proved previously, we have that $\exp(k(n)h_n) = (\exp(h_n))^{k(n)}$. But also $\exp(h_n) \in H \cap U \subset H$ and so $(\exp(h_n))^{k(n)} \in H$. Since H is closed, we have
$$\exp(tv) = \lim_{n \to \infty} (\exp(h_n))^{k(n)} \in H.$$

Claim. If W is the set of all sv where $s \in \mathbb{R}$ and v can be obtained as a limit $h_n / \|h_n\| \to v$ with $h_n \in \widetilde{H}$ and $h_n \to 0$, then W is a vector space.

5.4. Lie Algebras and Exponential Maps

Proof of the claim: We just need to show that if $h_n/\|h_n\| \to v$ and $h'_n/\|h'_n\| \to w$ with $h'_n, h_n \in \widetilde{H}$, then there is a sequence of elements h''_n from $\widetilde{H}$ with $h''_n \to 0$ such that

$$h''_n/\|h''_n\| \to \frac{v+w}{\|v+w\|}.$$

Using the previous claim, observe that

$$h(t) := \log_U(\exp(tv)\exp(tw)) = (\log_U \circ \mu)(\exp(tv), \exp(tw)).$$

Here μ is the group multiplication map. But, by the first claim, $\exp(tv)$ and $\exp(tw)$ are in H for all t, and so $h(t)$ is in $\widetilde{H}$ for small t. By Exercise 5.41 and the fact that $T_e \log = \mathrm{id}$, we have that

$$\lim_{t \downarrow 0} h(t)/t = h'(0) = v + w.$$

Thus,

$$\frac{h(t)}{\|h(t)\|} = \frac{h(t)/t}{\|h(t)/t\|} \to \frac{v+w}{\|v+w\|}.$$

Now just let $t_n \downarrow 0$ and let $h''_n := h(t_n)$. Notice that by the first claim, $\exp(W) \subset H$.

Claim. Let W be the set from the last claim. Then $\exp(W)$ contains an open neighborhood of e in H.

Proof of the claim: Let $W^\perp$ be the orthogonal complement of W with respect to the inner product chosen above. Then we have $T_e G = W^\perp \oplus W$. It is not difficult to show that the map $\Sigma : W \oplus W^\perp \to G$ defined by

$$w + x \mapsto \exp(w)\exp(x)$$

is a diffeomorphism in a neighborhood of the origin in $T_e G$. Denote this diffeomorphism by ψ. Now suppose that $\exp(W)$ does not contain an open neighborhood of e in H. Then we may choose a sequence $h_n \to e$ such that h_n is in H but not in $\exp(W)$. But this means that we can choose a corresponding sequence $(w_n, x_n) \in W \oplus W^\perp$ with $(w_n, x_n) \to 0$ and $\exp(w_n)\exp(x_n) \in H$ and yet, $x_n \neq 0$. The space $W^\perp$ is closed and the unit sphere in $W^\perp$ is compact. After passing to a subsequence, we may assume that $x_n/\|x_n\| \to x \in W^\perp$, and of course $\|x\| = 1$. Now $\exp(w_n) \in H$ since $\exp(W) \subset H$ and H is at least an algebraic subgroup, so we see that $\exp(w_n)\exp(x_n) \in H$. Thus it must be that $\exp(x_n) \in H$ also and so $x_n \in \widetilde{H}$. But $x_n \to 0$ and $x_n/\|x_n\| \to 0$, and so, since we now know that $x_n \in \widetilde{H}$, we have that $x \in W$ by definition. This contradicts the fact that $\|x\| = 1$ and $x \in W^\perp$. Thus $\exp(W)$ must contain a neighborhood of e in H.

Finally, we let $O \subset \exp(W)$ be a neighborhood of e in H. The set O must be of the form $O = H \cap V$ for some open V in $T_e G$ containing 0. By shrinking V further we obtain a diffeomorphism $\psi|_V$. The inverse of this diffeomorphism, $\phi : U \to V$, has the required properties by construction: $\phi(H \cap U) = O \cap W$. We have actually constructed a chart with values in $T_e G$, but this is clearly good enough since we can choose a basis of $T_e G$ adapted to W. □

5.5. The Adjoint Representation of a Lie Group

Definition 5.82. Fix an element $g \in G$. The map $C_g : G \to G$ defined by $C_g(x) = gxg^{-1}$ is a Lie group automorphism called the conjugation map, and the tangent map $T_e C_g : \mathfrak{g} \to \mathfrak{g}$, denoted Ad_g, is called the **adjoint map**.

Proposition 5.83. *The map $C_g : G \to G$ is a Lie group homomorphism. The map $C : g \mapsto C_g$ is a Lie group homomorphism $G \to \mathrm{Aut}(G)$.*

Proof. See Problem 4. □

Using Proposition 5.71, we get the following

Corollary 5.84. *The map $\mathrm{Ad}_g : \mathfrak{g} \to \mathfrak{g}$ is a Lie algebra homomorphism.*

Lemma 5.85. *Let $f : M \times N \to N$ be a smooth map and define the partial map at $x \in M$ by $f_x(y) = f(x,y)$. Suppose that for every $x \in M$ the point y_0 is fixed by f_x:*
$$f_x(y_0) = y_0 \text{ for all } x.$$
Then the map $A_{y_0} : x \mapsto T_{y_0} f_x$ is a smooth map from M to $\mathrm{GL}(T_{y_0} N)$.

Proof. It suffices to show that A_{y_0} composed with an arbitrary coordinate function from an atlas of charts on $\mathrm{GL}(T_{y_0} N)$ is smooth. But $\mathrm{GL}(T_{y_0} N)$ has an atlas consisting of a single chart. Namely, choose a basis $v_1, v_2, \ldots, v_n$ of $T_{y_0} N$ and let $v^1, v^2, \ldots, v^n$ be the dual basis of $T_{y_0}^* N$. Then $\chi_j^i : A \mapsto v^i(A v_j)$ is a typical coordinate function. Now we compose:
$$\chi_j^i \circ A_{y_0}(x) = v^i(A_{y_0}(x) v_j) = v^i(T_{y_0} f_x \cdot v_j).$$
It is enough to show that $T_{y_0} f_x \cdot v_j$ is smooth in x. But this is just the composition of the smooth maps $M \to TM \times TN \cong T(M \times N) \to TN$ given by
$$x \mapsto (0_x, v_j) \mapsto (\partial_1 f)(x, y_0) \cdot 0_x + (\partial_2 f)(x, y_0) \cdot v_j = T_{y_0} f_x \cdot v_j.$$
(Recall the discussion leading up to Lemma 2.28.) □

Proposition 5.86. *The map $\mathrm{Ad} : g \mapsto \mathrm{Ad}_g$ is a Lie group homomorphism $G \to \mathrm{GL}(\mathfrak{g})$, which is called the **adjoint representation** of G.*

5.5. The Adjoint Representation of a Lie Group

Proof. We have

$$\mathrm{Ad}(g_1 g_2) = T_e C_{g_1 g_2} = T_e(C_{g_1} \circ C_{g_2})$$
$$= T_e C_{g_1} \circ T_e C_{g_2} = \mathrm{Ad}_{g_1} \circ \mathrm{Ad}_{g_2},$$

which shows that Ad is a group homomorphism. The smoothness follows from the previous lemma applied to the map $C : (g, x) \mapsto C_g(x)$. $\square$

Recall that for $v \in \mathfrak{g}$ we have the associated left invariant vector field L^v as well as the right invariant field R^v. Using this notation, we have

Lemma 5.87. *Let $v \in \mathfrak{g}$. Then $L^v(x) = R^{\mathrm{Ad}_x v}$.*

Proof. We calculate as follows:

$$L^v(x) = T_e(L_x) \cdot v = T(R_x) T(R_{x^{-1}}) T_e(L_x) \cdot v$$
$$= T(R_x) T(R_{x^{-1}} \circ L_x) \cdot v = R^{\mathrm{Ad}(x)v}. \quad \square$$

We now go one step further and take the differential of Ad.

Definition 5.88. For a Lie group G with Lie algebra $\mathfrak{g}$, define the **adjoint representation of** $\mathfrak{g}$ as the map $\mathrm{ad} : \mathfrak{g} \to \mathfrak{gl}(\mathfrak{g})$ given by

$$\mathrm{ad} = T_e \mathrm{Ad} = d(\mathrm{Ad}).$$

Proposition 5.89. $\mathrm{ad}(v)w = [v, w]$ *for all $v, w \in \mathfrak{g}$.*

Proof. Let $v^1, \ldots, v^n$ be a basis for $\mathfrak{g}$ so that $\mathrm{Ad}(x)w = \sum a_i(x) v^i$ for some functions a_i. Let c be a curve with $\dot c(0) = v$. Then we have

$$\mathrm{ad}(v)w = \frac{d}{dt}\bigg|_{t=0} \mathrm{Ad}(c(t))w = \sum \frac{d}{dt}\bigg|_{t=0} a_i(c(t)) v^i$$
$$= \sum (v a_i) v^i.$$

On the other hand, by Lemma 5.87 we have

$$L^w(x) = R^{\mathrm{Ad}(x)w} = R\left(\sum a_i(x) v^i\right)$$
$$= \sum a_i(x) R^{v^i}(x).$$

From the fact that the bracket of any left invariant vector field with any right invariant vector field is zero, we have

$$[L^v, L^w] = \left[L^v, \sum a_i R^{v^i}\right] = 0 + \sum L^v(a_i) R^{v^i}.$$

Finally, using the equation for $\mathrm{ad}(v)w$ derived above, we have
$$\begin{aligned}[v,w] &= [L^v, L^w](e) \\ &= \sum L^v(a_i)(e)R^{v^i}(e) = \sum L^v(a_i)(e)v^i \\ &= \sum (va_i)v^i = \mathrm{ad}(v)w.\end{aligned}$$
$\square$

The map $\mathrm{ad} : \mathfrak{g} \to \mathfrak{gl}(\mathfrak{g}) = \mathrm{End}(T_e G)$ is given as the tangent map at the identity of the map Ad which is a Lie group homomorphism. Thus by Proposition 5.71 we obtain the following:

Proposition 5.90. $\mathrm{ad} : \mathfrak{g} \to \mathfrak{gl}(\mathfrak{g})$ *is a Lie algebra homomorphism.*

Since ad is defined as the Lie differential of Ad, Proposition 5.79 tells us that the following diagram commutes for any Lie group G:

$$\begin{array}{ccc} \mathfrak{g} & \xrightarrow{\mathrm{ad}} & \mathfrak{gl}(\mathfrak{g}) \\ {\scriptstyle\exp}\downarrow & & \downarrow{\scriptstyle\exp} \\ G & \xrightarrow{\mathrm{Ad}} & \mathrm{GL}(\mathfrak{g}) \end{array}$$

On the other hand, for any $g \in G$, the map $C_g : x \mapsto gxg^{-1}$ is also a homomorphism, and so Proposition 5.79 applies again, giving the following commutative diagram:

$$\begin{array}{ccc} \mathfrak{g} & \xrightarrow{\mathrm{Ad}_g} & \mathfrak{g} \\ {\scriptstyle\exp}\downarrow & & \downarrow{\scriptstyle\exp} \\ G & \xrightarrow{C_g} & G \end{array}$$

In other words,
$$\exp(t\,\mathrm{Ad}_g v) = g\exp(tv)g^{-1}$$
for any $g \in G, v \in \mathfrak{g}$ and $t \in \mathbb{R}$.

In the case of linear Lie groups $G \subset \mathrm{GL}(V)$, we have identified $\mathfrak{g}$ with a subspace of $\mathfrak{gl}(V)$, which is in turn identified with $L(V,V)$. In this case, the exponential map is given by the power series as explained above. It is easy to show from the power series that $B \circ \exp(tA) \circ B^{-1} = \exp(tB \circ A \circ B^{-1})$ for any $A \in \mathfrak{gl}(V)$ and $B \in \mathrm{GL}(V)$. In this special set of circumstances, we have
$$\mathrm{Ad}_B A = B \circ A \circ B^{-1}.$$

This is seen as follows:
$$\begin{aligned}\mathrm{Ad}_B A &= \left.\frac{d}{dt}\right|_{t=0} B \circ \exp(tA) \circ B^{-1} \\ &= \left.\frac{d}{dt}\right|_{t=0} \exp(tB \circ A \circ B^{-1}) = B \circ A \circ B^{-1}.\end{aligned}$$

5.5. The Adjoint Representation of a Lie Group

Earlier we noted that for a general Lie group we always have $\mathrm{Ad} \circ \exp = \exp \circ \mathrm{ad}$. In the current context of *linear* Lie groups, this can be written as

$$\exp(A) \circ B \circ \exp(-A) = \sum_{k=0}^{\infty} \frac{1}{k!} (\mathrm{ad}(A))^k B$$

for any $A \in \mathfrak{gl}(V)$ and any $B \in \mathrm{GL}(V)$.

We end this section with a statement of the useful **Campbell-Baker-Hausdorff** (CBH) formula. A proof may be found in [**Helg**] or [**Mich**]. First notice that if G is a Lie group with Lie algebra $\mathfrak{g}$, then for each $X \in \mathfrak{g}$ we have $\mathrm{ad}\, X \in L(\mathfrak{g}, \mathfrak{g})$. We choose a norm on $\mathfrak{g}$, and then we have a natural operator norm and consequent notion of convergence in $L(\mathfrak{g}, \mathfrak{g})$. The analytic function $\frac{\ln z}{z-1}$ is defined by the power series

$$\frac{\ln z}{z-1} := \sum_{n=0}^{\infty} \frac{(-1)^n}{n+1} (z-1)^n,$$

which converges in a small ball about $z = 1$. We may then use the corresponding power series to make sense of $(\ln A)(A-1)^{-1}$ for $A \in L(\mathfrak{g}, \mathfrak{g})$ sufficiently close to the identity map $I = \mathrm{id}_{\mathfrak{g}}$.

Theorem 5.91 (CBH Formula). *Let G be a Lie group and $\mathfrak{g}$ its Lie algebra. Let $f(z) = \frac{\ln z}{z-1}$ be defined by a power series as above. Then for sufficiently small $x, y \in \mathfrak{g}$,*

$$\exp(x)\exp(y) = \exp(C(x,y)),$$

where

$$C(x,y) = y + \int_0^1 f(e^{t\, \mathrm{ad}\, x} e^{t\, \mathrm{ad}\, y}) \cdot x\, dt$$

$$= x + y + \sum_{n=1}^{\infty} \frac{(-1)^n}{n+1} \int_0^1 \left(\sum_{k,l>0,\, k+l\geq 1} \frac{t^k}{k!l!} (\mathrm{ad}\, x)^k (\mathrm{ad}\, y)^l \right)^n x\, dt.$$

It follows from the above that

$$C(x,y) = \sum_{n=1}^{\infty} C_n(x,y),$$

where $C_1(x,y) = x+y$, $C_2(x,y) = \frac{1}{2}[x,y]$ and

$$C_3(x,y) = \frac{1}{12}\left([[x,y],y] + [[y,x],x]\right).$$

In particular,

$$\exp(tx)\exp(ty) = \exp\bigl(t(x+y) + O(t^2)\bigr)$$

for any x, y and sufficiently small t. One can show, using the above results, that G is abelian if and only if $\mathfrak{g}$ is abelian (i.e. $[x,y] = 0$ for all $x, y \in \mathfrak{g}$). The

CBH formula shows that the Lie algebra structure on $\mathfrak{g}$ locally determines the multiplicative structure on G.

5.6. The Maurer-Cartan Form

Let G be a Lie group, and for each $g \in G$, define $\omega_G(g) : T_g G \to \mathfrak{g}$ and $\omega_G^{\text{right}}(g) : T_g G \to \mathfrak{g}$ by

$$\omega_G(g)(X_g) = TL_{g^{-1}} \cdot X_g,$$

and

$$\omega_G^{\text{right}}(g)(X_g) = TR_{g^{-1}} \cdot X_g.$$

The maps $\omega_G : g \mapsto \omega_G(g)$ and $\omega_G^{\text{right}} : g \mapsto \omega_G^{\text{right}}(g)$ are $\mathfrak{g}$-valued 1-forms called the **left Maurer-Cartan form** and **right Maurer-Cartan form** respectively. We can view ω_G and ω_G^{right} as maps $TG \to \mathfrak{g}$. For example, $\omega_G(X_g) := \omega_G(g)(X_g)$.

As we have seen, $\text{GL}(n)$ is an open set in a vector space, and so its tangent bundle is trivial, $T\text{GL}(n) \cong \text{GL}(n) \times M_{n \times n}$ (recall Definition 2.58). A general abstract Lie group G is not an open subset of a vector space, but we are still able to show that TG is trivial. There are two such trivializations obtained from the Maurer-Cartan forms. These are $\text{triv}_L : TG \to G \times \mathfrak{g}$ and $\text{triv}_R : TG \to G \times \mathfrak{g}$ defined by

$$\text{triv}_L(v_g) = (g, \omega_G(v_g)),$$
$$\text{triv}_R(v_g) = (g, \omega_G^{\text{right}}(v_g))$$

for $v_g \in T_g G$. Observe that $\text{triv}_L^{-1}(g, v) = L^v(g)$ and $\text{triv}_R^{-1}(g, v) = R^v(g)$. It is easy to check that triv_L and triv_R are trivializations in the sense of Definition 2.58. Thus we have the following:

Proposition 5.92. *The tangent bundle of a Lie group is trivial: $TG \cong G \times \mathfrak{g}$.*

We will refer to triv_L and triv_R as the (left and right) **Maurer-Cartan trivializations**. How do these two trivializations compare? There is no special reason to prefer left multiplication. We could have used right invariant vector fields as our means of producing the Lie algebra, and the whole theory would work "on the other side", so to speak. The bridge between left and right is the adjoint map:

Lemma 5.93 (Left-right lemma). *For any $v \in \mathfrak{g}$, and $g \in G$ we have*

$$\text{triv}_R \circ \text{triv}_L^{-1}(g, v) = (g, \text{Ad}_g(v)).$$

5.6. The Maurer-Cartan Form

Proof. We compute:

$$\mathrm{triv}_R \circ \mathrm{triv}_L^{-1}(g, v)$$
$$= (g, TR_{g^{-1}}TL_g v) = (g, T(R_{g^{-1}}L_g) \cdot v)$$
$$= (g, TC_g \cdot v) = (g, \mathrm{Ad}_g(v)). \qquad \square$$

It is often convenient to identify the tangent bundle TG of a Lie group G with $G \times \mathfrak{g}$. Of course we must specify which of the two trivializations described above is being invoked. Unless indicated otherwise we shall use the "left version" described above: $v_g \mapsto (g, \omega_G(v_g)) = (g, TL_g^{-1}(v_g))$.

Warning: It must be realized that we now have three natural ways to trivialize the tangent bundle of the general linear group. In fact, the usual one which we introduced earlier is actually the restriction to $T\mathrm{GL}(n)$ of the Maurer-Cartan trivialization of the *abelian* Lie group $(M_{n \times n}, +)$.

In order to use the (left) Maurer-Cartan trivialization as an identification effectively, we need to find out how a few basic operations look when this identification is imposed.

The picture obtained from using the trivialization produced by the left Maurer-Cartan form:

(1) The tangent map of the left translation $TL_g : TG \to TG$ takes the form "TL_g" : $(x, v) \mapsto (gx, v)$. Indeed, the following diagram commutes:

$$\begin{array}{ccc} TG & \stackrel{TL_g}{\longrightarrow} & TG \\ \downarrow & & \downarrow \\ G \times \mathfrak{g} & \stackrel{``TL_g"}{\longrightarrow} & G \times \mathfrak{g} \end{array}$$

where elementwise we have

$$\begin{array}{ccc} v_x & \stackrel{TL_g}{\longmapsto} & TL_g \cdot v_x \\ \downarrow & & \downarrow \\ (x, TL_x^{-1} v_x) & \longmapsto & (gx, TL_{gx}^{-1} TL_g v_x) \\ = (x, v) & & = (gx, v) \end{array}$$

(2) The tangent map of multiplication: This time we will invoke two identifications. First, group multiplication is a map $\mu : G \times G \to G$, and so on the tangent level we have a map $T(G \times G) \to TG$. Recall that we have a natural isomorphism $T(G \times G) \cong TG \times TG$ given by $T\pi_1 \times T\pi_2 : (v_{(x,y)}) \mapsto (T\pi_1 \cdot v_{(x,y)}, T\pi_2 \cdot v_{(x,y)})$. If we also identify TG with $G \times \mathfrak{g}$, then $TG \times TG \cong (G \times \mathfrak{g}) \times (G \times \mathfrak{g})$, and we end

up with the following "version" of $T\mu$:
$$"T\mu" : (G \times \mathfrak{g}) \times (G \times \mathfrak{g}) \to G \times \mathfrak{g},$$
$$"T\mu" : ((x,v),(y,w)) \mapsto (xy, TR_y v + TL_x w)$$

(see Exercise 5.94).

(3) The (left) Maurer-Cartan form is a map $\omega_G : TG \to T_e G = \mathfrak{g}$, and so there must be a "version", "ω_G", that uses the identification $TG \cong G \times \mathfrak{g}$. In fact, the map we seek is just projection:
$$"\omega_G" : (x,v) \mapsto v.$$

(4) The right Maurer-Cartan form is a little more complicated since we are currently using the isomorphism $TG \cong G \times \mathfrak{g}$ obtained from the *left* Maurer-Cartan form. From Lemma 5.93 we obtain:
$$"\omega_G^{\text{right}}" : (x,v) \mapsto \text{Ad}_g(v).$$

The adjoint map is nearly the same thing as the right Maurer-Cartan form if we decide to use the left trivialization $TG \cong G \times \mathfrak{g}$ as an identification.

(5) A vector field $X \in \mathfrak{X}(G)$ should correspond to a section of the projection $G \times \mathfrak{g} \to G$ which must have the form $\overleftrightarrow{X} : x \mapsto (x, F^X(x))$ for some smooth $\mathfrak{g}$-valued function $F^X \in C^\infty(G; \mathfrak{g})$. It is an easy consequence of the definitions that $F^X(x) = \omega_G(X(x)) = TL_x^{-1} \cdot X(x)$. Under this identification, a left invariant vector field becomes a constant section of $G \times \mathfrak{g}$. For example, if X is left invariant, then the corresponding constant section is $x \mapsto (x, X(e))$.

Exercise 5.94. Refer to 2 above. Show that the map "$T\mu$" defined so that the diagram below commutes is $((x,v),(y,w)) \mapsto (xy, TR_y v + TL_x w)$.

$$\begin{array}{ccc} T(G \times G) & \xrightarrow{T\mu} & TG \\ \downarrow & & \downarrow \\ (G \times \mathfrak{g}) \times (G \times \mathfrak{g}) & \xrightarrow{"T\mu"} & G \times \mathfrak{g} \end{array}$$

We have already seen that using available identifications the Lie algebra of a matrix groups and associated formulas take a concrete form. We now consider the Maurer-Cartan form in the case of matrix groups. Suppose that G is a Lie subgroup of $\text{GL}(n)$ and consider the coordinate functions x^i_j on $\text{GL}(n)$ defined by $x^i_j(A) = a^i_j$ where $A = [a^i_j]$. We have the associated 1-forms dx^i_j. Both the functions x^i_j and the forms dx^i_j restrict to G. Denoting these restrictions by the same symbols, the Maurer-Cartan form can be expressed as
$$\omega_G = \left[x^i_j\right]^{-1} \left[dx^i_j\right].$$

5.6. The Maurer-Cartan Form

One sometimes sees the shorthand, $g^{-1}dg$, which is a bit cryptic. Our goal is to understand this expression better. First, from a practical point of view we think of it as follows: An element v_g of the tangent space T_gG is also an element of $T_g\mathrm{GL}(n)$ and so can be expressed in terms of the vectors $\partial/\partial x^i_j|_g$, say

$$v_g = \sum v^i_j \frac{\partial}{\partial x^i_j}\bigg|_g.$$

Then,

$$\omega_G(v_g) = [x^i_j]^{-1}[dx^i_j](v_g) = [g^i_j]^{-1}[v^i_j],$$

where $g = [g^i_j]$ and the matrix $[g^i_j]^{-1}[v^i_j]$ is interpreted as an element of the Lie algebra $\mathfrak{g}$. For instance, if $G = \mathrm{SO}(2)$, then the Maurer-Cartan form is given by

$$\begin{bmatrix} x^2_2 & -x^1_2 \\ -x^2_1 & x^1_1 \end{bmatrix} \begin{bmatrix} dx^1_1 & dx^1_2 \\ dx^2_1 & dx^2_2 \end{bmatrix} = \begin{bmatrix} x^2_2 dx^1_1 - x^1_2 dx^2_1 & x^2_2 dx^1_2 - x^1_2 dx^2_2 \\ -x^2_1 dx^1_1 + x^1_1 dx^2_1 & -x^2_1 dx^1_2 + x^1_1 dx^2_2 \end{bmatrix}$$

$$= \begin{bmatrix} x^1_1 dx^1_1 - x^1_2 dx^2_1 & x^1_1 dx^1_2 - x^1_2 dx^2_2 \\ x^1_2 dx^1_1 + x^1_1 dx^2_1 & x^1_2 dx^1_2 + x^1_1 dx^2_2 \end{bmatrix},$$

since on $\mathrm{SO}(2)$ we have $x^1_1 = x^2_2$, $x^1_2 = -x^2_1$ and $x^1_1 x^2_2 - x^1_2 x^2_1 = 1$. But this can be further simplified. If we let $v_g \in T_g\mathrm{SO}(2)$, then $\omega_G(v_g)$ is in $\mathfrak{so}(2)$ and so must be antisymmetric. We conclude that

$$\omega_{\mathrm{SO}(2)} = \begin{bmatrix} 0 & x^1_1 dx^1_2 - x^1_2 dx^2_2 \\ -x^1_1 dx^1_2 + x^1_2 dx^2_2 & 0 \end{bmatrix}.$$

Let us try to understand things a bit more thoroughly. If G is a subgroup of $\mathrm{GL}(n)$, then consider the inclusion map $j : G \hookrightarrow M_{n\times n}$, where $M_{n\times n}$ is the set of $n \times n$ matrices. Then we have the differential $dj : TG \to M_{n\times n}$ and the two left multiplications $L_g : G \to G$ and $\mathcal{L}_g : M_{n\times n} \to M_{n\times n}$ for $g \in G$. We have the following commutative diagrams:

$$\begin{array}{ccc} M_{n\times n} \xrightarrow{\mathcal{L}_g} M_{n\times n} & \qquad & M_{n\times n} \xrightarrow{\mathcal{L}_g} M_{n\times n} \\ j \uparrow \qquad \uparrow j & & dj|_h \uparrow \qquad \uparrow dj|_{gh} \\ G \xrightarrow{L_g} G & & T_hG \xrightarrow{T_hL_g} T_{gh}G \end{array}$$

for any $h \in G$. We have used the fact that since $\mathcal{L}_g$ is linear, $D\mathcal{L}_g(A) = \mathcal{L}_g$ for all $A \in M_{n\times n}$. Then for $v_g \in T_gG$ we have

$$dj|_e(\omega_G(v_g)) = dj|_e(TL_{g^{-1}}v_g) = \mathcal{L}_{g^{-1}} \, dj|_g(v_g)$$
$$= g^{-1} \, dj|_g(v_g) = (j \circ \pi(v_g))^{-1} \, dj(v_g),$$

where $\pi : TG \to G$ is the tangent bundle projection. Notice that the effect of $dj|_e$ is simply to interpret elements of T_eG as matrices, and so can be suppressed from the notation. Taking this into account and applying a reasonable abbreviation for $(j \circ \pi(v_g))^{-1} \, dj \, (v_g)$, we arrive at

$$\omega_G = j^{-1} \, dj.$$

But notice that for a matrix group, j is none other than the map $[x_j^i] : A \mapsto [x_j^i(A)] = [a_j^i]$, so that we are returned to the expression $\omega_G = [x_j^i]^{-1}[dx_j^i] =$ "$g^{-1}dg$". This tells us the meaning of g in the expression $g^{-1}dg$.

5.7. Lie Group Actions

The basic definitions for group actions were given earlier in Definitions 1.99 and 1.6.2. As before we give most of our definitions and results for left actions and ask the reader to notice that analogous statements can be made for right actions.

Definition 5.95. Let $l : G \times M \to M$ be a left action, where G is a Lie group and M is a smooth manifold. If l is a smooth map, then we say that l is a (smooth) Lie group action.

As before, we also use any of the notations gp, $g \cdot p$ or $l_g(p)$ for $l(g, p)$. We will need this notational flexibility. Recall that for $p \in M$, the orbit of p is denoted Gp or $G \cdot p$, and the action is **transitive** if $Gp = M$. Recall also that an action is **effective** if $l_g(p) = p$ for all p only if $g = e$. For right actions $r : M \times G \to M$, similar definitions apply and we write $pg = r_g(p) = r(p, g)$. A right action corresponds to a left action by the rule $gp := pg^{-1}$.

Definition 5.96. Let l be a Lie group action as above. For a fixed $p \in M$, the **isotropy group** of p is defined to be

$$G_p := \{g \in G : gp = p\}.$$

The isotropy group of p is also called the **stabilizer** of p.

Exercise 5.97. Show that G_p is a closed subset and an abstract subgroup of G. This means that G_p is a closed Lie subgroup.

Recalling the definition of a free action (Definition 1.100), it is easy to see that an action is free if and only if the isotropy subgroup of every point is the trivial subgroup consisting of the identity element alone.

Definition 5.98. Suppose that we have a Lie group action of G on M. If N is a subset of M and $Gx \subset N$ for all $x \in N$, then we say that N is an **invariant subset**. If N is also a submanifold, then it is called an **invariant submanifold**.

In this definition, we include the possibility that N is an open submanifold. If N is an invariant subset of M, then it is easy to see that $gN = N$,

5.7. Lie Group Actions

where $gN = l_g(N)$ for any g. Furthermore, if N is a submanifold, then the action restricts to a Lie group action $G \times N \to N$.

If G is zero-dimensional, then by definition it is just a group with discrete topology, and we recover the definition of a discrete group action. We have already seen several examples of discrete group actions, and now we list a few examples of more general Lie group actions.

Example 5.99. The maps $G \times G \to G$ given by $(g, x) \mapsto L_g x$ and $(g, x) \mapsto R_g x$ are Lie group actions.

Example 5.100. In case $M = \mathbb{R}^n$, the Lie group $\text{GL}(n, \mathbb{R})$ acts on $\mathbb{R}^n$ by matrix multiplication. Similarly, $\text{GL}(n, \mathbb{C})$ acts on $\mathbb{C}^n$. More abstractly, $\text{GL}(V)$ acts on the vector space V. This action is smooth since Ax depends smoothly (polynomially) on the components of A and on the components of $x \in \mathbb{R}^n$.

Example 5.101. Any Lie subgroup of $\text{GL}(n, \mathbb{R})$ acts on $\mathbb{R}^n$ also by matrix multiplication. For example, $\text{O}(n, \mathbb{R})$ acts on $\mathbb{R}^n$. For every $x \in \mathbb{R}^n$, the orbit of x is the sphere of radius $\|x\|$. This is trivially true if $x = 0$. In general, if $\|x\| \neq 0$, then $\|gx\| = \|x\|$ for any $g \in \text{O}(n, \mathbb{R})$. On the other hand, if $x, y \in \mathbb{R}^n$ and $\|x\| = \|y\| = r$, then let $\widehat{x} := x/r$ and $\widehat{y} := y/r$. Let $\widehat{x} := e_1$ and $\widehat{y} = f_1$ and then extend to orthonormal bases $(e_1, \ldots, e_n)$ and $(f_1, \ldots, f_n)$. Then there exists an orthogonal matrix S such that $S e_i = f_i$ for $i = 1, \ldots, n$. In particular, $S\widehat{x} = \widehat{y}$, and so $Sx = y$.

Exercise 5.102. From the last example we can restrict the action of $\text{O}(n, \mathbb{R})$ to a transitive action on S^{n-1}. Now $\text{SO}(n, \mathbb{R})$ also acts on S^{n-1} by restriction. Show that this action on S^{n-1} is transitive as long as $n > 1$.

Example 5.103. If H is a Lie subgroup of a Lie group G, then we can consider L_h for any $h \in H$ and thereby obtain a Lie group action of H on G.

Recall that a subgroup H of a group G is called a **normal subgroup** if $gkg^{-1} \in K$ for any $k \in H$ and all $g \in G$. In other words, H is normal if $gHg^{-1} \subset H$ for all $g \in G$, and it is easy to see that in this case we always have $gHg^{-1} = H$.

Example 5.104. If H is a normal Lie subgroup of G, then G acts on H by conjugation:
$$g \cdot h := C_g h = ghg^{-1}.$$
Notice that the notation $g \cdot h$ *cannot* reasonably be abbreviated to gh in this example.

Suppose that a Lie group G acts on smooth manifolds M and N. For simplicity, we take both actions to be left actions, which we denote by l and

λ, respectively. A map $f : M \to N$ such that $f \circ l_g = \lambda_g \circ f$ for all $g \in G$, is said to be an **equivariant map** (equivariant with respect to the given actions). This means that for all g the following diagram commutes:

(5.3)
$$\begin{array}{ccc} M & \xrightarrow{f} & N \\ l_g \downarrow & & \downarrow \lambda_g \\ M & \xrightarrow{f} & N \end{array}$$

If f is also a diffeomorphism, then f is an **equivalence** of Lie group actions.

Example 5.105. If $\phi : G \to H$ is a Lie group homomorphism, then we can define an action of G on H by $\lambda(g, h) = \lambda_g(h) = L_{\phi(g)} h$. We leave it to the reader to verify that this is indeed a Lie group action. In this situation, ϕ is equivariant with respect to the actions λ and L (left translation).

Example 5.106. Let $T^n = S^1 \times \cdots \times S^1$ be the n-torus, where we identify S^1 with the complex numbers of unit modulus. Fix $k = (k_1, \ldots, k_n) \in \mathbb{R}^n$. Then $\mathbb{R}$ acts on $\mathbb{R}^n$ by $\tau^k(t, x) = t \cdot x := x + tk$. On the other hand, $\mathbb{R}$ acts on T^n by $t \cdot (z^1, \ldots, z^n) = (e^{itk_1} z^1, \ldots, e^{itk_n} z^n)$. The map $\mathbb{R}^n \to T^n$ given by $(x^1, \ldots, x^n) \mapsto (e^{ix^1}, \ldots, e^{ix^n})$ is equivariant with respect to these actions.

Theorem 5.107 (Equivariant rank theorem). *Suppose that $f : M \to N$ is smooth and that a Lie group G acts on both M and N with the action on M being transitive. If f is equivariant, then it has constant rank. In particular, each level set of f is a closed regular submanifold.*

Proof. Let the actions on M and N be denoted by l and λ respectively as before. Pick any two points $p_1, p_2 \in M$. Since G acts transitively on M, there is a g with $l_g p_1 = p_2$. By hypothesis, we have $f \circ l_g = \lambda_g \circ f$, which corresponds to the commutative diagram (5.3). Upon application of the tangent functor we have the commutative diagram

$$\begin{array}{ccc} T_{p_1} M & \xrightarrow{T_{p_1} f} & T_{f(p_1)} N \\ T_{p_1} l_g \downarrow & & \downarrow T_{f(p_1)} \lambda_g \\ T_{p_2} M & \xrightarrow{T_{p_2} f} & T_{f(p_2)} N \end{array}$$

Since the maps $T_{p_1} l_g$ and $T_{f(p_1)} \lambda_g$ are linear isomorphisms, we see that $T_{p_1} f$ must have the same rank as $T_{p_2} f$. Since p_1 and p_2 were arbitrary, we see that the rank of f is constant on M. Apply Theorem C.5. □

There are several corollaries of this nice theorem. For example, we know that $O(n, \mathbb{R})$ is the level set $f^{-1}(I)$, where $f : \text{GL}(n, \mathbb{R}) \to \mathfrak{gl}(n, \mathbb{R}) = M_{n \times n}$ is given by $f(A) = A^T A$. The group $O(n, \mathbb{R})$ acts on itself via left translation, and we also let $O(n, \mathbb{R})$ act on $\mathfrak{gl}(n, \mathbb{R})$ by $Q \cdot A := Q^T A Q$ (adjoint

5.7. Lie Group Actions

action). One checks easily that f is equivariant with respect to these actions, and since the first action (left translation) is certainly transitive, we see that $\mathrm{O}(n, \mathbb{R})$ is a closed regular submanifold of $\mathrm{GL}(n, \mathbb{R})$. It follows from Proposition 5.9 that $\mathrm{O}(n, \mathbb{R})$ is a closed Lie subgroup of $\mathrm{GL}(n, \mathbb{R})$. Similar arguments apply for $\mathrm{U}(n, \mathbb{C}) \subset \mathrm{GL}(n, \mathbb{C})$ and other linear Lie groups. In fact, we have the following general corollary to Theorem 5.107 above.

Corollary 5.108. *If $\phi : G \to H$ is a Lie group homomorphism, then the kernel $\mathrm{Ker}(h)$ is a closed Lie subgroup of G.*

Proof. Let G act on itself and on H as in Example 5.105. Then ϕ is equivariant, and $\phi^{-1}(e) = \mathrm{Ker}(h)$ is a closed Lie subgroup by Theorem 5.107 and Proposition 5.9. □

Corollary 5.109. *Let $l : G \times M \to M$ be a Lie group action, and let G_p be the isotropy subgroup of some $p \in M$. Then G_p is a closed Lie subgroup of G.*

Proof. The orbit map $\theta_p : G \to M$ given by $\theta_p(g) = gp$ is equivariant with respect to left translation on G and the given action on M. Thus by the equivariant rank theorem, G_p is a regular submanifold of G, and then by Proposition 5.9 it is a closed Lie subgroup. □

Proper Actions and Quotients. At several points in this section, such as the proof of Proposition 5.111 below, we follow [**Lee, John**].

Definition 5.110. Let $l : G \times M \to M$ be a smooth (or merely continuous) group action. If the map $P : G \times M \to M \times M$ given by $(g, p) \mapsto (l_g p, p)$ is proper, we say that the action is a **proper action**.

It is important to notice that a proper action is *not* defined to be an action such that the defining map $l : G \times M \to M$ is proper. We now give a useful characterization of a proper action. For any subset $K \subset M$, let $g \cdot K := \{gx : x \in K\}$.

Proposition 5.111. *Let $l : G \times M \to M$ be a smooth (or merely continuous) group action. Then l is a proper action if and only if the set*
$$G_K := \{g \in G : (g \cdot K) \cap K \neq \emptyset\}$$
is compact whenever K is compact.

Proof. Suppose that l is proper so that the map P is a proper map. Let π_G be the first factor projection $G \times M \to G$. Then
$$\begin{aligned} G_K &= \{g : \text{there exists an } x \in K \text{ such that } gx \in K\} \\ &= \{g : \text{there exists an } x \in M \text{ such that } P(g, x) \in K \times K\} \\ &= \pi_G(P^{-1}(K \times K)), \end{aligned}$$
and so G_K is compact.

Next we assume that G_K is compact for all compact K. If $C \subset M \times M$ is compact, then letting $K = \pi_1(C) \cup \pi_2(C)$, where π_1 and π_2 are the first and second factor projections $M \times M \to M$ respectively, we have

$$P^{-1}(C) \subset P^{-1}(K \times K) \subset \{(g,x) : gx \in K \text{ and } x \in K\}$$
$$\subset G_K \times K.$$

Since $P^{-1}(C)$ is a closed subset of the compact set $G_K \times K$, it is compact. This means that P is proper since C was an arbitrary compact subset of $M \times M$. □

Using this proposition, one can show that Definition 1.106 for discrete actions is consistent with Definition 5.110 above.

Proposition 5.112. *If G is compact, then any smooth action $l : G \times M \to M$ is proper.*

Proof. Let $B \subset M \times M$ be compact. We find a compact subset $K \subset M$ such that $B \subset K \times K$ as in the proof of Proposition 5.111.

Claim: $P^{-1}(B)$ is compact. Indeed,

$$P^{-1}(B) \subset P^{-1}(K \times K) = \bigcup_{k \in K} P^{-1}(K \times \{k\})$$
$$= \bigcup_{k \in K} \{(g,p) : (gp,p) \in K \times \{k\}\}$$
$$= \bigcup_{k \in K} \{(g,k) : gp \in K\}$$
$$\subset \bigcup_{k \in K} (G \times \{k\}) = G \times K$$

Thus $P^{-1}(B)$ is a closed subset of the compact set $G \times K$ and hence is compact. □

Exercise 5.113. Prove the following:

(i) If $l : G \times M \to M$ is a proper action and $H \subset G$ is a closed subgroup, then the restricted action $H \times M \to M$ is proper.

(ii) If N is an invariant submanifold for a proper action $l : G \times M \to M$, then the restricted action $G \times N \to N$ is also proper.

Let us consider a Lie group action $l : G \times M \to M$ that is both *proper and free*. The **orbit map** at p is the map $\theta_p : G \to M$ given by $\theta_p(g) = g \cdot p$. It is easily seen to be smooth, and its image is obviously $G \cdot p$. In fact, since the action is free, each orbit map is injective. Also, θ_p is equivariant with respect to the left action of G on itself and the action l:

$$\theta_p(gx) = (gx) \cdot p = g \cdot (x \cdot p)$$
$$= g \cdot \theta_p(x)$$

for all $x, g \in G$. It follows from Theorem 5.107 (the equivariant rank theorem) that θ_p has constant rank, and since it is injective, it must be an

5.7. Lie Group Actions

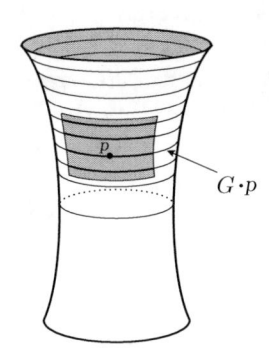

Figure 5.1. Action-adapted chart

immersion. Not only that, but it is a proper map. Indeed, for any compact $K \subset M$ the set $\theta_p^{-1}(K)$ is a closed subset of the set $G_{K \cup \{p\}}$, and since the latter set is compact by Theorem 5.111, $\theta_p^{-1}(K)$ is compact. By Exercise 3.9, θ_p is an embedding, so each orbit is a regular submanifold of M.

It will be very convenient to have charts on M which fit the action of G in a nice way. See Figure 5.1.

Definition 5.114. Let M be an n-manifold and G a Lie group of dimension k. If $l : G \times M \to M$ is a Lie group action, then an **action-adapted chart** on M is a chart $(U, \mathbf{x})$ such that

(i) $\mathbf{x}(U)$ is a product open set $V_1 \times V_2 \subset \mathbb{R}^k \times \mathbb{R}^{n-k} = \mathbb{R}^n$;

(ii) if an orbit has nonempty intersection with U, then that intersection has the form
$$\{x^{k+1} = c^1, \ldots, x^n = c^{n-k}\}$$
for some constants $c^1, \ldots, c^{n-k}$.

Theorem 5.115. *If $l : G \times M \to M$ is a free and proper Lie group action, then for every $p \in M$ there is an action-adapted chart centered at p.*

Proof. Let $p \in M$ be given. Since $G \cdot p$ is a regular submanifold, we may choose a regular submanifold chart $(W, \mathbf{y})$ centered at p so that $(G \cdot p) \cap W$ is exactly given by $y^{k+1} = \cdots = y^n = 0$ in W. Let S be the complementary slice in W given by $y^1 = \cdots = y^k = 0$. Note that S is a regular submanifold. The tangent space T_pM decomposes as
$$T_pM = T_p(G \cdot p) \oplus T_pS.$$
Let $\varphi : G \times S \to M$ be the restriction of the action l to the set $G \times S$. Also, let $i_p : G \to G \times S$ be the insertion map $g \mapsto (g, p)$ and let $j_e : S \to G \times S$ be the insertion map $s \mapsto (e, s)$. (See Figure 5.2.) These insertion maps

are embeddings, and we have $\theta_p = \varphi \circ i_p$ and also $\varphi \circ j_e = \iota$, where ι is the inclusion $S \hookrightarrow M$. Now $T_e\theta_p(T_eG) = T_p(G \cdot p)$ since θ_p is an embedding. On the other hand, $T\theta_p = T\varphi \circ Ti_p$, and so the image of $T_{(e,p)}\varphi$ must contain $T_p(G \cdot p)$. Similarly, from the composition $\varphi \circ j_e = \iota$ we see that the image of $T_{(e,p)}\varphi$ must contain T_pS. It follows that $T_{(e,p)}\varphi : T_{(e,p)}(G \times S) \to T_pM$ is surjective, and since $T_{(e,p)}(G \times S)$ and T_pM have the same dimension, it is also injective.

By the inverse mapping theorem, there is a neighborhood O of (e, p) such that $\varphi|_O$ is a diffeomorphism. By shrinking O further if necessary we may assume that $\varphi(O) \subset W$. We may also arrange that O has the form of a product $O = A \times B$ for A open in G and B open in S. In fact, we can assume that there are diffeomorphisms $\alpha : I^k \to A$ and $\beta : I^{n-k} \to B$, where I^k and I^{n-k} are the open cubes in $\mathbb{R}^k$ and $\mathbb{R}^{n-k}$ given respectively by $I^k = (-1,1)^k$ and $I^{n-k} = (-1,1)^{n-k}$ and where $\alpha^{-1}(e) = 0 \in \mathbb{R}^k$ and $\beta^{-1}(p) = 0 \in \mathbb{R}^{n-k}$. Let $U := \varphi(A \times B)$. The map $\varphi \circ (\alpha \times \beta) : I^k \times I^{n-k} \to U$ is a diffeomorphism, and so its inverse is a chart. We now show that B can be chosen small enough that the intersection of each orbit with B is either empty or a single point. If this were not true, then there would be a sequence of open sets $\{B_i\}$ with compact closure and $\overline{B}_{i+1} \subset B_i$ (and with corresponding diffeomorphisms $\beta_i : I^k \to B_i$ as above), such that for every i there is a pair of distinct points $p_i, p'_i \in B_i$ with $g_ip_i = p'_i$ for some sequence $\{g_i\} \subset G$. (We have used the fact that manifolds are first countable and normal.) This forces both p_i and $p'_i = g_ip_i$ to converge to p. From this we see that the set $K = \{(g_ip_i, p_i), (p, p)\} \subset M \times M$ is compact. Recall that by definition, the map $P : (g, x) \mapsto (gx, x)$ is proper. Since $(g_i, p_i) = P^{-1}(g_ip_i, p_i)$, we see that $\{(g_i, p_i)\}$ is a subset of the compact set $P^{-1}(K)$. Thus after passing to a subsequence, we have that (g_i, p_i) converges to (g, p) for some g and hence $g_i \to g$ and $g_ip_i \to gp$. But this means that we have

$$gp = \lim_{i \to \infty} g_ip_i = \lim_{i \to \infty} p'_i = p,$$

and since the action is free, we conclude that $g = e$. However, this is impossible since it would mean that for large enough i we would have $g_i \in A$, and this in turn would imply that

$$\varphi(g_i, p_i) = l_{g_i}(p_i) = p'_i = l_e(p'_i) = \varphi(e, p'_i)$$

contradicting the injectivity of φ on $A \times B$. Thus after shrinking B we may assume that the intersection of each orbit with B is either empty or a single point. One may now check that with $\mathbf{x} := (\varphi \circ (\alpha \times \beta))^{-1} : U \to I^k \times I^{n-k} \subset \mathbb{R}^n$, we obtain a chart $(U, \mathbf{x})$ with the desired properties. Write $\mathbf{x} = (x, y)$, where y takes values in $I^{n-k} \subset \mathbb{R}^{n-k}$ and x takes values in $I^k \subset \mathbb{R}^k$. Each $y = c$ slice is of the form $\varphi(A \times \{q\}) \subset Gq$ for $q \in B$ and so is contained in a single orbit. We see that the intersection of an orbit with U must be

5.7. Lie Group Actions 235

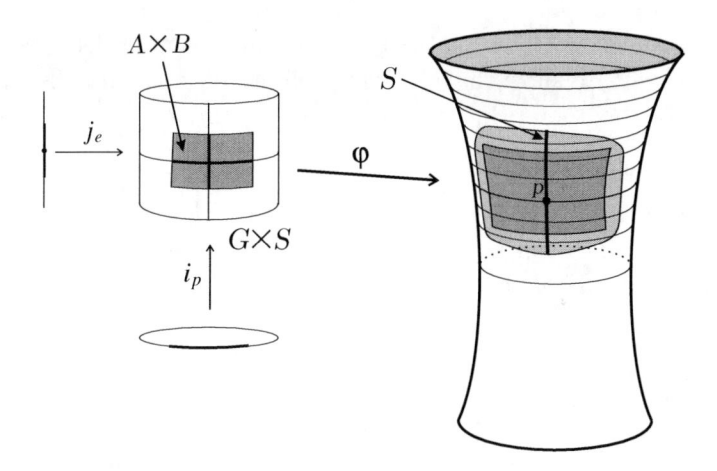

Figure 5.2. Construction of action-adapted charts

a union of such slices. But each orbit can only intersect B in at most one point, and so it is clear that each orbit intersects U in one slice or does not intersect U at all. □

Notice that we have actually constructed action-adapted charts that have image the cube I^n. For the next lemma, we continue with the convention that I is the interval $(-1, 1)$.

Lemma 5.116. *Let* $\mathbf{x} := (\varphi \circ (\alpha \times \beta))^{-1} : U \to I^k \times I^{n-k} = I^n \subset \mathbb{R}^n$ *be an action-adapted chart map obtained as in the proof of Theorem 5.115 above. Then given any $p_1 \in U$, there exists a diffeomorphism $\psi : I^n \to I^n$ such that $\psi \circ \mathbf{x}$ is an action-adapted chart centered at p_1 and with image I^n. Furthermore ψ can be decomposed as $(a, b) \mapsto (\psi_1(a), \psi_2(b))$, where $\psi_1 : I^k \to I^k$ and $\psi_2 : I^{n-k} \to I^{n-k}$ are diffeomorphisms.*

Proof. We need to show that for any $a \in I^n$, there is a diffeomorphism $\psi : I^n \to I^n$ such that $\psi(a) = 0$. Let a^i be the i-th component of a. Let $\psi_i : I \to I$ be defined by

$$\psi_i := \phi^{-1} \circ t_{-\phi(a_i)} \circ \phi,$$

where $t_{-c}(x) := x - c$ and $\phi : (-1, 1) \to \mathbb{R}$ is the diffeomorphism $\phi : x \mapsto \tan(\frac{\pi}{2}x)$. The diffeomorphism we want is $\psi(x) = (\psi_1(x^1), \ldots, \psi_1(x^n))$. The last part is obvious from the construction. □

We now discuss quotients. If $l : G \times M \to M$ is a Lie group action, then there is a natural equivalence relation on M whereby the equivalence classes are exactly the orbits of the action. The **quotient space** (or orbit space) is denoted $G \backslash M$, and we have the quotient map $\pi : M \to G \backslash M$. We put the quotient topology on $G \backslash M$ so that $A \subset G \backslash M$ is open if and only if

$\pi^{-1}(A)$ is open in M. The quotient map is also open. Indeed, let $U \subset M$ be open. We want to show that $\pi(U)$ is open, and for this it suffices to show that $\pi^{-1}(\pi(U))$ is open. But $\pi^{-1}(\pi(U))$ is the union $\bigcup_g l_g(U)$, and this is open since each $l_g(U)$ is open.

Proposition 5.117. *Let G act smoothly on M. Then $G\backslash M$ is a Hausdorff space if the set $\Gamma := \{(gp, p) : g \in G, p \in M\}$ is a closed subset of $M \times M$. If M is second countable then $G\backslash M$ is also.*

Proof. Let $\underline{p}, \underline{q} \in G\backslash M$ with $\pi(p) = \underline{p}$ and $\pi(q) = \underline{q}$. If $\underline{p} \neq \underline{q}$, then p and q are not in the same orbit. This means that $(p, q) \notin \Gamma$, and so there must be a product open set $U \times V$ such that $(p, q) \in U \times V$ and $U \times V$ is disjoint from Γ. This means that $\pi(U)$ and $\pi(V)$ are disjoint neighborhoods of $\underline{p}$ and $\underline{q}$ respectively. Finally, if $\{U_i\}$ is a countable basis for the topology on M, then $\{\pi(U_i)\}$ is a countable basis for the topology on $G\backslash M$. $\square$

Proposition 5.118. *If $l : G \times M \to M$ is a free and proper action, then $G\backslash M$ is Hausdorff.*

Proof. To show that $G\backslash M$ is Hausdorff, we use the previous lemma. We must show that Γ is closed. By Problem 2, proper continuous maps are closed. Thus $\Gamma = P(G \times M)$ is closed since P is proper. $\square$

We will shortly show that if the action is free and proper, then $G\backslash M$ has a smooth structure which makes the quotient map $\pi : M \to G\backslash M$ a submersion. Before coming to this let us note that if such a smooth structure exists, then it is unique and is determined by the smooth structure on M. Indeed, if $(G\backslash M)_{\mathcal{A}}$ is $G\backslash M$ with a smooth structure given by a maximal atlas $\mathcal{A}$ and similarly for $(G\backslash M)_{\mathcal{B}}$ for another atlas $\mathcal{B}$, then we have the following commutative diagram:

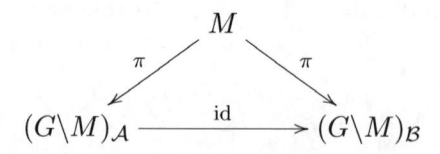

$$\begin{array}{ccc} & M & \\ {\pi}\swarrow & & \searrow{\pi} \\ (G\backslash M)_{\mathcal{A}} & \xrightarrow{\text{id}} & (G\backslash M)_{\mathcal{B}} \end{array}$$

Since π is a surjective submersion, Proposition 3.26 applies to show that $(G\backslash M)_{\mathcal{A}} \xrightarrow{\text{id}} (G\backslash M)_{\mathcal{B}}$ is smooth as is its inverse. This means that $\mathcal{A} = \mathcal{B}$.

Theorem 5.119. *If $l : G \times M \to M$ is a free and proper Lie group action, then there is a unique smooth structure on the quotient $G\backslash M$ such that*

    (i) *the induced topology is the quotient topology, and $G\backslash M$ is a smooth manifold;*

5.7. Lie Group Actions

(ii) the projection $\pi : M \to G\backslash M$ is a submersion;

(iii) $\dim(G\backslash M) = \dim(M) - \dim(G)$.

Proof. Let $\dim(M) = n$ and $\dim(G) = k$. We have already shown that $G\backslash M$ is a Hausdorff space. All that is left is to exhibit an atlas such that the charts are homeomorphisms with respect to this quotient topology. Let $q \in G\backslash M$ and choose p with $\pi(p) = q$. Let $(U, \mathbf{x})$ be an action-adapted chart centered at p and constructed exactly as in Theorem 5.115. Let $\pi(U) = V \subset G\backslash M$ and let B be the slice $x^1 = \cdots = x^k = 0$. By construction $\pi|_B : B \to V$ is a bijection. In fact, it is easy to check that $\pi|_B$ is a homeomorphism, and $\sigma := (\pi|_B)^{-1}$ is the corresponding local section. Consider the map $\mathbf{y} = \pi_2 \circ \mathbf{x} \circ \sigma$, where π_2 is the second factor projection $\pi_2 : \mathbb{R}^k \times \mathbb{R}^{n-k} \to \mathbb{R}^{n-k}$. This is a homeomorphism since $(\pi_2 \circ \mathbf{x})|_B$ is a homeomorphism and $\pi_2 \circ \mathbf{x} \circ \sigma = (\pi_2 \circ \mathbf{x})|_B \circ \sigma$. We now have a chart $(V, \mathbf{y})$.

Given two such charts $(V, \mathbf{y})$ and $(\bar{V}, \bar{\mathbf{y}})$, we must show that $\bar{\mathbf{y}} \circ \mathbf{y}^{-1}$ is smooth. The $(V, \mathbf{y})$ and $(\bar{V}, \bar{\mathbf{y}})$ are constructed from associated action-adapted charts $(U, \mathbf{x})$ and $(\bar{U}, \bar{\mathbf{x}})$ on M. Let $q \in V \cap \bar{V}$. As in the proof of Lemma 5.116, we may find diffeomorphisms ψ and $\bar{\psi}$ such that $(U, \psi \circ \mathbf{x})$ and $(\bar{U}, \bar{\psi} \circ \bar{\mathbf{x}})$ are action-adapted charts centered at points $p_1 \in \pi^{-1}(q)$ and $p_2 \in \pi^{-1}(q)$ respectively. Correspondingly, the charts $(V, \mathbf{y})$ and $(\bar{V}, \bar{\mathbf{y}})$ are modified to charts $(V, \mathbf{y}_\psi)$ and $(\bar{V}, \bar{\mathbf{y}}_{\bar{\psi}})$ centered at q, where

$$\mathbf{y}_\psi := \pi_2 \circ \mathbf{x}_\psi \circ \sigma',$$
$$\bar{\mathbf{y}}_{\bar{\psi}} := \pi_2 \circ \bar{\mathbf{x}}_{\bar{\psi}} \circ \bar{\sigma}',$$

with $\mathbf{x}_\psi := \psi \circ \mathbf{x}$ and similarly for $\bar{\mathbf{x}}_{\bar{\psi}}$. Also, the sections σ' and $\bar{\sigma}'$ are constructed to map into the zero slices of $\mathbf{x}_\psi$ and $\bar{\mathbf{x}}_{\bar{\psi}}$. However, it is not hard to see that $\mathbf{y}_\psi$ and $\bar{\mathbf{y}}_{\bar{\psi}}$ are unchanged if we replace σ' and $\bar{\sigma}'$ by the sections σ and $\bar{\sigma}$ corresponding to the zero slices of $\mathbf{x}$ and $\mathbf{y}$. Now, recall that ψ was chosen to have the form $(a,b) \mapsto (\psi_1(a), \psi_2(b))$. Using the above observations, one checks that $\mathbf{y}_\psi \circ \mathbf{y}^{-1} = \psi_2$ and similarly for $\bar{\mathbf{y}}_{\bar{\psi}} \circ \bar{\mathbf{y}}^{-1}$. From this it follows that the overlap map $\bar{\mathbf{y}}_{\bar{\psi}}^{-1} \circ \mathbf{y}_\psi$ will be smooth if and only if $\bar{\mathbf{y}}^{-1} \circ \mathbf{y}^{-1}$ is smooth. Thus we have reduced to the case where $(U, \mathbf{x})$ and $(\bar{U}, \bar{\mathbf{x}})$ are centered at $p_1 \in \pi^{-1}(q)$ and $p_2 \in \pi^{-1}(q)$ respectively. This entails that both $(V, \mathbf{y})$ and $(\bar{V}, \bar{\mathbf{y}})$ are centered at $q \in V \cap \bar{V}$. If we choose a $g \in G$ such that $l_g(p_1) = p_2$, then by composing with the diffeomorphism l_g we can reduce further to the case where $p_1 = p_2$. Here we use the fact that l_g takes the set of orbits to the set of orbits in a bijective manner and the special nature of our action-adapted charts with respect to these orbits. In this case, the overlap map $\bar{\mathbf{x}} \circ \mathbf{x}^{-1}$ must have the form $(a,b) \mapsto (f(a,b), g(b))$ for some smooth functions f and g. It follows that $\bar{\mathbf{y}} \circ \mathbf{y}^{-1}$ has the form $b \mapsto g(b)$.

Finally, we give an argument that $G\backslash M$ is paracompact. Since $G\backslash M$ is Hausdorff and locally Euclidean, we will be done once we show that every connected component of $G\backslash M$ is second countable (see Proposition B.5). Let Q_0 be any such connected component of $G\backslash M$. Let X_0 be a connected component of $\pi^{-1}(Q_0)$. Then X_0 is a second countable manifold and open in M. We now argue that $\pi(X_0) = Q_0$. To this end we show that the connected set $\pi(X_0)$ is open and closed in Q_0. It is open since π is an open map. Let $\bar{x}$ be in the closure of $\pi(X_0)$ and choose $x \in M$ with $\pi(x) = \bar{x}$. Let U be the domain of an action-adapted chart centered at x and let S be the slice of U that maps diffeomorphically onto an open neighborhood O of $\bar{x}$. Now we may find $\bar{y} \in O \cap \pi(X_0)$ and a corresponding $y \in S$ such that $\pi(y) = \bar{y}$. Since $\bar{y}$ is in the image of X_0, there is a $y' \in X_0$ such that $\pi(y') = \bar{y}$. But then there exists $g \in G$ such that $gy = y'$. The open set gU is diffeomorphic to U under l_g and hence is path connected. It contains y' and gx. Since X_0 is a path component, $gU \subset X_0$ and so $gx \in X_0$. But $\bar{x} = \pi(gx)$ so $\bar{x} \in \pi(X_0)$. We conclude that $\pi(X_0)$ is closed (and open) and connected. Hence $\pi(X_0) = Q_0$. Now since X_0 is a second countable manifold, we argue as in Proposition 5.117 that Q_0 is second countable. Conclusion: $G\backslash M$ is paracompact. $\square$

Similar results hold for right actions. Some of the most important examples of proper actions are usually presented as right actions. In fact, principal bundle actions studied in Chapter 6 are usually presented as right actions. We shall also encounter situations where there are both a right and a left action in play.

Example 5.120. Consider S^{2n-1} as the subset of $\mathbb{C}^n$ given by $S^{2n-1} = \{\xi \in \mathbb{C}^n : |\xi| = 1\}$. Here $\xi = (z^1, \ldots, z^n)$ and $|\xi| = \sum \bar{z}^i z^i$. Let S^1 act on S^{2n-1} by $(a, \xi) \mapsto a\xi = (az^1, \ldots, az^n)$. This action is free and proper. The quotient is the complex projective space $\mathbb{C}P^{n-1}$,

$$S^{2n-1}$$
$$\downarrow$$
$$\mathbb{C}P^{n-1}$$

These maps (one for each n) are called the **Hopf maps**. In this context, S^1 is usually denoted by $U(1)$.

In what follows, we will consider the similar right action $S^n \times U(1) \to S^n$. In this case, we think of $\mathbb{C}^{n+1}$ as a set of column vectors, and the action is given by $(\xi, a) \mapsto \xi a$. Of course, since $U(1)$ is abelian, this makes essentially no difference, but in the next example we consider the quaternionic analogue where keeping track of order is important.

5.7. Lie Group Actions

The quaternionic projective space $\mathbb{H}P^{n-1}$ is defined by analogy with $\mathbb{C}P^{n-1}$. The elements of $\mathbb{H}P^{n-1}$ are 1-dimensional subspaces of the right $\mathbb{H}$-vector space $\mathbb{H}^n$. Let us refer to these as $\mathbb{H}$-lines. Each of these are of real dimension 4. Each element of $\mathbb{H}^n\setminus\{0\}$ determines an $\mathbb{H}$-line and the $\mathbb{H}$-line determined by $(\xi^1,\ldots,\xi^n)^t$ will be the same as that determined by $(\widetilde{\xi}^1,\ldots,\widetilde{\xi}^n)^t$ if and only if there is a nonzero element $a \in \mathbb{H}$ such that $(\widetilde{\xi}^1,\ldots,\widetilde{\xi}^n)^t = (\xi^1,\ldots,\xi^n)^t a = (\xi^1 a,\ldots,\xi^n a)^t$. This defines an equivalence relation $\sim$ on $\mathbb{H}^n\setminus\{0\}$, and thus we may also think of $\mathbb{H}P^{n-1}$ as $(\mathbb{H}^n\setminus\{0\})/\sim$. The element of $\mathbb{H}P^{n-1}$ determined by $(\xi^1,\ldots,\xi^n)^t$ is denoted by $[\xi^1,\ldots,\xi^n]$. Notice that the subset $\{\xi \in \mathbb{H}^n : |\xi| = 1\}$ is S^{4n-1}. Just as for the complex projective spaces, we observe that all such $\mathbb{H}$-lines contain points of S^{4n-1}, and two points $\xi, \zeta \in S^{4n-1}$ determine the same $\mathbb{H}$-line if and only if $\xi = \zeta a$ for some a with $|a| = 1$. Thus we can think of $\mathbb{H}P^{n-1}$ as a quotient of S^{4n-1}. When viewed in this way, we also denote the equivalence class of $\xi = (\xi^1,\ldots,\xi^n)^t \in S^{4n-1}$ by $[\xi] = [\xi^1,\ldots,\xi^n]$. The equivalence classes are clearly the orbits of an action as described in the following example.

Example 5.121. Consider S^{4n-1} regarded as the subset of $\mathbb{H}^n$ given by $S^{4n-1} = \{\xi \in \mathbb{H}^n : |\xi| = 1\}$. Here $\xi = (\xi^1,\ldots,\xi^n)^t$ and $|\xi| = \sum \bar{\xi}^i \xi^i$. Now we define a *right* action of $U(1,\mathbb{H})$ on S^{4n-1} by $(\xi, a) \mapsto \xi a = (\xi^1 a,\ldots,\xi^n a)^t$. This action is free and proper. The quotient is the quaternionic projective space $\mathbb{H}P^{n-1}$, and we have the quotient map denoted by $\wp$,

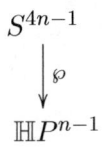

$$\begin{array}{c} S^{4n-1} \\ \downarrow \wp \\ \mathbb{H}P^{n-1} \end{array}$$

This map is also referred to as a **Hopf map**. Recall that $\mathbb{Z}_2 = \{1, -1\}$ acts on $S^{n-1} = \mathbb{R}^n$ on the right (or left) by multiplication, and the action is a (discrete) proper and free action with quotient $\mathbb{R}P^{n-1}$, and the examples above generalize this.

For completeness, we describe an atlas for $\mathbb{H}P^{n-1}$. View $\mathbb{H}P^{n-1}$ as the quotient $S^{4n-1}/\sim$ described above. Let

$$U_k := \{[\xi] \in S^{4n-1} \subset \mathbb{H}^n : \xi^k \neq 0\},$$

and define $\varphi_k : U_k \to \mathbb{H}^{n-1} \cong \mathbb{R}^{4n-1}$ by

$$\varphi_k([\xi]) = (\xi_1 \xi_k^{-1}, \ldots, \widehat{1}, \ldots, \xi_n \xi_k^{-1}),$$

where as before, the caret symbol ^ indicates that we have omitted the 1 in the k-th slot to obtain an element of $\mathbb{H}^{n-1}$. Notice that we insist that the ξ_k^{-1} in this expression multiply from the right. The general pattern for the overlap maps become clear from the example $\varphi_3 \circ \varphi_2^{-1}$. Here we have

$$\varphi_3 \circ \varphi_2^{-1}(y_1, y_3, \ldots, y_n) = \varphi_3([y_1, 1, y_3, \ldots, y_n])$$
$$= \left(y_1 y_3^{-1}, y_3^{-1}, y_4 y_3^{-1}, \ldots, y_n y_3^{-1}\right).$$

In the case $n = 1$, we have an atlas of just two charts $\{(U_1, \varphi_1), (U_2, \varphi_2)\}$. In close analogy with the complex case we have $U_1 \cap U_2 = \mathbb{H} \setminus \{0\}$ and $\varphi_1 \circ \varphi_2^{-1}(y) = \bar{y}^{-1} = \varphi_2 \circ \varphi_1^{-1}(y)$ for $y \in \mathbb{H} \setminus \{0\}$.

Exercise 5.122. Show that by identifying $\mathbb{H}$ with $\mathbb{R}^4$ and modifying the stereographic charts on $S^3 \subset \mathbb{R}^4$ we can obtain an atlas for S^3 with overlap maps of the same form as for $\mathbb{H}P^1$ given above. Use this to show that $\mathbb{H}P^1 \cong S^3$.

Combining the last exercise with previous results we have

$$\mathbb{R}P^1 \cong S^1, \quad \mathbb{C}P^1 \cong S^2, \quad \mathbb{H}P^1 \cong S^3.$$

5.8. Homogeneous Spaces

Let H be a closed Lie subgroup of a Lie group G. Then we have a *right* action of H on G given by right multiplication $r : G \times H \to G$. The orbits of this right action are exactly the left cosets of the quotient G/H. The action is clearly free, and we would like to show that it is also proper. Since we are now talking about a right action, and G is the manifold on which we are acting, we need to show that the map $P_{\text{right}} : G \times H \to G \times G$ given by $(p, h) \mapsto (p, ph)$ is a proper map. The characterization of proper action becomes the condition that

$$H_K := \{h \in H : (K \cdot h) \cap K \neq \emptyset\}$$

is a compact subset of H whenever K is compact in G. Let K be any compact subset of G. It will suffice to show that H_K is sequentially compact. To this end, let $(h_i)_{i \in \mathbb{N}}$ be a sequence in H_K. Then there must be sequences (a_i) and (b_i) in K such that $a_i h_i = b_i$. Since K is compact and hence sequentially compact, we can pass to subsequences $\left(a_{i(j)}\right)_{j \in \mathbb{N}}$ and $\left(b_{i(j)}\right)_{j \in \mathbb{N}}$ such that $\lim_{j \to \infty} a_{i(j)} = a$ and $\lim_{j \to \infty} b_{i(j)} = b$. Here $i \mapsto i(j)$ is a monotonic map on positive integers; $\mathbb{N} \to \mathbb{N}$. This means that $\lim_{j \to \infty} h_{i(j)} = \lim_{j \to \infty} a_{i(j)}^{-1} b_{i(j)} = a^{-1} b$. Thus the original sequence $\{h_i\}$ is shown to have a convergent subsequence. Since by Theorem 5.81, H is an embedded submanifold, this sequence converges in the topology of H. We conclude that the right action is proper. Using Theorem 5.119 (or its analogue for right actions), we obtain the following Proposition.

Proposition 5.123. *Let H be a closed Lie subgroup of a Lie group G. Then,*

5.8. Homogeneous Spaces

(i) *the right action $G \times H \to G$ is free and proper;*

(ii) *the orbit space is the left coset space G/H, and this has a unique smooth manifold structure such that the quotient map $\pi : G \to G/H$ is a surjection. Furthermore, $\dim(G/H) = \dim(G) - \dim(H)$.*

If K is a normal Lie subgroup of G, then the quotient is a group with multiplication defined by $[g_1][g_2] = (g_1 K)(g_2 K) = g_1 g_2 K$. In this case, we may ask whether G/K is a Lie group. If K is closed, then we know from the considerations above that G/K is a smooth manifold and that the quotient map is smooth. In fact, we have the following:

Proposition 5.124 (Quotient Lie groups). *If K is a closed normal subgroup of a Lie group G, then G/K is a Lie group and the quotient map $G \to G/K$ is a Lie group homomorphism. Furthermore, if $f : G \to H$ is a surjective Lie group homomorphism, then $\mathrm{Ker}(f)$ is a closed normal subgroup, and the induced map $\widetilde{f} : G/\mathrm{Ker}(f) \to H$ is a Lie group isomorphism.*

Proof. We have already observed that G/K is a smooth manifold and that the quotient map is smooth. After taking into account what we know from standard group theory, the only thing we need to prove for the first part, is that the multiplication and inversion in the quotient are smooth. It is an easy exercise using Corollary 3.27 to show that both of these maps are smooth.

Consider a Lie group homomorphism f as in the hypothesis of the proposition. It is standard that $\mathrm{Ker}(f)$ is a normal subgroup and it is clearly closed. It is also easy to verify fact that the induced $\widetilde{f}$ map is an isomorphism. One can then use Corollary 3.27 to show that the induced map $\widetilde{f}$ is smooth. $\square$

If a Lie group G acts smoothly and transitively on M (on the right or left), then M is called a **homogeneous space** with respect to that action. Of course it is possible that a single group G may act on M in more than one way and so M may be a homogeneous space in more than one way. We will give a few concrete examples shortly, but we already have an abstract example on hand.

Theorem 5.125. *If H is a closed Lie subgroup of a Lie group G, then the map $G \times G/H \to G/H$, given by $l : (g, g_1 H) \mapsto g g_1 H$, is a transitive Lie group action. Thus G/H is a homogeneous space with respect to this action.*

Proof. The fact that l is well-defined follows since if $g_1 H = g_2 H$, then $g_2^{-1} g_1 \in H$, and so $g g_2 H = g g_2 g_2^{-1} g_1 H = g g_1 H$. We already know that G/H is a smooth manifold and $\pi : G \to G/H$ is a surjective submersion.

We can form another submersion $\mathrm{id}_G \times \pi : G \times G \to G \times G/H$ making the following diagram commute:

$$\begin{array}{ccc} G \times G & \longrightarrow & G \\ \mathrm{id}_G \times \pi \downarrow & \searrow & \downarrow \pi \\ G \times G/H & \longrightarrow & G/H \end{array}$$

Here the upper horizontal map is group multiplication and the lower horizontal map is the action l. Since the diagonal map is smooth, it follows from Proposition 3.26 that l is smooth. We see that l is transitive by observing that if $g_1 H, g_2 H \in G/H$, then $l_{g_2 g_1^{-1}}(g_1 H) = g_2 H$. □

It turns out that up to appropriate equivalence, the examples of the above type account for all homogeneous spaces. Before proving this let us look at some concrete examples.

Example 5.126. Let $M = \mathbb{R}^n$ and let $G = \mathrm{Euc}(n, \mathbb{R})$ be the group of Euclidean motions. We realize $\mathrm{Euc}(n, \mathbb{R})$ as a matrix group

$$\mathrm{Euc}(n, \mathbb{R}) = \left\{ \begin{bmatrix} 1 & 0 \\ v & Q \end{bmatrix} : v \in \mathbb{R}^n \text{ and } Q \in \mathrm{O}(n) \right\}$$

The action of $\mathrm{Euc}(n, \mathbb{R})$ on $\mathbb{R}^n$ is given by the rule

$$\begin{bmatrix} 1 & 0 \\ v & Q \end{bmatrix} \cdot x = Qx + v,$$

where x is written as a column vector. Notice that this action is not given by a matrix multiplication, but one can use the trick of representing the points x of $\mathbb{R}^n$ by the $(n+1) \times 1$ column vectors $\begin{bmatrix} 1 \\ x \end{bmatrix}$, and then we have $\begin{bmatrix} 1 & 0 \\ v & Q \end{bmatrix} \begin{bmatrix} 1 \\ x \end{bmatrix} = \begin{bmatrix} 1 \\ Qx+v \end{bmatrix}$. The action is easily seen to be transitive.

Example 5.127. As in the previous example we take $M = \mathbb{R}^n$, but this time, the group acting is the affine group $\mathrm{Aff}(n, \mathbb{R})$ realized as a matrix group:

$$\mathrm{Aff}(n, \mathbb{R}) = \left\{ \begin{bmatrix} 1 & 0 \\ v & A \end{bmatrix} : v \in \mathbb{R}^n \text{ and } A \in \mathrm{GL}(n, \mathbb{R}) \right\}.$$

The action is

$$\begin{bmatrix} 1 & 0 \\ v & A \end{bmatrix} \cdot x = Ax + v,$$

and this is again a transitive action.

Comparing these first two examples, we see that we have made $\mathbb{R}^n$ into a homogeneous space in two different ways. It is sometimes desirable to give different names and/or notations for $\mathbb{R}^n$ to distinguish how we are acting on the space. In the first example we might write $\mathbb{E}^n$ (**Euclidean space**), and

5.8. Homogeneous Spaces

in the second case we write $\mathbb{A}^n$ and refer to it as **affine space**. Note that, roughly speaking, the action by $\text{Euc}(n,\mathbb{R})$ preserves all metric properties of figures such as curves defined in $\mathbb{E}^n$. On the other hand, $\text{Aff}(n,\mathbb{R})$ always sends lines to lines, planes to planes, etc.

Example 5.128. Let $M = \text{H} := \{z \in \mathbb{C} : \text{Im}\, z > 0\}$. This is the upper half-plane. The group acting on H will be $\text{SL}(2,\mathbb{R})$, and the action is given by
$$\begin{pmatrix} a & b \\ c & d \end{pmatrix} \cdot z = \frac{az+b}{cz+d}.$$
This action is transitive.

Example 5.129. We have already seen in Example 5.102 that both $\text{O}(n)$ and $\text{SO}(n)$ act transitively on the sphere $S^{n-1} \subset \mathbb{R}^n$, so S^{n-1} is a homogeneous space in at least two (slightly) different ways. Also, both $\text{SU}(n)$ and $\text{U}(n)$ act transitively on $S^{2n-1} \subset \mathbb{C}^n$.

Example 5.130. Let $V'_{n,k}$ denote the set of all k-frames for $\mathbb{R}^n$, where by a k-frame we mean an ordered set of k linearly independent vectors. Thus an n-frame is just an ordered basis for $\mathbb{R}^n$. This set can easily be given a smooth manifold structure. This manifold is called the (real) **Stiefel manifold of k-frames**. The Lie group $\text{GL}(n,\mathbb{R})$ acts (smoothly) on $V'_{n,k}$ by $g \cdot (e_1,\ldots,e_k) = (ge_1,\ldots,ge_k)$. To see that this action is transitive, let $(e_1,\ldots,e_k)$ and $(f_1,\ldots,f_k)$ be two k-frames. Extend each to n-frames $(e_1,\ldots,e_k,\ldots,e_n)$ and $(f_1,\ldots,f_k,\ldots,f_n)$. Since we consider elements of $\mathbb{R}^n$ as column vectors, these two n-frames can be viewed as invertible $n \times n$ matrices E and F. If we let $g := EF^{-1}$, then $gE = F$, or $g \cdot (e_1,\ldots,e_k) = (ge_1,\ldots,ge_k) = (f_1,\ldots,f_k)$.

Example 5.131. Let $V_{n,k}$ denote the set of all orthonormal k-frames for $\mathbb{R}^n$, where by an orthonormal k-frame we mean an ordered set of k orthonormal vectors. Thus an orthonormal n-frame is just an orthonormal basis for $\mathbb{R}^n$. This set can easily be given a smooth manifold structure and is called the **Stiefel manifold of orthonormal k-frames**. The group $\text{O}(n,\mathbb{R})$ acts transitively on $V_{n,k}$ for reasons similar to those given in the last example.

Theorem 5.132. *Let M be a homogeneous space via the transitive action $l : G \times M \to M$, and let G_p be the isotropy subgroup of a point $p \in M$. Recall that G acts on G/G_p. If G/G_p is second countable (in particular if G is second countable), then there is an equivariant diffeomorphism $\phi : G/G_p \to M$ such that $\phi(gG_p) = g \cdot p$.*

Proof. We want to define ϕ by the rule $\phi(gG_p) = g \cdot p$, but we must show that this is well-defined. This is a standard group theory argument; if $g_1 G_p = g_2 G_p$, then $g_1^{-1} g_2 \in G_p$, so that $(g_1^{-1} g_2) \cdot p = p$ or $g_1 \cdot p = g_2 \cdot p$.

This map is surjective by the transitivity of the action l. It is also injective since if $\phi(g_1 G_p) = \phi(g_2 G_p)$, then $g_1 \cdot p = g_2 \cdot p$ or $\left(g_1^{-1} g_2\right) \cdot p = p$, which by definition means that $g_1^{-1} g_2 \in G_p$ and $g_1 G_p = g_2 G_p$. Notice that the following diagram commutes:

$$\begin{array}{ccc} G & & \\ \downarrow & \searrow^{\theta_p} & \\ G/G_p & \xrightarrow{\phi} & M \end{array}$$

From Corollary 3.27 we see that ϕ is smooth.

To show that ϕ is a diffeomorphism, it suffices to show that the rank of ϕ is equal to $\dim M$ or in other words that ϕ is a submersion. Since $\phi(gg_1 G_p) = (gg_1) \cdot p = g\phi(g_1 G_p)$, the map ϕ is equivariant and so has constant rank. By Lemma 3.28, ϕ is a submersion and hence in the present case a diffeomorphism. $\square$

Without the technical assumption on second countability, the proof shows that we still have that $\phi : G/G_p \to M$ is a smooth equivariant bijection.

Exercise 5.133. Show that if instead of the hypothesis of second countability in the last theorem we assume that θ_p has full rank at the identity, then $\phi : G/G_p \to M$ is a diffeomorphism.

Let $l : G \times M \to M$ be a left Lie group action and fix $p_0 \in M$. Denote the projection onto cosets by π and also write $\theta_{p_0} : g \mapsto gp_0$ as before. Then we have the following equivalence of maps:

$$\begin{array}{ccc} G & = & G \\ \downarrow^{\pi} & & \downarrow^{\theta_{p_0}} \\ G/G_p & \cong & M \end{array}$$

Exercise 5.134. Let G act on M as above. Show that if $p_2 = gp_1$ for some $g \in G$ and $p_1, p_2 \in M$, then is a natural Lie group isomorphism $G_{p_1} \cong G_{p_2}$ and a natural equivariant diffeomorphism $G/G_{p_1} \cong G/G_{p_2}$.

We now look again at some of our examples of homogeneous spaces and apply the above theorem.

Example 5.135. Consider again Example 5.126. The isotropy group of the origin in $\mathbb{R}^n$ is the subgroup consisting of matrices of the form

$$\begin{pmatrix} 1 & 0 \\ 0 & Q \end{pmatrix},$$

5.8. Homogeneous Spaces

where $Q \in O(n)$. This group is clearly isomorphic to $O(n, \mathbb{R})$, and so by the above theorem we have an equivariant diffeomorphism

$$\mathbb{R}^n \cong \frac{\text{Euc}(n, \mathbb{R})}{O(n, \mathbb{R})}.$$

Example 5.136. Consider again Example 5.127. The isotropy group of the origin in $\mathbb{R}^n$ is the subgroup consisting of matrices of the form

$$\begin{pmatrix} 1 & 0 \\ 0 & A \end{pmatrix},$$

where $A \in \text{GL}(n, \mathbb{R})$. This group is clearly isomorphic to $\text{GL}(n, \mathbb{R})$, and so by the above theorem we have an equivariant diffeomorphism

$$\mathbb{R}^n \cong \frac{\text{Aff}(n, \mathbb{R})}{\text{GL}(n, \mathbb{R})}.$$

It is important to realize that there is an implied action on $\mathbb{R}^n$, which is different from that in the previous example.

Example 5.137. Consider the action of $\text{SL}(2, \mathbb{R})$ on the complex upper half-plane $H = \mathbb{C}_+$ as in Example 5.128. We determine the isotropy subgroup for the point $i = \sqrt{-1}$. A matrix $A = \begin{pmatrix} a & b \\ c & d \end{pmatrix}$ is in this subgroup if and only if

$$\frac{ai + b}{ci + d} = i.$$

This is true exactly if $bc - ad = 1$ and $bd + ac = 0$, and so the isotropy subgroup is $\text{SO}(2, \mathbb{R})$ ($\cong S^1 = U(1, \mathbb{C})$). Thus we have an equivariant diffeomorphism

$$H = \mathbb{C}_+ \cong \frac{\text{SL}(2, \mathbb{R})}{\text{SO}(2, \mathbb{R})}.$$

Example 5.138. From Example 5.129 we obtain

$$S^{n-1} \cong \frac{O(n)}{O(n-1)}, \qquad S^{n-1} \cong \frac{SO(n)}{SO(n-1)},$$

$$S^{2n-1} \cong \frac{U(n)}{U(n-1)}, \qquad S^{2n-1} \cong \frac{SU(n)}{SU(n-1)}.$$

Example 5.139. Let $(\mathbf{e}_1, \ldots, \mathbf{e}_n)$ be the standard basis for $\mathbb{R}^n$. Under the action of $\text{GL}(n, \mathbb{R})$ on $V'_{n,k}$ given in Example 5.130, the isotropy group of the k-frame $(\mathbf{e}_{k+1}, \ldots, \mathbf{e}_n)$ is the subgroup of $\text{GL}(n, \mathbb{R})$ of the form

$$\begin{pmatrix} A & 0 \\ 0 & I \end{pmatrix} \text{ for } A \in \text{GL}(n-k, \mathbb{R}).$$

We identify this group with $\text{GL}(n-k, \mathbb{R})$ and then we obtain

$$V'_{n,k} \cong \frac{\text{GL}(n, \mathbb{R})}{\text{GL}(n-k, \mathbb{R})}.$$

Example 5.140. A similar analysis leads to an equivariant diffeomorphism
$$V_{n,k} \cong \frac{\mathrm{O}(n,\mathbb{R})}{\mathrm{O}(n-k,\mathbb{R})},$$
where $V_{n,k}$ is the Stiefel manifold of *orthonormal* k-planes of Example 5.131. Notice that taking $k=1$ we recover Example 5.135.

Exercise 5.141. Show that if $k < n$, then we have $V_{n,k} \cong \frac{\mathrm{SO}(n,\mathbb{R})}{\mathrm{SO}(n-k,\mathbb{R})}$.

Next we introduce a couple of standard results concerning connectivity.

Proposition 5.142. *Let G be a Lie group acting smoothly on M. Let the action be a left (resp. right) action. If both G and $M\backslash G$ (resp. M/G) are connected, then M is connected.*

Proof. Assume for concreteness that the action is a left action and that G and $M\backslash G$ are connected. Suppose by way of contradiction that M is not connected. Then there are disjoint open sets U and V whose union is M. Each orbit $G \cdot p$ is the image of the connected space G under the orbit map $g \mapsto g \cdot p$ and so is connected. This means that each orbit must be contained in one and only one of U and V. Since the quotient map π is an open map, $\pi(U)$ and $\pi(V)$ are open, and from what we have just observed they must be disjoint and $\pi(U) \cup \pi(V) = M\backslash G$. This contradicts the assumption that $M\backslash G$ is connected. $\square$

Corollary 5.143. *Let H be a closed Lie subgroup of G. If both H and G/H are connected, then G is connected.*

Corollary 5.144. *For each $n \geq 1$, the groups $\mathrm{SO}(n)$, $\mathrm{SU}(n)$ and $\mathrm{U}(n)$ are connected while the group $\mathrm{O}(n)$ has exactly two components: $\mathrm{SO}(n)$ and the subset of $\mathrm{O}(n)$ consisting of elements with determinant -1.*

Proof. The groups $\mathrm{SO}(1)$ and $\mathrm{SU}(1)$ are both connected since they each contain only one element. The group $U(1)$ is the circle, and so it too is connected. We use induction. Suppose that $\mathrm{SO}(k)$, $\mathrm{SU}(k)$ are connected for $1 \leq k \leq n-1$. We show that this implies that $\mathrm{SO}(n)$, $\mathrm{SU}(n)$ and $\mathrm{U}(n)$ are connected. From Example 5.138 we know that $S^{n-1} = \mathrm{SO}(n)/\mathrm{SO}(n-1)$. Since S^{n-1} and $\mathrm{SO}(n-1)$ are connected (the second by the induction hypothesis), we see that $\mathrm{SO}(n)$ is connected. The same argument works for $\mathrm{SU}(n)$ and $\mathrm{U}(n)$.

Every element of $\mathrm{O}(n)$ has determinant either 1 or -1. The subset $\mathrm{SO}(n) \subset \mathrm{O}(n)$ is closed since it is exactly $\{g \in \mathrm{O}(n) : \det g = 1\}$. Fix an element a_0 with $\det a_0 = -1$. It is easy to show that $a_0 \mathrm{SO}(n)$ is exactly the set of elements of $\mathrm{O}(n)$ with determinant -1 so that $\mathrm{SO}(n) \cup a_0 \mathrm{SO}(n) = \mathrm{O}(n)$ and $\mathrm{SO}(n) \cap a_0 \mathrm{SO}(n) = \emptyset$. Indeed, by the multiplicative property of determinants, each element of $a_0 \mathrm{SO}(n)$ has determinant -1. But $a_0 \mathrm{SO}(n)$

5.8. Homogeneous Spaces

also contains every element of determinant -1 since for any such g we have $g = a_0 \left(a_0^{-1} g\right)$ and $a_0^{-1} g \in \mathrm{SO}(n)$. Since $\mathrm{SO}(n)$ and $a_0 \mathrm{SO}(n)$ are complements of each other, they are also both open. Both sets are connected, since $g \mapsto a_0 g$ is a diffeomorphism which maps the first to the second. Thus we see that $\mathrm{SO}(n)$ and $a_0 \mathrm{SO}(n)$ are the connected components of $\mathrm{O}(n)$. □

We close this chapter by relating the notion of a Lie group action with that of a Lie group representation. We give just a few basic definitions, some of which will be used in the next chapter.

Definition 5.145. A **linear action** of a Lie group G on a finite-dimensional vector space V is a left Lie group action $\lambda : G \times V \to V$ such that for each $g \in G$ the map $\lambda_g : v \mapsto \lambda(g, v)$ is linear.

The map $G \to \mathrm{GL}(V)$ given by $g \mapsto \lambda(g) := \lambda_g$ is a Lie group homomorphism and will be denoted by the same letter λ as the action so that $\lambda(g)v := \lambda(g, v)$. A Lie group homomorphism $\lambda : G \to \mathrm{GL}(V)$ is called a **representation** of G. Given such a representation, we obtain a linear action by letting $\lambda(g, v) := \lambda(g)v$. Thus a linear action of a Lie group is basically the same as a Lie group representation. The kernel of the action is the kernel of the associated homomorphism (representation). An effective linear action is one such that the associated homomorphism has trivial kernel, which, in turn, is the same as saying that the representation is **faithful**. Two representations $\lambda : G \to \mathrm{GL}(V)$ and $\lambda' : G \to \mathrm{GL}(V')$ are **equivalent** if there exists a linear isomorphism $T : V \to V'$ such that $T \circ \lambda_g = \lambda'_g \circ T$ for all g.

Exercise 5.146. Show that if $\lambda : G \times V \to V$ is a map such that $\lambda_g : v \mapsto \lambda(g, v)$ is linear for all g, then λ is smooth if and only if $\lambda_g : G \to \mathrm{GL}(V)$ is smooth for every $g \in G$. (Assume that V is finite-dimensional as usual.)

We have already seen one important example of a Lie group representation, namely, the adjoint representation. The adjoint representation came from first considering the action of G on itself given by conjugation which leaves the identity element fixed. The idea can be generalized:

Theorem 5.147. *Let* $l : G \times M \to M$ *be a (left) Lie group action. Suppose that* $p_0 \in M$ *is a fixed point of the action* ($l_g(p_0) = p_0$ *for all* g). *The map*

$$l^{(p_0)} : G \to \mathrm{GL}(T_{p_0} M)$$

given by

$$l^{(p_0)}(g) := T_{p_0} l_g$$

is a Lie group representation.

Proof. Since

$$l^{(p_0)}(g_1 g_2) = T_{p_0}(l_{g_1 g_2}) = T_{p_0}(l_{g_1} \circ l_{g_2})$$
$$= T_{p_0} l_{g_1} \circ T_{p_0} l_{g_2} = l^{(p_0)}(g_1) l^{(p_0)}(g_2),$$

we see that $l^{(p_0)}$ is a homomorphism. We must show that $l^{(p_0)}$ is smooth. It will be enough to show that $g \mapsto \alpha(T_{p_0} l_g \cdot v)$ is smooth for any $v \in T_{p_0} M$ and any $\alpha \in T_{p_0}^* M$. This will follow if we can show that for fixed $v_0 \in T_{p_0} M$, the map $G \to TM$ given by $g \mapsto T_{p_0} l_g \cdot v_0$ is smooth. This map is a composition

$$G \to TG \times TM \cong T(G \times M) \xrightarrow{Tl} TM,$$

where the first map is $g \mapsto (0_g, v_0)$, which is clearly smooth. By Exercise 5.146 this implies that the map $G \times T_{p_0} M \to T_{p_0} M$ given by $(g, v) \mapsto T_{p_0} l_g \cdot v$ is smooth. $\square$

Definition 5.148. For a Lie group action $l : G \times M \to M$ with fixed point p_0, the representation $l^{(p_0)}$ from the last theorem is called the **isotropy representation** for the fixed point.

Now let us consider a transitive Lie group action $l : G \times M \to M$ and a point p_0. For notational convenience, denote the isotropy subgroup G_{p_0} by H. Then H acts on M by restriction. We denote this action by $\lambda : H \times M \to M$,

$$\lambda : (h, p) \mapsto hp \text{ for } h \in H = G_{p_0}.$$

Notice that p_0 is a fixed point of this action, and so we have an isotropy representation $\lambda^{(p_0)} : H \times T_{p_0} M \to T_{p_0} M$. On the other hand, we have another action $C : H \times G \to G$, where $C_h : G \to G$ is given by $g \mapsto hgh^{-1}$ for $h \in H$. The Lie differential of C_h is the adjoint map $\mathrm{Ad}_h : \mathfrak{g} \to \mathfrak{g}$. The map C_h fixes H, and Ad_h fixes $\mathfrak{h}$. Thus the map $\mathrm{Ad}_h : \mathfrak{g} \to \mathfrak{g}$ descends to a map $\widetilde{\mathrm{Ad}}_h : \mathfrak{g}/\mathfrak{h} \to \mathfrak{g}/\mathfrak{h}$. We are going to show that there is a natural isomorphism $T_{p_0} M \cong \mathfrak{g}/\mathfrak{h}$ such that for each $h \in H$ the following diagram commutes:

(5.4)
$$\begin{array}{ccc} \mathfrak{g}/\mathfrak{h} & \xrightarrow{\widetilde{\mathrm{Ad}}_h} & \mathfrak{g}/\mathfrak{h} \\ \downarrow & & \downarrow \\ T_{p_0} M & \xrightarrow{\lambda_h^{(p_0)}} & T_{p_0} M \end{array}$$

One way to state the meaning of this result is to say that $h \mapsto \widetilde{\mathrm{Ad}}_h$ is a representation of H on the vector space $\mathfrak{g}/\mathfrak{h}$, which is equivalent to the linear isotropy representation. The isomorphism $T_{p_0} M \cong \mathfrak{g}/\mathfrak{h}$ is given in the

5.9. Combining Representations

following very natural way: Let $\xi \in \mathfrak{h}$ and consider $T_e\pi(\xi) \in T_{p_0}M$. We have

$$T_e\pi(\xi) = \left.\frac{d}{dt}\right|_{t=0} \pi(\exp \xi t) = 0$$

since $\exp \xi t \in \mathfrak{h}$ for all t. Thus $\mathfrak{h} \subset \mathrm{Ker}(T_e\pi)$. On the other hand, $\dim \mathfrak{h} = \dim H = \dim(\mathrm{Ker}(T_e\pi))$, so in fact $\mathfrak{h} = \mathrm{Ker}(T_e\pi)$ and we obtain an isomorphism $\mathfrak{g}/\mathfrak{h} \cong T_{p_0}M$ induced from $T_e\pi$. Let us see why the diagram (5.4) commutes. First, L_h is well-defined as a map from G/H to itself and the following diagram clearly commutes:

$$\begin{array}{ccc} G & \xrightarrow{C_h} & G \\ \pi \downarrow & & \downarrow \pi \\ G/H & \xrightarrow{L_h} & G/H \end{array}$$

Using our equivariant diffeomorphism $\phi : G/H \to M$, we obtain an equivalent commutative diagram

$$\begin{array}{ccc} G & \xrightarrow{C_h} & G \\ \theta_{p_0} \downarrow & & \downarrow \theta_{p_0} \\ M & \xrightarrow{\lambda_h} & M \end{array}$$

Applying these maps to $\exp t\xi$ for $\xi \in \mathfrak{g}$, we have

$$\begin{array}{ccc} \exp t\xi & \xmapsto{C_h} & h(\exp t\xi)h^{-1} \\ \theta_{p_0} \downarrow & & \downarrow \theta_{p_0} \\ (\exp t\xi)p_0 & \xmapsto{\lambda_h} & h(\exp t\xi)p_0 \end{array}$$

Applying the tangent functor (looking at the differential), we get the commutative diagram

$$\begin{array}{ccc} \mathfrak{g} & \xrightarrow{\mathrm{Ad}_h} & \mathfrak{g} \\ \downarrow & & \downarrow \\ T_{p_0}M & \xrightarrow{\lambda_h^{(p_0)}} & T_{p_0}M \end{array}$$

and, taking quotients, this gives the desired commutative diagram (5.4).

5.9. Combining Representations

We close this chapter with a bit about constructing new linear representations from old ones. Suppose that V is an $\mathbb{F}$-vector space and let $\mathcal{B} = (v_1, \ldots, v_n)$ be a basis for V. Then denoting the matrix representative of λ_g with respect to $\mathcal{B}$ by $[\lambda_g]_\mathcal{B}$ we obtain a homomorphism $G \to \mathrm{GL}(n, \mathbb{F})$

given by $g \mapsto [\lambda_g]_{\mathcal{B}}$. In general, a Lie group homomorphism of a Lie group G into $\mathrm{GL}(n,\mathbb{F})$ is called a matrix representation of G. We have already seen that any Lie subgroup of $\mathrm{GL}(\mathbb{R}^n)$ acts on $\mathbb{R}^n$ by matrix multiplication, and the corresponding homomorphism is the inclusion map $G \hookrightarrow \mathrm{GL}(\mathbb{R}^n)$. More generally, a Lie subgroup G of $\mathrm{GL}(\mathrm{V})$ acts on V in the obvious way simply by employing the definition of $\mathrm{GL}(\mathrm{V})$ as a set of linear transformations of V. We call this the **standard action** of the linear Lie subgroup of $\mathrm{GL}(\mathrm{V})$ on V, and the corresponding homomorphism is just the inclusion map $G \hookrightarrow \mathrm{GL}(\mathrm{V})$. Choosing a basis, the subgroup corresponds to a matrix group, and the standard action becomes matrix multiplication on the left of $\mathbb{F}^n$, where the latter is viewed as a space of column vectors. This action of a matrix group on column vectors is also referred to as a standard action.

Given a representation λ of G in a vector space V, we have a **dual representation** λ^* of G in the dual space V^* by defining $\lambda^*(g) := \left(\lambda(g^{-1})\right)^* : \mathrm{V}^* \to \mathrm{V}^*$. Recall that if $L : \mathrm{V} \to \mathrm{V}$ is linear, then $L^* : \mathrm{V}^* \to \mathrm{V}^*$ is defined by $L^*(\alpha)(v) = \alpha(Lv)$ for $\alpha \in \mathrm{V}^*$ and $v \in \mathrm{V}$. This dual representation is also sometimes called the **contragredient representation** (especially when $\mathbb{F} = \mathbb{R}$).

Now let λ^{V} and λ^{W} be representations of a lie group G in $\mathbb{F}$-vector spaces V and W respectively. We can then form the **direct sum representation** $\lambda^{\mathrm{V}} \oplus \lambda^{\mathrm{W}}$ by $\left(\lambda^{\mathrm{V}} \oplus \lambda^{\mathrm{W}}\right)_g := \lambda_g^{\mathrm{V}} \oplus \lambda_g^{\mathrm{W}}$ for $g \in G$, where we have $\left(\lambda_g^{\mathrm{V}} \oplus \lambda_g^{\mathrm{W}}\right)(v,w) = (\lambda_g^{\mathrm{V}} v, \lambda^{\mathrm{W}} w)$.

We will not pursue a serious study of Lie group representations but simply note that a major goal in the subject is the identification and classification of irreducible representations. A representation $\lambda : G \to \mathrm{GL}(\mathrm{V})$ is said to be **irreducible** if there is no nonzero proper subspace W of V such that $\lambda_g(\mathrm{W}) \subset \mathrm{W}$ for all g. A large class of Lie groups known as semisimple Lie groups have the property that their representations break into direct sums of irreducible representations.

Example 5.149. A homogeneous polynomial of degree d on $\mathbb{C}^2$ is a linear combination of monomials of total degree d. Let H_j denote the vector space of homogeneous polynomials of degree $2j$, where j is a nonnegative "half-integer" ($j = k/2$ for some nonnegative integer k). Define $\lambda_j : \mathrm{SU}(2) \to \mathrm{GL}(H_j)$ by $(\lambda_j(g)f)(z) := f(g^{-1}z)$ for $z = (z_1, z_2) \in \mathbb{C}^2$. Then λ_j is an irreducible representation called the **spin-j representation** of $\mathrm{SU}(2)$. The **spin**-1/2 representation turns out to be equivalent to the standard representation of $\mathrm{SU}(2)$ in $\mathbb{C}^2$. These spin representations play an important role in quantum physics.

One can also form the tensor product of representations. The definitions and basic facts about tensor products are given in the more general context of

5.9. Combining Representations

module theory in Appendix D. Here we give a quick recounting of the notion of a tensor product of vector spaces, and then we define tensor products of representations. Given two vector spaces V and W over some field $\mathbb{F}$, consider the class $\mathcal{C}_{V \times W}$ consisting of all bilinear maps $V \times W \to X$, where X varies over all $\mathbb{F}$-vector spaces, but V and W are fixed. We take members of $\mathcal{C}_{V \times W}$ as the objects of a category (see Appendix A). A morphism from, say, $\mu_1 : V \times W \to X$ to $\mu_2 : V \times W \to Y$ is defined to be a linear map $\ell : X \to Y$ such that the diagram

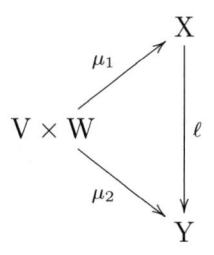

$$\begin{array}{ccc} & & X \\ & \nearrow^{\mu_1} & \\ V \times W & & \downarrow \ell \\ & \searrow_{\mu_2} & \\ & & Y \end{array}$$

commutes.

There exists a vector space $T_{V,W}$ together with a bilinear map $\otimes : V \times W \to T_{V,W}$ that has the following **universal property**: For every bilinear map $\mu : V \times W \to X$, there is a unique linear map $\widetilde{\mu} : T_{V,W} \to X$ such that the following diagram commutes:

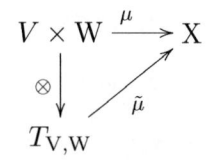

$$\begin{array}{ccc} V \times W & \xrightarrow{\mu} & X \\ \otimes \downarrow & \nearrow_{\widetilde{\mu}} & \\ T_{V,W} & & \end{array}$$

If such a pair $(T_{V,W}, \otimes)$ exists with this property, then it is unique up to isomorphism in $\mathcal{C}_{V \times W}$. In other words, if $\widehat{\otimes} : V \times W \to \widehat{T}_{V,W}$ also has this universal property, then there is a linear isomorphism $T_{V,W} \cong \widehat{T}_{V,W}$ such that the following diagram commutes:

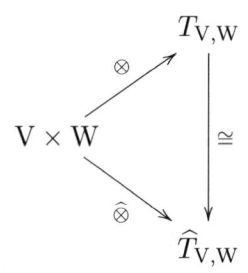

$$\begin{array}{ccc} & & T_{V,W} \\ & \nearrow^{\otimes} & \\ V \times W & & \downarrow \cong \\ & \searrow_{\widehat{\otimes}} & \\ & & \widehat{T}_{V,W} \end{array}$$

We refer to such universal object as a **tensor product** of V and W. We will indicate the construction of a specific tensor product that we denote by $V \otimes W$ with corresponding bilinear map $\otimes : V \times W \to V \otimes W$. The idea is simple: We let $V \otimes W$ be the set of all linear combinations of symbols of the form $v \otimes w$ for $v \in V$ and $w \in W$, subject to the relations

$$(v_1 + v_2) \otimes w = v_1 \otimes w + v_2 \otimes w,$$
$$v \otimes (w_1 + w_2) = v \otimes w_1 + v \otimes w_2,$$
$$r(v \otimes w) = rv \otimes w = v \otimes rw, \text{ for } r \in \mathbb{F}.$$

The map $\otimes$ is then simply $\otimes : (v, w) \to v \otimes w$. Somewhat more pedantically, let $F(V \times W)$ denote the free vector space generated by the *set* $V \times W$ (the elements of $V \times W$ are treated as a basis for the space, and so the free space has dimension equal to the cardinality of the set $V \times W$). Next consider the subspace R of $F(V \times W)$ generated by the set of all elements of the form

$$(av, w) - a(v, w),$$
$$(v, aw) - a(v, w),$$
$$(v_1 + v_2, w) - (v_1, w) + (v_2, w),$$
$$(v, w_1 + w_2) - (v, w_1) + (v, w_2),$$

for $v_1, v_2, v \in V$, $w_1, w_2, w \in W$, and $a \in \mathbb{F}$. Then we let $V \otimes W$ be defined as the quotient vector space $F(V \times W)/R$, and we have a corresponding quotient map $F(V \times W) \to V \otimes W$. The set $V \times W$ is contained in $F(V \times W)$, and the map $\otimes : V \times W \to V \otimes W$ is then defined to be the restriction of the quotient map to $V \times W$. The image of (v, w) under the quotient map is denoted by $v \otimes w$.

Tensor products of several vector spaces at a time are constructed similarly to be a universal space in a category of multilinear maps (Definition D.13). We may also form the tensor products two at a time and then use the easily proved fact that $(V \otimes W) \otimes U \cong V \otimes (W \otimes U)$, which is then denoted by $V \otimes W \otimes U$. Again the reader is referred to Appendix D for more about tensor products.

Elements of the form $v \otimes w$ generate $V \otimes W$, and in fact, if $(e_1, \ldots, e_r)$ is a basis for V and $(f_1, \ldots, f_s)$ is a basis for W, then

$$\{e_i \otimes f_j : 1 \leq i \leq r, 1 \leq j \leq s\}$$

is a basis for $V \otimes W$, which therefore has dimension $rs = \dim V \dim W$.

One more observation: If $A : V \to X$ and $B : W \to Y$ are linear maps, then we can define a linear map $A \otimes B : V \otimes W \to X \otimes Y$. First note that the map $(v, w) \mapsto Av \otimes Bw$ is bilinear. Thus, by the universal property, there is a unique map $A \otimes B$ such that

$$(A \otimes B)(v \otimes w) = Av \otimes Bw,$$

for all $v \in V, w \in W$.

Notice that if A and B are invertible, then $A \otimes B$ is invertible with $(A \otimes B)^{-1}(v \otimes w) = A^{-1}v \otimes B^{-1}w$. Pick bases for V and W as above and bases $\{e'_1, \ldots, e'_r\}$ and $\{f'_1, \ldots, f'_s\}$ for X and Y respectively. For $\tau \in V \otimes W$, we can write $\tau = \tau^{ij} e_i \otimes f_j$ using the Einstein summation convention. We have

$$\begin{aligned} A \otimes B(\tau) &= A \otimes B(\tau^{ij} e_i \otimes f_j) \\ &= \tau^{ij} A e_i \otimes B f_j \\ &= \tau^{ij} A_i^k e'_k \otimes B_j^l f'_l \\ &= \tau^{ij} A_i^k B_j^l \left(e'_k \otimes f'_l \right), \end{aligned}$$

so that the matrix of $A \otimes B$ is given by $(A \otimes B)_{ij}^{kl} = A_i^k B_j^l$.

Let λ^V and λ^W be representations of a Lie group G in $\mathbb{F}$-vector spaces V and W, respectively. We can form a representation of G in the tensor product space $V \otimes W$ by letting $\left(\lambda^V \otimes \lambda^W\right)_g := \lambda_g^V \otimes \lambda_g^W$ for all $g \in G$. There is a variation on the tensor product that is useful when we have two groups involved. If λ^V is a representation of a Lie group G_1 in the $\mathbb{F}$-vector space V and λ^W is a representation of a Lie group G_2 in the $\mathbb{F}$-vector space W, then we can form a representation of the Lie group $G_1 \times G_2$, also called the tensor product representation and denoted $\lambda^V \otimes \lambda^W$ as before. In this case, the definition is $\left(\lambda^V \otimes \lambda^W\right)_{(g_1, g_2)} := \lambda_{g_1}^V \otimes \lambda_{g_2}^W$. Of course if it happens that $G_1 = G_2$, then we have an ambiguity since $\lambda^V \otimes \lambda^W$ could be a representation of G or of $G \times G$. One can usually determine which version is meant from the context. Alternatively, one can use pairs to denote actions so that an action $\lambda : G \times V \to V$ is denoted (G, λ). Then the two tensor product representations would be $(G \times G, \lambda^V \otimes \lambda^W)$ and $(G, \lambda^V \otimes \lambda^W)$, respectively.

Problems

(1) Verify that each of the groups described in Section 5.2 is (isomorphic to) a Lie subgroup of an appropriate Lie group of linear automorphisms.

(2) Show that proper continuous maps are closed maps.

(3) Show that $SL(2, \mathbb{C})$ is simply connected and that $\wp : SL(2, \mathbb{C}) \to \text{Mob}$ is a universal covering homomorphism. See Example 5.50.

(4) Prove Proposition 5.83.

(5) Let $\mathfrak{g}$ be a Lie algebra of a Lie group G. Show that the set of all automorphisms of $\mathfrak{g}$, denoted $\text{Aut}(\mathfrak{g})$, forms a Lie group (actually a Lie subgroup of $GL(\mathfrak{g})$).

(6) Show that if we consider $\text{SL}(2,\mathbb{R})$ as a subset of $\text{SL}(2,\mathbb{C})$ in the obvious way, then $\text{SL}(2,\mathbb{R})$ is a Lie subgroup of $\text{SL}(2,\mathbb{C})$ and $\wp(\text{SL}(2,\mathbb{R}))$ is a Lie subgroup of Mob. Show that if $T \in \wp(\text{SL}(2,\mathbb{R}))$, then T maps the upper half-plane of $\mathbb{C}$ onto itself (bijectively).

(7) Show that for $v \in T_e G$, the field defined by $g \mapsto L^v(g) := TL_g \cdot v$ is automatically smooth.

(8) Determine explicitly the map $T_I \text{inv} : T_I \text{GL}(n,\mathbb{R}) \to T_I \text{GL}(n,\mathbb{R})$, where $\text{inv} : \text{GL}(n,\mathbb{R}) \to \text{GL}(n,\mathbb{R})$ is defined by $\text{inv}(A) = A^{-1}$.

(9) Let H be the set of real 3×3 matrices of the form

$$A = \begin{bmatrix} 1 & a & b \\ 0 & 1 & c \\ 0 & 0 & 1 \end{bmatrix}.$$

Find a global chart for H and show that this and the usual matrix multiplication gives H the structure of a Lie group.

(10) If G is a connected Lie group and $h : G \to H$ is a Lie group homomorphism with discrete kernel K, then $K \subset Z(G)$, where $Z(G) = \{x \in G : xg = gx \text{ for all } g \in G\}$ is the center of G.

(11) Show that for a Lie group G, the conjugation map $C_g : G \to G$ defined by $x \mapsto gxg^{-1}$ is a Lie group isomorphism. Show that the map $C : g \to \text{Diff}(G)$ is a group homomorphism. Note that we have not defined any Lie group structure on $\text{Diff}(G)$.

(12) Consider the map $T_e C_g : T_e G \to T_e G$. Show that $g \mapsto T_e C_g$ is a Lie group homomorphism from G into $\text{GL}(T_e G)$.

(13) Show that $\text{SO}(3)$ is the connected component of the identity in $\text{O}(3)$. Show that the special Lorentz group $\text{SO}(1,3)$ is not connected. Show that the first entry of elements of $\text{SO}(1,3)$ must have absolute value greater than or equal to 1. Define $\text{SO}(3,1)^\uparrow$ as the subset of $\text{SO}(1,3)$ consisting of matrices with positive first entry (which must be greater than 1). Show that $\text{SO}(3,1)^\uparrow$ is connected (and hence the connected component of the identity in $\text{O}(1,3)$).

(14) Let $A \in \mathfrak{gl}(V) = L(V,V)$ for some finite-dimensional vector space V. Show that if A has eigenvalues $\{\lambda_i\}_{i=1,\ldots,n}$, then $\text{ad}(A)$ has eigenvalues $\{\lambda_j - \lambda_k\}_{j,k=1,\ldots,n}$. Hint: Choose a basis for V such that A is represented by an upper triangular matrix. Show that this induces a basis for $\mathfrak{gl}(V)$ such that with the appropriate ordering, $\text{ad}(A)$ is upper triangular.

(15) Fix a nonzero vector $w \in \mathbb{R}^3$ with length $\theta = \|w\|$. Let $L_w : \mathbb{R}^3 \to \mathbb{R}^3$ denote the linear transformation $v \mapsto w \times v$, where $\times$ is the vector cross product. Show that for any right handed orthonormal basis $\{e_1, e_2, e_3\}$

Problems 255

with e_3 parallel to w we have

$$\exp(L_w)e_1 = \cos\theta\, e_1 + \sin\theta\, e_2,$$
$$\exp(L_w)e_2 = -\sin\theta\, e_1 + \cos\theta\, e_2,$$
$$\exp(L_w)e_3 = e_3.$$

(16) Let L_w be as in the previous problem. Show that

$$\exp L_w = I + \frac{\sin\theta}{\theta}L_w + \frac{1-\cos\theta}{\theta^2}L_w^2,$$

where $\frac{\sin\theta}{\theta}$ and $\frac{1-\cos\theta}{\theta^2}$ are defined in the obvious way using power series.

(17) Let $A, B \in \mathfrak{gl}(V)$, where V is a finite-dimensional vector space over the field $\mathbb{F} = \mathbb{R}$ or $\mathbb{C}$, and show that the following statements are equivalent:
 (a) $[A, B] = 0$.
 (b) $\exp sA$ and $\exp tB$ commute for all $s, t \in \mathbb{F}$.
 (c) $\exp(sA + tB) = \exp(sA)\exp(tB)$ for all $s, t \in \mathbb{F}$.

(18) Let V be a finite-dimensional normed space over the field $\mathbb{R}$ (resp. $\mathbb{C}$). Show that if $\sum_{n=0}^{\infty} a_n x^n$ is an absolutely convergent real (resp. complex) power series with radius of convergence R, then

$$\sum_{n=0}^{\infty} a_n A^n$$

converges (absolutely) in the normed space $\mathfrak{gl}(V)$ for $\|A\| < R$.

(19) Show that if G is a connected Lie group, then $\pi_1(G)$ is abelian.

(20) Let G be a Lie group and denote by $\mu : G \times G \to G$ the multiplication map.
 (a) Identify $T(G \times G)$ with $TG \times TG$ in the usual way. Show that the tangent map $T\mu : TG \times TG \to TG$ defines a Lie group structure on TG and show that if $(v_g, w_h) \in T_gG \times T_hG$, then

$$T\mu(v_g, w_h) = TR_h v_g + TL_g w_h,$$

 where R_h and L_g are right and left multiplications respectively.
 (b) Show that under the isomorphism $G \times \mathfrak{g}$ with TG, the Lie group multiplication takes the form

$$(g, A) \cdot (h, B) = (gh, \mathrm{Ad}_{h^{-1}} A + B).$$

(21) Let M be a smooth manifold, let G be a Lie group, and let $r : M \times G \to M$ be a right Lie group action. Recall from the previous problem that TG is naturally a Lie group.
 (a) Show that $Tr : TM \times TG \to TM$ defines a right Lie group action.

(b) For each $A \in \mathfrak{g} = T_e G$, define a vector field $\sigma(A)$ on M by $\sigma(A)(p) := \frac{d}{dt}\big|_{t=0} p \exp(At)$. Show that for $A, B \in \mathfrak{g}$, we have
$$\sigma([A, B]) = [\sigma(A), \sigma(B)].$$

(c) Show that if A^L denote the left invariant vector field generated by A, then for $v_p \in T_p M$,
$$Tr(v_p, A^L(g)) = TR_g(v_p) + \sigma(A)(pg).$$

Chapter 6

Fiber Bundles

The notion of a bundle is basic in both topology and geometry. The reader need not master everything in this chapter before going on to later chapters and should skip forward rather than become too bogged down. The definition and basic examples of a vector bundle are most important. In this chapter we also introduce the more advanced notion of a structure group for a bundle. There is more than one approach to structure groups. We start out with an approach that takes the notion of a G-atlas as basic. This is essentially the approach of Steenrod [**St**]. One may also approach G-bundle structures by first introducing the notion of a principal G-bundle (see [**Hus**]). We discuss principal bundles near the end of this chapter.

6.1. General Fiber Bundles

Definition 6.1. Let F, M, and E be C^r manifolds and let $\pi : E \to M$ be a C^r map. The quadruple (E, π, M, F) is called a (locally trivial) C^r **fiber bundle** if for each point $p \in M$ there is an open set U containing p and a C^r diffeomorphism $\phi : \pi^{-1}(U) \to U \times F$ such that the following diagram commutes:

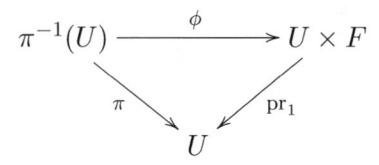

$$\begin{array}{ccc} \pi^{-1}(U) & \xrightarrow{\phi} & U \times F \\ & \searrow_{\pi} \quad \swarrow_{\mathrm{pr}_1} & \\ & U & \end{array}$$

In differential geometry, attention is usually focused on C^∞ fiber bundles (smooth fiber bundles), but the continuous case is also of interest. We will restrict ourselves to the smooth case, but the reader should keep in mind that most of the definitions and theorems have analogous C^0 versions where

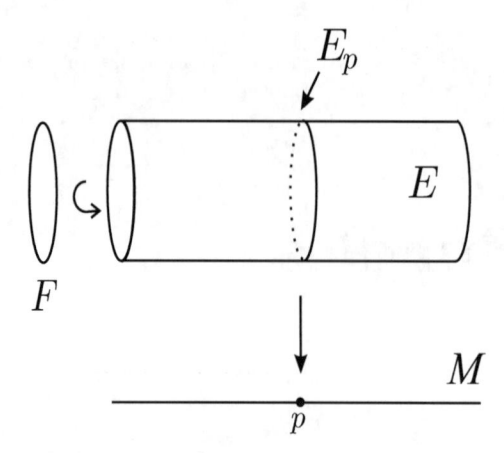

Figure 6.1. Schematic for fiber bundle

the spaces are assumed merely to be sufficiently nice topological spaces and the maps are only assumed to be continuous. In this chapter, all maps and spaces will be smooth unless otherwise indicated.

Definition 6.2. If (E, π, M, F) is a smooth fiber bundle, then E is called the **total space**, π is called the **bundle projection**, M is called the **base space** and F is called the **typical fiber**. For each $p \in M$, the set $E_p := \pi^{-1}(p)$ is called the **fiber** over p.

Because the quadruple notation is cumbersome, it is common to denote a fiber bundle by a single symbol. For example, we could write $\xi = (E, \pi, M, F)$. In the literature, it is common to see E refer both to the total space and to the fiber bundle itself (an abuse of notation). The map π is also a common way to reference the fiber bundle.

Example 6.3. For smooth manifolds M and F, we have the projection $\mathrm{pr}_1 : M \times F \to M$. Then, $(M \times F, \mathrm{pr}_1, M, F)$ is a fiber bundle called a **product bundle** (or trivial bundle).

Exercise 6.4. Show that if $\xi = (E, \pi, M, F)$ is a (smooth) fiber bundle, then $\pi : E \to M$ is a submersion and each fiber $\pi^{-1}(p)$ is a regular submanifold which is diffeomorphic to F. Show that if both F and M are connected, then E is connected.

There are various categories of bundles with corresponding notions of morphism. We give two very general definitions and modify them as needed.

Definition 6.5 (Bundle morphism (type I)). Let $\xi_1 = (E_1, \pi_1, M, F_1)$ and $\xi_2 = (E_2, \pi_2, M, F_2)$ be smooth fiber bundles with the same base space M. A **(type I) bundle morphism over** M from ξ_1 to ξ_2 is a smooth map

6.1. General Fiber Bundles

$h : E_1 \to E_2$ such that the following diagram commutes:

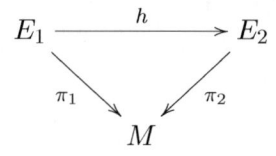

$$\begin{array}{ccc} E_1 & \xrightarrow{h} & E_2 \\ & \searrow{\pi_1} \quad \swarrow{\pi_2} & \\ & M & \end{array}$$

This type of morphism is also called an M-morphism or a morphism over M. If h is also a diffeomorphism, then h is called a **bundle isomorphism over** M and in this case the bundles are said to be isomorphic (over M) or **equivalent**. A bundle isomorphism from a bundle to itself is called a **bundle automorphism**.

Definition 6.6 (Bundle morphism (type II)). Let $\xi_1 = (E_1, \pi_1, M_1, F_1)$ and $\xi_2 = (E_2, \pi_2, M_2, F_2)$ be smooth fiber bundles. A **(type II) bundle morphism** from ξ_1 to ξ_2 is a pair of smooth maps $\widehat{f} : E_1 \to E_2$ and $f : M_1 \to M_2$ such that the following diagram commutes:

$$\begin{array}{ccc} E_1 & \xrightarrow{\widehat{f}} & E_2 \\ \downarrow{\pi_1} & & \downarrow{\pi_2} \\ M_1 & \xrightarrow{f} & M_2 \end{array}$$

We write $(\widehat{f}, f) : \xi_1 \to \xi_2$ and say that $\widehat{f}$ is a bundle morphism along f. If both $\widehat{f}$ and f are diffeomorphisms, then we call $(\widehat{f}, f)$ a bundle isomorphism. In this case, we say that the bundles are isomorphic over f.

Note that as a fiber preserving map, $\widehat{f}$ determines f and so it is also proper to refer to $\widehat{f}$ as the bundle morphism and we sometimes say that $\widehat{f}$ is a bundle morphism **along** (or over) f.

Warning: The definitions of bundle morphism above are quite relaxed. There are a variety of definitions in the literature that require more than the definitions above, especially when structure groups (discussed below) are emphasized.

Definition 6.7. A (global) **smooth section** of a fiber bundle $\xi = (E, \pi, M, F)$ is a smooth map $\sigma : M \to E$ such that $\pi \circ \sigma = \mathrm{id}_M$ (i.e., $\sigma(p) \in E_p$). A **local smooth section** over an open set U is a smooth map $\sigma : U \to E$ such that $\pi \circ \sigma = \mathrm{id}_U$. The set of smooth sections of ξ is denoted by $\Gamma(\xi)$ or sometimes by $\Gamma(E)$ or $\Gamma(\pi)$.

A very important point is that a fiber bundle may not have any global smooth sections. If two bundles are equivalent, via a bundle isomorphism h (of type I), then there is a natural bijection between the spaces of sections given by $\sigma \mapsto h \circ \sigma$. This means that one quick way to conclude that two bundles are *not* equivalent is by showing that one bundle has global sections while the other does not.

A bundle chart essentially gives a *local* type I bundle isomorphism, but it is sometimes more natural to consider bundle charts which are local type II isomorphisms. We will call these type II bundle charts. These are of the form $(\phi, \mathbf{x})$, where $\phi : \pi^{-1}U \to V \times F$ and $\mathbf{x} : U \to V$ are smooth diffeomorphisms such that the following diagram commutes:

$$\begin{array}{ccc} \pi^{-1}U & \stackrel{\phi}{\to} & V \times F \\ \downarrow & & \downarrow \\ U & \stackrel{\mathbf{x}}{\to} & V \end{array}$$

Usually, the pair $(U, \mathbf{x})$ is a chart on the base manifold. The two types of bundle charts are equivalent since one may always compose a type II chart with $(\mathbf{x}^{-1}, \mathrm{id}_F)$ to obtain a type I bundle chart.

More restricted notions of bundle morphism can be obtained by making requirements such as that the induced maps on fibers $\widehat{f}\big|_{\pi_1^{-1}(p)} : \pi_1^{-1}(p) \to \pi_2^{-1}(p)$ are C^∞ diffeomorphisms.

The maps $\phi : \pi^{-1}(U) \to U \times F$ occurring in the definition of a fiber bundle are said to be **local trivializations** of the bundle. It is easy to see that such a local trivialization must be a map of the form $\phi = (\pi|_{\pi^{-1}(U)}, \Phi)$ where $\Phi : \pi^{-1}(U) \to F$ is a smooth map with the property that $\Phi|_{E_p} : E_p \to F$ is a diffeomorphism. We will just write (π, Φ) instead of the more pedantic $(\pi|_{\pi^{-1}(U)}, \Phi)$. Thus

$$\phi(y) = (\pi, \Phi)(y) := (\pi(y), \Phi(y)).$$

The second component map Φ is called the **principal part** of the local trivialization. A pair (U, ϕ), where ϕ is a local trivialization over $U \subset M$, is called a **bundle chart**. (Clearly, a local trivialization and a bundle chart are essentially the same thing.) A family $\{(U_\alpha, \phi_\alpha)\}_{\alpha \in A}$ of bundle charts such that $\{U_\alpha\}_{\alpha \in A}$ is a cover of M is said to be a **bundle atlas**. Given two such bundle charts (U_α, ϕ_α) and (U_β, ϕ_β), we have

$$\phi_\alpha = (\pi, \Phi_\alpha) : \pi^{-1}(U_\alpha) \to U_\alpha \times F$$

and similarly for $\phi_\beta = (\pi, \Phi_\beta)$. If $U_\alpha \cap U_\beta$ is not empty, then $\pi^{-1}(U_\alpha) \cap \pi^{-1}(U_\beta) = \pi^{-1}(U_\alpha \cap U_\beta)$ is not empty and we have **overlap maps**

$$\phi_\alpha \circ \phi_\beta^{-1} : (U_\alpha \cap U_\beta) \times F \to (U_\alpha \cap U_\beta) \times F.$$

Since $\Phi_\alpha|_{E_p}$ is a diffeomorphism for each $p \in U_\alpha$, the map $\Phi_\alpha|_{E_p} \circ \Phi_\beta|_{E_p}^{-1} : F \to F$ is a diffeomorphism for all $p \in U_\alpha \cap U_\beta$. We then obtain a map $\Phi_{\alpha\beta} : U_\alpha \cap U_\beta \to \mathrm{Diff}(F)$ defined by

$$p \mapsto \Phi_{\alpha\beta}(p) = \Phi_\alpha|_{E_p} \circ \Phi_\beta|_{E_p}^{-1}.$$

6.1. General Fiber Bundles

It follows that
$$\phi_\alpha \circ \phi_\beta^{-1}(p, y) = (p, \Phi_{\alpha\beta}(p)(y)).$$
The functions $\Phi_{\alpha\beta} : U_\alpha \cap U_\beta \to \mathrm{Diff}(F)$ are called **transition maps** or **transition functions**. Given a bundle atlas, the corresponding transition functions clearly satisfy the following "cocycle conditions":

$$\begin{aligned}
&\Phi_{\alpha\alpha}(p) = e && \text{for } p \in U_\alpha, \\
&\Phi_{\alpha\beta}(p) = (\Phi_{\beta\alpha}(p))^{-1} && \text{for } p \in U_\alpha \cap U_\beta, \\
&\Phi_{\alpha\beta}(p) \circ \Phi_{\beta\gamma}(p) \circ \Phi_{\gamma\alpha}(p) = \mathrm{id} && \text{for } p \in U_\alpha \cap U_\beta \cap U_\gamma,
\end{aligned}$$

for all α, β, γ.

Notation 6.8. We will often denote $\Phi_{\alpha\beta}(p)(y)$ by $\Phi_{\alpha\beta}|_p(y)$, which is, in many contexts, more transparent.

$\mathrm{Diff}(F)$ is a group, and we have a group action $\mathrm{Diff}(F) \times F \to F$ given by $(\psi, y) \mapsto \psi(y)$. However, $\mathrm{Diff}(F)$ is too big for many purposes, and we have certainly not attempted to give $\mathrm{Diff}(F)$ a Lie group structure. Even if we were to somehow extend the notion of Lie group sufficiently to include $\mathrm{Diff}(F)$, it would be infinite-dimensional and thereby take us out of the circle of ideas we have been developing. Because of this, the transition functions $\Phi_{\alpha\beta}$ above, which could be called "**raw transition functions**", might not be appropriate for our needs. We remedy this below by bringing Lie groups into the picture. First we give a simple example of a nontrivial bundle.

Example 6.9. The circle S^1 can be considered as a quotient $\mathbb{R}/\sim$ where x is equivalent to y if and only if $x - y$ is an integer multiple of 2π. For this example, we put an equivalence relation on $\mathbb{R} \times (-1, 1)$ according to the prescription $(x, t) \sim (x + 2\pi n, (-1)^n t)$ for any integer n. The quotient $(\mathbb{R} \times (-1, 1))/\sim$ can easily be seen to be a smooth manifold and is none other than the familiar **Möbius band** which we denote by MB. Define a map $\pi \colon \mathrm{MB} \to \mathbb{R}/\sim \ = S^1$ by $\pi([x, t]) = [x]$. We show that this is a fiber bundle by exhibiting an atlas consisting of three bundle charts. We call it the **Möbius band bundle**. We use three bundle charts instead of two in order that the overlaps be connected sets. Let $U_1 = \{[x] \in \mathbb{R}/\sim \ : -2\pi/3 < x < 2\pi/3\}$ and $U_2 = \{[x] \in \mathbb{R}/\sim \ : 0 < x < 4\pi/3\}$ and $U_3 = \{[x] \in \mathbb{R}/\sim \ : 2\pi/3 < x < 2\pi\}$. Then $U_1 \cup U_2 \cup U_3 = \mathbb{R}/\sim \ = S^1$. For $i = 2, 3$ define $\phi_i : \pi^{-1}(U_i) \to U_i \times S^1$ by

$$\phi_i([x, t]) = ([x], t),$$

where (x, t) is the unique representative of $[x, t]$ in the set $(0, 2\pi) \times (-1, 1)$. For $\phi_1 : \pi^{-1}(U_1) \to U_1 \times S^1$, we define $\phi_1([x, t]) = ([x], t)$, where (x, t) is the unique representative of $[x, t]$ in the set $(-2\pi/3, 2\pi/3) \times (-1, 1)$. One can check that $\phi_2 \circ \phi_3^{-1} = \phi_3 \circ \phi_2^{-1} = \mathrm{id}$ on the overlap $\pi^{-1}(U_2) \cap \pi^{-1}(U_3)$. Now consider the overlap $\pi^{-1}(U_1) \cap \pi^{-1}(U_2)$. If $[x, t] \in \pi^{-1}(U_1) \cap \pi^{-1}(U_2)$, then

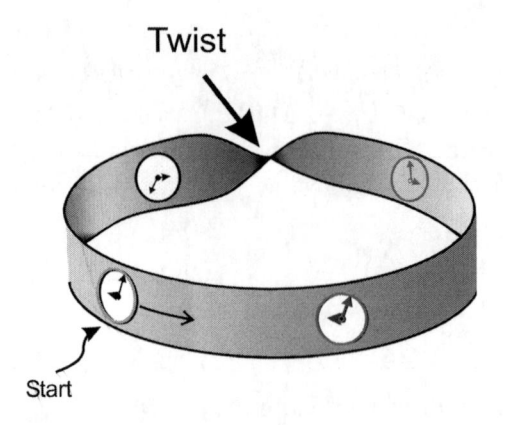

Figure 6.2. Möbius band

$[x,t]$ is uniquely represented by some $(x,t) \in (0, 2\pi/3) \times (-1,1)$ and in view of the definitions we see that $\phi_2 \circ \phi_3^{-1} = \mathrm{id}$ also. Finally we consider $\phi_1 \circ \phi_3^{-1}$. If $[x,t] \in \pi^{-1}(U_1) \cap \pi^{-1}(U_3)$, then it has a unique representative (x,t) in $(4\pi/3, 2\pi) \times (-1,1)$ and then $\phi_3^{-1}([x],t) = [x,t]$. For ϕ_1, we need to represent $[x,t]$ properly. We use the fact that $[x,t] = [x - 2\pi, -t]$ and $(x - 2\pi, -t) \in (-2\pi/3, 0) \times (-1,1)$ so that $\phi_1([x - 2\pi, -t]) = ([x - 2\pi], -t) = ([x], -t)$. In short, we have $\phi_1 \circ \phi_3^{-1}([x,t]) = ([x], -t)$. From these considerations and the fact that in general $\phi_\alpha \circ \phi_\beta^{-1}(p,y) = (p, \Phi_{\alpha\beta}(p)(y))$ we see that

$$\Phi_{12}(p) = \mathrm{id}_{(-1,1)} \in \mathrm{Diff}(-1,1) \text{ for } p \in U_1 \cap U_2,$$
$$\Phi_{23}(p) = \mathrm{id}_{(-1,1)} \in \mathrm{Diff}(-1,1) \text{ for } p \in U_2 \cap U_3,$$
$$\Phi_{13}(p) = -\mathrm{id}_{(-1,1)} \in \mathrm{Diff}(-1,1) \text{ for } p \in U_1 \cap U_3.$$

A "twist" occurs on the overlap $\pi^{-1}(U_1) \cap \pi^{-1}(U_3)$. There is no way to construct an atlas for this bundle without having such a twist on at least one of the overlaps.

Notice that if we define an action λ of $\mathbb{Z}_2 = \{1, -1\}$ on the interval $(-1,1)$ by $\lambda(g,x) \mapsto gx$, then we can describe the transition functions in the last example by
$$\Phi_{\alpha\beta}(p)(x) = \lambda(g_{\alpha\beta}(p), x),$$
where $g_{\alpha\beta} : U_\alpha \cap U_\beta \to \mathbb{Z}_2$ is given by

$$g_{12} = 1 \text{ on } U_1 \cap U_2,$$
$$g_{23} = 1 \text{ on } U_2 \cap U_3,$$
$$g_{13} = -1 \text{ on } U_1 \cap U_3,$$

and in this case the $g_{\alpha\beta}$ satisfy a cocycle condition like the $\Phi_{\alpha\beta}$. This is convenient since we understand $\mathbb{Z}_2$ very well. It is a zero-dimensional Lie group. Inspired by this, we seek to put Lie groups into the formalism. This

6.1. General Fiber Bundles

will alleviate our concerns about the group $\mathrm{Diff}(F)$ mentioned above. We are also led to the theory of G-bundles that involves the group in subtle ways.

Definition 6.10. Let $\{U_\alpha\}$ be an indexed open cover of a smooth manifold M and let G be a Lie group. A G-**cocycle** on $\{U_\alpha\}$ is the assignment of a smooth map $g_{\alpha\beta} : U_\alpha \cap U_\beta \to G$ to every nonempty intersection $U_\alpha \cap U_\beta$ such that the **cocycle conditions** hold:

$$g_{\alpha\alpha}(p) = e \text{ for } p \in U_\alpha,$$
$$g_{\alpha\beta}(p) = (g_{\beta\alpha}(p))^{-1} \text{ for } p \in U_\alpha \cap U_\beta$$
$$g_{\alpha\beta}(p)g_{\beta\gamma}(p)g_{\gamma\alpha}(p) = e \text{ for } p \in U_\alpha \cap U_\beta \cap U_\gamma,$$

where e is the identity in G. The family of maps $\{g_{\alpha\beta}\}$ forms a **cocycle**.

The idea that we wish to pursue is that of representing the action of the raw transition maps by using Lie group actions. There is a subtle point here that the reader should not miss. Consider the following fact: If $\lambda : G \times F \to F$ is a group action, then by letting $K = \{g : \lambda_g(p) = p \text{ for all } p \in F\}$ (the kernel of the action) we obtain an effective action of G/K on F. Things are not so simple on the global level of bundles as becomes clear when dealing with the notion of *spin structure* (See [**L-M**]). The best way to explain what is at stake is by the use of the notion of a principal bundle which we introduce later. Even before we get to that point, we will mention some things that will provide some idea as to why we need to be careful about ineffective actions.

We start out assuming that the action is effective (see Definition 1.100):

Definition 6.11. Let $\xi = (E, \pi, M, F)$ be a fiber bundle and G a Lie group. Suppose that we have an *effective* left action $\lambda : G \times F \to F$. Let $\{(\phi_\alpha, U_\alpha)\}$ be a bundle atlas for ξ. Suppose that for every nonempty intersection $U_\alpha \cap U_\beta$ there exists a smooth map $g_{\alpha\beta} : U_\alpha \cap U_\beta \to G$ such that $\lambda(g_{\alpha\beta}(p), y) = \Phi_{\alpha\beta}|_p (y)$ for all $p \in U_\alpha \cap U_\beta$ and $y \in F$. Then the atlas $\{(\phi_\alpha, U_\alpha)\}$ is called a (G, λ)-**bundle atlas**. If the action λ is understood or standard in some way, one also speaks of a G-**bundle atlas**.

Because the action λ in the above definition is assumed effective, it follows that the family $\{g_{\alpha\beta}\}$ satisfies the cocycle conditions of Definition 6.10. Notice that if we had not assume the action to be effective, then the maps $g_{\alpha\beta}$ would not be unique and may not satisfy a cocycle condition (although they would do so modulo the kernel of the action). Thus if we do not assume effectiveness, then we have to make the cocycle $\{g_{\alpha\beta}\}$ part of the definition. We return to this below.

The basic definition in the case of an effective action can be formulated as follows:

Definition 6.12. Let $\xi = (E, \pi, M, F)$ be a fiber bundle and G a Lie group. Suppose that we have an *effective* left action $\lambda : G \times F \to F$. Two (G, λ)-bundle atlases for ξ, say $\{(\phi_\alpha, U_\alpha)\}$ and $\{(\phi'_\alpha, U'_\alpha)\}$, are **strictly equivalent** if the union of the atlases is also a (G, λ)-bundle atlas. A *strict* equivalence class of atlases is referred to as an effective (G, λ)-**bundle structure** on ξ, and we say that ξ together with this (G, λ)-bundle structure is an effective (G, λ)-**bundle**. Again, if the action is standard or understood, then it is common to speak of a G-bundle structure and refer to ξ as a G-bundle.

Actually, there is a tiny point to be made. To keep things neat we should always arrange that the indexing map $\alpha \mapsto (\phi_\alpha, U_\alpha)$ for any atlas is injective. Thus when taking the union of two atlases per the definition of strict equivalence, one may need to reindex so that the $g_{\alpha\beta}$ are notationally unambiguous. For example, if we take the union of an atlas $\{(\phi_1, U_1), (\phi_2, U_2)\}$ with an atlas $\{(\psi_1, V_1), (\psi_2, V_2)\}$, then how should the transition maps for the bigger atlas be denoted? What would g_{11} mean? Sometimes the traditional indexing scheme is wisely dropped. Instead one uses the set of trivializing maps itself as the index set so that a chart is written as (ϕ, U_ϕ) or (ψ, U_ψ), and then one denotes transitions maps by $g_{\phi\psi}$ etc. The notion of **maximal** (G, λ)-**bundle atlas** is defined in the obvious way by direct analogy with the notion of maximal atlas for a smooth structure.

Our main emphasis will be on the effective (G, λ)-bundles, but as mentioned above, if we wish to allow ineffective actions, then the notion of atlas should include the cocycle as part of the data. But even then we have to be careful. Indeed, for an ineffective action it is conceivable that there could be a different cocycle $\{g'_{\alpha\beta}\}$ such that $\lambda(g'_{\alpha\beta}(p), y) = \Phi_{\alpha\beta}|_p(y)$ for all $y \in F$. Then $\{(\phi_\alpha, U_\alpha), (g_{\alpha\beta}), \lambda\}$ and $\{(\phi_\alpha, U_\alpha), (g'_{\alpha\beta}), \lambda\}$ would be different (but possibly equivalent) (G, λ)-bundle atlases. For $\{(\phi_\alpha, U_\alpha), (g_{\alpha\beta}), \lambda\}$ and $\{(\phi'_j, U'_j), (g'_{ij}), \lambda\}$ to define the same (G, λ)-**bundle** it must be the case that both of the cocycles are contained in a larger cocycle that gives the transitions for the atlas obtained as the union of the collection of charts from $\{(\phi_\alpha, U_\alpha), (g_{\alpha\beta}), \lambda\}$ and $\{(\phi'_j, U'_j), (g'_{ij}), \lambda\}$. We handle ineffective actions this way so as to keep aligned with the notion of *associated bundle* introduced later.

If there is no chance of confusion, we will drop the adjective *effective*. The reader is warned that some standard expositions on fiber bundles allow ineffective actions right from the start, but in some cases assertions are made that would only be true in the effective case! It is interesting to note that in his famous book on the subject [**St**], Norman Steenrod restricts himself

6.1. General Fiber Bundles

to effective actions, although he announces this restriction in one easily overlooked sentence early in the book.

Notice that an alternative way to say that $\lambda(g_{\alpha\beta}(p), y) = \Phi_{\alpha\beta}|_p(y)$ is $\phi_\alpha \circ \phi_\beta^{-1}(p, y) = (p, \lambda(g_{\alpha\beta}(p), y))$ or

$$\phi_\alpha \circ \phi_\beta^{-1}(p, y) = (p, g_{\alpha\beta}(p) \cdot y).$$

(Actually, we shall at first avoid the notation $g \cdot y$ for the action of a group element g on y since we do not want the beginner to forget the role of the choice of action.) The maps $g_{\alpha\beta}$ are also called **transition functions** for the (G, λ)-bundle atlas. If the G is literally a subgroup of $\mathrm{Diff}(F)$ and the action is simply $(\Phi, y) \mapsto \Phi(y)$, then the transition maps $g_{\alpha\beta}$ are simply the maps $\Phi_{\alpha\beta}$.

When dealing with (G, λ)-bundles there is a more specific notion of morphism and equivalence:

Definition 6.13. Let $\xi_1 = (E_1, \pi_1, M_1, F)$ be a (G, λ)-bundle with its (G, λ)-bundle structure determined by the strict equivalence class of the (G, λ)-atlas $\{(\varphi_\alpha, U_\alpha)\}_{\alpha \in A}$. Let $\xi_2 = (E_2, \pi_2, M_1, F)$ be a (G, λ)-bundle with its (G, λ)-bundle structure determined by the strict equivalence class of the (G, λ)-atlas $\{(\psi_\beta, V_\beta)\}_{\beta \in B}$. Then a type II bundle morphism $(\widehat{h}, h) : \xi_1 \to \xi_2$ is called a (G, λ)-**bundle morphism** along h if

(i) $\widehat{h}$ carries each fiber of E_1 diffeomorphically onto the corresponding fiber of E_2;

(ii) whenever $U_\alpha \cap h^{-1}(V_\beta)$ is not empty, there is a smooth map $h_{\alpha\beta} : U_\alpha \cap h^{-1}(V_\beta) \to G$ such that for each $p \in U_\alpha \cap h^{-1}(V_\beta)$ we have

$$\left(\Psi_\beta \circ \widehat{h} \circ \left(\Phi_\alpha|_{\pi_1^{-1}(p)}\right)^{-1}\right)(y) = \lambda(h_{\alpha\beta}(p), y) \text{ for all } y \in F,$$

where as usual $\varphi_\alpha = (\pi_1, \Phi_\alpha)$ and $\psi_\beta = (\pi_2, \Psi_\beta)$.

If $M_1 = M_2$ and $h = \mathrm{id}_M$, then we call $\widehat{h}$ a (G, λ)-**bundle equivalence** over M. (In this case, $\widehat{h}$ is a diffeomorphism.) Condition (ii) simply says that $\widehat{h}$ must be given by the action on each fiber when viewed in (G, λ)-charts. For this definition to be good it must be shown to be well-defined. That is, one must show that condition (ii) is independent of the choice of representatives $\{(\varphi_\alpha, U_\alpha)\}_{\alpha \in A}$ and $\{(\varphi_i, V_i)_{i \in J}\}$ of the *strict* equivalence classes of atlases that define the (G, λ)-bundle structures. We leave this as an exercise. Later we will discover another, perhaps better, way to talk about equivalence of (G, λ)-bundles.

The product bundle $\mathrm{pr}_1 : M \times F \to M$ has a trivial (G, λ)-bundle structure for any λ acting on F. Indeed, we just take the structure given by the single bundle chart $(\mathrm{id}_{M\times F}, M)$ where we have the resulting cocycle $\{g_{11}\}$, and where $g_{11}(x) := e$ for all x. In fact, if $\{U_\alpha\}_{\alpha \in A}$ is any open cover of M, then $\{(\mathrm{id}_{U_\alpha \times F}, U_\alpha)\}_{\alpha \in A}$ is a (G, λ)-atlas strictly equivalent to the atlas $\{(\mathrm{id}_{M\times F}, M)\}$ and so defining the same trivial (G, λ)-bundle. By **trivial (G, λ)-bundle** over M we will mean either this product (G, λ)-bundle or one that is (G, λ)-equivalent to it.

Remark 6.14. The special case of a (G, λ)-bundle equivalence in the case that $E_1 = E_2$ and $\pi_1 = \pi_2$ is interesting but easily misunderstood. Suppose that $\{(\varphi_\alpha, U_\alpha)\}_{\alpha \in A}$ determines a (G, λ)-bundle structure on (E, π, M, F) and suppose that $\{(\psi_\alpha, U_\alpha)\}_{\alpha \in A}$ determines another (G, λ)-bundle structure on (E, π, M, F). Then these two (G, λ)-bundle structures might be (G, λ)-bundle equivalent without being the very same. We would have (G, λ)-bundle equivalence in case the two atlases give "cohomologous" cocycles. This is *not* the same as strict equivalence despite what one occasionally finds stated in the literature. *Strict* equivalence is the notion that defines the (G, λ)-bundle structure and is analogous to equivalence of manifold atlases which defines the notion of differentiable structure on a manifold. The other, weaker, notion of equivalence above is analogous to diffeomorphism and we know that two atlases on a manifold may define different smooth structures, which may or may not be diffeomorphic. For a more detailed discussion of these issues see the online supplement [**Lee, Jeff**].

We have defined fiber bundles, (G, λ)-bundles and various notions of morphism and equivalence. The question of classifying fiber bundles generally, or (G, λ)-bundles more specifically, is a huge part of topology which we do not have the space to discuss. The interested reader should consult [**Hus**], [**Span**] and [**St**].

G-bundle structures vs. G-structures. We have deliberately opted to use the term "G-bundle structure" rather than simply "G-structure", which could reasonably be taken to mean the same thing. Perhaps the reader is aware that there is a theory of G-structures *on* a smooth manifold (see [**Stern**]). One may rightly ask whether a G-structure *on* M is nothing more than a G-bundle structure on the tangent bundle TM where G is a Lie subgroup of $GL(n)$ acting in the standard way. The answer is both yes and no. First, one could indeed say that a G-structure on M is a kind of G-bundle structure on TM even though the theory is usually fleshed out in terms of the frame bundle of M (defined below). However, the notion of equivalence of two G-structures on M is different from what we have given above. Roughly, G-structures on M are equivalent in this sense if there is a diffeomorphism ϕ such that $(T\phi, \phi)$ is a type II bundle isomorphism that is also a G-bundle morphism along ϕ.

6.1. General Fiber Bundles

Let $\xi = (E, \pi, M, F)$ have a (G, λ)-bundle structure given by a maximal (G, λ)-bundle atlas $\{(\phi_\alpha, U_\alpha)\}_{\alpha \in A}$. Suppose that there is a subatlas $\{(\phi_j, U_j)\}_{j \in J}$, $J \subset A$, such that the transition maps for the subatlas take values in a Lie subgroup H of G. Then clearly ξ has an $(H, \lambda|_H)$-bundle structure, and this is called a **reduction of the structure group**. We say that the structure group is **reducible**.

The following theorem shows how we can build a fiber bundle using a cocycle and a choice of action.

Theorem 6.15 (Fiber bundle construction theorem). *Let M and F be smooth manifolds and let G be a Lie group. Let $\{U_\alpha\}_{\alpha \in A}$ be a cover of M and $\{g_{\alpha\beta}\}$ a G-cocycle for the cover. For every action $\lambda : G \times F \to F$, there exists a fiber bundle with bundle-atlas $\{(U_\alpha, \phi_\alpha)\}$ satisfying $\phi_\alpha \circ \phi_\beta^{-1}(p, y) = (p, \lambda(g_{\alpha\beta}(p), y))$ on nonempty overlaps $U_\alpha \cap U_\beta$. Thus the resulting bundle has a (G, λ)-bundle structure.*

Proof. On the union $\Sigma := \bigcup_\alpha \{\alpha\} \times U_\alpha \times F$ define an equivalence relation such that $(\alpha, p, y) \in \{\alpha\} \times U_\alpha \times F$ is equivalent to $(\beta, p', y') \in \{\beta\} \times U_\beta \times F$ if and only if $p = p'$ and $y = g_{\alpha\beta}(p) \cdot y'$. Notice that $p = p'$ is possible only in the case $U_\alpha \cap U_\beta \neq \emptyset$. The first member of the triple is only needed to make the union above disjoint. The cocycle conditions ensure that the equivalence relation is well-defined.

The total space of our bundle is then $E := \Sigma/\sim$. The set Σ is essentially the disjoint union of the product spaces $U_\alpha \times F$ and so has an obvious topology. We then give $E := \Sigma/\sim$ the quotient topology. The bundle projection π is induced by $(\alpha, p, v) \mapsto p$. To get our trivializations, we define
$$\phi_\alpha(\epsilon) := (p, y) \text{ for } \epsilon \in \pi^{-1}(U_\alpha),$$
where (p, y) is the unique member of $U_\alpha \times F$ such that $(\alpha, p, y) \in \epsilon$ (recall that ϵ is an equivalence class). The point here is that $(\alpha, p_1, y_1) \sim (\alpha, p_2, y_2)$ only if $(p_1, y_1) = (p_2, y_2)$. Now suppose $U_\alpha \cap U_\beta \neq \emptyset$. Then for $p \in U_\alpha \cap U_\beta$, the element $\phi_\beta^{-1}(p, y)$ is in $\pi^{-1}(U_\alpha \cap U_\beta) = \pi^{-1}(U_\alpha) \cap \pi^{-1}(U_\beta)$ and so $\phi_\beta^{-1}(p, y) = [(\beta, p, y)]$. But since $\phi_\beta^{-1}(p, y)$ is in $\pi^{-1}(U_\alpha)$, it must be equal to $[(\alpha, q, y_2)]$ for some y_2. This means that $p = q$ and $y_2 = g_{\alpha\beta}(p) \cdot y = \lambda(g_{\alpha\beta}(p), y)$. Thus $\phi_\beta^{-1}(p, y) = [\alpha, p, g_{\alpha\beta}(p) \cdot y]$ and so
$$\phi_\alpha \circ \phi_\beta^{-1}(p, y) = (p, g_{\alpha\beta}(p) \cdot y).$$
We leave the existence of the smooth structure and the routine verification of the smoothness of these maps to the reader. That the topology induced is Hausdorff and paracompact is also easy to see. $\square$

Example 6.16. Let S^1 be the circle realized as the unit complex numbers. We will construct a fiber bundle with typical fiber $F := (-1, 1) \subset \mathbb{R}$.

Let $U_1 = \{e^{i\theta} \in S^1 : 0 < \theta < 2\pi\}$ and $U_2 = \{e^{i\theta} \in S^1 : -\pi < \theta < \pi\}$. Then $U_1 \cap U_2$ is a disjoint union of open sets V and W where $1 \in V$ and $-1 \in W$. Now we define maps with values in the multiplicative group of two elements $\{1, -1\} = \mathbb{Z}_2$. Let $g_{11}(x) = 1$ for all $x \in U_1$, $g_{22}(x) = 1$ for all $x \in U_1$, and then let g_{12} and g_{21} be defined on $U_1 \cap U_2$ by

$$g_{12}(x) := \begin{cases} 1 & \text{on } V, \\ -1 & \text{on } W, \end{cases}$$

and $g_{21} := g_{12}^{-1}$. Let $\mathbb{Z}_2$ act on the symmetric interval $(-1, 1) \subset \mathbb{R}$ by multiplication (so that $-1 \cdot x := -x$). Using the cocycle $\{g_{11}, g_{22}, g_{12}, g_{21}\}$ and this action on $\mathbb{R}$, Theorem 6.15 above gives a $\mathbb{Z}_2$-bundle which can be shown to be equivalent to the Möbius band bundle described in Example 6.9.

Example 6.17. Use the same cocycle on S^1 as in the last example but with typical fiber S^1 and action $\mathbb{Z}_2 \times S^1 \to S^1$ given by letting 1 act as the identity and -1 act by a rotation of S^1 by π, that is $-1 \cdot e^{i\theta} := e^{i(\theta + \pi)} = -e^{i\theta}$. Then the bundle obtained is a $\mathbb{Z}_2$-bundle, sometimes called the twisted torus. It is of the upmost importance to realize that there is a fiber preserving diffeomorphism between the twisted torus and the trivial bundle $S^1 \times S^1 \overset{\text{pr}_1}{\to} S^1$ and yet there is no $\mathbb{Z}_2$-bundle equivalence between the twisted torus and the trivial $\mathbb{Z}_2$-bundle $S^1 \times S^1 \overset{\text{pr}_1}{\to} S^1$. Thus the twisted torus is trivial as a general bundle but not as a $\mathbb{Z}_2$-bundle (See Problem 5). This shows how much the involvement of the group matters.

Notice that the previous theorem is true even without the assumption that the action is effective and we still end up with a genuine (ineffective) (G, λ)-bundle since there is no problem about the cocycle existing. A quotient by the kernel K of the action would give an effective structure group action. Thinking of the action as a homomorphism $G \to \text{Diff}(F)$, we have an induced homomorphism $G/K \to \text{Diff}(F)$ such that the following diagram commutes:

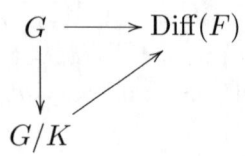

$$\begin{array}{ccc} G & \longrightarrow & \text{Diff}(F) \\ \downarrow & \nearrow & \\ G/K & & \end{array}$$

Definition 6.18. Let $\xi = (E, \pi, M, F)$ be a smooth fiber bundle and $f : N \to M$ a smooth map. The **pull-back bundle** $f^*\xi = (f^*E, \pi_1, M, F)$ (or **induced** bundle) is defined as follows: The total space f^*E is the set

$$f^*E := \{(q, \epsilon) \in N \times E : f(q) = \pi(\epsilon)\}.$$

Then we define π_1 as the restriction to f^*E of the projection $\text{pr}_1 : N \times E \to N$.

6.1. General Fiber Bundles

Notice that the second factor projection map $\mathrm{pr}_2 : N \times E \to E$ restricts to a map $\widetilde{f} : f^*E \to E$ which is a bundle morphism over the map f:

$$\begin{array}{ccc} f^*E & \xrightarrow{\widetilde{f}} & E \\ \downarrow & & \downarrow \\ N & \xrightarrow{f} & M \end{array}$$

The map $\widetilde{f}$ restricts to a diffeomorphism on each fiber. If $\phi = (\pi, \Phi)$ is a trivialization of the bundle ξ over the open set U, then $(\pi_1, \overline{\Phi}) := (\pi_1, \Phi \circ \widetilde{f})$ is a trivialization of $f^*\xi$ over the open set $f^{-1}(U)$. Thus a bundle atlas on ξ induces a bundle atlas on $f^*\xi$. If $\{\Phi_{\alpha\beta}\}$ are (raw) transition maps for ξ corresponding to a bundle atlas $\{(\pi, \Phi_\alpha)\}_{\alpha \in A}$, then $\{\Phi_{\alpha\beta} \circ f\}$ are transition maps for $f^*\xi$ corresponding to the atlas $\{(\pi_1, \Phi_\alpha \circ \widetilde{f})\}_{\alpha \in A}$. In fact,

$$\left.\overline{\Phi}_\alpha\right|_{q \times E_{f(q)}} (q, \epsilon) = \left.\Phi_\alpha \circ \widetilde{f}\right|_{q \times E_{f(q)}} (q, \epsilon) = \left.\Phi_\alpha\right|_{E_{f(q)}} (\epsilon),$$

so the inverse is

$$\left.\overline{\Phi}_\alpha\right|^{-1}_{q \times E_{f(q)}} : y \mapsto (q, \left.\Phi_\alpha\right|^{-1}_{E_{f(q)}} (y)).$$

Thus $\left.\overline{\Phi}_\alpha\right|_{q \times E_{f(q)}} \circ \left.\overline{\Phi}_\beta\right|^{-1}_{q \times E_{f(q)}}$ is given by

$$y \mapsto (q, \left.\Phi_\beta\right|^{-1}_{E_{f(q)}} (y)) \mapsto \left.\Phi_\alpha\right|_{E_{f(q)}} \circ \left.\Phi_\beta\right|^{-1}_{E_{f(q)}} (y),$$

and so

$$\overline{\Phi}_{\alpha\beta}(q) = \left.\overline{\Phi}_\alpha\right|_{q \times E_{f(q)}} \circ \left.\overline{\Phi}_\beta\right|^{-1}_{q \times E_{f(q)}}$$
$$= \left.\Phi_\alpha\right|_{E_{f(q)}} \circ \left.\Phi_\beta\right|^{-1}_{E_{f(q)}} = \Phi_{\alpha\beta} \circ f(q).$$

Furthermore, if $\{(\pi, \Phi_\alpha)\}_{\alpha \in A}$ is a (G, λ)-atlas for ξ, then since $\left.\Phi_{\alpha\beta}\right|_p (y) = \lambda(g_{\alpha\beta}(p), y)$, we have $\left.\overline{\Phi}_{\alpha\beta}\right|_q (y) = \Phi_{\alpha\beta}(f(q))(y) = \lambda(g_{\alpha\beta}(f(q)), y)$. We see that $f^*\xi$ has a (G, λ)-bundle structure with cocycles $g_{\alpha\beta} \circ f$. Note, however, that because of the composition with f, this structure may be reducible to a smaller group.

Definition 6.19. Let $\xi = (E, \pi, M, F)$ be a smooth fiber bundle and let $f : N \to M$ be a smooth map. A **section of ξ along** f is a map $\sigma : N \to E$ such that $\pi \circ \sigma = f$. The set of sections along f will be denoted $\Gamma_f(\xi)$ or $\Gamma_f(E)$.

If $\sigma : N \to E$ is a section of ξ along f, then the map $\sigma' : N \to f^*E$ given by $p \mapsto (p, \sigma(p))$ is a section of the pull-back bundle $f^*\xi$. It is not hard to show that *all* sections of $f^*\xi$ have this form.

Proposition 6.20. *Let $\xi = (E, \pi, M, F)$ be a smooth fiber bundle and $f : N \to M$ a smooth map. Then there is a natural bijection between $\Gamma_f(\xi)$ and $\Gamma(f^*\xi)$.*

Proof. Let $s \in \Gamma(f^*\xi)$. For each $p \in N$, we have $s(p) \in (f^*E)_p$, which must have the form (p, y) for some $y \in E_{f(p)}$. Therefore the smooth map $\sigma := \mathrm{pr}_2 \circ s : N \to E$ has the property that $\pi \circ \sigma = f$. Thus we obtain a map $\Gamma(f^*\xi) \to \Gamma_f(\xi)$ given by $s \mapsto \sigma$. But the inverse of this map is clearly $\sigma \mapsto (\mathrm{id}_N, \sigma)$. $\square$

6.2. Vector Bundles

The tangent and cotangent bundles are examples of a general type of fiber bundle called a vector bundle. Roughly speaking, a vector bundle is a parametrized family of vector spaces. We shall need both complex vector bundles and real vector bundles, and so to facilitate definitions we let $\mathbb{F}$ denote either $\mathbb{R}$ or $\mathbb{C}$. Let V be a finite-dimensional $\mathbb{F}$-vector space. The simplest examples of vector bundles over a manifold M are the **product vector bundles** which consist of a Cartesian product $M \times V$ together with the projection onto the first factor $\mathrm{pr}_1 : M \times V \to M$. Each set of the form $\{x\} \times V \subset M \times V$ inherits an $\mathbb{F}$-vector space structure from that of V in the obvious way: $a(p, v) + b(p, w) := (p, av + bw)$. We think of $M \times V$ as copies of V parametrized by M.

Definition 6.21. Let V be a finite-dimensional $\mathbb{F}$-vector space. A smooth $\mathbb{F}$-**vector bundle** with typical fiber V is a fiber bundle (E, π, M, V) such that:

(i) for each $x \in M$ the set $E_x := \pi^{-1}(x)$ has the structure of a vector space over the field $\mathbb{F}$, isomorphic to the fixed vector space V;

(ii) every $p \in M$ is in the domain of some bundle chart (U, ϕ), such that for each $x \in U$ the map $\Phi|_{E_x} : E_x \to V$ is a vector space isomorphism where $\phi = (\pi, \Phi)$.

Definition 6.22 (Terminology). We refer to a vector bundle as a complex vector bundle (resp. real vector bundle) if $\mathbb{F} = \mathbb{C}$ (resp. $\mathbb{F} = \mathbb{R}$). "Vector bundle" shall mean either real or complex vector bundle as determined by the context.

A bundle chart (U, ϕ) of the sort described in the definition is called a **vector bundle chart** (VB-chart) or a local vector bundle trivialization (over U). In the setting of vector bundles, a local trivialization is assumed to be linear on fibers. A family $\{(U_\alpha, \phi_\alpha)\}$ of vector bundle charts such that $\{U_\alpha\}$ is an open cover of M is called a **vector bundle atlas** for $\pi : E \to M$. The definition of a vector bundle guarantees that such an atlas exists. The dimension of the typical fiber V is called the **rank** of the vector bundle.

Remark 6.23. By choosing a basis for V, one gets an isomorphism with $\mathbb{C}^k$ or $\mathbb{R}^k$ as the case may be. Composing with this isomorphism we can

6.2. Vector Bundles

convert the V-valued VB-charts into $\mathbb{C}^k$- or $\mathbb{R}^k$-valued VB-charts. Thus we could have assumed from the start that we were dealing with one of these standard vector spaces, but it is not always natural to do so since our vector space may arise in a specific way (it could be a Lie algebra or perhaps a space of algebraic tensors) and may not have a preferred choice of basis.

Exercise 6.24. Show that the tangent and cotangent bundles of an n-manifold are vector bundles with typical fiber $\mathbb{R}^n$ (the cotangent bundle may be viewed as having typical fiber $(\mathbb{R}^n)^*$).

Let $\mathbb{F}$ be $\mathbb{R}$ or $\mathbb{C}$ as above. The space of smooth sections $\Gamma(\xi)$ of an $\mathbb{F}$-vector bundle $\xi = (E, \pi, M, V)$ has the structure of a module over the ring $C^\infty(M; \mathbb{F})$; for $\sigma, \sigma_1, \sigma_2 \in \Gamma(\xi)$ and $f \in C^\infty(M; \mathbb{F})$, we define

$$(\sigma_1 + \sigma_2)(p) := \sigma_1(p) + \sigma_2(p) \text{ for all } p \in M,$$
$$f\sigma(p) := f(p)\sigma(p) \text{ for all } p \in M.$$

Definition 6.25. Let ξ_1 and ξ_2 be $\mathbb{F}$-vector bundles with respective bundle projections π_1 and π_2. A bundle morphism $(\widehat{f}, f) : \xi_1 \to \xi_2$ is called a **vector bundle morphism** if the restrictions to fibers, $\widehat{f}|_{\pi_1^{-1}(p)} \pi_1^{-1}(p) \to \pi_2^{-1}(f(p))$, are $\mathbb{F}$-linear. If ξ_1 and ξ_2 have the same base space M, then we obtain the definition of a vector bundle morphism over M by specializing to the case $f = \mathrm{id}_M$. We then also have the corresponding notions of **vector bundle isomorphism** and **automorphism** (for both type I and II bundle morphisms).

A vector bundle (E, π, M, V) is said to be **trivial** if it is vector bundle isomorphic to the product vector bundle $\mathrm{pr}_1 : M \times V \to M$. This happens exactly when there is a vector bundle trivialization over the entire manifold M, which we call a global vector bundle trivialization (a notion already introduced for tangent bundles).

Definition 6.26. Let (E, π, M, V) be a rank k vector bundle with typical fiber V and fix an l-dimensional subspace V′ of V. If $E' \subset E$ is a submanifold with the property that for every $p \in M$ there is a VB-chart (U, ϕ) such that

$$\phi(\pi^{-1}(U) \cap E') = U \times V' \subset U \times V,$$

then $(E', \pi|_{E'}, M, V')$ is called a rank l **vector subbundle** of (E, π, M, V). Charts with this property are said to be adapted to the subbundle.

The triple $(E', \pi|_{E'}, M, V')$ is a vector bundle and every adapted VB-chart (U, ϕ) on E gives rise to a chart on $(E', \pi|_{E'}, M, V')$; namely, (U, ϕ'), where ϕ' is the restriction of ϕ to $\pi^{-1}(U) \cap E' = \pi|_{E'}^{-1}(U)$. By picking a basis for V′ and extending to a basis for V one may take V to be $\mathbb{R}^k$ and V′ to be $\mathbb{R}^l$ embedded in $\mathbb{R}^k$ as $\mathbb{R}^l \times \{0\} \subset \mathbb{R}^k$.

Exercise 6.27. Let $E \to M$ be a vector bundle as above. Suppose that a subspace E'_p of E_p is given for each $p \in M$ and consider the set $E' = \bigcup_{p \in M} E'_p$. Show that E' is the total space of a rank l vector subbundle if and only if for each $p \in M$, there is an open neighborhood U of p on which smooth sections $\sigma_1, \ldots, \sigma_l$ are defined such that for each $q \in U$ the set $\{\sigma_1(q), \ldots, \sigma_l(q)\}$ is a basis of the subspace E'_q.

If $h : E_1 \to E_2$ is a vector bundle morphism over M, then
$$\operatorname{Ker} h := \bigcup_{p \in M} \operatorname{Ker} h|_{E_{1p}}$$
is a *subset* of E_1. This subset is not necessarily (the total space of) a subbundle, at least in the ordinary sense. However, if the rank l of $h|_{E_{1p}}$ is independent of p, then we say that the bundle map has rank l and, in this case, $\operatorname{Ker} h$ is a vector subbundle. Similarly, if h has constant rank in this sense, then the image $\operatorname{Im} h$ is a vector subbundle of E_2. Both of these facts follow from

Proposition 6.28. *Suppose that $h : E_1 \to E_2$ is a vector bundle morphism over M of constant rank r and that E_1 and E_2 have typical fibers V_1 and V_2 respectively. Fix a rank r linear map $A : V_1 \to V_2$. Then for every $p \in M$ there is a VB-chart (U, ϕ) for E_1 with $p \in U$ and a VB-chart (U, ψ) for E_2 such that $\psi \circ h \circ \phi^{-1} : U \times V_1 \to U \times V_2$ has the form*
$$(p, v) \mapsto (p, Av).$$

It follows that $\operatorname{Ker} h$ is a vector subbundle with typical fiber $\operatorname{Ker} A$ and $\operatorname{Im} h$ is a vector subbundle of E_2 with typical fiber $\operatorname{Im} A$.

Proof. Let us first make an observation. Notice that only the rank of the linear map $A : V_1 \to V_2$ in this last proposition is important and we may replace A by any linear map of the same rank. The reason for this is that if $B : V_1 \to V_2$ is any other linear map with the same rank as A, then there exist linear isomorphisms α and β such that $B = \beta A \alpha^{-1}$. In particular, if one has chosen bases and identified V_1 with $\mathbb{R}^{k_1}$ and V_2 with $\mathbb{R}^{k_2}$, then we may take A to be a map of the form
$$(x^1, \ldots, x^{k_1}) \mapsto (x^1, \ldots, x^r, 0, \ldots, 0),$$
so that $\operatorname{Ker} A$ is a copy of $\mathbb{R}^{k_1 - r}$ and $\operatorname{Im} h$ is a copy of $\mathbb{R}^r$.

What we need to prove is entirely local. Thus our task is to show that for any smooth map $h : U \times \mathbb{F}^{k_1} \to U \times \mathbb{F}^{k_2}$ of the form $(p, v) \to (p, h_p v)$, with h_p a linear map of rank r and where $p \mapsto h_p$ is smooth, we may find maps ψ and ϕ such that $\psi \circ h \circ \phi^{-1} : U \times \mathbb{F}^{k_1} \to U \times \mathbb{F}^{k_2}$ is given by $(p, x^1, \ldots, x^{k_1}) \mapsto (p, x^1, \ldots, x^r, 0, \ldots, 0)$. Fix $p_0 \in U$. There exist linear

6.2. Vector Bundles

isomorphisms $\alpha : \mathbb{F}^{k_1} \to \mathbb{F}^{k_1}$ and $\beta : \mathbb{F}^{k_2} \to \mathbb{F}^{k_2}$ such that $\beta \circ h_p \circ \alpha^{-1}$ is given by a $k_2 \times k_1$ matrix of the form

$$\begin{bmatrix} A_{11}(p) & A_{12}(p) \\ A_{21}(p) & A_{22}(p) \end{bmatrix},$$

and where $A_{11}(p_0)$ is an $r \times r$ matrix which is invertible. By shrinking U if needed, we may assume that $A_{11}(p)$ is invertible for all $p \in U$. Thus we may as well assume from the start that h_p is represented by a matrix of this form. Now consider the map $\phi_p : \mathbb{F}^{k_1} \to \mathbb{F}^{k_1}$ whose matrix is given by

$$\begin{bmatrix} A_{11}(p) & A_{12}(p) \\ 0 & I_{(k_1-r) \times (k_1-r)} \end{bmatrix}_{k_1 \times k_1}.$$

Then $h_p \circ \phi_p^{-1}$ has a matrix of the form

$$\begin{bmatrix} I_{r \times r} & 0 \\ A_{21}(p) A_{11}^{-1}(p) & C \end{bmatrix}_{k_2 \times k_1},$$

and since this matrix must have rank r, we see that $C = 0$. Let $M_p := A_{21}(p) A_{11}^{-1}(p)$ and let ψ_p be the linear map $\mathbb{R}^{k_2} \to \mathbb{R}^{k_2}$ with matrix

$$\begin{bmatrix} I_{r \times r} & 0 \\ -M_p & I_{(k_2-r) \times (k_2-r)} \end{bmatrix}.$$

Then $\psi_p \circ h_p \circ \phi_p^{-1}$ has the form $\begin{bmatrix} I_{r \times r} & 0 \\ 0 & 0 \end{bmatrix}$. Now define $\phi(p, v) = (p, \phi_p v)$, $h(p, x) := (p, h_p x)$ and $\psi(p, v) := (p, \psi_p v)$ for $p \in U$, $x \in \mathbb{F}^{k_1}$, and $v \in \mathbb{F}^{k_2}$. Notice that ψ_p, h_p and ϕ_p^{-1} each depend smoothly on p. The map $\psi \circ h \circ \phi^{-1}$ has the required form. $\square$

Proposition 6.29. *Let $\lambda_0 : \mathrm{GL}(V) \times V \to V$ be the standard action of $\mathrm{GL}(V)$ on the $\mathbb{F}$-vector space V. A fiber bundle with typical fiber V has an $\mathbb{F}$-vector bundle structure if and only if it admits a λ_0-bundle atlas (a $\mathrm{GL}(V)$-bundle atlas). Furthermore, if $\lambda : \mathrm{GL}(V) \times V \to V$ is any effective action which acts linearly, then any fiber bundle (E, π, M, V) that has a λ-atlas is a vector bundle in a natural way.*

Proof. That a vector bundle has a $\mathrm{GL}(V)$-bundle structure follows directly from the definition. All that remains to show is the second part of the theorem, since this will imply the remainder of the first part. Let $\lambda : G \times V \to V$ be any effective Lie group action which acts linearly and suppose that (E, π, M, V) has a λ-bundle structure. Let (U_α, ϕ_α) and (U_β, ϕ_β) be λ-compatible bundle charts and let $\phi_\alpha = (\pi, \Phi_\alpha)$ and $\phi_\beta = (\pi, \Phi_\beta)$. Fix

$p \in U_\alpha \cap U_\beta$. For $v, w \in V$, and $a, b \in \mathbb{F}$, we have

$$\Phi_{\alpha\beta}(p)(av + bw) = \lambda(g_{\alpha\beta}(p), av + bw)$$
$$= a\lambda(g_{\alpha\beta}(p), v) + b\lambda(g_{\alpha\beta}(p), w)$$
$$= a\Phi_{\alpha\beta}(p)(v) + b\Phi_{\alpha\beta}(p)(w),$$

which shows that $\Phi_{\alpha\beta}(p) \in \mathrm{GL}(V)$ for all $p \in U_\alpha \cap U_\beta$. We transfer the vector space structure from V to E_p via $\Phi_\alpha|_{E_p}^{-1}$ and note that this is well-defined by Proposition 2.3. With this linear structure on the fibers it is easy to verify that (E, π, M, V) is a vector bundle. $\square$

Theorem 6.30 (Vector bundle construction theorem). *Let $\{U_\alpha\}_{\alpha \in A}$ be a cover of M and let $\{g_{\alpha\beta}\}$ be a G-cocycle for a Lie group G. If G acts linearly on the vector space V (by say λ), then there exists a vector bundle over M with a VB-atlas $\{(U_\alpha, \phi_\alpha)\}$ satisfying $\phi_\alpha \circ \phi_\beta^{-1}(p, v) = (p, g_{\alpha\beta}(p) \cdot v))$ on nonempty overlaps $U_\alpha \cap U_\beta$. In other words, there exists a vector bundle with (G, λ)-atlas.*

Proof. This is essentially a special case of Theorem 6.15. One only needs to check linearity of the ϕ_α on fibers. $\square$

Perhaps some clarification is in order. In the case of a vector bundle, the raw transition maps $\Phi_{\alpha\beta}$ take values in the general linear group $\mathrm{GL}(V)$, which is a Lie group. They correspond to a λ_0-bundle structure where λ_0 is the standard linear action of $\mathrm{GL}(V)$ on V (the standard representation), and they automatically satisfy the cocycle condition. The more general transition maps that define a (G, λ)-bundle structure (G-bundle structure) are G-valued. It is important to note that G may be small compared to $\mathrm{GL}(V)$ and certainly need not be thought of as a subset of $\mathrm{GL}(V)$. For example, the tensor bundles have (possibly ineffective) $\mathrm{GL}(V)$-bundle structures coming from tensor representations, but the tensor bundles themselves generally have rank greater than $k = \dim(V)$. Since in the vector bundle case, the $\Phi_{\alpha\beta}$ arise directly from a VB-atlas and act by the standard action, we will call these **standard transition maps**, and the corresponding $\mathrm{GL}(V)$-bundle structure will be called the **standard $\mathrm{GL}(V)$-bundle structure**. *The standard $\mathrm{GL}(V)$-bundle structure is the structure that a vector bundle has simply by virtue of being an $\mathbb{F}$-vector bundle with typical fiber V.*

Remark 6.31. We have previously mentioned that the notion of a representation is equivalent to that of a left linear action. When dealing with vector bundles it is perhaps more common to use the representation terminology and notation and this we shall do as convenient. So if λ is a left linear action, then the map $G \to \mathrm{GL}(V)$ given by $g \mapsto \lambda(g) := \lambda_g$ is a **representation** of G. Conversely, if λ is such a representation, we obtain a linear action by letting $\lambda(g, v) := \lambda(g)v$.

6.2. Vector Bundles

We already know what it means for two vector bundles over M to be equivalent. Of course any two vector bundles that are equivalent in a natural way can be thought of as the same. Since we can and often do construct our bundles according to the above recipe, it will pay to know something about when two vector bundles over M are isomorphic, based on their respective transition functions. Notice that the standard transition functions are easily recovered from every (G, λ)-atlas by the formula $\lambda\left(g_{\alpha\beta}(p)\right) y = \left.\Phi_{\alpha\beta}\right|_p (y)$.

Proposition 6.32. *Two vector bundles $\pi : E \to M$ and $\pi' : E' \to M$ with standard transition maps $\{\Phi_{\alpha\beta} : U_\alpha \cap U_\beta \to \mathrm{GL}(V)\}$ and $\{\Phi'_{\alpha\beta} : U_\alpha \cap U_\beta \to \mathrm{GL}(V)\}$ over the same cover $\{U_\alpha\}$ are isomorphic (over M) if and only if there are $\mathrm{GL}(V)$-valued functions f_α defined on each U_a such that*

$$(6.1) \qquad \Phi'_{\alpha\beta}(x) = f_\alpha(x) \Phi_{\alpha\beta}(x) f_\beta^{-1}(x) \text{ for } x \in U_\alpha \cap U_\beta.$$

Proof. (Sketch) Given a vector bundle isomorphism $f : E \to E'$ over M, let $f_\alpha(x) := \Phi'_\alpha \circ f \circ \left.\Phi_\alpha\right|_{E_x}$. Check that this works. Conversely, given functions f_α satisfying equations (6.1), define $\widetilde{f}_\alpha : U_\alpha \times V \to U_\alpha \times V$ by $(x, v) \mapsto (x, f_\alpha(x) v)$. We define $f : E \to E'$ by

$$f(\epsilon) := \left(\left(\phi'_\alpha\right)^{-1} \circ \widetilde{f}_\alpha \circ \phi_\alpha \right)(\epsilon) \text{ for } \epsilon \in \left.E\right|_{U_a}.$$

The conditions (6.1) insure that f is well-defined on the overlaps $\left.E\right|_{U_a} \cap \left.E\right|_{U_\beta} = \left.E\right|_{U_a \cap U_\beta}$. One easily checks that this is a vector bundle isomorphism. $\square$

We can use this construction to arrive at several common vector bundles.

Example 6.33. Given an atlas $\{(U_\alpha, \mathbf{x}_\alpha)\}$ for a smooth manifold M, we let $g_{\alpha\beta}(p) = T_p \mathbf{x}_\alpha \circ (T_p \mathbf{x}_\beta)^{-1}$ for all $p \in U_\alpha \cap U_\beta$. The bundle constructed according to the recipe of Theorem 6.30 is a vector bundle which is (naturally isomorphic to) the tangent bundle TM. If we let $g^*_{\alpha\beta}(p) = (T_p \mathbf{x}_\beta \circ (T_p \mathbf{x}_\alpha)^{-1})^*$, then we obtain the cotangent bundle $T^* M$.

Proposition 6.34. *Let $\pi : E \to M$ be an $\mathbb{F}$-vector bundle with typical fiber V and with VB-atlas $\{(U_\alpha, \phi_\alpha)\}$. Let $s_\alpha : U_\alpha \to V$ be a collection of maps such that whenever $U_\alpha \cap U_\beta \ne \emptyset$, we have $s_\alpha(p) = \Phi_{\alpha\beta}(p) s_\beta(p)$ for all $p \in U_\alpha \cap U_\beta$. Then there is a global section s such that $\left.s\right|_{U_\alpha} = s_\alpha$ for all α.*

Proof. Let $\phi_\alpha : \left.E\right|_{U_\alpha} \to U_\alpha \times V$ be the trivializations that give rise to the cocycle $\{\Phi_{\alpha\beta}\}$. Let $\gamma_\alpha(p) := (p, s_\alpha(p))$ for $p \in U_\alpha$ and let $\left.s\right|_{U_\alpha} := \phi_\alpha^{-1} \circ \gamma_\alpha$.

This gives a well-defined section s because for $x \in U_\alpha \cap U_\beta$ we have

$$\phi_\alpha^{-1} \circ \gamma_\alpha(p) = \phi_\alpha^{-1}(p, s_\alpha(p))$$
$$= \phi_\alpha^{-1}(p, \Phi_{\alpha\beta}(p) s_\beta(p))$$
$$= \phi_\alpha^{-1} \circ \phi_\alpha \circ \phi_\beta^{-1}(p, s_\beta(p))$$
$$= \phi_\beta^{-1}(p, s_\beta(p)) = \phi_\beta^{-1} \circ \gamma_\beta(p). \qquad \square$$

Suppose we have two vector bundles, $\pi_1 : E_1 \to M$ and $\pi_2 : E_2 \to M$. We give two constructions of the **Whitney sum bundle** $\pi_1 \oplus \pi_2 : E_1 \oplus E_2 \to M$. This is a globalization of the direct sum construction of vector spaces. In fact, the first construction simply takes $E_1 \oplus E_2 = \bigcup_{p \in M} E_{1p} \oplus E_{2p}$. Now, we have a vector bundle atlas $\{(\phi_\alpha, U_\alpha)\}$ for π_1 and a vector bundle atlas $\{(\psi_\alpha, U_\alpha)\}$ for π_2. Assume that both atlases have the same family of open sets (we can arrange this by taking a common refinement). Now let $\phi_\alpha \oplus \psi_\alpha : (v_p, w_p) \mapsto (p, \mathrm{pr}_2 \circ \phi_\alpha(v_p), \mathrm{pr}_2 \circ \psi_\alpha(w_p))$ for all $(v_p, w_p) \in (E_1 \oplus E_2)|_{U_\alpha}$. Then $\{(\phi_\alpha \oplus \psi_\alpha, U_\alpha)\}$ is a VB-atlas for $\pi_1 \oplus \pi_2 : E_1 \oplus E_2 \to M$.

Another method of constructing this bundle is to take the cocycle $\{g_{\alpha\beta}\}$ for π_1 and the cocycle $\{h_{\alpha\beta}\}$ for π_2 and then let $g_{\alpha\beta} \oplus h_{\alpha\beta} : U_\alpha \cap U_\beta \to \mathrm{GL}(\mathbb{F}^{k_1} \times \mathbb{F}^{k_2})$ be defined by $(g_{\alpha\beta} \oplus h_{\alpha\beta})(x) = g_{\alpha\beta}(x) \oplus h_{\alpha\beta}(x) : (v, w) \mapsto (g_{\alpha\beta}(x)v, h_{\alpha\beta}(x)w)$. The maps $g_{\alpha\beta} \oplus h_{\alpha\beta}$ form a cocycle which determines a bundle by the construction of Proposition 6.30, which is (isomorphic to) $\pi_1 \oplus \pi_2 : E_1 \oplus E_2 \to M$.

The pull-back of a vector bundle $\pi : E \to M$ by a smooth map $f : N \to M$ is naturally a vector bundle whose linear structure on each fiber $(f^*E)_q = \{q\} \times E_p$ is the obvious one induced from E_p. Put another way, we give the unique linear structure to each fiber that makes the bundle map $\widetilde{f} : f^*E \to E$ linear on fibers. When given this vector bundle structure, we call f^*E the **pull-back vector bundle**.

Example 6.35. Let $\pi_1 : E_1 \to M$ and $\pi_2 : E_2 \to M$ be vector bundles and let $\triangle : M \to M \times M$ be the diagonal map $x \mapsto (x, x)$. From π_1 and π_2 one can construct a bundle $\pi_{E_1 \times E_2} : E_1 \times E_2 \to M \times M$ by $\pi_{E_1 \times E_2}(\epsilon_1, \epsilon_2) := (\pi_1(\epsilon_1), \pi_2(\epsilon_2))$. The Whitney sum bundle $E_1 \oplus E_2$ defined previously is naturally isomorphic to the pull-back $\triangle^* \pi_{E_1 \times E_2} : \triangle^*(E_1 \times E_2) \to M$ (Problem 10).

Exercise 6.36. Recall the space $\Gamma_f(\xi)$ from Definition 6.19. Show that if $\pi : E \to M$ is an $\mathbb{F}$-vector bundle and $f : N \to M$ is a smooth map, then both $\Gamma_f(\xi)$ and $\Gamma(f^*\xi)$ are modules over $C^\infty(N; \mathbb{F})$, and that the natural correspondence between $\Gamma_f(\xi)$ and $\Gamma(f^*\xi)$ is a module isomorphism.

Every vector bundle has global sections. An obvious example is the **zero section** which maps each $x \in M$ to the zero element 0_x of the fiber

6.2. Vector Bundles

E_x. The image of the zero section is also referred to as the **zero section** and is often identified with M. (Of course, the image of any global section is a submanifold diffeomorphic to the base manifold.) We have the following simple analogue of Lemma 2.65:

Lemma 6.37. *Let $\pi : E \to M$ be an $\mathbb{F}$-vector bundle with typical fiber V. If $v \in \pi^{-1}(p)$ then there exists a global section $\sigma \in \Gamma(\xi)$ such that $\sigma(p) = v$. Furthermore, if s is a local section defined on U, and V is an open set with compact closure with $V \subset \overline{V} \subset U$, then there is a section $\sigma \in \Gamma(\xi)$ such that $\sigma = s$ on V.*

Proof. Using a local trivialization one can easily get a local section σ_{loc} defined near p such that $\sigma(p) = v$. Now just use a cut-off function as in the proof of Lemma 2.65. For the second part we just choose a cut-off function β with support in U and such that $\beta = 1$ on $\overline{V}$. Then βs extends by zero to the desired global section. $\square$

If a section of a vector bundle takes the zero value in some fiber we say that it vanishes at that point. Global smooth sections that never vanish do not always exist; such sections are called **nowhere vanishing** or **nonvanishing**. However, there is one case where it is easy to see that nonvanishing smooth sections exist:

Proposition 6.38. *Any bundle equivalent to a (trivial) product bundle must have a nowhere vanishing smooth global section.*

It is a fact that the tangent bundle of S^2 does not have any such nowhere vanishing smooth sections. In other words, all smooth (or even continuous) vector fields on S^2 must vanish at some point. This is a result from algebraic topology called the "hairy sphere theorem" (see Theorem 10.15). If one fancifully imagines a vector field on a sphere to be hair, then the theorem suggests that one cannot comb the hair neatly "flat" without creating a cowlick somewhere. More generally, the analogous result holds for S^{2n} if $n \geq 1$.

Exercise 6.39. Modify either the construction of Example 6.9 or Example 6.16 to obtain a rank one vector bundle version of the Möbius band and give an argument proving that every global continuous section of this bundle must vanish somewhere.

Definition 6.40. If $\xi = (E, \pi, M, V)$ is a vector bundle and $p \in M$, then a vector space basis for the fiber E_p is called a **frame** at p.

Definition 6.41. Let $\pi : E \to M$ be a rank k vector bundle. A k-tuple $\sigma = (\sigma_1, \ldots, \sigma_k)$ of sections of E over an open set U is called a (local) **frame field** over U if for all $p \in U$, $(\sigma_1(p), \ldots, \sigma_k(p))$ is a frame at p.

If we choose a fixed basis $\{\mathbf{e}_i\}_{i=1,\ldots,k}$ for the typical fiber V, then a choice of a local frame field over an open set $U \subset M$ is equivalent to a local trivialization (a vector bundle chart). Namely, if ϕ is such a trivialization over U, then defining $\sigma_i(p) = \phi^{-1}(p, \mathbf{e}_i)$, we have that $\sigma_\phi = (\sigma_1, \ldots, \sigma_k)$ is a local frame over U. Conversely, if $\sigma_\phi = (\sigma_1, \ldots, \sigma_k)$ is a local frame over U, then every $v \in \pi^{-1}(U)$ has the form $v = \sum v^i \sigma_i(p)$ for a unique p and unique numbers $v^i(p)$. Then the map $f : U \times V \to \pi^{-1}(U)$ defined by $(p, v) \mapsto \sum v^i \sigma_i(p)$ is a diffeomorphism and its inverse $\phi = f^{-1}$ is a trivialization. Thus if there is a global frame field, then the vector bundle is trivial. A manifold M is said to be **parallelizable** if $TM \to M$ has a global frame field; i.e. if the tangent bundle is trivial. For example, the hairy sphere theorem mentioned above implies that S^2 is not parallelizable. On the other hand, $S^2 \times \mathbb{R}$ *is* parallelizable. It is easy to show that the torus $T^2 = S^1 \times S^1$ and its higher-dimensional analogues $T^n = S^1 \times \cdots \times S^1$ are also parallelizable.

If G is a Lie subgroup of GL(V) and G acts in the standard way on V, that is, if the action is $\lambda_0|_G$, the restriction of the standard action, then a $\lambda_0|_G$-bundle structure is a reduction to the group G. Put another way, one has achieved such a reduction if one can find a cocycle of standard transition maps $\{\Phi_{\alpha\beta}\}$ arising from a vector bundle atlas which take values in G (acting in the standard way on V). By a slight extension, if λ is a linear action on V, then an effective (G, λ)-bundle structure on E can be considered as a reduction of the standard GL(V)-bundle structure.

Let $\sigma_i(p) = \phi_\alpha^{-1}(p, \mathbf{e}_i)$ and $\sigma'_i(p) = \phi_\beta^{-1}(p, \mathbf{e}_i)$ be frame fields corresponding to VB-charts (U_α, ϕ_α) and (U_β, ϕ_β) that lie in a (G, λ)-atlas and where $U_\alpha \cap U_\beta$ is nonempty. Then $\phi_\alpha \circ \phi_\beta^{-1}(p, v) = (p, \lambda(g_{\alpha\beta}(p))(v))$ for transition functions $g_{\alpha\beta} : U_\alpha \cap U_\beta \to G$. For each $g \in G$, there is a matrix $(\lambda_i^j(g))$ which represents λ_g with respect to $(\mathbf{e}_1, \ldots, \mathbf{e}_k)$. Unwinding definitions, we see that ϕ_α^{-1} is linear on the vector space $\{p\} \times V$. Using this we have

$$\sigma'_i(p) = \phi_\beta^{-1}(p, \mathbf{e}_i) = \phi_\alpha^{-1}(p, \lambda(g_{\alpha\beta}(p))(\mathbf{e}_i))$$
$$= \phi_\alpha^{-1}\left(p, \sum_j \lambda_i^j(g_{\alpha\beta}(p))\mathbf{e}_j\right) = \sum_j \lambda_i^j(g_{\alpha\beta}(p))\phi_\alpha^{-1}(p, \mathbf{e}_j)$$
$$= \sum_j \lambda_i^j(g_{\alpha\beta}(p))\sigma_j(p).$$

Thus the smooth matrix-valued function $(\lambda_i^j \circ g_{\alpha\beta})$ defined on $U_\alpha \cap U_\beta$ gives the change of frame and embodies the transition map on $U_\alpha \cap U_\beta$. In practice, this is often a good way to look at things. Consider the common situation where $V = \mathbb{R}^k$ and where λ_0 is the standard representation of GL($\mathbb{R}^k$). If

6.2. Vector Bundles

we identify GL($\mathbb{R}^k$) with the matrix group GL(k), then $((\lambda_0)_i^j \circ g_{\alpha\beta}(p))$ is just the matrix $g_{\alpha\beta}(p)$ itself.

Metric differential geometry begins if we have a scalar product on the fibers of a vector bundle. We introduce the concept at this point so as to have an example of a reduction of the structure group.

Definition 6.42. A **Riemannian metric** on a real vector bundle $\pi : E \to M$ is a map $p \mapsto g_p(\cdot, \cdot)$ which assigns to each $p \in M$ a positive definite scalar product $g_p(\cdot, \cdot)$ on the fiber E_p that is smooth in the sense that $p \mapsto g_p(s_1(p), s_2(p))$ is smooth for all smooth sections s_1 and s_2. A real vector bundle together with a Riemannian metric is referred to as a **Riemannian vector bundle**.

For example, a Riemannian metric on the tangent bundle of a smooth manifold is what one means by a Riemannian metric *on the manifold*. A smooth manifold with a Riemannian metric is called a Riemannian manifold and such will be studied later in this book. If a rank k real vector bundle $\pi : E \to M$ has a Riemannian metric, then it is convenient to assume that a fixed inner product is chosen on the typical fiber V and that a distinguished orthonormal basis $(\mathbf{e}_1, \ldots, \mathbf{e}_k)$ has been chosen. In most applications, V is $\mathbb{R}^k$, the inner product is the standard dot product, and the distinguished basis is the usual standard basis. Recall that the orthogonal group O(V) is the subgroup of GL(V) consisting of elements that preserve the inner product. Once the inner product and distinguished orthonormal basis are fixed, every choice of Riemannian metric on E corresponds to a reduction of the standard structure group GL(V) to the subgroup O(V) as follows. Let us first show how a metric leads to a reduction. We start with an arbitrary VB-atlas $\{(U_\alpha, \phi_\alpha)\}$. Since we have fixed a basis for V, each chart (U_α, ϕ_α) defines a frame field $(\sigma_1^\alpha, \ldots, \sigma_k^\alpha)$ on U_α by $\sigma_i^\alpha(p) := \phi_\alpha^{-1}(p, \mathbf{e}_i)$ as explained above. One can perform a Gram-Schmidt process on the basis $(\sigma_1^\alpha(p), \ldots, \sigma_k^\alpha(p))$ simultaneously for all $p \in U_\alpha$ so that we have a new orthonormal basis $(e_1^\alpha(p), \ldots, e_k^\alpha(p))$ for each p, where $e_j^\alpha(p) = \sum A_j^i(p) \sigma_i^\alpha(p)$ for all $p \in U_\alpha$ and the matrix entries $A_j^i(p)$ depend smoothly on p. Thus $(e_1^\alpha, \ldots, e_k^\alpha)$ is a (smooth) local frame field called an **orthonormal frame field**. One then replaces the original chart (U_α, ϕ_α) by a new chart (U_α, ϕ_α') that is the inverse of the map $(p, v) \mapsto \sum v^i e_i(p)$, where $v = \sum v^i \mathbf{e}_i(p)$ in V. In other words, $\phi_\alpha^{-1} : (p, v) \mapsto \sum v^i \sigma_i(p)$ is replaced by

$$\phi_\alpha'^{-1} : (p, v) \mapsto \sum v^i e_i(p).$$

Make this replacement for each (U_α, ϕ_α) to obtain a new atlas $\{(U_\alpha, \phi_\alpha')\}$. Any two of these orthonormal frame fields, say $(e_1^\alpha, \ldots, e_k^\alpha)$ and $(e_1^\beta, \ldots e_k^\beta)$,

are related by
$$e_j^\alpha(p) = \sum Q_j^i(p) e_i^\beta(p)$$
for some smooth orthogonal matrix function Q_j^i. One now checks that the transition maps for this new atlas (which is still a subatlas for the maximal VB-atlas) take values in O(V). Indeed,

$$(p, \Phi'_{\alpha\beta}(p)(v)) = \phi'_\beta \circ \phi'^{-1}_\alpha(p,v) = \phi'_\beta\left(\sum v^j e_j^\alpha(p)\right)$$
$$= \phi'_\beta\left(\sum v^j Q_j^i(p) e_i^\beta(p)\right) = \left(p, \Phi'_\beta\big|_p \sum v^j Q_j^i(p) e_i^\beta(p)\right)$$
$$= \left(p, \sum v^j Q_j^i(p) \, \Phi'_\beta\big|_p e_i^\beta(p)\right) = \left(p, \sum v^j Q_j^i(p) \mathbf{e}_i\right),$$

from which we see that $\Phi'_{\alpha\beta}(p)(v) = \Phi'_{\alpha\beta}(p)(\sum v^j \mathbf{e}_j) = \sum v^j Q_j^i(p) \mathbf{e}_i$. Since $(Q_j^i(p))$ is an orthogonal matrix for all p, we have $\Phi'_{\alpha\beta}(p) \in O(V)$ for all p. The converse is also true. Namely, a reduction to the structure group O(V) (acting in the standard way) is tantamount to the introduction of a Riemannian metric. The correspondence presumes the prior choice of inner product and distinguished orthonormal basis on V.

Exercise 6.43. Prove the converse statement referred to above.

Exercise 6.44. Let E be a complex vector bundle of rank k. Define by analogy with Riemannian metric, the notion of a **Hermitian metric** on E and show that every Hermitian metric on E corresponds to a reduction of the standard $GL(k, \mathbb{C})$-bundle structure to a $U(n)$-bundle structure.

Proposition 6.45. *On every real vector bundle E there can be defined a Riemannian metric. Similarly, on any complex vector bundle there exists a Hermitian metric.*

Proof. We prove the Riemannian case; the Hermitian case is entirely analogous. The proof uses the fact that a strict convex combination of positive definite scalar products is a positive definite scalar product. This allows us to use a partition of unity argument. Endow V with an inner product. Let $\{(U_\alpha, \phi_\alpha)\}$ be a VB-atlas and let (U_α, ϕ_α) be a given VB-chart. On the trivial bundle $U_\alpha \times V \to U_\alpha$ there certainly exists a Riemannian metric given on each fiber by $\langle (p,v), (p,w) \rangle_\alpha = \langle v, w \rangle$. We may transfer this to the bundle $\pi^{-1}(U_\alpha) \to U_\alpha$ by using the map ϕ_α^{-1}, thus obtaining a metric g_α on this restricted bundle over U_α. We do this for every VB-chart in the atlas. The trick is to piece these together in a smooth way. For that, we take a smooth partition of unity (U_α, ρ_α) subordinate to the cover $\{U_\alpha\}$. Let

$$g(p) = \sum \rho_\alpha(p) g_\alpha(p).$$

6.2. Vector Bundles

The sum is finite at each $p \in M$ since the partition of unity is locally finite and the functions $\rho_\alpha g_\alpha$ are extended to be zero outside of the corresponding U_α. The fact that $\rho_\alpha \geq 0$ and $\rho_\alpha > 0$ at p for at least one α easily gives the result that g is positive definite at each p and so it is a Riemannian metric on E. $\square$

Example 6.46 (Tautological line bundle). Recall that $\mathbb{R}P^n$ is the set of all lines through the origin in $\mathbb{R}^{n+1}$. Define the subset $\mathbb{L}(\mathbb{R}P^n)$ of $\mathbb{R}P^n \times \mathbb{R}^{n+1}$ consisting of all pairs (l, v) such that $v \in l$ (think about this). This set together with the map $\pi_{\mathbb{R}P^n} : \mathbb{L}(\mathbb{R}P^n) \to \mathbb{R}P^n$ given by $(l, v) \mapsto l$, is a rank one vector bundle.

Example 6.47 (Tautological bundle). Let $G(n, k)$ denote the Grassmann manifold of k-planes in $\mathbb{R}^n$. Let $\gamma_{n,k}$ be the subset of $G(n, k) \times \mathbb{R}^n$ consisting of pairs (P, v) where P is a k-plane (k-dimensional subspace) and v is a vector in the plane P. The projection $\pi_{n,k} : \gamma_{n,k} \to G(n, k)$ is simply $(P, v) \mapsto P$. The result is a vector bundle $(\gamma_{n,k}, \pi_{n,k}, G(n, k), \mathbb{R}^k)$. We leave it to the reader to discover an appropriate VB-atlas (see Problem 12).

These tautological vector bundles are not just trivial bundles, and in fact their topology or twistedness (for large n) is of the utmost importance for classifying vector bundles (see [**Bo-Tu**]). One may take the inclusions $\mathbb{R}^n \subset \mathbb{R}^{n+1} \subset \cdots \subset \mathbb{R}^\infty$ to construct inclusions $G(n, k) \subset G(n+1, k) \subset \cdots$ and $\gamma_{n,k} \subset \gamma_{n+1,k}$. Given a rank k vector bundle $\pi : E \to M$, there is an n such that $\pi : E \to M$ is (isomorphic to) the pull-back of $\gamma_{n,k}$ by some map $f : M \to G(n, k)$:

$$\begin{array}{ccc} E \cong f^*\gamma_{n,k} & \longrightarrow & \gamma_{n,k} \\ \downarrow & & \downarrow \\ M & \xrightarrow{f} & G(n, k) \end{array}$$

Exercise 6.48. To each point on a unit sphere in $\mathbb{R}^n$, attach the space of all vectors normal to the sphere at that point. Show that this *normal bundle* is in fact a (smooth) vector bundle. Generalize to define the normal bundle of a hypersurface in $\mathbb{R}^n$. When is such a normal bundle trivial?

Exercise 6.49. Fix a nonnegative integer j. Let $Y = \mathbb{R} \times (-1, 1)$ and let $(x_1, y_1) \sim (x_2, y_2)$ if and only if $x_1 = x_2 + jk$ and $y_1 = (-1)^{jk} y_2$ for some integer k. Show that $E := Y/\sim$ is a vector bundle of rank 1 that is trivial if and only if j is even. Prove or at least convince yourself that this is the Möbius band when j is odd.

6.3. Tensor Products of Vector Bundles

Given two vector bundles $\pi_1 : E_1 \to M$ and $\pi_2 : E_2 \to M$ with respective typical fibers V_1 and V_2, we let

$$E_1 \otimes E_2 := \bigcup_{p \in M} E_{1p} \otimes E_{2p} \quad \text{(a disjoint union)}.$$

Then we have a projection map $\pi : E_1 \otimes E_2 \to M$ given by mapping any element in a fiber $E_{1p} \otimes E_{2p}$ to the base point p. We show how to construct a VB-atlas for $E_1 \otimes E_2$ from an atlas on each of E_1 and E_2. The smooth structure and topology can be derived from the atlas as usual in such a way as to make all the relevant maps smooth. We leave the verification of this to the reader. The resulting bundle is the **tensor product bundle**. As usual we can assume that the atlases are based on the same open cover. Thus suppose that $\{(U_\alpha, \phi_\alpha)\}$ is a VB-atlas for E_1 while $\{(U_\alpha, \psi_\alpha)\}$ is a VB-atlas for E_2. Now let $\Phi_\alpha \otimes \Psi_\alpha : (E_1 \otimes E_2)|_{U_\alpha} \to V_1 \otimes V_2$ be defined by $(\Phi_\alpha \otimes \Psi_\alpha)|_{E_{1p} \otimes E_{2p}} := \Phi_\alpha|_{E_{1p}} \otimes \Psi_\alpha|_{E_{2p}}$ for $p \in U_\alpha$. Then let

$$\phi_\alpha \otimes \psi_\alpha : (E_1 \otimes E_2)|_{U_\alpha} \to U_\alpha \times (V_1 \otimes V_2)$$

be defined by $\phi_\alpha \otimes \psi_\alpha := (\pi, \Phi_\alpha \otimes \Psi_\alpha)$. To clarify, the map $\Phi_\alpha|_{E_{1p}} \otimes \Psi_\alpha|_{E_{2p}} : E_{1p} \otimes E_{2p} \to V_1 \otimes V_2$ is the tensor product map of two linear maps as described at the end of Chapter 5. To see what the transition maps look like, we compute;

$$(\Phi_\alpha \otimes \Psi_\alpha)|_{E_{1p} \otimes E_{2p}} \circ (\Phi_\beta \otimes \Psi_\beta)|_{E_{1p} \otimes E_{2p}}^{-1}$$
$$= \Phi_\alpha|_{E_{1p}} \otimes \Psi_\alpha|_{E_{2p}} \circ \Phi_\beta|_{E_{1p}}^{-1} \otimes \Psi_\beta|_{E_{2p}}^{-1}$$
$$= \left(\Phi_\alpha|_{E_{1p}} \circ \Phi_\beta|_{E_{1p}}^{-1}\right) \otimes \left(\Psi_\alpha|_{E_{1p}} \circ \Psi_\beta|_{E_{1p}}^{-1}\right)$$
$$= \Phi_{\alpha\beta}(p) \otimes \Psi_{\alpha\beta}(p).$$

Thus the transition maps are given by $p \to \Phi_{\alpha\beta}(p) \otimes \Psi_{\alpha\beta}(p)$, which is a map from U_α to $\mathrm{GL}(V_1 \otimes V_2)$. The group $\mathrm{GL}(V_1 \otimes V_2)$ acts on $V_1 \otimes V_2$ in a standard way, and this is the standard effective structure group of the bundle as we have just seen. However, it is also true that the bundle $E_1 \otimes E_2$ has (ineffective) structure group $\mathrm{GL}(V_1) \times \mathrm{GL}(V_2)$ via a tensor product representation. Indeed, if ι_1 denotes the standard representation of $\mathrm{GL}(V_1)$ in V_1 and ι_2 denotes the standard representation of $\mathrm{GL}(V_2)$ in V_2, then we have a tensor product representation $\iota_1 \otimes \iota_2$ of $\mathrm{GL}(V_1) \times \mathrm{GL}(V_2)$ in $V_1 \otimes V_2$. This is usually not a faithful representation. Using the $\mathrm{GL}(V_1) \times \mathrm{GL}(V_2)$-valued cocycle $p \mapsto h_{\alpha\beta}(p) := (\Phi_{\alpha\beta}(p), \Psi_{\alpha\beta}(p))$, together with $\iota_1 \otimes \iota_2$, we see that by definition

$$(\iota_1 \otimes \iota_2)_{h_{\alpha\beta}(p)}(\tau) = (\Phi_{\alpha\beta}(p) \otimes \Psi_{\alpha\beta}(p))(\tau).$$

Furthermore, if $V_1 = V_2 = V$, then the tensor product representation is usually defined as a representation of $GL(V)$ rather than $GL(V) \times GL(V)$, and so $E_1 \otimes E_2$ would have a $(GL(V), \iota \otimes \iota)$-bundle structure where ι is the standard representation. In this case $\iota \otimes \iota$ is still not a faithful representation since $-\operatorname{id}_V$ is in the kernel. We can reconstruct the same vector bundle using any of these representation-cocycle pairs via Lemma 6.30. In fact, it is quite common that we have different representations by *one group*. Suppose that we have two faithful representations λ_1 and λ_2 of a Lie group G acting on V_1 and V_2 respectively. If $\{g_{\alpha\beta}\}$ is a cocycle of transition maps, then we can use the pair $\{g_{\alpha\beta}, \lambda_1\}$ in Lemma 6.30 to form a vector bundle E_1 that has a (G, λ_1)-bundle structure by construction. Similarly, we can construct a vector bundle E_2 with (G, λ_2)-bundle structure. If we use $\lambda_1 \otimes \lambda_2$ and the same cocycle $\{g_{\alpha\beta}\}$, then we obtain a bundle which, as a vector bundle, is $E_1 \otimes E_2$. But by construction, it has a $(G, \lambda_1 \otimes \lambda_2)$-structure (possibly ineffective). This is the case in the following exercise:

Exercise 6.50. Suppose that E is a vector bundle with a (G, λ)-bundle structure given by a (G, λ)-atlas with a corresponding cocycle of transition functions. Show how one may use Theorem 6.30 to construct bundles isomorphic to E^*, $E \otimes E$ and $E \otimes E^*$ which will have a (G, λ^*)-bundle structure, a $(G, \lambda \otimes \lambda)$-bundle structure and a $(G, \lambda \otimes \lambda^*)$-bundle structure respectively.

6.4. Smooth Functors

We have seen that various new vector bundles can be constructed starting with one or more vector bundles. Most of the operations of linear algebra extend to the vector bundle category. We can unify our thinking on these matters by introducing the notion of a C^∞ functor (or **smooth functor**). With $\mathbb{F} = \mathbb{R}$ or $\mathbb{C}$, the set of all $\mathbb{F}$-vector spaces together with linear maps is a category that we denote by $\operatorname{Lin}(\mathbb{F})$. The set of morphisms from V to W is the space of $\mathbb{F}$-linear maps $L(V, W)$ (also denoted $\operatorname{Hom}(V, W)$).

Definition 6.51. A **covariant C^∞ functor** $\mathcal{F}$ of one variable on $\operatorname{Lin}(\mathbb{F})$ consists of a map, denoted again by $\mathcal{F}$, that assigns to every $\mathbb{F}$-vector space V an $\mathbb{F}$-vector space $\mathcal{F}V$, and a map, also denoted by $\mathcal{F}$, which assigns to every linear map $A \in L(V, W)$, a linear map $\mathcal{F}A \in L(\mathcal{F}V, \mathcal{F}W)$ such that

(i) $\mathcal{F}: L(V, W) \to L(\mathcal{F}V, \mathcal{F}W)$ is smooth;
(ii) $\mathcal{F}(\operatorname{id}_V) = \operatorname{id}_{\mathcal{F}V}$ for all $\mathbb{F}$-vector spaces V;
(iii) $\mathcal{F}(A \circ B) = \mathcal{F}A \circ \mathcal{F}B$ for all $A \in L(U, V)$ and $B \in L(V, W)$ and vector spaces U, V and W.

As an example we have the C^∞ functor which assigns to each V the k-fold direct sum $\bigoplus_k V = V \oplus \cdots \oplus V$ and to each linear map $A \in L(V, W)$

the map
$$\bigoplus_k A : \bigoplus_k V \to \bigoplus_k W$$
given by $\bigoplus_k A(v_1,\ldots,v_k) := (Av_1,\ldots,Av_k)$. Similarly there is the functor which assigns to each V the k-fold tensor product $\bigotimes^k V = V \otimes \cdots \otimes V$ and to each $A \in L(V,W)$ the map $\bigotimes^k A : \bigotimes^k V \to \bigotimes^k W$ given on homogeneous elements by $(\bigotimes^k A)(v_1 \otimes \cdots \otimes v_k) := Av_1 \otimes \cdots \otimes Av_k$.

One can also consider C^∞ covariant functors of several variables. For example, we may assign to each pair of vector spaces (V, W) the tensor product $V \otimes W$, and to each pair $(A, B) \in L(V, V') \times L(W, W')$, the map $A \otimes B : V \otimes W \to V' \otimes W'$.

There is also a similar notion of *contravariant* C^∞ functor:

Definition 6.52. A **contravariant C^∞ functor** $\mathcal{F}$ of one variable on Lin($\mathbb{F}$) consists of a map, denoted again by $\mathcal{F}$, which assigns to every $\mathbb{F}$-vector space V an $\mathbb{F}$-vector space $\mathcal{F}V$, and a map, also denoted by $\mathcal{F}$, which assigns to every linear map $A \in L(V,W)$ a linear map $\mathcal{F}A \in L(\mathcal{F}W, \mathcal{F}V)$ (notice the reversal) such that

(i) $\mathcal{F} : L(V, W) \to L(\mathcal{F}W, \mathcal{F}V)$ is smooth;

(ii) $\mathcal{F}(\mathrm{id}_V) = \mathrm{id}_{\mathcal{F}V}$ for all $\mathbb{F}$-vector spaces V;

(iii) $\mathcal{F}(A \circ B) = \mathcal{F}B \circ \mathcal{F}A$ for all $A \in L(U, V)$ and $B \in L(V, W)$ and vector spaces U, V and W.

The map that assigns to each vector space its dual and to each map its dual map (transpose) is a contravariant C^∞ functor $\mathcal{F}$. One may define the notion of a C^∞ functor of several variables which may be covariant in some variables and contravariant in others. For example, consider the functor of two variables that assigns to each pair (V, W) the space $V \otimes W^*$ and to each pair $(A, B) \in L(V_1, V_2) \times L(W_1, W_2)$ the map $A \otimes B^* : V_1 \otimes W_2^* \to V_2 \otimes W_1^*$.

Theorem 6.53. *Let $\mathcal{F}$ be a C^∞ functor of m variables on* Lin($\mathbb{F}$) *and let $E_1, \ldots, E_m$ be $\mathbb{F}$-vector bundles with respective typical fibers $V_1, \ldots, V_m$. Then the set*
$$E := \mathcal{F}(E_1,\ldots,E_m) := \bigcup_p \mathcal{F}(E_1|_p,\ldots,E_m|_p)$$
together with the map $\pi : E \to M$ which takes elements of $\mathcal{F}(E_1|_p,\ldots,E_m|_p)$ to p is naturally a vector bundle with typical fiber $\mathcal{F}(V_1,\ldots,V_m)$.

Proof. We will only prove the case of $m = 2$ with covariant first variable and contravariant second variable. This should make it clear how the general case would go while keeping the notational complexity under control.

Given vector bundles $\pi_1 : E_1 \to M$ and $\pi_2 : E_2 \to M$, the total space of the constructed bundle is $\bigcup_p \mathcal{F}(E_1|_p, E_2|_p)$ with the obvious projection which we call π. Let (ϕ_α, U_α) be a VB-atlas for E_1 and (ψ_α, U_α) a VB-atlas for E_2 (we have arranged that both atlases use the same cover by going to a common refinement as usual). For each p, let E_p denote the fiber $\mathcal{F}(E_1|_p, E_2|_p)$. Fix α and for each $p \in U_\alpha$ define $\Theta_\alpha|_p \in L(E_p, \mathcal{F}(V_1, V_2))$ by
$$\Theta_\alpha|_p := \mathcal{F}(\Phi_\alpha|_p, \Psi_\alpha|_p^{-1}),$$
where $\phi_\alpha = (\pi_1, \Phi_\alpha)$ and $\psi_\alpha = (\pi_2, \Psi_\alpha)$. Then define $\Theta_\alpha : \pi^{-1}(U_\alpha) \to \mathcal{F}(V_1, V_2)$ by $\Theta_\alpha(\epsilon) = \Theta_\alpha|_p(\epsilon)$ whenever $\epsilon \in \mathcal{F}(E_1|_p, E_2|_p)$. Next define
$$\theta_\alpha = (\pi, \Theta_\alpha) : \pi^{-1} U_\alpha \to U_\alpha \times \mathcal{F}(V_1, V_2).$$
The family $\{(\theta_\alpha, U_\alpha)\}$ is to be a VB-atlas for E. We check the transition maps:
$$\begin{aligned}\Theta_{\alpha\beta}(p) &= \Theta_\alpha|_p \circ \Theta_\beta^{-1}|_p \\ &= \mathcal{F}(\Phi_\alpha|_p, \Psi_\alpha|_p^{-1}) \circ \mathcal{F}(\Phi_\beta|_p, \Psi_\beta|_p^{-1})^{-1} \\ &= \mathcal{F}(\Phi_\alpha|_p, \Psi_\alpha|_p^{-1}) \circ \mathcal{F}(\Phi_\beta|_p^{-1}, \Psi_\beta|_p) \\ &= \mathcal{F}(\Phi_\alpha|_p \circ \Phi_\beta|_p^{-1}, \Psi_\beta|_p \circ \Psi_\alpha|_p^{-1}) \\ &= \mathcal{F}(\Phi_{\alpha\beta}(p), \Psi_{\beta\alpha}(p)).\end{aligned}$$
(Remember that the functor is contravariant in the second variable.) Now we can see from the properties of $\Phi_{\alpha\beta}, \Psi_{\beta\alpha}$ and the definition of C^∞ functor that $\mathcal{F}(\Phi_{\alpha\beta}(p), \Psi_{\beta\alpha}(p)) \in \mathrm{GL}(\mathcal{F}(V_1, V_2))$ and the maps $\Theta_{\alpha\beta} : U_\alpha \cap U_\beta \to \mathrm{GL}(\mathcal{F}(V_1, V_2))$ are smooth. $\square$

6.5. Hom

Let $\xi_1 := (E_1, \pi_1, M, V)$ and $\xi_2 := (E_2, \pi_2, M, V)$ be smooth $\mathbb{F}$-vector bundles. The bundle whose fiber over $p \in M$ is $L(E_{1p}, E_{2p}) = \mathrm{Hom}(E_{1p}, E_{2p})$ is denoted by $\mathrm{Hom}(\xi_1, \xi_2)$ or less precisely, by referring to the total space $\mathrm{Hom}(E_1, E_2)$. Here $\mathrm{Hom}(E_{1p}, E_{2p})$ denotes $\mathbb{F}$-linear maps. If $f : E_1 \to E_2$ is vector bundle homomorphism over M, then we may obtain a section s of $\mathrm{Hom}(E_1, E_2)$ by defining $s : p \mapsto f|_{E_{1p}}$. Conversely, given $s \in \Gamma(\mathrm{Hom}(E_1, E_2))$ we define $f : E_1 \to E_2$ by requiring that
$$f|_{E_{1p}} = s(p).$$
Thus every element of $\mathrm{Hom}(E_1, E_2)$ can be identified with a vector bundle homomorphism over M.

Exercise 6.54. Let $E_1 \to M_1$ and $E_2 \to M_2$ be smooth vector bundles. Show that the set of vector bundle homomorphisms along a smooth map $g : M_1 \to M_2$ is in natural bijection with the sections of the bundle $\mathrm{Hom}(E_1, g^*E_2)$.

Since $\Gamma(E_1)$ and $\Gamma(E_2)$ are $C^\infty(M, \mathbb{F})$ modules, we can look at the $C^\infty(M, \mathbb{F})$ module $\mathrm{Hom}(\Gamma(E_1), \Gamma(E_2))$. Then we have

Proposition 6.55. *Let $E_1 \to M$ and $E_2 \to M$ be smooth $\mathbb{F}$-vector bundles. Then $\Gamma(\mathrm{Hom}(E_1, E_2))$ and $\mathrm{Hom}(\Gamma(E_1), \Gamma(E_2))$ are naturally isomorphic as $C^\infty(M, \mathbb{F})$ modules.*

Proof. To each section $s \in \Gamma(\mathrm{Hom}(E_1, E_2))$ we assign a map $\phi_s : \Gamma(E_1) \to \Gamma(E_2)$ defined by the formula

$$\phi_s(\sigma)(p) = s(p)\sigma(p) \text{ for } \sigma \in \Gamma(E_1).$$

Then we obtain a map $\Phi : \Gamma(\mathrm{Hom}(E_1, E_2)) \to \mathrm{Hom}(\Gamma(E_1), \Gamma(E_2))$ which is defined by $\Phi : s \mapsto \phi_s$. The smoothness of s and σ implies the smoothness of $\phi_s(\sigma)$ and then the smoothness of ϕ_s. The map ϕ_s is clearly in $\mathrm{Hom}(\Gamma(E_1), \Gamma(E_2))$, and it is not difficult to check that Φ is also a module homomorphism.

Now suppose we are given a module homomorphism $\phi : \Gamma(E_1) \to \Gamma(E_2)$. Define $s \in \Gamma(\mathrm{Hom}(E_1, E_2))$ by

$$s(p)(v_p) := \phi(\sigma)(p) \text{ for } v_p \in E_{1p},$$

where σ is any section in $\Gamma(E_1)$ such that $\sigma(p) = v_p$. We need to show that this is well-defined. It suffices to show that if $\sigma(p) = 0$ then $\phi(\sigma)(p) = 0$. Let $(\chi_1, \ldots, \chi_k)$ be a local frame field for E_1 defined over an open set U. Choose $g \in C^\infty(M)$ with support in U and $g(p) = 1$. Define fields $X_i := g\chi_i$ and extend by zero outside of U. Then there exist functions f^i such that

$$g\sigma = \sum_{i=1}^{k} f^i X_i,$$

and since $\sigma(p) = 0$, we must have $f^i(p) = 0$ for all i. Then we have

$$(\phi\sigma)(p) = g(p)(\phi(\sigma))(p) = (g\phi(\sigma))(p)$$
$$= \phi(g\sigma)(p) = \phi\left(\sum_{i=1}^{k} f^i X_i\right)(p)$$
$$= \sum_{i=1}^{k} (f^i \phi(X_i))(p) = \sum_{i=1}^{k} f^i(p)\phi(X_i)(p) = 0.$$

The constructed map is easily checked to be the inverse of the map $\Phi : s \mapsto \phi_s$. $\square$

If $(\sigma_1, \ldots, \sigma_{k_1})$ is a local frame field for $E_1 \to M$ over an open set U, and if $(\phi_1, \ldots, \phi_{k_2})$ a local frame field for $E_2 \to M$ also over U, then we may construct a local frame field for $\mathrm{Hom}(E_1, E_2)$. The frame field is $\{e_i^j\}$ where $e_i^j(\sigma_k) = \delta_{kj}\phi_i$. If $\mathrm{Hom}(E_1, E_2)$ is identified with $E_2 \otimes E_1^*$, then e_i^j is $\phi_i \otimes \sigma^j$ where $(\sigma^1, \ldots, \sigma^{k_1})$ is the dual frame field to $(\sigma_1, \ldots, \sigma_{k_1})$. Suppose that over an open set V, we have frame fields $(\widetilde{\sigma}_1, \ldots, \widetilde{\sigma}_{k_1})$ and $(\widetilde{\phi}_1, \ldots, \widetilde{\phi}_{k_2})$. If $U \cap V$ is nonempty, then

$$(\widetilde{\sigma}_1, \ldots, \widetilde{\sigma}_{k_1}) = (\sigma_1, \ldots, \sigma_{k_1})C,$$
$$(\widetilde{\phi}_1, \ldots, \widetilde{\phi}_{k_2}) = (\phi_1, \ldots, \phi_{k_2})D$$

for smooth matrix-valued functions C and D defined on $U \cap V$. We obtain $\widetilde{e}_i^j = \widetilde{\phi}_i \otimes \widetilde{\sigma}^j = \sum_{r,s}(C^{-1})_s^j e_r^s D_i^r$. If $A \in \Gamma\left(\mathrm{Hom}(E_1, E_2)\right)$ and on $U \cap V$ we have

$$A = \sum \widetilde{A}_j^i \widetilde{e}_i^j = \sum A_j^i e_i^j,$$

then $\widetilde{A}_j^i = \sum_{r,s} \left(D^{-1}\right)_r^i A_s^r C_j^s$ on $U \cap V$.

Of special importance is $\mathrm{End}(E) := \mathrm{Hom}(E, E) \to M$ for a given vector bundle $E \to M$. Here if $(\widetilde{\sigma}_1, \ldots, \widetilde{\sigma}_k) = (\sigma_1, \ldots, \sigma_k)C$ is a change of frame field, then we take $\phi_i = \sigma_i$ and $\widetilde{\phi}_i = \widetilde{\sigma}_i$ so that the above rule specializes to

(6.2) $$\widetilde{A}_j^i = \sum_{r,s} \left(C^{-1}\right)_r^i A_s^r C_j^s \quad \text{on } U \cap V,$$

which is the signature transformation law for sections of $\mathrm{End}(E)$. This bundle is important in the study of covariant derivatives and curvature. Notice how everything is formally similar to operations in linear algebra, but here we are dealing with fields and functions (often only locally defined).

6.6. Algebra Bundles

Let $\mathbb{F}$ be $\mathbb{C}$ or $\mathbb{R}$. Recall that an $\mathbb{F}$-algebra is a vector space V with a bilinear map $V \times V \to V$ giving a product on V. Such a bilinear map is uniquely specified by the associated linear map $V \otimes V \to V$ (see Definition D.17).

Definition 6.56. Let V be an $\mathbb{F}$-algebra. An $\mathbb{F}$-vector bundle $\xi = (E, \pi, M, V)$ is called an $\mathbb{F}$-**algebra bundle** if each fiber E_p has an $\mathbb{F}$-algebra structure in such a way that the associated maps $E_p \otimes E_p \to E_p$ combine to give a vector bundle homomorphism $E \otimes E \to E$ and ξ has a VB-atlas $\{(U_\alpha, \phi_\alpha)\}$ such that for each α, $\Phi_\alpha|_{E_p} : E_p \to V$ is an algebra isomorphism for all $p \in U_\alpha$. Here $\phi_\alpha = (\pi, \Phi_\alpha)$ as usual.

Example 6.57. The endomorphism bundle $\mathrm{End}(E) \to M$ of a vector bundle (E, π, M, V) is the bundle whose fiber at p is $\mathrm{End}(E_p) = \mathrm{Hom}(E_p, E_p)$. The space $\mathrm{End}(E_p)$ has an $\mathbb{F}$-algebra structure given by composition of linear maps.

Example 6.58. Let $\xi = (E, \pi, M, V)$ be an $\mathbb{F}$-vector bundle. The corresponding **general linear Lie algebra bundle** is the bundle whose fiber at p is $L(E_p, E_p)$ but with the product being $(f, g) \mapsto [f, g] := f \circ g - g \circ f$. When given this product, the fiber is denoted $\mathfrak{gl}(E_p)$ and is the Lie algebra of $\mathrm{GL}(E_p)$. The total space of the general linear Lie algebra bundle is then denoted by $\mathfrak{gl}(E)$. This provides an example of a *Lie algebra bundle*.

Recall that when a direct sum has an infinite number of summands, we require that each element have finite support. In other words, $\bigoplus_{i=1}^{\infty} V_i$ is the vector space of all formal sums $\sum_{i=1}^{\infty} v_i$, where $v_i \in V_i$ and all but finitely many of the v_i's are zero.

Example 6.59. The definition of tensor algebra is given as Definition D.46 of Appendix D. Let $\xi = (E, \pi, M, V)$ be an $\mathbb{F}$-vector bundle and consider the bundle $\bigotimes E \to M$ whose fiber at p is the $\mathbb{F}$-tensor algebra $\bigotimes E_p = \mathbb{R} \oplus E_p \oplus (E_p \otimes E_p) \oplus \cdots$. If we consider the disjoint union

$$\bigotimes E = \bigcup_{p \in M} \bigotimes E_p,$$

with the obvious projection $\bigotimes E \to M$, then it seems that we have a bundle with algebra fibers. However, each fiber $\bigotimes E_p$ is infinite-dimensional. On the other hand, we do at least have a nested sequence of vector bundles $\cdots \subset \bigotimes^{\leq k} E \subset \bigotimes^{\leq k+1} E \subset \cdots$, where

$$\bigotimes^{\leq k} E = \mathbb{R} \oplus E \oplus (E \otimes E) \oplus \cdots \oplus \left(\bigotimes^{k} E\right).$$

6.7. Sheaves

Let $E \to M$ be a vector bundle. We have seen that $\Gamma(M, E)$ is a module over the smooth functions $C^\infty(M)$. It is important to realize that having a vector bundle at hand not only provides a module, but a *family* of modules $\{\Gamma(U, E)\}$ parametrized by the open subsets U of M. How are these modules related to each other?

Consider a section $\sigma : M \to E$. Given any open set $U \subset M$, we may always produce the restricted section $\sigma|_U : U \to E$. This gives us a family of sections; one for each open set U. Conversely, suppose that we have a family of sections $\sigma_U : U \to E$ where U varies over the open sets (or just a cover of M). When is it the case that such a family is just the family of restrictions of some section $\sigma : M \to E$? To help with these kinds of questions, and to provide a language that will occasionally be convenient, we will introduce another formalism. This is the formalism of sheaves and presheaves. The formalism of sheaf theory is used for studying the interplay between the local and the global, for example, using sheaf cohomology theory. It is especially

6.7. Sheaves

useful in complex geometry. Sheaf theory also provides a very good framework within which to develop the foundations of supergeometry, which is an extension of differential geometry that incorporates the important notion of "fermionic variables". A deep understanding of sheaf theory is not necessary for what we do here and it would be enough to acquire a basic familiarity with the definitions since we only want the convenience of the language.

Definition 6.60. A **presheaf of abelian groups** (resp. **rings**, etc.) on a manifold (or more generally a topological space) M assigns to each open set an abelian group (resp. ring, etc.) $\mathcal{M}(U)$ and assigns to each nested pair $V \subset U$ of open sets a homomorphism $r_V^U : \mathcal{M}(U) \to \mathcal{M}(V)$ of abelian groups (resp. rings, etc.) such that

(i) $r_W^V \circ r_V^U = r_W^U$ whenever $W \subset V \subset U$;
(ii) $r_V^V = \text{id}_V$ for all open $V \subset M$.

Definition 6.61. Let $\mathcal{M}$ be a presheaf and $\mathcal{R}$ a presheaf of rings over M. If for each open $U \subset M$ we have that $\mathcal{M}(U)$ is a module over the ring $\mathcal{R}(U)$, and if multiplication commutes with restriction, that is, if the following diagram commutes for each nested pair $V \subset U$,

$$\begin{array}{ccc} \mathcal{R}(U) \times \mathcal{M}(U) & \longrightarrow & \mathcal{M}(U) \\ {\scriptstyle r_V^U \times r_V^U} \downarrow & & \downarrow {\scriptstyle r_V^U} \\ \mathcal{R}(V) \times \mathcal{M}(V) & \longrightarrow & \mathcal{M}(V) \end{array}$$

then we say that $\mathcal{M}$ is a **presheaf of modules over** $\mathcal{R}$.

Definition 6.62. Let $\mathcal{M}_1$ and $\mathcal{M}_2$ be presheaves over M. A **presheaf morphism** $h : \mathcal{M}_1 \to \mathcal{M}_2$ over M is a collection of morphisms, $h_U : \mathcal{M}_1(U) \to \mathcal{M}_2(U)$, one for each open set and such that whenever $V \subset U$, the following diagram commutes:

$$\begin{array}{ccc} \mathcal{M}_1(U) & \xrightarrow{h_U} & \mathcal{M}_2(U) \\ {\scriptstyle r_V^U} \downarrow & & \downarrow {\scriptstyle r_V^U} \\ \mathcal{M}_1(V) & \xrightarrow{h_V} & \mathcal{M}_2(V) \end{array}$$

Note that we have used the same notation for the restriction maps of both presheaves.

Definition 6.63. Let $\mathcal{M}$ be a presheaf. A family $\{s_\alpha\}$ with $s_\alpha \in \mathcal{M}(U_\alpha)$ is called **consistent** if $r_{U_\alpha \cap U_\beta}^U s_\alpha = r_{U_\alpha \cap U_\beta}^U s_\beta$ whenever $U_\alpha \cap U_\beta \neq \emptyset$.

Definition 6.64. We will call a presheaf $\mathcal{M}$ a **sheaf** if the following properties hold whenever $U = \bigcup_{U_\alpha \in \mathcal{U}} U_\alpha$ for some collection of open sets $\mathcal{U}$.

(i) If $s_1, s_2 \in \mathcal{M}(U)$ and $r_{U_\alpha}^U s_1 = r_{U_\alpha}^U s_2$ for all $U_\alpha \in \mathcal{U}$, then $s_1 = s_2$.

(ii) Given a *consistent* family $\{s_\alpha : s_\alpha \in \mathcal{M}(U_\alpha)\}$, there exists $s \in \mathcal{M}(U)$ such that $r_{U_\alpha}^U s = s_\alpha$.

(iii) $\mathcal{M}(\emptyset)$ is the trivial group, (resp. ring, etc.).

If we need to indicate the space M involved, we will write $\mathcal{M}_M$ instead of $\mathcal{M}$.

Definition 6.65. A **morphism of sheaves** is a morphism of the underlying presheaf.

The assignment $C^\infty(\cdot) : U \mapsto C^\infty(U)$ is a sheaf of rings. This sheaf will also be denoted by C_M^∞. The best and most important example of a sheaf of modules over $C^\infty(\cdot)$ is the assignment $\Gamma(\cdot, E) : U \mapsto \Gamma(U, E)$ for some vector bundle $E \to M$, where by definition $r_V^U(s) = s|_V$ for $s \in \Gamma(U, E)$. In other words, r_V^U is just the restriction map. Let us denote this (pre)sheaf by $\Gamma_E : U \mapsto \Gamma_E(U) := \Gamma(U, E)$.

Exercise 6.66. For each open set U in a manifold M, let $B(U)$ denote the ring of bounded smooth functions defined on U. Show that $U \mapsto B(U)$ defines a presheaf that is not a sheaf.

Many if not most of the constructions and operations we introduce for sections of vector bundles are really also operations appropriate to the (pre)sheaf category. Naturality with respect to restrictions is one of the features that is often not even mentioned (precisely because it seems obvious). This is the inspiration for a slight twist on our notation.

	Global	Local	Sheaf
Functions on M	$C^\infty(M)$	$C^\infty(U)$	C_M^∞
Vector fields on M	$\mathfrak{X}(M)$	$\mathfrak{X}(U)$	$\mathfrak{X}_M$
Sections of E	$\Gamma(E)$	$\Gamma(U, E)$	Γ_E

where $C_M^\infty : U \mapsto C_M^\infty(U) := C^\infty(U)$, $\mathfrak{X}_M : U \mapsto \mathfrak{X}_M(U) := \mathfrak{X}(U)$ and so on.

Notation 6.67. For example, when we say that $D : C_M^\infty \to C_M^\infty$ is a derivation we mean that D is actually a family of algebra derivations $D_U : C_M^\infty(U) \to C_M^\infty(U)$ indexed by open sets U such that we have naturality with respect to restrictions; i.e., diagrams of the form below for $V \subset U$ commute:

$$\begin{array}{ccc} C_M^\infty(U) & \xrightarrow{D_U} & C_M^\infty(U) \\ {\scriptstyle r_V^U}\downarrow & & \downarrow{\scriptstyle r_V^U} \\ C_M^\infty(V) & \xrightarrow{D_V} & C_M^\infty(V) \end{array}$$

It is easy to see that all of the following examples are sheaves. In each case, the maps r_V^U are just the restriction maps.

Example 6.68 (Sheaf of holomorphic functions). Sheaf theory really shows its strength in complex analysis. This example is one of the most studied. However, we have not studied the notion of a complex manifold, and so this example is for those readers with some exposure to complex manifolds (see the online supplement). Let M be a complex manifold and let $\mathcal{O}_M(U)$ be the algebra of holomorphic functions defined on U. Here too, $\mathcal{O}_M$ is a sheaf of modules over itself. Whereas the sheaf C_M^∞ always has global sections, the same is not true for $\mathcal{O}_M$. The sheaf-theoretic approach to the study of obstructions to the existence of global holomorphic functions has been very successful.

For a bit more on sheaves, see the online supplement [**Lee, Jeff**].

6.8. Principal and Associated Bundles

Let $\pi : E \to M$ be a vector bundle with typical fiber V and for every $p \in M$ let $\mathrm{GL}(V, E_p)$ denote the set of linear isomorphisms from V to E_p. If we choose a fixed basis $(\mathbf{e}_1, \ldots, \mathbf{e}_k)$ for V, then each frame $(u_1, \ldots, u_k)$ at p gives an element $u \in \mathrm{GL}(V, E_p)$ defined by

$$u(v) := \sum v^i u_i,$$

where $v = \sum v^i \mathbf{e}_i$. We identify u with $(u_1, \ldots, u_k)$ and refer to it as a frame. With this identification, notice that if $\sigma_\alpha := \sigma_{\phi_\alpha}$ is the local frame field coming from a VB-chart (U_α, ϕ_α) as described above, then we have

$$\sigma_\alpha(p) = \Phi_\alpha|_{E_p} \text{ for } p \in U_\alpha.$$

Now let

$$F(E) := \bigcup_{p \in M} \mathrm{GL}(V, E_p) \quad \text{(disjoint union)}.$$

It will shortly be clear that $F(E)$ is a smooth manifold and the total space of a fiber bundle. Let $\wp : F(E) \to M$ be the projection map defined by $\wp(u) = p$ for $u \in \mathrm{GL}(V, E_p)$. Observe that $\mathrm{GL}(V)$ acts on the right of the set $F(E)$; the action $F(E) \times \mathrm{GL}(V) \to F(E)$ is given by $r : (u, g) \mapsto ug = u \circ g$. If we pick a fixed basis for V as above, then we may view g as a matrix and an element $u \in \mathrm{GL}(V, E_p)$ as a basis $(u_1, \ldots, u_k)$. In this case, we have

$$ug = \left(\sum u_i g_1^i, \ldots, \sum u_i g_k^i\right).$$

It is easy to see that the orbit of a frame at p is exactly the set $\wp^{-1}(p) = \mathrm{GL}(V, E_p)$ and that the action is free. For each VB-chart (U, ϕ) for E, let σ_ϕ be the associated frame field. Define $f_\phi : U \times \mathrm{GL}(V) \to \wp^{-1}(U)$

by $f_\phi(p,g) = \sigma_\phi(p)g$. It is easy to check that this is a bijection. Let $\widetilde{\phi} : \wp^{-1}(U) \to U \times \mathrm{GL}(V)$ be the *inverse* of this map. We have $\widetilde{\phi} = (\wp, \widetilde{\Phi})$, where $\widetilde{\Phi}$ is uniquely determined by $\widetilde{\phi}$. Starting with a VB-atlas $\{(U_\alpha, \phi_\alpha)\}$ for E, we obtain a family $\{\widetilde{\phi}_\alpha : \wp^{-1}(U_\alpha) \to U_\alpha \times \mathrm{GL}(V)\}$ of trivializations which gives a fiber bundle atlas $\{(U_\alpha, \widetilde{\phi}_\alpha)\}$ for $F(E) \to M$ and simultaneously induces the smooth structure.

Definition 6.69. Let $\pi : E \to M$ be a vector bundle with typical fiber V. The fiber bundle $(F(E), \wp, M, \mathrm{GL}(V))$ constructed above is called the **linear frame bundle** of E and is usually denoted simply by $F(E)$. The frame bundle for the tangent bundle of a manifold M is often denoted by $F(M)$ rather than by $F(TM)$.

Notation 6.70. It would be perhaps more appropriate to refer to the frame bundle of a vector bundle $\xi = (E, \pi, M, \mathrm{V})$ as $F(\xi)$ since the notation $F(E)$ is inconsistent with the notation $F(M)$ above. After all, E is itself a manifold. Despite this we will continue with the dangerous notation.

Notice that any VB-atlas for E induces an atlas on $F(E)$ according to our considerations above. We have

$$\widetilde{\phi}_\alpha \circ \widetilde{\phi}_\beta^{-1}(p,g) = \widetilde{\phi}_\alpha(\sigma_\beta(p)g) = \widetilde{\phi}_\alpha(\Phi_\beta|_{E_p} g) = (p, \Phi_{\alpha\beta}(p)g).$$

Thus the transition functions of $F(E)$ are given by the standard transition functions of E acting by *left* multiplication on $\mathrm{GL}(V)$. The cocycle corresponding to the bundle atlas for $F(E)$ that we constructed from a VB-atlas $\{(U_\alpha, \phi_\alpha)\}$ for the original vector bundle E, is the very cocycle $\Phi_{\alpha\beta}$ deriving from this atlas on E.

We need to make one more observation concerning the right action of $\mathrm{GL}(V)$ on $F(E)$. Take a VB-chart for E, say (U, ϕ), and let us look again at the associated chart $(U, \widetilde{\phi})$ for $F(E)$. First, consider the trivial bundle $\mathrm{pr}_1 : U \times \mathrm{GL}(V) \to U$ and define the obvious *right* action on the total space $\widetilde{\phi}(\wp^{-1}(U)) = U \times \mathrm{GL}(V)$ by $((p, g_1), g) \mapsto (p, g_1 g) := (p, g_1) \cdot g$. Then this action is transitive on the fibers of this trivial bundle. Of course since $\mathrm{GL}(V)$ acts on $F(E)$ and preserves fibers, it also acts by restriction on $\wp^{-1}(U)$.

Proposition 6.71. *The bundle map $\widetilde{\phi} : \wp^{-1}(U) \to U \times \mathrm{GL}(V)$ is equivariant with respect to the right actions described above.*

Proof. We first look at the inverse:

$$\widetilde{\phi}^{-1}(p, g_1)g = \sigma_\phi(p)g_1 g = \widetilde{\phi}^{-1}(p, g_1 g).$$

To see this from the point of view of $\widetilde{\phi}$ rather than its inverse, take $u \in \wp^{-1}(U) \subset F(E)$ and let (p, g_1) be the unique pair such that $u = \widetilde{\phi}^{-1}(p, g_1)$.

6.8. Principal and Associated Bundles

Then
$$\widetilde{\phi}(ug) = \widetilde{\phi}\left(\widetilde{\phi}^{-1}(p, g_1)g\right) = \widetilde{\phi}\widetilde{\phi}^{-1}(p, g_1 g) = (p, g_1 g) = (p, g_1) \cdot g = \widetilde{\phi}(u) \cdot g. \quad \square$$

A section of $F(E)$ over an open set U in M is just a frame field over U. A global frame field is a global section of $F(E)$, and clearly a global section exists if and only if E is trivial.

For a frame bundle $F(E)$, the following things stand out: The typical fiber is the structure group $\mathrm{GL}(V)$, and we constructed an atlas which showed that $F(E)$ has a $\mathrm{GL}(V)$-bundle structure where the action is left multiplication. Furthermore there is a right action of $\mathrm{GL}(V)$ on the total space $F(E)$ which has the fibers as orbits. The charts constructed were of the form $(U, \widetilde{\phi})$, where $\widetilde{\phi}$ is equivariant in the sense that $\widetilde{\phi}(ug) = \widetilde{\phi}(u)g$, where if $\widetilde{\phi}(u) = (p, g_1)$, then $(p, g_1)g := (p, g_1 g)$ by definition. These facts motivate the concept of a principal bundle:

Definition 6.72. Let $\wp : P \to M$ be a smooth fiber bundle with typical fiber a Lie group G. The bundle $(P, \wp, M, G)$ is called a **principal G-bundle** if there is a smooth free right action of G on P such that

(i) The action preserves fibers; $\wp(ug) = \wp(u)$ for all $u \in P$ and $g \in G$.

(ii) For each $p \in M$, there exists a bundle chart (U, ϕ) with $p \in U$ and such that if $\phi = (\wp, \Phi)$, then
$$\Phi(ug) = \Phi(u)g$$
for all $u \in \wp^{-1}(U)$ and $g \in G$.

If the group G is understood, then we may refer to $(P, \wp, M, G)$ simply as a **principal bundle**. Define a right action on $U \times G$ by $(p, g_1)g = (p, g_1 g)$. Then $U \times G \to U$ is a trivial principal bundle. If $\phi = (\wp, \Phi)$, then using this right action on $U \times G$ we have that $\phi(ug) = \phi(u)g$ if and only if $\Phi(ug) = \Phi(u)g$. Charts of the form described in (ii) of the definition are called **principal bundle charts**, and an atlas consisting of principal bundle charts is called a **principal bundle atlas**.

Proposition 6.73. If $(P, \wp, M, G)$ is a principal G-bundle, then the fibers are exactly the orbits of the right G-action.

Proof. The definition makes it clear that each orbit is contained in some fiber. Suppose that u_1 and u_2 are in the same fiber so that $\wp(u_1) = \wp(u_2)$. We wish to find a $g \in G$ such that $u_1 = u_2 g$. Let $g := \Phi(u_2)^{-1}\Phi(u_1)$. Then $\Phi(u_1) = \Phi(u_2)g$ and so
$$\phi(u_1) = (\wp(u_1), \Phi(u_1)) = (\wp(u_2), \Phi(u_2)g)$$
$$= (\wp(u_2 g), \Phi(u_2 g)) = \phi(u_2 g).$$

Since ϕ is bijective, we see that $u_1 = u_2 g$. The conclusion can be expressed by saying that the action is **transitive on fibers**. □

The definition of a principal G-bundle above does not use the notion of a G-atlas or strict equivalence. The definition of principal bundle atlas is given *after* the notion of a principal bundle is already defined. However, once we have singled out this type of atlas, we can find out what the transition functions are and thereby make connection with earlier developments. Notice that if (ϕ_α, U_α) and (ϕ_β, U_β) are overlapping principal bundle charts with $\phi_\alpha = (\wp, \Phi_\alpha)$ and $\phi_\beta = (\wp, \Phi_\beta)$, then

$$\Phi_\alpha(ug)\Phi_\beta(ug)^{-1} = \Phi_\alpha(u)gg^{-1}\Phi_\beta(u)^{-1} = \Phi_\alpha(u)\Phi_\beta(u)^{-1},$$

so that the map $u \mapsto \Phi_\alpha(u)\Phi_\beta(u)^{-1}$ is constant on fibers. This means that there is a smooth function $g_{\alpha\beta} : U_\alpha \cap U_\beta \to G$ such that

(6.3) $$g_{\alpha\beta}(p) = \Phi_\alpha(u)\Phi_\beta(u)^{-1},$$

where u is any element in the fiber at p.

Lemma 6.74. *Let (ϕ_α, U_α) and (ϕ_β, U_β) be overlapping principal bundle charts. For each $p \in U_\alpha \cap U_\beta$,*

$$\Phi_\alpha|_{\wp^{-1}(p)} \circ \left(\Phi_\beta|_{\wp^{-1}(p)} \right)^{-1}(g) = g_{\alpha\beta}(p)g,$$

where the $g_{\alpha\beta}$ are given as above.

Proof. Let $(\Phi_\beta|_{\wp^{-1}(p)})^{-1}(g) = u$. Then $g = \Phi_\beta(u)$ and so $\Phi_\alpha|_{\wp^{-1}(p)} \circ \Phi_\beta|_{\wp^{-1}(p)}(g) = \Phi_\alpha(u)$. On the other hand, $u \in \wp^{-1}(p)$ and so

$$g_{\alpha\beta}(p)g = \Phi_\alpha(u)\Phi_\beta(u)^{-1}g = \Phi_\alpha(u)\Phi_\beta(u)^{-1}\Phi_\beta(u)$$
$$= \Phi_\alpha(u) = \Phi_\alpha|_{\wp^{-1}(p)} \circ \Phi_\beta|_{\wp^{-1}(p)}(g). \quad \square$$

From this lemma we see that the structure group of a principal bundle is G acting on itself by left translation. Conversely if $(P, \wp, M, G)$ is a fiber bundle with a G-atlas with G acting by left translation, then $(P, \wp, M, G)$ is a principal bundle. To see this, we only need to exhibit the free right action. Let $u \in P$ and choose a chart (ϕ_α, U_α) from the G-atlas. Then let $ug := \phi_\alpha^{-1}(p, \Phi_\alpha(u)g)$ where $p = \wp(u)$. We need to show that this is well-defined, so let (ϕ_β, U_β) be another such bundle chart with $p = \wp(u) \in U_\beta \cap U_\alpha$. Then if $u_1 := \phi_\beta^{-1}(p, \Phi_\beta(u)g)$, we have

$$\phi_\alpha(u_1) = \phi_\alpha \phi_\beta^{-1}(p, \Phi_\beta(u)g) = (p, g_{\alpha\beta}(p)\Phi_\beta(u)g) = (p, \Phi_\alpha(u)g)$$

so that $u_1 = \phi_\alpha^{-1}(p, \Phi_\alpha(u)g) = ug$. It is easy to see that this action is free. Furthermore, since

$$\phi_\alpha^{-1}(p, \Phi_\alpha(ug)) = \phi_\alpha^{-1} \circ \phi_\alpha(ug) = ug := \phi_\alpha^{-1}(p, \Phi_\alpha(u)g),$$

6.8. Principal and Associated Bundles

we see that $\Phi_\alpha(ug) = \Phi_\alpha(u)g$ as required by the definition of principal bundle. Obviously the frame bundles of vector bundles are examples of principal bundles.

Definition 6.75. Suppose that $\pi : E \to M$ is a rank k vector bundle with typical fiber V. Let G be a Lie subgroup of GL(V) and suppose that $\pi : E \to M$ has a G-bundle structure where G acts on V by the standard action as a subgroup of GL(V). Thus we have a reduction of the structure group to G. Let $\mathcal{A}_G = \{(U_\alpha, \phi_\alpha)\}$ be the maximal G-atlas that defines this structure. For each $p \in M$, let

$$F_G(E_p) := \{u \in \mathrm{GL}(\mathrm{V}, E_p) : u = \phi^{-1}(p, \cdot) \text{ for some } (U, \phi) \in \mathcal{A}_G\}.$$

Elements of $F_G(E_p)$ are called G-**frames at** p (associated to $\mathcal{A}_G$).

The group G acts on the right on $F_G(E_p)$ by $(u, g) \mapsto u \circ g$, and letting $F_G(E) := \bigcup_{p \in M} F_G(E_p)$ we see that G acts on the right on $F_G(E)$. It is not hard to show that $F_G(E)$ has the structure of a principal G-bundle with this action. It is a subprincipal bundle of the frame bundle.

Definition 6.76. The bundle $F_G(E)$ is called the **bundle of** G-**frames** associated to the G-bundle structure on $\pi : E \to M$.

Actually, G acts on the whole frame bundle of E by the same formula and on the set of all frames $F(E_p)$ at a fixed p. Then, $F_G(E_p)$ is an orbit of this action on $F(E)$. In fact, one may simply *define* a G-bundle structure on a vector bundle to be a subbundle of the frame bundle such that G acts transitively on each fiber. It is not hard to see that the notion of a bundle of G-frames gives us another way to describe the notion of reduction of the structure group. It is also important to notice that since there may be more than one G-bundle structure on E, the notation $F_G(E)$ is ambiguous. If one must deal with two different G-bundle structures on E, then one must resort to another notation such as $F_G^1(E)$ and $F_G^2(E)$.

Exercise 6.77. Show that a choice of metric on a vector bundle $E \to M$ is equivalent to a specification of a subbundle of the frame bundle such that $O(k)$ acts transitively on each fiber of the subbundle.

As another example of a principal bundle we also have the Hopf bundles described in the next example and the following exercise.

Example 6.78 (Hopf bundles). Recall the Hopf map $\wp : S^{2n-1} \to \mathbb{C}P^{n-1}$ defined in Example 5.120. The quadruple $(S^{2n-1}, \wp, \mathbb{C}P^{n-1}, \mathrm{U}(1))$ is a principal fiber bundle. We have already defined the left action of U(1) on S^{2n-1} in Example 5.120. Since U(1) is abelian, we may take this action to also be a right action. Recall that in this context, we have $S^{2n-1} = \{\xi \in \mathbb{C}^n : |\xi| = 1\}$, where for $\xi = (z^1, \ldots, z^n)$, we have $|\xi|^2 = \sum \bar{z}^i z^i$. The right action of $\mathrm{U}(1) = S^1$ on S^{2n-1} is $(\xi, g) \mapsto \xi g = (z^1 g, \ldots, z^n g)$. It is clear that $\wp(\xi g) = \wp(\xi)$. To finish the verification that $(S^{2n-1}, \wp, \mathbb{C}P^{n-1}, \mathrm{U}(1))$ is a principal bundle, we exhibit appropriate principal bundle charts. For each $k = 1, 2, \ldots, n$, we let $U_k := \{[z^1, \ldots, z^n] \in \mathbb{C}P^{n-1} : z^k \neq 0\}$ and we let $\psi_k : \wp^{-1}(U_k) \to U_k \times \mathrm{U}(1)$ be defined by $\psi_k := (\wp, \Psi_k)$, where

$$\Psi_k(\xi) = \Psi_k(z^1, \ldots, z^n) := |z^k|^{-1} z^k.$$

We leave it to the reader to show that $\psi_k := (\wp, \Psi_k)$ is a diffeomorphism. For $g \in \mathrm{U}(1)$, we have

$$\Psi_k(\xi g) = |z^k g|^{-1}(z^k g) = |z^k|^{-1}(z^k g) = (|z^k|^{-1} z^k) g = \Psi_k(\xi) g,$$

as desired. Let us compute the transition cocycle $\{g_{ij}\}$. For $p = [\xi] \in U_i \cap U_j$, we have

$$g_{ij}(p) = \Psi_i(\xi) \Psi_j(\xi)^{-1} = \left|z^i\right|^{-1} z^i \left(z^j\right)^{-1} \left|z^j\right| \in \mathrm{U}(1).$$

Exercise 6.79. By analogy with the above example, show that we have principal bundles $(S^{n-1}, \wp, \mathbb{R}P^{n-1}, \mathbb{Z}_2)$ and $(S^{4n-1}, \wp, \mathbb{H}P^{n-1}, \mathrm{U}(1, \mathbb{H}))$. Show that in the quaternionic case $g_{ij}(p) = \left|q^i\right|^{-1} q^i \left(q^j\right)^{-1} \left|q^j\right|$ for $p = [q^1, \ldots, q^n]$ and that the order matters in this case.

If (U, ϕ) is a principal bundle chart for a principal bundle $(P, \wp, M, G)$, then for each fixed $g \in G$, the map $\sigma_{\phi, g} : p \mapsto \phi^{-1}(p, g)$ is a smooth local section. The canonical choice for the fixed element is the identity e. Thus to each principal bundle chart (U, ϕ) we associate the local section $\sigma_\phi : p \mapsto \phi^{-1}(p, e)$. Conversely if $\sigma : U \to P$ is a smooth local section, we let $f_\sigma : U \times G \to \wp^{-1}(U)$ be defined by $f_\sigma(p, g) = \sigma(p) g$. The proof of the next proposition shows that f_σ is a diffeomorphism and its inverse $\phi : \wp^{-1}(U) \to U \times G$ defines a principal bundle chart. This is a principle worth emphasizing: *Local sections of a principal bundle give rise to associated principal bundle charts, and conversely as described above.*

Proposition 6.80. *If $\wp : P \to M$ is a surjective submersion and a Lie group G acts freely on P so that for each $p \in M$ the orbit of p is exactly $\wp^{-1}(p)$, then $(P, \wp, M, G)$ is a principal bundle.*

Proof. Let us assume (without loss of generality) that the action is a right action since it can always be converted into such by group inversion if needed. We use Proposition 3.25: For each point $p \in M$, there is a local section $\sigma : U \to P$ on some neighborhood U containing p. Consider the map $f_\sigma : U \times G \to \wp^{-1}(U)$ given by $f_\sigma(p, g) = \sigma(p) g$. One can check that this map is injective and has an invertible tangent map at each point of U. Now

6.8. Principal and Associated Bundles

let $\phi := f_\sigma^{-1}$. Then we have $\phi = (\wp, \Phi)$ for a uniquely determined smooth map $\Phi : U \to G$. If $p = \wp(u)$, we have $\phi(ug) = (p, \Phi(ug))$ and so

$$ug = \phi^{-1}(p, \Phi(ug)),$$

while

$$\begin{aligned}\phi^{-1}(p, \Phi(u)g) &= f_\sigma(p, \Phi(u)g) \\ &= \sigma(p)(\Phi(u)g) = (\sigma(p)\Phi(u))g \\ &= f_\sigma(p, \Phi(u))g = \phi^{-1}(p, \Phi(u))g \\ &= ug = \phi^{-1}(p, \Phi(ug)).\end{aligned}$$

Since ϕ^{-1} is a bijection, we have $\Phi(ug) = \Phi(u)g$. Thus the section σ gives rise to a principal bundle chart (U, ϕ), where $\phi = (\pi, \Phi)$. $\square$

Combining this with our results on proper free actions from Chapter 5, we obtain the following corollary:

Corollary 6.81. *If a Lie group G acts properly and freely on M (on the right), then $(M, \pi, M/G, G)$ is a principal bundle. In particular, if H is a closed subgroup of a Lie group G, then $(G, \pi, G/H, H)$ is a principal bundle (with structure group H).*

Definition 6.82. Let $(P_1, \wp_1, M_1, G)$ and $(P_2, \wp_2, M_2, G)$ be two principal G-bundles. A (type II) bundle morphism $\widetilde{f} : P_1 \to P_2$ along a smooth map $f : M_1 \to M_2$ is called a **principal G-bundle morphism** (along f) if

$$\widetilde{f}(u \cdot g) = \widetilde{f}(u) \cdot g$$

for all $g \in G$ and $u \in P$. If $M_1 = M_2$ and $f = \mathrm{id}_M$, then we say that $\widetilde{f}$ is a **principal G-bundle morphism over M**.

Exercise 6.83. Show that if $(P_1, \wp_1, M_1, G)$ and $(P_2, \wp_2, M_2, G)$ are principal G-bundles and $\widetilde{f} : P_1 \to P_2$ is a principal G-bundle morphism along a diffeomorphism f, then $\widetilde{f}$ is a diffeomorphism.

If $\widetilde{f}$ is a principal bundle morphism over M, then it is a diffeomorphism and hence a bundle equivalence (or bundle isomorphism over M) with the property $\widetilde{f}(u \cdot g) = \widetilde{f}(u) \cdot g$ for all $g \in G$ and $u \in P$. In this case, we call $\widetilde{f}$ a **principal G-bundle equivalence** and the two bundles are **equivalent** principal G-bundles over M. A principal G-bundle equivalence from a principal bundle to itself is called a **principal bundle automorphism** or also a **(global) gauge transformation**.

The classification problem for principal bundles (in the topological category) is regrettably beyond the scope of this volume and can be found in [**Hus**]. We can only offer the following comments: In the topological category, the notion of a Lie group is replaced by that of a topological group, but the reader may continue to think of Lie groups. For every topological group G, there is a principal bundle $\xi(G) = ((E(G), \wp_\infty, B(G), G)$ called a **universal bundle** with the property that all the homotopy groups of $E(G)$ are trivial. There is then a classification theorem that states that the equivalence classes of principal bundles with a fixed sufficiently nice[1] base space M are in one-to-one correspondence with the set of homotopy classes of maps from M to $B(G)$. The correspondence is given by assigning to the homotopy class $[f]$, the pull-back principal bundle $f^*\xi(G)$.

The notion of a principal bundle morphism over M can be generalized to the situation where we have two groups in play.

Definition 6.84. Let $(P_1, \wp_1, M, G_1)$ and $(P_2, \wp_2, M, G_2)$ be principal bundles and let $h: G_1 \to G_2$ be a Lie group homomorphism. A bundle morphism over M is called a principal bundle homomorphism with respect to h if

$$\widetilde{f}(u \cdot g) = \widetilde{f}(u) \cdot h(g).$$

If $G_1 \subset G_2$ and the homomorphism h is the inclusion, then we call $\widetilde{f}$ a **reduction** of $(P_2, \wp_2, M, G_2)$ to $(P_1, \wp_1, M, G_1)$.

In the most common case of a reduction, P_1 is a submanifold of P_2 and $\widetilde{f}$ is the inclusion $P_1 \hookrightarrow P_2$. For example, this is the case when one chooses a metric on a vector bundle and thereby obtains a bundle of orthonormal frames $F_{O(n)}(E)$. The inclusion $F_{O(n)}(E) \hookrightarrow F(E)$ is then a reduction, and we just say that $F_{O(n)}(E)$ is a reduction of the frame bundle $F(E)$.

We have seen that a principal G-bundle atlas $\{(U_\alpha, \phi_\alpha)\}$ is associated to a cocycle $\{g_{\alpha\beta}\}$. From this cocycle and the left action of G on itself we may construct a bundle which has $\{g_{\alpha\beta}\}$ as a transition cocycle. In fact, recall that in the construction we formed the total space by putting an equivalence relation on the set $\Sigma := \bigcup_\alpha \{\alpha\} \times U_\alpha \times G$, where $(\alpha, p, g) \in \{\alpha\} \times U_\alpha \times G$ is equivalent to $(\beta, p', g') \in \{\beta\} \times U_\beta \times G$ if and only if $p = p'$ and $g' = g_{\beta\alpha}(p) \cdot g$. If we define a right action on the total space of the constructed bundle by $[\alpha, p, g_1] \cdot g = [\alpha, p, g_1 g]$, then this is well-defined, smooth, and makes the constructed bundle a principal G-bundle equivalent to the original principal G-bundle.

Exercise 6.85. Prove the last assertion above.

Thus we see that G-cocycles on a smooth manifold M give rise to principal G-bundles and conversely. If we start with two G-cocycles on M, then we may ask whether the principal G-bundles constructed from these cocycles are equivalent or not. First notice that the constructed bundles will have principal bundle atlases with the respective original transition cocycles. Thus we are led to the following related question: What conditions on

---

[1] M should be a CW-complex.

6.8. Principal and Associated Bundles

the transition cocycles arising from principal bundle atlases on two principal G-bundles will ensure that the bundles are equivalent principal G-bundles? By restricting the trivializing maps to open sets of a common refinement, we obtain new atlases and so we may as well assume from the start that the respective principal bundle atlases are defined on the same cover of M.

Theorem 6.86. *Let $(P_1, \wp_1, M, G)$ and $(P_2, \wp_2, M, G)$ be principal G-bundles with principal bundle atlases $\{(\phi_\alpha, U_\alpha)\}$ and $\{(\phi'_\alpha, U_\alpha)\}$ respectively. Then $(P_1, \wp_1, M, G)$ is equivalent to $(P_2, \wp_2, M, G)$ if and only if there exists a family of (smooth) maps $\tau_\alpha : U_\alpha \to G$ such that $g'_{\alpha\beta}(p) = (\tau_\alpha(p))^{-1} g_{\alpha\beta}(p) \tau_\beta(p)$ for all $p \in U_\alpha \cap U_\beta$ and for all nonempty intersections $U_\alpha \cap U_\beta$. (Here $\{g_{\alpha\beta}\}$ is the cocycle associated to $\{(\phi_\alpha, U_\alpha)\}$ and $\{g'_{\alpha\beta}\}$ is the cocycle associated to $\{(\phi'_\alpha, U_\alpha)\}$.)*

Sketch of proof. First suppose that P_1 and P_2 are equivalent principal G-bundles and let $\widetilde{f} : P_1 \to P_2$ be an equivalence. Let $p \in U_\alpha$ and choose some $u \in \wp_1^{-1}(p)$, so $\widetilde{f}(u) \in \wp_2^{-1}(p)$. Write $\phi_\alpha = (\wp_1, \Phi_\alpha)$ and $\phi'_\alpha = (\wp_2, \Phi'_\alpha)$. One can easily show that $\Phi_\alpha(u)(\Phi'_\alpha(\widetilde{f}(u)))^{-1}$ is an element of G that is independent of the choice of $u \in \wp_1^{-1}(p)$. For each α, define $\tau_\alpha : U_\alpha \to G$ by

$$\tau_\alpha(p) := \Phi_\alpha(u)(\Phi'_\alpha(\widetilde{f}(u)))^{-1},$$

where $u \in \wp_1^{-1}(p)$. Suppose that $p \in U_\alpha \cap U_\beta$. Then we have $(\tau_\alpha(p))^{-1} = \Phi'_\alpha(\widetilde{f}(u))(\Phi_\alpha(u))^{-1}$. Using the definitions of $g_{\alpha\beta}$ and $g'_{\alpha\beta}$ (see equation (6.3)), we immediately have

$$g'_{\alpha\beta}(p) = (\tau_\alpha(p))^{-1} g_{\alpha\beta}(p) \tau_\beta(p).$$

Conversely, given the maps $\tau_\alpha : U_\alpha \to G$ satisfying

$$g'_{\alpha\beta}(p) = (\tau_\alpha(p))^{-1} g_{\alpha\beta}(p) \tau_\beta(p),$$

we define, for each α, a map $f_\alpha : \wp_1^{-1}(U_\alpha) \to \wp_2^{-1}(U_\alpha)$ by

$$f_\alpha(u) := (\phi'_\alpha)^{-1} \left(p, (\tau_\alpha(p))^{-1} \Phi_\alpha(u) \right).$$

Check that $f_\alpha(u) = f_\beta(u)$ when $\wp_1(u) \in U_\alpha \cap U_\beta$ so that there is a well-defined map $\widetilde{f} : P_1 \to P_2$ such that $f_\alpha(u) = \widetilde{f}(u)$ whenever $\wp_1(u) \in U_\alpha$. Finally, check that $\widetilde{f}(u \cdot g) = \widetilde{f}(u) \cdot g$. □

Let $\wp : P \to M$ be a principal G-bundle and suppose that we are given a smooth left action $\lambda : G \times F \to F$ on some smooth manifold F. Define a right action of G on $P \times F$ according to

$$(u, y) \cdot g := (ug, g^{-1}y) = (ug, \lambda(g^{-1}, y)).$$

Denote the orbit space of this action by $P \times_\lambda F$ (or $P \times_G F$) and let $\widetilde{\wp}$ denote the quotient map. Also denote the equivalence class of (u, y) by $[u, y]$ so

$\widetilde{\wp}(u,y) = [u,y]$. One may check that there is a unique map $\pi : P \times_\lambda F \to M$ such that $\pi([u,y]) = \wp(u)$, and so we have a commutative diagram:

$$\begin{array}{ccc} P \times F & \xrightarrow{\mathrm{pr}_1} & P \\ \downarrow \widetilde{\wp} & & \downarrow \wp \\ P \times_\lambda F & \xrightarrow{\pi} & M \end{array}$$

Next we show that $(P \times_\lambda F, \pi, M, F)$ is a fiber bundle (a (G, λ)-bundle). It is said to be **associated** to the principal bundle P. Bundles constructed in this way are called **associated bundles**. More precisely, if the action λ is *not* effective, we should say that $P \times_\lambda F$ is **weakly associated** to P. This is what lies behind our previously introduced notion of "ineffective (G, λ)-bundle".

Theorem 6.87. *Referring to the above diagram and notations, $P \times_\lambda F$ is a smooth manifold and the following hold:*

(i) *$(P \times_\lambda F, \pi, M, F)$ is a fiber bundle, and for every principal bundle atlas $\{(U_\alpha, \phi_\alpha)\}$, there is a corresponding bundle atlas $\{(U_\alpha, \widetilde{\phi}_\alpha)\}$ for $P \times_\lambda F$ such that*

$$\widetilde{\phi}_\alpha \circ \widetilde{\phi}_\beta^{-1}(p, y) = (p, \lambda(g_{\alpha\beta}(p), y)) \text{ if } p \in U_\alpha \cap U_\beta \text{ and } y \in F,$$

where the $g_{\alpha\beta}$ are defined by equation (6.3).

(ii) *$(P \times F, \widetilde{\wp}, P \times_\lambda F, G)$ is a principal bundle with the right action given by*

$$(u, y) \cdot g := (ug, g^{-1}y).$$

(iii) *$P \times F \xrightarrow{\mathrm{pr}_1} P$ is a principal bundle morphism along π.*

Proof. Let $\{(U_\alpha, \phi_\alpha)\}$ be a principal bundle atlas for $\wp : P \to M$. Note that $\widetilde{\wp}\left(\wp^{-1}(U_\alpha) \times F\right) = \pi^{-1}(U_\alpha)$. For each α, define $\widetilde{\Phi}_\alpha : \pi^{-1}(U_\alpha) \to F$ by requiring that $\widetilde{\Phi}_\alpha \circ \widetilde{\wp}(u, y) = \Phi_\alpha(u) \cdot y$ for all $(u, y) \in \wp^{-1}(U_\alpha) \times F$ and then let $\widetilde{\phi}_\alpha := (\pi, \widetilde{\Phi}_\alpha)$ on $\pi^{-1}(U_\alpha)$. We want to show that $\widetilde{\phi}_\alpha$ is bijective by defining an inverse for $\widetilde{\phi}_\alpha$. For every $p \in U_\alpha$, let $\sigma_\alpha(p) := \phi_\alpha^{-1}(p, e)$, where e is the identity element in G. Then we have

$$\sigma_\alpha(p) \cdot \Phi_\alpha(u) = \phi_\alpha^{-1}(p, e) \cdot \Phi_\alpha(u) = \phi_\alpha^{-1}(p, \Phi_\alpha(u)) = u.$$

Define $\eta_\alpha : U_\alpha \times F \to \pi^{-1}(U_\alpha)$ by $\eta_\alpha(p, y) := \widetilde{\wp}(\sigma_\alpha(p), y)$. We have

$$\eta_\alpha \circ \widetilde{\phi}_\alpha(\widetilde{\wp}(u, y)) = \eta_\alpha(p, \Phi_\alpha(u) \cdot y) = \widetilde{\wp}(\sigma_\alpha(p), \Phi_\alpha(u) \cdot y)$$
$$= \widetilde{\wp}(\sigma_\alpha(p) \cdot \Phi_\alpha(u), y) = \widetilde{\wp}(u, y).$$

Thus η_α is a right inverse for $\widetilde{\phi}_\alpha$ and so $\widetilde{\phi}_\alpha$ is injective. It is easily checked that η_α is also a left inverse for $\widetilde{\phi}_\alpha$. To see this first note that $(p, \Phi_\alpha(\sigma_\alpha(p))) =$

6.8. Principal and Associated Bundles

$\phi_\alpha(\sigma_a(p)) = (p, e)$ so $\Phi_\alpha(\sigma_\alpha(p)) = e$. Thus we have

$$\widetilde{\phi}_\alpha \circ \eta_\alpha(p, y) = \widetilde{\phi}_\alpha \left(\widetilde{\wp}(\sigma_\alpha(p)), y) \right) = (p, \widetilde{\Phi}_\alpha(\widetilde{\wp}(\sigma_\alpha(p)), y)))$$
$$= (p, \Phi_\alpha(\sigma_\alpha(p)) \cdot y) = (p, y).$$

Thus $\widetilde{\phi}_\alpha$ is a bijection. Next we check the overlaps. We use Lemma 6.74;

$$\widetilde{\phi}_\alpha \circ \widetilde{\phi}_\beta^{-1}(p, y) = \widetilde{\phi}_\alpha \circ \eta_\beta(p, y) = \widetilde{\phi}_\alpha \left(\widetilde{\wp}(\sigma_\beta(p)), y \right)$$
$$= (p, \Phi_\alpha(\sigma_\beta(p)) \cdot y) = (p, \Phi_\alpha(\phi_\beta^{-1}(p, e)) \cdot y)$$
$$= (p, \Phi_\alpha|_p \circ \Phi_\beta|_p^{-1}(e)) \cdot y)$$
$$= (p, g_{\alpha\beta}(p) \cdot e \cdot y) = (p, g_{\alpha\beta}(p)y).$$

This shows that the transitions mappings have the stated form and that the overlap maps $\widetilde{\phi}_\alpha \circ \widetilde{\phi}_\beta^{-1}$ are smooth. The family $\{(U_\alpha, \phi_\alpha)\}$ provides the induced smooth structure and is also a bundle atlas. Since $\phi_\alpha \circ \widetilde{\wp}(u, p) = (\pi, \widetilde{\Phi}_\alpha) \circ \widetilde{\wp}(u, p) = (\wp(u), \Phi_\alpha(u)y)$ in the domain of every bundle chart (U_α, ϕ_α), it follows that $\widetilde{\wp}$ is smooth.

We leave it to the reader to verify that $(P \times F, \widetilde{\wp}, P \times_\lambda F, G)$ is a principal G-bundle. Notice that while the map $\mathrm{pr}_1 : P \times F \to P$ is clearly a bundle map along π, we also have

$$\mathrm{pr}_1\left((u, y) \cdot g\right) = \mathrm{pr}_1\left((u \cdot g, g^{-1}y)\right) = u \cdot g = \mathrm{pr}_1(u, y) \cdot g,$$

and so pr_1 is in fact a principal bundle morphism. $\square$

Clearly what we have is another way of looking at bundle construction. The principal bundle takes the place of the cocycle of transition maps.

Exercise 6.88. Construct a principal $\mathbb{Z}_2$-bundle P and left actions λ_1 and λ_2 of $\mathbb{Z}_2$ on S^1 and $\mathbb{R}$ respectively, such that $P \times_{\lambda_1} S^1$ is the twisted torus and $P \times_{\lambda_2} \mathbb{R}$ is the Mobius band line bundle.

We have seen that given a principal G-bundle, one may construct various fiber bundles with G-bundle structures. Let us look at the converse situation. Suppose that (E, π, M, F) is a fiber bundle. Suppose that this bundle has a (G, λ)-atlas $\{(U_\alpha, \phi_\alpha)\}$ with associated G-valued cocycle of transition functions $\{g_{\alpha\beta}\}$. Using Theorem 6.15, one may construct a bundle with typical fiber G by using left translation as the action. The resulting bundle is then a principal bundle $(P, \wp, M, G)$, and it turns out that $P \times_\lambda F$ is equivalent to the original bundle E.

If (E, π, M, V) is a vector bundle and we use the standard $\mathrm{GL}(\mathrm{V})$-cocycle $\{\Phi_{\alpha\beta}\}$ associated to a VB-atlas, then the principal bundle obtained by the above construction is (equivalent to) the linear frame bundle $F(E)$. Letting $\mathrm{GL}(\mathrm{V})$ act on V according to the standard action we have $F(E) \times_{\mathrm{GL(V)}} \mathrm{V}$, which is equivalent to the original bundle (E, π, M, V). More generally, if

$\lambda : G \to \mathrm{GL}(\mathrm{V})$ is a Lie group representation, then by treating λ as a linear action we can form $P \times_\lambda \mathrm{V}$.

Proposition 6.89. *Let P be a principal G-bundle and let $\lambda : G \to \mathrm{GL}(\mathrm{V})$ be a representation. Then $P \times_\lambda \mathrm{V}$ has a natural vector bundle structure with typical fiber V.*

Proof. This follows from Theorem 6.87, but we can argue more directly. Let us denote the total space of $P \times_\lambda \mathrm{V}$ by B and let B_p be the fiber over some point $p \in M$. Then for each $u \in P_p$ there is a map $\psi_u := [u,\cdot] : \mathrm{V} \to B_p$ given by $v \mapsto [u,v]$. We compare ψ_u with ψ_{ug} for $g \in G$ and $u \in P_p$. Since $[ug,v] = [u,\lambda(g)v]$ for all $v \in \mathrm{V}$, the following diagram commutes:

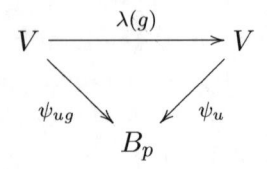

$$\begin{array}{ccc} \mathrm{V} & \xrightarrow{\lambda(g)} & \mathrm{V} \\ & \searrow\psi_{ug} \quad \psi_u \swarrow & \\ & B_p & \end{array}$$

From this it follows that ψ_u transfers the linear structure of V to B_p independently of the choice of $u \in P_p$. We leave it to the reader to show that the local trivializations of $P \times_\lambda \mathrm{V}$ constructed as in the proof of Theorem 6.87 are linear on each fiber. $\square$

Example 6.90. Let M be an n-manifold and let $F(M)$ be the frame bundle of M. Then, if λ_0 is the standard action of $\mathrm{GL}(n,\mathbb{R})$ on $\mathbb{R}^n$, we have the following vector bundle isomorphisms:

$$F(M) \times_{\lambda_0} \mathbb{R}^n \cong TM,$$
$$F(M) \times_{\lambda_0^*} \mathbb{R}^n \cong T^*M,$$
$$F(M) \times_{\lambda_0 \otimes \lambda_0^*} \mathbb{R}^n \cong TM \otimes T^*M.$$

If $E = P \times_\lambda \mathrm{V}$ is an associated vector bundle for λ a representation, then we can map P into the frame bundle of E. Indeed, the map is just $\psi : u \mapsto \psi(u) = \psi_u$, where $\psi_u := [u,\cdot]$ as above. Furthermore, $\psi(ug) = \psi(u) \circ \lambda(g)$ and so we have a principal bundle morphism with respect to the homomorphism λ:

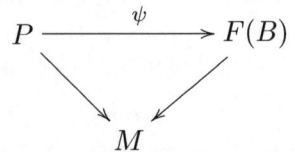

$$\begin{array}{ccc} P & \xrightarrow{\psi} & F(B) \\ & \searrow \quad \swarrow & \\ & M & \end{array}$$

The map $\psi : P \to F(E)$ is only injective if the action λ is effective.

Based on what we have seen above we can say that the theory of principal bundles and associated bundles is an alternative and "invariant" approach to G-bundles. By "invariant" we mean that the foundations can be laid out without recourse to strict equivalence classes of G-atlases or the use of cocycles (of course, these notions can be brought in as convenient). According to this approach, the central notion is the principal bundle, and one recovers the other G-bundles of interest as associated bundles. Developing the theory in this way has the advantage that much can be accomplished without the direct need of bundle atlases. It is a more "intrinsic" approach. This approach seems to have originated with Ehresmann and is the approach followed by [**Hus**].

Problems

(1) Show that $S^n \times \mathbb{R}$ and $S^n \times S^1$ are parallelizable.

(2) Let $X := [0,1] \times \mathbb{R}^n$. Fix a linear isomorphism $L : \mathbb{R}^n \to \mathbb{R}^n$ and consider the quotient space $E = X/\sim$, where the equivalence relation is given by $(0,v) \sim (1,Lv)$. Show that E is the total space of a smooth vector bundle over the circle S^1.

(3) Exhibit the vector bundle charts for the pull-back bundle construction of Definition 6.18.

(4) Let $\xi = (E, \pi, M, F)$ be a G-bundle. Let $g_{\alpha\beta}$ be cocycles associated to a G-altas $\{(U_\alpha, \phi_\alpha)\}$ for ξ. Show that ξ is G-equivalent to a product bundle if and only if there exist functions $\lambda_\alpha : U_\alpha \to G$ such that
$$g_{\beta\alpha}(x) = \lambda_\beta(x)\lambda_\alpha^{-1}(x) \quad \text{for all } x \in U_\alpha \cap U_\beta$$
and all α, β.

(5) Show that the twisted torus of Example 6.17 is trivial as a fiber bundle but not trivial as a $\mathbb{Z}_2$-bundle. (Use Problem 4.)

(6) Show that the space of sections of a vector bundle over a compact base is a finitely generated module. Show that if the bundle is trivial, then the space of sections is a finitely generated free module.

(7) Let $E_1 \to M_1$ and $E_2 \to M_2$ be smooth vector bundles. Show that if $F : E_1 \to E_2$ is a vector bundle homomorphism along a map $f : M_1 \to M_2$ such that F is an isomorphism of fibers, then $E_1 \to M_1$ is isomorphic to the pull-back bundle $f^*E_2 \to M_1$.

(8) Suppose that $\xi = (E, \pi, M)$ is a vector bundle with a positive definite metric. Show that the metric induces a vector bundle isomorphism $E \cong E^*$.

(9) Show that the tangent bundle of the real projective plane is a vector bundle isomorphic to $\text{Hom}(\mathbb{L}(\mathbb{R}P^n), \mathbb{L}(\mathbb{R}P^n)^\perp)$, where $\mathbb{L}(\mathbb{R}P^n) \to \mathbb{R}P^n$

is the tautological line bundle and $\mathbb{L}(\mathbb{R}P^n)^\perp \to \mathbb{R}P^n$ is the rank n vector bundle whose fiber at $l \in \mathbb{R}P^n$ is $\{(l,v) \in \mathbb{R}P^n \times \mathbb{R}^{n+1} : v \perp l\}$.

(10) Recall Example 6.35. Show that the Whitney sum bundle $E_1 \oplus E_2$ is naturally isomorphic to the pull-back $\triangle^* \pi_{E_1 \times E_2} : \triangle^*(E_1 \times E_2) \to M$.

(11) (a) Let $\pi : E \to M$ be an $\mathbb{F}$-vector bundle. We wish to show that $T\pi : TE \to TM$ is naturally a vector bundle. Consider the maps $\alpha : E \oplus E \to E$ and $\mu_s : E \to E$ for each $s \in \mathbb{F}$ given by

$$\alpha(v_p, w_p) := v_p + w_p \text{ for } v_p, w_p \in E_p$$
$$\mu_s(e_p) := se_p \text{ for } e_p \in E_p$$

Show that we may identify $T(E \oplus E)$ with the submanifold of $TE \times TE$ given by

$$\{(v, w) \in TE \times TE : T\pi \cdot v = T\pi \cdot w\}$$

Now suppose that for $v, w \in TE$ with $T\pi \cdot v = T\pi \cdot w$ we define $v \boxplus w := T\alpha \cdot (v, w)$ and for $s \in \mathbb{F}$ and $v \in TE$ we define $s \odot v := T\mu_s \cdot v$. Show that with these definitions of addition and scalar multiplication, $T\pi : TE \to TM$ is indeed an $\mathbb{F}$-vector bundle.

(b) Let E be as above but assume for simplicity that $\mathbb{F} = \mathbb{R}$. Let $x^1, \ldots, x^n$ be coordinates on $U \subset M$. Suppose that $e_1, \ldots, e_k$ is a frame field over U. Let $\xi^1, \ldots, \xi^n$ be defined on $E|_U$ by $y = \sum \xi^i(y) e^i(\pi(y))$ for any $y \in E$. Then, identifying x^i with $x^i \circ \pi$, the functions $x^1, \ldots, x^n, \xi^1, \ldots, \xi^n$ are a coordinate system for E defined on $E|_U$ and such that the $\frac{\partial}{\partial \xi^i}$ are in the kernel of $T\pi$. Now if $v, w \in TE$ are such that $T\pi \cdot v = T\pi \cdot w$, then we may express v and w as

$$v = \sum_i a^i \left.\frac{\partial}{\partial x^i}\right|_y + \sum_\alpha b^\alpha \left.\frac{\partial}{\partial \xi^\alpha}\right|_y$$

and

$$w = \sum_i a^i \left.\frac{\partial}{\partial x^i}\right|_{\widetilde{y}} + \sum_\alpha \widetilde{b}^\alpha \left.\frac{\partial}{\partial \xi^\alpha}\right|_{\widetilde{y}}.$$

Here y and $\widetilde{y}$ are the base points of v and w, and the fact that the a's are the same for both v and w is a result of the condition $T\pi \cdot v = T\pi \cdot w$. Show that

$$v \boxplus w = \sum_i a^i \left.\frac{\partial}{\partial x^i}\right|_{y+\widetilde{y}} + \sum_\alpha \left(b^\alpha + \widetilde{b}^\alpha\right) \left.\frac{\partial}{\partial \xi^\alpha}\right|_{y+\widetilde{y}},$$

where in $v \boxplus w$, the $\boxplus$ refers to the addition described in part (a).

(12) Exhibit a VB-atlas for the tautological bundle of Example 6.47.

(13) Show that the tautological bundle over $\mathbb{R}P^1$ is a Möbius band.

(14) Let $P \to M$ be a principal bundle with group G. If H is a Lie subgroup of G, then the quotient P/H is an H-principal bundle. Show that $P/H \to M$ admits a global section if and only if the structure group of $P \to M$ is reducible to H.

(15) Show that the notions of smooth fiber bundle and vector bundle make sense when the base space is allowed to be a manifold with boundary. What issues arise if one considers allowing both the base space and typical fiber to have boundary?

Chapter 7

Tensors

In this chapter we shall employ the *Einstein summation convention*. For example, $\tau_j^{ik}\alpha^j v_k$ is taken to be shorthand for

$$\sum_{k=1}^{m}\sum_{j=1}^{n} \tau_j^{ik}\alpha^j v_k,$$

where the range of summation is understood from the context. Normally, the repeated indices that are summed over occur once as a subscript and once as a superscript. For example, if $A = (a_j^i)$ is an $n \times m$ matrix and $B = (b_j^i)$ is an $m \times k$ matrix, where in this case we use upper indices to indicate rows and lower indices for columns, then $C = AB$ corresponds to

$$c_j^i = \sum_{l=1}^{m} a_l^i b_j^l.$$

This is reduced by the summation convention to $c_j^i = a_l^i b_j^l$. We will occasionally include the summation symbol $\sum$ for emphasis, or to meet the demands of clarity.

Tensor fields (often referred to simply as tensors) can be introduced in a rough and ready way by describing their local expressions in charts and then going on to explain how such expressions are transformed under a change of coordinates. With this approach one can gain proficiency with tensor calculations in short order, and this is usually the way physicists and engineers are introduced to tensors. However, since this approach hides much of the underlying algebraic and geometric structure, we will not pursue it here. Instead, we present tensors in terms of multilinear maps.

7.1. Some Multilinear Algebra

It will be convenient to define the notion of an algebraic tensor on a vector space or module. The reader who has looked over the material in Appendix D will find this chapter easier to understand. In particular, we assume the definition of "multilinear" (Definition D.13). In this chapter, if we say that a module is finite-dimensional,[1] we mean that it is free and finitely generated and thus has a basis. All modules in this chapter are assumed to be over a commutative ring with unity.

Definition 7.1. Let V and W be modules over a commutative ring R with unity. Then, an algebraic W-valued **tensor** on V is a multilinear mapping of the form
$$\tau : V_1 \times V_2 \times \cdots \times V_m \to W,$$
where each factor V_i is either V or V*. If the number of V* factors occurring is r and the number of V factors is s, then we say that the tensor is r-**contravariant** and s-**covariant**. We also say that the tensor is of **total type** $\binom{r}{s}$.

The most common situation is where W is the ring R itself, in which case we often drop the adjective "R-valued". Notice that if $\tau : V^* \times V \times V^* \times V \to R$ is a tensor, then we can define a tensor $\tilde{\tau} : V^* \times V \times V \times V^* \to R$ by
$$\tilde{\tau}(\alpha_1, v_1, v_2, \alpha_2) := \tau(\alpha_1, v_1, \alpha_2, v_2).$$
Although these two tensors clearly contain the same information, they are nevertheless different. We indicate this with a more specific notation. We say that τ is a tensor of type $\binom{1\ \ 1}{\ 1\ \ 1}$, while $\tilde{\tau}$ is of type $\binom{1\ \ \ \ 1}{\ 2\ }$. More generally, a tensor might, for example, be specified to be of type
$$\begin{pmatrix} r_1 & r_2 & \cdots & r_a \\ s_1 & s_2 & & s_b \end{pmatrix} \text{ or } \begin{pmatrix} r_1 & & \cdots & r_a \\ & s_1 & s_2 & & s_b \end{pmatrix}.$$
The general pattern should be clear. If $r = r_1 + \cdots + r_a$ and $s = s_1 + \cdots + s_b$, then the tensor would be of **total type** $\binom{r}{s}$, which we also write as (r,s). The set of all tensors of fixed type (as above) is easily seen to be an R-module with the scalar multiplication and addition defined as is usual for spaces of functions. As another example, a multilinear map
$$\Upsilon : V \times V^* \times V \times V^* \times V^* \to W$$
is a W-tensor which is of type $\binom{\ \ 1\ \ \ 2}{1\ \ 1\ \ \ }$ and total type $\binom{3}{2}$. The set of all W-valued tensors on V of type $\binom{\ \ 1\ \ \ 2}{1\ \ 1\ \ \ }$ is denoted $T_{\ 1\ \ 1\ \ }^{\ \ 1\ \ \ 2}(V; W)$, and we have analogous notations for other types. In many, if not most, circumstances we agree to associate to each tensor of total type $\binom{r}{s}$, a unique element of $T^r_s(V; W)$ by simply keeping the relative order among the V variables and

---
[1] For modules, what we mean by dimension is what is usually called the rank.

among the V* variables separately, but shifting all V variables to the right of the V* variables. Following this procedure, we have, for example, the map
$$T_1{}^1{}_1{}^2(V;W) \to T^3_2(V;W).$$
Maps like this will be called **consolidation maps** or **consolidation isomorphisms**.

Definition 7.2. A tensor
$$\tau : \underbrace{V^* \times V^* \times \cdots \times V^*}_{r \text{ times}} \times \underbrace{V \times V \times \cdots \times V}_{s \text{ times}} \to W,$$
where all the V factors occur last, is said to be in **consolidated form**. The set of all such (consolidated) W-valued tensors on V will be denoted $T^r_s(V;W)$. As a special case we have $T^0_1(V;R) = V^*$. We will often abbreviate $T^r_s(V;R)$ to $T^r_s(V)$.

For example, elements of $T^3_2(V;W)$ are in consolidated form, while tensors from $T_1{}^1{}_1{}^2(V;W)$ are unconsolidated.

Remark 7.3. Some authors consolidate by putting all V arguments first. Also, sometimes it is appropriate to forgo the consolidation especially in connection with the "type changing" operations introduced later. Our policy will be to work with tensors in consolidated form whenever convenient.

Example 7.4. One always has the special tensor $\delta \in T^1_1(V;R)$ defined by
$$\delta(a,v) = a(v)$$
for $a \in V^*$ and $v \in V$. This tensor is sometimes referred to as the **Kronecker delta tensor**.

There is a natural map from V to V** given by $v \mapsto \tilde{v}$, where $\tilde{v} : \alpha \mapsto \alpha(v)$. If this map is an isomorphism, we say that V is a **reflexive** module and we identify V with V**. Finite-dimensional vector spaces are reflexive.

Exercise 7.5. Show that the $C^\infty(M)$ module of sections of a vector bundle $E \to M$ is a reflexive module. (It is important here that we are only considering vector bundles with finite-dimensional fibers.)

We now consider the relationship between tensors as defined above and the abstract tensor product spaces described in Appendix D. We restrict our discussion to tensors in consolidated form since the implications for the general situation will be obvious. We specialize to the case of R-valued tensors where R is the ring. Recall that the k-th tensor power of an R-module V is denoted by $\bigotimes^k V := V \otimes \cdots \otimes V$. We always have a module homomorphism

(7.1) $$\left(\bigotimes^r V\right) \otimes \left(\bigotimes^s V^*\right) \to T^r_s(V;R),$$

whereby an element $u_1 \otimes \cdots \otimes u_r \otimes \beta^1 \otimes \cdots \otimes \beta^s \in (\bigotimes^r V) \otimes (\bigotimes^s V^*)$ corresponds to the multilinear map given by

$$(\alpha^1, \ldots, \alpha^r, v_1, \ldots, v_s) \mapsto \alpha^1(u_1) \cdots \alpha^r(u_r) \beta^1(v_1) \cdots \beta^s(v_s).$$

We will identify $u_1 \otimes \cdots \otimes u_r \otimes \beta^1 \otimes \cdots \otimes \beta^s$ with this multilinear map. In particular, this entails identifying $v \in V$ with the element $\tilde{v} \in V^{**}$ where $\tilde{v}: \alpha \mapsto \alpha(v)$. If V is a finite-dimensional vector space, then the map (7.1) is an isomorphism. In fact, it is also true that if V is the space of sections of some vector bundle over M (with finite-dimensional fibers), then V is a $C^\infty(M)$-module and the map (7.1) is still an isomorphism. A tensor which can be written in the form $u_1 \otimes \cdots \otimes \beta^s$ is called a **simple** or **decomposable** tensor. Note well that not all tensors are simple.

Remark 7.6. The reader should take careful notice of how we treat the orders of the factors: An element of $V \otimes V^* \otimes V^*$ corresponds to a multilinear map $V^* \times V \times V \to R$ and **not** to a map $V \times V^* \times V^* \to R$.

Since the map (7.1) is not always an isomorphism for general modules, and since no analogous isomorphism exists in the case of tangent spaces to infinite-dimensional manifolds such as those discussed in [**L1**], it becomes important to ask to what extent the map (7.1) is needed in differential geometry. Serge Lang has written a very fine differential geometry book for manifolds modeled on Banach spaces [**L1**] without the help of such an isomorphism. In any case, we still can and will consider $u_1 \otimes \cdots \otimes u_r \otimes \beta^1 \otimes \cdots \otimes \beta^s$ to be an element of $T^r_s(V)$ as described above. Another thing to notice is that if (7.1) is an isomorphism for all r and s, then in particular $V \cong V^{**}$, that is, V must be reflexive. Corollary D.33 of Appendix D states that for a finitely generated free module, being reflexive is enough to insure that (7.1) is an isomorphism for all r and s. In the latter case, the consolidation maps introduced earlier can be described in terms of simple tensors. For example, the consolidation map

$$T_1{}^2{}_2(V) \to T^2{}_3(V)$$

is given on simple tensors by

$$\alpha \otimes v \otimes w \otimes \beta \otimes \gamma \to v \otimes w \otimes \alpha \otimes \beta \otimes \gamma.$$

Now let us consider the spaces $V \otimes V^*$ and $V \otimes V^* \otimes V^*$. By a straightforward argument using the universal property of tensor product spaces, one can construct a bilinear map

$$(V \otimes V^*) \times (V \otimes V^* \otimes V^*) \to V \otimes V^* \otimes V \otimes V^* \otimes V^*$$

such that $(v \otimes \alpha, w \otimes \beta_1 \otimes \beta_2)$ is mapped to $v \otimes \alpha \otimes w \otimes \beta_1 \otimes \beta_2$. This corresponds to a product map

$$\otimes : T^1{}_1(V) \times T^1{}_2(V) \to T^1{}_1{}^1{}_2(V)$$

7.1. Some Multilinear Algebra

such that $(S, T) \to S \otimes T$, where

$$(S \otimes T)(\alpha_1, v_1, \alpha_2, v_2, v_3) = S(\alpha_1, v_1) T(\alpha_2, v_2, v_3).$$

The general pattern should be clear, but writing down the general case is notationally onerous. This product is the **(unconsolidated) tensor product** of tensors. Note carefully the order of the factors. To simplify the notation, the tensor product is often defined in a slightly different way when dealing with tensors which are in consolidated form:

Definition 7.7. For tensors $S \in T^{r_1}_{s_1}(V)$ and $T \in T^{r_2}_{s_2}(V)$, we define the **(consolidated) tensor product** $S \otimes T \in T^{r_1+r_2}{}_{s_1+s_2}(V)$ by

$$S \otimes T(\theta^1, \ldots, \theta^{r_1+r_2}, v_1, \ldots, v_{s_1+s_2})$$
$$:= S(\theta^1, \ldots, \theta^{r_1}, v_1, \ldots, v_{s_1}) T(\theta^{r_1+1}, \ldots, \theta^{r_1+r_2}, v_{s_1+1}, \ldots, v_{s_1+s_2}).$$

Whether or not a tensor product is the consolidated version will normally be clear from the context, and so we will drop the word "consolidated". We can also extend to products of several tensors at a time. While it is easy to see that the tensor product defined above is associative, it is *not* commutative since the order of the slots is an issue.

Let $T^*(V)$ denote the direct sum of all spaces of the form $T^r_0(V)$, where we take $T^0_0(V) := \mathsf{R}$. The tensor product gives $T^*(V)$ the structure of an algebra over R as long as we make the definition that $r \otimes A := rA$ for $r \in \mathsf{R}$.

Proposition 7.8. *Let* V *be a free* R*-module with basis* $(e_1, \ldots, e_n)$ *and corresponding dual basis* $(e^1, \ldots, e^n)$ *for* V^*. *Then the indexed set*

$$\{e_{i_1} \otimes \cdots \otimes e_{i_r} \otimes e^{j_1} \otimes \cdots \otimes e^{j_s} : i_1, \ldots, i_r, j_1, \ldots, j_s = 1, \ldots, n\}$$

is a basis for $T^r_s(V)$. *If* $\tau \in T^r_s(V)$, *then*

$$\tau = \tau^{i_1 \ldots i_r}{}_{j_1 \ldots j_s} e_{i_1} \otimes \cdots \otimes e_{i_r} \otimes e^{j_1} \otimes \cdots \otimes e^{j_s} \quad \text{(summation!)}$$

where $\tau^{i_1 \ldots i_r}{}_{j_1 \ldots j_s} = \tau(e^{i_1}, \ldots, e^{i_r}, e_{j_1}, \ldots, e_{j_s})$.

Proof. If $\tau \in T^r_s(V; \mathsf{R})$ and we *define* $\tau^{i_1 \ldots i_r}{}_{j_1 \ldots j_s} = \tau(e^{i_1}, \ldots, e^{i_r}, e_{j_1}, \ldots, e_{j_s})$, then it is easy to check that

$$\tau = \tau^{i_1 \ldots i_r}{}_{j_1 \ldots j_s} e_{i_1} \otimes \cdots \otimes e_{i_r} \otimes e^{j_1} \otimes \cdots \otimes e^{j_s},$$

and so, in particular, our indexed set spans $T^r_s(V; \mathsf{R})$. Indeed, if we denote the right hand side of the above equation by τ', we obtain (using the summation convention throughout)

$$\tau'(e^{k_1}, \ldots, e^{k_r}, e_{l_1}, \ldots, e_{l_s})$$
$$= \tau^{i_1 \ldots i_r}{}_{j_1 \ldots j_s} e_{i_1}(e^{k_1}) \cdots e_{i_r}(e^{k_r}) e^{j_1}(e_{l_1}) \cdots e^{j_s}(e_{l_s})$$
$$= \tau^{i_1 \ldots i_r}{}_{j_1 \ldots j_s} \delta^{k_1}_{i_1} \cdots \delta^{k_r}_{i_r} \delta^{j_1}_{l_1} \cdots \delta^{j_s}_{l_s} = \tau^{k_1 \ldots k_r}{}_{l_1 \ldots l_s}$$
$$= \tau(e^{k_1}, \ldots, e^{k_r}, e_{l_1}, \ldots, e_{l_s}).$$

Thus τ' and τ agree on basis elements, and by multilinearity $\tau' = \tau$.

For independence, suppose $\tau^{i_1 \ldots i_r}{}_{j_1 \ldots j_s} e_{i_1} \otimes \cdots \otimes e_{i_r} \otimes e^{j_1} \otimes \cdots \otimes e^{j_s} = 0$ for some n^{r+s} elements $\tau^{i_1 \ldots i_r}{}_{j_1 \ldots j_s}$ of R. This is an equality of multilinear maps, and if we apply both sides to $(e^{k_1}, \ldots, e^{k_r}, e_{l_1}, \ldots, e_{l_s})$, then we obtain $\tau^{k_1 \ldots k_r}{}_{l_1 \ldots l_s} = 0$. Since our choices were arbitrary, we see that all n^{r+s} elements $\tau^{i_1 \ldots i_r}{}_{j_1 \ldots j_s}$ are equal to 0. $\square$

As a special case we see that if $A \in T^1_1(V; R)$, then $A = A^i{}_j \, e_i \otimes e^j$, where $A^i{}_j = A(e^i, e_j)$. This theorem is a special case of Theorem D.29 of Appendix D, which we will also invoke below for spaces like $W \otimes V^*$.

If we are dealing with tensors that are not in consolidated form, it should still be clear how to obtain a basis. For example, $\{e_i \otimes e^j \otimes e_k\}$ is a basis for $T^1{}_1{}^1(V; R)$, and a typical element A would have an expansion

$$A = A^i{}_j{}^k e_i \otimes e^j \otimes e_k.$$

Notice the purposeful staggered positioning of the indices in $A^i{}_j{}^k$.

Definition 7.9. The elements $A^{i_1 \ldots i_r}{}_{j_1 \ldots j_s}$ from the previous proposition are called the **components** of τ with respect to the basis $e_1, \ldots, e_n$.

Example 7.10. If $V = T_pM$ for some smooth manifold M, then we can use any basis of T_pM we please. That said, we realize that if p is in a coordinate chart $(U, \mathbf{x})$, then the vectors $\frac{\partial}{\partial x^1}\big|_p, \ldots, \frac{\partial}{\partial x^n}\big|_p$ form a basis for T_pM, and we may form a basis for $T^r{}_s(T_pM)$ consisting of all tensors of the form

$$\frac{\partial}{\partial x^{i_1}}\bigg|_p \otimes \cdots \otimes \frac{\partial}{\partial x^{i_r}}\bigg|_p \otimes dx^{j_1}\big|_p \otimes \cdots \otimes dx^{j_s}\big|_p.$$

For example, an element A_p of $T^1{}_1(T_pM)$ can be expressed in coordinate form as

$$A_p = A^i{}_j \frac{\partial}{\partial x^i}\bigg|_p \otimes dx^j\big|_p.$$

An element A_p of $T^r{}_s(T_pM)$ can be written as

$$A_p = A^{i_1 \ldots i_r}{}_{j_1 \ldots j_s} \frac{\partial}{\partial x^{i_1}}\bigg|_p \otimes \cdots \otimes \frac{\partial}{\partial x^{i_r}}\bigg|_p \otimes dx^{j_1}\big|_p \otimes \cdots \otimes dx^{j_s}\big|_p,$$

and this is called the coordinate expression for A_p.

7.1. Some Multilinear Algebra

The components of a tensor depend on the basis chosen, and a different choice will give new components related to the first by a transformation law. This is the content of the following exercise:

Exercise 7.11. Let $e_1, \ldots, e_n$ be a basis for V and let $e^1, \ldots, e^n$ be the corresponding dual basis for V*. If $\bar{e}_1, \ldots, \bar{e}_n$ is another basis for V with

$$\bar{e}_i = C_i^k e_k,$$

then the dual basis $\bar{e}^1, \ldots, \bar{e}^n$ is related to $e^1, \ldots, e^n$ by $\bar{e}^i = \left(C^{-1}\right)^i_k e^k$, where $C = (C_j^i)$. Show that if $\tau^i{}_{jk}$ are the components of τ with respect to the first basis (and its dual) and if $\bar{\tau}^i{}_{jk}$ are the components with respect to the second basis, then

$$\bar{\tau}^i{}_{jk} = \tau^a{}_{bc}\, C_j^b C_k^c \left(C^{-1}\right)^i_a \quad \text{(sum over } a, b, c\text{)}.$$

This is a **transformation law**. What is the analogous statement for $\tau \in T^r_s(V; \mathsf{R})$?

Example 7.12. It is easy to show that for any basis (with corresponding dual basis) as above, the Kronecker delta tensor δ has components δ^i_j, where $\delta^i_j = 0$ if $i \neq j$ and $\delta^i_i = 1$.

It is easy to show that if $S \in T^1_2(V)$ and $T \in T^2_2(V)$, then $S \otimes T$ has components given by

$$(S \otimes T)^{abc}{}_{defg} = S^a{}_{de}\, T^{bc}{}_{fg}.$$

More generally, if $S \in T^{r_1}_{s_1}(V)$ and $T \in T^{r_2}_{s_2}(V)$, then

$$(7.2) \quad (S \otimes T)^{a_1 \ldots a_{r_1} \alpha_1 \ldots \alpha_{r_2}}{}_{b_1 \ldots b_{s_1} \beta_1 \ldots \beta_{s_2}} = S^{a_1 \ldots a_{r_1}}{}_{b_1 \ldots b_{s_1}}\, T^{\alpha_1 \ldots \alpha_{r_2}}{}_{\beta_1 \ldots \beta_{s_2}}.$$

Notice the consolidation. If we choose not to employ consolidation, then the (unconsolidated) tensor product would be expressed differently in component form. For example,

$$(S \otimes T)^a{}_{de}{}^{bc}{}_{fg} = S^a{}_{de}\, T^{bc}{}_{fg}.$$

This way of treating the position of the indices is a convention that is called (naturally enough) "positional notation". For more on positional notation, see [**Pe**], [**Dod-Pos**] and [**Stern**].

If V is a finitely generated free module, then we have a natural isomorphism $V \otimes V^* \cong L(V, V)$. We can be a bit more general. Let W be another finitely generated free module. Consider the map $\Psi : W \times V^* \to L(V, W)$ given by $(w, \alpha) \mapsto \Psi(w, \alpha)$, where

$$\Psi(w, \alpha)(v) := \alpha(v) w.$$

This map is multilinear, and so using the universal property of tensor products we get a map
$$\widehat{\Psi} : W \otimes V^* \to L(V, W)$$
such that $\widehat{\Psi} : w \otimes \alpha \mapsto \Psi(w, \alpha) \in L(V, W)$. Let us show that this map is an isomorphism. A given element $A \in W \otimes V^*$ may be written as $A = \sum w^i \otimes \alpha_i$, where the w^i are linearly independent. Indeed, we just write A in terms of a basis and collect terms. Then if $\widehat{\Psi}(A) = 0$, we have $\sum \alpha_i(v) w^i = 0$ for all v. Thus $\alpha_i(v) = 0$ for all v. It follows that $A = 0$. We see that $\widehat{\Psi}$ is injective. Both V and W are finite-dimensional, and $W \otimes V^*$ and $L(V, W)$ have the same dimension. Indeed, $L(V, W)$ is isomorphic to the space of $m \times n$ matrices with entries from R, where m and n are the dimensions (or ranks) of W and V respectively. If V and W were vector spaces, this would imply that $\widehat{\Psi}$ is onto. However, it remains true that $\widehat{\Psi}$ is onto in the case where W and V are finite-dimensional free modules, but we must argue differently. In Problem 2 we ask the reader to show that if $e_1, \ldots, e_n$ is a basis for V and $f_1, \ldots, f_m$ is a basis for W, then $\{\widehat{\Psi}(f_i \otimes e^j)\}$ is a basis for $L(V, W)$. Thus $\widehat{\Psi} : W \otimes V^* \to L(V, W)$ is an isomorphism. If $\tau = \tau^i_j f_i \otimes e^j$, then
$$\widehat{\Psi}(\tau) : v \mapsto \tau^i_j e^j(v) f_i = \tau^i_j v^j f_i.$$
The components τ^i_j are exactly the entries of the matrix that represents $\widehat{\Psi}(\tau)$. When the above conditions hold, so that $\widehat{\Psi}$ is an isomorphism, we often identify $W \otimes V^*$ with $L(V, W)$ and write τ even when we mean $\widehat{\Psi}(\tau)$. We say that τ has two "interpretations". Under this identification
$$(w \otimes \alpha)(v) = \alpha(v) w.$$
In component form, the two interpretations of τ show up as
$$v \mapsto w, \text{ where } w^i = \tau^i_j v^j, \text{ and}$$
$$(\alpha, v) \mapsto \tau(\alpha, v), \text{ where } \tau(\alpha, v) = \tau^i_j v^j \alpha_i.$$
In particular, we identify $T^1_1(V)$ with $L(V, V)$.

Remark 7.13 (Tensions of conventions). Both $W \otimes V^*$ and $V^* \otimes W$ can be identified with $L(V, V)$. For example, we may also interpret $\alpha \otimes w \in V^* \otimes W$ as the map $(\alpha \otimes w)(v) = \alpha(v) w$. If the reader looks at how we have consolidated the spaces in the definition of $T^r_s(V)$, it will be apparent that we have preferred $W \otimes V^*$ over $V^* \otimes W$. However, if one considers the case where the underlying ring is not commutative, it becomes clear that the isomorphism $V^* \otimes W \cong L(V, W)$ is correct for left modules while the other is correct for right modules (V^* is a right module if V is a left module, and vice versa). On the other hand, the identification $W \otimes V^* \cong L(V, W)$ is more natural for the conventions of matrix multiplication. For example, if we think of elements of $\mathbb{R}^n$ as column vectors and elements of $(\mathbb{R}^n)^*$ as row

7.1. Some Multilinear Algebra

vectors, then for $w \in \mathbb{R}^n$ and $\alpha \in (\mathbb{R}^n)^*$, the linear map $\alpha \otimes w$ is indeed given by the matrix αw. The tension between standard matrix conventions and left modules is well known to algebraists. We think of modules over commutative rings as simultaneously both left and right modules.

If V is finite-dimensional, then one can make various other reinterpretations of tensors:

Example 7.14. Suppose that V is finite-dimensional. Then, elements of $T^r_s(V)$ can be interpreted as members of
$$T^0_s(V; T^r_0(V))$$
according to the prescription that $\tau(v_1, \ldots, v_s)$ acts by
$$\tau(v_1, \ldots, v_s)(\alpha^1, \ldots, \alpha^r) := \tau(\alpha^1, \ldots, \alpha^r, v_1, \ldots, v_s).$$
Similarly, elements of $T^r_s(V)$ can be interpreted as members of $T^r_0(V; T^0_s(V))$.

Example 7.15. Let V be as in the previous example. Elements of $T^0_{s_1+s_2}(V)$ can be interpreted as members of
$$T^0_{s_1}(V; T^0_{s_2}(V))$$
by the prescription
$$\tau(v_1, \ldots, v_{s_1})(u_1, \ldots, u_{s_2}) := \tau(v_1, \ldots, v_{s_1}, u_1, \ldots, u_{s_2}).$$

One can easily see from the above examples that many reinterpretations are possible. One of the most common is where one interprets elements of $T^0_2(V, \mathbb{R})$ as elements of $L(V, V^*)$ according to
$$\tau(v)(u) = \tau(v, u) \text{ for } u, v \in V.$$

Exercise 7.16. Show that under the identification of $T^1_1(V, \mathbb{R})$ with $L(V, V)$ we can interpret the Kronecker delta tensor as the identity map.

Definition 7.17. A covariant tensor $\tau \in T^0_s(V, W)$ is said to be **symmetric** if
$$\tau(v_1, \ldots, v_s) = \tau(v_{\sigma(1)}, \ldots, v_{\sigma(s)})$$
for all $v_1, \ldots, v_s$ and all permutations σ of the letters $\{1, 2, \ldots, s\}$. We define a **symmetric contravariant tensor** similarly.

Definition 7.18. A covariant tensor $\tau \in T^0_s(V, W)$ is said to be **alternating** if
$$\tau(v_1, \ldots, v_s) = \text{sgn}(\sigma) \tau(v_{\sigma(1)}, \ldots, v_{\sigma(s)})$$
for all $v_1, \ldots, v_s$ and all permutations σ of the letters $\{1, 2, \ldots, s\}$, where $\text{sgn}(\sigma) = 1$ if σ is an even permutation and -1 if it is an odd permutation. $\text{sgn}(\sigma)$ is called the sign of the permutation σ; We define an **alternating contravariant tensor** similarly.

If $\ell : U \to V$ is a linear map, then the map $\ell^* : T^0_s(V;W) \to T^0_s(U;W)$ is defined by
$$(\ell^*\tau)(u_1,\ldots,u_s) := \tau(\ell(u_1),\ldots,\ell(u_s)).$$
$\ell^*\tau$ is called the **pull-back** of τ by ℓ. It is easy to show that ℓ^* is linear. If we have linear maps $\ell : U_1 \to U_2$ and $\lambda : U_2 \to V$, then
$$(\lambda \circ \ell)^* : T^0_s(V;W) \to T^0_s(U_1;W)$$
and
$$(\lambda \circ \ell)^* = \ell^* \circ \lambda^*.$$
Thus the pull-back $\ell \to \ell^*$ defines a *contravariant* functor in the category of W-valued *covariant* tensors. (Because of this, one might wish that covariant tensors were called contravariant and vice versa, and indeed some authors have reversed the traditional terminology.) Suppose that $(e_1,\ldots,e_n)$ is a basis for V and that $(f_1,\ldots,f_m)$ is a basis for W. If $\ell(e_i) = \sum \ell_i^k f_k$, then we have
$$(\ell^*\tau)_{i_1\ldots i_s} = (\ell^*\tau)(e_{i_1},\ldots,e_{i_s}) = \tau(\ell(e_{i_1}),\ldots,\ell(e_{i_s}))$$
$$= \tau(\ell_{i_1}^{k_1} f_{k_1},\ldots,\ell_{i_s}^{k_s} f_{k_s}) = \ell_{i_1}^{k_1}\cdots\ell_{i_s}^{k_s}\tau(f_{k_1},\ldots,f_{k_s})$$
$$= \tau_{k_1\ldots k_s}\ell_{i_1}^{k_1}\cdots\ell_{i_s}^{k_s},$$
which gives the component form of the pull-back operation in terms of the matrix (ℓ_i^k).

Proposition 7.19. *If $\lambda \in L(V,V)$, $\alpha \in T^0_{s_1}(V)$ and $\beta \in T^0_{s_2}(V)$, then*
$$\lambda^*(\alpha \otimes \beta) = \lambda^*\alpha \otimes \lambda^*\beta.$$

Proof. We have
$$\lambda^*(\alpha \otimes \beta)(u_1,\ldots,u_{s_1},u_{s_1+1},\ldots,u_{s_1+s_2})$$
$$= \alpha \otimes \beta(\lambda u_1,\ldots,\lambda u_{s_1},\lambda u_{s_1+1},\ldots,\lambda u_{s_1+s_2})$$
$$= \alpha(\lambda u_1,\ldots,\lambda u_{s_1})\beta(\lambda u_{s_1+1},\ldots,\lambda u_{s_1+s_2})$$
$$= \lambda^*\alpha(u_1,\ldots,u_{s_1})\lambda^*\beta(u_{s_1+1},\ldots,u_{s_1+s_2})$$
$$= \lambda^*\alpha \otimes \lambda^*\beta(u_1,\ldots,u_{s_1},u_{s_1+1},\ldots,u_{s_1+s_2}). \qquad \square$$

Before going on to study tensor *fields*, we introduce one more notion from multilinear algebra referred to as contraction.

Definition 7.20. Let $(e_1,\ldots,e_n)$ be a basis for V and $(e^1,\ldots,e^n)$ the dual basis. If $\tau \in T^r_s(V)$, then for $k \leq r$ and $l \leq s$, we define $C_l^k\tau \in T^{r-1}_{s-1}(V)$ by
$$C_l^k\tau(\theta^1,\ldots,\theta^{r-1},w_1,\ldots,w_{s-1})$$
$$:= \sum_{a=1}^n \tau(\theta^1,\ldots,\underbrace{e^a}_{k\text{-th position}},\ldots,\theta^{r-1},w_1,\ldots,\underbrace{e_a}_{l\text{-th position}},\ldots,w_{s-1}).$$

7.1. Some Multilinear Algebra

This processes is called **contraction**. Write the components of τ with respect to our basis as $\tau^{i_1...i_r}{}_{j_1...j_s}$. If we pick out an upper index, say i_k, and also a lower index, say j_l, then we obtain the components of the contracted tensor $C_l^k \tau$ by:

$$\left(C_l^k \tau\right)^{i_1...\widehat{i_k}...i_r}{}_{j_1...\widehat{j_l}...j_s} := \tau^{i...a...i_r}{}_{j_1...a...j_s} \quad \text{(sum over } a\text{)}.$$

Here the caret means omission. In practice, one often just writes

$$\tau^{i_1...\widehat{i_k}...i_r}{}_{j_1...\widehat{j_l}...j_s}$$

instead of $\left(C_l^k \tau\right)^{i_1...\widehat{i_k}...i_r}{}_{j_1...\widehat{j_l}...j_s}$ as long as it has been made clear how the contraction was carried out. For example, one often sees expressions like $R_{ij} := R^r{}_{irj}$. Notice that we always contract an upper index with a lower index. The map C_l^k contracts the k-th upper index with the l-th lower index.

Consider a tensor of the form $v \otimes w \otimes \eta \otimes \theta \in T_2^2(V)$. One can show that

$$C_1^1(v \otimes w \otimes \eta \otimes \theta) = \eta(v) w \otimes \theta.$$

Similarly,

$$C_2^1(v \otimes w \otimes \eta \otimes \theta) = \theta(v) w \otimes \eta.$$

In general, C_l^k acts on simple tensors $v_1 \otimes v_2 \otimes \cdots \otimes v_r \otimes \eta^1 \otimes \eta^2 \otimes \cdots \otimes \eta^s$ by an obvious extension of the above. Universal mapping properties can be invoked to give a basis free definition of contraction. Contraction generalizes the notion of the trace of a linear transformation.

A common use of contraction involves first taking the tensor product of two tensors, and then performing a contraction of a contravariant slot of one with a covariant slot of the other. One often performs several contractions. For example, we may form a tensor that is given in components as

$$\tau^{ac}{}_{efg} = S^a{}_{ke} T^{kc}{}_{fg} \quad \text{(sum over } k\text{)}.$$

Evaluation of a tensor on its arguments is the result of a repeated contraction. For example, let V have a basis $e_1, \ldots, e_n$ and let $e^1, \ldots, e^n$ be the dual basis for V^* as above. If $v = v^i e_i$, $w = w^i e_i$, and $\alpha = \alpha_i e^i$, then for $\tau \in T_2^1(V)$ we easily deduce that

(7.3) $$\tau(\alpha, v, w) = \tau^i{}_{jk} \alpha_i v^j w^k,$$

which is the result of a repeated contraction on the tensor $\tau \otimes \alpha \otimes v \otimes w$. More generally, if we express elements $v_1, \ldots, v_s \in V$ and $\alpha_1, \ldots, \alpha_r \in V^*$ in terms of our basis and its dual, then for $\tau \in T^r{}_s(V)$, we have an analogous general expression for $\tau(\alpha_1, \ldots, \alpha_r, v_1, \ldots, v_s)$ in terms of the components of the tensor and its arguments.

7.2. Bottom-Up Approach to Tensor Fields

There are two approaches to tensor fields on smooth manifolds that turn out to be equivalent (at least for finite-dimensional manifolds). We start with the "bottom-up" approach where we apply multilinear algebra first to individual tangent spaces. The second approach directly defines tensors on M as tensors on the module $\mathfrak{X}(M)$.

Roughly speaking, a smooth (r,s)-tensor field on a manifold M assigns to each $p \in M$ an element of $T^r{}_s(T_pM)$ in a smooth way. We are interested in making sense of smoothness for tensor fields, so we wish to view a tensor field as a section of an appropriate vector bundle (a tensor bundle). Let us start out being a bit more general by considering a real rank k vector bundle $\xi = (E, \pi, M)$. For convenience, we take the typical fiber to be $\mathbb{R}^k$. Let $T^r{}_s(E) = \bigcup_{p \in M} T^r{}_s(E_p)$. We wish to construct a bundle $T^r{}_s(\xi)$ which has $T^r{}_s(E)$ as total space, M as base space, and $T^r{}_s(E_p)$ as fiber over p. If (U, ϕ) is a VB-chart for ξ, then we construct a VB-chart for $T^r{}_s(\xi)$ in the following way: Recall that ϕ has the form $\phi = (\pi, \Phi)$, where $\Phi : \pi^{-1}U \to \mathbb{R}^k$ and where $\Phi_p := \Phi|_{E_p} : E_p \to \mathbb{R}^k$ is a linear isomorphism for each p. We obtain a map $\Phi_p^{r,s} : T^r{}_s(E_p) \to T^r{}_s(\mathbb{R}^k)$ by

$$(\Phi_p^{r,s}\tau_p)(\alpha_1, \ldots, \alpha_r, v_1, \ldots, v_s)$$
$$:= \tau_p((\Phi_p)^*\alpha_1, \ldots, (\Phi_p)^*\alpha_r, \Phi_p^{-1}v_1, \ldots, \Phi_p^{-1}v_s).$$

These maps combine to give a map $\Phi^{r,s} : \pi^{-1}U \to T^r{}_s(\mathbb{R}^k)$ which is smooth (exercise). Our chart for $T^r{}_s(\xi)$ is

$$\phi^{r,s} := (\pi, \Phi^{r,s}) : \pi^{-1}U \to U \times T^r{}_s(\mathbb{R}^k).$$

If desired, one can choose, once and for all, an isomorphism $T^r{}_s(\mathbb{R}^k) \cong \mathbb{R}^{k^{r+s}}$. A VB-atlas $\{(U_\alpha, \phi_\alpha)\}$ for $\xi = (E, \pi, M)$ gives a VB-atlas $\{(U_\alpha, \phi_\alpha^{r,s})\}$ for $T^r{}_s(\xi)$.

Exercise 7.21. Show that there is a natural vector bundle isomorphism

$$T^r{}_s(\xi) \cong (\otimes^r E) \otimes (\otimes^s E^*).$$

We leave it to the interested reader to prove the following useful theorem.

Proposition 7.22. *Let $\xi = (E, \pi, M)$ be a vector bundle as above and let $\Upsilon : M \to T^r{}_s(E)$ be a map which assigns to each $p \in M$ an element of $T^r{}_s(E_p)$. Then Υ is smooth if and only if*

$$p \mapsto \Upsilon(p)(\alpha_1(p), \ldots, \alpha_r(p), X_1(p), \ldots, X_s(p))$$

is smooth for all smooth sections $p \mapsto \alpha_i(p)$ and $p \mapsto X_i(p)$ of $E^* \to M$ and $E \to M$ respectively. The same statement is true if we use local sections.

7.2. Bottom-Up Approach to Tensor Fields

The set of smooth sections of $T^r_s(\xi)$) is denoted $\Gamma(T^r_s(\xi))$. If $\Upsilon \in \Gamma(T^r_s(\xi))$, then for $X_1, \ldots, X_s$ any smooth sections of $E \to M$ and $\alpha_1, \ldots, \alpha_r$ smooth sections of $E^* \to M$, define $\Upsilon(\alpha_1, \ldots, \alpha_r, X_1, \ldots, X_s) \in C^\infty(M)$ by

$$\Upsilon(\alpha_1, \ldots, \alpha_r, X_1, \ldots, X_s)(p) := \Upsilon_p(\alpha_1(p), \ldots, \alpha_r(p), X_1(p), \ldots, X_s(p)).$$

Now we have a map $\Upsilon : (\Gamma E^*)^k \times (\Gamma E)^l \to C^\infty(M)$. This map is clearly multilinear over $C^\infty(M)$, and we see that we can interpret elements of $\Gamma(T^r_s(\xi))$ as such maps when convenient. This extends the idea of thinking of a 1-form α as a $C^\infty(M)$ linear map $\mathfrak{X}(M) \to C^\infty(M)$.

Like most linear algebraic structures existing at the level of a single fiber E_p, the notion of tensor product is easily extended to the level of sections: For $\tau \in \Gamma(T^{r_1}_{s_1}(\xi))$ and $\eta \in \Gamma(T^{r_2}_{s_2}(\xi))$, we *define* the (consolidated) **tensor product** $\tau \otimes \eta \in \Gamma(T^{r_1+r_2}_{s_1+s_2}(\xi))$ by $(\tau \otimes \eta)(p) := \tau_p \otimes \eta_p$. Thus

$$(\tau \otimes \eta)(p)(\alpha^1, \ldots, \alpha^{r_1+r_2}, v_1, \ldots, v_{s_1+s_2})$$
$$= \tau(\alpha^1, \ldots, \alpha^{r_1}, v_1, \ldots, v_{s_1})\eta(\alpha^{r_1+1}, \ldots, \alpha^{r_1+r_2}, v_{s_1+1}, \ldots, v_{s_1+s_2})$$

for all $\alpha^i \in E^*_p$ and $v_i \in E_p$.

Let $(s_1, \ldots, s_k)$ be a local frame field for ξ over an open set U and let $\sigma^1, \ldots, \sigma^k$ be the dual frame field of the dual bundle $E^* \to M$ so that $\sigma^i(s_j) = \delta^i_j$. Consider the set

$$\{\sigma^{i_1} \otimes \cdots \otimes \sigma^{i_r} \otimes s_{j_1} \otimes \cdots \otimes s_{j_s} : i_1, \ldots, i_r, j_1, \ldots, j_s = 1, \ldots, k\}.$$

If $\tau \in \Gamma(T^r_s(\xi))$, then we have functions $\tau^{i_1 \ldots i_r}{}_{j_1 \ldots j_s} \in C^\infty(U)$ defined by $\tau^{i_1 \ldots i_r}{}_{j_1 \ldots j_s} = \tau(\sigma^{i_1}, \ldots, \sigma^{i_r}, s_{j_1}, \ldots, s_{j_s})$. It follows from Proposition 7.8 that τ (restricted to U) has the expansion

$$\tau = \tau^{i_1 \ldots i_r}{}_{j_1 \ldots j_s} \sigma^{i_1} \otimes \cdots \otimes \sigma^{i_r} \otimes s_{j_1} \otimes \cdots \otimes s_{j_s}.$$

Also, applying equation (7.2) in each fiber E_p, we see that the component functions for $\tau \otimes \eta$ are given by

$$(\tau \otimes \eta)^{i_1 \ldots i_{r_1+r_2}}{}_{j_1 \ldots j_{s_1+s_2}}$$
$$= \tau^{i_1 \ldots i_{r_1}}{}_{j_1 \ldots j_{s_1}} \eta^{i_{r_1+1} \ldots i_{r_2}}{}_{j_{s_1+1} \ldots j_{s_2}}.$$

Here, and wherever convenient, we use the consolidated tensor product.

Notation 7.23. Whenever there is no chance of confusion, we will refer to $T^r_s(\xi)$ by $T^r_s(E) \to M$ or even just $T^r_s(E)$ (the latter is the notation for the total space of the bundle).

In the case of the tangent bundle TM, we have special terminology and notation:

Definition 7.24. The bundle $T^r_s(TM) \to M$ is called the (r,s)-**tensor bundle on** M.

By Exercise 7.21, $T^r{}_s(TM) \cong (\bigotimes^r TM) \otimes (\bigotimes^s T^*M)$ and this natural isomorphism is taken as an identification so the latter bundle is also referred to as a tensor bundle. *We now restrict ourselves to the case of the tangent bundle of a manifold but note that much of what follows makes sense for general vector bundles.*

Definition 7.25. The space of sections $\Gamma(T^r{}_s(TM))$ is denoted by $\mathcal{T}^r_s(M)$ and its elements are referred to as r-contravariant s-covariant **tensor fields** or just **type** (r,s)-**tensor fields**. The space $\mathcal{T}^r_0(M)$ is denoted by $\mathcal{T}^r(M)$ and $\mathcal{T}^0_s(M)$ by $\mathcal{T}_s(M)$.

In summary, a smooth tensor field A is a smooth assignment of a multilinear map on each tangent space of the manifold. Thus for each p, $A(p)$ is a multilinear map

$$A(p) : (T_p^*M)^r \times (T_pM)^s \to \mathbb{R},$$

or in other words, an element of $T^r{}_s(T_pM)$. Elements of $T^r{}_s(T_pM)$ are called tensors *at p*. We also write A_p for $A(p)$.

Example 7.26. In Definition 6.42, we introduced the notion of a Riemannian metric on a real vector bundle. We saw that such metrics always exist. The most important case is where the bundle is the tangent bundle TM of a manifold M. In this case, we say that we have a **Riemannian metric** on M. Thus a Riemannian metric on M is an element of $\mathcal{T}_2(M)$ which is symmetric and positive definite at each point.

Of course the manifold in question could be an open submanifold U of M so we have $C^\infty(U)$-module (r,s)-tensor fields over that set denoted $\mathcal{T}^r_s(U)$. The open subsets are partially ordered by inclusion $V \subset U$ and the tensor fields on these are related by restriction. Let $r^U_V : \mathcal{T}^r_s(U) \to \mathcal{T}^r_s(V)$ denote the restriction map. The assignment $U \to \mathcal{T}^r_s(U)$ is an example of a presheaf and in fact a sheaf.

We will also sometimes deal with tensors with values in TM (or in T^*M). First note that the space $T^r{}_s(T_pM; T_pM)$ of all multilinear maps $(T_p^*M)^r \times (T_pM)^s \to T_pM$ is a vector space. The set $T^r{}_s(TM; TM) := \bigcup_p T^r{}_s(T_pM; T_pM)$ can be given a smooth vector bundle structure in a way that is closely analogous to $T^r{}_s(TM) \to M$.

Definition 7.27. The space of sections $\Gamma(T^r{}_s(TM; TM))$ is denoted by $\mathcal{T}^r_s(M; TM)$ and its elements are referred to as r-contravariant s-covariant TM-valued **tensor fields**. Similarly, we may define T^*M-valued **tensor fields**.

Note that TM-valued tensor fields can be associated in a natural manner with ordinary tensor fields. For example, if $A \in T^0{}_2(TM; TM)$, then using

7.2. Bottom-Up Approach to Tensor Fields

the same letter A by abuse of notation, we may define an element $A \in T^1_2(TM)$ by

$$A_p(\theta_p, v_p, w_p) = \theta_p\left(A_p(v_p, w_p)\right) \text{ for } \theta_p \in T^*_p M \text{ and } v_p, w_p \in T_p M,$$

Many such reinterpretations are possible. For this reason, we shall stick to studying ordinary tensors and tensor fields in what follows.

We shall define several operations on spaces of tensor fields. We would like each of these to be natural with respect to restriction. We already have one such operation: the tensor product. If $A \in \mathcal{T}^{r_1}_{s_1}(U)$ and $B \in \mathcal{T}^{r_2}_{s_2}(U)$ and $V \subset U$, then $r^U_V(A \otimes B) = r^U_V A \otimes r^U_V B$. A (k,l)-tensor field A may generally be expressed in a chart $(U, \mathbf{x})$ as

$$(7.4) \qquad A = A^{i_1 \ldots i_r}{}_{j_1 \ldots j_s} \frac{\partial}{\partial x^{i_1}} \otimes \cdots \otimes \frac{\partial}{\partial x^{i_r}} \otimes dx^{j_1} \otimes \cdots \otimes dx^{j_s},$$

where $A^{i_1 \ldots i_r}{}_{j_1 \ldots j_s}$ are functions, $\frac{\partial}{\partial x^i} \in \mathfrak{X}(U)$ and $dx^j \in \mathfrak{X}^*(U)$. Actually, it is the restriction of τ to U that can be written in this way, but because of the naturality of all the operations we introduce, it is generally safe to use the same letter to denote a tensor field and its restriction to an open set such as a chart domain. It is easy to show that

$$A^{i_1 \ldots i_r}{}_{j_1 \ldots j_s} = A\left(dx^{i_1}, \ldots, dx^{i_r}, \frac{\partial}{\partial x^{j_1}}, \ldots, \frac{\partial}{\partial x^{j_s}}\right),$$

and so the components of a smooth tensor field are C^∞ for every choice of coordinates $(U, \mathbf{x})$ by Proposition 7.22. Conversely, one can obviously define tensors that are not necessarily smooth sections of the appropriate tensor bundle, and then a tensor will be smooth exactly when its components with respect to every chart in an atlas are smooth. Evaluating the expression (7.4) above at a point $p \in U$ results in an expression such as that given at the end of Example 7.10.

Exercise 7.28 (Transformation laws). Suppose that we have two charts $(U, \mathbf{x})$ and $(V, \bar{\mathbf{x}})$. If $A \in \mathcal{T}^1_2(M)$ has components A^i_{jk} in the first chart and $\overline{A}^i_{jk}$ in the second chart, then on the overlap $U \cap V$ we have

$$\overline{A}^i_{jk} = A^a_{bc} \frac{\partial \bar{x}^i}{\partial x^a} \frac{\partial x^b}{\partial \bar{x}^j} \frac{\partial x^c}{\partial \bar{x}^k},$$

where

$$d\bar{x}^i = \frac{\partial \bar{x}^i}{\partial x^a} dx^a \text{ and } \frac{\partial}{\partial \bar{x}^i} = \frac{\partial x^b}{\partial \bar{x}^i} \frac{\partial}{\partial x^b}.$$

This last exercise reveals the transformation law for tensor fields in $\mathcal{T}^1_2(M)$, and there is obviously an analogous law for tensor fields from $\mathcal{T}^r_s(M)$ for any values of r and s. In some presentations, tensor fields are defined in terms of such transformations laws (see [**L-R**] for this approach). It should be emphasized again that there are two slightly different ways of reading local expressions like the above. We may think of all of these functions as living on the manifold in the domain $U \cap V$. In this interpretation, we read the above as

$$\overline{A}^i_{jk}(p) = A^l_{ab}(p)\frac{\partial x^a}{\partial \bar{x}^j}(p)\frac{\partial x^b}{\partial \bar{x}^k}(p)\frac{\partial \bar{x}^i}{\partial x^l}(p) \text{ for each } p \in U \cap V.$$

This is the default modern viewpoint. Alternatively, we could take $\frac{\partial x^j}{\partial \bar{x}^m}$ to be functions on $\bar{\mathbf{x}}(U \cap V)$ and write $\frac{\partial x^j}{\partial \bar{x}^m}(\bar{x}^1, \ldots, \bar{x}^n)$. Then, $\frac{\partial \bar{x}^i}{\partial x^l}$ would refer to $\frac{\partial \bar{x}^i}{\partial x^l} \circ \mathbf{x} \circ \bar{\mathbf{x}}^{-1}(\bar{x}^1, \ldots, \bar{x}^n)$ so that both sides of the equation are functions of variables which we abusively write as $(\bar{x}^1, \ldots, \bar{x}^n)$. The first version seems theoretically pleasing, but for specific calculations that use familiar coordinates such as polar coordinates, the second version is often convenient. For example, suppose that a tensor τ has components with respect to rectangular coordinates on $\mathbb{R}^2$ given by A_{jk} and we wish to find the components in polar coordinates. For indexing purposes, we take $(x,y) = (u^1, u^2)$ and $(r, \theta) = (v^1, v^2)$. Then we have

(7.5) $$\overline{A}_{jk} = A_{ab}\frac{\partial u^a}{\partial v^j}\frac{\partial u^b}{\partial v^k},$$

which can be read so that both sides are functions of (v^1, v^2) by writing u^1 and u^2 as a function of (v^1, v^2), etc. Of course, the charts are there to "identify" open sets in Euclidean space with open sets on the manifold so these viewpoints are really somehow the same after all. Using (x, y) and (r, θ), the transformation (7.5) is given in matrix form as

$$\begin{bmatrix} \overline{A}_{11} & \overline{A}_{12} \\ \overline{A}_{12} & \overline{A}_{22} \end{bmatrix} = \begin{bmatrix} \cos\theta & -r\sin\theta \\ \sin\theta & r\cos\theta \end{bmatrix} \begin{bmatrix} A_{11} & A_{12} \\ A_{12} & A_{22} \end{bmatrix} \begin{bmatrix} \cos\theta & \sin\theta \\ -r\sin\theta & r\cos\theta \end{bmatrix}.$$

We now introduce the pull-back of a covariant tensor field, which will play a big role in the next chapter.

Definition 7.29. If $f : M \to N$ is a smooth map and $\tau \in \mathcal{T}_s(N)$, then we define the **pull-back** $f^*\tau \in \mathcal{T}_s(M)$ by

$$f^*\tau(v_1, \ldots, v_s)(p) = \tau(Tf \cdot v_1, \ldots, Tf \cdot v_s)$$

for all $v_1, \ldots, v_s \in T_pM$ and any $p \in M$.

Notice the connection of this with the pull-back defined earlier in a purely algebraic context. It is not hard to see that $f^* : \mathcal{T}_s(N) \to \mathcal{T}_s(M)$ is linear over $\mathbb{R}$, and for any $h \in C^\infty(N)$ and $\tau \in \mathcal{T}_s(N)$ we have $f^*(h\tau) = (h \circ f) f^*\tau$. If $f : M \to N$ and $g : N \to P$ are smooth maps, then of course $(g \circ f)^* = f^* \circ g^*$.

Let us discover the local expression for pull-back. Choose a chart $(U, \mathbf{x})$ on M and a chart $(V, \mathbf{y})$ on N and assume that $f(U) \subset V$. Let us denote

$\frac{\partial (y^i \circ f)}{\partial x^j}$ by $\frac{\partial y^i}{\partial x^j}$ for simplicity. We have $T_p f \cdot \frac{\partial}{\partial x^i}\big|_p = \sum \frac{\partial y^k}{\partial x^i}(p) \frac{\partial}{\partial y^k}\big|_{f(p)}$ and

$$(f^*\tau)_{i_1\ldots i_s}(p) = (f^*\tau)\left(\frac{\partial}{\partial x^{i_1}}\bigg|_p, \ldots, \frac{\partial}{\partial x^{i_s}}\bigg|_p\right)$$

$$= \tau\left(Tf \frac{\partial}{\partial x^{i_1}}\bigg|_p, \ldots, Tf \frac{\partial}{\partial x^{i_s}}\bigg|_p\right)$$

$$= \tau\left(\frac{\partial y^{k_1}}{\partial x^{i_1}}(p) \frac{\partial}{\partial y^{k_1}}\bigg|_{f(p)}, \ldots, \frac{\partial y^{k_s}}{\partial x^{i_s}}(p) \frac{\partial}{\partial y^{k_s}}\bigg|_{f(p)}\right)$$

$$= \tau\left(\frac{\partial}{\partial y^{k_1}}\bigg|_{f(p)}, \ldots, \frac{\partial}{\partial y^{k_s}}\bigg|_{f(p)}\right) \frac{\partial y^{k_1}}{\partial x^{i_1}}(p) \cdots \frac{\partial y^{k_s}}{\partial x^{i_s}}(p)$$

$$= \tau_{k_1\ldots k_s}(f(p)) \frac{\partial y^{k_1}}{\partial x^{i_1}}(p) \cdots \frac{\partial y^{k_s}}{\partial x^{i_s}}(p).$$

Thus we have

$$(f^*\tau)_{i_1\ldots i_s} = (\tau_{k_1\ldots k_s} \circ f) \frac{\partial y^{k_1}}{\partial x^{i_1}} \cdots \frac{\partial y^{k_s}}{\partial x^{i_s}}.$$

This looks similar to a transformation law for a tensor, but here f is not a change of coordinates and need not even be a diffeomorphism. Pull-back respects tensor products:

Exercise 7.30. Let $f : M \to N$ be as above. Show that for $\tau_1 \in \mathcal{T}_{s_1}(N)$ and $\tau_2 \in \mathcal{T}_{s_2}(N)$ we have $f^*(\tau_1 \otimes \tau_2) = f^*\tau_1 \otimes f^*\tau_2$.

In the case that $f : M \to N$ is a diffeomorphism, the notion of pull-back can be extended to contravariant tensors and tensors of mixed covariance. For such a diffeomorphism, let $(Tf^{-1})^* : T_p^*M \to T_p^*N$ denote the dual of the map $Tf^{-1} : T_pN \to T_pM$.

Definition 7.31. If $f : M \to N$ is a diffeomorphism and τ is an (r,s)-tensor field on N, then define the **pull-back** $f^*\tau \in T^r{}_s(M)$ by

$$f^*\tau(a_1, \ldots, a_r, v_1, \ldots, v_s)(p)$$
$$:= \tau\big((Tf^{-1})^* a_1, \ldots, (Tf^{-1})^* a_r, Tf \cdot v_1, \ldots, Tf \cdot v_s\big)$$

for all $v_1, \ldots, v_s \in T_pM$ and $a_1, \ldots, a_r \in T_p^*M$ and any $p \in M$. The **push-forward** is then defined for $\tau \in T^r{}_s(M)$ as $f_*\tau := (f^{-1})^*\tau$.

7.3. Top-Down Approach to Tensor Fields

Specializing what we learned from the discussion following Proposition 7.22 to the case of the tangent bundle, we see that a tensor field gives us a $C^\infty(M)$-multilinear map based on the module $\mathfrak{X}(M)$. This observation leads to an alternative definition of a tensor field over M. In this "top-down" view, we simply *define* an (r,s)-tensor field to be a $C^\infty(M)$-multilinear map

$$\mathfrak{X}^*(M)^r \times \mathfrak{X}(M)^s \to C^\infty(M).$$

In this view, a tensor field is an element of $T^r{}_s(\mathfrak{X}(M))$. For example, a global covariant 2-tensor field on a manifold M is a map $\tau : \mathfrak{X}(M) \times \mathfrak{X}(M) \to C^\infty(M)$ such that

$$\tau(f_1 X_1 + f_2 X_2, Y) = f_1 \tau(X_1, Y) + f_2 \tau(X_2, Y),$$
$$\tau(Y, f_1 X_1 + f_2 X_2) = f_1 \tau(Y, X_1) + f_2 \tau(Y, X_2)$$

for all $f_1, f_2 \in C^\infty(M)$ and all $X_1, X_2, Y \in \mathfrak{X}(M)$. As we shall see, it turns out that such $C^\infty(M)$-multilinear maps determine tensor fields in the sense of the previous section.

If we take a top-down approach to tensor fields, then we must work to recover the presheaf/sheaf aspects. Indeed, it is not obvious what is the relation between $T^r{}_s(\mathfrak{X}(M))$ and $T^r{}_s(\mathfrak{X}(U))$ for some proper open subset $U \subset M$. Indeed, thinking purely in terms of modules makes the issue clear. The module $\mathfrak{X}(M)$ is not the same module as $\mathfrak{X}(U)$ unless $U = M$. A priori, there is no immediate reason to think that a multilinear map with arguments from the module $\mathfrak{X}(M)$ should be able to take elements of $\mathfrak{X}(U)$ as arguments! For instance, from the top-down viewpoint, how can we insert coordinate fields $\frac{\partial}{\partial x^i}$ and dx^i into an element of $T^r{}_s(\mathfrak{X}(M))$ to get coordinate expressions if the chart domain is not all of M? We address this in the next section indirectly by showing how the top-down approach gives back tensors as sections (the bottom-up approach). Another comment is that both $\mathfrak{X}(U)$ and $T^r{}_s(\mathfrak{X}(U))$ are finite-dimensional *free* modules over the ring $C^\infty(U)$ whenever U is a chart domain or, more generally, the domain of a frame field. The reason is that a local frame field and its dual frame field provide a module basis for $\mathfrak{X}(U)$ and $\mathfrak{X}^*(U)$ and the latter really is the dual of the first in the module sense. On the other hand, the $C^\infty(M)$-modules $\mathfrak{X}(M)$ and $T^r{}_s(\mathfrak{X}(M))$ are not generally free unless M is parallelizable.

7.4. Matching the Two Approaches to Tensor Fields

If we define a tensor field as we first did, that is, as a field of tensors in tangent spaces, then we immediately obtain a tensor as defined in the top-down approach. On the other hand, if τ is initially defined as a $C^\infty(M)$-multilinear map, then how should we recover a field of tensors on the tangent spaces?[2] Answering this is our next goal.

---

[2]This is exactly where things might not go so well if the manifold is not finite-dimensional. What we need is the existence of smooth cut-off functions. Some Banach manifolds support cut-off functions but not all do.

7.4. Matching the Two Approaches to Tensor Fields

Proposition 7.32. *Let $p \in M$ and $\tau \in T^r_s(\mathfrak{X}(M))$. Let $\theta_1, \ldots, \theta_r$ and $\bar{\theta}_1, \ldots, \bar{\theta}_r$ be smooth 1-forms such that $\theta_i(p) = \bar{\theta}_i(p)$ for $1 \leq i \leq r$; also let $X_1, \ldots, X_s$ and $\bar{X}_1, \ldots, \bar{X}_s$ be smooth vector fields such that $X_i(p) = \bar{X}_i(p)$ for $1 \leq i \leq s$. Then we have that*

$$\tau(\theta_1, \ldots, \theta_r, X_1, \ldots, X_s)(p) = \tau(\bar{\theta}_1, \ldots, \bar{\theta}_r, \bar{X}_1, \ldots, \bar{X}_s)(p).$$

Proof. The result will follow easily if we can show that

$$\tau(\theta_1, \ldots, \theta_r, X_1, \ldots, X_s)(p) = 0$$

whenever one of $\theta_1(p), \ldots, \theta_r(p), X_1(p), \ldots, X_s(p)$ is zero. We shall assume for simplicity of notation that $r = 1$ and $s = 2$. Now suppose that $X_1(p) = 0$. If $(U, \mathbf{x})$, with $\mathbf{x} = (x^1, \ldots, x^n)$, is a chart with $p \in U$, then $X_1|_U = \sum \xi^i \frac{\partial}{\partial x^i}$ for some smooth functions $\xi^i \in C^\infty(U)$. Let β be a cut-off function with support in U and $\beta(p) = 1$. Then for any smooth vector field X defined on U we can consider both βX and $\beta^2 X$ to be globally defined and zero outside of U. Similarly, if f is a smooth function defined on U, then βf can be taken to be globally defined on M and zero outside of U. Now $\beta^2 X_1 = \sum (\beta \xi^i)(\beta \frac{\partial}{\partial x^i})$. (Notice that in this last expression we have used β to extend both the functions ξ^i and the coordinate fields $\frac{\partial}{\partial x^i}$, which is why we used β^2 rather than just β.) Thus

$$\beta^2 \tau(\theta_1, X_1, X_2) = \tau(\theta_1, \beta^2 X_1, X_2)$$
$$= \tau\left(\theta_1, \beta^2 \xi^i \frac{\partial}{\partial x^i}, X_2\right)$$
$$= \tau\left(\theta_1, (\beta \xi^i) \beta \frac{\partial}{\partial x^i}, X_2\right)$$
$$= \beta \xi^i \tau\left(\theta_1, \beta \frac{\partial}{\partial x^i}, X_2\right).$$

(Notice that the point of the above expression is that, at this moment, τ is defined only for *global* sections and is linear over *global* functions. For example, $\frac{\partial}{\partial x^i}$ is not a global section while $\beta \frac{\partial}{\partial x^i}$ is a global section.) Since $X_1(p) = 0$, we must have $\xi^i(p) = 0$ for all i. Also recall that $\beta(p) = 1$. Plugging p into the formula above we obtain

$$\tau(\theta_1, X_1, X_2)(p) = 0.$$

A similar argument holds when $X_2(p) = 0$ or $\theta_1(p) = 0$.

Assume that $\theta_1(p) = \bar{\theta}_1(p)$, $X_1(p) = \bar{X}_1(p)$ and $X_2(p) = \bar{X}_2(p)$. Then we have

$$\tau(\bar{\theta}_1, \bar{X}_1, \bar{X}_2) - \tau(\theta_1, X_1, X_2)$$
$$= \tau(\bar{\theta}_1 - \theta_1, \bar{X}_1, \bar{X}_2) + \tau(\theta_1, \bar{X}_1 - X_1, \bar{X}_2) + \tau(\theta_1, X_1, \bar{X}_2 - X_2).$$

Since $\bar{\theta}_1 - \theta_1$, $\bar{X}_1 - X_1$, and $\bar{X}_2 - X_2$ are all zero at p, we obtain the result that $\tau(\bar{\theta}_1, \bar{X}_1, \bar{X}_2)(p) = \tau(\theta_1, X_1, X_2)(p)$. $\square$

Thus we have a natural correspondence between $T^r_s(\mathfrak{X}(M))$ and $\mathcal{T}^r_s(M)$ (the latter being smooth sections of the bundle $T^r{}_s(TM) \to M$). For example, if $A \in \mathcal{T}^1_3(M)$, then we obtain an element of $T^1_3(\mathfrak{X}(M))$, also denoted by A, by defining a smooth function $A(\theta, X, Y, Z)$ for given fields (θ, X, Y, Z) by

$$A(\theta, X, Y, Z)(p) := A(p)(\theta(p), X(p), Y(p), Z(p)).$$

Conversely, if $A \in T^1_3(\mathfrak{X}(M))$, then we can use the above proposition to define an element of $\mathcal{T}^1_3(M)$, which we denote by the same letter. Given $A \in T^1_3(\mathfrak{X}(M))$, define $A(p) \in T^1_3(T_pM)$ for each p as follows: For $X_p, Y_p, Z_p \in T_pM$ and $\theta_p \in T^*_pM$ we let

$$A(p)(\theta_p, X_p, Y_p, Z_p) := A(\theta, X, Y, Z)(p),$$

where θ, X, Y, Z are any fields chosen so that $\theta(p) = \theta_p$, $X(p) = X_p$, $Y(p) = Y_p$, and $Z(p) = Z_p$. By Lemma 6.37 we can always find such extensions, and by Proposition 7.32 above, A is well-defined. That A so defined is smooth follows from Proposition 7.22. The general case should be clear and, all said, we end up with a natural isomorphism of $C^\infty(M)$-modules:

$$T^r_s(\mathfrak{X}(M)) \cong \mathcal{T}^r_s(M).$$

Similar reasoning shows that there is a correspondence between fields of TM-valued tensors and $\mathfrak{X}(M)$-valued tensors on $\mathfrak{X}(M)$. For example, we have

(7.6) $$T^0_2(\mathfrak{X}(M); \mathfrak{X}(M)) \cong \mathcal{T}_2(M; TM).$$

Elements of $T^0_2(\mathfrak{X}(M); \mathfrak{X}(M))$ are $C^\infty(M)$-bilinear maps $\mathfrak{X}(M) \times \mathfrak{X}(M) \to \mathfrak{X}(M)$, while elements of $\mathcal{T}_2(M; TM)$ are sections of the bundle whose fiber at p is $T^0_2(T_pM; T_pM)$. So, if $A \in \mathcal{T}_2(M; TM)$, then for each p, $A(p)$ is an $\mathbb{R}$-multilinear map $T_pM \times T_pM \to T_pM$. Similarly, we have

(7.7) $$T^0_3(\mathfrak{X}(M); \mathfrak{X}(M)) \cong \mathcal{T}_3(M; TM).$$

In fact, later, when we define the curvature tensor on a semi-Riemannian manifold, it will initially be given as a multilinear map on modules of fields with values in $\mathfrak{X}(M)$. The correspondence is then invoked to get a tensor field (as defined in the bottom-up approach) with values in TM. It is just as easy to give a similar correspondence between $T^r_s(\Gamma(\xi))$ and $\Gamma(T^r_s(\xi))$ for some vector bundle $\xi = (E, \pi, M)$, where we view $\Gamma(\xi)$ as a $C^\infty(M)$ module.

Exercise 7.33. Exhibit the isomorphism (7.7) in detail.

Exercise 7.34. Suppose that S and T are tensors of the same type and we wish to show that they are equal. Then it is enough to check equality under the assumption that the vector fields inserted into the slots of S and T are locally defined and have vanishing Lie brackets. Hint: Think about coordinate vector fields.

We end this section with some warnings. It may seem that there is a simple way to obtain a pull-back by a smooth map $f : M \to N$ entirely from the top-down or module-theoretic view. In fact, one often sees expressions like

$$f^*\tau(X_1, \ldots, X_s) = \tau(f_*X_1, \ldots, f_*X_s) \quad \text{(problematic expression!)}.$$

This looks cute, but invites misunderstanding. The left hand side takes fields $X_1, \ldots, X_s$ as arguments, while on the right hand side, if we consider τ as a multilinear map $\mathfrak{X}(N) \times \cdots \times \mathfrak{X}(N) \to C^\infty(N)$, then f_*X_i must be fields. But the push-forward map f_* is generally not defined on fields, and even if it were, the above expression would seem to be an equality of a function on M with a function on N. Note that $\mathfrak{X}(M)$ is a $C^\infty(M)$-module, while $\mathfrak{X}(N)$ is a $C^\infty(N)$-module. The above expression may be taken to mean something like $f^*\tau(X_1, \ldots, X_s)(p) = \tau(Tf \cdot X_1(p), \ldots, Tf \cdot X_s(p))$, but now the right hand side has tangent vectors as arguments, and we are back to the bottom-up approach! A correct statement is the following:

Proposition 7.35. *Let $f : M \to N$ be a smooth map. Let τ be a $(0, s)$-tensor field. If τ and $f^*\tau$ are interpreted as elements of $T^0_s(\mathfrak{X}(N))$ and $T^0_s(\mathfrak{X}(M))$ respectively, then*

$$f^*\tau(X_1, \ldots, X_s) = \tau(Y_1, \ldots, Y_s) \circ f$$

whenever Y_i is f-related to X_i for $i = 1, \ldots, s$.

Of course, we can use Definition 7.31 to make sense of both push-forward and pull-back in the case that f is a diffeomorphism.

7.5. Tensor Derivations

We would like to be able to differentiate tensor fields. In particular, we would like to extend the Lie derivative to tensor fields. For this purpose we introduce the following definition, which will be useful not only for extending the Lie derivative, but also in several other contexts. Recall the presheaf of tensor fields $U \mapsto \mathcal{T}^r_s(U)$ on a manifold M.

Definition 7.36. A **tensor derivation** is a collection of maps $\mathcal{D}^r_s|_U : \mathcal{T}^r_s(U) \to \mathcal{T}^r_s(U)$, all denoted by $\mathcal{D}$ for convenience, such that

(1) $\mathcal{D}$ is a presheaf map for $\mathcal{T}_s^r$ considered as a presheaf of vector spaces over $\mathbb{R}$. In particular, for all open U and V with $V \subset U$ we have
$$\mathcal{D}A|_V = \mathcal{D}\left(A|_V\right)$$
for all $A \in \mathcal{T}_s^r(U)$, i.e., the restriction of $\mathcal{D}A$ to V is just $\mathcal{D}\left(A|_V\right)$.

(2) $\mathcal{D}$ commutes with contractions.

(3) $\mathcal{D}$ satisfies a derivation law. Specifically, for $A \in \mathcal{T}_s^r(U)$ and $B \in \mathcal{T}_k^j(U)$ we have
$$\mathcal{D}(A \otimes B) = \mathcal{D}A \otimes B + A \otimes \mathcal{D}B.$$

For smooth n-manifolds, the conditions (2) and (3) imply that for $A \in \mathcal{T}_s^r(U)$, $\alpha_1, \ldots, \alpha_r \in \mathfrak{X}^*(U)$ and $X_1, \ldots, X_s \in \mathfrak{X}(U)$, we have

(7.8)
$$\begin{aligned}&\mathcal{D}(A(\alpha_1, \ldots, \alpha_r, X_1, \ldots, X_s))\\ &= (\mathcal{D}A)(\alpha_1, \ldots, \alpha_r, X_1, \ldots, X_s)\\ &\quad + \sum_{i=1}^r A(\alpha_1, \ldots, \mathcal{D}\alpha_i, \ldots, \alpha_r, X_1, \ldots, X_s)\\ &\quad + \sum_{i=1}^s A(\alpha_1, \ldots, \alpha_r, X_1, \ldots, \mathcal{D}X_i, \ldots, X_s).\end{aligned}$$

This follows by noticing that
$$A(\alpha_1, \ldots, \alpha_r, X_1, \ldots, X_s) = C(A \otimes (\alpha_1 \otimes \cdots \otimes \alpha_r \otimes X_1 \otimes \cdots \otimes X_s))$$
(where C is the repeated contraction) and then applying (2) and (3).

Note that $\mathcal{D}$ stands for a family of maps whose domains $\mathcal{T}_s^r(U)$ depend not only on r and s, but also on U. The next proposition considers the situation where we only have derivations defined for $U = M$ (the global case).

Proposition 7.37. *Let M be a smooth manifold and suppose we have a map on globally defined tensor fields $\mathcal{D} : \mathcal{T}_s^r(M) \to \mathcal{T}_s^r(M)$ for all nonnegative integers r, s such that (2) and (3) above hold for the case $U = M$. Then there is a unique induced tensor derivation that agrees with $\mathcal{D}$ on global sections, that is, on the various $\mathcal{T}_s^r(M)$.*

Proof. We need to define $\mathcal{D} : \mathcal{T}_s^r(U) \to \mathcal{T}_s^r(U)$ for arbitrary open U as a derivation. Let δ be a function in $C^\infty(U)$ that vanishes on a neighborhood V of $p \in U$. We claim that $(\mathcal{D}\delta)(p) = 0$. To see this, let β be a cut-off function equal to 1 on a neighborhood of p and zero outside of V. Then $\delta = (1 - \beta)\delta$ and so
$$\begin{aligned}\mathcal{D}\delta(p) &= \mathcal{D}((1-\beta)\delta)(p)\\ &= \delta(p)\mathcal{D}(1-\beta)(p) + (1-\beta(p))\mathcal{D}\delta(p) = 0.\end{aligned}$$

7.5. Tensor Derivations

Now given $\tau \in \mathcal{T}_s^r(U)$, let β be a cut-off function with support in U and equal to 1 on a neighborhood of $p \in U$. Then $\beta\tau \in \mathcal{T}_s^r(M)$ after extending by zero. Define $(\mathcal{D}\tau)(p) := \mathcal{D}(\beta\tau)(p)$. To show that this is well-defined let β_2 be any other cut-off function with support in U and equal to 1 on a neighborhood of $p \in U$. Then we have

$$\mathcal{D}(\beta\tau)(p) - \mathcal{D}(\beta_2\tau)(p)$$
$$= (\mathcal{D}(\beta\tau) - \mathcal{D}(\beta_2\tau))(p) = \mathcal{D}((\beta - \beta_2)\tau)(p) = 0,$$

where the last equality follows from our claim above with $\delta = \beta - \beta_2$. Thus $\mathcal{D}$ is well-defined on $\mathcal{T}_s^r(U)$. We now show that $\mathcal{D}\tau$ so defined is an element of $\mathcal{T}_s^r(U)$. Let $(U', \mathbf{x})$ be a chart with $p \in U' \subset U$. Then we can write $\tau|_{U'} \in \mathcal{T}_s^r(U')$ as

$$\tau_{U'} = \tau_{i_1,\ldots,i_s}^{j_1,\ldots,j_r} dx^{i_1} \otimes \cdots \otimes dx^{i_s} \otimes \frac{\partial}{\partial x^{j_1}} \otimes \cdots \otimes \frac{\partial}{\partial x^{j_r}}.$$

We can use this to show that $\mathcal{D}\tau$ as defined agrees with a global section in a neighborhood O of p and so must be a smooth section itself since the choice of $p \in U$ was arbitrary. To save on notation, let us take the case $r = 1$, $s = 1$. Then $\tau_{U_\alpha} = \tau_j^i dx^j \otimes \frac{\partial}{\partial x^i}$. Let β be a cut-off function equal to 1 in the neighborhood O of p and zero outside of U'. Extend each of the sections $\beta\tau_j^i$, βdx^j, and $\beta\frac{\partial}{\partial x^i}$ to global sections and apply $\mathcal{D}$ to $\beta^3\tau = (\beta\tau_j^i)(\beta dx^j) \otimes (\beta\frac{\partial}{\partial x^i})$ to get

$$\mathcal{D}(\beta^3\tau) = \mathcal{D}\left(\beta\tau_j^i \beta dx^j \otimes \beta\frac{\partial}{\partial x^i}\right)$$
$$= \mathcal{D}(\beta\tau_j^i)\beta dx^j \otimes \beta\frac{\partial}{\partial x^i} + \beta\tau_j^i \mathcal{D}(\beta dx^j) \otimes \beta\frac{\partial}{\partial x^i}$$
$$+ \beta\tau_j^i \beta dx^j \otimes \mathcal{D}\left(\beta\frac{\partial}{\partial x^i}\right).$$

By assumption, $\mathcal{D}$ takes smooth global sections to smooth global sections, so both sides of the above equation are smooth. On the other hand, we have $\mathcal{D}(\beta^3\tau)(q) = \mathcal{D}(\tau)(q)$ by definition and valid for all $q \in O$. Thus $\mathcal{D}(\tau)$ is smooth and is the restriction of a smooth global section. This gives a unique derivation $\mathcal{D} : \mathcal{T}_s^r(U) \to \mathcal{T}_s^r(U)$ for all U satisfying the naturality conditions (1), (2) and (3). We leave it to the reader to check this last statement. $\square$

Exercise 7.38. Let $\mathcal{D}_1$ and $\mathcal{D}_2$ be two tensor derivations (so satisfying conditions (1), (2) and 3 of Definition 7.36) that agree on functions and vector fields. Then $\mathcal{D}_1 = \mathcal{D}_2$. [Hint: If $\alpha \in \mathfrak{X}^*(U) = \mathcal{T}_1^0(U)$, we must have $(\mathcal{D}_i\alpha)(X) = \mathcal{D}_i(\alpha(X)) - \alpha(\mathcal{D}_iX)$ for $i = 1, 2$. Then both $\mathcal{D}_1$ and $\mathcal{D}_2$ must obey formula (7.8) above.]

Theorem 7.39. *If $\mathcal{D}_U$ can be defined on $C^\infty(U)$ and $\mathfrak{X}(U)$ for each open $U \subset M$ so that*

(1) $\mathcal{D}_U(fg) = (\mathcal{D}_U f)g + f\mathcal{D}_U g$ *for all* $f, g \in C^\infty(U)$,
(2) $(\mathcal{D}_M f)|_U = \mathcal{D}_U f|_U$ *for each* $f \in C^\infty(M)$,
(3) $\mathcal{D}_U(fX) = (\mathcal{D}_U f)X + f\mathcal{D}_U X$ *for all* $f \in C^\infty(U)$ *and* $X \in \mathfrak{X}(U)$,
(4) $(\mathcal{D}_M X)|_U = \mathcal{D}_U X|_U$ *for each* $X \in \mathfrak{X}(M)$,

then there is a unique tensor derivation $\mathcal{D}$ that is equal to $\mathcal{D}_U$ on $C^\infty(U)$ and $\mathfrak{X}(U)$ for all U.

Sketch of proof. We wish to define $\mathcal{D}$ on $\mathfrak{X}^*(U)$ so that

(7.9) $$\mathcal{D}_U(\alpha \otimes X) = \mathcal{D}_U \alpha \otimes X + \alpha \otimes \mathcal{D}_U X.$$

By contraction we see that we must have $(\mathcal{D}_U \alpha)(X) = \mathcal{D}_U(\alpha(X)) - \alpha(\mathcal{D}_U X)$, which we take as the definition. Then check that (7.9) holds. Now define $\mathcal{D}_U$ by formula (7.8) and verify that we really have a map $\mathcal{T}_s^r(U) \to \mathcal{T}_s^r(U)$. Check that $\mathcal{D}_U$ commutes with contraction $C : \mathcal{T}_1^1(U) \to C^\infty(U)$ for simple tensors $\alpha \otimes X \in \mathcal{T}_1^1(U)$. Use the fact that, locally, every element of $\mathcal{T}_1^1$ can be written as a sum of simple tensors. Next extend to $\mathcal{T}_s^r$ along the lines exemplified by the case of $\mathcal{T}_2^1(U)$ and the contraction C_2^1 as follows: For $\tau \in \mathcal{T}_2^1(U)$, we have

$$\begin{aligned}
(\mathcal{D}_U C_2^1 \tau)(X) &= \mathcal{D}_U \left((C_2^1 \tau)(X) \right) - (C_2^1 \tau) \mathcal{D}_U X \\
&= \mathcal{D}_U \left(C(\tau(\cdot, X, \cdot)) \right) - C(\tau(\cdot, \mathcal{D}_U X, \cdot)) \\
&= C\mathcal{D}_U \left(\tau(\cdot, X, \cdot) - \tau(\cdot, \mathcal{D}_U X, \cdot) \right) \\
&= C\left((\mathcal{D}_U \tau)(\cdot, X, \cdot) \right) = \left(C_2^1 \mathcal{D}_U \tau \right)(X).
\end{aligned}$$

The general case would involve an inconvenient profusion of parentheses. Uniqueness follows from Exercise 7.38. Finally check by direct calculation that (3) of Definition 7.36 holds. □

Corollary 7.40. *The Lie derivative $\mathcal{L}_X$ can be extended to a tensor derivation for any $X \in \mathfrak{X}(M)$.*

The last corollary extends the Lie derivative to tensor fields. It follows from formula (7.8) that we have

(7.10) $$(\mathcal{L}_X S)(Y_1, \ldots, Y_s) = X(S(Y_1, \ldots, Y_s)) \\ - \sum_{i=1}^{s} S(Y_1, \ldots, Y_{i-1}, \mathcal{L}_X Y_i, Y_{i+1}, \ldots, Y_s).$$

We now present a different way of extending the Lie derivative to tensor fields that is equivalent to what we have just done. First let $A \in \mathcal{T}_s^r(M)$

7.6. Metric Tensors

and recall that if $f : M \to M$ is a *diffeomorphism*, then we can define $f^*A \in \mathcal{T}_s^r(M)$ by

$$(f^*A)(p)(\alpha^1, \ldots, \alpha^r, v_1, \ldots, v_s)$$
$$= A(f(p))\left(\left(T_p f^{-1}\right)^*(\alpha^1), \ldots, \left(T_p f^{-1}\right)^*(\alpha^r), T_p f(v_1), \ldots, T_p f(v_s)\right)$$

for all $\alpha^1, \ldots, \alpha^r \in (T_p M)^*$ and $v_1, \ldots, v_s \in T_p M$. If X is a vector field on M (possibly locally defined), we can define

$$(7.11) \qquad (\mathcal{L}_X A)(p) = \left.\frac{d}{dt}\right|_0 \left(\left(\varphi_t^X\right)^* A\right)(p),$$

just as we did for vector fields. We leave it as a project for the reader to show that this definition agrees with our first definition of the Lie derivative of a tensor field.

The Lie derivative on tensor fields is natural with respect to diffeomorphisms in the sense that for any diffeomorphism $f : M \to N$ and any vector field X we have

$$(7.12) \qquad \mathcal{L}_{f_*X} f_* A = f_* \mathcal{L}_X A.$$

This property is not shared by some other important derivations such as the covariant derivative, which we define later in this book.

Exercise 7.41. Show that the Lie derivative on tensor fields is natural with respect to diffeomorphisms in the above sense of equation (7.12) by using the fact that it is natural on functions and vector fields.

7.6. Metric Tensors

We start out again considering some linear algebra that we wish to globalize. Thus the vector space V that we discuss next should be thought of as a tangent space of a manifold or a fiber of some vector bundle.

We recall the following definitions: A symmetric bilinear form g on a finite-dimensional vector space V is **nondegenerate** if and only if $g(v, w) = 0$ for all $w \in V$ implies that $v = 0$. A (real) **scalar product** on a (real) finite-dimensional vector space V is a nondegenerate symmetric bilinear form $g : V \times V \to \mathbb{R}$. A **scalar product space** is a pair (V, g) where V is a vector space and g is a scalar product. We say that g is **positive** (resp. negative) **definite** if $g(v, v) \geq 0$ (resp. $g(v, v) \leq 0$) for all $v \in V$ and $g(v, v) = 0 \Longrightarrow v = 0$. In case the scalar product is positive definite, we also refer to it as an **inner product** and the pair (V, g) as an inner product space. Otherwise we say that the scalar product is **indefinite**. A scalar product on V is sometimes called a **metric tensor** on V. We now need to introduce quite a few more definitions.

Definition 7.42. The **index** of a symmetric bilinear form g on V is the dimension of the largest subspace $W \subset V$ such that the restriction $g|_W$ is negative definite. The index is denoted $\text{ind}(g)$.

Definition 7.43. Let (V, g) be a scalar product space. We say that v and w are mutually orthogonal if and only if $g(v, w) = 0$. Furthermore, given two subspaces W_1 and W_2 of V we say that W_1 is orthogonal to W_2 and write $W_1 \perp W_2$ if and only if every element of W_1 is orthogonal to every element of W_2.

Since, in general, g is not necessarily positive definite or negative definite, there may be nonzero elements that are orthogonal to themselves.

Definition 7.44. Given a subspace W of a scalar product space V, we define the **orthogonal complement** as $W^\perp = \{v \in V : g(v, w) = 0 \text{ for all } w \in W\}$.

Exercise 7.45. We always have $\dim(W) + \dim(W^\perp) = \dim(V)$, but unless g is definite, we may *not* have $W \cap W^\perp = \{0\}$.

Definition 7.46. A subspace W of a scalar product space (V, g) is called **nondegenerate** if $g|_W$ is nondegenerate.

Lemma 7.47. *A subspace* $W \subset (V, g)$ *is nondegenerate if and only if* $V = W \oplus W^\perp$ *(inner direct sum).*

Proof. This an easy exercise. One uses the standard fact that
$$\dim W + \dim W^\perp = \dim(W + W^\perp) + \dim(W \cap W^\perp). \qquad \square$$

It is a standard fact from linear algebra, already mentioned in Chapter 5, that if g is a scalar product, then there exists a basis $e_1, \ldots, e_n$ for V such that the matrix representative of g with respect to this basis is a diagonal matrix with ones or minus ones along the diagonal. Such a basis is called an orthonormal basis for (V, g). The number of minus ones appearing is the index $\text{ind}(g)$ and so is independent of the orthonormal basis chosen. It is easy to see that the index $\text{ind}(g)$ is zero if and only if g is positive definite.

Definition 7.48. For each $v \in V$ with $\langle v, v \rangle \neq 0$, let $\epsilon(v) := \text{sgn}\langle v, v \rangle$. Then if $e_1, \ldots, e_n$ are orthonormal, we have $\epsilon_i = \epsilon(i) := \epsilon(e_i)$.

Thus if $e_1, \ldots, e_n$ is an orthonormal basis for (V, g), then $g(e_i, e_j) = \epsilon_i \delta_{ij}$, where $\epsilon_i = g(e_i, e_i) = \pm 1$ are the entries of the diagonal matrix $\text{ind}(g)$ of which are equal to -1 and the remaining are equal to 1. Let us refer to the list of ± 1's given by $(\epsilon_1, \ldots, \epsilon_n)$ as the **signature**. We may arrange for the -1's to come first by permuting the elements of the basis. For example, if $(-1, -1, 1, 1)$ is the signature, then the index is 2.

7.6. Metric Tensors

Remark 7.49. *From now on, whenever context allows, we shall always assume that by "orthonormal basis" we mean an orthonormal basis that is arranged so that the -1's come first as described above.* The convention of putting the minus signs first is not universal, and in fact we used the opposite convention in Chapter 5. The negative signs first convention is popular in relativity theory and semi-Riemannian geometry, but the reverse convention is perhaps more common in Lie group theory and quantum field theory. It makes no difference in the final analysis as long as one is consistent, but it can be confusing when comparing references in the literature.

Another difference between the theory of positive definite scalar products and indefinite scalar products is the appearance of the ϵ_i's from the signature in formulas that would be familiar in the positive definite case. For example, we have the following:

Proposition 7.50. *Let $e_1, \ldots, e_n$ be an orthonormal basis for (V, g). For any $v \in V$, we have a unique expansion given by $v = \sum_i \epsilon_i \langle v, e_i \rangle e_i$.*

Proof. The usual proof works. One just has to notice the appearance of the ϵ_i's. □

Definition 7.51. If $v \in V$, then let $\|v\|$ denote the nonnegative number $|g(v,v)|^{1/2}$ and call this the (absolute or positive) **length** or **norm** of v.

Some authors call $g(v,v)$ or $g(v,v)^{1/2}$ the norm, which would make it possible for the norm to be negative or even complex-valued. We will avoid this.

Just as for positive definite inner product spaces, we call a linear isomorphism $\Phi : (V_1, g_1) \to (V_2, g_2)$ from one scalar product space to another an **isometry** if $g_1(v, w) = g_2(\Phi v, \Phi w)$. It is not hard to show that if such an isometry exists, then g_1 and g_2 have the same index and signature.

Let (V_i, g_i) be scalar product spaces for $i = 1, \ldots, k$. By Corollary D.35 of Appendix D, there is a unique bilinear form

$$\varphi : \bigotimes_{i=1}^k V_i \times \bigotimes_{i=1}^k V_i \to \mathbb{R}$$

such that for $v_i \in V_i$ and $w_i \in W_i$,

$$\varphi(v_1 \otimes \cdots \otimes v_k, w_1 \otimes \cdots \otimes w_k) = g_1(v_1, w_1) \cdots g_k(v_k, w_k).$$

The form φ is clearly symmetric and by Problem 3, it is a scalar product on $\bigotimes_{i=1}^k V_i$ that is positive definite if each g_i is positive definite. If (V, g) is a scalar product space, then we use the above to endow each of the various tensor spaces $T^r_s(V)$ with a scalar product. First consider V^*. Since g is nondegenerate, there is a linear isomorphism $g_\flat : V \to V^*$ defined by

$$g_\flat(v)(w) = g(v, w).$$

Denote the inverse by $g^\sharp : V^* \to V$. We force this to be an isometry by defining the scalar product on V^* to be

$$g^*(\alpha, \beta) = g(g^\sharp(\alpha), g^\sharp(\beta)).$$

Under this prescription, the dual basis $(e^1, \ldots, e^n)$ corresponding to an orthonormal basis $(e_1, \ldots, e_n)$ for V will also be orthonormal. The signatures (and hence the indexes) of g^* and g are the same.

Notation 7.52. When convenient, we shall also denote $g_\flat(v)$ by either $\flat v$ or $v^\flat$ and similarly for $g^\sharp(\alpha)$.

The above procedure now applies to give a scalar product on any tensor space $T^r_s(V)$. For example, consider $T^1_1(V) = V \otimes V^*$. Then there is a unique scalar product g^1_1 on $V \otimes V^*$ such that for $v_1 \otimes \alpha_1$ and $v_2 \otimes \alpha_2 \in V \otimes V^*$ we have

$$g^1_1(v_1 \otimes \alpha_1, v_2 \otimes \alpha_2) = g(v_1, v_2) g^*(\alpha_1, \alpha_2).$$

One can then see that for orthonormal $e_1, \ldots, e_n$ we have that

$$\{e_i \otimes e^j\}_{1 \le i,j \le n}$$

is an orthonormal basis for $(T^1_1(V), g^1_1)$. In general, if we endow $T^r_s(V)$ with a scalar product as above, then the natural basis for $T^r_s(V)$ formed from the orthonormal basis $(e_1, \ldots, e_n)$ (and its dual $(e^1, \ldots, e^n)$) will also be orthonormal.

Notation 7.53. In order to reduce notational clutter, let us reserve the option to denote all these scalar products coming from g by the same letter g or, even more conveniently, by $\langle \cdot, \cdot \rangle$. So, for example, $\langle v_1 \otimes \alpha_1, v_2 \otimes \alpha_2 \rangle = \langle v_1, v_2 \rangle \langle \alpha_1, \alpha_2 \rangle$ by definition.

Exercise 7.54. Show that under the natural identification of $V \otimes V^*$ with $L(V, V)$, the scalar product of linear transformations A and B is given by $\langle A, B \rangle = \mathrm{trace}(A^t B)$.

The maps $g_\flat$ and $g^\sharp$ are called musical isomorphisms. Let us see how things look in terms of components. Let $f_1, \ldots, f_n$ be an arbitrary basis of V and let $f^1, \ldots, f^n$ be the dual basis for V^*. The components of g are given by $g_{ij} := g(f_i, f_j)$. So if $v = v^i f_i$ and $w = w^i f_i$, then

$$g(v, w) = g(v^i f_i, w^j f_j) = v^i w^j g(f_i, f_j) = g_{ij} v^i w^j,$$

where we continue to use the Einstein summation convention. There must be a matrix (A_{ij}) such that $\flat f_i = A_{ki} f^k$. On the other hand,

$$g_{ij} = g(f_i, f_j) = (\flat f_i)(f_j) = A_{ki} f^k(f_j)$$
$$= A_{ki} \delta^k_j = A_{ji}.$$

7.6. Metric Tensors

So we have
$$\flat f_i = g_{ik} f^k.$$
Thus if $v = v^i f_i$, then $\flat v = v^j \flat f_j = v^j g_{ji} f^i$, so the components of $\flat v$ are $(\flat v)_i = v^j g_{ji} = g_{ij} v^j$. It is a common convention that if v^i are the components of v with respect to a basis, then the components of $\flat v$ are denoted simply by lowering the index:
$$v_i = g_{ij} v^j.$$
The map $g_\flat$ is called the **flatting operator** and the effect of this operator is sometimes described as "**index lowering**". If we write $\sharp f^i = g^{ij} f_j$ for some matrix (g^{ij}), then $(g_{ij})^{-1} = (g^{ij})$ so that
$$g^{ik} g_{kj} = \delta^k_j.$$
This follows from $f^i = \flat \sharp f^i = g^{ik} \flat f_k = g^{ik} g_{kj} f^j$. If $\omega \in V^*$ is written as $\omega = \omega_i f^i$, then an easy calculation shows that the components of $\sharp \omega$ are given by
$$\omega^i := (\sharp \omega)^i = g^{ik} \omega_k.$$
The isomorphism $g^\sharp$ is called the **sharping operator** and its effect is referred to as "**index raising**". The scalar product g_* on V^* introduced above has components g^{ij} with respect to the dual basis. Indeed,
$$g^*(f^i, f^j) = g(\sharp f^i, \sharp f^j) = g(g^{ik} f_k, g^{jl} f_l)$$
$$= g^{ik} g^{jl} g_{kl} = g^{jl} \delta^i_l = g^{ji} = g^{ij}.$$

Next we see how to extend the notions of index raising and lowering to tensors. Suppose we have a tensor $A \in T^2_2(V)$ and we wish to obtain a new related tensor A' whose final slot takes elements of V^* rather than elements of V. Thus we want A' be of type $\binom{2\ \ 1}{1}$. The trick is to define A' using $\sharp$ as follows:
$$A'(\omega, \eta, v, \alpha) := A(\omega, \eta, v, \sharp \alpha).$$
Let us compute the components of A' with respect to our basis. We have
$$(A')^{ij}{}_k{}^\ell = A'(f^i, f^j, f_k, f^\ell) = A(f^i, f^j, f_k, \sharp f^\ell)$$
$$= A(f^i, f^j, f_k, g^{\ell r} f_{kr}) = g^{\ell r} A^{ij}{}_{kr}.$$
It is common to write $A^{ij}{}_k{}^\ell$ in place of $(A')^{ij}{}_k{}^\ell$ when the context makes the meaning clear. In other words, we use the same root symbol but reposition the indices. This is an instance of index raising. Similarly we might use the flatting operation to obtain a new tensor from A. For $\omega \in V^*$ and $u, v, w \in V$ we could define A' by
$$A'(u, \omega, v, w) := A(\flat u, \omega, v, w)$$

and the components of A' would be given by

$$(A')_i{}^j{}_{k\ell} := g_{ir} A^{rj}{}_{k\ell},$$

and again it is common to see simply $A_i{}^j{}_{k\ell}$. This process of raising and lowering indices is called **type changing**. Notice that we often obtain tensors that are not in consolidated form. However, one may simply apply consolidation as desired. But, notice that the staggering of position makes the relation of $A_i{}^j{}_{k\ell}$ to the original tensor $A^{ij}{}_{k\ell}$ clear. This is where positional index notation excels.

The above can be approached in a slightly different way. We can take tensor products of various combinations of $g_\flat$, $g^\sharp$ and the identity maps id_V and id_{V^*}. For example,

$$g_\flat \otimes \mathrm{id}_V \otimes g_\flat \otimes g^\sharp \otimes \mathrm{id}_{V^*} : V \otimes V \otimes V \otimes V^* \otimes V^* \to V^* \otimes V \otimes V^* \otimes V \otimes V^*.$$

Depending on convenience, this map might then be followed by the consolidation isomorphism

$$V^* \otimes V \otimes V^* \otimes V \otimes V^* \to V^* \otimes V^* \otimes V^* \otimes V \otimes V.$$

Exercise 7.55. Show that the map $g_\flat \otimes \mathrm{id}_V \otimes g_\flat \otimes g^\sharp \otimes \mathrm{id}_{V^*}$ effects three iterated type changes and is given in component form by

$$A^{ijk}{}_{\ell m} \mapsto A_i{}^j{}_k{}^\ell{}_m := g_{ia} g_{kb} g^{\ell c} A^{ajb}{}_{cm}.$$

In the presence of a scalar product, a type-changed tensor is considered to be just a different manifestation of the original tensor. We say that it is **metrically equivalent**. The reader should expect to see some slight variability with regard to how index positioning and order of slots is handled when type changing is done. For example, there is nothing stopping us from raising the a-th lower index into the b-th upper position while keeping everything in consolidated form:

$$(A')^{i_1...i_{b-1} j\, i_{b+1}...i_r}{}_{j_1...j_{a-1} j_{a+1}...j_s} := A^{i_1...i_r}{}_{j_1...j_{a-1}\, m\, j_{a+1}...j_s}\, g^{mj}.$$

Invariantly, this is described as

$$A'(\alpha^1, \ldots, \alpha^{r+1}, v_1, \ldots, v_{s-1})$$
$$:= A(\alpha^1, \ldots, \widehat{\alpha^b}, \ldots, \alpha^{r+1}, v_1, \ldots, v_{a-1}, g^\sharp(\alpha^b), v_{a+1}, \ldots, v_s).$$

Notice that if we raise all the lower indices and lower all the upper indices on a tensor, then we can "completely contract" against another tensor of the original type. We leave it to the reader to show that the result is the scalar product of tensors defined earlier. For example, let $\chi = \sum \chi_{ij} f^i \otimes f^j$ and $\tau = \sum \tau_{ij} f^i \otimes f^j$. We may apply two type changes to τ that are given

7.6. Metric Tensors

in component form as $\tau_{ij} \mapsto \tau^i{}_j \mapsto \tau^{ij}$. In other words, $\tau^{ij} = g^{ik}g^{jl}\tau_{kl}$. Then we have
$$\langle \chi, \tau \rangle = \chi_{ij}\tau^{ij}.$$
See Problem 18.

7.6.1. Metrics on manifolds. If $g \in \mathcal{T}_2(M)$ is nondegenerate, symmetric and positive definite at every tangent space, we call g a **Riemannian metric** (tensor). If g is a Riemannian metric, then we call the pair (M,g) a **Riemannian manifold**. For example, in Chapter 4, we saw how a hypersurface in $\mathbb{R}^n$ inherits a Riemannian metric. This works just as well for a regular submanifold $M \subset \mathbb{R}^n$ of arbitrary codimension.

In Riemannian geometry, it is the metric that is the basis for generalizations of length, volume and so on. Motivated by a desire to generalize and to include the mathematics needed for general relativity, we also allow the metric to be indefinite. In this case, some nonzero tangent vectors v might have zero or negative self-scalar product $\langle v, v \rangle$. If $g \in \mathcal{T}_2(M)$ is a symmetric tensor field, then we say that it is nondegenerate if g_p is nondegenerate on T_pM for every p. If furthermore g_p has the same index for all p, then we say it has constant index.

Definition 7.56. If $g \in \mathcal{T}_2(M)$ is symmetric nondegenerate and has constant index on M, then we call g a semi-Riemannian metric and (M,g) a **semi-Riemannian manifold** or **pseudo-Riemannian manifold**. The index is called the **index** of (M,g) and denoted $\mathrm{ind}(g)$ or $\mathrm{ind}(M)$. The signature is also constant and so the manifold has a signature also. If the signature of a semi-Riemannian manifold (with $\dim(M) \geq 2$) is $(-1,+1,+1,+1,\ldots)$ (or according to some conventions $(1,-1,-1,-1,\ldots)$), then the manifold is called a **Lorentz manifold**.

The simplest semi-Riemannian manifolds are the spaces $\mathbb{R}^n_\nu$, which are the spaces $\mathbb{R}^n$ endowed with the scalar products given by
$$\langle x, y \rangle_\nu = -\sum_{i=1}^{\nu} x^i y^i + \sum_{i=\nu+1}^{n} x^i y^i.$$
Since ordinary Euclidean geometry does not use indefinite scalar products, we shall call the spaces $\mathbb{R}^n_\nu$ **semi-Euclidean** spaces when the index $\nu = \mathrm{ind}(g)$ is not zero. If we write just $\mathbb{R}^n$, then either we are not concerned with a scalar product at all, or the scalar product is assumed to be the usual inner product ($\nu = 0$). Thus a Riemannian metric is just the special case of index 0. The space $\mathbb{R}^4_1$ is called the **Minkowski** space.

We will usually write $\langle X_p, Y_p \rangle$ or $g(X_p, Y_p)$ in place of $g(p)(X_p, X_p)$. Also, for a pair of vector fields X and Y, we define the function $\langle X, Y \rangle$ which is given by $\langle X, Y \rangle(p) = \langle X_p, Y_p \rangle$. In local coordinates $(x^1, \ldots, x^n)$ on $U \subset M$, we have that $g|_U = g_{ij}dx^i \otimes dx^j$, where $g_{ij} = \langle \frac{\partial}{\partial x^i}, \frac{\partial}{\partial x^j} \rangle$. Thus if $X = X^i \frac{\partial}{\partial x^i}$ and $Y = Y^i \frac{\partial}{\partial x^i}$ on U, then

(7.13) $$\langle X, Y \rangle = g_{ij} X^i Y^i,$$

which is a smooth function defined on U. The expression $\langle X, Y \rangle = g_{ij} X^i Y^i$ means that for all $p \in U$ we have $\langle X(p), Y(p) \rangle = g_{ij}(p) X^i(p) Y^i(p)$. As we know, the functions X^i and Y^i are given by $X^i = dx^i(X)$ and $Y^i = dx^i(Y)$.

On a semi-Riemannian manifold, the musical isomorphisms are globalized in the obvious way to act on tensor fields. We simply apply the type change at each point in the domain of a given tensor field. For example, if A is a tensor field of type $(2,2)$, then we may obtain a new metrically equivalent tensor field A' of type $(1,3)$ by the rule that for any $p \in M$, we have $A'(p)(\alpha, u, v, w) := A(p)(\alpha, \flat u, v, w)$ for $\alpha \in T_p^*M$ and $u, v, w \in T_pM$. If we choose a chart $(U, \mathbf{x})$, then in terms of the coordinate frames and the corresponding g_{ij}, we have

(7.14) $$A' = A^i{}_{jkl} \frac{\partial}{\partial x^i} \otimes dx^j \otimes dx^k \otimes dx^l,$$

where

$$A^i{}_{jkl} = g_{ja} A^{ia}{}_{kl}.$$

Of course, each of the possible conventions for consolidation and index position globalize accordingly. The reader is invited to compare our treatment with those found in [**Pe**], [**Dod-Pos**], [**ON1**] and [**Stern**].

We have been using coordinate frame fields, but there is nothing preventing us from giving local components of tensor fields with respect to arbitrary smooth frame fields. For example, if we choose a frame field $(E_1, \ldots, E_n)$ and the corresponding dual frame field, then we may define $g_{ij} := \langle E_i, E_j \rangle$ with the corresponding g^{ij}, and then local expressions analogous to those above hold with the $\frac{\partial}{\partial x^i}$'s and dx^j's replaced by the E_i's and E_j's, where the components of the tensor are obtained by evaluating on these frame fields. In particular, if the frame field is an orthonormal frame field, then $g_{ij} = \pm 1$ for $i = j$ and $g_{ij} = 0$ for $i \neq j$. This can result in a good deal of simplification.

We now say a few words about the appropriate notion of equivalence of semi-Riemannian manifolds.

Definition 7.57. Let (M, g) and (N, h) be two semi-Riemannian manifolds. A diffeomorphism $\Phi : M \to N$ is called an **isometry** if $\Phi^* h = g$. Thus for an isometry $\Phi : M \to N$ we have $g(v, w) = h(T\Phi \cdot v, T\Phi \cdot w)$ for all $v, w \in TM$. If $\Phi : M \to N$ is a local diffeomorphism such that $\Phi^* h = g$, then Φ is called a **local isometry**. If there is an isometry $\Phi : M \to N$, then we say that (M, g) and (N, h) are **isometric**.

7.6. Metric Tensors

Definition 7.58. The set of all isometries of a semi-Riemannian manifold M to itself is a group called the isometry group. It is denoted by $\text{Isom}(M)$.

The isometry group of a generic manifold is most likely trivial, but examples of manifolds with relatively large isometry groups are easy to find using Lie group theory. Also, Myers and Steenrod showed that the isometry group of a compact Riemannian manifold is a Lie group (see [**My-St**]). Recall from Chapter 5 that associated to $\mathbb{R}^n_\nu$ we have the matrix groups $\text{O}(\nu, n - \nu)$ and $\text{SO}(\nu, n - \nu)$. The isometry group of $\mathbb{R}^n_\nu$ is given by

$$\text{Iso}(\nu, n - \nu) = \{L : L(x) = Qx + x_0$$
$$\text{for some } Q \in \text{O}(\nu, n - \nu) \text{ and } x_0 \in \mathbb{R}^n_\nu\}.$$

This is the **group of semi-Euclidean motions.**

Example 7.59. We have seen that a regular submanifold of a Euclidean space $\mathbb{R}^n$ is a Riemannian manifold with the metric inherited from $\mathbb{R}^n$. In particular, the sphere $S^{n-1} \subset \mathbb{R}^n$ is a Riemannian manifold. Every isometry of S^{n-1} is the restriction to S^{n-1} of an isometry of $\mathbb{R}^n$ that fixes the origin.

Definition 7.60. Let $\widetilde{M}$ and M be semi-Riemannian manifolds. If $\wp : \widetilde{M} \to M$ is a covering map such that $\wp$ is a local isometry, we call $\wp : \widetilde{M} \to M$ a **semi-Riemannian** covering.

If we have a local isometry $\phi : N \to M$, then any lift $\tilde{\phi} : N \to \widetilde{M}$ is also a local isometry (Problem 16). Deck transformations are lifts of the identity map $M \to M$, and so are diffeomorphisms which are local isometries. Thus deck transformations are, in fact, isometries. We conclude that the group of deck transformations of a semi-Riemannian cover is a subgroup of the group of isometries $\text{Isom}(\widetilde{M})$.

Let us consider here the case of a discrete group G and a discrete group action $\lambda : G \times M \to M$ that is smooth, proper, and free. We have already seen that the quotient space M/G has a unique structure as a smooth manifold such that the projection $\kappa : M \to M/G$ is a covering. Let us now assume that G acts by isometries so that $\lambda_g^* \langle \cdot, \cdot \rangle = \langle \cdot, \cdot \rangle$ for all $g \in G$. The tangent map $T\kappa : T_pM \to T_{\kappa(p)}(M/G)$ is onto. For $x \in M/G$, let $\bar{v}_1, \bar{v}_2 \in T_x(M/G)$. Define $h_x(\bar{v}_1, \bar{v}_2) = \langle v_1, v_2 \rangle$, where v_1 and v_2 are chosen at the same point and such that $T\kappa \cdot v_i = \bar{v}_i$. We wish to show that this is well-defined. Indeed, if $v_i \in T_pM$ and $w_i \in T_qM$ are such that $T_p\kappa \cdot v_i = T_q\kappa \cdot w_i = \bar{v}_i$ for $i = 1, 2$, then there is an isometry λ_g with $\lambda_g p = q$. Furthermore, since λ_g is a deck transformation and curves representing v_i and w_i must be related by this deck transformation, we also have $T_p\lambda_g \cdot v_i = w_i$. Thus

$$\langle v_1, v_2 \rangle = \langle T_p\lambda_g v_1, T_p\lambda_g v_2 \rangle = \langle w_1, w_2 \rangle,$$

which means h_x is well-defined. It is easy to show that $x \mapsto h_x$ is smooth and defines a metric on M/G with the same signature as that of $\langle \cdot, \cdot \rangle$ and that further, $\kappa^* h = \langle \cdot, \cdot \rangle$. In fact, we will use the same notation for either the metric on M/G or on M.

Definition 7.61. A lattice of rank k in $\mathbb{R}^n$ is a set of the form
$$\Gamma := \{x \in \mathbb{R}^n : x = \sum n_i f_i \text{ where } n_i \in \mathbb{Z}\},$$
where $f_1, \ldots, f_k$ are linearly independent elements of $\mathbb{R}^n$. The $f_1, \ldots, f_k$ are called the generators of the lattice.

The lattice $\mathbb{Z}^n \subset \mathbb{R}^n$ is the standard rank n lattice, and it is generated by the standard basis. A lattice is a subgroup of $\mathbb{R}^n$ and so acts on $\mathbb{R}^n$ by $a \mapsto a + v$ for $v \in \Gamma$. This is a discrete, free and proper action, and so the quotient $\mathbb{R}^n/\Gamma$ provides a simple example of the above construction and so has a metric induced from $\mathbb{R}^n$. If the lattice has full rank n, then $\mathbb{R}^n/\Gamma$ is called a flat torus (or flat n-torus) and is diffeomorphic to the product of n copies of the circle S^1. Each of these n-dimensional flat tori is locally isometric, but may not be globally isometric. To be more precise, suppose that $f_1, f_2, \ldots, f_n$ is a basis for $\mathbb{R}^n$ which is not necessarily orthonormal. Let Γ_f be the lattice consisting of integer linear combinations of $f_1, f_2, \ldots, f_n$. Now suppose we have two such lattices Γ_f and $\Gamma_{\bar{f}}$. When is $\mathbb{R}^n/\Gamma_f$ isometric to $\mathbb{R}^n/\Gamma_{\bar{f}}$? It may seem that, since these are clearly diffeomorphic and since they are locally isometric, they must be (globally) isometric. But this is *not* the case (see Problem 17). The study of the global geometry of flat tori is quite interesting and even has deep connections with fields outside of geometry such as arithmetic, which we shall not have the space to pursue.

We know from Chapter 4 that there are surfaces in $\mathbb{R}^3$ that are diffeomorphic to a torus $S^1 \times S^1$. Such surfaces inherit a metric from the ambient space, but it turns out that the Riemannian surface obtained in this way cannot be isometric to one of the flat 2-tori introduced here. In Chapter 13 we will see how each metric on a manifold gives rise to an associated curvature tensor. The reason the tori just introduced are referred to as flat is because (being locally isometric to some $\mathbb{R}^n$) they have vanishing curvature tensor.

If we have semi-Riemannian manifolds (M, g) and (N, h), then we can consider the product manifold $M \times N$ and the projections $\mathrm{pr}_1 : M \times N \to M$ and $\mathrm{pr}_2 : M \times N \to N$. The tensor $g \times h = \mathrm{pr}_1^* g + \mathrm{pr}_2^* h$ provides a semi-Riemannian metric on the manifold $M \times N$, which is then called the **semi-Riemannian product** of (M, g) and (N, h). Let $(U_1, \mathbf{x})$ and $(U_2, \mathbf{y})$ denote charts on M and N respectively. Then we may form a product chart for $M \times N$ defined on $U_1 \times U_2$. The coordinate functions of this chart are given by $\widetilde{x}^i = x^i \circ \mathrm{pr}_1$ and $\widetilde{y}^i = y^i \circ \mathrm{pr}_2$. We have the associated frame fields $\frac{\partial}{\partial \widetilde{x}^i}$

and $\frac{\partial}{\partial \widetilde{y}^i}$. The components of $g \times h = \mathrm{pr}_1^* g + \mathrm{pr}_2^* h$ in these coordinates are discovered by choosing a point $(p_1, p_2) \in U_1 \times U_2$ and then calculating. We have

$$\begin{aligned}
&g \times h\Big(\frac{\partial}{\partial \widetilde{x}^i}\Big|_{(p,q)}, \frac{\partial}{\partial \widetilde{y}^j}\Big|_{(p,q)}\Big) \\
&= \mathrm{pr}_1^* g\Big(\frac{\partial}{\partial \widetilde{x}^i}\Big|_{(p,q)}, \frac{\partial}{\partial \widetilde{y}^j}\Big|_{(p,q)}\Big) + \mathrm{pr}_2^* h\Big(\frac{\partial}{\partial \widetilde{x}^i}\Big|_{(p,q)}, \frac{\partial}{\partial \widetilde{y}^j}\Big|_{(p,q)}\Big) \\
&= g\Big(T\mathrm{pr}_1 \frac{\partial}{\partial \widetilde{x}^i}\Big|_{(p,q)}, T\mathrm{pr}_1 \frac{\partial}{\partial \widetilde{y}^j}\Big|_{(p,q)}\Big) + h\Big(T\mathrm{pr}_2 \frac{\partial}{\partial \widetilde{x}^i}\Big|_{(p,q)}, T\mathrm{pr}_2 \frac{\partial}{\partial \widetilde{y}^j}\Big|_{(p,q)}\Big) \\
&= g\Big(\frac{\partial}{\partial x^i}\Big|_p, 0_p\Big) + h\Big(0_q, \frac{\partial}{\partial y^j}\Big|_q\Big) \\
&= 0 + 0 = 0,
\end{aligned}$$

and (abbreviating a bit)

$$g \times h\Big(\frac{\partial}{\partial \widetilde{x}^i}\Big|_{(p,q)}, \frac{\partial}{\partial \widetilde{x}^j}\Big|_{(p,q)}\Big) = g\Big(\frac{\partial}{\partial x^i}\Big|_p, \frac{\partial}{\partial x^i}\Big|_p\Big) + h(0_q, 0_q) = g_{ij}(p).$$

Similarly $g \times h(\frac{\partial}{\partial \widetilde{y}^i}, \frac{\partial}{\partial \widetilde{y}^j})(p,q) = h_{ij}(q)$. In practice, the coordinate functions constructed above are often abusively denoted by $(x^1, \ldots, x^{n_1}, y^1, \ldots, y^{n_2})$ and the frame field,s by $\frac{\partial}{\partial x^1}, \ldots, \frac{\partial}{\partial x^{n_1}}, \frac{\partial}{\partial y^1}, \ldots, \frac{\partial}{\partial y^{n_2}}$. So with respect to these coordinates, the matrix of $g \times h$ is of the form

$$\begin{pmatrix} G & 0 \\ 0 & H \end{pmatrix},$$

where $G = (g_{ij} \circ \mathrm{pr}_1)$ and $H = (h_{ij} \circ \mathrm{pr}_2)$.

Notation 7.62. The product metric is often denoted by $g + h$ or in coordinates by $ds^2 = g_{ij}dx^i dx^j + h_{kl}dy^k dy^l$.

Every smooth manifold that admits partitions of unity also admits at least one (in fact infinitely many) Riemannian metric. This includes all (finite-dimensional) paracompact manifolds. The reason for this is that the set of all Riemannian metric tensors is, in an appropriate sense, convex. We record this as a proposition.

Proposition 7.63. *Every smooth (paracompact) manifold admits a Riemannian metric.*

Proof. This is a special case of Proposition 6.45. □

If M is a regular submanifold of a Riemannian manifold (N, h), then M inherits a Riemannian metric $g := \imath^* h$, where $\imath : M \hookrightarrow N$ is the inclusion map. We have already used this idea for submanifolds of the Euclidean space $\mathbb{R}^d$. More generally, if $f : M \to N$ is an immersion, then (M, f^*h) is a Riemannian manifold. In particular, if $f : M \to \mathbb{R}^d$ is an immersion, then we obtain a Riemannian metric on M. It turns out that every Riemannian metric on M can be obtained in this way. Actually, more is true! For any Riemannian manifold (M, g) there is an *embedding* $f : M \to \mathbb{R}^d$, for some d, such that $g := \imath^* g_0$, where g_0 denotes the standard metric on $\mathbb{R}^d$. What this means is that $f(M)$ is a regular submanifold, and if we give $f(M)$ the metric induced from the ambient space $\mathbb{R}^d$, then f becomes an isometry when viewed as a map into $f(M)$. We say that such an f is an **isometric embedding** of M into $\mathbb{R}^d$. In short, the result is that every Riemannian manifold can be isometrically embedded into some Euclidean space of sufficiently high dimension. This difficult theorem is called the **Nash embedding theorem** and is due to John Forbes Nash (see [**Nash1**] and [**Nash2**]). Note that d must be quite large in general ($d = (\dim M)^2 + 5(\dim M) + 3$ is sufficient). For an indefinite semi-Riemannian manifold (N, h), the pull-back f^*h by a smooth map $f : M \to N$ may not be a metric because there may be points $p \in M$ such that $T_p f(T_p M)$ is a degenerate subspace of $T_{f(p)} N$. In particular, not every embedding of a manifold M into a semi-Euclidean space $\mathbb{R}^d_\nu$ (with $1 < \nu < d$) induces a metric on M. Nevertheless, every metric on M of any index can be obtained using an appropriate embedding into some $\mathbb{R}^d_\nu$ (see [**Clark**]).

Problems

(1) Show that if $\tau \in V \otimes V^*$ has the same components τ^i_j with respect to every basis, then $\tau^i_j = a \delta^i_j$ for some $a \in \mathbb{R}$.

(2) If $e_1, \ldots, e_n$ is a basis for V and $f_1, \ldots, f_m$ is a basis for W, then $\{E^i_j\}_{i=1,\ldots,n\ j=1,\ldots,m}$ is a basis for $L(V, W)$, where $E^i_j(v) := e^i(v) f_j$. Show this directly without assuming the isomorphism of $W \otimes V^*$ with $L(V, W)$.

(3) Let (V_i, g_i) be scalar product spaces for $i = 1, \ldots, k$. By Corollary D.35 of Appendix D, there is a unique bilinear form

$$\varphi : \bigotimes_{i=1}^k V_i \times \bigotimes_{i=1}^k V_i \to \mathbb{R}$$

such that for $v_i \in V_i$ and $w_i \in W_i$,

$$\varphi(v_1 \otimes \cdots \otimes v_k, w_1 \otimes \cdots \otimes w_k) = \varphi_1(v_1, w_1) \cdots \varphi_k(v_k, w_k).$$

Show that φ is nondegenerate and that it is positive definite if each g_i is positive definite.

Problems

(4) Define $\tau : \mathfrak{X}(M) \times \mathfrak{X}(M) \to C^\infty(M)$ by $\tau(X, Y) = XYf$. Show that τ does not define a tensor field.

(5) Let b_{ij} and b'_{ij} be the components of a bilinear form b with respect to bases $e_1, \ldots, e_n$ and $e'_1, \ldots, e'_n$ respectively. Show that in general $\det(b_{ij})$ does not equal $\det(b'_{ij})$. Show that if $\det(b_{ij})$ is nonzero, then the same is true of $\det(b'_{ij})$.

(6) Show that while a single algebraic tensor τ_p at a point on a manifold can always be extended to a smooth tensor field, it is not the case that one may always extend a (smooth) tensor field defined on an open subset to a smooth tensor field on the whole manifold.

(7) Let $\phi : \mathbb{R}^2 \to \mathbb{R}^2$ be defined by $(x, y) \mapsto (x + 2y, y)$. Let $\tau := x \frac{\partial}{\partial x} \otimes dy + \frac{\partial}{\partial y} \otimes dy$. Compute $\phi_*\tau$ and $\phi^*\tau$.

(8) Prove Proposition 7.35.

(9) Let $\mathcal{D}$ be a tensor derivation on M and suppose that in a local chart we have $\mathcal{D}(\frac{\partial}{\partial x^i}) = \sum D_i^j \frac{\partial}{\partial x^j}$ for smooth functions D_i^j. Show that $\mathcal{D}(dx^j) = -\sum D_i^j dx^i$. Let X be a fixed vector field with components X^i in our chart. Find the D_i^j in the case that $\mathcal{D} = \mathcal{L}_X$.

(10) Let $A \in \mathcal{T}_0^2(M)$. Show that the component form of the Lie derivative with respect to a chart is given as
$$(\mathcal{L}_X A)^{ab} = \frac{\partial A^{ab}}{\partial x^h} X^h - \frac{\partial X^a}{\partial x^h} A^{hb} - \frac{\partial X^b}{\partial x^h} A^{ah}$$
(where we use the Einstein summation convention). Show that if $A \in \mathcal{T}_2^0(M)$, then the formula becomes
$$(\mathcal{L}_X A) = \frac{\partial A_{ab}}{\partial x^h} X^h + \frac{\partial X^h}{\partial x^a} A_{hb} + \frac{\partial X^h}{\partial x^b} A_{ah}.$$
Find a formula for $A \in \mathcal{T}_2^2(M)$.

(11) Show that our two definitions of the Lie derivative of a tensor field agree with each other.

(12) In some chart $(U, (x, y))$ on a 2-manifold, let $A = x \frac{\partial}{\partial y} \otimes dx \otimes dy + \frac{\partial}{\partial x} \otimes dy \otimes dy$ and let $X = \frac{\partial}{\partial x} + x \frac{\partial}{\partial y}$. Compute the coordinate expression for $\mathcal{L}_X A$.

(13) Suppose that for every chart $(U, \mathbf{x})$ in an atlas for a smooth n-manifold M we have assigned n^3 smooth functions Γ_{ij}^k, which we call Christoffel symbols. Suppose that rather than obeying the transformation law expected for a tensor, we have the following horrible formula relating the Christoffel symbols Γ'^k_{ij} on a chart $(U', \mathbf{y})$ to the symbols Γ^k_{ij}:
$$\Gamma'^k_{ij} = \frac{\partial^2 x^l}{\partial y^i \partial y^j} \frac{\partial y^k}{\partial x^l} + \Gamma^t_{rs} \frac{\partial x^r}{\partial y^i} \frac{\partial x^s}{\partial y^j} \frac{\partial y^k}{\partial x^t} \quad \text{(sum)}.$$

Assume that such a transformation law holds between the Christoffel symbol functions for all pairs of intersecting charts. For any pair of vector fields $X, Y \in \mathfrak{X}(M)$, consider the functions $(D_X Y)^k$ given in every chart by the formula $(D_X Y)^k := \frac{\partial Y^k}{\partial x^h} X^h + \Gamma^k_{ij} X^i Y^j$. Show that the local vector fields of the form $(D_X Y)^k \frac{\partial}{\partial x^k}$, defined on each chart, are the restrictions of a single global vector field $D_X Y$. Show that $D_X : Y \mapsto D_X Y$ is a derivation of $\mathfrak{X}(M)$ and that with $D_X f := Xf$ for smooth functions, we may extend to a tensor derivation; D_X is called a covariant derivative with respect to X. There are many possible covariant derivatives.

(14) Continuing on the last problem, show that $D_{fX+gY}\Upsilon = f D_X \Upsilon + g D_Y \Upsilon$ for all $f, g \in C^\infty(M)$ and $X, Y \in \mathfrak{X}(M)$ and $\Upsilon \in \mathcal{T}^r_s(M)$.

(15) Show that if $\frac{\partial}{\partial x^1}, \ldots, \frac{\partial}{\partial x^n}$ are coordinate vector fields from some chart, then $[\frac{\partial}{\partial x^i}, \frac{\partial}{\partial x^j}] \equiv 0$. Consider the vector fields $\frac{\partial}{\partial x}$ and $\frac{\partial}{\partial y}$ arising from standard coordinates on $\mathbb{R}^2$ and also the $\frac{\partial}{\partial r}$ and $\frac{\partial}{\partial \theta}$ from polar coordinates. Show that $[\frac{\partial}{\partial x}, \frac{\partial}{\partial r}]$ is not identically zero by explicit computation.

(16) Let $\widetilde{M} \to M$ be a semi-Riemannian cover. Prove that if we have a local isometry $\phi : N \to M$, then any lift $\phi : N \to \widetilde{M}$ is also a local isometry.

(17) Suppose we have two lattices Γ_f and $\Gamma_{\bar{f}}$ in $\mathbb{R}^n$. Let $\mathbb{R}^n$ have the standard metric. Describe the induced metrics on $\mathbb{R}^n / \Gamma_f$ and $\mathbb{R}^n / \Gamma_{\bar{f}}$ and provide a necessary and sufficient condition for the existence of an isometry $\mathbb{R}^n / \Gamma_f \to \mathbb{R}^n / \Gamma_{\bar{f}}$.

(18) Show that if $\alpha, \beta \in T_k(V)$ where V is a scalar product space, then the scalar product on $T_k(V)$ is given in terms of index raising and contraction by
$$\langle \alpha, \beta \rangle = \sum \alpha_{i_1 \ldots i_k} \beta^{i_1 \ldots i_k}.$$

Chapter 8

Differential Forms

In one guise, a differential form is nothing but an alternating (antisymmetric) tensor field. What is new is the introduction of an antisymmetrized version of the tensor product and also a natural differential operator called the exterior derivative. We start off with some more multilinear algebra.

8.1. More Multilinear Algebra

Definition 8.1. Let V and W be real finite-dimensional vector spaces. A k-multilinear map $\alpha : V \times \cdots \times V \to W$ is called **alternating** if $\alpha(v_1, \ldots, v_k) = 0$ whenever $v_i = v_j$ for some $i \neq j$. The space of all **alternating** k-multilinear maps into W will be denoted by $L_{\text{alt}}^k(V; W)$ or by $L_{\text{alt}}^k(V)$ if $W = \mathbb{R}$. By convention, $L_{\text{alt}}^0(V; W)$ is taken to be W and in particular, $L_{\text{alt}}^0(V) = \mathbb{R}$.

Since we are dealing with the field $\mathbb{R}$ (which has characteristic zero), it is easy to see that **alternating** k-multilinear maps are the same as (completely) **antisymmetric** k-multilinear maps which are defined by the property that for any permutation σ of the letters $1, 2, \ldots, k$ we have

$$\omega(v_1, v_2, \ldots, v_k) = \text{sgn}(\sigma)\omega(v_{\sigma(1)}, v_{\sigma(2)}, \ldots, v_{\sigma(k)}).$$

Let us denote the group of permutations of the k letters $1, 2, \ldots, k$ by S_k. In what follows, we will occasionally write σ_i in place of $\sigma(i)$.

Definition 8.2. The **antisymmetrization map** $\text{Alt}^k : T_k^0(V) \to L_{\text{alt}}^k(V)$ is defined by

$$\text{Alt}^k(\omega)(v_1, v_2, \ldots, v_k) := \frac{1}{k!} \sum_{\sigma \in S_k} \text{sgn}(\sigma)\omega(v_{\sigma_1}, v_{\sigma_2}, \ldots, v_{\sigma_k}).$$

Lemma 8.3. For $\alpha \in T^0_{k_1}(V)$ and $\beta \in T^0_{k_2}(V)$, we have

$$\operatorname{Alt}^{k_1+k_2}(\operatorname{Alt}^{k_1}\alpha \otimes \beta) = \operatorname{Alt}^{k_1+k_2}(\alpha \otimes \beta),$$
$$\operatorname{Alt}^{k_1+k_2}(\alpha \otimes \operatorname{Alt}^{k_2}\beta) = \operatorname{Alt}^{k_1+k_2}(\alpha \otimes \beta),$$

and

$$\operatorname{Alt}^{k_1+k_2}(\operatorname{Alt}^{k_1}(\alpha) \otimes \operatorname{Alt}^{k_2}(\beta)) = \operatorname{Alt}^{k_1+k_2}(\alpha \otimes \beta).$$

Proof. For a permutation $\sigma \in S_k$ and any $T \in T^0_k(V)$, let σT denote the element of $T^0_k(V)$ given by $(\sigma T)(v_1, \ldots, v_k) := T(v_{\sigma(1)}, \ldots, v_{\sigma(k)})$. We then have $\operatorname{Alt}^k(\sigma T) = \operatorname{sgn}(\sigma)\operatorname{Alt}^k(T)$ as may easily be checked. Also, by definition $\operatorname{Alt}^k(T) = \sum \operatorname{sgn}(\sigma)\sigma T$. We have

$$\operatorname{Alt}^{k_1+k_2}(\operatorname{Alt}^{k_1}(\alpha) \otimes \beta)$$
$$= \operatorname{Alt}^{k_1+k_2}\left(\left(\frac{1}{k_1!}\sum_{\sigma \in S_{k_1}} \operatorname{sgn}\sigma\, (\sigma\alpha)\right) \otimes \beta\right)$$
$$= \operatorname{Alt}^{k_1+k_2}\left(\frac{1}{k_1!}\sum_{\sigma \in S_{k_1}} \operatorname{sgn}\sigma\, (\sigma\alpha \otimes \beta)\right)$$
$$= \frac{1}{k_1!}\sum_{\sigma \in S_{k_1}} \operatorname{sgn}\sigma\, \operatorname{Alt}^{k_1+k_2}(\sigma\alpha \otimes \beta).$$

Let us examine the expression $\operatorname{sgn}\sigma\, \operatorname{Alt}^{k_1+k_2}(\sigma\alpha \otimes \beta)$. If we extend each $\sigma \in S_{k_1}$ to a corresponding element $\sigma' \in S_{k_1+k_2}$ by letting $\sigma'(i) = \sigma(i)$ for $i \leq k_1$ and $\sigma'(i) = i$ for $i > k_1$, then we have $\sigma\alpha \otimes \beta = \sigma'(\alpha \otimes \beta)$ and also $\operatorname{sgn}(\sigma) = \operatorname{sgn}(\sigma')$. Thus $\operatorname{sgn}\sigma\, \operatorname{Alt}^{k_1+k_2}(\sigma\alpha \otimes \beta) = \operatorname{sgn}\sigma'\, \operatorname{Alt}^{k_1+k_2}\sigma'(\alpha \otimes \beta)$ and so

$$\operatorname{Alt}^{k_1+k_2}(\operatorname{Alt}^{k_1}\alpha \otimes \beta)$$
$$= \frac{1}{k_1!}\sum_{\sigma \in S_{k_1}} \operatorname{sgn}\sigma'\operatorname{Alt}^{k_1+k_2}\left(\sigma'(\alpha \otimes \beta)\right)$$
$$= \frac{1}{k_1!}\sum_{\sigma \in S_{k_1}} \operatorname{sgn}\sigma'\operatorname{sgn}\sigma'\operatorname{Alt}^{k_1+k_2}(\alpha \otimes \beta)$$
$$= \operatorname{Alt}^{k_1+k_2}(\alpha \otimes \beta)\frac{1}{k_1!}\sum_{\sigma \in S_{k_1}} 1 = \operatorname{Alt}^{k_1+k_2}(\alpha \otimes \beta).$$

We arrive at $\operatorname{Alt}^{k_1+k_2}(\operatorname{Alt}^{k_1}(\alpha) \otimes \beta) = \operatorname{Alt}^{k_1+k_2}(\alpha \otimes \beta)$. In a similar way, $\operatorname{Alt}^{k_1+k_2}(\alpha \otimes \operatorname{Alt}^{k_2}\beta) = \operatorname{Alt}^{k_1+k_2}(\alpha \otimes \beta)$, and so the last part of the theorem follows. $\square$

8.1. More Multilinear Algebra

Given $\omega \in L_{\text{alt}}^{k_1}(V)$ and $\eta \in L_{\text{alt}}^{k_2}(V)$, we define their **exterior product** or **wedge product** $\omega \wedge \eta \in L_{\text{alt}}^{k_1+k_2}(V)$ by the formula

$$\omega \wedge \eta := \frac{(k_1+k_2)!}{k_1! k_2!} \text{Alt}^{k_1+k_2}(\omega \otimes \eta).$$

Written out, this is

(8.1)
$$\omega \wedge \eta(v_1, \ldots, v_{k_1}, v_{k_1+1}, \ldots, v_{k_1+k_2})$$
$$= \frac{1}{k_1! k_2!} \sum_{\sigma \in S_{k_1+k_2}} \text{sgn}(\sigma) \omega(v_{\sigma_1}, \ldots, v_{\sigma_{k_1}}) \eta(v_{\sigma_{k_1+1}}, \ldots, v_{\sigma_{k_1+k_2}}).$$

Warning: The factor in front of Alt in the definition of the **exterior product** is a convention but not the only convention in use. This choice has an effect on many of the formulas to follow which differ by a factor from the corresponding formulas written by authors following other conventions.

It is an exercise in combinatorics that we also have

$$(\omega \wedge \eta)(v_1, \ldots, v_{k_1}, v_{k_1+1}, \ldots, v_{k_1+k_2})$$
$$= \sum_{(k_1,k_2)\text{-shuffles } \sigma} \text{sgn}(\sigma) \omega(v_{\sigma_1}, \ldots, v_{\sigma_{k_1}}) \eta(v_{\sigma_{k_1+1}}, \ldots, v_{\sigma_{k_1+k_2}}).$$

In the latter formula, we sum over all permutations such that $\sigma(1) < \sigma(2) < \ldots < \sigma(k_1)$ and $\sigma(k_1+1) < \sigma(k_1+2) < \cdots < \sigma(k_1+k_2)$. This kind of permutation is called a (k_1, k_2)-**shuffle** as indicated in the summation.

The most important case of (8.1) is for $\omega, \eta \in L_{\text{alt}}^1(V)$, in which case

$$(\omega \wedge \eta)(v, w) = \omega(v)\eta(w) - \omega(w)\eta(v).$$

This clearly defines an antisymmetric bilinear map.

Proposition 8.4. *For $\alpha \in L_{\text{alt}}^{k_1}(V)$, $\beta \in L_{\text{alt}}^{k_2}(V)$, and $\gamma \in L_{\text{alt}}^{k_3}(V)$, we have*

(i) $\wedge : L_{\text{alt}}^{k_1}(V) \times L_{\text{alt}}^{k_2}(V) \to L_{\text{alt}}^{k_1+k_2}(V)$ *is $\mathbb{R}$-bilinear;*

(ii) $\alpha \wedge \beta = (-1)^{k_1 k_2} \beta \wedge \alpha$;

(iii) $\alpha \wedge (\beta \wedge \gamma) = (\alpha \wedge \beta) \wedge \gamma$.

Proof. We leave the proof of (i) as an easy exercise. For (ii), we consider the special permutation f given by $(f(1), f(2), \ldots, f(k_1+k_2)) = (k_1+1, \ldots, k_1+k_2, 1, \ldots, k_1)$. We have that $\alpha \otimes \beta = f(\beta \otimes \alpha)$. Also $\text{sgn}(f) = (-1)^{k_1 k_2}$. So we have

$$\text{Alt}^{k_1+k_2}(\alpha \otimes \beta) = \text{Alt}^{k_1+k_2}(f(\beta \otimes \alpha)) = (-1)^{k_1 k_2} \text{Alt}^{k_1+k_2}(\beta \otimes \alpha),$$

which gives (ii).

For (iii), we compute

$$\alpha \wedge (\beta \wedge \gamma) = \frac{(k_1 + k_2 + k_3)!}{k_1!\,(k_2 + k_3)!} \operatorname{Alt}(\alpha \otimes (\beta \wedge \gamma))$$

$$= \frac{(k_1 + k_2 + k_3)!}{k_1!\,(k_2 + k_3)!} \frac{(k_2 + k_3)!}{k_2!k_3!} \operatorname{Alt}(\alpha \otimes \operatorname{Alt}(\beta \otimes \gamma))$$

$$= \frac{(k_1 + k_2 + k_3)!}{k_1!k_2!k_3!} \operatorname{Alt}(\alpha \otimes \operatorname{Alt}(\beta \otimes \gamma)).$$

By Lemma 8.3, we know that $\operatorname{Alt}(\alpha \otimes \operatorname{Alt}(\beta \otimes \gamma)) = \operatorname{Alt}(\alpha \otimes (\beta \otimes \gamma))$, and so we arrive at

$$\alpha \wedge (\beta \wedge \gamma) = \frac{(k_1 + k_2 + k_3)!}{k_1!k_2!k_3!} \operatorname{Alt}(\alpha \otimes (\beta \otimes \gamma)).$$

By a symmetric computation, we also have

$$(\alpha \wedge \beta) \wedge \gamma = \frac{(k_1 + k_2 + k_3)!}{k_1!k_2!k_3!} \operatorname{Alt}((\alpha \otimes \beta) \otimes \gamma),$$

and so by the associativity of the tensor product we obtain the result. $\square$

Example 8.5. Let V have a basis e_1, e_2, e_3 with dual basis e^1, e^2, e^3. Let $\alpha = 2e^1 \wedge e^2 + e^1 \wedge e^2$ and $\beta = e^1 - e^3$. Then as a sample calculation we have

$$\alpha \wedge \beta = \left(2e^1 \wedge e^2 + e^1 \wedge e^3\right) \wedge (e^1 - e^3)$$

$$= 2e^1 \wedge e^2 \wedge e^1 + e^1 \wedge e^3 \wedge e^1$$

$$= -2e^1 \wedge e^2 \wedge e^3 - e^1 \wedge e^3 \wedge e^3$$

$$= -2e^1 \wedge e^2 \wedge e^3,$$

where we have used that $e^1 \wedge e^2 \wedge e^1 = -e^1 \wedge e^1 \wedge e^2 = 0$, etc.

Lemma 8.6. *Let $\alpha^1, \ldots, \alpha^k$ be elements of $V^* = L^1_{\operatorname{alt}}(V)$ and let $v_1, \ldots, v_k \in V$. Then we have*

$$\alpha^1 \wedge \cdots \wedge \alpha^k (v_1, \ldots, v_k) = \det A,$$

where $A = (a^i_j)$ is the $k \times k$ matrix whose ij-th entry is $a^i_j = \alpha^i(v_j)$.

Proof. From the proof of the last theorem we have

$$\alpha \wedge (\beta \wedge \gamma) = \frac{(k_1 + k_2 + k_3)!}{k_1!k_2!k_3!} \operatorname{Alt}(\alpha \otimes (\beta \otimes \gamma)).$$

By inductive application of this we have

$$\alpha^1 \wedge \cdots \wedge \alpha^k = k!\,\operatorname{Alt}(\alpha^1 \otimes \cdots \otimes \alpha^k).$$

Thus

$$\alpha^1 \wedge \cdots \wedge \alpha^k(v_1, \ldots, v_k) = \sum_\sigma \operatorname{sgn}(\sigma) \alpha^1(v_{\sigma_1}) \cdots \alpha^k(v_{\sigma_k}) = \det A. \quad \square$$

8.1. More Multilinear Algebra

Let us define

$$(8.2) \quad \epsilon^{i_1\ldots i_k}_{j_1\ldots j_k} = \begin{cases} 1 & \text{if } j_1,\ldots,j_k \text{ is an even permutation of } i_1,\ldots,i_k, \\ -1 & \text{if } j_1,\ldots,j_k \text{ is an odd permutation of } i_1,\ldots,i_k, \\ 0 & \text{otherwise.} \end{cases}$$

Then we have

Corollary 8.7. *Let $e_1,\ldots,e_n$ be a basis for V and $e^1, e^2,\ldots, e^n$ the dual basis for V*. Then we have*

$$e^{i_1} \wedge \cdots \wedge e^{i_k}(e_{j_1},\ldots,e_{j_k}) = \epsilon^{i_1\ldots i_k}_{j_1\ldots j_k}.$$

Since any $\alpha \in L^k_{\text{alt}}(V)$ is also a member of $T^0_k(V)$, we may write

$$\alpha = \sum \alpha_{i_1\ldots i_k} e^{i_1} \otimes \cdots \otimes e^{i_k},$$

where $\alpha_{i_1\ldots i_k} = \alpha(e_{i_1},\ldots,e_{i_k})$. By $\text{Alt}(\alpha) = \alpha$ and the linearity of Alt we have

$$\alpha = \sum \alpha_{i_1\ldots i_k} \text{Alt}\left(e^{i_1} \otimes \cdots \otimes e^{i_k}\right) = \frac{1}{k!} \sum \alpha_{i_1\ldots i_k} e^{i_1} \wedge \cdots \wedge e^{i_k}.$$

We conclude that the set of elements of the form $e^{i_1} \wedge \cdots \wedge e^{i_k}$ spans $L^{k_1}_{\text{alt}}(V)$. Furthermore, if we use the fact that both $\alpha_{i_{\sigma_1}\ldots i_{\sigma_k}} = \text{sgn}\,\sigma\, \alpha_{i_1\ldots i_k}$ and $e^{i_{\sigma_1}} \wedge \cdots \wedge e^{i_{\sigma_k}} = \text{sgn}\,\sigma\, e^{i_1} \wedge \cdots \wedge e^{i_k}$ for any permutation $\sigma \in S_k$, we see that we can permute the indices into increasing order and collect terms to get

$$\alpha = \frac{1}{k!} \sum \alpha_{i_1\ldots i_k} e^{i_1} \wedge \cdots \wedge e^{i_k} = \sum_{i_1 < i_2 < \ldots < i_k} a_{i_1 i_2,\ldots,i_k} e^{i_1} \wedge e^{i_2} \wedge \cdots \wedge e^{i_k},$$

where in the last expression we sum only over strictly increasing indices. We can check that the set of $\binom{n}{k}$ elements of the form $e^{i_1} \wedge e^{i_2} \wedge \cdots \wedge e^{i_k}$ with $1 \leq i_1 < i_2 < \cdots < i_k \leq n$ is linearly independent as follows: Suppose $\alpha = \sum_{i_1 < i_2 < \cdots < i_k} a_{i_1 i_2,\ldots,i_k} e^{i_1} \wedge e^{i_2} \wedge \cdots \wedge e^{i_k} = 0$. Let $j_1 < j_2 < \cdots < j_k$. Then we have

$$0 = \alpha(e_{j_1},\ldots,e_{j_k})$$
$$= \sum_{i_1 < i_2 < \cdots < i_k} a_{i_1 i_2\ldots i_k} e^{i_1} \wedge e^{i_2} \wedge \cdots \wedge e^{i_k}(e_{j_1},\ldots,e_{j_k})$$
$$= \sum_{i_1 < i_2 < \cdots < i_k} a_{i_1 i_2\ldots i_k} \epsilon^{i_1\ldots i_k}_{j_1\ldots j_k} = a_{j_1 j_2\ldots j_k}.$$

We have used that $\epsilon^{i_1\ldots i_k}_{j_1\ldots j_k}$ is zero unless $j_1\ldots j_k$ is a permutation of $i_1\ldots i_k$, and then in this case, since both are increasing, we must have $i_r = j_r$ for $r = 1,\ldots,k$. Thus we get $0 = a_{j_1 j_2\ldots j_k}$, and since the choice of j's was arbitrary, we have shown independence. Thus, we have proved the following theorem.

Theorem 8.8. *If $(e^1, e^2, \ldots, e^n)$ is a basis for V^*, then the set of elements*
$$\{e^{i_1} \wedge e^{i_2} \wedge \cdots \wedge e^{i_k} : 1 \leq i_1 < i_2 < \cdots < i_k \leq n\}$$
is a basis for $L_{\text{alt}}^k(V)$. Thus $\dim(L_{\text{alt}}^k(V)) = \binom{n}{k} = \frac{n!}{k!(n-k)!}$. In particular, $L_{\text{alt}}^k(V) = 0$ for $k > n$.

Corollary 8.9. *If $\alpha^1, \ldots, \alpha^k \in V^*$, then $\alpha^1, \ldots, \alpha^k$ are linearly independent if and only if*
$$\alpha^1 \wedge \cdots \wedge \alpha^k \neq 0.$$

Proof. If $\alpha^1, \ldots, \alpha^k$ are independent, then there are elements $\alpha^{k+1}, \ldots, \alpha^n$ such that $\alpha^1, \ldots, \alpha^n$ is a basis for V^*. Let $v_1, \ldots, v_n$ be the basis for V dual to the above basis. Then since $\alpha^1 \wedge \cdots \wedge \alpha^k$ is a basis element for $L_{\text{alt}}^k(V)$, it cannot be zero.

For the other direction, suppose that, after rearranging if needed, we have
$$\alpha^1 = c_2 \alpha^2 + \cdots + c_k \alpha^k.$$
Then we would have
$$\alpha^1 \wedge \cdots \wedge \alpha^k = \left(c_2 \alpha^2 + \cdots + c_k \alpha^k \right) \wedge \alpha^2 \wedge \cdots \wedge \alpha^k = 0.$$
Thus we see that $\alpha^1 \wedge \cdots \wedge \alpha^k \neq 0$ implies that $\alpha^1, \ldots, \alpha^k$ are linearly independent. $\square$

Notation 8.10. In order to facilitate notation, we will sometimes abbreviate a sequence of k integers, say $i_1, i_2, \ldots, i_k$, from the set $\{1, 2, \ldots, \dim(V)\}$ as I, and $e^{i_1} \wedge e^{i_2} \wedge \cdots \wedge e^{i_k}$ will be written as e^I and $\epsilon_L^I = \epsilon_{\ell_1 \ldots \ell_k}^{i_1 \ldots i_k}$. Also, if we require that $i_1 < i_2 < \cdots < i_k$, then we will write $\vec{I}$. We will freely use similar self-explanatory notation as we go along without further comment. For example, we may write
$$\alpha = \sum a_{\vec{I}} e^{\vec{I}}$$
to mean $\alpha = \sum_{i_1 < i_2 < \cdots < i_k} a_{i_1 i_2 \ldots i_k} e^{i_1} \wedge e^{i_2} \wedge \cdots \wedge e^{i_k}$.

Whenever we have $\alpha = \sum a_{\vec{I}} e^{\vec{I}}$, where $a_{\vec{I}} = a_{i_1 \ldots i_k}$ and $i_1 < \cdots < i_k$, we can define a_J for any k-tuple of indices $J = (j_1, \ldots, j_k)$ by requiring that $a_J = \sum \epsilon_J^{\vec{I}} a_{\vec{I}}$. Then we have $a_J = 0$ when the entries of J are indices that are not distinct. Otherwise, $a_J = \epsilon_J^{\vec{J}} a_{\vec{J}}$ where $\vec{J}$ is J rearranged in increasing order. But, it is also easy to see that $\alpha(e_{j_1}, \ldots, e_{j_k}) = \sum \epsilon_J^{\vec{I}} a_{\vec{I}}$. We then have
$$\alpha = \sum a_{\vec{I}} e^{\vec{I}} = \frac{1}{k!} \sum a_I e^I = \frac{1}{k!} \sum a_{i_1 \ldots i_k} e^{i_1} \wedge \cdots \wedge e^{i_k}$$
$$= \sum a_{i_1 \ldots i_k} e^{i_1} \otimes \cdots \otimes e^{i_k}.$$

8.1. More Multilinear Algebra

Exercise 8.11. Show that for $\alpha = \frac{1}{k!}\sum a_I e^I$ and $\beta = \frac{1}{\ell!}\sum b_J e^J$ with I and J as above, we have

$$\alpha \wedge \beta = \frac{1}{(k+\ell)!}\sum_{i_1\ldots i_{k+\ell}} (\alpha \wedge \beta)_{i_1\ldots i_{k+\ell}}\, e^{i_1} \wedge \cdots \wedge e^{i_{k+\ell}},$$

where

(8.3) $$(\alpha \wedge \beta)_{i_1\ldots i_{k+\ell}} = \frac{1}{k!\ell!}\sum a_{j_1\ldots j_k} b_{j_{k+1}\ldots j_{k+\ell}} \epsilon^{j_1\ldots j_{k+\ell}}_{i_1\ldots i_{k+\ell}}.$$

Definition 8.12. Let $v \in V$ and $\omega \in L^k_{\text{alt}}(V)$. Define $i_v \omega \in L^{k-1}_{\text{alt}}(V)$ by

$$i_v\omega(v_1,\ldots,v_{k-1}) := \omega(v, v_1,\ldots,v_{k-1});$$

$i_v\omega$ is called the **interior product** of v with ω or the **contraction** of ω by v. By convention $i_v a = 0$ for $a \in L^0_{\text{alt}}(V) := \mathbb{R}$. Thus we obtain a linear map $i_v : L^k_{\text{alt}}(V) \to L^{k-1}_{\text{alt}}(V)$.

It is clear that i_v depends linearly on v and that $i_v i_w \omega = -i_w i_v \omega$ for all $v, w \in V$ and hence $i_v \circ i_v = 0$.

With $L^1_{\text{alt}}(V) = V^*$ and $L^0_{\text{alt}}(V) = \mathbb{R}$, the sum

$$L_{\text{alt}}(V) = \bigoplus_{k=0}^{\dim(V)} L^k_{\text{alt}}(V)$$

is made into an $\mathbb{R}$-algebra via the exterior product just defined. Since $\alpha \wedge \beta \in L^{k_1+k_2}_{\text{alt}}(V)$ whenever $\alpha \in L^{k_1}_{\text{alt}}(V)$ and $\beta \in L^{k_2}_{\text{alt}}(V)$, the algebra $L_{\text{alt}}(V)$ is a graded algebra (see Definition D.43).

The contraction map i_v for $v \in V$ extends to a map $L_{\text{alt}}(V) \to L_{\text{alt}}(V)$. Since $i_v(L^k_{\text{alt}}(V)) \subset L^{k-1}_{\text{alt}}(V)$, we say that i_v is a map of degree -1.

Proposition 8.13. *For $v \in V$, the contraction map i_v satisfies the product rule*

$$i_v(\alpha \wedge \beta) = (i_v\alpha) \wedge \beta + (-1)^k \alpha \wedge (i_v \beta) \text{ for } \alpha \in L^k_{\text{alt}}(V).$$

The map $i_v : L_{\text{alt}}(V) \to L_{\text{alt}}(V)$ is the unique degree -1 map satisfying the above product rule and satisfying $i_v a = 0$ for $a \in \mathbb{R}$ and

$$i_v \theta = \theta(v) \text{ for } \theta \in V^* = L^1_{\text{alt}}(V) \text{ and } v \in V.$$

Proof. Let $\alpha \in L^k_{\text{alt}}(V)(M)$ and $\beta \in L^\ell_{\text{alt}}(V)(M)$. In the following computation we use the permutation $\widetilde{\sigma}$ which is given by

$$(2,3,\ldots,k+1,1,k+2,\ldots,k+\ell) \mapsto (1,2,\ldots,k+\ell).$$

The sign of $\tilde{\sigma}$ is $(-1)^k$. We compute as follows:

$(\imath_v \alpha \wedge \beta + (-1)^k \alpha \wedge \imath_v \beta)(v_2, \ldots, v_{k+\ell})$

$= \dfrac{(k+\ell-1)!}{(k-1)!\ell!} \mathrm{Alt}(\imath_v \alpha \otimes \beta)(v_2, \ldots, v_{k+\ell})$

$\quad + (-1)^k \dfrac{(k+\ell-1)!}{k!(\ell-1)!} \mathrm{Alt}(\alpha \otimes \imath_v \beta)(v_2, \ldots, v_{k+1})$

$= \displaystyle\sum_\sigma \dfrac{\mathrm{sgn}(\sigma)}{(k-1)!\ell!} \alpha(v, v_{\sigma_2}, \ldots, v_{\sigma_k}) \beta(v_{\sigma_{k+1}}, \ldots, v_{\sigma_{k+\ell}})$

$\quad + (-1)^k \dfrac{1}{k!(\ell-1)!} \displaystyle\sum_\sigma \mathrm{sgn}(\sigma) \alpha(v_{\sigma_2}, \ldots, v_{\sigma_{(k+1)}}) \beta(v, v_{\sigma_{k+2}}, \ldots, v_{\sigma_{k+\ell}})$

$= \dfrac{1}{(k-1)!\ell!} \displaystyle\sum_\sigma \mathrm{sgn}(\sigma) \alpha(v, v_{\sigma_2}, \ldots, v_{\sigma_k}) \beta(v_{\sigma_{k+1}}, \ldots, v_{\sigma_{k+\ell}})$

$\quad + (-1)^k \dfrac{1}{k!(\ell-1)!} \displaystyle\sum_\sigma \mathrm{sgn}(\sigma\tilde{\sigma}) \alpha(v, v_{\sigma_2}, \ldots, v_{\sigma_{k+1}}) \beta(v_{\sigma_{k+1}}, \ldots, v_{\sigma_{k+\ell}})$

$= \dfrac{1}{(k-1)!\ell!} \displaystyle\sum_\sigma \mathrm{sgn}(\sigma) \alpha(v, v_{\sigma_2}, \ldots, v_{\sigma_k}) \beta(v_{\sigma_{k+1}}, \ldots, v_{\sigma_{k+\ell}})$

$\quad + \dfrac{1}{k!(\ell-1)!} \displaystyle\sum_\sigma \mathrm{sgn}(\sigma) \alpha(v, v_{\sigma_2}, \ldots, v_{\sigma_{k+1}}) \beta(v_{\sigma_{k+1}}, \ldots, v_{\sigma_{k+\ell}})$

$= \dfrac{1}{(k-1)!\ell!} \displaystyle\sum_\sigma \alpha(v, v_{\sigma_2}, \ldots, v_{\sigma_{(k+1)}}) \beta(v_{\sigma_{k+1}}, \ldots, v_{\sigma_{k+\ell}})$

$= \alpha \wedge \beta(v, v_2, \ldots, v_{k+1}) = \imath_v(\alpha \wedge \beta)(v_2, \ldots, v_{k+1}).$

We leave the proof of the remaining statements of the theorem to the reader (see Problem 14). $\square$

It follows from the above that if $\theta_1, \ldots, \theta_k \in V^*$ and $v \in V$, then

$$\imath_v(\theta_1 \wedge \cdots \wedge \theta_k) = \sum_{\ell=1}^{k} (-1)^{k+1} \theta_\ell(v) \theta_1 \wedge \cdots \wedge \widehat{\theta_\ell} \wedge \cdots \wedge \theta_k,$$

where the caret over θ_ℓ denotes omission.

Proposition 8.14. *If $\lambda \in L(V, V)$, $\alpha \in L_{\mathrm{alt}}^{k_1}(V)$ and $\beta \in L_{\mathrm{alt}}^{k_2}(V)$, then*

$$\lambda^*(\alpha \wedge \beta) = \lambda^*\alpha \wedge \lambda^*\beta.$$

Proof. This follows from Proposition 1.83 and the definition of the exterior product. $\square$

We look more closely at $L_{\mathrm{alt}}^n(V)$ where $n = \dim V$. The dimension of $L_{\mathrm{alt}}^n(V)$ is one, and any nonzero element of $L_{\mathrm{alt}}^n(V)$ provides a basis. If $\lambda \in L(V, V)$, then $\lambda^* : L_{\mathrm{alt}}^n(V) \to L_{\mathrm{alt}}^n(V)$ is a linear transformation

8.1. More Multilinear Algebra

between one-dimensional vector spaces, and so it must be multiplication by an element of $\mathbb{R}$. Thus there is a unique number $\det(\lambda) \in \mathbb{R}$ such that

$$\lambda^*\omega = \det(\lambda)\omega$$

for any $\omega \in L^n_{\text{alt}}(V)$. This number is called the **determinant** of λ. This provides a definition of determinant that does not involve a choice of basis. We will show that if $A = (a^i_j)$ represents $\lambda \in L(V, V)$ with respect to a basis for V, then $\det(\lambda) = \det(A)$, where the determinant of a matrix is given by the standard definition. Let $\lambda \in L(V, V)$ and suppose that $\lambda(e_i) = \sum a^j_i e_j$ for some basis $(e_1, \ldots, e_n)$ with dual $(e^1, \ldots, e^n)$. Then $A = (a^i_j)$ represents λ, and we have

$$\lambda^*(e^1 \wedge \cdots \wedge e^n)(e_1, \ldots, e_n) = \left(e^1 \wedge \cdots \wedge e^n\right)(\lambda e_1, \ldots, \lambda e_n)$$
$$= \det(e^i(\lambda e_j)) = \det A.$$

On the other hand, $\lambda^*(e^1 \wedge \cdots \wedge e^n) = (\det \lambda)(e^1 \wedge \cdots \wedge e^n)$, and since $e^1 \wedge \cdots \wedge e^n(e_1, \ldots, e_n) = 1$, it must be that $\det(\lambda) = \det(A)$.

Exercise 8.15. Show (without using a basis) that if $\ell, \lambda \in L(V, V)$, then

(i) $\det(\ell \circ \lambda) = \det \ell \det \lambda$;

(ii) $\det(\text{id}) = 1$;

(iii) $\lambda \in \text{GL}(V)$ if and only if $\det \lambda \neq 0$;

(iv) if $\lambda \in \text{GL}(V)$, then $\det(\lambda^{-1}) = (\det \lambda)^{-1}$.

We now briefly discuss orientations of real vector spaces. In what follows, let V be a *real* vector space with $n = \dim V$. The nonzero elements of $L^n_{\text{alt}}(V)$ are sometimes referred to as volume elements (although this term will also apply to a global object later on). Two nonzero elements ω_1 and ω_2 of $L^n_{\text{alt}}(V)$ are said to be **equivalent** if there is a scalar $c > 0$ such that $\omega_1 = c\omega_2$. The equivalence class of ω will be denoted $[\omega]$. There are clearly exactly two equivalence classes. An equivalence class is referred to as an **orientation** for V. An **oriented vector space** is a vector space with a choice of orientation and is sometimes written as a pair $(V, [\omega])$.

An ordered basis $(e_1, \ldots, e_n)$ for an oriented real vector space $(V, [\omega])$ is said to be positive with respect to the orientation if $\omega(e_1, \ldots, e_n) > 0$ for some and hence any $\omega \in [\omega]$. Equivalently, $(e_1, \ldots, e_n)$ is positive if $e^1 \wedge \cdots \wedge e^n \in [\omega]$. Actually a choice of ordered basis for a real vector space determines an orientation with respect to which it is positive. Indeed, if $e^1, \ldots, e^n$ is dual to such a basis, then choose $[\omega]$ where $\omega = e^1 \wedge \cdots \wedge e^n$.

Definition 8.16. Let $(V_1, [\omega_1])$ and $(V_2, [\omega_2])$ be oriented real vector spaces. A linear isomorphism $\lambda : V_1 \to V_2$ is said to be orientation preserving if $\lambda^*\omega_2 = c\omega_1$ for some $c > 0$ and some, and hence any, choices $\omega_1 \in [\omega_1]$ and $\omega_2 \in [\omega_2]$.

When one talks about an element $\lambda \in \mathrm{GL}(V)$ being orientation preserving, one means that $\det \lambda > 0$ and this is tantamount to $\lambda : (V, [\omega]) \to (V, [\omega])$ being orientation preserving for any choice of orientation $[\omega]$.

8.1.1. The abstract Grassmann algebra. We now take a very abstract approach to constructing an algebra that will be seen to be isomorphic to $L_{\mathrm{alt}}(V)$. We wish to construct a space that is universal with respect to alternating multilinear maps. We work in the category of real vector spaces, although much of what we do here makes sense for modules. Consider the tensor space $T^k(V) := \bigotimes^k V$ (take any realization of the abstract tensor product as in Definition D.17). Let A be the submodule of $T^k(V)$ generated by elements of the form

$$v_1 \otimes \cdots v_i \otimes \cdots \otimes v_i \cdots \otimes v_k.$$

In other words, A is generated by simple tensors with two (or more) equal factors. Recall that associated to $T^k(V)$ we have the canonical map $\otimes : V \times \cdots \times V \to T^k(V)$ defined so that $\otimes(v_1, \ldots, v_k) = v_1 \otimes \cdots \otimes v_k$. We define the space of k-vectors to be

$$V \wedge \cdots \wedge V := \bigwedge\nolimits^k V := T^k(V)/A.$$

Let $\wedge_k : V \times \cdots \times V \to \bigwedge^k V$ be the composition of the canonical map $\otimes$ with quotient map of $T^k(V)$ onto $\bigwedge^k V$. This map turns out to be an alternating multilinear map. We will denote $\wedge_k(v_1, \ldots, v_k)$ by $v_1 \wedge \cdots \wedge v_k$. Using the universal property of $T^k(V)$ as described in Appendix D, one can show that the pair $(\bigwedge^k V, \wedge_k)$ is universal with respect to alternating k-multilinear maps: That is, given any alternating k-multilinear map $\alpha : V \times \cdots \times V \to W$, there is a unique linear map $\alpha_\wedge : \bigwedge^k V \to W$ such that $\alpha = \alpha_\wedge \circ \wedge_k$; that is, the following diagram commutes:

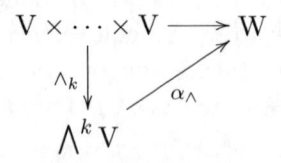

$$\begin{array}{ccc} V \times \cdots \times V & \longrightarrow & W \\ {\scriptstyle \wedge_k} \downarrow & \nearrow {\scriptstyle \alpha_\wedge} & \\ \bigwedge^k V & & \end{array}$$

Notice that we also have that $v_1 \wedge \cdots \wedge v_k$ is the image of $v_1 \otimes \cdots \otimes v_k$ under the quotient map $T^k(V) \to \bigwedge^k V$. Next we define $\bigwedge V := \bigoplus_{k=0}^\infty \bigwedge^k V$, which is a direct sum, and we take $\bigwedge^0 V := \mathbb{R}$. We impose on $\bigwedge V$ the multiplication generated by the rule

$$(v_1 \wedge \cdots \wedge v_i) \times (v'_1 \wedge \cdots \wedge v'_j) \mapsto v_1 \wedge \cdots \wedge v_i \wedge v'_1 \wedge \cdots \wedge v'_j \in \bigwedge\nolimits^{i+j} V.$$

The resulting graded algebra is called the **Grassmann algebra** or **exterior algebra**. (Of course, the definition of $\wedge$ here is different from what we defined previously.) If we need to have a $\mathbb{Z}$ grading rather than an $\mathbb{N}$ grading,

8.1. More Multilinear Algebra

we may define $\bigwedge^k V := 0$ for $k < 0$ and extend the multiplication in the obvious way. Elements of $\bigwedge V$ are called **multivectors** and specifically, elements of $\bigwedge^k V$ are called k-**multivectors**.

Notice that since $(v+w) \wedge (w+v) = 0$, it follows that $v \wedge w = -w \wedge v$. In fact, any transposition of the factors in a simple element such as $v_1 \wedge \cdots \wedge v_k$, introduces a change of sign:

$$v_1 \wedge \cdots \wedge v_i \wedge \cdots \wedge v_j \wedge \cdots \wedge v_k$$
$$= -v_1 \wedge \cdots \wedge v_j \wedge \cdots \wedge v_i \wedge \cdots \wedge v_k.$$

Lemma 8.17. *If V has dimension n, then $\bigwedge^k V = 0$ for $k > n$. If $e_1, \ldots, e_n$ is a basis for V, then the set*

$$\{e_{i_1} \wedge \cdots \wedge e_{i_k} : 1 \leq i_1 < \cdots < i_k \leq n\}$$

is a basis for $\bigwedge^k V$ where we agree that $e_{i_1} \wedge \cdots \wedge e_{i_k} = 1$ if $k = 0$.

Proof. The first statement is easy and we leave it to the reader. We will show that the set above is indeed a basis. First note that $\bigwedge^n V$ is spanned by $e_1 \wedge \cdots \wedge e_n$. To see that $e_1 \wedge \cdots \wedge e_n$ is not zero we let

$$\det : V \times \cdots \times V \to \mathbb{R}$$

be the multilinear map given by representing the arguments as column vectors of components with respect to the given basis and then taking the determinant of the $n \times n$ matrix built from these column vectors. Then $\det(e_1, \ldots, e_n) = 1$. But by the universal property above there is a linear map $\det_\wedge$ such that $\det = \det_\wedge \circ \wedge_k$ and so

$$\det_\wedge(e_1 \wedge \cdots \wedge e_n) = \det_\wedge \circ \wedge_k (e_1, \ldots, e_n)$$
$$= \det(e_1, \ldots, e_n) = 1;$$

thus, we conclude that $e_1 \wedge \cdots \wedge e_n$ is not zero (and is a basis for $\bigwedge^n V$). Now it is easy to see that the elements of the form $e_{i_1} \wedge \cdots \wedge e_{i_k}$ span $\bigwedge^k V$. To see that we have linear independence, suppose that

$$\sum_{1 \leq i_1 < \cdots < i_k \leq n} a_{i_1 \ldots i_k} e_{i_1} \wedge \cdots \wedge e_{i_k} = 0.$$

Now multiply both sides by $e_{j_1} \wedge \cdots \wedge e_{j_{n-k}}$, where $\{j_1, \ldots, j_{n-k}\}$ is the complement of $\{i_1, \ldots, i_k\}$. Then we obtain

$$\pm a_{i_1 \ldots i_k} e_1 \wedge \cdots \wedge e_n = 0,$$

from which we conclude that $a_{i_1 \ldots i_k} = 0$, and since $i_1, \ldots, i_k$ was arbitrary, we are done. $\square$

Exercise 8.18. Show that $v_1, \ldots, v_k \in V$ are linearly independent if and only if $v_1 \wedge \cdots \wedge v_k \neq 0$. Compare this with Corollary 8.9.

Definition 8.19. An element $\xi \in \bigwedge^k V$ is called **decomposable** if $\xi = v_1 \wedge \cdots \wedge v_k$ for some $v_1, \ldots, v_k \in V$.

Let us gain a little practice dealing with multivectors by proving the following proposition:

Proposition 8.20. *Let $\xi \in \bigwedge^2 V$ with $\xi \neq 0$. Then there is a basis $v_1, \ldots, v_n$ of V such that*

$$\xi = v_1 \wedge v_2 + v_3 \wedge v_4 + \cdots + v_{2r-1} \wedge v_{2r}.$$

Furthermore, in this case the r-fold product $\xi \wedge \cdots \wedge \xi$ is nonzero and decomposable while the $r+1$-fold product is zero.

Proof. First we prove that there exists such a decomposition for ξ. Let $e_1, \ldots, e_n$ be a basis for V. We have

$$\xi = \sum_{i<j} a_{ij} e_i \wedge e_j = a_{12} e_1 \wedge e_2 + a_{13} e_1 \wedge e_2 + \cdots + a_{1n} e_1 \wedge e_n$$

$$+ a_{23} e_2 \wedge e_3 + a_{24} e_2 \wedge e_4 + \cdots + a_{2n} e_2 \wedge e_n + \xi',$$

where ξ' does not involve e_1 or e_2. By renumbering if necessary we may assume that a_{12} is not zero. But then if

$$v_1 := e_1 + (a_{23}/a_{12}) e_3 + \cdots + (a_{2n}/a_{12}) e_n$$
$$v_2 := a_{12} e_2 + a_{13} e_3 + \cdots + a_{1n} e_n,$$

we have that $v_1, v_2, e_3, \ldots, e_n$ are linearly independent and

$$\xi = v_1 \wedge v_2 + v,$$

where v does not involve e_1 or e_2. If $v = 0$, then we are done. Otherwise we may repeat the process with

$$v = \sum_{\substack{i<j \\ i,j \geq 3}} b_{ij} e_i \wedge e_j.$$

Clearly a simple induction gives

$$\xi = v_1 \wedge v_2 + v_3 \wedge v_4 + \cdots + v_{2r-1} \wedge v_{2r}$$

for some nonzero $v_1, \ldots, v_{2r}$ such that $v_1, \ldots, v_{2r}, e_{2r+1}, \ldots, e_n$ is the desired basis.

Next we consider the r-fold product. If we set $\phi_k := v_{2k-1} \wedge v_{2k}$ for $k = 1, \ldots, r$, then

$$\xi = \sum_{i=1}^r \phi_i \text{ and } \phi_i \wedge \phi_i = 0.$$

8.1. More Multilinear Algebra

On the other hand, if $i \neq j$, then $\phi_i \wedge \phi_j = \phi_j \wedge \phi_i$ is plus or minus one times one of the elements in the linearly independent set

$$\{2v_{\ell_1} \wedge v_{\ell_2} \wedge v_{\ell_3} \wedge v_{\ell_4} : \ell_1 < \ell_2 < \ell_3 < \ell_4 \text{ and } \ell_i \in \{1, \ldots, 2r\}\}.$$

Thus
$$\xi^2 = \xi \wedge \xi = 2 \sum_{1 \leq i < j \leq r} \phi_i \wedge \phi_i.$$

In general, it is easy to see that since all of the ϕ_i commute with each other, we have

$$\xi^k = \sum_{1 \leq i_1, \ldots, i_k \leq r} \phi_{i_1} \wedge \phi_{i_2} \wedge \cdots \wedge \phi_{i_k} = k! \sum_{1 \leq i_1 < \cdots < i_k \leq r} \phi_{i_1} \wedge \phi_{i_2} \wedge \cdots \wedge \phi_{i_k}.$$

Then, it is easy to see that
$$\xi^r = r! \phi_1 \wedge \phi_2 \wedge \cdots \wedge \phi_r = r! v_1 \wedge v_2 \wedge \cdots \wedge v_{2r} \neq 0,$$

while $\xi^{r+1} = 0$. $\square$

The number r from the previous proposition is called the rank of the element $\xi \in \bigwedge^2 V$ and the definition works just as well for elements of $\bigwedge^2 V^*$. It can be shown that if

$$\xi = \sum_{i<j} a_{ij} e_i \wedge e_j = \frac{1}{2} \sum_{i,j} a_{ij} e_i \wedge e_j,$$

where (a_{ij}) is an antisymmetric matrix, then the rank of ξ is the rank of the matrix (a_{ij}).

The following proposition follows easily from the universal properties of the exterior product.

Remark 8.21. There is a natural isomorphism
$$L_{\text{alt}}^k(V; W) \cong L\left(\bigwedge^k V; W\right).$$

Lemma 8.22. *In particular,*
$$L_{\text{alt}}^k(V) \cong \left(\bigwedge^k V\right)^*.$$

We would now like to embed $\bigwedge^k V^*$ into $\bigotimes^k V^*$, and this involves a choice. For each k, let $A_k : V^* \times \cdots \times V^* \to \bigotimes^k V^*$ be defined by

$$A_k(\alpha_1, \ldots, \alpha_k) := \sum_\sigma \text{sgn}(\sigma) \alpha_{\sigma_1} \otimes \cdots \otimes \alpha_{\sigma_k}.$$

By the universal property of $\bigwedge^k V^*$ we obtain an induced map
$$\widetilde{A_k} : \bigwedge^k V^* \to \bigotimes^k V^*.$$

If we identify $\bigotimes^k V^*$ with $T_k(V)$, then we get a map $\widetilde{A}_k : \bigwedge^k V^* \to T_k(V)$.

Proposition 8.23. *The image of the map $\widetilde{A}_k : \bigwedge^k V^* \to T_k(V)$ is a linear isomorphism with image equal to $L_{\mathrm{alt}}^k(V)$ such that*
$$\widetilde{A}_k(\alpha_1 \wedge \cdots \wedge \alpha_k)(v_1, \ldots, v_k) = \det(\alpha_i(v_j)).$$

Proof. We leave the proof as Problem 5. $\square$

We combine these maps to obtain a linear isomorphism
$$\widetilde{A} : \bigwedge V^* \to L_{\mathrm{alt}}(V).$$
Now both $L_{\mathrm{alt}}(V)$ and $\bigwedge V^*$ have already independently been given the exterior algebra structures via their respective wedge products. One may check that $\widetilde{A}$ has been defined in such a way as to be an isomorphism of these algebras:
$$\bigwedge V^* \cong L_{\mathrm{alt}}(V) \text{ (as exterior algebras)}.$$
Also notice that
$$\bigwedge\nolimits^k V^* \cong L_{\mathrm{alt}}^k(V) \cong \left(\bigwedge\nolimits^k V\right)^*,$$
which allows us to think of $\bigwedge^k V^*$ as dual to $\bigwedge^k V$ in such a way that
$$(\alpha_1 \wedge \cdots \wedge \alpha_k)(v_1 \wedge \cdots \wedge v_k) = \det(\alpha_i(v_j)).$$

Remark 8.24 (An identification). In what follows, we will identify $\bigwedge^k V^*$ with $L_{\mathrm{alt}}^k(V)$ and hence $\bigwedge V^*$ with $L_{\mathrm{alt}}(V)$ whenever convenient. In other words, we freely treat elements of $\bigwedge^k V^*$ as alternating multilinear forms.

8.2. Differential Forms

Let M be an n-manifold. We now bundle together the various spaces $L_{\mathrm{alt}}^k(T_pM)$. That is, we form the natural bundle $L_{\mathrm{alt}}^k(TM)$ that has as its fiber at p, the space $L_{\mathrm{alt}}^k(T_pM)$. Thus $L_{\mathrm{alt}}^k(TM) = \bigcup_{p \in M} L_{\mathrm{alt}}^k(T_pM)$.

Exercise 8.25. Exhibit the smooth structure and vector bundle structure on $L_{\mathrm{alt}}^k(TM) = \bigcup_{p \in M} L_{\mathrm{alt}}^k(T_pM)$. [Hint: Let $(U, \mathbf{x})$ be a chart of M and $x^1, \ldots, x^n$ the coordinate functions. Let $\widetilde{U} := \bigcup_{p \in U} L_{\mathrm{alt}}^k(T_pM)$ and let $d = \binom{n}{k}$. Then we have a map $\widetilde{U} \to U \times \mathbb{R}^d$ given by $\alpha_p \mapsto (p, a)$ where a is the d-tuple of components (in some fixed order) of α_p given by its local coordinate representation.]

Let the smooth sections of this bundle be denoted by
$$(8.4) \qquad \Omega^k(M) = \Gamma(M; L_{\mathrm{alt}}^k(TM)),$$
and sections over $U \subset M$ by $\Omega_M^k(U)$.

8.2. Differential Forms

Definition 8.26. Elements of $\Omega^k(M)$ are called **differential k-forms** or just k-forms.

The space $\Omega^k(M)$ is a module over the algebra of smooth functions $C^\infty(M) = \mathcal{F}(M)$. If $n = \dim M$ then we have the *direct* sum

$$\Omega(M) = \bigoplus_{k=0}^{n} \Omega^k(M),$$

with a similar decomposition for any open $U \subset M$.

Exercise 8.27. Show that there is a module isomorphism

$$\bigoplus_{k=0}^{n} \Omega^k(M) \cong \Gamma\left(\bigoplus_{k=0}^{n} L^k_{\text{alt}}(TM)\right).$$

Definition 8.28. Let M be an n-manifold. The elements of $\Omega(M)$ are called **differential forms** on M. We identify $\Omega^k(M)$ with the obvious subspace of $\Omega(M) = \bigoplus_k \Omega^k(M)$. A differential form in $\Omega^k(M)$ is said to be **homogeneous** of **degree** k.

If $\omega \in \Omega(M)$, then we can uniquely write $\omega = \sum_{k=1}^{n} \omega_k$ where $\omega_k \in \Omega^k(M)$ and the ω_k are called **homogeneous components** of ω.

Definition 8.29. For $\omega \in \Omega^{k_1}(M)$, and $\eta \in \Omega^{k_2}(M)$, we define the **exterior product** $\omega \wedge \eta \in \Omega^{k_1+k_2}(M)$ by

$$(\omega \wedge \eta)(p) := \omega(p) \wedge \eta(p).$$

It is easy to see that $\Omega(M)$ is a ring under the wedge product and, in fact, a $C^\infty(M)$-algebra. Whenever convenient, we may extend this to a sum over all $n \in \mathbb{Z}$ by defining (as before) $\Omega^k(M) := 0$ for $k < 0$ and $\Omega^k(M) := 0$ if $k > \dim(M)$. We have made the trivial extension of $\wedge$ to a $\mathbb{Z}$-**graded algebra** by declaring that $\omega \wedge \eta = 0$ if either η or ω is homogeneous of negative degree. Thus $\Omega(M)$ is a graded algebra and is said to be **graded commutative** because $\alpha \wedge \beta = (-1)^{k\ell} \beta \wedge \alpha$ for $\alpha \in \Omega^k(M)$ and $\beta \in \Omega^\ell(M)$. Sometimes one see this notion referred to as skew-commutativity.

Just as a tangent vector is the infinitesimal version of a (parametrized) curve through a point $p \in M$, so a covector at $p \in M$ is the infinitesimal version of a function defined near p. At this point one must be careful. It is true that for any single covector $\alpha_p \in T_pM$ there always exists a function f such $df_p = \alpha_p$. But as we saw in Chapter 2, if $\alpha \in \Omega^1(M)$, then it is not necessarily true that there is a function $f \in C^\infty(M)$ such that $df = \alpha$. If $f_1, f_2, \ldots, f_k$ are smooth functions, then one way to picture $df_1 \wedge \cdots \wedge df_k$ is by thinking of the intersecting family of level sets of the functions $f_1, f_2, \ldots, f_k$, which in some cases can be pictured as a sort of "egg crate", structure. For

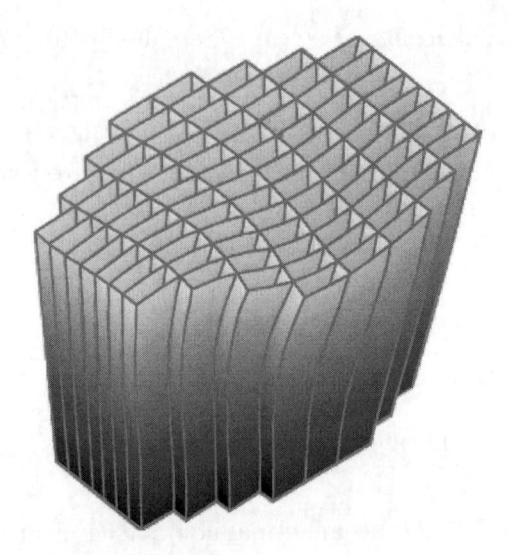

Figure 8.1. 2-form as "flux tubes"

a 2-form in a 3-manifold, one obtains "flux tubes" as shown in Figure 8.1. The infinitesimal version of this is a sort of straightened out "linear egg crate structure," which may be thought of as existing in the tangent space at a point. This is the rough intuition for $df_1|_p \wedge \cdots \wedge df_k|_p$, and the k-form $df_1 \wedge \cdots \wedge df_k$ is a field of such structures which somehow fit the level sets of the family $f_1, f_2, \ldots, f_k$. Of course, $df_1 \wedge \cdots \wedge df_k$ is a very special kind of k-form. In general, a k-form over U may not arise from a family of functions.

In local coordinates, calculation is often quite easy and formal. For example, in $\mathbb{R}^3$ with standard coordinates x, y, z, a simple wedge product calculation is as follows:

$$(xy dx + z dy + dz) \wedge (x dy + z dz)$$
$$= xy dx \wedge x dy + xy dx \wedge z dz + z dy \wedge x dy$$
$$\quad + z dy \wedge z dz + dz \wedge x dy + dz \wedge z dz$$
$$= x^2 y dx \wedge dy + xyz dx \wedge dz + z^2 dy \wedge dz + x dz \wedge dy$$
$$= x^2 y dx \wedge dy + xyz dx \wedge dz + (z^2 - x) dy \wedge dz.$$

An equally trivial calculation shows that

$$(xyz^2 dx \wedge dy + dy \wedge dz) \wedge (dx + x dy + z dz) = (xyz^3 + 1) dx \wedge dy \wedge dz.$$

8.2.1. Pull-back of a differential form. Since we treat differential forms as alternating covariant tensor fields, we have a notion of pull-back already defined. It is easy to see that the pull-back of an alternating tensor field is also an alternating tensor field, and so given any smooth map $f : M \to N$,

8.2. Differential Forms

we get a map $f^* : \Omega^k(N) \to \Omega^k(M)$. We recall here the definition:

$$(f^*\eta)(p)(v_1,\ldots,v_k) = \eta_{f(p)}(T_pf \cdot v_1,\ldots,T_pf \cdot v_k),$$

for tangent vectors $v_1,\ldots,v_k \in T_pM$. The pull-back extends in the obvious way to a map $f^* : \Omega(N) \to \Omega(M)$.

Proposition 8.30. *Let $f : M \to N$ be a smooth map and let $\eta_1, \eta_2 \in \Omega(N)$. Then we have*

$$f^*(\eta_1 \wedge \eta_2) = f^*\eta_1 \wedge f^*\eta_2.$$

Proof. This follows directly from Proposition 8.14. □

Proposition 8.31. *Let $f : M \to N$ and $g : N \to P$ be smooth maps. Then for any smooth differential form $\eta \in \Omega(P)$ we have $(f \circ g)^*\eta = g^*(f^*\eta)$. Thus $(f \circ g)^* = g^* \circ f^*$.*

Proof. We prove only the case $\eta \in \Omega^1(N)$. The general case is entirely similar and is left to the reader. For $v \in T_pM$, we have

$$(f \circ g)^*\eta(v) = \eta(T(f \circ g)v) = \eta(Tf \circ Tg(v))$$
$$= f^*\eta(Tg \cdot v) = g^*(f^*\eta)(v),$$

which completes the proof for the considered case. □

From the above propositions we see that we have a contravariant functor from the category of smooth manifolds and smooth maps to the category of rings which assigns to each smooth manifold M the space $\Omega(M)$ and to each smooth map f the ring homomorphism f^*.

In case S is a regular submanifold of M, we have the inclusion map $\iota : S \hookrightarrow M$, which maps $p \in S$ to the very same point $p \in M$. As mentioned before, it is natural to identify T_pS with $T_p\iota(T_pS)$ for any $p \in S$. In other words, we often do not distinguish between a vector v_p and $T_p\iota(v_p)$. Thus we view the tangent bundle of S as a subset of TM. With this in mind we must realize that for any $\alpha \in \Omega^k(M)$ the form $\iota^*\alpha$ is just the restriction of α to vectors tangent to S. In particular, if $U \subset M$ is open and $\iota : U \hookrightarrow M$, then $\iota^*\alpha = \alpha|_U$.

The local expression for the pull-back is described as follows. Let us abbreviate: $\epsilon_L^I = \epsilon_{\ell_1\ldots\ell_k}^{i_1\ldots i_k}$ (recall the definition given by equation (8.2)). Let $(U, \mathbf{x})$ be a chart on M and $(V, \mathbf{y})$ a chart on N with $f(U) \subset V$. Then, writing $\eta = \sum b_{\vec{J}} dy^{\vec{J}}$ and abbreviating $\frac{\partial (y^j \circ f)}{\partial x^i}$ to simply $\frac{\partial y^j}{\partial x^i}$, etc. we have

$$f^*\eta = \sum b_{\vec{J}} \circ f \, d\left(y^{j_1} \circ f\right) \wedge \cdots \wedge d\left(y^{j_k} \circ f\right)$$

$$= \sum_{\vec{J}} \left(b_{\vec{J}} \circ f\right) \left(\sum_{i_1} \frac{\partial y^{j_1}}{\partial x^{i_1}} dx^{i_1}\right) \wedge \cdots \wedge \left(\sum_{i_k} \frac{\partial y^{j_k}}{\partial x^{i_k}} dx^{i_k}\right)$$

$$= \sum_{\vec{J}} \left(b_{\vec{J}} \circ f\right) \sum_{I} \frac{\partial y^{j_1}}{\partial x^{i_1}} \cdots \frac{\partial y^{j_k}}{\partial x^{i_k}} dx^{i_1} \wedge \cdots \wedge dx^{i_k}$$

$$= \sum_{\vec{J}} \left(b_{\vec{J}} \circ f\right) \sum_{\vec{L}} \left(\sum_{I} \epsilon_{\vec{L}}^{I} \frac{\partial y^{j_1}}{\partial x^{i_1}} \cdots \frac{\partial y^{j_k}}{\partial x^{i_1}}\right) dx^{\ell_1} \wedge \cdots \wedge dx^{\ell_k}$$

$$= \sum_{\vec{J}} \sum_{\vec{L}} b_{\vec{J}} \circ f \, \frac{\partial (y^{j_1}, \ldots, y^{j_k})}{\partial (x^{\ell_1}, \ldots, x^{\ell_k})} dx^{\ell_1} \wedge \cdots \wedge dx^{\ell_k},$$

where

$$\frac{\partial y^J}{\partial x^L} := \frac{\partial (y^{j_1}, \ldots, y^{j_k})}{\partial (x^{\ell_1}, \ldots, x^{\ell_k})} = \det \begin{bmatrix} \frac{\partial y^{j_1}}{\partial x^{\ell_1}} & \cdots & \frac{\partial y^{j_1}}{\partial x^{\ell_k}} \\ \vdots & & \vdots \\ \frac{\partial y^{j_k}}{\partial x^{\ell_1}} & \cdots & \frac{\partial y^{j_k}}{\partial x^{\ell_k}} \end{bmatrix}.$$

Since the above formula is a bit intimidating at first sight, we work out the case where $\dim M = 2$, $\dim N = 3$ and $k = 2$. As a warm up, notice that since $dx^i \wedge dx^i = 0$, we have

$$d\left(y^2 \circ f\right) \wedge d\left(y^3 \circ f\right) = \sum \frac{\partial y^2}{\partial x^i} dx^i \wedge \frac{\partial y^3}{\partial x^j} dx^j = \sum \frac{\partial y^2}{\partial x^i} \frac{\partial y^3}{\partial x^j} dx^i \wedge dx^j$$

$$= \frac{\partial y^2}{\partial x^1} \frac{\partial y^3}{\partial x^2} dx^1 \wedge dx^2 + \frac{\partial y^2}{\partial x^2} \frac{\partial y^3}{\partial x^1} dx^2 \wedge dx^1$$

$$= \left(\frac{\partial y^2}{\partial x^1} \frac{\partial y^3}{\partial x^2} - \frac{\partial y^2}{\partial x^2} \frac{\partial y^3}{\partial x^1}\right) dx^1 \wedge dx^2$$

$$= \frac{\partial (y^2, y^3)}{\partial (x^1, x^2)} dx^1 \wedge dx^2.$$

Using similar expressions, we have

$$f^*\eta = f^*\left(b_{23} dy^2 \wedge dy^3 + b_{13} dy^1 \wedge dy^3 + b_{12} dy^1 \wedge dy^2\right)$$

$$= b_{23} \circ f \sum \frac{\partial y^2}{\partial x^i} dx^i \wedge \frac{\partial y^3}{\partial x^j} dx^j + b_{13} \circ f \sum \frac{\partial y^1}{\partial x^i} dx^i \wedge \frac{\partial y^3}{\partial x^j} dx^j$$

$$+ b_{12} \circ f \sum \frac{\partial y^1}{\partial x^i} dx^i \wedge \frac{\partial y^2}{\partial x^j} dx^j$$

$$= \left(b_{23} \circ f \frac{\partial (y^2, y^3)}{\partial (x^1, x^2)} + b_{13} \circ f \frac{\partial (y^1, y^3)}{\partial (x^1, x^2)} + b_{12} \circ f \frac{\partial (y^1, y^2)}{\partial (x^1, x^2)}\right) dx^1 \wedge dx^2.$$

8.3. Exterior Derivative

Remark 8.32. Notice that the space $\Omega^0(M)$ is just the space of smooth functions $C^\infty(M)$, and so unfortunately we now have several notations for the same space: $C^\infty(M) = \Omega^0(M) = \mathcal{T}^0_0(M)$.

8.3. Exterior Derivative

Here we will define and study the exterior derivative d. For 0-forms, exterior differentiation is just the operation of taking the differential: $f \mapsto df$. Let us start right out with giving an idea of what the exterior derivative looks like for k-forms defined on an open set in $U \subset \mathbb{R}^n$. Using standard rectangular coordinates x^i, all k-forms can be written as sums of terms of the form $f dx^{i_1} \wedge \cdots \wedge dx^{i_k}$ for some $f \in C^\infty(U)$. We know that the differential of a 0-form is a 1-form: $d : f \mapsto \frac{\partial f}{\partial x^i} dx^i$. We inductively extend the definition of d to an operator that takes k-forms to $k+1$ forms. We declare d to be linear over real numbers and then define $d(f dx^{i_1} \wedge \cdots \wedge dx^{i_k}) = df \wedge dx^{i_1} \wedge \cdots \wedge dx^{i_k}$. It is an easy exercise to show that if the latter formula holds for increasing indices $i_1 < \cdots < i_k$, then it holds for all choices of indices. For example, if in $\mathbb{R}^2$ we have a 1-form $\alpha = x^2 dx + xy\, dy$, then

$$d\alpha = d(x^2 dx + xy\, dy)$$
$$= d(x^2) \wedge dx + d(xy) \wedge dy$$
$$= 2x\, dx \wedge dx + (y\, dx + x\, dy) \wedge dy$$
$$= y\, dx \wedge dy.$$

We now develop the general theory on manifolds. For the next theorem we think of Ω_M as an assignment $\Omega_M : U \mapsto \Omega_M(U) = \Omega(U)$ for open $U \subset M$. Thus we are simply thinking in terms of (pre)sheaves. We will drop the subscript M on Ω_M when there is no chance of confusion.

Definition 8.33. A (natural) **graded derivation** of degree r on Ω_M is a family of maps, one for each open set $U \subset M$, denoted $\mathcal{D}_U : \Omega_M(U) \to \Omega_M(U)$, such that for each $U \subset M$,

$$\mathcal{D}_U : \Omega_M^k(U) \to \Omega_M^{k+r}(U)$$

and such that

(1) $\mathcal{D}_U$ is $\mathbb{R}$ linear;
(2) $\mathcal{D}_U(\alpha \wedge \beta) = \mathcal{D}_U \alpha \wedge \beta + (-1)^{kr} \alpha \wedge \mathcal{D}_U \beta$ for $\alpha \in \Omega^k(U)$ and $\beta \in \Omega_M(U)$;
(3) $\mathcal{D}_U$ is natural with respect to restriction:

$$\begin{array}{ccc} \Omega^k(U) & \xrightarrow{\mathcal{D}_U} & \Omega^{k+r}(U) \\ \downarrow & & \downarrow \\ \Omega^k(V) & \xrightarrow{\mathcal{D}_V} & \Omega^{k+r}(V) \end{array}$$

As usual we will denote all of the maps by a single symbol $\mathcal{D}$. In summary, we have a map of (pre)sheaves $\mathcal{D} : \Omega_M \to \Omega_M$. Along the lines similar to our study of tensor derivations, one can show that a graded derivation of Ω_M is completely determined by, and can be defined by its action on 0-forms (functions) and 1-forms. In fact, since every form can be locally built out of functions and exact 1-forms, i.e. differentials, we only need to know the action on 0-forms and exact 1-form to determine the graded derivation. Recall that an element of $\Omega^1(U)$ is said to be exact if it is the differential of a smooth function.

Remark 8.34. If one has a map $\mathcal{D} : \Omega(M) \to \Omega(M)$ that satisfies (1) and (2) of the previous definition, then it is just called a graded derivation of degree k. But, when we meet derivations below, they will also be defined on open submanifolds and will all give natural derivations.

Proposition 8.35. *If $\mathcal{D}_1$ and $\mathcal{D}_2$ are (natural) graded derivations of degrees r_1 and r_2 respectively then the operator*

$$[\mathcal{D}_1, \mathcal{D}_2] := \mathcal{D}_1 \circ \mathcal{D}_2 - (-1)^{r_1 r_2} \mathcal{D}_2 \circ \mathcal{D}_1$$

is a (natural) graded derivation of degree $r_1 + r_2$.

Proof. See Problem 15. □

Lemma 8.36. *Suppose $\mathcal{D}_1 : \Omega_M^k(U) \to \Omega_M^{k+r}(U)$ and $\mathcal{D}_2 : \Omega_M^k(U) \to \Omega_M^{k+r}(U)$ are defined for each open set $U \subset M$ and both satisfy (1), (2) and (3) of Definition 8.33 above. If $\mathcal{D}_1$ and $\mathcal{D}_2$ agree when applied to functions and exact forms, then $\mathcal{D}_1 = \mathcal{D}_2$.*

Proof. By (3), if $\mathcal{D}_1$ and $\mathcal{D}_2$ agree on chart domains, then they agree globally. Let $x^1, \ldots, x^n$ be local coordinates on U. Then every element of $\Omega_M^k(U)$ is a sum of elements of the form $f dx^{i_1} \wedge \cdots \wedge dx^{i_k}$. But by (2) we have

$$\mathcal{D}_1 \left(f dx^{i_1} \wedge \cdots \wedge dx^{i_k} \right) = \mathcal{D}_1 f \wedge dx^{i_1} \wedge \cdots \wedge dx^{i_k} \pm f \mathcal{D}_1 \left(dx^{i_1} \wedge \cdots \wedge dx^{i_k} \right)$$
$$= \mathcal{D}_2 f \wedge dx^{i_1} \wedge \cdots \wedge dx^{i_k} \pm f \mathcal{D}_1 \left(dx^{i_1} \wedge \cdots \wedge dx^{i_k} \right).$$

The last term can be expanded using (2), and then the elements $\mathcal{D}_1 dx^{i_j}$ can be replaced by $\mathcal{D}_2 dx^{i_j}$. The result is equal to $\mathcal{D}_2 \left(f dx^{i_1} \wedge \cdots \wedge dx^{i_k} \right)$. □

The differential d defined by

(8.5) $\qquad df(X) = Xf$ for $X \in \mathfrak{X}_M(U)$ and $f \in C^\infty(U)$

gives a map $\Omega_M^0 \to \Omega_M^1$. Next we show that this map can be extended to a degree one graded derivation.

8.3. Exterior Derivative

Theorem 8.37. *Let M be a smooth manifold. There is a unique degree one graded derivation $d : \Omega_M \to \Omega_M$ such that*

$$d \circ d = 0$$

and such that for each open $U \subset M$ and $f \in C^\infty(U) = \Omega_M^0(U)$, the 1-form df coincides with the usual differential. Furthermore, for any chart $(U, \mathbf{x})$ for M, we have the following local formula:

$$d \sum \alpha_{\vec{I}} dx^{\vec{I}} = \sum_{\vec{I}} (d\alpha_{\vec{I}}) \wedge dx^{\vec{I}}.$$

Proof. We define an operator $d_\mathbf{x}$ for each chart $(U, \mathbf{x})$. For a 0-form on U (i.e. a smooth function), we just define $d_\mathbf{x} f$ to be the usual differential given by $df = \sum \frac{\partial f}{\partial x^i} dx^i$. For $\alpha \in \Omega_M^k(U)$, we have $\alpha = \sum \alpha_{\vec{I}} dx^{\vec{I}}$ and we define $d_\mathbf{x} \alpha = \sum d\alpha_{\vec{I}} \wedge dx^{\vec{I}}$. To show the product rule ((2) of Definition 8.33), consider $\alpha = \sum \alpha_{\vec{I}} dx^{\vec{I}} \in \Omega_M^k(U)$ and $\beta = \sum \beta_{\vec{J}} dx^{\vec{J}} \in \Omega_M^\ell(U)$. Then

$$\begin{aligned} d_\mathbf{x}(\alpha \wedge \beta) &= d_\mathbf{x} \left(\sum \alpha_{\vec{I}} dx^{\vec{I}} \wedge \sum \beta_{\vec{J}} dx^{\vec{J}} \right) \\ &= d_\mathbf{x} \left(\sum \alpha_{\vec{I}} \beta_{\vec{J}} dx^{\vec{I}} \wedge dx^{\vec{J}} \right) \\ &= \sum \left((d\alpha_{\vec{I}}) \beta_{\vec{J}} + \alpha_{\vec{I}} (d\beta_{\vec{J}}) \right) dx^{\vec{I}} \wedge dx^{\vec{J}} \\ &= \left(\sum_{\vec{I}} d\alpha_{\vec{I}} \wedge dx^{\vec{I}} \right) \wedge \sum_{\vec{J}} \beta_{\vec{J}} dx^{\vec{J}} \\ &\quad + \sum_{\vec{I}} \alpha_{\vec{I}} dx^{\vec{I}} \wedge \left((-1)^k \sum_{\vec{J}} d\beta_{\vec{J}} \wedge dx^{\vec{J}} \right), \end{aligned}$$

since $d\beta_{\vec{J}} \wedge dx^{\vec{I}} = (-1)^k dx^{\vec{I}} \wedge d\beta_{\vec{J}}$ due to the k interchanges of the basic differentials dx^i. This means that the product rule holds for each $d_\mathbf{x}$. For any function f, we have $d_\mathbf{x} d_\mathbf{x} f = d_\mathbf{x} df = \sum_{ij} \left(\frac{\partial^2 f}{\partial x^i \partial x^j} \right) dx^i \wedge dx^j = 0$ since $\frac{\partial^2 f}{\partial x^i \partial x^j}$ is symmetric in i, j and $dx^i \wedge dx^j$ is antisymmetric in i, j. More generally, for any functions $f, g \in C^\infty(U)$ we have $d_\mathbf{x}(df \wedge dg) = 0$ because of the graded commutativity. Inductively we get $d_\mathbf{x}(df_1 \wedge df_2 \wedge \cdots \wedge df_k) = 0$ for any functions $f_i \in C^\infty(U)$. From this it follows that for any $\alpha = \sum \alpha_{\vec{I}} dx^{\vec{I}} \in \Omega_M^k(U)$ we have $d_\mathbf{x} d_\mathbf{x} \sum \alpha_{\vec{I}} dx^{\vec{I}} = d_\mathbf{x} \sum d\alpha_{\vec{I}} \wedge dx^{\vec{I}} = 0$ since $d_\mathbf{x} d\alpha_{\vec{I}} = 0$ and $d_\mathbf{x} dx^{\vec{I}} = 0$. We have now defined, for each coordinate chart $(U, \mathbf{x})$, an operator $d_\mathbf{x}$ that clearly has the desired properties on that chart. Consider two different charts $(U, \mathbf{x})$ and $(V, \mathbf{y})$ such that $U \cap V \neq \emptyset$. We need to show that $d_\mathbf{x}$ restricted to $U \cap V$ coincides with $d_\mathbf{y}$ restricted to $U \cap V$, but it is clear that these restrictions of $d_{\mathbf{x}}$ and $d_{\mathbf{y}}$ satisfy the hypothesis of Lemma 8.36 and so they must agree on $U \cap V$.

It is now clear that the individual operators on coordinate charts fit together to give a well-defined operator with the desired properties. □

Definition 8.38. The degree one graded derivation just introduced is called the **exterior derivative**.

Another approach to the existence of the exterior derivative is to exhibit a global coordinate free formula. Let $\omega \in \Omega^k(M)$ and view ω as an alternating multilinear map on $\mathfrak{X}(M)$. Then for $X_0, X_1, \ldots, X_k \in \mathfrak{X}(M)$, define

$$d\omega(X_0, X_1, \ldots, X_k)$$
$$= \sum_{0 \leq i \leq k} (-1)^i X_i(\omega(X_0, \ldots, \widehat{X_i}, \ldots, X_k))$$
$$+ \sum_{0 \leq i < j \leq k} (-1)^{i+j} \omega([X_i, X_j], X_0, \ldots, \widehat{X_i}, \ldots, \widehat{X_j}, \ldots, X_k).$$

One can check that $d\omega$ is an alternating $C^\infty(M)$-multilinear map on $\mathfrak{X}(M)$ and so defines a differential form of degree $k+1$. By applying this formula to coordinate fields one obtains the same local operator defined previously.

Lemma 8.39. *Given any smooth map $f : M \to N$, we have that d is natural with respect to the pull-back:*

$$f^*(d\eta) = d(f^*\eta).$$

Proof. By Lemma 2.119 we know the result is true if η is a 1-form. Because d is natural with respect to restriction, we need only prove the formula for a differential form defined in the domain of a chart $(U, \mathbf{x})$. By linearity we may assume that $\eta = g\, dx^{i_1} \wedge \cdots \wedge dx^{i_k}$ since an arbitrary form on U is a sum of forms of this type:

$$f^*(d\eta) = f^*(d(g\, dx^{i_1} \wedge \cdots \wedge dx^{i_k})) = f^*(dg \wedge dx^{i_1} \wedge \cdots \wedge dx^{i_k})$$
$$= d(f^*g) \wedge d(f^*x^{i_1}) \wedge \cdots \wedge d(f^*x^{i_k})$$
$$= d(g \circ f) \wedge d(x^{i_1} \circ f) \wedge \cdots \wedge d(x^{i_k} \circ f)$$
$$= d((g \circ f)\, d(x^{i_1} \circ f) \wedge \cdots \wedge d(x^{i_k} \circ f)) = d(f^*\eta). \quad \square$$

Definition 8.40. A smooth differential form α is called **closed** if $d\alpha = 0$ and **exact** if $\alpha = d\beta$ for some differential form β.

Notice that since $d \circ d = 0$, *every exact form is closed*. In general, the converse is not true. The extent to which the converse fails is a topological property of the manifold. This is the point of the de Rham[1] cohomology

---
[1] Georges de Rham 1903–1990.

to be studied in detail in Chapter 10. Here we just give the following basic definitions. The set of closed forms of degree k on a smooth manifold M is the kernel of $d : \Omega^k(M) \to \Omega^{k+1}(M)$ and is denoted $Z^k(M)$. The set of exact k-forms is the image of the map $d : \Omega^{k-1}(M) \to \Omega^k(M)$ and is denoted $B^k(M)$. Since $d \circ d = 0$, we have $B^k(M) \subset Z^k(M)$.

Definition 8.41. The k-th **de Rham cohomology group** (actually a vector space) is given by

$$(8.6) \qquad H^k(M) = \frac{Z^k(M)}{B^k(M)}.$$

In other words, we look at closed forms and identify any two whose difference is an exact form.

If $\alpha \in Z^k(M) \subset \Omega^k(M)$, then the equivalence class that contains α is denoted $[\alpha]$ and called the **cohomology class** of α. If $f : M \to N$ is smooth and $[\beta] \in H^k(N)$, then it is easy to see that $f^*\beta$ is closed since β is closed. Thus we obtain a cohomology class $[f^*\beta]$. Also, $[f^*\beta]$ depends only on the equivalence class of β. Indeed, if $\beta - \beta' = d\eta$, then $f^*\beta - f^*\beta' = df^*\eta$. Thus we may define a linear map $f^* : H^k(N) \to H^k(M)$ by $f^*[\beta] := [f^*\beta]$. We return to this topic later.

8.4. Vector-Valued and Algebra-Valued Forms

Now we consider a straightforward generalization. Let V and W be real vector spaces so that we have $L_{\text{alt}}^k(V; W)$ (Definition 8.1). For $\omega \in L_{\text{alt}}^k(V; W)$ and $\eta \in L_{\text{alt}}^\ell(V; W)$, we define the exterior product using the same formula as before except that we use the tensor product so that $\omega \wedge \eta$ is an element of $L_{\text{alt}}^{k+\ell}(V; W \otimes W)$:

$$(\omega \wedge \eta)(v_1, v_2, \ldots, v_k, v_{k+1}, v_{k+2}, \ldots, v_{k+\ell})$$
$$= \frac{1}{k_1! k_2!} \sum_{\sigma \in S_{k_1+k_2}} \text{sgn}(\sigma) \omega(v_{\sigma_1}, v_{\sigma_2}, \ldots, v_{\sigma_k}) \otimes \eta(v_{\sigma(k+1)}, v_{\sigma(k+2)}, \ldots, v_{\sigma(k+\ell)}).$$

We want to globalize this algebra. Let M be a smooth n-manifold and consider the set

$$L_{\text{alt}}^k(TM; W) = \bigcup_{p \in M} L_{\text{alt}}^k(T_p M; W).$$

This set can easily be given a rather obvious vector bundle structure. In this setting, it is convenient to identify $L_{\text{alt}}^k(T_p M; W)$ with $W \otimes (\bigwedge^k T_p^* M)$, so that our bundle be identified with the vector bundle $W \otimes (\bigwedge^k T^* M)$ whose fiber at p is $W \otimes (\bigwedge^k T_p^* M)$. The $C^\infty(M)$-module of sections of this bundle is denoted $\Omega^k(M, W)$. Elements of $\Omega^k(M, W)$ are called (smooth) W-**valued** k-**forms**.

We obtain an exterior product $\Omega^k(M,\mathrm{W}) \times \Omega^\ell(M,\mathrm{W}) \to \Omega^{k+\ell}(M,\mathrm{W} \otimes \mathrm{W})$ as usual by $(\alpha \wedge \beta)(p) := \alpha(p) \wedge \beta(p)$.

We give an alternative definition of $\wedge$ and leave it to the assiduous reader to show that the result is the same. Let $\mathrm{w}_1, \ldots, \mathrm{w}_m$ be a basis for W and $\omega \in \Omega^k(M,\mathrm{W})$ and $\eta \in \Omega^\ell(M,\mathrm{W})$. We have

$$\omega = \sum_{i=1}^{m} \mathrm{w}_i \omega^i \text{ and } \eta = \sum_{j=1}^{m} \mathrm{w}_j \eta^j$$

for some $\omega^i \in \Omega^k(M)$ and $\eta^j \in \Omega^\ell(M)$, where we write $\mathrm{w}_i \omega^i$ rather than $\mathrm{w}_i \otimes \omega^i$, etc. Then,

$$\omega \wedge \eta = \sum_{i=1}^{m} \sum_{j=1}^{m} \mathrm{w}_i \otimes \mathrm{w}_j \, \omega^i \wedge \eta^j.$$

One can show that this definition is independent of the choices and is consistent with the basis free definition.

We still have a pull-back operation defined as before so that if $f : N \to M$ is smooth and $\omega = \sum_{i=1}^{m} \mathrm{w}_i \omega^i \in \Omega^k(M,\mathrm{W})$ for smooth k-forms ω^i, then

$$f^*\omega = \sum_{i=1}^{m} f^*\left(\mathrm{w}_i \omega^i\right) = \sum_{i=1}^{m} \mathrm{w}_i f^* \omega^i$$

is an element of $\Omega^k(N,\mathrm{W})$. We also define a natural exterior derivative

$$d : \Omega^k(M,\mathrm{W}) \to \Omega^{k+1}(M,\mathrm{W})$$

as follows: If $\omega \in \Omega^k(M,\mathrm{W})$, then choose a basis as above and write $\omega = \sum_{i=1}^{m} \mathrm{w}_i \omega^i$. Then

$$d\omega := \sum_{i=1}^{m} \mathrm{w}_i d\omega^i.$$

If $\omega = \sum_{i=1}^{m} \widetilde{\mathrm{w}}_i \widetilde{\omega}^i$ where $\widetilde{\mathrm{w}}_i = \sum \mathrm{w}_j A_i^j$, then we must have

$$\sum_{i=1}^{m} \widetilde{\mathrm{w}}_i \widetilde{\omega}^i = \sum_{i=1}^{m} \left(\sum_{j=1}^{m} \mathrm{w}_j A_i^j\right) \widetilde{\omega}^i = \sum_{j=1}^{m} \mathrm{w}_j \left(\sum_{i=1}^{m} A_i^j \widetilde{\omega}^i\right),$$

from which it follows that

$$\omega^j = \sum_{i=1}^{m} A_i^j \widetilde{\omega}^i,$$

and so

$$\sum_{j=1}^{m} \mathrm{w}_j d\omega^j = \sum_{j=1}^{m} \mathrm{w}_j \sum_{i=1}^{m} A_i^j d\widetilde{\omega}^i = \sum_{j=1}^{m} \widetilde{\mathrm{w}}_i d\widetilde{\omega}^i.$$

8.4. Vector-Valued and Algebra-Valued Forms

Thus the definition does not depend on the choices. It turns out that the invariant definition is

$$\begin{aligned}
d\omega&(X_0, X_1, \ldots, X_k) \\
&= \sum_{0 \le i \le k} (-1)^i X_i(\omega(X_0, \ldots, \widehat{X_i}, \ldots, X_k)) \\
&\quad + \sum_{0 \le i < j \le k} (-1)^{i+j} \omega([X_i, X_j], X_0, \ldots, \widehat{X_i}, \ldots, \widehat{X_j}, \ldots, X_k),
\end{aligned}$$

where now $\omega(X_0, \ldots, \widehat{X_i}, \ldots, X_k)$ is a W-valued function. (A vector field X acts on a W-valued function as $Xf := df(X)$.)

If W happens to be an algebra, then the algebra product $W \times W \to W$ is bilinear, and so it gives rise to a linear map $m \colon W \otimes W \to W$. We compose the exterior product with this map to get an exterior product $\overset{m}{\wedge} \colon \Omega^k(M, W) \times \Omega^\ell(M, W) \to \Omega^{k+\ell}(M, W)$. If $\omega \in \Omega^k(M, W)$ and $\eta \in \Omega^\ell(M, W)$, then as above we may write

$$\omega = \sum_{i=1}^m \mathrm{w}_i \omega^i \text{ and } \eta = \sum_{j=1}^m \mathrm{w}_j \eta^j,$$

for some $\omega^i \in \Omega^k(M)$ and $\eta^j \in \Omega^\ell(M)$ and

$$\omega \overset{m}{\wedge} \eta = \sum_{i=1}^m \sum_{j=1}^m m(\mathrm{w}_i \otimes \mathrm{w}_j) \omega^i \wedge \eta^j.$$

Using a dot for the multiplication, we also have

$$(\omega \overset{m}{\wedge} \eta)(X_1, X_2, \ldots, X_r, X_{r+1}, X_{r+2}, \ldots, X_{r+s})$$
$$= \frac{1}{k_1! k_2!} \sum_{\sigma \in S_{k_1+k_2}} \mathrm{sgn}(\sigma) \omega(X_{\sigma_1}, X_{\sigma_2}, \ldots, X_{\sigma_r}) \cdot \eta(X_{\sigma_{r+1}}, X_{\sigma_{r+2}}, \ldots, X_{\sigma_{r+s}}).$$

In this case d is also defined as before. A particularly important case is when W is a Lie algebra $\mathfrak{g}$ with bracket $[\cdot, \cdot]$. Then we write the resulting product $\overset{m}{\wedge}$ as $[\cdot, \cdot]^\wedge$ or just $[\cdot, \cdot]$ when there is no risk of confusion. Thus if $\omega, \eta \in \Omega^1(U, \mathfrak{g})$ are Lie algebra-valued 1-forms, then

$$[\omega, \eta]^\wedge(X, Y) = [\omega(X), \eta(Y)] + [\eta(X), \omega(Y)].$$

In particular, $\frac{1}{2}[\omega, \omega]^\wedge(X, Y) = [\omega(X), \omega(Y)]$, which might not be zero in general!

Example 8.42. The Maurer-Cartan forms are $\mathfrak{g}$-valued 1-forms.

8.5. Bundle-Valued Forms

It is convenient in several contexts to have on hand the notion of a differential form with values in a vector bundle. Let $\xi = (E, \pi, M)$ be a smooth real vector bundle of rank r. We can consider the vector bundle $L_{\text{alt}}^k(TM, E)$ over M whose fiber at p is $L_{\text{alt}}^k(T_pM, E_p)$. We identify $L_{\text{alt}}^k(T_pM, E_p)$ with $E_p \otimes \bigwedge^k T_p^* M$ and thus the bundle is identified with $E \otimes \bigwedge^k T^* M$.

Definition 8.43. Let $\xi = (E, \pi, M)$ be a smooth vector bundle. A smooth differential k-form with values in ξ (or values in E) is a smooth section of the bundle $E \otimes \bigwedge^k T^* M$. These are denoted by $\Omega^k(M; E)$.

Remark 8.44. The reader should avoid confusion between $\Omega^k(M; E)$ and the space of sections $\Gamma(M, \bigwedge^k E)$.

Theorem 8.45. *There is a natural $C^\infty(M)$ module isomorphism $\Omega^k(M; E) \cong L_{\text{alt}}^k(\mathfrak{X}(M), \Gamma(E))$. If this isomorphism is taken as an identification, then*
$$\mu(X_1, \ldots, X_k)(p) = \mu(X_1(p), \ldots, X_k(p))$$
for $\mu \in \Omega^k(M; E)$ and $X_1, \ldots, X_k \in L_{\text{alt}}^k(\mathfrak{X}(M), \Gamma(E))$.

Proof. We use Proposition 6.55. The reader should check each of the following:
$$\Omega^k(M; E) \cong \Gamma(E \otimes \bigwedge^k T^* M) \cong \Gamma\left(L_{\text{alt}}^k(TM, E)\right)$$
$$\cong \Gamma\left(\text{Hom}(\bigwedge^k TM, E)\right) \cong \text{Hom}\left(\Gamma\left(\bigwedge^k TM\right), \Gamma(E)\right)$$
$$\cong \text{Hom}\left(\bigwedge^k \mathfrak{X}(M), \Gamma(E)\right) \cong L_{\text{alt}}^k(\mathfrak{X}(M), \Gamma(E)). \qquad \square$$

In order to get a grip on the meaning of $\Omega^k(M; E)$, let us exhibit transition functions. For a vector bundle, knowing the transition functions is tantamount to knowing how local expressions with respect to a frame transform as we change frames. A local frame for $E \otimes \bigwedge^k T^* M$ can be given by combining a local frame for E with a local frame for $\bigwedge^k T^* M$. Let $(e_1, \ldots, e_r)$ be a frame field for E defined on an open set U. We may as well take U to also be a chart domain for the manifold M. Then any local section of $\Omega^k(M; E)$ defined on U has the form
$$s = \sum a_{\vec{I}}^j e_j \otimes dx^{\vec{I}}$$
for some smooth functions $a_{\vec{I}}^j = a_{i_1 \ldots i_k}^j$ defined in U. Then for a new local set up with frames $(f_1, \ldots, f_r)$ and $dy^{\vec{I}} = dy^{i_1} \wedge \cdots \wedge dy^{i_k}$ ($i_1 < \cdots < i_k$) we have
$$s = \sum \widetilde{a}_{\vec{I}}^j f_j \otimes dy^{\vec{I}}$$

8.5. Bundle-Valued Forms

for some $\widetilde{a}_{\vec{I}}^j$ and the transformation law

$$\widetilde{a}_{\vec{I}}^j = \sum a_{\vec{J}}^i C_i^j \frac{\partial x^{\vec{J}}}{\partial y^{\vec{I}}},$$

where C_i^j is defined by $f_s C_j^s = e_j$.

Exercise 8.46. Derive the above transformation law.

Note that if we write $\omega^j = \sum a_{\vec{I}}^j dx^{\vec{I}}$, then we may write s in U as

$$s = \sum e_j \otimes \omega^j \text{ for } \omega^j \in \Omega^k(U).$$

Example 8.47. If E is a trivial product bundle $M \times V \to M$, then $\Omega^k(M; E)$ is canonically isomorphic to $\Omega^k(M; V)$.

Example 8.48. For an n-manifold M, the bundle map $I : TM \to TM$ which is the identity on each fiber can be interpreted as an element of $\Omega^1(M; TM)$.

Example 8.49. If $f : M \to N$ is a smooth map, then the tangent map $Tf : TM \to TN$ can be interpreted as an element of $\Omega^1(M; f^*TN)$, where f^*TN is the pull-back bundle (Definition 6.18).

Now we want to define an important graded module structure on the direct sum

$$\Omega(M; E) = \bigoplus_k \Omega^k(M; E).$$

This will be a module over the graded algebra $\Omega(M)$. The action of $\Omega(M)$ on $\Omega(M; E)$ is given by maps $\wedge : \Omega^k(M) \times \Omega^\ell(M; E) \to \Omega^{k+\ell}(M; E)$, which in turn are defined by extending the following rule linearly:

$$\omega^1 \wedge (s \otimes \omega^2) := s \otimes \omega^1 \wedge \omega^2 \text{ for } \omega^1 \in \Omega^k(M), \omega^2 \in \Omega^\ell(M) \text{ and } s \in \Gamma(E)$$

(with the same formula for local sections). Actually we can also define the analogous right multiplication $\wedge : \Omega^\ell(M; E) \times \Omega^k(M) \to \Omega^{k+\ell}(M; E)$,

$$(s \otimes \omega^1) \wedge \omega^2 := s \otimes \omega^1 \wedge \omega^2,$$

and then we have

$$\eta \wedge \mu = (-1)^{k\ell} \mu \wedge \eta \text{ for } \mu \in \Omega^\ell(M; E), \eta \in \Omega^k(M).$$

If the vector bundle is actually an *algebra bundle*, say $\mathcal{A} \to M$, then we may turn $\mathcal{A} \otimes \wedge T^*M := \sum_{p=0}^n \mathcal{A} \otimes \bigwedge^p T^*M$ into an algebra bundle whose sections can be multiplied: For $\omega^1 \in \Omega^k(M)$, $\omega^2 \in \Omega^\ell(M)$, and $s_1, s_2 \in \Gamma(\mathcal{A})$, define

$$(s_1 \otimes \omega^1) \circledast (s_2 \otimes \omega^2) := s_1 \cdot s_2 \otimes \omega^1 \wedge \omega^2,$$

where $\cdot$ is the product in $\mathcal{A}$. This extends linearly on (possibly locally defined) sections:

$$\left(\sum a^i_j s_i \otimes \omega^j\right) \circledast \left(\sum b^k_\ell s_k \otimes \omega^\ell\right) := \sum a^i_j b^k_\ell s_i s_k \otimes \omega^j \wedge \omega^\ell,$$

where a^i_j and b^k_ℓ are smooth functions. We obtain a product that is natural with respect to restriction to open sets. From this the sections $\Omega(M,\mathcal{A}) = \Gamma(M, \mathcal{A} \otimes \wedge T^*M)$ become an algebra over the ring of smooth functions. We can think of elements of $\Omega^k(M, \mathcal{A})$ as $C^\infty(M)$ multilinear maps on $\mathfrak{X}(M)$. Then the invariant formula for $\phi \in \Omega^k(M,\mathcal{A})$ and $\psi \in \Omega^\ell(M,\mathcal{A})$ is

$$\phi \circledast \psi(X_1, \ldots, X_{k+\ell})$$
$$= \frac{1}{k!\ell!} \sum_\sigma \mathrm{sgn}(\sigma) \phi(X_{\sigma_1}, \ldots, X_{\sigma_k}) \cdot \psi(X_{\sigma(k+1)}, \ldots, X_{\sigma(k+\ell)})$$

for $X_1, \ldots, X_{k+\ell} \in \mathfrak{X}(M)$. Depending on the context, the symbol may be chosen to be the same as that for the product in $\mathcal{A}$, or it may just be $\wedge$ (especially if $\mathcal{A}$ is commutative), or it may be some hybrid symbol.

Another important example is where $\mathcal{A} = \mathrm{End}(E)$. Locally, say on U, sections μ_1 and μ_2 of $\Omega(M, \mathrm{End}(E))$ take the form $\mu_1 = \sum A_i \otimes \alpha^i$ and $\mu_2 = \sum B_i \otimes \beta^i$, where A_i and B_i are maps $U \to \mathrm{End}(E)$. Thus for each $x \in U$, the A_i and B_i evaluate to give $A_i(x), B_i(x) \in \mathrm{End}(E_x)$. The multiplication is then

$$\left(\sum A_i \otimes \alpha^i\right) \wedge \left(\sum B_j \otimes \beta^j\right) = \sum_{i,j} A_i B_j \otimes \alpha^i \wedge \beta^j,$$

where the $A_i B_j : U \to \mathrm{End}(E)$ are local sections given by composition:

$$A_i B_j : x \mapsto A_i(x) \circ B_j(x).$$

Exercise 8.50. Show that $\Omega(M, \mathrm{End}(E))$ acts on $\Omega(M, E)$ making $\Omega(M, E)$ a module over the algebra $\Omega(M, \mathrm{End}(E))$.

Perhaps it would help to think as follows: We have a cover of a manifold M by open sets $\{U_\alpha\}$ that simultaneously locally trivialize both E and T^*M. Then these also give local trivializations over these open sets of the bundles $\mathrm{End}(E)$ and $\bigwedge T^*M$. Associated with each local trivialization is a frame field for $E \to M$, say $(e_1, \ldots, e_r)$, which allows us to associate with each section $\sigma \in \Omega^k(M, E)$ an r-tuple of k-forms $\sigma_U = (\sigma^i_U)$ for each U such that $\sigma = \sum \sigma^i_U e_i$. Similarly, a section $A \in \Omega^\ell(M, \mathrm{End}(E))$ is equivalent to assigning to each open set $U \in \{U_\alpha\}$ a matrix of ℓ-forms A_U. The algebra structure on $\Omega(M, \mathrm{End}(E))$ is then just matrix multiplication, where the entries are multiplied using the exterior product $A_U \wedge B_U$,

$$(A_U \wedge B_U)^i_j = \sum A^i_s \wedge B^s_j.$$

The module structure of the above exercise is given locally by $\sigma_U \mapsto A_U \wedge \sigma_U$. Where did the bundle go? The global topology is now encoded in the transformation laws, which tell us what the same section looks like when we change to a new frame field on an overlap $U_\alpha \cap U_\beta$. In this sense, the bundle is a combinatorial recipe for pasting together local objects.

Recall that with the product $[A, B] := A \circ B - B \circ A$, the bundle $\mathrm{Hom}(E, E)$ is written as $\mathfrak{gl}(E)$ rather than $\mathrm{End}(E)$, and then our general construction gives $\Omega(M, \mathfrak{gl}(E))$ an algebra structure, whose product is denoted $[\phi, \psi]^\wedge$ or something similar.

8.6. Operator Interactions

The Lie derivative acts on differential forms since the latter are, from one viewpoint, alternating tensor fields. When we apply the Lie derivative to a differential form, we get a differential form, so we should think about the Lie derivative in the context of differential forms.

Lemma 8.51. *For any $X \in \mathfrak{X}(M)$ and any $f \in \Omega^0(M)$, we have $\mathcal{L}_X df = d\mathcal{L}_X f$.*

Proof. For a function f, we compute as

$$(\mathcal{L}_X df)(Y) = \left(\frac{d}{dt}\bigg|_0 (\varphi_t^X)^* df\right)(Y) = \frac{d}{dt}\bigg|_0 df(T\varphi_t^X \cdot Y) = \frac{d}{dt}\bigg|_0 Y((\varphi_t^X)^* f)$$

$$= Y\left(\frac{d}{dt}\bigg|_0 (\varphi_t^X)^* f\right) = Y(\mathcal{L}_X f) = d(\mathcal{L}_X f)(Y),$$

where $Y \in \mathfrak{X}(M)$ is arbitrary. $\square$

We now have two ways to differentiate sections in $\Omega(M)$. First, there is the Lie derivative $\mathcal{L}_X : \Omega^i(M) \to \Omega^i(M)$, which turns out to be a graded derivation of degree zero,

(8.7) $$\mathcal{L}_X(\alpha \wedge \beta) = \mathcal{L}_X \alpha \wedge \beta + \alpha \wedge \mathcal{L}_X \beta.$$

We may apply $\mathcal{L}_X$ to elements of $\Omega(U)$ for $U \subset M$, and it is easy to see that we obtain a natural derivation in the sense of Definition 8.33.

Exercise 8.52. Prove the above product rule.

Second, there is the exterior derivative d which is a graded derivation of degree 1. In order to relate the two operations, we need a third map, which, like the Lie derivative, is taken with respect to a given field $X \in \mathfrak{X}(M)$. This map is defined using the interior product given in Definition 8.12 by letting

(8.8) $$i_X \omega(X_1, \ldots, X_{i-1})(p) := i_{X_p} \omega_p(X_1(p), \ldots, X_{i-1}(p)).$$

Alternatively, if $\omega \in \Omega^i(M)$ is viewed as a skew-symmetric multilinear map from $\mathfrak{X}(M) \times \cdots \times \mathfrak{X}(M)$ to $C^\infty(M)$, then we simply define

$$i_X \omega(X_1, \ldots, X_{i-1}) := \omega(X, X_1, \ldots, X_{i-1}).$$

By convention $i_X f = 0$ for $f \in C^\infty(M)$. We will call this operator the **interior product** or **contraction operator**. This operator is clearly linear over $\mathbb{R}$. Notice that for any $f \in C^\infty(M)$ we have

$$i_{fX}\omega = f i_X \omega, \text{ and } i_X df = df(X) = \mathcal{L}_X f.$$

Proposition 8.53. i_X is a graded derivation of $\Omega(M)$ of degree -1:

$$i_X(\alpha \wedge \beta) = (i_X \alpha) \wedge \beta + (-1)^k \alpha \wedge (i_X \beta) \text{ for } \alpha \in \Omega^k(M).$$

It is the unique degree -1 graded derivation of $\Omega(M)$ such that $i_X f = 0$ for $f \in \Omega^0(M)$ and

$$i_X \theta = \theta(X) \text{ for } \theta \in \Omega^1(M) \text{ and } X \in \mathfrak{X}(M).$$

Proof. This follows from Proposition 8.13. □

Actually, i_X is natural with respect to restriction, so it is a natural graded derivation in the sense of Definition 8.33. Formulas developed for the interior product in the vector space category also hold for vector fields and differential forms. For example, if $\theta_1, \ldots, \theta_k \in \Omega^1(M)$ and $X \in \mathfrak{X}(M)$, then

$$i_X(\theta_1 \wedge \cdots \wedge \theta_k) = \sum_{\ell=1}^{k} (-1)^{k+1} \theta_\ell(X) \theta_1 \wedge \cdots \wedge \widehat{\theta_\ell} \wedge \cdots \wedge \theta_k.$$

Notation 8.54. Other notations for $i_X \omega$ include $X \rfloor \omega$ and $\langle X, \omega \rangle$. These notations make the following theorem look more natural:

Theorem 8.55. The Lie derivative is a derivation with respect to the pairing $(X, \omega) \mapsto \langle X, \omega \rangle$. That is,

$$\mathcal{L}_X(i_Y \omega) = i_{\mathcal{L}_X Y} \omega + i_Y \mathcal{L}_X \omega,$$

or in alternative notations,

$$\mathcal{L}_X(Y \rfloor \omega) = (\mathcal{L}_X Y) \rfloor \omega + Y \rfloor (\mathcal{L}_X \omega),$$
$$\mathcal{L}_X \langle Y, \omega \rangle = \langle \mathcal{L}_X Y, \omega \rangle + \langle Y, \mathcal{L}_X \omega \rangle.$$

Proof. Exercise. □

Now we can relate the Lie derivative, the exterior derivative and the contraction operator.

Theorem 8.56. Let $X \in \mathfrak{X}_M$. Then we have **Cartan's formula**,

(8.9) $$\mathcal{L}_X = d \circ i_X + i_X \circ d.$$

Proof. Both sides of the equation define derivations of degree zero (use Proposition 8.35). So by Lemma 8.36 we just have to check that they agree on functions and exact 1-forms. On functions we have $i_X f = 0$ and $i_X df = Xf = \mathcal{L}_X f$ so the formula holds. On differentials of functions we have

$$(d \circ i_X + i_X \circ d)df = (d \circ i_X)df = d\mathcal{L}_X f = \mathcal{L}_X df,$$

where we have used Lemma 8.51 in the last step. $\square$

As a corollary, we can extend Lemma 8.51:

Corollary 8.57. $d \circ \mathcal{L}_X = \mathcal{L}_X \circ d$.

Proof. We have

$$d\mathcal{L}_X \alpha = d(di_X + i_X d)(\alpha) = di_X d\alpha$$
$$= di_X d\alpha + i_X dd\alpha = (\mathcal{L}_X \circ d)\alpha. \quad \square$$

Corollary 8.58. *We have the following formulas:*

(i) $i_{[X,Y]} = \mathcal{L}_X \circ i_Y - i_Y \circ \mathcal{L}_X$;

(ii) $\mathcal{L}_{fX}\omega = f\mathcal{L}_X \omega + df \wedge i_X \omega$ for all $\omega \in \Omega(M)$.

Proof. We leave (i) as Problem 9. For (ii), we compute:

$$\mathcal{L}_{fX}\omega = i_{fX}d\omega + d(i_{fX}\omega) = i_{fX}d\omega + d(f i_X \omega)$$
$$= f i_X d\omega + df \wedge i_X \omega + f d(i_X \omega)$$
$$= f(i_X d\omega + d(i_X \omega)) + df \wedge i_X \omega = f\mathcal{L}_X \omega + df \wedge i_X \omega. \quad \square$$

8.7. Orientation

A vector bundle $E \to M$ is called **oriented** if every fiber E_p is given a smooth choice of orientation. There are several equivalent ways to make a rigorous definition:

Proposition 8.59. *Let $E \to M$ be a rank k real vector bundle with typical fiber V. The following are equivalent:*

(i) *There is a smooth global section ω of the bundle $\bigwedge^k E^* \cong L_{\text{alt}}^k(E) \to M$ such that ω is nowhere vanishing.*

(ii) *There is a smooth global section s of the bundle $\bigwedge^k E \to M$ such that s is nowhere vanishing.*

(iii) *The vector bundle has an atlas of VB-charts (local trivializations) such that the corresponding transition maps take values in $\mathrm{GL}^+(\mathrm{V})$, the group of positive determinant elements of $\mathrm{GL}(\mathrm{V})$. This means that the standard $\mathrm{GL}(\mathrm{V})$-structure on $E \to M$ can be reduced to a $\mathrm{GL}^+(\mathrm{V})$-structure (refer to Chapter 6).*

Proof. We show that (i) is equivalent to (iii) and leave the rest as an easy exercise. Suppose that (i) holds and that ω is a nonvanishing section of $L_{\text{alt}}^k(E)$. Now fix a basis $(\mathbf{e}_1, \ldots, \mathbf{e}_k)$ on V and recall that with this basis fixed, each VB-chart (U, ϕ) for $E \to M$ corresponds to a local frame field. Indeed, we let $e_i(p) := \phi^{-1}(p, \mathbf{e}_i)$. The transition maps between two charts will have values in $\text{GL}^+(\text{V})$ exactly when the matrix function that relates the corresponding frame fields has positive determinant (check this). Given a VB-atlas we construct a new atlas. We retain those VB-charts (U, ϕ) whose corresponding frame field $e_1, \ldots, e_k$ satisfies $\omega(e_1, \ldots, e_k) > 0$. For the charts for which $\omega(e_1, \ldots, e_k) < 0$, we replace $e_1, \ldots, e_k$ by $-e_1, \ldots, e_k$, and the resulting chart will be included in our new atlas. Now if $e_1, \ldots, e_k$ and $f_1, \ldots, f_k$ are two frame fields coming from this atlas, then $f_i = \sum C_i^j e_j$ and
$$\omega(f_1, \ldots, f_k) = (\det C)\,\omega(e_1, \ldots, e_k).$$
We conclude that $\det C > 0$.

Conversely, suppose the vector bundle has an atlas $\{(U_\alpha, \phi_\alpha)\}$ taking values in $\text{GL}^+(\text{V})$. We will use the frame fields coming from this atlas to construct a nowhere vanishing section of $L_{\text{alt}}^k(E) \to M$. If $e_1, \ldots, e_k$ and $f_1, \ldots, f_k$ are two frame fields coming from this atlas, then let $f^1, \ldots, f^k$ be dual to $f_1, \ldots, f_k$ and consider $f^1 \wedge \cdots \wedge f^k$. We have $f_i = \sum C_i^j e_j$ and
$$\left(f^1 \wedge \cdots \wedge f^k\right)(e_1, \ldots, e_k) = \det C > 0.$$
For each chart (U_α, ϕ_α) in our VB-atlas, let $f_1^\alpha, \ldots, f_k^\alpha$ be the corresponding frame field and let $\left(f_\alpha^1, \ldots, f_\alpha^k\right)$ be the dual frame field. Then let $\{\rho_\alpha\}$ be a partition of unity subordinate to the cover $\{U_\alpha\}$. Let
$$\omega := \sum_\alpha \rho_\alpha f_\alpha^1 \wedge \cdots \wedge f_\alpha^k.$$
Then, ω is nowhere vanishing. To see this let $p \in M$ and suppose that $p \in U_\beta$ for some chart (U_β, ϕ_β) from the $\text{GL}^+(\text{V})$-valued atlas. Then
$$\omega(f_1^\beta, \ldots, f_k^\beta)(p) = \sum_\alpha \rho_\alpha(p) \det C_{\beta\alpha}(p) > 0,$$
where $C_{\beta\alpha}$ is the matrix that relates $f_1^\beta, \ldots, f_k^\beta$ and $f_1^\alpha, \ldots, f_k^\alpha$. □

Definition 8.60. If any one (and hence all) of the conditions in Proposition 8.59 hold, then $E \to M$ is said to be **orientable**. A VB-atlas that satisfies (iii) will be called an oriented atlas.

If $E \to M$ is orientable as above, then two nowhere vanishing sections of $L_{\text{alt}}^k(E)$, say ω_1 and ω_2, are said to be equivalent if $\omega_1 = f\omega_2$, where f is a smooth positive function. We denote the equivalence class of such a nowhere vanishing ω by $[\omega]$.

8.7. Orientation

Definition 8.61. An **orientation** for an orientable vector bundle of rank k is an equivalence class $[\omega]$ of nowhere vanishing sections of $\bigwedge^k E^*$. If such an orientation is chosen, then the vector bundle is said to be **oriented** by $[\omega]$.

Notice that if we have two oriented VB-atlases on a vector bundle, then we know what it means for them to determine the same $GL^+(V)$-structure. This was the notion of strict equivalence from Chapter 6. The next exercise shows that the notion of a reduction to a $GL^+(V)$-structure is equivalent to the notion of an orientation as we have defined it.

Exercise 8.62. Recall the construction of a nowhere vanishing section ω in the proof of Proposition 8.59. Show that the class $[\omega]$ does not depend on the partition of unity used in the construction. Show that if two oriented VB-atlases determine the same $GL^+(V)$-structure, then the constructed sections are equivalent and so determine the same orientation. Conversely, show that an orientation as we have defined it determines a unique reduction to a $GL^+(V)$-structure on the vector bundle.

Let $E \to M$ be oriented by $[\omega]$. A frame $(v_1, \ldots, v_k)$ of fiber E_p is **positively oriented** (or just positive) with respect to $[\omega]$ if and only if $\omega(p)(v_1, \ldots, v_k) > 0$. This condition is independent of the choice of representative ω for the class $[\omega]$.

Definition 8.63. Let $\pi : E \to M$ be an oriented vector bundle. A frame field $(f_1, \ldots, f_k)$ over an open set U is called a **positively oriented frame field** if $(f_1(p), \ldots, f_k(p))$ is a positively oriented basis of E_p for each $p \in U$.

Exercise 8.64. Let $\pi : E \to M$ be an oriented vector bundle. Show that if M is connected, then there are exactly two possible orientations for the vector bundle.

Exercise 8.65. If $\pi_1 : E_1 \to M$ and $\pi_2 : E_2 \to M$ are orientable, then so is the Whitney sum $\pi_1 \oplus \pi_2 : E_1 \oplus E_2 \to M$.

Definition 8.66. A smooth manifold M is said to be **orientable** if TM is orientable. An orientation for the vector bundle TM is also called an **orientation** for M. A manifold M together with an orientation for M is said to be an **oriented manifold**.

Definition 8.67. An atlas $\{(U_\alpha, \mathbf{x}_\alpha)\}$ for M is said to be an **oriented atlas** if the associated frame fields are positively oriented. If this atlas is positively oriented with respect to an orientation on M (an orientation $[\omega]$ of TM), then we call $\{(U_\alpha, \mathbf{x}_\alpha)\}$ a **positively oriented atlas**.

It follows from the definitions that an oriented atlas induces an orientation for which it is a positively oriented atlas. For this reason, a choice of oriented atlas is equivalent to a choice of orientation, and so one often sees an orientation of an orientable manifold defined as simply being given by a choice of oriented atlas.

Now let M be an n-manifold. Consider a **top form**, i.e. an n-form $\varpi \in \Omega^n(M)$, and assume that ϖ is nowhere vanishing. Thus M must be orientable. We call such a nonvanishing ϖ a **volume form** for M, and every such volume form obviously determines an orientation for M. If $\varphi : M \to M$ is a diffeomorphism, then we must have that $\varphi^*\varpi = \delta\varpi$ for some $\delta \in C^\infty(M)$, which we will call the **Jacobian determinant** of φ with respect to the volume element ϖ:

$$\varphi^*\varpi = J_\varpi(\varphi)\varpi.$$

Clearly $J_\varpi(\varphi)$ is a nowhere vanishing smooth function.

Proposition 8.68. *Let $(M, [\varpi])$ be an oriented n-manifold. The sign of $J_\varpi(\varphi)$ is independent of the choice of volume form ϖ in the orientation class $[\varpi]$.*

Proof. Let $\varpi' \in \Omega^n(M)$. We have $\varpi = a\varpi'$ for some function a that is never zero on U. Furthermore,

$$J(\varphi)\varpi = (\varphi^*\varpi) = (a \circ \varphi)(\varphi^*\varpi') = (a \circ \varphi)J_{\varpi'}(\varphi)\varpi' = \frac{a \circ \varphi}{a}\varpi,$$

and since $\frac{a \circ \varphi}{a} > 0$ and ϖ is nonzero, the conclusion follows. □

Let us consider a very important special case of this: Suppose that $\varphi : U \to U$ is a diffeomorphism and $U \subset \mathbb{R}^n$. Then letting $\varpi_0 = du^1 \wedge \cdots \wedge du^n$ we have for any $x \in U$,

$$\varphi^*\varpi_0(x) = \varphi^*du^1 \wedge \cdots \wedge \varphi^*du^n(x)$$
$$= \left(\sum \frac{\partial(u^1 \circ \varphi)}{\partial u^{i_1}}\bigg|_x du^{i_1}\right) \wedge \cdots \wedge \left(\sum \frac{\partial(u^n \circ \varphi)}{\partial u^{i_n}}\bigg|_x du^{i_n}\right)$$
$$= \det\left(\frac{\partial(u^i \circ \varphi)}{\partial u^j}(x)\right)\varpi_0(x) = J\varphi(x)\varpi_0(x).$$

So in this case, $J_{\varpi_0}(\varphi)$ is just the usual Jacobian determinant of φ. More generally, let a nonvanishing top form ϖ be defined on M and let ϖ' be another such form defined on N. Then we say that a diffeomorphism $\varphi : M \to N$ is **orientation preserving** (or positive) with respect to the orientations determined by ϖ and ϖ' if the unique function $J_{\varpi,\varpi'}$ such that $\varphi^*\varpi' = J_{\varpi,\varpi'}\varpi$ is strictly positive on M.

Exercise 8.69. An open subset of an oriented manifold M inherits the orientation from M since we can just restrict a defining volume form. Show that a chart $(U, \mathbf{x})$ on an oriented manifold is positive if and only if

8.7. Orientation

$\mathbf{x} : U \to \mathbf{x}(U)$ is orientation preserving. Here, $\mathbf{x}(U)$ inherits its orientation from the ambient Euclidean space with its standard orientation.

We now construct a **two-fold covering manifold** M^{or} for any manifold M. The orientation cover will itself always be orientable. Recall that the zero section of a vector bundle over M is a submanifold of the total space diffeomorphic to M. Consider the vector bundle whose total space is $\bigwedge^n T^*M$ and remove the zero section to obtain

$$\left(\bigwedge^n T^*M\right)^\times := \left(\bigwedge^n T^*M\right) \setminus \{\text{zero section}\}.$$

Define an equivalence relation on $(\bigwedge^n T^*M)^\times$ by declaring $\nu_1 \sim \nu_2$ if and only if ν_1 and ν_2 are in the same fiber and if $\nu_1 = a\nu_2$ with $a > 0$. The space of equivalence classes is denoted M^{or} and we will show that it is a smooth manifold. Let $\mathsf{q} : (\bigwedge^n T^*M)^\times \to M^{\mathrm{or}}$ be the quotient map and give M^{or} the quotient topology. There is a unique smooth map π_{or} making the following diagram commute:

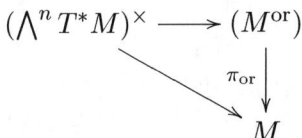

$$(\bigwedge^n T^*M)^\times \longrightarrow (M^{\mathrm{or}})$$
$$\searrow \quad \pi_{\mathrm{or}} \downarrow$$
$$M$$

It is easy to see that for each $p \in M$, the set $\pi_{\mathrm{or}}^{-1}(p)$ contains exactly two elements, which are the two orientations of T_pM. We give the set M^{or} a smooth structure. First let $[\mu_0]$ be the standard orientation of $\mathbb{R}^n$ and choose a fixed orientation reversing linear involution $r_0 : \mathbb{R}^n \to \mathbb{R}^n$. Let $\{(U_\alpha, \mathbf{x}_\alpha)\}$ be an atlas for M. By composing some of the charts with r_0 and adding the resulting charts to the atlas, we may suppose that $\{(U_\alpha, \mathbf{x}_\alpha)\}$ has the property that for every chart $(U, \mathbf{x})$ in the atlas there is a chart $(U, \mathbf{y})$ in the atlas with the same domain such that $\mathbf{y} \circ \mathbf{x}^{-1}$ is orientation reversing. Let us say that such an atlas is "balanced". (The maximal atlas is obviously balanced.) Now for each chart $(U, \mathbf{x})$, where $\mathbf{x} = (x^1, \ldots, x^n)$, define a map $\phi_\mathbf{x} : \mathbf{x}(U) \to M^{\mathrm{or}}$ by

$$\phi_\mathbf{x}(u) := [dx^1(\mathbf{x}^{-1}(u)) \wedge \cdots \wedge dx^n(\mathbf{x}^{-1}(u))] \in M^{\mathrm{or}}.$$

Because we have assumed that the atlas is balanced, it is easy to see that each element of M^{or} is in the image of some $\phi_\mathbf{x}$. The $\phi_\mathbf{x}$ are injective and so are bijections onto their images. These maps are local parameterizations of M^{or} and the inverses of these maps are charts on M^{or}. Let us denote the chart arising from $(U, \mathbf{x})$ by $(\widetilde{U}, \widetilde{\mathbf{x}})$. Thus $\widetilde{U} = \phi_\mathbf{x}(U)$ and $\widetilde{\mathbf{x}} = \phi_\mathbf{x}^{-1}$. In this way the balanced atlas $\{(U_\alpha, \mathbf{x}_\alpha)\}$ gives an atlas on M^{or}. We must check that the overlaps are smooth. So let $\widetilde{U}_\alpha \cap \widetilde{U}_\beta \neq \emptyset$ and consider $\widetilde{\mathbf{x}}_\beta \circ \widetilde{\mathbf{x}}_\alpha^{-1} = \widetilde{\mathbf{x}}_\beta \circ \phi_{\mathbf{x}_\alpha}$. For $u \in \mathbf{x}_\alpha(U_\alpha \cap U_\beta)$ and $w = \mathbf{x}_\beta \circ \mathbf{x}_\alpha^{-1}(u)$ we have

$$dx_\alpha^1(\mathbf{x}_\alpha^{-1}(u)) \wedge \cdots \wedge dx_\alpha^n(\mathbf{x}_\alpha^{-1}(u)) = \lambda \, dx_\beta^1(\mathbf{x}_\beta^{-1}(w)) \wedge \cdots \wedge dx_\beta^n(\mathbf{x}_\beta^{-1}(w))$$

for some $\lambda > 0$. Indeed, we must have $\lambda = \det(D(\mathbf{x}_\alpha \circ \mathbf{x}_\beta^{-1})(u))$. Then,

$$\widetilde{\mathbf{x}}_\beta \circ \phi_{\mathbf{x}_\alpha}(u) = \widetilde{\mathbf{x}}_\beta \left([dx_\alpha^1(\mathbf{x}_\alpha(u)) \wedge \cdots \wedge dx_\alpha^n(\mathbf{x}_\alpha(u))]\right)$$
$$= \widetilde{\mathbf{x}}_\beta \left([dx_\beta^1(\mathbf{x}_\beta(w)) \wedge \cdots \wedge dx_\beta^n(\mathbf{x}_\beta(w))]\right)$$
$$= w = \mathbf{x}_\beta \circ \mathbf{x}_\alpha^{-1}(u).$$

Thus $\widetilde{\mathbf{x}}_\beta \circ \widetilde{\mathbf{x}}_\alpha^{-1} = \mathbf{x}_\beta \circ \mathbf{x}_\alpha^{-1}$ on $\widetilde{\mathbf{x}}_\alpha(\widetilde{U}_\alpha \cap \widetilde{U}_\beta)$. Note that for each α and β the set $\widetilde{\mathbf{x}}_\alpha(\widetilde{U}_\alpha \cap \widetilde{U}_\beta)$ consists of exactly those connected components of $\mathbf{x}_\alpha(U_\alpha \cap U_\beta)$ on which $\det\left(D(\mathbf{x}_\beta \circ \mathbf{x}_\alpha^{-1})\right)$ is positive. Thus $\widetilde{\mathbf{x}}_\alpha(\widetilde{U}_\alpha \cap \widetilde{U}_\beta)$ is open. One may check that the topology induced by this atlas coincides with the quotient topology and that the quotient map is smooth. Also, note that if $\widetilde{U}_\alpha \cap \widetilde{U}_\beta \neq \emptyset$, then $\det\left(D(\widetilde{\mathbf{x}}_\beta \circ \widetilde{\mathbf{x}}_\alpha^{-1})\right) > 0$ and so our atlas is oriented! We thus have a canonical orientation of M^{or}.

For each admissible chart $(U, \mathbf{x})$, we obtain a section $U \to M^{\mathrm{or}}$ given by

$$p \mapsto [dx^1(p) \wedge \cdots \wedge dx^n(p)].$$

It is easy to see that such sections are smooth. From the existence of these sections, one surmises that the connected components of the chart domains are evenly covered and so $\pi_{\mathrm{or}} : M^{\mathrm{or}} \to M$ is a two-fold covering map (but M^{or} may not be connected). This two-fold covering map is called the **orientation covering map**. The space M^{or} itself is called the **orientation (double) cover**.

Exercise 8.70. Let R^+ be the multiplicative Lie group of positive real numbers. Show that R^+ acts freely and properly on $(\bigwedge^n T^*M)^\times$ and that the quotient manifold is M^{or}. Show that the canonical orientation on M^{or} can be described as follows: Each $y \in M^{\mathrm{or}}$ is an orientation of $T_{\pi_{\mathrm{or}}(y)}M$. Since $T_y\pi_{\mathrm{or}} : T_yM^{\mathrm{or}} \to T_{\pi_{\mathrm{or}}(y)}M$ is an isomorphism, we may transfer the orientation y on $T_{\pi_{\mathrm{or}}(y)}M$ to an orientation for T_yM^{or}. This gives a canonical orientation on each fiber of TM^{or} which agrees with the orientation derived from the oriented atlas described above. How can we get a smooth global nonvanishing top form that induced this orientation?

Exercise 8.71. A manifold M is orientable if and only if M^{or} is disconnected. If M is oriented, then M^{or} has exactly two components and an orientation of M corresponds to a choice of connected component of M^{or}.

8.7.1. Orientation induced on a boundary. Now let M be a manifold with boundary. According to Problem 15, we can consider vector bundles over M. The definitions of *orientable* and *orientation* make sense for M. Here we wish to consider orientations of ∂M. If $[\omega]$ is an orientation for an orientable vector bundle $\pi : E \to M$, then $[\omega|_{\partial M}]$ is an orientation on $E|_{\partial M} \to \partial M$ where $E|_{\partial M} = \pi^{-1}(\partial M)$. In particular, if M is oriented by

8.7. Orientation

$[\omega]$, then we may obtain an orientation on $TM|_{\partial M} \to \partial M$. But what we would really like to obtain at this time is an orientation on ∂M. In other words, we need an orientation on the bundle $T(\partial M)$. First notice that by our conventions, an atlas for M consists of charts that take values in half-spaces of the form $\mathbb{R}^n_{\lambda \geq c}$.

Exercise 8.72. Consider $v_p \in T_p M$ for some boundary point $p \in \partial M$. Let $(U, \mathbf{x})$ be a half-space chart containing p and taking values in $\mathbb{R}^n_{\lambda \geq c}$ and let $(V, \mathbf{y})$ be another chart containing p, but taking values in $\mathbb{R}^n_{\mu \geq d}$. Then $d\mathbf{x} \circ d\mathbf{y}^{-1}$ maps $\mathbb{R}^n_{\mu < 0}$ diffeomorphically onto $\mathbb{R}^n_{\lambda < 0}$. Thus $\lambda \circ d\mathbf{x}(v_p) < 0$ if and only if $\mu \circ d\mathbf{y}(v_p) < 0$. Similarly, $\lambda \circ d\mathbf{x}(v_p) > 0$ if and only if $\mu \circ d\mathbf{y}(v_p) > 0$.

In light of the exercise above, the following definition makes sense:

Definition 8.73. Let $p \in \partial M$. A vector $v_p \in TM_p \subset TM|_{\partial M}$ is called **outward pointing** if in some $\mathbb{R}^n_{\lambda \geq c}$-valued chart $(U, \mathbf{x})$ we have $\lambda \circ d\mathbf{x}(v_p) < 0$. A smooth section χ of $TM|_{\partial M}$ is called **outward pointing** if $\chi(p)$ is outward pointing for each p. **Inward pointing** is defined analogously.

To clarify the situation, let us consider the special case of an $\mathbb{R}^n_{u^k \leq 0}$-valued chart $(U, \mathbf{x})$ for fixed $k \in \{1, 2, \ldots, n\}$ (thus $\lambda = -u^k$). If $p \in U$, then v_p is outward pointing exactly when $dx^k(v_p) > 0$. In particular, $\frac{\partial}{\partial x^k}\big|_p$ is outward pointing. For an $\mathbb{R}^n_{u^k \geq 0}$-valued chart $(U, \mathbf{x})$, the reverse is true; v_p is outward pointing exactly when $dx^k(v_p) < 0$. We shall see that $\mathbb{R}^n_{u^1 \leq 0}$ is special, so keep the corresponding criterion $dx^1(v_p) > 0$ in mind.

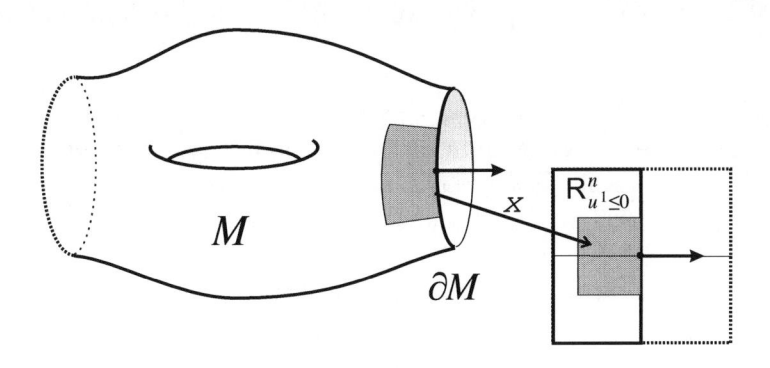

Figure 8.2. Outward pointing

Notice that the notion of outward pointing on ∂M does not depend on M being orientable (see Figure 8.3). In fact, we have the following:

Lemma 8.74. *Outward pointing sections of $TM|_{\partial M}$ always exist.*

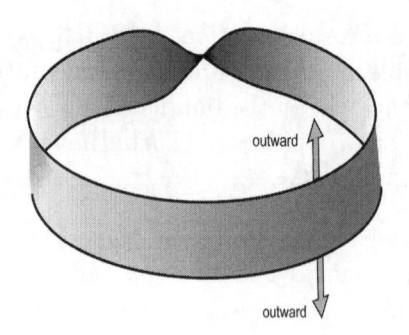

Figure 8.3. Outward at boundary of Möbius band

Proof. We may assume that we are dealing with an $\mathbb{R}^n_{u^1 \leq 0}$-valued atlas $\{(U_\alpha, \mathbf{x}_\alpha)\}$ for M. Let $\{\rho_\alpha\}$ be a partition of unity subordinate to $\{U_\alpha\}$. Define
$$\chi := \sum_\alpha \rho_\alpha \frac{\partial}{\partial x^1_\alpha}.$$
Then for $p \in \partial M$ we have
$$\chi(p) = \sum_\alpha \rho_\alpha(p) \left.\frac{\partial}{\partial x^1_\alpha}\right|_p.$$
To see if $\chi(p)$ is outward pointing, let $(U_\beta, \mathbf{x}_\beta)$ be a given $\mathbb{R}^n_{u^1 \leq 0}$-valued chart from the atlas. Then
$$dy^1(\chi(p)) = \sum_\alpha \rho_\alpha(p) \, dy^1\left(\left.\frac{\partial}{\partial x^1_\alpha}\right|_p\right) > 0,$$
since $\rho_\alpha(p) \geq 0$ for all α, we have $\rho_\alpha(p) > 0$ for at least one α, and $dy^1\left(\frac{\partial}{\partial x^1_\alpha}(p)\right) > 0$ for all α. $\square$

In the case that M is oriented we can use the orientation plus the notion of outward to define an orientation on the boundary. Let $(M, [\omega])$ be an oriented smooth manifold with boundary and suppose that χ is an outward pointing section of $TM|_{\partial M}$. Let $\omega \in [\omega]$; then we define $i_\chi \omega \in \Omega^{n-1}(\partial M)$ by
$$i_\chi \omega(p)(v_2, \ldots, v_n) = \omega(p)(\chi(p), v_2, \ldots, v_n).$$
It is clear that $i_\chi \omega$ is nowhere vanishing. If also $\omega_1 \in [\omega]$, then it is easy to see that $i_\chi \omega = f i_\chi \omega_1$ for a positive function f. Thus $[i_\chi \omega]$ depends only on $[\omega]$ and provides an orientation on ∂M.

Definition 8.75. The orientation on ∂M defined above is called the **induced orientation**.

It follows that if $(f_1, f_2, \ldots, f_n)$ is a positively oriented frame field on an open set $U \subset M$ with nonempty intersection with ∂M and if $f_1(p)$ is

8.7. Orientation

outward pointing for all $p \in U \cap \partial M$, then $(f_2, \ldots, f_n)$ restricted to $U \cap \partial M$ is a positively oriented frame with respect to the induced orientation on ∂M.

The case $n = 1$ needs some interpretation. Here ∂M is a discrete set of points $\partial M = \{p_1, \ldots, p_k\}$, and an orientation is an assignment of $+1$ or -1 to each point. For $p_i \in \partial M$, we assign $+1$ if $\omega(p_i)(\chi(p)) > 0$ for some outward pointing vector $\chi(p)$.

If every chart in an atlas takes values in a fixed half-space $\mathbb{R}^n_{\lambda \geq c}$, then we say the atlas is $\mathbb{R}^n_{\lambda \geq c}$-valued. It is not true generally that an oriented n-manifold has an atlas of positively oriented charts with values in a fixed half-space. However, it is almost true:

Lemma 8.76. *If M is an oriented n-manifold with nonempty boundary and $n \geq 2$, then there is an atlas of positively oriented charts.*

Proof. If $(U, \mathbf{x})$ is not positively oriented, then replace $\mathbf{x} = (x^1, x^2, \ldots, x^n)$ by $\mathbf{y} = (y^1, y^2, \ldots, y^n) := (x^1, -x^2, \ldots, x^n)$ to obtain a positively oriented chart. Notice that this does not work if $n = 1$. $\square$

Exercise 8.77. Show that although, as a manifold with boundary, the interval $M = [0, 1]$ is oriented in the standard way, there is no atlas consisting of positively oriented $\mathbb{R}^1_{u^1 \leq 0}$-valued charts. Trace the problem to the fact that we have defined an atlas for a manifold with boundary as having charts with values in a *fixed* half-space (in this case $\mathbb{R}^1_{u^1 \leq 0}$). Hint: There are no positively oriented charts containing 0. Changing to $\mathbb{R}^1_{u^1 \geq 0}$ pushes the problem to the other endpoint.

This last exercise exhibits a fact that is an annoyance if one wants to work with an atlas taking values in a fixed half-space such as the ever popular upper half-space $\mathbb{R}^n_{u^n \geq 0}$. It is an issue often overlooked in the literature.

Definition 8.78. A **nice chart** on a smooth manifold (possibly with boundary) is a chart $(U, \mathbf{x})$ such that $\mathbf{x}(U) = \mathbb{R}^n_{u^1 \leq 0}$ if $U \cap \partial M \neq \emptyset$ or where $\mathbf{x}(U) = \mathbb{R}^n$ if $U \cap \partial M = \emptyset$.

Lemma 8.79. *The following assertions hold:*

  (i) *Every smooth manifold has an atlas consisting of nice charts.*

  (ii) *Every oriented smooth manifold without boundary has an atlas consisting of positively oriented nice charts.*

  (iii) *Every oriented smooth manifold with boundary of dimension $n \geq 2$ has an atlas consisting of positively oriented nice charts.*

Proof. (i) If $(U, \mathbf{x})$ is a chart with range in the interior of the left half-space $\mathbb{R}^n_{u^1 \leq 0}$, then we can find a ball B inside $\mathbf{x}(U)$ in $\mathbb{R}^n_{u^1 < 0}$ and then we form a new chart on $\mathbf{x}^{-1}(B)$ with range B. But a ball is diffeomorphic to $\mathbb{R}^n$ (by an orientation preserving diffeomorphism). If $(U, \mathbf{x})$ is a chart with range meeting the boundary of the left half-space $\mathbb{R}^n_{u^1 \leq 0}$, then we can find a half-ball B_- in $\mathbb{R}^n_{u^1 \leq 0}$ with center on $\mathbb{R}^n_{u^1 = 0}$. Reduce the chart domain as before to have range equal to this half-ball. But every half-ball is diffeomorphic to the half-space $\mathbb{R}^n_{u^1 \leq 0}$ so we can proceed by composition as before.

(ii) For this, we repeat the procedure of (i) and notice that since $\mathbb{R}^n$ is diffeomorphic to B by an orientation preserving diffeomorphism, the new nice chart will be positively oriented if $(U, \mathbf{x})$ is positive.

(iii) If $(U, \mathbf{x})$ is a chart with range in the interior of the left half-space $\mathbb{R}^n_{u^1 \leq 0}$, we proceed as in (ii). If $(U, \mathbf{x})$ is a chart with range meeting the boundary of the left half-space $\mathbb{R}^n_{u^1 \leq 0}$ that is already positively oriented, then we proceed as in (i). If $(U, \mathbf{x})$ is not positively oriented, then replace $\mathbf{x} = (x^1, x^2, \ldots, x^n)$ by $\mathbf{y} = (y^1, y^2, \ldots, y^n) := (x^1, -x^2, \ldots, x^n)$ as in the proof of Lemma 8.76 and proceed as before. $\square$

Exercise 8.80. If $\mathbf{x} = (x^1, \ldots, x^n)$ is an oriented $\mathbb{R}^n_{u^1 \leq 0}$-valued chart on a neighborhood of p on the boundary of an oriented manifold with boundary, then the vectors $\frac{\partial}{\partial x^2}, \ldots, \frac{\partial}{\partial x^n}$ form a positive basis for $T_p \partial M$ with respect to the induced orientation on ∂M and thus $(x^2, \ldots, x^n)$ restricts to ∂M to give a positively oriented chart.

8.8. Invariant Forms

Throughout this section G is an n-dimensional Lie group.

Definition 8.81. An element $\omega \in \Omega^k(G)$ is called **left invariant** if $L_g^* \omega = \omega$ for all $g \in G$.

It is easy to see that a left invariant k-form is determined by its value $\omega(e)$ at the identity element. Now suppose that $\omega^1, \ldots, \omega^n$ are invariant 1-forms such that $\omega^1(e), \ldots, \omega^n(e)$ is a basis of $T_e^* G$. The $\omega^1, \ldots, \omega^n$ are independent everywhere (why?). If $X_1, \ldots, X_n$ is the frame field dual to $\omega^1, \ldots, \omega^n$, then each X_i is a left invariant vector field as may be easily checked. By definition, a left invariant form satisfies

$$\omega(x)(v_1, \ldots, v_k) = \omega(gx)(TL_g v_1, \ldots, TL_g v_k)$$

for all $x \in M$ and $g \in G$. This would make sense even if ω were not smooth, but $\omega(X_1, \ldots, X_k)$ is not only smooth but constant, so any left invariant form must be smooth after all. Every left invariant k-form ω can be written as

$$(8.10) \qquad \omega = \sum_{1 \leq i_1 < \cdots < i_k \leq n} a_{i_1 \ldots i_k} \omega^{i_1} \wedge \cdots \wedge \omega^{i_k}$$

for unique constants $a_{i_1 \ldots i_k}$.

8.8. Invariant Forms

The exterior derivative of a left invariant form is left invariant since for any $g \in G$, we have
$$L_g^* d\omega = dL_g^*\omega = d\omega.$$
Furthermore for any left invariant vector fields X, Y we have

(8.11) $\qquad d\omega(X, Y) = X\omega(Y) - Y\omega(X) - \omega([X, Y])$
(8.12) $\qquad\qquad\quad = -\omega([X, Y]),$

and so in particular

(8.13) $\qquad\qquad d\omega|_e(v, w) = -\omega_e([v, w]),$

for any $v, w \in \mathfrak{g}$.

A form is called right invariant if $R_g^*\omega = \omega$ for all $g \in G$. Of course, if ω is right invariant, then so is $d\omega$. If $\mathrm{inv}: x \mapsto x^{-1}$ denotes the inversion map on G, then
$$\mathrm{inv} \circ R_g = L_{g^{-1}} \circ \mathrm{inv},$$
and so
$$R_g^* \circ \mathrm{inv}^* = \mathrm{inv}^* \circ L_{g^{-1}}^*.$$
As a consequence, if ω is left invariant, then
$$R_g^* \circ \mathrm{inv}^*\omega = \mathrm{inv}^* \circ L_{g^{-1}}^*\omega = \mathrm{inv}^*\omega,$$
so $\mathrm{inv}^*\omega$ is right invariant. Similarly, if ω is right invariant, then $\mathrm{inv}^*\omega$ is left invariant.

Lemma 8.82. $T_e\mathrm{inv} = -\mathrm{id} : \mathfrak{g} \to \mathfrak{g}$.

Proof. Let $v \in \mathfrak{g}$. Then the curve $t \mapsto \exp(tv)$ has tangent v. Thus $(\exp(tv))^{-1}$ has tangent $T_e\mathrm{inv} \cdot v$. But $(\exp(tv))^{-1} = \exp(-tv)$ so we must have $T_e\mathrm{inv} \cdot v = -v$. $\square$

Proposition 8.83. *Let* $\mathrm{inv} : x \mapsto x^{-1}$ *be the inversion map on* G.

(i) *If* $\omega \in \Omega^k(G)$, *then* $(\mathrm{inv}^*\omega)_e = (-1)^k \omega_e$.

(ii) *If* ω *is left and right invariant, then* $d\omega = 0$.

Proof. (i) It suffices to assume that $\omega = f\omega^{i_1} \wedge \cdots \wedge \omega^{i_k}$ for certain 1-forms $\omega^{i_1}, \ldots, \omega^{i_k}$ that may be taken to be left invariant. But the result will follow if we can show that $(\mathrm{inv}^*\omega)_e = -1\omega_e$ for any 1-form. So let $v \in \mathfrak{g}$. Then, by Lemma 8.82,
$$(\mathrm{inv}^*\omega)_e(v) = \omega_e(T_e\mathrm{inv} \cdot v) = -\omega_e(v).$$

(ii) From (i) we have
$$(\mathrm{inv}^*\omega)_e = (-1)^k \omega_e.$$

If ω is left and right invariant, then both inv$^*\omega$ and ω are left invariant and this continues to hold globally:
$$\text{inv}^*\omega = (-1)^k\omega;$$
$d\omega$ is also left and right invariant, so
$$\text{inv}^*d\omega = (-1)^{k+1}d\omega.$$
On the other hand,
$$\text{inv}^*d\omega = d\,\text{inv}^*\omega = (-1)^k d\omega,$$
so $(-1)^{k+1}d\omega = (-1)^k d\omega$ and then $d\omega = 0$. $\square$

Corollary 8.84. *If G is abelian, then $\mathfrak{g}$ is abelian.*

Proof. If G is abelian, then every left invariant form is also right invariant, so $d\omega = 0$ for all left invariant forms ω. But then by equation (8.13), for any $v, w \in \mathfrak{g}$ we have $\omega_e([v, w]) = 0$ for all $\omega_e \in \mathfrak{g}^*$. Thus $[v, w] = 0$ for any v, w. $\square$

If $X_1, \ldots, X_n$ are left invariant vector fields with $X_1(e), \ldots, X_n(e)$ dual to the basis $\omega^1(e), \ldots, \omega^n(e)$, then
$$[X_i(e), X_j(e)] = \sum_k c_{ij}^k X_k(e) \quad \text{for } 1 \leq i, j \leq n,$$
where c_{ij}^k are the structure constants from Definition 5.56. We also have $[X_i, X_j] = \sum_k c_{ij}^k X_k$ for $1 \leq i, j \leq n$, so by the equation above and equation (8.13) we have for $i < j$,
$$d\omega^k(X_i, X_j) = -\omega^k([X_i, X_j]) = -c_{ij}^k.$$
Since $d\omega^k(X_i, X_j)$ gives the components of $d\omega^k$, we have
$$(8.14) \qquad d\omega^k = -\sum_{i<j} c_{ij}^k \omega^i \wedge \omega^j = -\frac{1}{2}\sum_{i,j} c_{ij}^k \omega^i \wedge \omega^j,$$
which are called the **equations of structure,** or the **structural equations.**

We now use the concept of Lie algebra-valued forms and the product $[\,,\,]^\wedge$ introduced earlier. Recall the left Maurer-Cartan form ω_G. If $\omega^1, \ldots, \omega^n$ is a basis of left invariant 1-forms dual to $X_1, \ldots, X_n$, then (omitting tensor product signs),
$$\omega_G = \sum_{i=1}^n X_i(e)\omega^i.$$

8.8. Invariant Forms

Indeed, if $X_g = \sum_{i=1}^n a^i X_i(g) \in T_g G$, then $X = \sum_{i=1}^n a^i X_i$ is the unique left invariant vector field such that $X(g) = X_g$ and we have

$$\left(\sum_{i=1}^n X_i(e)\omega^i\right)(X_g) = \sum_{i=1}^n a^i X_i(e)$$
$$= X(e) = TL_{g^{-1}} \cdot X(g) = \omega_G(X_g).$$

Moreover,

$$d\omega_G = \sum_{k=1}^n X_k(e)d\omega^k = -\sum_{k=1}^n X_k(e)\left(\sum_{i<j} c_{ij}^k \omega^i \wedge \omega^j\right)$$
$$= -\sum_{k=1}^n \sum_{i<j} c_{ij}^k X_k(e) \omega^i \wedge \omega^j.$$

On the other hand we have

$$[\omega_G, \omega_G]^\wedge = \left[\sum_{i=1}^n X_i(e)\omega^i, \sum_{i=1}^n X_j(e)\omega^j\right]^\wedge$$
$$= \sum_{i,j=1}^n [X_i(e), X_j(e)]\,\omega^i \wedge \omega^j$$
$$= \sum_{i,j=1}^n \sum_k c_{ij}^k X_k \omega^i \wedge \omega^j = \frac{1}{2}\sum_k \sum_{i<j} c_{ij}^k X_k \omega^i \wedge \omega^j,$$

so that an alternative and concise form of the equations of structure is the single equation

(8.15) $$d\omega_G = -\frac{1}{2}[\omega_G, \omega_G]^\wedge.$$

Using what we learned at the end of Section 8.4, this equation may be written as

$$d\omega_G(X, Y) = -[\omega_G(X), \omega_G(Y)]$$

for any $X, Y \in \mathfrak{X}(G)$. If G is a matrix group, then the structural equation can be written as

$$d\omega_G = -\omega_G \wedge \omega_G.$$

Indeed, if we abbreviate ω_G to just ω, then we have $(d\omega)_j^i = d\omega_j^i$ and

$$[\omega_G(X), \omega_G(Y)]_j^i = (\omega(X)\omega(Y) - \omega(Y)\omega(X))_j^i$$
$$= \omega(X)_k^i \omega(Y)_j^k - \omega(Y)_k^i \omega(X)_j^k$$
$$= \left(\omega_k^i \wedge \omega_j^k\right)(X, Y).$$

Equation (8.15) is also called the **Maurer-Cartan equation**.

Problems

(1) Let S_N denote the group of permutations of the set $\{1, 2, \ldots, N\}$. Now let $G_{k,\ell}$ be the subgroup of $S_{k+\ell}$ which consists of permutations that leave the sets $\{1, \ldots, k\}$ and $\{k+1, \ldots, k+\ell\}$ each invariant. A **cross section** of $G_{k,\ell}$ is a subset K of $S_{k+\ell}$ that contains exactly one element from each coset in $S_{k+\ell}/G_{k,\ell}$. Show that for any such cross section we have

$$\omega \wedge \eta(v_1, \ldots, v_k, v_{k+1}, \ldots, v_{k+\ell})$$
$$= \sum_{\sigma \in K} \operatorname{sgn}(\sigma) \omega(v_{\sigma_1}, \ldots, v_{\sigma_k}) \eta(v_{\sigma_{k+1}}, \ldots, v_{\sigma_{k+\ell}}).$$

Also, show that the set of all (k, ℓ)-shuffle permutations is a cross section of $G_{k,\ell}$.

(2) Show that if $v_1, \ldots, v_k \in V$, then $v_1 \wedge \cdots \wedge v_k \neq 0$ if and only if $v_1, \ldots, v_k$ are linearly independent.

(3) Let $f_1, \ldots, f_n$ be smooth functions on an open set in an n-manifold. Let p be in their common domain. Then $df_1 \wedge \cdots \wedge df_n$ is nonzero at p if and only if $f_1, \ldots, f_n$ agree with the coordinate functions of a chart whose domain is a neighborhood of p.

(4) (a) Let $v \in \bigwedge^k V$. Show that if $v \wedge w = 0$ for all $w \in \bigwedge^{n-k} V$, where $n = \dim V$, then $v = 0$.
(b) Using part (a), show that more generally, if $v \wedge w = 0$ for all $w \in \bigwedge^m V$, where $m \leq n - k$, then $v = 0$. [Hints: If $v \wedge w = 0$ for all $w \in \bigwedge^m V$, then $v \wedge (w \wedge x) = 0$ for all $w \in \bigwedge^m V$ and all $x \in \bigwedge^{n-k-m} V$. Elements of the form $w \wedge x$ as above span $\bigwedge^{n-k} V$.]

(5) Prove Proposition 8.23.

(6) Show that the sphere is orientable.

(7) Prove (i) of Proposition 8.4.

(8) Prove Proposition 8.30.

(9) Prove equation (i) of Corollary 8.58.

(10) Prove Cartan's lemma: Let $k \leq n = \dim M$ and let $\omega_1, \ldots, \omega_k$ be 1-forms on M which are linearly independent at each point. Suppose that there are 1-forms $\theta_1, \ldots, \theta_k$ such that

$$\sum_{i=1}^{k} \theta_i \wedge \omega_i = 0 \text{ (identically)}.$$

Then there exists a symmetric $k \times k$ matrix of smooth functions (A_{ij}) such that
$$\theta_i = \sum_{j=1}^{k} A_{ij}\omega_j \text{ for } i = 1, \ldots, k.$$

(11) Let $M = \mathbb{R}^3 \setminus \{0\}$ and let
$$\omega = \frac{x\,dy \wedge dz + y\,dz \wedge dx + z\,dx \wedge dy}{(x^2 + y^2 + z^2)^{3/2}}.$$
Find $d\omega$ and determine whether ω is closed and if so, whether it is exact. Find the expression for ω in spherical coordinates.

(12) Show that every simply connected manifold is orientable.

(13) (a) Let V and W be vector spaces with $\dim V = n$ and $\dim W = m$. Show that if $A \in L(V, W)$, then there is a unique map $\wedge_A : \bigwedge^k V \to \bigwedge^k W$ such that $\wedge_A(v_1 \wedge \cdots \wedge v_k) = Av_1 \wedge \cdots \wedge Av_k$ whenever $v_1, \ldots, v_k \in V$.

(b) Show that if $e_1, \ldots, e_n$ is a basis for V and $f_1, \ldots, f_m$ is a basis for W and if $Ae_i = \sum a_i^j f_j$, then for $1 \leq i_1 < \cdots < i_k \leq n$ we have
$$\wedge_A(e_{i_1} \wedge \cdots \wedge e_{i_k}) = \sum_{1 \leq j_1 < \cdots < j_k \leq m} a_{i_1 \ldots i_k}^{j_1 \ldots j_k} f_{j_1} \wedge \cdots \wedge f_{j_k},$$
where
$$a_{i_1 \ldots i_k}^{j_1 \ldots j_k}$$
is the $k \times k$ minor determinant of the matrix (a_j^i) given by
$$a_{i_1 \ldots i_k}^{j_1 \ldots j_k} = \sum_{\sigma \in S_k} \mathrm{sgn}(\sigma) a_{i_1}^{\sigma(j_1)} \cdots a_{i_k}^{\sigma(j_k)}.$$

(14) Finish the proof of Proposition 8.13.

(15) Prove Proposition 8.35.

Chapter 9

Integration and Stokes' Theorem

Let M be a smooth n-manifold possibly with boundary ∂M and assume that M is oriented and that ∂M has the induced orientation (Definition 8.75).

Definition 9.1. The **support** of a differential form $\alpha \in \Omega(M)$ is the *closure* of the set $\{p \in M : \alpha(p) \neq 0\}$ and is denoted by $\operatorname{supp}(\alpha)$. The set of all k-forms that have compact support is denoted $\Omega_c^k(M)$, and the set of all k-forms with compact support contained in $U \subset M$ is denoted by $\Omega_c^k(U)$.

Let us consider the case of an n-form α on an open subset U of $\mathbb{R}^n$. Let $(u^1, \ldots, u^n)$ be standard coordinates on U. We may write $\alpha = a\, du^1 \wedge \cdots \wedge du^n$ for some function a. If α has *compact support* in U, we may define the integral $\int_U \alpha$ by

$$\int_U \alpha = \int_U a\, du^1 \wedge \cdots \wedge du^n$$
$$:= \int_U a(u)\, du^1 \cdots du^n,$$

where this latter integral is the Riemann (or Lebesgue) integral of $a(\cdot)$. For this we extend $a(\cdot)$ by zero to all of $\mathbb{R}^n$ and integrate over a sufficiently large closed n-cube containing the support of $a(\cdot)$ in its interior. We could have written $|du^1 \cdots du^n|$ instead of $du^1 \cdots du^n$ to emphasize that the order of the du^i does not matter as it does for $du^1 \wedge \cdots \wedge du^n$. If U is an open subset of the half-space $\mathbb{R}^n_{\lambda \geq c}$, then we define $\int_U \alpha$ by the same formula. If $\phi : V \to U$ is an orientation preserving diffeomorphism of open sets in $\mathbb{R}^n$, then $\det D\phi > 0$. Let $u^1, \ldots, u^n$ denote standard coordinates on U, and let

391

$v^1, \ldots, v^n$ denote standard coordinates on V. Then by the classical change of variable formula,

$$\int_U \alpha = \int_U a(u)\, du^1 \cdots du^n = \int_V a \circ \phi(v) \, |\det D\phi| \, dv^1 \cdots dv^n$$
$$= \int_V a \circ \phi(v) \det D\phi \, dv^1 \cdots dv^n$$
$$= \int_V a \circ \phi(v) \det D\phi \, dv^1 \wedge \cdots \wedge dv^n = \int_V \phi^* \alpha.$$

So

(9.1) $$\int_U \alpha = \int_V \phi^* \alpha.$$

Next consider an oriented n-manifold M without boundary and let $\alpha \in \Omega^n(M)$. If α has compact support inside U for some positively oriented chart $(U, \mathbf{x})$, then $\mathbf{x}^{-1} : \mathbf{x}(U) \to U$ and $(\mathbf{x}^{-1})^*\alpha$ has compact support in $\mathbf{x}(U) \subset \mathbb{R}^n$. We define

$$\int_U \alpha := \int_{\mathbf{x}(U)} (\mathbf{x}^{-1})^* \alpha.$$

The change of variables formula (9.1) shows that this definition is independent of the positively oriented chart chosen. In fact, if $(\mathbf{y}, V)$ is another chart and α has support inside $U \cap V$, then $(\mathbf{x}^{-1})^*\alpha$ has support inside $\mathbf{x}(U \cap V)$ and $(\mathbf{y}^{-1})^*\alpha$ has support in $\mathbf{y}(U \cap V)$. Thus since $\mathbf{x} \circ \mathbf{y}^{-1}$ is orientation preserving, we have

$$\int_{\mathbf{x}(U)} (\mathbf{x}^{-1})^*\alpha = \int_{\mathbf{x}(U \cap V)} (\mathbf{x}^{-1})^*\alpha = \int_{\mathbf{y}(U \cap V)} (\mathbf{x} \circ \mathbf{y}^{-1})^* (\mathbf{x}^{-1})^*\alpha$$
$$= \int_{\mathbf{y}(U \cap V)} (\mathbf{y}^{-1})^*\alpha = \int_{\mathbf{y}(U)} (\mathbf{y}^{-1})^*\alpha.$$

The same definition works fine in case M has nonempty boundary, but there is a small technicality. Namely, suppose we wish to work only with positive charts taking values in a *fixed* half-space $\mathbb{R}^n_{\lambda \geq c}$. Then as long as $n \geq 2$ there is no problem, but if $n = 1$, we are faced with the fact that there may be no positively oriented $\mathbb{R}^1_{\lambda \geq c}$-valued atlas at all even if M is orientable. Some authors define manifold with boundary completely in terms of a fixed half-space and seem unaware of this little glitch. Since we allow multiple half-spaces, this is not a problem for us, but in any case, we could modify the definition slightly in a way that works in all dimensions and for any chart. Let M be an oriented manifold with boundary. If α has compact support inside U for some chart $(U, \mathbf{x})$, then

(9.2) $$\int_U \alpha = \text{sgn}(\mathbf{x}) \int_{\mathbf{x}(U)} (\mathbf{x}^{-1})^*\alpha,$$

9. Integration and Stokes' Theorem

where

$$\text{sgn}(\mathbf{x}) = \begin{cases} 1 & \text{if } (U, \mathbf{x}) \text{ is positively oriented,} \\ -1 & \text{if } (U, \mathbf{x}) \text{ is not positively oriented.} \end{cases}$$

This could be taken as a definition. Once again, the standard change of variables formula shows that this definition is independent of the chart chosen.

Remark 9.2. Because we have used the $\text{sgn}(\mathbf{x})$ factor in the definition, we can use arbitrary charts. But the manifold still must be oriented so that $\text{sgn}(\mathbf{x})$ makes sense!

If $\alpha \in \Omega_c^n(M)$ has compact support but does not have support contained in some chart domain, then we choose a positively oriented atlas $\{(\mathbf{x}_i, U_i)\}$ for M and a smooth partition of unity $\{(\rho_i, U_i)\}$ subordinate to the atlas, and consider the sum

$$(9.3) \qquad \sum_i \int_{U_i} \rho_i \alpha = \sum_i \int_{\mathbf{x}_i(U_i)} (\mathbf{x}_i^{-1})^*(\rho_i \alpha).$$

Proposition 9.3. *In the sum above, only a finite number of terms are nonzero. The sum is independent of the choice of atlas and smooth partition of unity* $\{(\rho_i, U_i)\}$.

Proof. First, for any $p \in M$ there is an open set O containing p such that only a finite number of ρ_i are nonzero on O. But α has compact support, and so a finite number of such open sets cover the support. This means that only a finite number of the ρ_i are nonzero on this support. Now let $\{(\bar{\mathbf{x}}_i, V_i)\}$ be another positive atlas and $\bar{\rho}_i$ a partition of unity subordinate to it. Then we have

$$\sum_i \int_{U_i} \rho_i \alpha = \sum_i \int_{U_i} \left(\rho_i \sum_j \bar{\rho}_j \alpha \right)$$

$$= \sum_i \sum_j \int_{U_i \cap V_j} \rho_i \bar{\rho}_j \alpha = \sum_j \int_{V_j} \left(\bar{\rho}_j \sum_i \rho_i \alpha \right)$$

$$= \sum_j \int_{V_j} \bar{\rho}_j \alpha. \qquad \square$$

Since the sum (9.3) above is the same independently of the allowed choices, we make the following definition:

Definition 9.4. Let $(M, [\varpi])$ be an oriented smooth manifold with or without boundary. Let $\alpha \in \Omega^n(M)$ have compact support. Choose a positively oriented atlas $\{(\mathbf{x}_i, U_i)\}$ and a smooth partition of unity $\{(\rho_i, U_i)\}$ subordinate to $\{U_i\}$. Then we define

$$\int_{(M,[\varpi])} \alpha := \sum_i \int_{U_i} \rho_i \alpha.$$

We usually omit the explicit reference to the orientation $[\varpi]$ and simply write $\int_M \alpha$.

Remark 9.5. Of course if we take (9.2) as a definition, then we may take

$$\sum_i \int_{U_i} \rho_i \alpha = \sum_i \mathrm{sgn}(\mathbf{x}_i) \int_{\mathbf{x}_i(U_i)} (\mathbf{x}_i^{-1})^*(\rho_i \alpha)$$

and $\int_{(M,[\varpi])} \alpha = \sum_i \int_{U_i} \rho_i \alpha$ even for an arbitrary unoriented atlas. Once again, we still need M to be oriented. In the online supplement we introduce twisted n-forms, and these may be integrated even on nonorientable manifolds!

In case M is zero-dimensional, and therefore a discrete set of points $M = \{p_1, p_2, \ldots\}$, an orientation $[\varpi]$ is an assignment of $+1$ or -1 to each point. Then, if $\alpha = f \in \Omega^0(M)$, we have

$$\int_{(M,[\varpi])} f = \sum \pm f(p_i),$$

where we choose the $\pm$ according to the orientation at the point.

9.1. Stokes' Theorem

In this section we take up the main theorem of the chapter. It is the fundamental theorem of exterior calculus known as Stokes' theorem. Our definition of integration works for any (positive) atlas, but we can use a specific atlas for theoretical purposes. We will employ $\mathbb{R}^n_{u^1 \leq 0}$-valued atlases in this section. This is $\mathbb{R}^n_{\lambda \geq c}$ where $c = 0$ and $\lambda = -u^1$. Let us consider two special cases of integration.

Case 1. This is the case of a compactly supported $(n-1)$-form on $\mathbb{R}^n$. Let $\omega_j = f du^1 \wedge \cdots \wedge \widehat{du^j} \wedge \cdots \wedge du^n$ be a smooth $(n-1)$-form with compact support in $\mathbb{R}^n$, where the caret symbol over the du^j means that this j-th factor is omitted. Then we have

$$\int_{\mathbb{R}^n_{u^1\leq 0}} d\omega_j = \int_{\mathbb{R}^n_{u^1\leq 0}} d(fdu^1 \wedge \cdots \wedge \widehat{du^j} \wedge \cdots \wedge du^n)$$

$$= \int_{\mathbb{R}^n_{u^1\leq 0}} (df \wedge du^1 \wedge \cdots \wedge \widehat{du^j} \wedge \cdots \wedge du^n)$$

$$= \int_{\mathbb{R}^n_{u^1\leq 0}} \left(\sum_k \frac{\partial f}{\partial u^k} du^k \wedge du^1 \wedge \cdots \wedge \widehat{du^j} \wedge \cdots \wedge du^n \right)$$

$$= \int_{\mathbb{R}^n_{u^1\leq 0}} (-1)^{j-1} \frac{\partial f}{\partial u^j} du^1 \wedge \cdots \wedge du^n$$

$$= \int_{\mathbb{R}^n} (-1)^{j-1} \frac{\partial f}{\partial u^j} du^1 \cdots du^n = 0$$

by the fundamental theorem of calculus and the fact that f has compact support. All compactly supported $(n-1)$-forms ω on $\mathbb{R}^n$ are sums of forms of this type and so we have

$$\int_{\mathbb{R}^n} d\omega = \int_{\partial \mathbb{R}^n} \omega.$$

Case 2. This is the case of a compactly supported $(n-1)$-form on $\mathbb{R}^n_{u^1\leq 0}$. Let $\omega_j = fdu^1 \wedge \cdots \wedge \widehat{du^j} \wedge \cdots \wedge du^n$ be a smooth $(n-1)$-form with compact support meeting $\partial \mathbb{R}^n_{u^1\leq 0} = \mathbb{R}^n_{u^1=0} = \{0\} \times \mathbb{R}^{n-1}$. Then, if $j \neq 1$,

$$\int_{\mathbb{R}^n_{u^1\leq 0}} d\omega_j = \int_{\mathbb{R}^n_{u^1\leq 0}} d(fdu^1 \wedge \cdots \wedge \widehat{du^j} \wedge \cdots \wedge du^n)$$

$$= \int_{\mathbb{R}^{n-1}} (-1)^{j-1} \left(\int_{-\infty}^{\infty} \frac{\partial f}{\partial u^j} du^j \right) du^1 \cdots \widehat{du^j} \cdots du^n = 0.$$

If $j = 1$ we have

$$\int_{\mathbb{R}^n_{u^1\leq 0}} d\omega_1 = \int_{\mathbb{R}^{n-1}} (-1)^{j-1} \left(\int_{-\infty}^0 \frac{\partial f}{\partial u^1} du^1 \right) du^2 \wedge \cdots \wedge du^n$$

$$= \int_{\mathbb{R}^{n-1}} f(0, u^2, \ldots, u^n) du^2 \cdots du^n$$

$$= \int_{\partial \mathbb{R}^n_{u^1\leq 0}} f(0, u^2, \ldots, u^n) du^2 \wedge \cdots \wedge du^n = \int_{\partial \mathbb{R}^n_{u^1\leq 0}} \omega_1.$$

For this last equality, it is important that we have set things up so that $du^2 \wedge \cdots \wedge du^n$ be positive on $\partial \mathbb{R}^n_{u^1\leq 0}$ with the induced orientation. Since clearly $\int_{\partial \mathbb{R}^n_{u^1\leq 0}} \omega_j = 0$ if $j \neq 1$ or if ω_j has support that does not meet $\partial \mathbb{R}^n_{u^1\leq 0}$, we see that in any case $\int_{\mathbb{R}^n_{u^1\leq 0}} d\omega_j = \int_{\partial \mathbb{R}^n_{u^1\leq 0}} \omega_j$. All compactly supported $(n-1)$-forms ω on $\mathbb{R}^n_{u^1 \leq 0}$ are sums of forms of this type, and so summing we have

$$\int_{\mathbb{R}^n_{u^1 \leq 0}} d\omega = \int_{\partial \mathbb{R}^n_{u^1 \leq 0}} \omega.$$

Let us assume that M is an oriented smooth manifold of dimension $n \geq 2$. Then there is a positively oriented atlas $\{(U_\alpha, \mathbf{x}_\alpha)\}_{\alpha \in A}$ consisting of nice charts so either $\mathbf{x}_\alpha : U_\alpha \cong \mathbb{R}^n$ or $\mathbf{x}_\alpha : U_\alpha \cong \mathbb{R}^n_{u^1 \leq 0}$. Let $\{\rho_\alpha\}$ be a smooth partition of unity subordinate to $\{U_\alpha\}$. Notice that $\{\rho_\alpha|_{U_\alpha \cap \partial M}\}$ is a partition of unity for the cover $\{U_\alpha \cap \partial M\}$ of ∂M. Next we apply the special cases above. For $\omega \in \Omega^{n-1}(M)$ with compact support, we have that

$$\int_M d\omega = \int_{U_\alpha} \sum_\alpha d(\rho_\alpha \omega) = \sum_\alpha \int_{U_\alpha} d(\rho_\alpha \omega)$$

$$= \sum_\alpha \int_{\mathbf{x}_\alpha(U_\alpha)} (\mathbf{x}_\alpha^{-1})^* d(\rho_\alpha \omega) = \sum_\alpha \int_{\mathbf{x}_\alpha(U_\alpha)} d((\mathbf{x}_\alpha^{-1})^* \rho_\alpha \omega)$$

$$= \sum_\alpha \int_{\partial\{\mathbf{x}_\alpha(U_\alpha)\}} ((\mathbf{x}_\alpha^{-1})^* \rho_\alpha \omega) = \sum_\alpha \int_{\partial U_\alpha} \rho_\alpha \omega = \int_{\partial M} \omega,$$

where we have used the fact that $\mathbf{x}_\alpha^{-1} : \mathbf{x}_\alpha(\partial U_\alpha) \to \partial U_\alpha$ is orientation preserving. We have now proved Stokes' theorem stated below for the $n > 1$ case. The $n = 1$ case is easily proved directly and amounts to the familiar fundamental theorem of calculus.

Theorem 9.6 (Stokes' theorem). *Let M be an oriented smooth manifold with boundary (possibly empty) and give ∂M the induced orientation. Then for any $\omega \in \Omega_c^{n-1}(M)$ (i.e. with compact support) we have*

$$\int_M d\omega = \int_{\partial M} \omega.$$

Note that $\int_{\partial M} \omega := 0$ if $\partial M = \emptyset$. It can be shown that if $(U, \mathbf{x})$ is a chart for a manifold such that $M \backslash U$ has measure zero, then

$$\int_M \omega := \int_{\mathbf{x}(U)} (\mathbf{x}^{-1})^* \omega.$$

The definition of integration using partitions of unity is fine for the theoretical purposes we intend to pursue (such as cohomology), but for actually calculating integrals it is nearly useless. Consider the task of integrating the form $\omega := z\,dy \wedge dz + x\,dx \wedge dy$ over the sphere $S^2 \subset \mathbb{R}^3$. Technically speaking, we are integrating the restriction of ω to S^2. Let $\sigma(u, v) = (\cos u \cos v, \sin u \cos v, \sin v)$ for $(u, v) \in (0, 2\pi) \times (-\pi, \pi)$. This gives a parametrization of a portion of the sphere. Thus σ plays the role of $\mathbf{x}^{-1}$. The image of our parametrization is the domain of the chart $\mathbf{x}$, and the complement of the domain of the chart has measure zero in S^2.

9.2. Differentiating Integral Expressions; Divergence

Because of this it seems plausible that we would get the correct answer by calculating $\int_{(0,2\pi)\times(-\pi,\pi)} \sigma^*\omega$. But $\sigma^*\omega$ does not have compact support in $(0, 2\pi) \times (-\pi, \pi)$, and so this is not justified in terms of the theory we have developed. Let us try it anyway. We pull the form back to the u, v space and then integrate just as one normally would in calculus of several variables:

$$\int_{S^2} \omega := \int_{(0,2\pi]\times(-\pi,\pi)} \sigma^*\omega$$

$$= \int_{(0,2\pi]\times(-\pi,\pi)} \left(z(u,v) \frac{\partial(y,z)}{\partial(u,v)} + x(u,v) \frac{\partial(x,y)}{\partial(u,v)} \right) du\, dv$$

$$= \int_{-\pi}^{\pi} \int_{0}^{2\pi} \cos^2 v \cos u \sin v\, du\, dv = 0.$$

We are led to a nice integral in this case, but in general, proceeding in this way may lead to improper Riemann integrals, and anyway, it is not immediately clear how to connect this with our theory given in terms of partitions of unity. That the answer we obtained above is indeed the correct answer can be seen by applying Stokes' theorem:

$$\int_{S^2} \omega \stackrel{\text{Stokes}}{=} \int_B d\omega = \int_B 0 = 0.$$

Here B is the unit ball and we used the fact that $d\omega = 0$.

A practical way of calculating integrals of forms is to break up the manifold in a nice way and add up the integrals over the pieces. In Problem 10 we ask the reader to prove the following theorem:

Theorem 9.7. *Let M be an oriented n-manifold with possibly nonempty boundary. Suppose that there are subsets $D_i \subset \mathbb{R}^n$ for $i = 1, \ldots, K$ and smooth maps $\phi_i : D_i \to M$ such that the following assertions hold:*

(i) *Each D_i is compact and has a boundary of measure zero.*

(ii) *Each ϕ_i restricts to an orientation preserving diffeomorphism of the interior of D_i to the interior of $\phi_i(D_i)$.*

(iii) *$\phi_i(D_i) \cap \phi_j(D_j)$ is either empty or contains only boundary points of $\phi_i(D_i)$ and $\phi_j(D_j)$.*

Then, for any smooth n-form ω with support in $\bigcup \phi_i(D_i)$, we have

$$\int_M \omega = \sum_{i=1}^{K} \int_{D_i} \phi_i^*\omega.$$

9.2. Differentiating Integral Expressions; Divergence

Suppose that $S \subset M$ is a k-dimensional regular submanifold with boundary ∂S (possibly empty) and Φ_t is the flow of some complete vector field $X \in$

$\mathfrak{X}(M)$. In this case, $\Phi_t(S)$ is also a regular submanifold with boundary. We then consider $\frac{d}{dt}\int_S \Phi_t^*\eta$ for some k-form η with compact support. We have

$$\frac{d}{dt}\int_S \Phi_t^*\eta = \lim_{h\to 0}\frac{1}{h}\left[\int_S \Phi_{t+h}^*\eta - \int_S \Phi_t^*\eta\right]$$

$$= \lim_{h\to 0}\left[\int_S \frac{1}{h}\Phi_t^*\left(\Phi_h^*\eta - \eta\right)\right]$$

$$= \left[\int_{\Phi_t S}\lim_{h\to 0}\frac{1}{h}\left(\Phi_h^*\eta - \eta\right)\right] = \int_{\Phi_t S}\mathcal{L}_X\eta.$$

But also $\int_S \Phi_t^*\eta = \int_{\Phi_t S}\eta$. Thus we obtain the useful formula

$$\frac{d}{dt}\int_{\Phi_t S}\eta = \int_{\Phi_t S}\mathcal{L}_X\eta.$$

As a special case ($t=0$), we have $\frac{d}{dt}\big|_{t=0}\int_{\Phi_t S}\eta = \int_S \mathcal{L}_X\eta$. We can go further using Cartan's formula $\mathcal{L}_X = i_X\circ d + d\circ i_X$. We get

$$\frac{d}{dt}\int_{\Phi_t S}\eta = \frac{d}{dt}\int_S \Phi_t^*\eta = \int_{\Phi_t S}\mathcal{L}_X\eta = \int_{\Phi_t S}i_X d\eta + \int_{\Phi_t S}d i_X\eta$$

$$= \int_{\Phi_t S}i_X d\eta + \int_{\partial(\Phi_t S)}i_X\eta.$$

This formula is particularly interesting in the case when $S = \Omega$ is an open submanifold of M with compact closure and smooth boundary $\partial\Omega$, and where $\eta = \mu$ is a volume form on M. Let $\Omega_t := \Phi_t\Omega$. We have $\frac{d}{dt}\int_{\Omega_t}\mu = \int_{\partial\Omega_t}i_X\mu$ and then

$$\frac{d}{dt}\bigg|_{t=0}\int_{\Omega_t}\mu = \int_{\partial\Omega}i_X\mu.$$

Definition 9.8. If μ is a volume form orienting a manifold M, then $\mathcal{L}_X\mu = (\operatorname{div} X)\mu$ for a unique function $\operatorname{div} X$ called the **divergence** of X with respect to the volume form μ.

We have

$$\frac{d}{dt}\bigg|_{t=0}\int_{\Omega_t}\mu = \int_{\partial\Omega}i_X\mu$$

and

$$\frac{d}{dt}\bigg|_{t=0}\int_{\Omega_t}\mu = \int_\Omega \mathcal{L}_X\mu = \int_\Omega (\operatorname{div} X)\mu.$$

From this we conclude that

$$\int_\Omega (\operatorname{div} X)\mu = \int_{\partial\Omega}i_X\mu.$$

This formula helps to give geometric meaning to $\operatorname{div} X$. The above formula is a version of **Gauss' theorem** and can be easily proven not just for domains

9.2. Differentiating Integral Expressions; Divergence

in a manifold with a volume form, but for a general manifold with boundary with a volume form:

Theorem 9.9. *Let M be a manifold with boundary and let μ be a volume form on M. In particular, M is oriented by $[\mu]$. Then we have*

$$\int_M (\operatorname{div} X)\,\mu = \int_{\partial M} \iota_X \mu$$

for any compactly supported smooth vector field X.

Proof. The proof uses the trick we used above:

$$\begin{aligned}
\int_M (\operatorname{div} X)\,\mu &= \int_M \mathcal{L}_X \mu \\
&= \int_M d\iota_X \mu + \int_M \iota_X d\mu \\
&= \int_M d\iota_X \mu = \int_{\partial M} \iota_X \mu \quad \text{(Stokes' theorem)}. \qquad \square
\end{aligned}$$

Now let $E_1, \ldots, E_n$ be a local frame field on $U \subset M$ and $\theta^1, \ldots, \theta^n$ the dual frame field. Then for some smooth function ρ we have

$$\mu = \rho\,\theta^1 \wedge \cdots \wedge \theta^n,$$

and a simple calculation gives

$$\iota_X \left(\rho\,\theta^1 \wedge \cdots \wedge \theta^n \right) = \sum_{k=1}^n (-1)^{j-1} \rho X^k \theta^1 \wedge \cdots \wedge \widehat{\theta^k} \wedge \cdots \wedge \theta^n.$$

Then we have

$$\begin{aligned}
\mathcal{L}_X \mu &= \mathcal{L}_X \left(\rho\,\theta^1 \wedge \cdots \wedge \theta^n \right) = d\iota_X \left(\rho\,\theta^1 \wedge \cdots \wedge \theta^n \right) \\
&= d \sum_{k=1}^n (-1)^{j-1} \rho X^k \theta^1 \wedge \cdots \wedge \widehat{\theta^k} \wedge \cdots \wedge \theta^n \\
&= \sum_{k=1}^n (-1)^{j-1} d(\rho X^k) \wedge \theta^1 \wedge \cdots \wedge \widehat{\theta^k} \wedge \cdots \wedge \theta^n \\
&= \sum_{k=1}^n (-1)^{j-1} \sum_{i=1}^n (\rho X^k)_i \theta^i \wedge \theta^1 \wedge \cdots \wedge \widehat{\theta^k} \wedge \cdots \wedge \theta^n \\
&= \sum_{k=1}^n \left(\frac{1}{\rho}(\rho X^k)_k \right) \rho\,\theta^1 \wedge \cdots \wedge \theta^n = \sum_{k=1}^n \left(\frac{1}{\rho}(\rho X^k)_k \right) \mu,
\end{aligned}$$

where $(\rho X^k)_k := d(\rho X^k)(E_k)$. Thus we end up with a nice formula

$$\operatorname{div} X = \sum_{k=1}^n \frac{1}{\rho}(\rho X^k)_k.$$

In particular, if $E_k = \frac{\partial}{\partial x^k}$ for some chart $(U, \mathbf{x}) = (x^1, \ldots, x^n)$, then

(9.4) $$\operatorname{div} X = \sum_{k=1}^{n} \frac{1}{\rho} \frac{\partial}{\partial x^k}(\rho X^k).$$

If we were to replace the volume form μ by $-\mu$, then divergence with respect to that volume form would be given locally by $\sum_{k=1}^{n} \frac{1}{-\rho} \frac{\partial}{\partial x^k}(-\rho X^k) = \sum_{k=1}^{n} \frac{1}{\rho} \frac{\partial}{\partial x^k}(\rho X^k)$ and so the orientation seems superfluous! What if M is not even orientable? In fact, since divergence is a local concept and orientation is a global concept, it seems that we should replace the volume form in the definition by something else that makes sense even on a nonorientable manifold. On nonorientable manifolds, *pseudoforms* can be used in place of forms for many purposes. Alternatively, many situations can be handled by going to the two-fold orientation cover. See the online supplement [**Lee, Jeff**] for a discussion.

9.3. Stokes' Theorem for Chains

Here we give another version of Stokes' theorem that is very useful in connection with topology. We say that distinct points $p_0, \ldots, p_{k+1} \in \mathbb{R}^n$ are in **general position** if they are not contained in any k-dimensional affine subspace of $\mathbb{R}^n$. Equivalently, $p_0, \ldots, p_k$ are in general position if the vectors $p_1 - p_0, \ldots, p_k - p_0$ are linearly independent. A set of the form

$$\left\{ \sum_{i=0}^{p} t_i p_i : 0 \leq t^i \leq 1 \text{ and } \sum_{i=0}^{p} t_i = 1 \right\}$$

for $p_0, \ldots, p_k$ in general position is called a **geometric k-simplex**. The set above is a closed convex set and is the smallest such set containing the points $p_0, \ldots, p_k$. The integer k is the dimension of the simplex, and a geometric 0-simplex is just a point. Let us denote the geometric k-simplex determined by the points $p_0, \ldots, p_k$ by $\langle p_0, \ldots, p_k \rangle$. If $\langle p_0, \ldots, p_k \rangle$ is such a geometric k-simplex and $\{p_{i_1}, \ldots, p_{i_j}\} \subset \{p_0, \ldots, p_{k+1}\}$, then $p_{i_1}, \ldots, p_{i_j}$ are in general position and the geometric j-simplex $\langle p_{i_1}, \ldots, p_{i_j} \rangle$ is called a face of $\langle p_0, \ldots, p_k \rangle$. In particular, the geometric $(k-1)$-simplices obtained by omitting one of $p_0, \ldots, p_k$ are called boundary faces. The i-th boundary face of $\langle p_0, \ldots, p_k \rangle$ is the simplex obtained by omitting the p_i and is denoted $\partial_i \langle p_0, \ldots, p_k \rangle = \langle p_0, \ldots, \widehat{p_i}, \ldots, p_k \rangle$. Geometric simplices can be combined to create geometric simplicial spaces, but we take a slightly more flexible approach.

Definition 9.10. For $k \geq 0$, the set $\Delta^k := \{a \in \mathbb{R}^k : a^i \geq 0 \text{ and } \sum a^i \leq 1\}$ is called the **standard k-simplex** in $\mathbb{R}^k$. Note that $\Delta^0 = \mathbb{R}^0$ consists of just a single point denoted 0. In other words, Δ^k is $\langle e_0, \ldots, e_k \rangle$, where $e_0 = (0, \ldots, 0), e_1 = (1, 0, \ldots, 0), e_2 = (0, 1, \ldots, 0), \ldots$, etc.

9.3. Stokes' Theorem for Chains

Definition 9.11. Let M be an n-manifold. A C^r map $\sigma : \Delta^k \to M$ is called a C^r **singular k-simplex**. Let G be an abelian group. A formal sum $c = \sum_\sigma c_\sigma \sigma$, where the sum is over all singular simplices σ, $c_\sigma \in G$, and $c_\sigma = 0$ for all but finitely many simplices σ, is called a smooth **singular k-chain** with coefficients in G. A 0-simplex $\sigma : \Delta^0 \to M$ is often identified with the image $\sigma(0) = p \in M$.

By definition, $c_\sigma = 0$ for all but finitely many simplices; we say that a **singular k-chain has finite support**. The set of all C^r k-chains with coefficients in G is an abelian group denoted $C_k(M,G)^r$, where we use additive notation for the group operation. Addition is given by

$$\left(\sum_\sigma c_\sigma \sigma\right) + \left(\sum_\sigma c'_\sigma \sigma\right) = \sum_{i=0}^{\infty} (c_\sigma + c'_\sigma) \sigma.$$

The cases $r = 0$ and $r = \infty$ are the most important. We will restrict to the $r = \infty$ case and drop the superscript in $C_k(M,G)^\infty$. We now define the notion of the i-th face of a singular simplex and then the boundary of a simplex. First, if $q_0, \ldots, q_k$ are distinct points in $\mathbb{R}^n$, then there is a unique affine map $\mathbb{R}^k \to \mathbb{R}^n$ which maps e_i to q_i for $0 \leq i \leq k$. This map restricts to a map from Δ^k onto the convex hull of $q_0, \ldots, q_k$, which will be a geometric k-simplex if $q_0, \ldots, q_k$ are in general position. For given $q_0, \ldots, q_k$, denote this map by $\alpha(q_0, \ldots, q_k)$. For $0 \leq i \leq k$, let $f_i^k : \Delta^{k-1} \to \Delta^k$ be $\alpha(e_0, \ldots, \widehat{e_i}, \ldots, e_k)$. Explicitly, $f_0^1(0) = 1$, $f_1^1(0) = 0$ and for $k > 0$,

$$f_0^k(a^1, \ldots, a^{k-1}) = \left(1 - \sum_{i=1}^{k-1} a^i, \ a^1, \ldots, a^{k-1}\right),$$

$$f_i^k(a^1, \ldots, a^{k-1}) = \left(a^1, \ldots, a^{i-1}, 0, a^i, \ldots, a^{k-1}\right).$$

Thus f_i^k is a homeomorphism of Δ^{k-1} onto its image, which is the i-th boundary face of Δ^k. Thus it parametrizes the face of Δ^k opposite e_i while keeping track of the order of the vertices.

The i-th face of a singular k-simplex σ is the singular $(k-1)$-simplex σ^i given by

$$\sigma^i := \sigma \circ f_i^k.$$

Definition 9.12. The boundary operator $\partial_k : C_k(M,G) \to C_{k-1}(M,G)$ is defined on a simplex σ by

$$\partial_k \sigma := \sum_{i=0}^{k} (-1)^i \sigma^i.$$

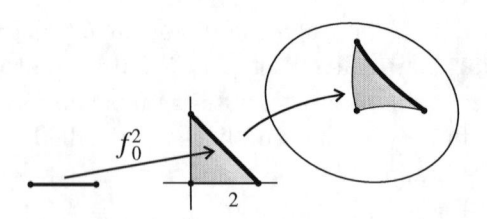

Figure 9.1. A face of σ

and then extended to be a group homomorphism $\partial_k \sum_\sigma c_\sigma \sigma = \sum_\sigma c_\sigma \partial_k \sigma$. If σ is a 0-simplex, then we define $\partial_0 \sigma = 0$ (the group identity) so that for a 0-chain $\sum_\sigma c_\sigma \sigma$ we have $\partial_k \sum_\sigma c_\sigma \sigma = 0$.

We will denote all the maps ∂_k simply by ∂ for all k. It can be shown that

(9.5) $$\partial \circ \partial = 0.$$

Exercise 9.13. Prove that $\partial \circ \partial = 0$. Hint: First show that $f_i^k \circ f_j^{k-1} = f_j^k \circ f_{i-1}^{k-1}$ for $k \geq 1$ and $i > j$. The maps f_i^k are defined above.

It is convenient to define $C_k(M, G)$ to be the trivial group $\{0\}$ for all $k < 0$ and define $\partial_k = 0$ for all $k < 0$. Then $\partial \circ \partial = 0$ remains true. The sequence of spaces and maps

$$\cdots \xrightarrow{\partial} C_{k+1}(M, G) \xrightarrow{\partial} C_k(M, G) \xrightarrow{\partial} C_{k-1}(M, G) \xrightarrow{\partial} \cdots$$

is called the singular chain complex with coefficients in G.

Let $Z_k(M, G) = \operatorname{Ker} \partial_k$ and $B_k(M, G) := \operatorname{Im} \partial_{k-1}$. Then because of equation (9.5) we have $B_k(M, G) \subset Z_k(M, G)$. The k-**th singular homology group** $H_k(M; G)$ with coefficients in G is defined by

$$H_k(M; G) = \frac{Z_k(M, G)}{B_k(M, G)}.$$

If $c \in Z_k(M, G)$, then the equivalence class of c is denoted $[c]$. If c_1 and c_2 are k-chains in the same class, then $c_1 = c_2 + \partial c$ for some $c \in B_k(M, G)$, and in this case we say that c_1 and c_2 are **homologous** (or in the same homology class). The most important choices for G are $\mathbb{R}$ and $\mathbb{Z}$.

Exercise 9.14. Check that if $G = \mathbb{R}$, then $C_k(M, \mathbb{R}), Z_k(M, \mathbb{R}), B_k(M, \mathbb{R})$ and $H_k(M; \mathbb{R})$ are all vector spaces in a natural way and the boundary maps ∂ extend to linear maps.

We define the integral of a k-form α over a smooth singular k-simplex σ as

$$\int_\sigma \alpha := \int_{\Delta^k} \phi^* \alpha,$$

9.3. Stokes' Theorem for Chains

and then for a chain $c = \sum_\alpha c_\sigma \sigma \in C_k(M, \mathbb{R})$ we can define

$$\int_c \alpha = \sum_\sigma c_\sigma \int_\sigma \alpha.$$

If σ is a 0-simplex and if $f \in \Omega^0(M) = C^\infty(M)$, then $\int_\sigma f = f(0)$. We state without proof the following version of Stokes' theorem:

Theorem 9.15 (Stokes' theorem for chains). *Let M be a smooth manifold. For $c \in C_{k+1}(M, \mathbb{R})$, and α a k-form on M, we have*

$$\int_c d\alpha = \int_{\partial c} \alpha.$$

For a proof see [**War**]. Notice that in this version, M is not assumed to be orientable. Also, α need not have compact support since the chain c has finite support. If σ is a 1-simplex and $f \in C^\infty(M)$, then the above reduces to

$$\int_\sigma df = f(\sigma(1)) - f(\sigma(0)).$$

Now we define the de Rham map. First, if $\alpha \in Z^k(M) \subset \Omega^k(M)$, then we can define an element $I\alpha$ of $H^k(M; \mathbb{R})$, the dual space of $H_k(M; \mathbb{R})$: For $[c] \in H_k(M; \mathbb{R})$ represented by $c \in Z_k(M; \mathbb{R})$ we define

$$I\alpha([c]) = \int_c \alpha.$$

This is well-defined since if $c + \partial c' \in [c]$, then

$$I\alpha(c + \partial c') = \int_{c + \partial c'} \alpha$$
$$= \int_c \alpha + \int_{\partial c'} \alpha = \int_c \alpha + \int_{c'} d\alpha \quad \text{(Stokes)}$$
$$= \int_c \alpha \quad (\text{since } d\alpha = 0).$$

This gives a linear map $Z^k(M) \to H^k(M; \mathbb{R})$. Now if $\alpha \in B^k(M)$ (image of d), then $\alpha = d\beta$ and

$$I\alpha([c]) = \int_c \alpha = \int_c d\beta = \int_{\partial c} \beta = 0 \text{ (since } \partial c = 0\text{)}.$$

Thus we obtain a linear map called the **de Rham map**

$$H_{\text{deR}}^k(M) \to H^k(M; \mathbb{R}),$$

where $H_{\text{deR}}^k(M) = Z^k(M)/B^k(M)$ is the de Rham cohomology defined earlier. The content of the celebrated de Rham theorem is in part that this map is an isomorphism. The theorem is fairly difficult to prove.

Theorem 9.16 (de Rham). *The de Rham map defined above is an isomorphism,*
$$H^k_{\text{deR}}(M) \cong H^k(M; \mathbb{R}).$$

Proof. See [**Bo-Tu**] or [**War**]. □

We have defined $H^k(M; \mathbb{R})$ using smooth singular chains, but we could have used continuous chains. The result is isomorphic to $H^k(M; \mathbb{R})$ as we have defined it (see [**War**] or [**Bo-Tu**]).

9.4. Differential Forms and Metrics

Let (V, g) be a real scalar product space (not necessarily positive definite). We wish to induce a scalar product on $L^k_{\text{alt}}(V) \cong \bigwedge^k(V^*)$. Even though elements of $L^k_{\text{alt}}(V) \cong \bigwedge^k(V^*)$ can be thought of as tensors of type $(0, k)$ that just happen to be alternating, we will give a scalar product to this space in such a way that the basis

(9.6) $$\{e^{i_1} \wedge \cdots \wedge e^{i_k}\}_{i_1 < \cdots < i_k} = \{e^{\vec{I}}\}$$

is orthonormal if $e^1, \ldots, e^n$ is orthonormal. Recall that if $\alpha, \beta \in V^*$, then by definition $\langle \alpha, \beta \rangle = g(\alpha^\sharp, \beta^\sharp)$. Now suppose that $\alpha = \alpha^1 \wedge \alpha^2 \wedge \cdots \wedge \alpha^k$ and $\beta = \beta^1 \wedge \beta^2 \wedge \cdots \wedge \beta^k$, where the α^i and β^i are 1-forms. Then we want to have

$$\langle \alpha | \beta \rangle = \langle \alpha^1 \wedge \alpha^2 \wedge \cdots \wedge \alpha^k \mid \beta^1 \wedge \beta^2 \wedge \cdots \wedge \beta^k \rangle$$
$$= \det \left[\langle \alpha^i, \beta^j \rangle \right],$$

where $[\langle \alpha^i, \beta^j \rangle]$ is an $n \times n$ matrix. Notice that we use $\langle \alpha | \beta \rangle$ rather than $\langle \alpha, \beta \rangle$ since the latter could be taken to be the natural inner product of α and β as tensors—the latter differs from the first by a factor. We want to extend this bilinearly to all k-forms. We could just declare the basis (9.6) above to be orthonormal and thus define a scalar product. Of course, one must then show that this scalar product does not depend on the choice of basis. For completeness, we now show how to arrive at the appropriate scalar product using universal mapping properties and obtain some formulas. Fix $\beta^1, \beta^2, \ldots, \beta^k \in V^*$ and consider the map

$$\mu_{\beta^1, \beta^2, \ldots, \beta^k} : V^* \times \cdots \times V^* \to \mathbb{R}$$

given by

$$\mu_{\beta^1, \beta^2, \ldots, \beta^k} : (\alpha^1, \alpha^2, \ldots, \alpha^k) \mapsto \det \left[\langle \alpha^i, \beta^j \rangle \right].$$

Since this is an alternating multilinear map, we can use the universal property of $\bigwedge^k(V^*)$ to see that this map defines a unique linear map

$$\widetilde{\mu}_{\beta^1, \beta^2, \ldots, \beta^k} : \bigwedge^k(V^*) \to \mathbb{R}$$

9.4. Differential Forms and Metrics

such that
$$\widetilde{\mu}_{\beta^1,\beta^2,\ldots,\beta^k}\left(\alpha^1 \wedge \alpha^2 \wedge \cdots \wedge \alpha^k\right) = \det\left[\langle \alpha^i, \beta^j \rangle\right].$$

Similarly, for fixed $\alpha \in \bigwedge^k(V^*)$, the map $\widetilde{m}_\alpha \colon (\beta^1, \beta^2, \ldots, \beta^k) \mapsto \widetilde{\mu}_{\beta^1,\beta^2,\ldots,\beta^k}(\alpha)$ is alternating multilinear and so gives rise to a linear map

$$\widetilde{m}_\alpha : \bigwedge\nolimits^k(V^*) \to \mathbb{R}$$

such that
$$\widetilde{m}_\alpha(\beta^1 \wedge \beta^2 \wedge \cdots \wedge \beta^k) = \widetilde{\mu}_{\beta^1,\beta^2,\ldots,\beta^k}(\alpha).$$

Lemma 9.17. *The map*
$$\langle \cdot | \cdot \rangle : \bigwedge\nolimits^k(V^*) \times \bigwedge\nolimits^k(V^*) \to \mathbb{R}$$

defined by
$$\langle \alpha | \beta \rangle := \widetilde{m}_\alpha(\beta)$$

is symmetric and bilinear. We have
$$\langle \alpha^1 \wedge \alpha^2 \wedge \cdots \wedge \alpha^k \mid \beta^1 \wedge \beta^2 \wedge \cdots \wedge \beta^k \rangle$$
$$= \det\left[\langle \alpha^i, \beta^j \rangle\right]$$

for 1-forms $\alpha^1, \alpha^2, \ldots, \alpha^k, \beta^1, \beta^2, \ldots, \beta^k$.

Proof. By construction, the map is linear in the second slot for each fixed α. Fix β and write β as a sum $\beta = \sum b_{i_1 \ldots i_k} \beta^{i_1} \wedge \cdots \wedge \beta^{i_k}$ in any way. Then

$$\widetilde{m}_\alpha(\beta) = \sum b_{i_1 \ldots i_k} \widetilde{m}_\alpha\left(\beta^{i_1} \wedge \cdots \wedge \beta^{i_k}\right)$$
$$= \sum b_{i_1 \ldots i_k} \widetilde{\mu}_{\beta^1,\beta^2,\ldots,\beta^k}(\alpha).$$

But each map $\alpha \mapsto \widetilde{\mu}_{\beta^1,\beta^2,\ldots,\beta^k}(\alpha)$ is linear and so $\alpha \mapsto \widetilde{m}_\alpha(\beta)$ is also linear.

Now let $\varphi(\cdot,\cdot) : W \times W \to \mathbb{R}$ be any bilinear map on a real vector space W. If $S \subset W$ spans W and if $\varphi(s_1, s_2) = \varphi(s_2, s_1)$ for all $s_1, s_2 \in S$, then φ is symmetric. The set of all elements $\beta \in \bigwedge^k(V^*)$ of the form $\beta = \beta^1 \wedge \cdots \wedge \beta^k$ for 1-forms $\beta^1, \ldots, \beta^k$, is a spanning set. Since

$$\langle \alpha^1 \wedge \alpha^2 \wedge \cdots \wedge \alpha^k \mid \beta^1 \wedge \beta^2 \wedge \cdots \wedge \beta^k \rangle$$
$$= \det\left[\langle \alpha^i, \beta^j \rangle\right] = \det\left[\langle \beta^j, \alpha^i \rangle\right]$$
$$= \langle \beta^1 \wedge \beta^2 \wedge \cdots \wedge \beta^k \mid \alpha^1 \wedge \alpha^2 \wedge \cdots \wedge \alpha^k \rangle,$$

we conclude that $\langle \cdot | \cdot \rangle$ is symmetric. $\square$

The bilinear map defined in the previous lemma is a scalar product for the vector space $\bigwedge^k(V^*)$. Notice the vertical bar rather than a comma in the notation. If $e^1, \ldots, e^n$ is an orthonormal basis for V*, then $\{e^{i_1} \wedge \cdots \wedge e^{i_k}\}_{i_1 < \cdots < i_k}$ is orthonormal since

$$\langle e^{i_1} \wedge \cdots \wedge e^{i_k} | e^{j_1} \wedge \cdots \wedge e^{j_k} \rangle = \begin{vmatrix} \langle e^{i_1}, e^{j_1} \rangle & \langle e^{i_1}, e^{j_2} \rangle & \cdots \\ \langle e^{i_2}, e^{j_1} \rangle & \langle e^{i_2}, e^{j_2} \rangle & \cdots \\ \vdots & \vdots & \ddots \end{vmatrix},$$

which is zero if $i_1, \ldots, i_k$ is not a permutation of $j_1, \ldots, j_k$, while on the other hand, if $(j_1, \ldots, j_k) = (\sigma(i_1), \ldots, \sigma(i_k))$, then every column of the above determinant has exactly one nonzero entry, which is either 1 or -1. Abbreviating $e^{i_1} \wedge \cdots \wedge e^{i_k}$ to $e^{\vec{I}}$ and so on, we have in the orthonormal case

$$\langle e^{\vec{I}} | e^{\vec{I}} \rangle = \pm 1.$$

So, if $\alpha = \sum \alpha_{\vec{I}} e^{\vec{I}}$, then $\langle e^{\vec{I}} | \alpha \rangle = \langle e^{\vec{I}} | e^{\vec{I}} \rangle \alpha_{\vec{I}}$ and

$$\alpha_{\vec{I}} = \langle e^{\vec{I}} | e^{\vec{I}} \rangle \langle e^{\vec{I}} | \alpha \rangle = \pm \langle e^{\vec{I}} | \alpha \rangle.$$

We see that the scalar product need not be positive definite. Unless we give a specific linear order to the basis $\{e^{\vec{I}}\}$, the signature is perhaps best thought of as the indexed set $\{\epsilon(\vec{I})\}_{\vec{I}}$ so that

$$\alpha = \sum_{\vec{I}} \epsilon(\vec{I}) \langle e^{\vec{I}} | \alpha \rangle e^{\vec{I}}.$$

We are very much interested in $\bigwedge^k(V^*)$, but note that $\bigwedge^k(V)$ is also a scalar product space in the analogous way so that

$$\langle v_1 \wedge v_2 \wedge \cdots \wedge v_k | w_1 \wedge w_2 \wedge \cdots \wedge w_k \rangle = \det [\langle v_i, w_j \rangle].$$

Exercise 9.18. Let $\mathbb{R}^4_1$ denote Minkowski space. Determine the index of the scalar product on $\bigwedge^2(\mathbb{R}^4_1)$ described above.

We already have a scalar product denoted $\langle \cdot, \cdot \rangle$ for tensors. We wish to compare it with the scalar product $\langle \cdot | \cdot \rangle$ just defined on forms. For simplicity we consider the positive definite case. If $\alpha = \sum \alpha_{\vec{I}} e^{\vec{I}}$ and $\beta = \sum \beta_{\vec{I}} e^{\vec{I}}$ are k-forms considered as covariant tensor fields, then by Problem 18 of Chapter 7 we have

$$\langle \alpha, \beta \rangle = \sum \alpha_{i_1 \ldots i_k} \beta^{i_1 \ldots i_k} = \sum \alpha_I \beta^I.$$

Now let $e^1, \ldots, e^n$ be an *orthonormal* basis for V*. Then

$$\langle \alpha, \beta \rangle = \sum \alpha_I \beta^I = \sum \alpha_I \beta_I,$$

9.4. Differential Forms and Metrics

and we have

$$\langle \alpha | \beta \rangle = \sum_{\vec{I}, \vec{J}} \langle \alpha_{\vec{I}} e^{\vec{I}} | \beta_{\vec{J}} e^{\vec{J}} \rangle = \sum_{\vec{I}, \vec{J}} \alpha_{\vec{I}} \beta_{\vec{J}} \langle e^{\vec{I}} | e^{\vec{J}} \rangle$$

$$= \sum_{\vec{I}} \alpha_{\vec{I}} \beta_{\vec{I}} = \frac{1}{k!} \sum_{I} \alpha_{I} \beta_{I} = \frac{1}{k!} \langle \alpha, \beta \rangle.$$

We conclude that

$$\langle \alpha | \beta \rangle = \frac{1}{k!} \langle \alpha, \beta \rangle,$$

so the two scalar products differ by a factor of $k!$. Now let $\theta^1, \ldots, \theta^n$ be *any* basis for V^* (not necessarily orthonormal). For a given $\alpha \in \bigwedge^k(V^*)$, we can also write $\alpha = \frac{1}{k!} \sum \alpha_I e^I$ and then as a tensor

$$\alpha = \frac{1}{k!} \sum \alpha_{i_1 \ldots i_k} \theta^{i_1} \wedge \cdots \wedge \theta^{i_k} = \sum \alpha_{i_1 \ldots i_k} \theta^{i_1} \otimes \cdots \otimes \theta^{i_k}.$$

But when $\alpha = \sum a_{\vec{I}} \theta^{\vec{I}}$ and $\beta = \sum b_{\vec{I}} \theta^{\vec{I}}$ are viewed as k-forms, we must have

$$\langle \alpha | \beta \rangle = \frac{1}{k!} \langle \alpha, \beta \rangle = \frac{1}{k!} \sum a_I b^I = \sum a_{\vec{I}} b^{\vec{I}}.$$

Definition 9.19. We defined the scalar product on $\bigwedge^k V^* \cong L^k_{\text{alt}}(V)$ by first using the above formula for exterior products of 1-forms and then extending (bi)linearly to all of $\bigwedge^k V^*$. We can also extend to the whole Grassmann algebra $\bigwedge V^* = \bigoplus \bigwedge^k V^*$ by declaring forms of different degree to be orthogonal. We also have the obvious similar definition for $\bigwedge^k V$ and $\bigwedge V$.

If we have an orthonormal basis $e^1, \ldots, e^n$ for V^*, then $e^1 \wedge \cdots \wedge e^n \in \bigwedge^n V^*$. But $\bigwedge^n V^*$ is one-dimensional, and if $\ell : V \to V$ is any isometry of (V^*, g^*), then the dual map $\ell^* : V^* \to V^*$ is an isometry of (V, g). Thus $\ell^* e^1 \wedge \cdots \wedge \ell^* e^n = \pm e^1 \wedge \cdots \wedge e^n$. In particular, for any permutation σ of the letters $\{1, 2, \ldots, n\}$ we have $e^1 \wedge \cdots \wedge e^n = \text{sgn}(\sigma) e^{\sigma(1)} \wedge \cdots \wedge e^{\sigma(n)}$.

For a given ordered basis $(e_1, \ldots, e_n)$ for V, with dual basis $(e^1, \ldots, e^n)$, we have

$$\langle e^1 \wedge \cdots \wedge e^n | e^1 \wedge \cdots \wedge e^n \rangle = \epsilon_1 \epsilon_2 \cdots \epsilon_n = \pm 1.$$

Exercise 9.20. Let $(e_1, \ldots, e_n)$ be orthonormal. Show that the only elements ω of $\bigwedge^n V^*$ with $|\langle \omega, \omega \rangle| = 1$ are $e^1 \wedge \cdots \wedge e^n$ and $-e^1 \wedge \cdots \wedge e^n$.

Given a fixed orthonormal basis $(e_1, \ldots, e_n)$, all ordered orthonormal bases for V fall into two classes. Namely, those bases $(f_1, \ldots, f_n)$ for which $f^1 \wedge \cdots \wedge f^n = e^1 \wedge \cdots \wedge e^n$ and those for which $f^1 \wedge \cdots \wedge f^n = -e^1 \wedge \cdots \wedge e^n$. Thus for each orientation of V there is a corresponding element of $\bigwedge^n V^*$ called a **metric volume element** for (V, g). The metric volume element corresponding to a basis $(e_1, \ldots, e_n)$ is just $e^1 \wedge \cdots \wedge e^n$. On the other hand, we have seen that *any* nonzero top form ω determines an orientation. If we have an orientation given by a top form ω, then obviously, $(e_1, \ldots, e_n)$ determines the same orientation if and only if $\omega(e_1, \ldots, e_n) > 0$ since this means that $\omega = ce^1 \wedge \cdots \wedge e^n$ for some $c > 0$.

Proposition 9.21. *Let an orientation be chosen on the scalar product space* (V, g) *and let* $\mathcal{E} = (e_1, \ldots, e_n)$ *be an oriented orthonormal frame so that* $\mathrm{vol} := e^1 \wedge \cdots \wedge e^n$ *is the corresponding volume element. Then if* $\mathcal{F} = (f_1, \ldots, f_n)$ *is a positively oriented basis for* V *with dual basis* $\mathcal{F}^* = (f^1, \ldots, f^n)$*, then*

$$\mathrm{vol} = \sqrt{|\det(g_{ij})|} f^1 \wedge \cdots \wedge f^n,$$

where $g_{ij} = \langle f_i, f_j \rangle$*. If* g *is positive definite, then* $\det(g_{ij}) > 0$ *and so*

$$\mathrm{vol} = \sqrt{\det(g_{ij})} f^1 \wedge \cdots \wedge f^n.$$

Proof. Let $e^i = \sum a^i_j f^j$. Then we have

$$\epsilon_i \delta^{ij} = \pm \delta^{ij} = \langle e^i, e^j \rangle = \left\langle \sum a^i_k f^k, \sum a^j_m f^m \right\rangle$$
$$= \sum_{k,m} a^i_k a^j_m \langle f^k, f^m \rangle = \sum_{k,m} a^i_k a^j_m g^{km},$$

so that $\pm 1 = \det([a^i_k])^2 \det([g^{km}]) = (\det([a^i_k]))^2 (\det(g_{ij}))^{-1}$. Thus,

$$\pm \sqrt{|\det(g_{ij})|} = \det([a^i_k]).$$

But since $\mathcal{E}$ and $\mathcal{F}$ are both positively oriented, we must have $\det([a^i_k]) > 0$ and so

$$\sqrt{|\det(g_{ij})|} = \det([a^i_k]).$$

On the other hand,

$$\mathrm{vol} = e^1 \wedge \cdots \wedge e^n$$
$$= \left(\sum a^1_{k_1} f^{k_1} \right) \wedge \cdots \wedge \left(\sum a^n_{k_1} f^{k_1} \right) = \det([a^i_k]) f^1 \wedge \cdots \wedge f^n,$$

and the result follows. If g is positive definite, then $\det([a^i_k])^2 \det([g^{km}]) = 1$ so $\det(g_{ij}) > 0$. $\square$

Fix an orientation and let $(e_1, \ldots, e_n)$ be an orthonormal basis in that orientation class. Then we have the corresponding volume element $\mathrm{vol} = e^1 \wedge \cdots \wedge e^n$. Now we define the **Hodge star operator** $* : \bigwedge^k V^* \to \bigwedge^{n-k} V^*$ for $1 \leq k \leq n$, where $n = \dim(V)$.

Theorem 9.22. *Let* (V, g) *be a scalar product space with* $\dim(V) = n$ *and the corresponding volume element* vol. *For each* k*, there is a unique linear*

9.4. Differential Forms and Metrics

isomorphism $* : \bigwedge^k V^* \to \bigwedge^{n-k} V^*$ *such that* $\alpha \wedge *\beta = \langle \alpha | \beta \rangle \operatorname{vol}$ *for all* $\alpha, \beta \in \bigwedge^k V^*$.

Proof. Given $\gamma \in \bigwedge^{n-k} V^*$, define a linear map $L_\gamma : \bigwedge^k V^* \to \mathbb{R}$ by requiring that $L_\gamma(\alpha) \operatorname{vol} = \alpha \wedge \gamma$. By Problem 3, if $L_\gamma(\alpha) = 0$ for all α, then $\gamma = 0$ and $\gamma \mapsto L_\gamma$ gives a one-to-one linear map $\bigwedge^k V^* \to (\bigwedge^k V^*)^*$. But since $\dim \bigwedge^k V^* = \dim(\bigwedge^k V^*)^*$, this must be a linear isomorphism. Thus, for each $\beta \in \bigwedge^k V^*$, there is a unique element $*\beta$ such that

$$L_{*\beta}(\alpha) = \langle \alpha | \beta \rangle$$

for all $\alpha \in \bigwedge^k V^*$. This defines a map $* : \bigwedge^k V^* \to \bigwedge^{n-k} V^*$ such that

$$\alpha \wedge *\beta = L_{*\beta}(\alpha) \operatorname{vol} = \langle \alpha | \beta \rangle \operatorname{vol}.$$

This map is easily seen to be linear. The equation above also shows that it is one-to-one and hence an isomorphism. $\square$

Proposition 9.23. *Let* (V, g) *be a scalar product space with* $\dim(V) = n$ *and corresponding volume element* vol. *Let* $(e_1, \ldots, e_n)$ *be a positively oriented orthonormal basis with dual* $(e^1, \ldots, e^n)$. *Let* σ *be a permutation of* $(1, 2, \ldots, n)$. *On the basis elements* $e^{\sigma(1)} \wedge \cdots \wedge e^{\sigma(k)}$ *for* $\bigwedge^k V^*$ *we have*

$$(9.7) \qquad *(e^{\sigma(1)} \wedge \cdots \wedge e^{\sigma(k)}) = \epsilon_{\sigma_1} \epsilon_{\sigma_2} \cdots \epsilon_{\sigma_k} \operatorname{sgn}(\sigma) e^{\sigma(k+1)} \wedge \cdots \wedge e^{\sigma(n)}.$$

In other words, if we let $\{i_{k+1}, \ldots, i_n\} = \{1, 2, \ldots, n\} \setminus \{i_1, \ldots, i_k\}$, *then*

$$*(e^{i_1} \wedge \cdots \wedge e^{i_k}) = \pm (\epsilon_{i_1} \epsilon_{i_2} \cdots \epsilon_{i_k}) e^{i_{k+1}} \wedge \cdots \wedge e^{i_n},$$

where we take the $+$ *sign if and only if* $e^{i_1} \wedge \cdots \wedge e^{i_k} \wedge e^{i_{k+1}} \wedge \cdots \wedge e^{i_n} = e^1 \wedge \cdots \wedge e^n$.

Proof. Formula (9.7) above actually defines a map $\bigwedge^k V^* \to \bigwedge^{n-k} V^*$. For this proof we denote this map by $*$ and show it satisfies the same defining equations as the Hodge star. It is enough to check that the defining formula $\alpha \wedge *\beta = \langle \alpha | \beta \rangle \operatorname{vol}$ holds for typical (orthonormal) basis elements $\alpha = e^{i_1} \wedge \cdots \wedge e^{i_k}$ and $\beta = e^{m_1} \wedge \cdots \wedge e^{m_k}$. We have

$$(9.8) \qquad (e^{m_1} \wedge \cdots \wedge e^{m_k}) \wedge *(e^{i_1} \wedge \cdots \wedge e^{i_k})$$
$$= e^{m_1} \wedge \cdots \wedge e^{m_k} \wedge (\pm e^{i_{k+1}} \wedge \cdots \wedge e^{i_{n_1}}).$$

The last expression is zero unless $\{m_1, \ldots, m_k\} \cup \{i_{k+1}, \ldots, i_n\} = \{1, 2, \ldots, n\}$, or in other words, unless $\{i_1, \ldots, i_k\} = \{m_1, \ldots, m_k\}$. But this is also true for

$$(9.9) \qquad \langle e^{m_1} \wedge \cdots \wedge e^{m_k} | e^{i_1} \wedge \cdots \wedge e^{i_k} \rangle \operatorname{vol}.$$

On the other hand if $\{i_1, \ldots, i_k\} = \{m_1, \ldots, m_k\}$, then both (9.8) and (9.9) give $\pm \operatorname{vol}$. So the lemma is proved up to a sign. We leave it to the reader to show that the definitions are such that the signs match. $\square$

Remark 9.24. In case the scalar product is positive definite, $\epsilon_1 = \epsilon_2 = \cdots = \epsilon_n = 1$ and so the formulas are a bit less cluttered.

The Hodge star operators for each k can be combined to give a Hodge star operator $\bigwedge V^* \to \bigwedge V^*$. We also note that the scalar product on $\bigwedge^0 V^* = \mathbb{R}$ is just $\langle a|b\rangle = ab$. Then, we still have the formula $\alpha \wedge *\beta = \langle \alpha|\beta\rangle$ vol for any α, β in the algebra $\bigwedge V^*$.

Proposition 9.25. *Let V be as above with* $\dim(V) = n$. *The following identities hold for the star operator:*

(1) $*1 = \mathrm{vol}$;
(2) $*\mathrm{vol} = (-1)^{\mathrm{ind}(g)}$;
(3) $**\alpha = (-1)^{\mathrm{ind}(g)}(-1)^{k(n-k)}\alpha$ *for* $\alpha \in \bigwedge^k V^*$;
(4) $\langle *\alpha | *\beta\rangle = (-1)^{\mathrm{ind}(g)}\langle \alpha|\beta\rangle$.

Proof. (1) and (2) follow directly from the definitions. For (3), it suffices to let $\alpha = e^{\sigma(1)} \wedge \cdots \wedge e^{\sigma(k)}$ for some permutation $\sigma \in S_n$. We first compute $*(e^{\sigma(k+1)}\wedge \cdots \wedge e^{\sigma(n)})$. We must have $*(e^{\sigma(k+1)}\wedge \cdots \wedge e^{\sigma(n)}) = c e^{\sigma(1)}\wedge \cdots \wedge e^{\sigma(k)}$ for some constant c. On the other hand,

$$\begin{aligned}\epsilon_{\sigma(k+1)}\cdots \epsilon_{\sigma(n)}\mathrm{vol} &= \langle e^{\sigma(k+1)}\wedge \cdots \wedge e^{\sigma(n)} | e^{\sigma(k+1)}\wedge \cdots \wedge e^{\sigma(n)}\rangle \mathrm{vol}\\ &= \left(e^{\sigma(k+1)}\wedge \cdots \wedge e^{\sigma(n)}\right)\wedge *\left(e^{\sigma(k+1)}\wedge \cdots \wedge e^{\sigma(n)}\right)\\ &= \left(e^{\sigma(k+1)}\wedge \cdots \wedge e^{\sigma(n)}\right)\wedge c e^{\sigma(1)}\wedge \cdots \wedge e^{\sigma(k)}\\ &= (-1)^{k(n-k)}c\,\mathrm{sgn}(\sigma)\,\mathrm{vol}\end{aligned}$$

so that $c = \epsilon_{\sigma(k+1)}\cdots \epsilon_{\sigma(n)}(-1)^{k(n-k)}\mathrm{sgn}(\sigma)$. Using this and equation (9.7), we have

$$\begin{aligned}**\left(e^{\sigma(1)}\wedge \cdots \wedge e^{\sigma(k)}\right) &= *\epsilon_{\sigma(1)}\epsilon_{\sigma(2)}\cdots \epsilon_{\sigma(k)}\,\mathrm{sgn}(\sigma)\,e^{\sigma(k+1)}\wedge \cdots \wedge e^{\sigma(n)}\\ &= \epsilon_{\sigma(1)}\epsilon_{\sigma(2)}\cdots \epsilon_{\sigma(k)}\epsilon_{\sigma(k+1)}\cdots \epsilon_{\sigma(n)}\,(\mathrm{sgn}(\sigma))^2\\ &\quad \times (-1)^{k(n-k)}e^{\sigma(1)}\wedge \cdots \wedge e^{\sigma(k)}\\ &= (-1)^{\mathrm{ind}(g)}(-1)^{k(n-k)}e^{\sigma(1)}\wedge \cdots \wedge e^{\sigma(k)},\end{aligned}$$

which implies the result. For (4) we compute as follows:

$$\begin{aligned}\langle *\alpha | *\beta\rangle\mathrm{vol} &= *\alpha \wedge **\beta = (-1)^{\mathrm{ind}(g)}(-1)^{k(n-k)}*\alpha \wedge \beta\\ &= (-1)^{\mathrm{ind}(g)}\beta \wedge *\alpha = (-1)^{\mathrm{ind}(g)}\langle \alpha|\beta\rangle\mathrm{vol}.\end{aligned}\qquad\square$$

We obtain a formula for the star operator that uses a basis that is not necessarily orthonormal. Recall $\epsilon^{i_1\ldots i_k}_{\ell_1\ldots \ell_k}$ defined by equation (8.2). If we let $\epsilon_{i_1\ldots i_n} := \epsilon^{1\,2\ldots n}_{i_1\ldots i_n}$, then $\epsilon_{i_1\ldots i_n}$ is just the sign of the permutation $\binom{1\,2\ldots n}{i_1\ldots i_n}$. Using this notation, we have the following theorem.

9.4. Differential Forms and Metrics

Theorem 9.26. *Let (V, g) be an oriented scalar product space with corresponding volume element* vol. *Let $e^1, \ldots, e^n$ be a positively oriented basis of V^*. If $\omega = \frac{1}{k!} \sum \omega_{i_1 \ldots i_k} e^{i_1} \wedge \cdots \wedge e^{i_k}$ (or $\omega = \sum_{i_1 < i_2 \cdots < i_k} \omega_{i_1 \ldots i_k} e^{i_1} \wedge \cdots \wedge e^{i_k}$), then*

$$(9.10) \quad *\omega = \sqrt{|\det[g_{ij}]|} \sum_{j_{k+1} < \cdots < j_n} \omega^{j_1 \ldots j_k} \epsilon_{j_1 \ldots j_k j_{k+1} \ldots j_n} e^{j_{k+1}} \wedge \cdots \wedge e^{j_n}.$$

Proof. Let us use the common abbreviation $g := \det[g_{ij}]$. Let ♣ denote the operator defined so that ♣ω is given by the right hand side of equation (9.10) above. Our task is to show that ♣ $= *$. For this we need to show that $\alpha \wedge$ ♣$\beta = \langle \alpha | \beta \rangle$ vol for all $\alpha, \beta \in \bigwedge^k V^*$. First note that for fixed $j_1, \ldots, j_n$ we have $\epsilon_{j_1 \ldots j_n} \epsilon^{j_1 \ldots j_n}_{i_1 \ldots i_n} = \epsilon_{i_1 \ldots i_n}$ (no sum). We compute the coefficients of $\alpha \wedge$ ♣β using formula (8.3):

$$\begin{aligned}(\alpha \wedge \clubsuit\beta)_{i_1 \ldots i_n} &= \frac{1}{k!(n-k)!} \sum \alpha_{j_1 \ldots j_k} (\clubsuit\beta)_{j_{k+1} \ldots j_n} \epsilon^{j_1 \ldots j_n}_{i_1 \ldots i_n} \\ &= \frac{1}{k!(n-k)!} \frac{1}{k!} \sum |g|^{1/2} \alpha_{j_1 \ldots j_k} \beta^{m_1 \ldots m_k} \epsilon_{m_1 \ldots m_k j_{k+1} \ldots j_n} \epsilon^{j_1 \ldots j_n}_{i_1 \ldots i_n} \\ &= |g|^{1/2} \frac{1}{k!(n-k)!} \sum \alpha_{j_1 \ldots j_k} \beta^{j_1 \ldots j_k} \epsilon_{j_1 \ldots j_k j_{k+1} \ldots j_n} \epsilon^{j_1 \ldots j_n}_{i_1 \ldots i_n} \\ &= |g|^{1/2} \frac{1}{k!} \sum \alpha_{j_1 \ldots j_k} \beta^{j_1 \ldots j_k} \epsilon_{i_1 \ldots i_n} = (\langle \alpha | \beta \rangle \text{ vol})_{i_1 \ldots i_n}\end{aligned}$$

and so $\alpha \wedge$ ♣$\beta = \langle \alpha | \beta \rangle$ vol; thus by uniqueness we have ♣ $= *$. □

The definitions we gave for volume element, star operator, and the musical isomorphisms all apply to each tangent space T_pM of a semi-Riemannian manifold and these concepts globalize to the whole of TM accordingly.

Let M be oriented. The **metric volume form** induced by the metric tensor g is defined to be the n-form vol such that vol_p is the metric volume form on T_pM matching the orientation. If $(U, \mathbf{x})$ is a positively oriented chart on M, then we have

$$\text{vol}|_U = \sqrt{|\det[g_{ij}]|} dx^1 \wedge \cdots \wedge dx^n.$$

If f is a smooth function with compact support, then we may integrate f over M by using the volume form:

$$\int_M f \text{ vol}.$$

The volume of a compact oriented Riemannian manifold is

$$\text{vol}(M) := \int_M \text{vol}.$$

The volume of an open set $D \subset M$ is defined as

$$\mathrm{vol}(D) := \sup\left\{ \int_M f\, \mathrm{vol} : \mathrm{supp}(f) \subset D \text{ with } f \in C_c^\infty(M) \text{ and } 0 \le f \le 1 \right\},$$

where $C_c^\infty(M)$ denotes the space of smooth functions with compact support.

Example 9.27. The metric volume form $\mathrm{vol}_{\mathbb{R}^n}$ on $\mathbb{R}^n$ is $dx^1 \wedge \cdots \wedge dx^n$, where $x^1, \ldots, x^n$ are standard coordinates. Also,

$$\mathrm{vol}_{\mathbb{R}^n}(v_{1p}, \ldots, v_{np}) = \det(v_1, \ldots, v_n),$$

where $v_{ip} = (p, v_i) \in T_p\mathbb{R}^n = \{p\} \times \mathbb{R}^n$.

Once again assume that M is oriented by a metric volume form "vol". The star operator gives two types of maps, which are both denoted by $*$. Namely, the bundle maps $* : \bigwedge^k T^*M \to \bigwedge^{n-k} T^*M$, which have the obvious definition in terms of the star operators on each fiber, and the maps on forms $* : \Omega^k(M) \to \Omega^{n-k}(M)$, which are also induced in the obvious way. The definitions are set up so that $*$ is natural with respect to restriction and so that for any oriented orthonormal frame field $\{E_1, \ldots, E_n\}$ with dual $\{\theta^1, \ldots, \theta^n\}$ we have $*(\theta^{i_1} \wedge \cdots \wedge \theta^{i_k}) = \pm(\epsilon_{i_1}\epsilon_{i_2}\cdots\epsilon_{i_k})\theta^{j_1} \wedge \cdots \wedge \theta^{j_{n-k}}$, where, as before, we use the $+$ sign if and only if $\theta^{i_1} \wedge \cdots \wedge \theta^{i_k} \wedge \theta^{j_1} \wedge \cdots \wedge \theta^{j_{n-k}} = \theta^1 \wedge \cdots \wedge \theta^n = \mathrm{vol}$. As expected we have

$$*1 = \mathrm{vol},$$
$$*\,\mathrm{vol} = (-1)^{\mathrm{ind}(g)},$$
$$**\alpha = (-1)^{\mathrm{ind}(g)}(-1)^{k(n-k)}\alpha$$

for $\alpha \in \Omega^k(M)$.

Example 9.28. On open sets in $\mathbb{R}^3$ the star operator associated with the standard metric and volume is given by

$$\begin{aligned} f &\mapsto f\,dx \wedge dy \wedge dz, \\ f_1 dx + f_2 dy + f_3 dz &\mapsto f_1 dy \wedge dz + f_2 dz \wedge dx + f_3 dx \wedge dy, \\ g_1 dy \wedge dz + g_2 dz \wedge dx + g_3 dx \wedge dy &\mapsto g_1 dx + g_2 dy + g_3 dz, \\ f\,dx \wedge dy \wedge dz &\mapsto f. \end{aligned}$$

For an oriented Riemannian manifold, div will always be the divergence with respect to the metric volume form (giving the orientation). From equation (9.4) we see that the local coordinate expression for the divergence of $X = \sum X^k \frac{\partial}{\partial x^k}$ is

$$\mathrm{div}\, X = \sum_{k=1}^n \frac{1}{\sqrt{\det g}} \frac{\partial}{\partial x^k}(\sqrt{\det g}\, X^k),$$

where $\det g = \det(g_{ij})$.

9.4. Differential Forms and Metrics

Definition 9.29. The gradient of a smooth function f on a Riemannian manifold (M, g) is defined to be

$$\operatorname{grad}(f) := \sharp df,$$

and the **Laplacian** Δ is defined by

$$\Delta f := -\operatorname{div}(\operatorname{grad}(f))$$

if M is oriented. If $\Delta f = 0$, then f is called a **harmonic function**.

The minus sign in the above definition is a matter of convention, and this choice of sign is popular with differential geometers. Ironically, this choice of sign makes the Laplacian a positive operator.

Lemma 9.30. *If M is an oriented semi-Riemannian manifold with metric volume element* vol, *then* $i_X \operatorname{vol} = *\flat X$ *for* $X \in \mathfrak{X}(M)$.

Proof. We show that for fixed $p \in M$ we have $i_{v_1} \operatorname{vol} = *\flat v_1$ for all $v_1 \in T_pM$. First consider the case where v is not a null vector and $\langle v_1, v_1 \rangle = \epsilon_1 = \pm 1$. Since the maps $(v_2, \ldots, v_n) \mapsto i_{v_1} \operatorname{vol}(v_2, \ldots, v_n)$ and $(v_2, \ldots, v_n) \mapsto (*\flat v_1)(v_2, \ldots, v_n)$ are both alternating multilinear maps, it suffices to check that they are equal on some basis. We extend to an orthonormal basis $(v_1, v_2, \ldots, v_n)$ with dual basis $e^1, \ldots, e^n$. Then we wish to show that $i_{v_1} \operatorname{vol}(v_{i_1}, \ldots, v_{i_{n-1}}) = (*\flat v_1)(v_{i_1}, \ldots, v_{i_{n-1}})$ for $1 \le i_1 < \cdots < i_{n-1} \le n$. But if $v_{i_1} = v_1$, then

$$i_{v_1} \operatorname{vol}(v_{i_1}, \ldots, v_{i_{n-1}}) = \operatorname{vol}(v_1, v_1, \ldots) = 0$$

and

$$(*\flat v_1)(v_{i_1}, \ldots, v_{i_{n-1}}) = \epsilon_1 (*e^1)(v_{i_1}, \ldots, v_{i_{n-1}})$$
$$= \epsilon_1^2 e^2 \wedge \cdots \wedge e^n (v_1, \ldots) = 0.$$

On the other hand, if $v_{i_1} \ne v_1$, then $(v_{i_1}, \ldots, v_{i_{n-1}}) = (v_2, \ldots, v_n)$ and

$$i_{v_1} \operatorname{vol}(v_2, \ldots, v_n) = \operatorname{vol}(v_1, v_2, \ldots, v_n) = 1$$

while

$$(*\flat v_1)(v_2, \ldots, v_n) = \epsilon_1 (*e^1)(v_2, \ldots, v_n)$$
$$= \epsilon_1^2 e^2 \wedge \cdots \wedge e^n (v_2, \ldots, v_n) = 1.$$

Thus $i_{v_1} \operatorname{vol} = *\flat v_1$ for all nonnull vectors v_1. But since both sides are linear in v_1, this establishes the result. $\square$

Proposition 9.31. *If M is an oriented semi-Riemannian manifold with metric volume element* vol, *then* $\operatorname{div} X = (-1)^{\operatorname{ind} g} * d * \flat X$ *for* $X \in \mathfrak{X}(M)$.

Proof. By Lemma 9.30 and Cartan's formula (8.56), we have $\mathcal{L}_X \text{vol} = d\, i_X \text{vol} = d * \flat X$ so that $* \operatorname{div} X\, \text{vol} = *d * \flat X$ or

$$\operatorname{div} X * \text{vol} = *d * \flat X$$

or

$$(-1)^{\operatorname{ind} g} \operatorname{div} X = *d * \flat X. \qquad \square$$

Proposition 9.32. *If M is an oriented semi-Riemannian manifold with metric volume element* vol, *then*

$$\operatorname{div} fX = \langle \operatorname{grad} f, X \rangle + f \operatorname{div} X.$$

Proof. Write $\text{vol} = \mu$ and compute as follows:

$$\begin{aligned}
(\operatorname{div} fX)\mu &= \mathcal{L}_{fX} \mu = d i_{fX} \mu + i_{fX} d\mu \\
&= d(f i_X \mu) = df \wedge i_X \mu + f d i_X \mu \\
&= df \wedge i_X \mu + f (\operatorname{div} X) \mu \\
&= -i_X (df \wedge \mu) + i_X df \wedge \mu + f (\operatorname{div} X) \mu \\
&= df(X) \mu + f (\operatorname{div} X) \mu \\
&= (\langle X, \operatorname{grad} f \rangle + f (\operatorname{div} X)) \mu,
\end{aligned}$$

where in the third line we use the fact that i_X is a graded derivation and $df \wedge \mu = 0$. $\qquad \square$

Exercise 9.33. Show that the local expression for Δf is

$$\Delta f = -\frac{1}{\sqrt{\det g}} \sum_{j,k} \frac{\partial}{\partial x^j} \left(g^{jk} \sqrt{\det g}\, \frac{\partial f}{\partial x^k} \right).$$

9.5. Integral Formulas

Definition 9.34. Let S be a regular codimension one submanifold of an n-dimensional Riemannian manifold (M, g). A smooth vector field N along S is called a **unit normal field** if $\langle N, N \rangle = 1$ and if $N(p) \perp T_p S$ for every $p \in S$.

Let (M, g) be oriented by a metric volume form vol and suppose that a regular $(n-1)$-dimensional submanifold S of M has a (global) unit normal field N along S. If N is extended to a field $\widetilde{N}$ on a neighborhood of S, then $i_{\widetilde{N}} \text{vol}$ is an $(n-1)$-form on this neighborhood. Clearly, the restriction of $i_{\widetilde{N}} \text{vol}$ to S is independent of this extension, and so we can denote this restriction by $i_N \text{vol}|_S$. Then we have

$$(i_N \text{vol}|_S)_p (v_1, \ldots, v_{n-1}) = \text{vol}_p(N(p), v_1, \ldots, v_{n-1})$$

for $v_1, \ldots, v_{n-1} \in T_p S$. This is a volume form on S which orients S and is clearly the metric volume form for the induced metric $\iota^* g$ on S. Note that

9.5. Integral Formulas

$i_N \operatorname{vol}|_S$ is sometimes denoted by $i_N \operatorname{vol}$, but this is an abuse of notation since $i_N \operatorname{vol}(p)(v_1, \ldots, v_{n-1})$ makes sense for $v_1, \ldots, v_{n-1} \in T_pM$, but this is not what we want. On the other hand, $\int_S i_N \operatorname{vol}|_S$ is $\int_S \iota^*\left(i_{\widetilde{N}} \operatorname{vol}\right)$, where $\iota : S \hookrightarrow M$ is the inclusion map. In turn, this means the same thing as $\int_S i_N \operatorname{vol}$ by convention.

Proposition 9.35. *Let S be a regular codimension one submanifold of an n-dimensional oriented Riemannian manifold (M, g) and let $\iota : S \hookrightarrow M$ be the inclusion. Let N be a smooth unit normal vector field along S. Let vol_M be the volume form of M and vol_S the volume form of S with respect to the orientation induced by N and the metric ι^*g. If X is a smooth vector field along S, then*

$$i_X \operatorname{vol}_M = \langle X, N \rangle \operatorname{vol}_S.$$

Proof. Let $X^\top := X - \langle X, N \rangle N$;

$$i_X \operatorname{vol}_M = i_{X^\top} \operatorname{vol}_M + i_{\langle X, N \rangle N} \operatorname{vol}_M$$
$$= i_{X^\top} \operatorname{vol}_M + \langle X, N \rangle i_N \operatorname{vol}_M,$$

and since $i_N \operatorname{vol}_M|_{\partial M} = \operatorname{vol}_S$, it remains to show that $i_{X^\top} \operatorname{vol}_M = 0$. But for $v_1, \ldots, v_{n-1} \in T_pS$,

$$i_{X^\top} \operatorname{vol}_M(v_1, \ldots, v_{n-1}) = \operatorname{vol}_M(X^\top, v_1, \ldots, v_{n-1}) = 0$$

since $X^\top, v_1, \ldots, v_{n-1}$ cannot be linearly independent. $\square$

Now suppose that an oriented (M, g) has a boundary ∂M. Since ∂M is basically a codimension one regular submanifold of M, the above constructions can be applied to ∂M so that ∂M is Riemannian with induced metric. Since ∂M has a global outward pointing vector field, we may normalize to obtain a global outward pointing *unit* normal field N along ∂M. The orientation induced on ∂M by the metric volume form $i_N \operatorname{vol}_M|_{\partial M}$ is exactly the induced orientation on ∂M described previously. We will denote $i_N \operatorname{vol}_M|_{\partial M}$ by $\operatorname{vol}_{\partial M}$ when the orientation and normal field are understood.

Exercise 9.36. Let $\mathbf{x} : V \to U \subset M$ be a parametrization of an open subset of a surface S in $\mathbb{R}^3$ and $N := (\mathbf{x}_{u^1} \times \mathbf{x}_{u^2})/\|\mathbf{x}_{u^1} \times \mathbf{x}_{u^2}\|$. Show that if $dS := i_N \operatorname{vol}|_S$, where vol is the usual volume form on $\mathbb{R}^3$, then $\mathbf{x}^*dS = \|\mathbf{x}_{u^1} \times \mathbf{x}_{u^2}\| du^1 \wedge du^2$.

We now introduce a special $(n-1)$-form σ on $\mathbb{R}^n$ (we use the same symbol for all n if no confusion arises). This form is used in many constructions. If $x^1, \ldots, x^n$ denote standard coordinates, then

(9.11) $$\sigma := \sum_{i=1}^n (-1)^{i-1} x^i \, dx^1 \wedge \cdots \wedge \widehat{dx^i} \wedge \cdots \wedge dx^n,$$

where, as usual, the caret denotes omission. For example, on $\mathbb{R}^3$
$$\sigma = x\, dy \wedge dz - y\, dx \wedge dz + z\, dx \wedge dy.$$

Proposition 9.37. *If we use the outward pointing normal field N on S^{n-1} and the metric induced from $\mathbb{R}^n$, then the induced volume form on S^{n-1} (the area form) is given by*
$$\mathrm{vol}_{S^{n-1}} = \iota^*\sigma,$$
where $\iota : S^{n-1} \hookrightarrow \mathbb{R}^n$ is inclusion.

Proof. Let $v_i = \sum v_i^j\, \partial/\partial x^j\big|_x \in T_x S^{n-1}$. Identify elements of $T_x S^{n-1}$ with vectors in $\mathbb{R}^n$ so that v_i is the column vector $(v_i^1, \ldots, v_i^n)^T$. Then we use expansion along the first column by cofactors to obtain
$$\mathrm{vol}_{S^{n-1}}(v_1, \ldots, v_{n-1}) = \det(x, v_1, \ldots, v_{n-1})$$
$$= \sum_{i=1}^n (-1)^{i-1} x^i \det(M_i)$$
$$= \sum_{i=1}^n (-1)^{i-1} x^i \left(dx^1 \wedge \cdots \wedge \widehat{dx^i} \wedge \cdots \wedge dx^n \right)(v_1, \ldots, v_{n-1}),$$

where M_i is the submatrix obtained by deleting the first column and i-th row. $\square$

Exercise 9.38. Show that the volume form on $S^{n-1}(R) = \{\sum_{i=1}^n (x^i)^2 = R^2\}$ is $\frac{1}{R}\iota^*\sigma$, where $\iota : S^{n-1}(R) \hookrightarrow \mathbb{R}^n$ is the inclusion map.

Theorem 9.39. *Let M be an oriented Riemannian n-manifold with boundary. For any $X \in \mathfrak{X}(M)$ with compact support,*
$$\int_M \mathrm{div}\, X\, \mathrm{vol}_M = \int_{\partial M} \langle X, N \rangle\, \mathrm{vol}_{\partial M},$$
where N is the outward pointing unit normal field on ∂M.

Proof. We have $\int_M \mathrm{div}\, X\, \mathrm{vol}_M = \int_M d(i_X\, \mathrm{vol}_M) = \int_{\partial M} i_X\, \mathrm{vol}_M$ which is equal to $\int_{\partial M} \langle X, N \rangle\, \mathrm{vol}_{\partial M}$ by Proposition 9.35. $\square$

Corollary 9.40. *Let M be a compact oriented Riemannian n-manifold with boundary and $X \in \mathfrak{X}(M)$. Then for $f \in C^\infty(M)$ we have*
$$\int_M \langle X, \mathrm{grad}\, f \rangle\, \mathrm{vol}_M + \int_M f\, \mathrm{div}\, X\, \mathrm{vol}_M = \int_{\partial M} f\, \langle X, N \rangle\, \mathrm{vol}_{\partial M}.$$

Proof. We have
$$\int_M (\langle \mathrm{grad}\, f, X \rangle + f\, \mathrm{div}\, X)\, \mathrm{vol}_M = \int_M \mathrm{div}\, fX = \int_{\partial M} f\, \langle X, N \rangle\, \mathrm{vol}_{\partial M}. \quad \square$$

9.5. Integral Formulas

Theorem 9.41. *Let M be an oriented Riemannian n-manifold with boundary and $f, g \in C^\infty(M)$. Then*

$$\int_M \langle \operatorname{grad} f, \operatorname{grad} g \rangle \operatorname{vol}_M - \int_M f \Delta g \operatorname{vol}_M = \int_{\partial M} f \langle \operatorname{grad} g, N \rangle \operatorname{vol}_{\partial M}$$

and

$$\int_M (-f \Delta g + g \Delta f) \operatorname{vol}_M = \int_{\partial M} [f \langle \operatorname{grad} g, N \rangle - g \langle \operatorname{grad} f, N \rangle] \operatorname{vol}_{\partial M}$$

where N is the outward unit normal field.

Proof.

$$-\int_M f \Delta g \operatorname{vol}_M = \int_M f \operatorname{div}(\operatorname{grad} g) \operatorname{vol}_M$$
$$= \int_{\partial M} f \langle \operatorname{grad} g, N \rangle \operatorname{vol}_{\partial M} - \int_M \langle \operatorname{grad} g, \operatorname{grad} f \rangle \operatorname{vol}_M$$

by Corollary 9.40. The second formula follows from the first by interchanging the roles of f and g and subtracting. $\square$

The integral equations of the previous theorem are called Green's first and second formulas respectively (when comparing with other versions in the literature, do not forget our convention: $\Delta f := -\operatorname{div}\operatorname{grad} f$). If the boundary of M is empty, then the boundary terms are zero, so that for instance $\int_M f \Delta g \operatorname{vol}_M = \int_M g \Delta f \operatorname{vol}_M = 0$. We close this section with an application of Green's formulas.

Theorem 9.42 (Hopf). *Let (M, g) be a compact, connected and oriented Riemannian n-manifold without boundary. If $f \in C^\infty(M)$ and $\Delta f \geq 0$ (or $\Delta f \leq 0$), then f is a constant function.*

Proof. Suppose $\Delta f \geq 0$. Then integration gives

$$0 \leq \int_M 1 \Delta f \operatorname{vol}_M = \int_M f \Delta 1 \operatorname{vol}_M = 0,$$

which means that $\Delta f \equiv 0$ on M. Now use Green's second formula with $f = g$ to get

$$\int_M \|\operatorname{grad} f\|^2 \operatorname{vol}_M = \int_M f \Delta f \operatorname{vol}_M = 0,$$

so that $\|\operatorname{grad} f\|^2 \equiv 0$ on M. Thus f is constant since M is connected. $\square$

9.6. The Hodge Decomposition

In this section we follow [**War**]. Let (M, g) be an oriented semi-Riemannian n-manifold. (However, we soon restrict to the compact Riemannian case.) Let vol denote the metric volume form. For each k, the scalar products induced by g_p on $\bigwedge^k T_p^* M$ for each p combine to give a symmetric $C^\infty(M)$-bilinear map

$$\langle \cdot | \cdot \rangle : \Omega^k(M) \times \Omega^k(M) \to C^\infty(M)$$

with $\langle \alpha | \beta \rangle(p) := \langle \alpha(p) | \beta(p) \rangle_p = g_p(\alpha(p), \beta(p))$. We see that $\langle \alpha | \beta \rangle$ vol $= \alpha \wedge *\beta$. We can put a scalar product $(\cdot | \cdot)$ on the space $\Omega(M) = \sum_k \Omega^k(M)$ by letting $(\alpha | \beta) = 0$ for $\alpha \in \Omega^{k_1}(M)$ and $\beta \in \Omega^{k_2}(M)$ with $k_1 \neq k_2$ and letting

$$(\alpha | \beta) = \int_M \alpha \wedge *\beta = \int_M \langle \alpha | \beta \rangle \text{ vol } \text{ if } \alpha, \beta \in \Omega^k(M).$$

Definition 9.43. Let (M, g) be an oriented semi-Riemannian n-manifold. For each k with $0 \le k \le n = \dim M$, the **codifferential** $\delta : \Omega^k(M) \to \Omega^{k-1}(M)$ is defined by

$$\delta := (-1)^{\text{ind}(g)} (-1)^{n(k+1)+1} *d*$$

and $\delta = 0$ on $\Omega^0(M)$. These operators combine to give a linear map $\Omega(M) \to \Omega(M)$, which is also denoted δ.

Remark 9.44. Notice that if M is nonorientable, then $*$ is still defined locally by choosing an orientation valid on a chart domain. But $*$ occurs twice in the formula for δ, so any sign ambiguities cancel and thus δ is globally well-defined even if M is nonorientable!

Proposition 9.45. *The codifferential δ is the formal adjoint of the exterior derivative on $\Omega(M)$. That is,*

$$(d\alpha | \beta) = (\alpha | \delta \beta)$$

for all α, β in $\Omega(M)$ with compact support in the interior of M.

Proof. It suffices to check that $(d\alpha | \beta) = (\alpha | \delta \beta)$ for α a $(k-1)$-form and β a k-form. In this case, $d*\beta$ is an $(n-k+1)$-form and so

$$*\delta \beta = *\left((-1)^{\text{ind}(g)} (-1)^{n(k+1)+1} * d * \beta \right)$$
$$= (-1)^{\text{ind}(g)} (-1)^{n(k+1)+1} ** d * \beta$$
$$= (-1)^{n(k+1)+1} (-1)^{(n-k+1)(k-1)} d*\beta = (-1)^{k^2} d*\beta,$$

9.6. The Hodge Decomposition

where we have used that $n(k+1)+1+(n-k+1)(k-1) = 2k+2kn-k^2 \equiv k^2 \bmod 2$. Then we have

$$d(\alpha \wedge *\beta) = d\alpha \wedge *\beta + (-1)^{k-1}\alpha \wedge d*\beta$$
$$= d\alpha \wedge *\beta + (-1)^{k-1}(-1)^{k^2}\alpha \wedge *\delta\beta$$
$$= d\alpha \wedge *\beta - \alpha \wedge *\delta\beta$$

since $k-1+k^2 = k(k+1)-1$, which is always odd. Now we integrate and use Stokes' theorem to get the desired result:

$$\int_M d\alpha \wedge *\beta = \int_M \alpha \wedge *\delta\beta. \qquad \square$$

Obviously, δ is defined on $\Omega^k(U)$ for open sets U in M and δ is natural with respect to restriction.

Definition 9.46. For $0 \leq k \leq n$, the differential operator $\Delta : \Omega(M) \to \Omega(M)$ defined by $\Delta = \delta d + d\delta$ is called the **Laplace-Beltrami operator**. For each k, the restriction of Δ to $\Omega^k(M)$ is also called the Laplace-Beltrami operator (on k-forms). If $\Delta\omega = 0$, we call ω a **harmonic** form.

Notice that by Remark 9.44 the Laplace-Beltrami operator is defined whether or not M is orientable. On $\Omega^0(M) = C^\infty(M)$ the operator Δ reduces to the Laplacian defined earlier.

Proposition 9.47. *The following assertions hold:*

(i) *For all $\alpha, \beta \in \Omega(M)$, we have*

$$(\Delta\alpha|\beta) = (d\alpha|d\beta) + (\delta\alpha|\delta\beta).$$

(ii) *Δ is formally self-adjoint. That is $(\Delta\alpha|\beta) = (\alpha|\Delta\beta)$ for all $\alpha, \beta \in \Omega(M)$ with compact support in the interior of M.*

Proof. By Proposition 9.45 we have

$$(\Delta\alpha|\beta) = (\delta d\alpha + d\delta\alpha|\beta) = (\delta d\alpha|\beta) + (d\delta\alpha|\beta) = (d\alpha|d\beta) + (\delta\alpha|\delta\beta)$$
$$= (\alpha|\delta d\beta + d\delta\beta) = (\alpha|\Delta\beta),$$

which proves (i). But by symmetry $(\Delta\alpha|\beta) = (d\alpha|d\beta) + (\delta\alpha|\delta\beta) = (\alpha|\Delta\beta)$, which proves (ii). $\qquad \square$

In the remainder of this section, we restrict attention to an oriented compact *Riemannian* n-manifold (M, g) without boundary. In this case, $(\alpha|\beta) = \int_M \alpha \wedge *\beta$ defines a *positive definite* inner product on $\Omega(M)$. We write $\|\alpha\| := (\alpha|\alpha)^{1/2}$.

Proposition 9.48. *If (M, g) is compact, oriented and Riemannian, then $\Delta\omega = 0$ if and only if $d\omega = 0$ and $\delta\omega = 0$.*

Proof. This follows from $(\Delta\omega|\omega) = (d\omega|d\omega) + (\delta\omega|\delta\omega)$ since $(\cdot|\cdot)$ is positive definite in the Riemannian case. $\square$

Now we wish to consider the equation $\Delta\omega = \alpha$ for $\alpha \in \Omega^k(M)$. First notice that if $\Delta\omega = \alpha$ holds, then for any $\beta \in \Omega^k(M)$ we have $(\Delta\omega|\beta) = (\omega|\Delta\beta) = (\alpha|\beta)$. Then since $|(\omega|\Delta\beta)| = |(\alpha|\beta)| \le \|\alpha\|\,\|\beta\|$, the map $\ell_\omega : \beta \mapsto (\omega|\Delta\beta)$ is a bounded linear functional. Employing a common idea from the theory of differential equations, we make the following definition:

Definition 9.49. A bounded linear functional $\ell : \Omega^k(M) \to \mathbb{R}$ is called a **weak solution** of the equation $\Delta\omega = \alpha$ if

$$\ell(\Delta\beta) = (\alpha|\beta)$$

for all $\beta \in \Omega^k(M)$.

Notice that if a weak solution ℓ is represented by ω, then it must be an ordinary solution since in that case we have

$$(\Delta\omega|\beta) = (\omega|\Delta\beta) = \ell(\Delta\beta) = (\alpha|\beta)$$

for all β, which means that $\Delta\omega = \alpha$. We will use the following powerful regularity result (see [**War**] for a proof):

Theorem 9.50. *Let $\alpha \in \Omega^k(M)$. If ℓ is a weak solution of $\Delta\omega = \alpha$, then there exists $\omega \in \Omega^k(M)$ such that*

$$\ell(\beta) = (\omega|\beta)$$

for all $\beta \in \Omega^k(M)$ and hence

$$\Delta\omega = \alpha.$$

We will also need a rather technical result whose proof can also be found in [**War**].

Proposition 9.51. *Let $\{\alpha_i\}_1^\infty$ be a sequence in $\Omega^k(M)$. Suppose that for some $C > 0$ we have*

$$\|\alpha_i\| \le C \text{ and } \|\Delta\alpha_i\| \le C$$

for all i. Then $\{\alpha_i\}_1^\infty$ has a Cauchy subsequence.

Let $\mathcal{H}^k = \{\omega \in \Omega^k(M) : \Delta\omega = 0\}$ and let $(\mathcal{H}^k)^\perp$ denote $\{\beta \in \Omega^k(M) : (\beta|\omega) = 0 \text{ for all } \omega \in \mathcal{H}^k\}$.

Lemma 9.52. *There exists a constant $C > 0$ such that $\|\beta\| \le C \|\Delta\beta\|$ for all $\beta \in (\mathcal{H}^k)^\perp$.*

9.6. The Hodge Decomposition

Proof. Suppose there is no such constant C. Then we may find a sequence $\{\beta_i\}_1^\infty \subset (\mathcal{H}^k)^\perp$ such that $\|\beta_i\| = 1$ while $\lim_{i\to\infty} \|\Delta\beta_i\| = 0$. By Proposition 9.51, $\{\beta_i\}_1^\infty$ has a Cauchy subsequence, which we may as well assume to be $\{\beta_i\}_1^\infty$. Hence we have that for any fixed $\theta \in \Omega^k(M)$, the sequence $\{(\beta_i|\theta)\}$ is Cauchy in $\mathbb{R}$ and so converges to some number. Now we define $\ell : \Omega^k(M) \to \mathbb{R}$ by

$$\ell(\theta) := \lim_{i\to\infty}(\beta_i|\theta).$$

Then,

$$\ell(\Delta\theta) = \lim_{i\to\infty}(\beta_i|\Delta\theta) = \lim_{i\to\infty}(\Delta\beta_i|\theta) = 0,$$

and since ℓ is clearly bounded, it is a weak solution to $\Delta\omega = 0$. By Theorem 9.50, there must exist $\beta \in \Omega^k(M)$ such that $\ell(\theta) = (\beta|\theta)$ for all $\theta \in \Omega^k(M)$. It follows that $(\beta|\theta) = \lim_{i\to\infty}(\beta_i|\theta)$ for all θ and so $\beta_i \to \beta$ since

$$\|\beta_i - \beta\|^2 = (\beta_i - \beta|\beta_i - \beta)$$
$$= (\beta_i|\beta_i) - 2(\beta_i|\beta) + (\beta|\beta) \to 0.$$

Since $\|\beta_i\| = 1$, we must have $\|\beta\| = 1$ and of course $\beta \in (\mathcal{H}^k)^\perp$. But by Theorem 9.50, $\Delta\beta = 0$ so $\beta \in \mathcal{H}^k \cap (\mathcal{H}^k)^\perp = 0$, which contradicts $\|\beta\| = 1$. We conclude that C exists after all. □

Theorem 9.53 (Hodge decomposition). *Let (M, g) be compact and oriented (and without boundary). For each k with $0 \leq k \leq n = \dim M$, the space of harmonic k-forms $\mathcal{H}^k$ is finite-dimensional. Furthermore, we have an orthogonal decomposition of $\Omega^k(M)$,*

$$\Omega^k(M) = \Delta\left(\Omega^k(M)\right) \oplus \mathcal{H}^k$$
$$= d\delta\left(\Omega^k(M)\right) \oplus \delta d\left(\Omega^k(M)\right) \oplus \mathcal{H}^k$$
$$= d\left(\Omega^{k-1}(M)\right) \oplus \delta\left(\Omega^{k+1}(M)\right) \oplus \mathcal{H}^k.$$

Proof. If $\mathcal{H}^k$ were infinite-dimensional, then it would contain an infinite orthonormal sequence $\{\omega_i\}_1^\infty$. In this case, we would have

$$\|\omega_i - \omega_j\|^2 = 2$$

for all i, j with $i \neq j$. By Proposition 9.51, this sequence would contain a Cauchy subsequence. But this contradicts the above equation. Thus $\mathcal{H}^k$ must be finite-dimensional.

We now prove the orthogonal decomposition $\Omega^k(M) = \Delta\left(\Omega^k(M)\right) \oplus \mathcal{H}^k$. The other two decompositions can be derived from the first, and we leave this as a problem for the reader (Problem 9). Choose an orthonormal basis $\omega_1, \ldots, \omega_d$ for $\mathcal{H}^k$. If $\alpha \in \Omega^k(M)$, then we may write

$$\alpha = \beta + \sum_{i=1}^{d} (\omega_i|\alpha)\omega_i$$

where $\beta \in \left(\mathcal{H}^k\right)^\perp$. It is easy to show that this decomposition is unique, so we have the orthogonal decomposition $\Omega^k(M) = \left(\mathcal{H}^k\right)^\perp \oplus \mathcal{H}^k$ and our task is to show that $\left(\mathcal{H}^k\right)^\perp = \Delta\left(\Omega^k(M)\right)$. Since $(\Delta\alpha|\omega) = (\alpha|\Delta\omega) = 0$ whenever $\omega \in \mathcal{H}^k$, we see that $\Delta\left(\Omega^k(M)\right) \subset \left(\mathcal{H}^k\right)^\perp$. Now let $\alpha \in \left(\mathcal{H}^k\right)^\perp$ and define a linear functional ℓ on $\Delta\left(\Omega^k(M)\right)$ by

$$\ell(\Delta\theta) := (\alpha|\theta).$$

This is well-defined since if $\Delta\theta_1 = \Delta\theta_2$, then $\theta_1 - \theta_2 \in \mathcal{H}^k$ and so $(\alpha|\theta_1) - (\alpha|\theta_2) = (\alpha|\theta_1 - \theta_2) = 0$. We show that ℓ is bounded. Let $\phi := \theta - H(\theta)$, where $H_k : \Omega^k(M) \to \mathcal{H}^k$ is the orthogonal projection. Then using Lemma 9.52 we have

$$|\ell(\Delta\theta)| = |\ell(\Delta\phi)| = |(\alpha|\phi)| \leq \|\alpha\| \|\phi\|$$
$$\leq C \|\alpha\| \|\Delta\phi\| = C \|\alpha\| \|\Delta\theta\|.$$

By the Hahn-Banach theorem, the functional ℓ extends to a bounded functional $\widetilde{\ell}$ defined on all of $\Omega^k(M)$, which is then a weak solution of $\Delta\omega = \alpha$. By Theorem 9.50, there is an $\omega \in \Omega^k(M)$ with $\Delta\omega = \alpha$ so $\left(\mathcal{H}^k\right)^\perp \subset \Delta\left(\Omega^k(M)\right)$ and so $\left(\mathcal{H}^k\right)^\perp = \Delta\left(\Omega^k(M)\right)$. $\square$

In order to take full advantage of the Hodge decomposition, we now introduce a so-called "Green's operator" $G_k : \Omega^k(M) \to \left(\mathcal{H}^k\right)^\perp$. We simply define $G_k(\alpha)$ to be the unique solution of $\Delta\omega = \alpha - H_k(\alpha)$ where $H_k : \Omega^k(M) \to \mathcal{H}^k$ is the orthogonal projection as above. The G_k combine to give a map $G : \Omega(M) \to \bigoplus_k \left(\mathcal{H}^k\right)^\perp$ also called the Green's operator.

Lemma 9.54. *Let $G_k : \Omega^k(M) \to \left(\mathcal{H}^k\right)^\perp$ be the Green's operator defined above for each k. Then*

(i) *G_k is formally self-adjoint;*

(ii) *if $L : \Omega^k(M) \to \Omega^r(M)$ is linear and commutes with Δ, then G commutes with L (that is, $L \circ G_k = G_r \circ L$). In particular, G commutes with d and δ.*

Proof. We have

$$(G_k(\alpha)|\beta) = (G_k(\alpha)|\beta - H_k(\beta)) = (G_k(\alpha)|\Delta G_k(\beta)) = (\Delta G_k(\alpha)|G_k(\beta))$$
$$= (\alpha - H_k(\alpha)|G_k(\beta)) = (\alpha|G_k(\beta)),$$

9.6. The Hodge Decomposition

so G is self-adjoint. For each j, let $\pi_j : \Omega^j(M) \to (\mathcal{H}^j)^\perp$ denote orthogonal projection (thus $\pi_j + H_j = \mathrm{id}_{\Omega^j}$). Now suppose that $L : \Omega^k(M) \to \Omega^r(M)$ is linear and commutes with Δ. Notice that by definition we have $G_k = (\Delta|\,(\mathcal{H}^k)^\perp)^{-1} \circ \pi_k$. The fact that $L\Delta = \Delta L$ implies that $L(\mathcal{H}^k) \subset \mathcal{H}^r$. Also, since $\Delta(\Omega^k(M)) = (\mathcal{H}^k)^\perp$, we have $L((\mathcal{H}^k)^\perp) \subset (\mathcal{H}^r)^\perp$. Thus

$$L \circ \pi_k = \pi_r \circ L,$$

$$L \circ (\Delta|\,(\mathcal{H}^k)^\perp) = (\Delta|\,(\mathcal{H}^r)^\perp) \circ L|_{(\mathcal{H}^r)^\perp},$$

and

$$(\Delta|\,(\mathcal{H}^r)^\perp)^{-1} \circ L = L|_{(\mathcal{H}^r)^\perp} \circ (\Delta|\,(\mathcal{H}^k)^\perp)^{-1}.$$

It follows that G commutes with L. $\square$

We note in passing that G maps bounded sequences into sequences that have Cauchy subsequences. Indeed, suppose that $\{\alpha_i\}_1^\infty \subset \Omega^k(M)$ is a sequence with $\|\alpha_i\| \leq C$. If $\beta_i := G(\alpha_i)$, then using Lemma 9.52 we have

$$\|\beta_i\| \leq \|\Delta\beta_i\| = \|\alpha_i - H(\alpha_i)\| \leq \|\alpha_i\| \leq C,$$

and so by Proposition 9.51, $\{\beta_i\}_1^\infty$ has a Cauchy subsequence.

Theorem 9.55. *Let (M,g) be a compact Riemannian manifold (without boundary). Then each de Rham cohomology class contains a unique harmonic representative.*

Proof. First assume that M is orientable and fix an orientation. Let $\alpha \in \Omega^k(M)$ and use the Hodge decomposition and the definition of G to obtain

$$\alpha = \Delta G_k(\alpha) + H_k(\alpha) = d\delta G_k(\alpha) + \delta d G_k(\alpha) + H_k(\alpha).$$

Then since G commutes with d, we have

$$\alpha = d\delta G_k(\alpha) + \delta G_{k+1}(d\alpha) + H_k(\alpha).$$

So if $d\alpha = 0$, then $\alpha - H_k(\alpha) = d\delta G_k(\alpha)$, and so $H_k(\alpha)$ represents the same cohomology class as α. To show uniqueness, suppose that α_1 and α_2 are both harmonic and in the same class so that $\alpha_2 - \alpha_1 = d\beta$ for some β. Then we have $0 = d\beta + (\alpha_1 - \alpha_2)$. But $\alpha_1 - \alpha_2$ is orthogonal to $d\beta$ since by Proposition 9.48

$$(d\beta\,|\,\alpha_1 - \alpha_2) = (\beta\,|\,\delta\alpha_1 - \delta\alpha_2) = (\beta\,|\,0) = 0.$$

Next suppose that M is nonorientable. If $\pi : M^{\mathrm{or}} \to M$ is the 2-fold orientation cover of M (Section 8.7), then there is a unique metric on M^{or} such that π restricts to an isometry on sufficiently small open sets (π is a Riemannian covering). Now suppose that $c \in H^k(M)$ is a de Rham cohomology class and consider the class $\pi^* c \in H^k(M^{\mathrm{or}})$ (see the discussion after Definition 8.41). Since M^{or} is orientable, $\pi^* c$ is uniquely represented by a harmonic form $\widetilde{\eta}$ on M^{or}. If $\tau : M^{\text{or}} \to M^{\text{or}}$ is the involution which transposes the two points in each fiber, then it is easy to see that τ is an isometry so that $\tau^*\widetilde{\eta}$ is also harmonic. It follows from the identity $\pi \circ \tau = \pi$ that $\tau^*\widetilde{\eta}$ is also a representative of the cohomology class π^*c. Indeed, $[\tau^*\widetilde{\eta}] = \tau^*[\widetilde{\eta}] = \tau^*\pi^*c = \pi^*c$. So by uniqueness, we must have $\tau^*\widetilde{\eta} = \widetilde{\eta}$, which is exactly the condition that guarantees that $\widetilde{\eta} = \pi^*\eta$ for some $\eta \in \Omega^k(M)$. But π is a local isometry and so by Problem 5, η must be harmonic. We show that η represents c. Suppose that $c = [\mu]$ for $\mu \in \Omega^k(M)$. Then both $\pi^*\mu$ and $\pi^*\eta$ represent the class π^*c, so $\pi^*(\eta - \mu) = d\beta$ for some β. But then, as for $\tau^*\widetilde{\eta}$ and $\widetilde{\eta}$ above, $\tau^*\beta$ is in the same class as β, which means that $d\beta = d\tau^*\beta$. Then

$$\pi^*(\eta - \mu) = d\beta = \frac{1}{2}(d\beta + d\tau^*\beta) = d\left(\frac{1}{2}(\beta + \tau^*\beta)\right).$$

Since τ is an involution, $\beta + \tau^*\beta$ is τ-invariant, that is, $\tau^*(\beta + \tau^*\beta) = (\beta + \tau^*\beta)$, so there exists $\alpha \in \Omega^k(M)$ with $\pi^*\alpha = \frac{1}{2}(\beta + \tau^*\beta)$. Thus

$$\pi^*(\eta - \mu) = d\pi^*\alpha = \pi^*d\alpha,$$

and since π^* is a local diffeomorphism, we must have $\eta - \mu = d\alpha$, which means that $[\eta] = [\mu] = c$. The uniqueness of the harmonic representative η is clear. $\square$

Notice that if τ is the involution of M^{or} introduced in the above proof, then $\pi^*\alpha = \tau^*\pi^*\alpha$ for any $\alpha \in \Omega^k(M)$.

Corollary 9.56. *If M is compact, then its cohomology spaces $H^k(M)$ are finite-dimensional.*

Proof. Any manifold can be given a Riemannian metric, and so if M is compact and oriented, then Theorems 9.53 and 9.55 combine to give the result. Now suppose that M is not orientable. Let $\alpha \in \Omega^k(M)$ and suppose that $\pi^*\alpha = d\beta$ so that $\pi^*[\alpha] = 0$. Then $\pi^*\alpha = \tau^*\pi^*\alpha = \tau^*d\beta = d\tau^*\beta$. Now $\pi^*\theta = \frac{1}{2}(\beta + \tau^*\beta)$ for some θ and so

$$\pi^*\alpha = d\left(\frac{1}{2}(\beta + \tau^*\beta)\right) = d\pi^*\theta = \pi^*d\theta.$$

Since π is a local diffeomorphism, we have $\alpha = d\theta$ or $[\alpha] = 0$. Thus the map $\pi^* : H^k(M) \to H^k(M^{\text{or}})$ has trivial kernel and is injective. The finite-dimensionality of $H^k(M)$ now follows from that of $H^k(M^{\text{or}})$. $\square$

The results above allow us to give a quick proof of Poincaré duality for de Rham cohomology. Choose an orientation for M. We define a bilinear pairing $H^k(M) \times H^{n-k}(M) \to \mathbb{R}$ as follows:

$$(([\omega], [\eta])) = \int_M \omega \wedge \eta.$$

This is well-defined since if $\omega_1 = \omega + d\alpha$ and $\eta_1 = \eta + d\beta$ (with ω, η closed), then by Stokes' theorem

$$\int_M \omega_1 \wedge \eta_1 = \int_M \omega \wedge \eta + \int_M d\alpha \wedge \eta + \int_M \omega \wedge d\beta + \int_M d\alpha \wedge d\beta$$
$$= \int_M \omega \wedge \eta + \int_M d(\alpha \wedge \eta) - \int_M d(\omega \wedge \beta) + \int_M d(\alpha \wedge d\beta)$$
$$= \int_M \omega \wedge \eta.$$

We wish to show that the pairing defined above is nondegenerate. Thus, given any nonzero $[\omega] \in H^k(M)$ we wish to produce a $[\eta] \in H^{n-k}(M)$ such that $(([\omega], [\eta])) \neq 0$. Choose a Riemannian metric and metric volume element for M. We may assume that ω is the harmonic representative for $[\omega]$. But it is easily checked that Δ commutes with $*$ and so $*\omega$ is also harmonic. In particular, $*\omega$ is closed and so represents a cohomology class $[*\omega]$. Then $[*\omega]$ is the desired class since

$$(([\omega], [*\omega])) = \int_M \omega \wedge *\omega = \int_M \langle \omega | *\omega \rangle \, \text{vol} = \|\omega\|^2 > 0.$$

For each fixed $[\omega] \in H^k(M)$, we have the linear map $\ell_{[\omega]} \in \left(H^{n-k}(M)\right)^*$ given by $\ell_{[\omega]}([\eta]) := (([\omega], [\eta]))$. Since the pairing is nondegenerate, the map $[\omega] \mapsto \ell_{[\omega]}$ defines an isomorphism from $H^k(M)$ to $\left(H^{n-k}(M)\right)^*$. Thus we have proved the following:

Theorem 9.57 (Poincaré duality). *If M is an orientable compact n-manifold without boundary, then we have an isomorphism*

$$H^k(M) \cong \left(H^{n-k}(M)\right)^*$$

coming from the pairing defined above.

We will take up Poincaré duality again in Chapter 10.

9.7. Vector Analysis on $\mathbb{R}^3$

In $\mathbb{R}^3$, the 1-forms may all be written (even globally) in the form $\theta = f_1 dx + f_2 dy + f_3 dz$ for some smooth functions f_1, f_2 and f_3 and all 2-forms β may be written $\beta = g_1 dy \wedge dz + g_2 dz \wedge dx + g_3 dx \wedge dy$. The forms $dy \wedge dz, dz \wedge dx, dx \wedge dy$ constitute a basis (in the module sense) for the space of 2-forms on $\mathbb{R}^3$ just as dx, dy, dz form a basis for the 1-forms. The single form $dx \wedge dy \wedge dz$ provides a module basis for the 3-forms in $\mathbb{R}^3$. Suppose that $\mathbf{x}(u, v)$ parametrizes a surface $S \subset \mathbb{R}^3$ so that we have a map $\mathbf{x} : U \to \mathbb{R}^3$. Then the surface is oriented by this parametrization, and the integral of β over S is

$$\int_S \beta = \int_S g_1 dy \wedge dz + g_2 dz \wedge dx + g_3 dx \wedge dy$$

$$= \int_U \Big(g_1(\mathbf{x}(u,v)) \frac{\partial(y,z)}{\partial(u,v)} + g_2(\mathbf{x}(u,v)) \frac{\partial(z,x)}{\partial(u,v)}$$

$$+ g_3(\mathbf{x}(u,v)) \frac{\partial(x,y)}{\partial(u,v)}\Big) du\, dv.$$

Here and in the following we disregard technical issues about integration (but recall Theorem 9.7).

Exercise 9.58. Find the integral of $\beta = x\, dy \wedge dz + dz \wedge dx + xz\, dx \wedge dy$ over the sphere oriented by the parametrization given by the usual spherical coordinates ϕ, θ, ρ.

If $\omega = h\, dx \wedge dy \wedge dz$ has support in a bounded open subset $U \subset \mathbb{R}^3$ which we may take to be given the usual orientation implied by the rectangular coordinates x, y, z, then

$$\int_U \omega = \int_U h\, dx \wedge dy \wedge dz = \int_U h\, dx\, dy\, dz.$$

In order to relate differential forms on $\mathbb{R}^3$ to vector calculus on $\mathbb{R}^3$, we will need some ways to relate forms to vector fields. To a 1-form $\theta = f_1 dx + f_2 dy + f_3 dz$, we can obviously associate the vector field $\sharp\theta = f_1 \mathbf{i} + f_2 \mathbf{j} + f_3 \mathbf{k}$. But recall that this association depends on the notion of orthonormality provided by the dot product. If θ is expressed in say spherical coordinates $\theta = \widetilde{f_1} d\rho + \widetilde{f_2} d\theta + \widetilde{f_3} d\phi$, then it is **not** true that $\sharp\theta = \widetilde{f_1}\mathbf{i} + \widetilde{f_2}\mathbf{j} + \widetilde{f_3}\mathbf{k}$. Neither is it generally true that $\sharp\theta = \widetilde{f_1}\widehat{\rho} + \widetilde{f_2}\widehat{\theta} + \widetilde{f_3}\widehat{\phi}$, where $\widehat{\rho}, \widehat{\theta}, \widehat{\phi}$ are unit vector fields in the coordinate directions[1] and where the $\widetilde{f_i}$ are just the f_i expressed in polar coordinates. Rather, our general formulas give

$$\sharp\theta = \widetilde{f_1}\widehat{\rho} + \widetilde{f_2}\frac{1}{\rho}\widehat{\theta} + \widetilde{f_3}\frac{1}{\rho\sin\theta}\widehat{\phi}.$$

In rectangular coordinates x, y, z, we have

$$\sharp: \quad dx \;\mapsto\; \mathbf{i},$$
$$\sharp: \quad dy \;\mapsto\; \mathbf{j},$$
$$\sharp: \quad dz \;\mapsto\; \mathbf{k},$$

while in spherical coordinates we have

$$\sharp: \quad d\rho \;\mapsto\; \widehat{\rho},$$
$$\sharp: \quad \rho\, d\theta \;\mapsto\; \widehat{\theta},$$
$$\sharp: \quad \rho\sin\theta\, d\phi \;\mapsto\; \widehat{\phi}.$$

---

[1] Here θ is the polar angle ranging from 0 to π.

9.7. Vector Analysis on $\mathbb{R}^3$

As an example, we can derive the familiar formula for the gradient in spherical coordinates by first just writing f in the new coordinates $f(\rho, \theta, \phi) := f(x(\rho, \theta, \phi), y(\rho, \theta, \phi), z(\rho, \theta, \phi))$ and then sharpening the differential

$$df = \frac{\partial f}{\partial \rho}d\rho + \frac{\partial f}{\partial \theta}d\theta + \frac{\partial f}{\partial \phi}d\phi$$

to get

$$\operatorname{grad} f = \sharp df = \frac{\partial f}{\partial \rho}\frac{\partial}{\partial \rho} + \frac{1}{\rho}\frac{\partial f}{\partial \theta}\frac{\partial}{\partial \theta} + \frac{1}{\rho \sin \theta}\frac{\partial f}{\partial \phi}\frac{\partial}{\partial \phi},$$

where we have used

$$(g^{ij}) = \begin{bmatrix} 1 & 0 & 0 \\ 0 & \rho & 0 \\ 0 & 0 & \rho \sin \theta \end{bmatrix}^{-1} = \begin{bmatrix} 1 & 0 & 0 \\ 0 & \frac{1}{\rho} & 0 \\ 0 & 0 & \frac{1}{\rho \sin \theta} \end{bmatrix}.$$

In order to proceed to the point of including the curl and divergence of traditional vector calculus, we need a way to relate 2-forms with vector fields. This part definitely depends on the fact that we are talking about forms in $\mathbb{R}^3$. We associate to a 2-form η the vector fields $\sharp(*\eta)$. Thus $g_1 dy \wedge dz + g_2 dz \wedge dx + g_3 dx \wedge dy$ gives the vector field $X = g_1\mathbf{i} + g_2\mathbf{j} + g_3\mathbf{k}$.

Now we can see how the usual divergence of a vector field comes about. First flat the vector field, say $X = f_1\mathbf{i} + f_2\mathbf{j} + f_3\mathbf{k}$, to obtain $\flat X = f_1 dx + f_2 dy + f_3 dz$ and then apply the star operator to obtain $f_1 dy \wedge dz + f_2 dz \wedge dx + f_3 dx \wedge dy$. Finally, we apply exterior differentiation to obtain

$$\begin{aligned}
&d(f_1 dy \wedge dz + f_2 dz \wedge dx + f_3 dx \wedge dy) \\
&= df_1 \wedge dy \wedge dz + df_2 \wedge dz \wedge dx + df_3 \wedge dx \wedge dy \\
&= \left(\frac{\partial f_1}{\partial x}dx + \frac{\partial f_1}{\partial y}dy + \frac{\partial f_1}{\partial z}dz\right) \wedge dy \wedge dz + \text{the other two terms} \\
&= \frac{\partial f_1}{\partial x}dx \wedge dy \wedge dz + \frac{\partial f_2}{\partial x}dx \wedge dy \wedge dz + \frac{\partial f_3}{\partial x}dx \wedge dy \wedge dz \\
&= \left(\frac{\partial f_1}{\partial x} + \frac{\partial f_2}{\partial x} + \frac{\partial f_3}{\partial x}\right)dx \wedge dy \wedge dz.
\end{aligned}$$

Now we see the divergence appearing. In fact, if we apply the star operator one more time, we get the function $\operatorname{div} X = \frac{\partial f_1}{\partial x} + \frac{\partial f_2}{\partial x} + \frac{\partial f_3}{\partial x}$. We are thus led to $*d*(\flat X) = \operatorname{div} X$ which agrees with the definition of divergence given earlier for a general semi-Riemannian manifold.

What about the curl? For this, we just take $d\,(\flat X)$ to get

$$d\,(f_1 dx + f_2 dy + f_3 dz) = df_1 \wedge dx + df_2 \wedge dy + df_3 \wedge dz$$

$$= \left(\frac{\partial f_1}{\partial x} dx + \frac{\partial f_2}{\partial y} dy + \frac{\partial f_3}{\partial z} dz\right) \wedge dx + \text{the obvious other two terms}$$

$$= \left(\frac{\partial f_2}{\partial y} - \frac{\partial f_3}{\partial z}\right) dy \wedge dz + \left(\frac{\partial f_3}{\partial z} - \frac{\partial f_1}{\partial x}\right) dz \wedge dx + \left(\frac{\partial f_1}{\partial x} - \frac{\partial f_2}{\partial y}\right) dx \wedge dy$$

and then apply the star operator and sharping to get back to vector fields obtaining

$$\left(\frac{\partial f_2}{\partial y} - \frac{\partial f_3}{\partial z}\right)\mathbf{i} + \left(\frac{\partial f_3}{\partial z} - \frac{\partial f_1}{\partial x}\right)\mathbf{j} + \left(\frac{\partial f_1}{\partial x} - \frac{\partial f_2}{\partial y}\right)\mathbf{k} = \operatorname{curl} X.$$

In short, we have

$$\sharp * d\,(\flat X) = \operatorname{curl} X.$$

Exercise 9.59. Show that the fact that $dd = 0$ leads to both of the following familiar facts:

$$\operatorname{curl}(\operatorname{grad} f) = 0,$$
$$\operatorname{div}(\operatorname{curl} X) = 0.$$

The 3-form $dx \wedge dy \wedge dz$ is the (oriented) volume element of $\mathbb{R}^3$. Every 3-form is a function times this volume form, and integration of a 3-form over a sufficiently nice subset (say a compact region) is given by $\int_D \omega = \int_D f\,dx \wedge dy \wedge dz = \int_D f\,dx\,dy\,dz$ (usual Riemann integral). Let us denote $dx \wedge dy \wedge dz$ by dV. Of course, dV is *not* to be considered as the exterior derivative of some object V. Let u^1, u^2, u^3 be curvilinear coordinates on an open set $U \subset \mathbb{R}^3$ and let g denote the determinant of the matrix $[g_{ij}]$ where $g_{ij} = \langle \frac{\partial}{\partial u^i}, \frac{\partial}{\partial u^j}\rangle$. Then $dV = \sqrt{g}\,du^1 \wedge du^2 \wedge du^3$. A familiar example is the case when (u^1, u^2, u^3) are spherical coordinates ρ, θ, ϕ, in which case

$$dV = \rho^2 \sin\theta\,d\rho \wedge d\theta \wedge d\phi,$$

and if $D \subset \mathbb{R}^3$ is parametrized by these coordinates, then

$$\int_D f\,dV = \int_D f(\rho, \theta, \phi)\rho^2 \sin\theta\,d\rho \wedge d\theta \wedge d\phi$$
$$= \int_D f(\rho, \theta, \phi)\rho^2 \sin\theta\,d\rho\,d\theta\,d\phi.$$

If we go to the trouble of writing Stokes' theorem for curves, surfaces and domains in $\mathbb{R}^3$ in terms of vector fields associated to the forms in an appropriate way, we obtain the following familiar theorems (using standard

notation):
$$\int_c \nabla f \cdot d\mathbf{r} = f(\mathbf{r}(b)) - f(\mathbf{r}(a)),$$
$$\iint_S \text{curl}(\mathbf{X}) \times d\mathbf{S} = \oint_c X \cdot dr \text{ (Stokes' theorem)},$$
$$\iiint_D \text{div}(\mathbf{X})\, dV = \iint_S \mathbf{X} \cdot d\mathbf{S} \text{ (Divergence theorem)}.$$

Similar and simpler things can be done in $\mathbb{R}^2$ leading for example to the following version of Green's theorem for a planar domain D with (oriented) boundary $c = \partial D$:
$$\int_D \left(\frac{\partial M}{\partial y} - \frac{\partial M}{\partial y}\right) dx \wedge dy = \int_D d(M\, dx + N\, dy) = \int_c M\, dx + N\, dy.$$

All of the standard integral theorems from vector calculus are special cases of the general Stokes' theorem.

9.8. Electromagnetism

In this subsection we take a short trip into physics. Consider Maxwell's equations[2]:
$$\nabla \cdot \mathbf{B} = 0,$$
$$\nabla \times \mathbf{E} + \frac{\partial \mathbf{B}}{\partial t} = 0,$$
$$\nabla \cdot \mathbf{E} = \varrho,$$
$$\nabla \times \mathbf{B} - \frac{\partial \mathbf{E}}{\partial t} = \mathbf{j}.$$

Here $\mathbf{E}$ and $\mathbf{B}$, the electric and magnetic fields, are functions of space and time. We write $\mathbf{E} = \mathbf{E}(t, \mathbf{x})$, $\mathbf{B} = \mathbf{B}(t, \mathbf{x})$. The notation suggests that we have conceptually separated space and time as if we were stuck in the conceptual framework of the Galilean spacetime. Our purpose is to slowly discover how much better the theory becomes when we combine space and time in Minkowski spacetime $\mathbb{R}_1^4$. Recall that $\mathbb{R}_1^4$ is treated as a semi-Riemannian manifold, which is $\mathbb{R}^4$ endowed with the indefinite metric $\langle x, y \rangle_\nu = -x^0 y^0 + \sum_{i=1}^3 x^i y^i$. Here the standard coordinates are conventionally denoted by (x^0, x^1, x^2, x^3), and x^0 is to be thought of as a time coordinate and is also denoted by t (we take units so that the speed of light c is unity). In what follows, we let $\mathbf{r} = (x^1, x^2, x^3)$, so that $(x^0, x^1, x^2, x^3) = (t, \mathbf{r})$.

---

[2] Actually, this is the form of Maxwell's equations after a certain convenient choice of units, and we are ignoring the somewhat subtle distinction between the two types of electric fields E and D and the two types of magnetic fields B and H and their relation in terms of dielectric constants.

The electric field **E** is produced by the presence of charged particles. Under normal conditions a generic material is composed of a large number of atoms. To simplify matters, we will think of the atoms as being composed of just three types of particle; electrons, protons and neutrons. Protons carry a positive charge, electrons carry a negative charge and neutrons carry no charge. Normally, each atom will have a zero net charge since it will have an equal number of electrons and protons. If a relatively small percent of the electrons in a material body are stripped from their atoms and conducted away, then there will be a net positive charge on the body. In the vicinity of the body, there will be an electric field which exerts a force on charged bodies. Let us assume for simplicity that the charged body which has the larger, positive charge, is a point particle and stationary at $\mathbf{r}_0$ with respect to a rigid rectangular coordinate system that is stationary with respect to the laboratory. We must assume that our test particle carries a sufficiently small charge, so that the electric field that it creates contributes negligibly to the field we are trying to detect (think of a single electron). Let the test particle be located at $\mathbf{r}$. Careful experiments show that when both charges are positive, the force experienced by the test particle is directly away from the charged body located at $\mathbf{r}_0$ and has magnitude proportional to qe/r^2, where $r = |\mathbf{r} - \mathbf{r}_0|$ is the distance between the charged body and the test particle, and where q and e are positive numbers which represent the amount of charge carried by the stationary body and the test particle respectively. If the units are chosen in an appropriate way, we can say that the force F is given by

$$F = qe \frac{\mathbf{r} - \mathbf{r}_0}{|\mathbf{r} - \mathbf{r}_0|^3}.$$

By definition, the electric field at the location $\mathbf{r}$ of the test particle is

(9.12) $$\mathbf{E} = q \frac{\mathbf{r} - \mathbf{r}_0}{|\mathbf{r} - \mathbf{r}_0|^3}.$$

If the test particle has charge opposite to that of the source body, then one of q or e will be negative and the force is directed toward the source. The test particle could have been placed anywhere in space, and so the electric field is implicitly defined at each point in space and so gives a vector field on $\mathbb{R}^3$. If the charge is modeled as a smoothly distributed charge density ρ which is nonzero in some region $U \subset \mathbb{R}^3$, then the total charge is given by integration $Q = \int_U \rho(t, \mathbf{r}) \, dV_{\mathbf{r}}$ and the field at $\mathbf{r}$ is now given by $\mathbf{E}(t, \mathbf{r}) = \int_U \rho(t, \mathbf{y}) \frac{\mathbf{r} - \mathbf{y}}{|\mathbf{r} - \mathbf{y}|^3} \, dV_{\mathbf{y}}$. Since the source particle is stationary at $\mathbf{r}_0$, the electric field will be independent of time t. A magnetic field is produced by circulating charge (a current). If charge e is located at $\mathbf{r}$ and moving with velocity $\mathbf{v}$ in a magnetic field $\mathbf{B}(t, \mathbf{r})$, then the force felt by the charge is $\mathbf{F} = e\mathbf{E} + \frac{e}{c}\mathbf{v} \times \mathbf{B}$, where $\mathbf{v}$ is the velocity of the test particle. The test

9.8. Electromagnetism

particle has to be moving to feel the magnetic part of the field! At this point it is worth pointing out that from the point of view of spacetime, we are not staying true to the spirit of differential geometry since a vector field should have a geometric reality that is independent of its expression in a coordinate system. But a change of inertial frame can make **B** zero. Only by treating **E** and **B** together as aspects of a single field can we obtain the proper view.

Our next task is to write Maxwell's equations in terms of differential forms. We already have a way to convert (time dependent) vector fields **E** and **B** on $\mathbb{R}^3$ into (time dependent) differential forms on $\mathbb{R}^3$. Namely, we use the flatting operation with respect to the standard metric on $\mathbb{R}^3$. For the electric field we have

$$\mathbf{E} = E^x \partial_x + E^y \partial_y + E^z \partial_z \mapsto \mathcal{E} = E_x\, dx + E_y\, dy + E_z\, dz.$$

For the magnetic field we do something a bit different. Namely, we flat and then apply the star operator. In rectangular coordinates, we have

$$\mathbf{B} = B^x \partial_x + B^y \partial_y + B^z \partial_z \mapsto \mathcal{B} = B_x\, dy \wedge dz + B_y\, dz \wedge dx + B_z\, dx \wedge dy.$$

If we stick to rectangular coordinates (as we have been), the matrix of the standard metric is just $I = (\delta_{ij})$, and so we see that the above operations do not numerically change the components of the fields. Thus in any rectangular coordinate system we have

$$E^x = E_x, \quad E^y = E_y, \quad E_z = E^z$$

and similarly for the B's. It is not hard to check that in the static case where $\mathcal{E}$ and $\mathcal{B}$ are time independent, the first pair of (static) Maxwell's equations are equivalent to

$$d\mathcal{E} = 0 \text{ and } d\mathcal{B} = 0.$$

This is nice, but if we put time dependence back into the picture, we need to do a couple more things to get a nice viewpoint. So assume now that **E** and **B** and hence the forms $\mathcal{E}$ and $\mathcal{B}$ are time dependent, and let us view these as differential forms on spacetime $\mathbb{R}_1^4$. In fact, let us combine $\mathcal{E}$ and $\mathcal{B}$ into a single 2-form on $\mathbb{R}_1^4$ by setting

$$\mathcal{F} = \mathcal{B} + \mathcal{E} \wedge dt.$$

Since $\mathcal{F}$ is a 2-form, it can be written in the form $\mathcal{F} = \frac{1}{2} F_{\mu\nu} dx^\mu \wedge dx^\nu$, where $F_{\mu\nu} = -F_{\nu\mu}$ and where the Greek indices are summed over $\{0, 1, 2, 3\}$. It is traditional in physics to let the Greek indices run over this set and to let Latin indices run over just the "space indices" $1, 2, 3$. We will follow this convention for a while. If we compare $\mathcal{F} = \mathcal{B} + \mathcal{E} \wedge dt$ with $\frac{1}{2} F_{\mu\nu} dx^\mu \wedge dx^\nu$, we see that the $F_{\mu\nu}$ form an antisymmetric matrix which is none other than

$$\begin{bmatrix} 0 & -E_x & -E_y & -E_z \\ E_x & 0 & B_z & -B_y \\ E_y & -B_z & 0 & B_x \\ E_z & B_y & -B_x & 0 \end{bmatrix}.$$

Our goal now is to show that the first pair of Maxwell's equations are equivalent to the single differential form equation

$$d\mathcal{F} = 0.$$

Let N be an n-manifold and let $M = (a,b) \times N$ for some interval (a,b). Let the coordinate on (a,b) be $t = x^0$ (time). Let $(x^1, \ldots, x^n)$ be a coordinate system on N. With the usual abuse of notation, $(x^0, x^1, \ldots, x^n)$ is a coordinate system on $(a,b) \times N$. One can easily show that the local expression $d\omega = \partial_\mu f_{\mu_1 \ldots \mu_k} \wedge dx^\mu \wedge dx^{\mu_1} \wedge \cdots \wedge dx^{\mu_k}$ for the exterior derivative of a form $\omega = f_{\mu_1 \ldots \mu_k} dx^{\mu_1} \wedge \cdots \wedge dx^{\mu_k}$ can be written as

$$(9.13) \quad d\omega = \sum_{i=1}^{3} \partial_i \omega_{\mu_1 \ldots \mu_k} \wedge dx^i \wedge dx^{\mu_1} \wedge \cdots \wedge dx^{\mu_k}$$
$$+ \partial_0 \omega_{\mu_1 \ldots \mu_k} \wedge dx^0 \wedge dx^{\mu_1} \wedge \cdots \wedge dx^{\mu_k},$$

where the μ_i sum over $\{0, 1, 2, \ldots, n\}$. Thus we may consider the spatial part d_S of the exterior derivative operator d on $(a,b) \times S = M$. That is, we think of a given form ω on $(a,b) \times S$ as a time dependent form on N so that $d_S \omega$ is exactly the first term in the expression (9.13) above. Then we may write $d\omega = d_S \omega + dt \wedge \partial_t \omega$ as a compact version of the expression (9.13). The part $d_S \omega$ contains no dt's. By definition $\mathcal{F} = \mathcal{B} + \mathcal{E} \wedge dt$ on $\mathbb{R} \times \mathbb{R}^3 = \mathbb{R}^4_1$, and so

$$d\mathcal{F} = d\mathcal{B} + d(\mathcal{E} \wedge dt)$$
$$= d_S \mathcal{B} + dt \wedge \partial_t \mathcal{B} + (d_S \mathcal{E} + dt \wedge \partial_t \mathcal{E}) \wedge dt$$
$$= d_S \mathcal{B} + (\partial_t \mathcal{B} + d_S \mathcal{E}) \wedge dt.$$

The part $d_S \mathcal{B}$ is the spatial part and contains no dt's. It follows that $d\mathcal{F}$ is zero if and only if both $d_S \mathcal{B}$ and $\partial_t \mathcal{B} + d_S \mathcal{E}$ are zero. Unraveling the definitions shows that the pair of equations $d_S \mathcal{B} = 0$ and $\partial_t \mathcal{B} + d_S \mathcal{E} = 0$ (which we just showed to be equivalent to $d\mathcal{F} = 0$) are Maxwell's first two equations disguised in a new notation. In summary, we have

$$d\mathcal{F} = 0 \quad \Longleftrightarrow \quad \begin{array}{c} d_S \mathcal{B} = 0 \\ \partial_t \mathcal{B} + d_S \mathcal{E} = 0 \end{array} \quad \Longleftrightarrow \quad \begin{array}{c} \nabla \cdot \mathbf{B} = 0 \\ \nabla \times \mathbf{E} + \frac{\partial \mathbf{B}}{\partial t} = 0 \end{array}$$

Below we rewrite the last pair of Maxwell's equations, where the advantage of combining time and space together manifests itself to an even greater degree. Let us first pause to notice an interesting aspect of the

9.8. Electromagnetism

first pair. Suppose that the electric and magnetic fields were really all along most properly thought of as differential forms. Then we see that the equation $d\mathcal{F} = 0$ has nothing to do with the metric on Minkowski space at all. In fact, if $\phi : \mathbb{R}_1^4 \to \mathbb{R}_1^4$ is any diffeomorphism at all, we have $d\mathcal{F} = 0$ if and only if $d(\phi^*\mathcal{F}) = 0$, and so the truth of the equation $d\mathcal{F} = 0$ is really a differential topological fact; a certain form $\mathcal{F}$ is closed. The metric structure of Minkowski space is irrelevant. The same will not be true for the second pair. Even if we start out with the form $\mathcal{F}$ on spacetime it will turn out that the metric will necessarily be implicit in the differential forms version of the second pair of Maxwell's equations. In fact, what we will show is that if we use the star operator for the Minkowski metric, then the second pair can be rewritten as the single equation $*d*\mathcal{F} = \mathcal{J}$, where $\mathcal{J}$ is formed from $\mathbf{j} = (j^1, j^2, j^3)$ and ρ as follows: First we form the 4-vector field $J = \rho\partial_t + j^1\partial_x + j^2\partial_y + j^3\partial_z$ (called the 4-current) and then using the flatting operation we produce $\mathcal{J} = -\rho\, dt + j^1\, dx + j^2\, dy + j^3\, dz = J_0\, dt + J_1\, dx + J_2\, dy + J_3\, dz$, which is the covariant form of the 4-current.

We will only outline the passage from $*d*\mathcal{F} = \mathcal{J}$ to the pair $\nabla \cdot \mathbf{E} = \varrho$ and $\nabla \times \mathbf{B} - \frac{\partial \mathbf{E}}{\partial t} = \mathbf{j}$. Let $*_S$ be the operator one gets by viewing differential forms on $\mathbb{R}_1^4$ as time dependent forms on $\mathbb{R}^3$ and then acting by the star operator with respect to the standard metric on $\mathbb{R}^3$. The first step is to verify that the pair $\nabla \cdot \mathbf{E} = \varrho$ and $\nabla \times \mathbf{B} - \frac{\partial \mathbf{E}}{\partial t} = \mathbf{j}$ is equivalent to the pair $*_S d_S *_S \mathcal{E} = \varrho$ and $-\partial_t \mathcal{E} + *_S d_S *_S \mathcal{B} = \jmath$, where $\jmath := j^1 dx + j^2 dy + j^3 dz$ and $\mathcal{B}$ and $\mathcal{E}$ are as before. Next we verify that

$$*\mathcal{F} = *_S \mathcal{E} - *_S \mathcal{B} \wedge dt.$$

So the next goal is to get from $*d*\mathcal{F} = *\mathcal{J}$ to the pair $*_S d_S *_S \mathcal{E} = \varrho$ and $-\partial_t \mathcal{E} + *_S d_S *_S \mathcal{B} = \jmath$. The following exercise finishes things off.

Exercise 9.60. Show that $*d*\mathcal{F} = -\partial_t \mathcal{E} - *_S d_S *_S \mathcal{E} \wedge dt + *_S d_S *_S \mathcal{B}$ and then use this and what we did above to show that $*d*\mathcal{F} = \mathcal{J}$ is equivalent to the pair $*_S d_S *_S \mathcal{E} = \varrho$ and $-\partial_t \mathcal{E} + *_S d_S *_S \mathcal{B} = \jmath$.

We have arrived at the following formulation of Maxwell's equations:

$$d\mathcal{F} = 0,$$
$$*d*\mathcal{F} = \mathcal{J}.$$

If we just think of this as a pair of equations to be satisfied by a 2-form $\mathcal{F}$ where the 1-form $\mathcal{J}$ is given, then this last version of Maxwell's equations makes sense on any semi-Riemannian manifold. In fact, on a Lorentz manifold that can be written as $(a, b) \times S = M$ with the product metric $-dt \otimes dt \times g$, for some Riemannian metric g on S, we can write $\mathcal{F} = \mathcal{B} + \mathcal{E} \wedge dt$, which allows us to identify the electric and magnetic fields in their covariant form.

9.9. Surface Theory Redux

Consider a surface $M \subset \mathbb{R}^3$. In what follows, we take advantage of the natural identifications of the tangent spaces of $\mathbb{R}^3$ with $\mathbb{R}^3$ itself. Let e_1, e_2 be an oriented orthonormal frame field defined on some open subset U of M and let $e_3 = e_1 \times e_2$. We can think of each e_1, e_2 and e_3 as vector fields *along* U. Using the identifications mentioned above, we also think of them as $\mathbb{R}^3$-valued 0-forms.

Note that the identity map $\mathrm{id}_{\mathbb{R}^3}$ may be considered as an $\mathbb{R}^3$-valued 0-form on $\mathbb{R}^3$. Let $I := \iota^*(\mathrm{id}_{\mathbb{R}^3})$ where $\iota : U \subset M \hookrightarrow \mathbb{R}^3$ is inclusion. Then $dI := \iota^* d(\mathrm{id}_{\mathbb{R}^3}) = 0$ since $\mathrm{id}_{\mathbb{R}^3}$ is constant. Thus, I is an $\mathbb{R}^3$-valued 0-form, and we may write
$$I = e_1 \theta^1 + e_2 \theta^2,$$
where (θ^1, θ^2) is the frame field dual to (e_1, e_2). Note that $\mathrm{vol}_M = \theta^1 \wedge \theta^2$. We have the $\mathbb{R}^3$-valued 1-forms de_j and
$$de_j = \sum_{k=1}^{3} e_k \omega_j^k$$
for some matrix of 1-forms (ω_j^i). Note that if $v \in T_p M$, then $de_j(v) = \overline{\nabla}_v e_j$, where $\overline{\nabla}$ is the flat Levi-Civita derivative on $\mathbb{R}^3$. (In fact, we should point out that if $\mathbf{i}, \mathbf{j}, \mathbf{k}$ is the standard basis on $\mathbb{R}^3$, then any field X along U may be written as $X = f_1 \mathbf{i} + f_2 \mathbf{j} + f_3 \mathbf{k}$ for some smooth functions f_1, f_2, f_3 defined on U and may be considered as an $\mathbb{R}^3$-valued 1-form. Then $dX(v) = \overline{\nabla}_v X = df_1(v)\mathbf{i} + df_2(v)\mathbf{j} + df_3(v)\mathbf{k}$.) For an arbitrary tangent vector v we have
$$\omega_j^i(v) = \langle de_j(v), e_i \rangle = -\langle e_j, de_i(v) \rangle = -\omega_i^j(v),$$
and so it follows that the matrix (ω_j^i) is antisymmetric. We write $\mathbb{R}^3$-valued forms on U as $\sum_{k=1}^{3} e_k \eta^k$ where $\eta^k \in \Omega(U)$. This is to conform with the order of matrix multiplication.

Theorem 9.61. *Let $M \subset \mathbb{R}^3$, e_1, e_2, θ^1, θ^2 and (ω_j^i) be as above. Then the following structure equations hold:*
$$d\theta^i = -\sum_{k=1}^{2} \omega_k^i \wedge \theta^k,$$
$$\omega_1^3 \wedge \theta^1 + \omega_2^3 \wedge \theta^2 = 0,$$
$$d\omega_j^i = -\sum_{k=1}^{3} \omega_k^i \wedge \omega_j^k.$$

9.9. Surface Theory Redux

Proof. We calculate as follows:

$$0 = dI = d\sum_{k=1}^{2} e_k \theta^k = \sum_{k=1}^{2} de_k \wedge \theta^k + e_k \wedge d\theta^k$$

$$= \sum_{k=1}^{2}\left(\sum_{j=1}^{3} \omega_k^j e_j\right) \wedge \theta^k + \sum_{k=1}^{2} e_k \wedge d\theta^k$$

$$= \sum_{k=1}^{2}\left(\sum_{j=1}^{2} e_j \omega_k^j + e_3 \omega_k^3\right) \wedge \theta^k + \sum_{j=1}^{2} e_k \wedge d\theta^k$$

$$= \sum_{j=1}^{2}\sum_{k=1}^{2} e_j \omega_k^j \wedge \theta^k + \sum_{j=1}^{2} e_j \wedge d\theta^j + \sum_{k=1}^{2} e_3 \omega_k^3 \wedge \theta^k$$

$$= \sum_{j=1}^{2} e_j \left(d\theta^j + \sum_{k=1}^{2} \omega_k^j \wedge \theta^k\right) + e_3 \left(\sum_{k=1}^{2} \omega_k^3 \wedge \theta^k\right),$$

and it follows that $d\theta^i = -\sum_{k=1}^{2} \omega_k^i \wedge \theta^k$ and $\omega_1^3 \wedge \theta^1 + \omega_2^3 \wedge \theta^2 = 0$.

Also,

$$0 = dde_j = d\sum_{k=1}^{3} e_k \omega_j^k$$

$$= \sum_{k=1}^{3} de_k \wedge \omega_j^k + \sum_{k=1}^{3} e_k \wedge d\omega_j^k$$

$$= \sum_{k=1}^{3}\left(\sum_{i=1}^{3} e_i \omega_k^i\right) \wedge \omega_j^k + \sum_{i=1}^{3} e_i \wedge d\omega_j^i$$

$$= \sum_{i=1}^{3} e_i \left(\sum_{k=1}^{3} \omega_k^i \wedge \omega_j^k + d\omega_j^i\right),$$

and so $d\omega_j^i = -\sum_{k=1}^{3} \omega_k^i \wedge \omega_j^k$. $\square$

Notice that for $v \in T_pM$ we have $de_3(v) = \overline{\nabla}_v e_3 = -S(v)$ and so $II(v,w) = \langle -de_3(v), w \rangle$ (recall Definition 4.20). Therefore,

$$II(e_i, e_j) = \langle -de_3(e_i), e_j \rangle = \left\langle -\sum_{k=1}^{3} e_k \omega_3^k(e_i), e_j \right\rangle = -\omega_3^j(e_i)$$

and so

$$\begin{bmatrix} II(e_1, e_1) & II(e_1, e_2) \\ II(e_2, e_1,) & II(e_2, e_2) \end{bmatrix} = -\begin{bmatrix} \omega_3^1(e_1) & \omega_3^2(e_1) \\ \omega_3^1(e_2) & \omega_3^2(e_2) \end{bmatrix}.$$

Since (e_1, e_2) is an orthonormal frame field, the matrix of the first fundamental form is the identity matrix. It follows that the matrix which represents the shape operator is the same as that which represents the second fundamental form. Thus

$$K = \det \begin{bmatrix} II(e_1,e_1) & II(e_1,e_2) \\ II(e_2,e_1,) & II(e_2,e_2) \end{bmatrix} = -\det \begin{bmatrix} \omega_3^1(e_1) & \omega_3^2(e_1) \\ \omega_3^1(e_2) & \omega_3^2(e_2) \end{bmatrix}.$$

Notice that $de_3 = \sum_{k=1}^3 e_k \omega_3^k$ reduces to $de_3 = e_1 \omega_3^1 + e_2 \omega_3^2$ since $\omega_3^3 = 0$. We have

$$\omega_3^1 = \omega_3^1(e_1)\theta^1 + \omega_3^1(e_2)\theta^2,$$
$$\omega_3^2 = \omega_3^2(e_1)\theta^1 + \omega_3^2(e_2)\theta^2,$$

so that

$$\omega_3^1 \wedge \omega_3^2 = -K\,\theta^1 \wedge \theta^2,$$
$$\omega_3^1 \wedge \omega_2^3 = K\,\theta^1 \wedge \theta^2,$$
$$d\omega_2^1 = K\,\theta^1 \wedge \theta^2.$$

Recall that for $v, w \in T_x S^2$ we have $\operatorname{vol}_{S^2}(v,w) = \langle v \times w, x \rangle$. If we consider e_3 as a map $e_3 : U \subset M \to S^2$, then it is called the **Gauss map**. If M is assumed oriented, then we may take e_3 to be equal to a global normal field. If $v, w \in T_p M$, then we have

$$e_3^* \operatorname{vol}_{S^2}(v,w) = \langle de_3(v) \times de_3(w), e_3 \rangle$$
$$= \langle (\omega_3^1(v)e_1 + \omega_3^2(v)e_2) \times (\omega_3^1(w)e_1 + \omega_3^2(w)e_2), e_3 \rangle$$
$$= \omega_3^1 \wedge \omega_2^3(v,w) = K\,\theta^1 \wedge \theta^2(v,w) = K \operatorname{vol}_M(v,w).$$

Thus we obtain

$$e_3^* \operatorname{vol}_{S^2} = K \operatorname{vol}_M.$$

This shows that the Gauss curvature is a measure of distortion of the signed volume under the Gauss map. In particular, if $e_3 : U \subset M \to S^2$ is orientation reversing at p, then the curvature at p is negative. Let $p \in M$ with p in the domain U of e_3. If A is a nice domain in U, then e_3 maps A to a set in S^2. Without worrying about measure-theoretic technicalities, we have

$$\operatorname{vol}(e_3(A)) = \int_{e_3(A)} \operatorname{vol}_{S^2}$$
$$= \int_A e_3^* \operatorname{vol}_{S^2} = \int_A K \operatorname{vol}_M.$$

One can get an idea of the curvature near a point on a surface by visualizing the Gauss map (see Figure 9.2). For example, it is clear that the right circular cylinder has Gauss curvature zero since it maps every region of the cylinder onto a set with zero area. It is also fairly clear that the saddle surface in the diagram has negative curvature.

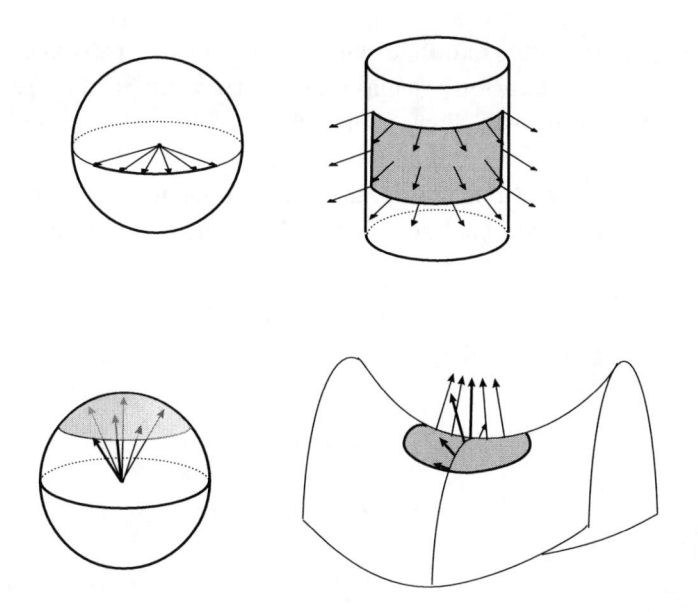

Figure 9.2. Gauss map

Problems

(1) Let $\iota : S^2 \hookrightarrow \mathbb{R}^3\backslash\{0\}$ be the inclusion map. Let $\tau = \iota^*\omega$ where
$$\omega = \frac{x\,dy \wedge dz + y\,dz \wedge dx + z\,dx \wedge dy}{(x^2 + y^2 + z^2)^{3/2}}.$$
Compute $\int_{S^2} \tau$ where S^2 is given the orientation induced by τ itself.

(2) Let M be an oriented smooth compact manifold with boundary ∂M and suppose that ∂M has two connected components N_0 and N_1. Let $\iota_i : N_i \hookrightarrow M$ be the inclusion map for $i = 0, 1$. Suppose that α is a p-form with $\iota_0^*\alpha = 0$ and β an $(n-p-1)$-form with $\iota_1^*\beta = 0$. Prove that in this case
$$\int_M d\alpha \wedge \beta = (-1)^{p+1} \int_M \alpha \wedge d\beta.$$

(3) Consider the set up in the proof of Theorem 9.22 where $\gamma \in \bigwedge^{n-k}$ and $L_\gamma : \bigwedge^k V^* \to \mathbb{R}$ is defined so that $L_\gamma(\alpha)\,\mathrm{vol} = \alpha \wedge \gamma$. Show that $\gamma \mapsto L_\gamma \in (\bigwedge^k V^*)^*$ is linear. Show that if $L_\gamma(\alpha) = 0$ for all $\alpha \in \bigwedge^k V^*$, then $\gamma = 0$. Thus $\gamma \mapsto L_\gamma$ is injective (and hence an isomorphism). See the related Problem 4.

(4) Recall the notion of a manifold with corners as in Problem 21 in Chapter 1. Define orientation on manifolds with corners. Let M be an n-manifold with corners. Show that if CM is the set of corner points of M, then $M\backslash CM$ is a manifold with boundary. Develop integration theory and Stokes' theorem for manifolds with corners. [Hint: If ω is an $(n-1)$-form with compact support in the domain of a chart $(U,\mathbf{x})$, then define
$$\int_{\partial M} \omega := \sum_{i=1}^{n} \int_{F_i} (\mathbf{x}^{-1})^* \omega,$$
where
$$F_i := \{x \in \overline{\mathbb{R}_+^n} : x^i = 0\}$$
is given the induced orientation as a subset of the boundary of $\{x \in \mathbb{R}^n : x^i \geq 0\}$.]

(5) Show that if $f : (M,g) \to (N,h)$ is an isometry or a local isometry and ω is a k-form on N, then $f^*\omega$ is harmonic if and only if ω is harmonic.

(6) Let M be a connected oriented compact Riemannian manifold with Laplace operator Δ. A smooth nonzero function f is called an eigenfunction for Δ with eigenvalue λ if $\Delta f = \lambda f$.
   (a) Show that zero is an eigenvalue and that all other eigenvalues are strictly positive.
   (b) Show that if $\Delta f_1 = \lambda_1 f_1$ and $\Delta f_2 = \lambda_2 f_2$ for $\lambda_1 \neq \lambda_2$, then $(f_1|f_2) = \int_M f_1 f_2 \, \text{vol}_M = 0$.
   (a) Show that if $\wp : \widetilde{M} \to M$ is a smooth covering space of multiplicity m, then for any compactly supported $\omega \in \Omega^n(M)$ we have
$$\int_{\widetilde{M}} \wp^*\omega = m \int_M \omega.$$

(7) Let $M \subset \mathbb{R}^{n+1}$ be an oriented hypersurface. If N is a positively oriented normal field along M with $N = \sum N^i \partial/\partial x^i$, then the following formula gives the volume form corresponding to the induced metric on M:
$$\text{vol}_M := \sum_i (-1)^{i-1} N^i dx^1 \wedge \cdots \wedge \widehat{dx^i} \wedge \cdots \wedge dx^{n+1},$$
where $x^1, \ldots, x^{n+1}$ are the standard coordinates on $\mathbb{R}^{n+1}$ and the dx^i are restricted to M.

(8) Let U be a starshaped open set in $\mathbb{R}^n$ and let $x^1, \ldots, x^n$ be standard coordinates. Given
$$\omega = \sum_{i_1 < \cdots < i_k} \omega_{i_1,\ldots,i_k} dx^{i_1} \wedge \cdots \wedge dx^{i_k},$$

define
$$I\omega := \sum_{i_1<\cdots<i_k} \sum_{\alpha=1}^{k} (-1)^{\alpha-1} \left(\int_0^1 t^{k-1} \omega_{i_1,\ldots,i_k}(tx) dt \right)$$
$$\times x^{i_\alpha} dx^{i_1} \wedge \cdots \wedge \widehat{dx^{i_\alpha}} \wedge \cdots \wedge dx^{i_k},$$
where the caret means omission. Show that if $d\omega = 0$, then $dI\omega = 0$.

(9) Derive the decompositions
$$\Omega^k(M) = d\delta\left(\Omega^k(M)\right) \oplus \delta d\left(\Omega^k(M)\right) \oplus \mathcal{H}^k$$
and
$$\Omega^k(M) = d\left(\Omega^{k-1}(M)\right) \oplus \delta\left(\Omega^{k+1}(M)\right) \oplus \mathcal{H}^k$$
from
$$\Omega^k(M) = \Delta\left(\Omega^k(M)\right) \oplus \mathcal{H}^k.$$

(10) Prove Theorem 9.7.

Chapter 10

De Rham Cohomology

We have already given the definition of de Rham cohomology (8.41) and the de Rham theorem which says that for a compact oriented smooth manifold M, the de Rham cohomology is isomorphic with the singular cohomology. In this chapter we introduce just a few of the most basic tools proper to the subject such as the Mayer-Vietoris sequence. We shall also introduce the compactly supported de Rham cohomology and revisit Poincaré duality. *In this chapter all manifolds will be without boundary.*

For a given n-manifold M, we have the sequence of maps called the **de Rham complex**

$$0 \xrightarrow{d} \Omega^0(M) \xrightarrow{d} \Omega^1(M) \xrightarrow{d} \cdots \xrightarrow{d} \Omega^n(M) \xrightarrow{d} 0,$$

and we have defined the de Rham cohomology groups (actually vector spaces) as the quotients

$$H^k(M) = \frac{Z^k(M)}{B^k(M)},$$

where $Z^k(M) := \mathrm{Ker}(d : \Omega^k(M) \to \Omega^{k+1}(M))$ and $B^k(M) := \mathrm{Im}(d : \Omega^{k-1}(M) \to \Omega^k(M))$. The elements of $Z^k(M)$ are called **closed** k-forms or k-**cocycles**, and the elements of $B^k(M)$ are called **exact** k-forms or k-**coboundaries**. If $f : M \to N$ is smooth and $[\beta] \in H^k(N)$, then since d commutes with pull-back, it is easy to see that $f^*\beta$ is closed because β is closed. Thus we obtain a cohomology class $[f^*\beta]$. Also, $[f^*\beta]$ depends only on the equivalence class of β. Indeed, if $\beta - \beta' = d\eta$, then $f^*\beta - f^*\beta' = df^*\eta$. Thus we may define a linear map $f^* : H^k(N) \to H^k(M)$ by $f^*[\beta] := [f^*\beta]$.

Let us immediately review a simple situation from Section 2.10, which will help the reader better see what the de Rham cohomologies are all about.

Let $M = \mathbb{R}^2 \setminus \{0\}$ and consider the 1-form
$$\vartheta := \frac{xdy - ydx}{x^2 + y^2}.$$
We got this 1-form by taking the exterior derivative of $\theta = \arctan(y/x)$. This function is not defined as a single-valued smooth function on all of $\mathbb{R}^2 \setminus \{0\}$, but ϑ *is* well-defined on *all* of $\mathbb{R}^2 \setminus \{0\}$. One can extend θ to be defined on the plane minus the ray $\{x \leq 0, y = 0\}$ as follows:

(10.1) $\theta = \begin{cases} \arctan(y/x) & \text{if } x > 0, \\ \arccos(x/\sqrt{x^2+y^2}) & \text{if } x \leq 0 \text{ and } y > 0, \\ -\arccos(x/\sqrt{x^2+y^2}) & \text{if } x \leq 0 \text{ and } y < 0. \end{cases}$

See Problem 1. However, this is still not defined on all of $\mathbb{R}^2 \setminus \{0\}$. One may also check that $d\vartheta = 0$ and so ϑ is closed. Using Proposition 2.125 and Example 2.126 of Section 2.10, it is easy to see that we have the following situation:

(1) $\vartheta := \frac{xdy-ydx}{x^2+y^2}$ is a smooth 1-form on $\mathbb{R}^2 \setminus \{0\}$ with $d\vartheta = 0$.

(2) There is no function f defined on (all of) $\mathbb{R}^2 \setminus \{0\}$ such that $\vartheta = df$.

(3) From Section 2.10 we know that for any ball $B(p, \varepsilon)$ in $\mathbb{R}^2 \setminus \{0\}$ there is a function $f \in C^\infty(B(p, \varepsilon))$ such that $\vartheta|_{B(p,\varepsilon)} = df$.

Assertion (1) says that ϑ is globally well-defined and closed, while (2) says that ϑ is not exact. Assertion (3) says that ϑ is what we might call locally exact or locally conservative. What prevents us from finding a (global) function with $\vartheta = df$? Could the same kind of situation occur if the manifold is $\mathbb{R}^2$? The answer is no, and the difference between $\mathbb{R}^2$ and $\mathbb{R}^2 \setminus \{0\}$ is that $H^1(\mathbb{R}^2) = 0$ while $H^1(\mathbb{R}^2 \setminus \{0\}) \neq 0$.

Exercise 10.1. Verify (1) and (3) above.

We recall from Chapter 2 that this example has something to do with path independence. In fact, if we could show that for a given 1-form α, the path integral $\int_c \alpha$ only depended on the beginning and ending points of the curve c, then we could define a function $f(x) := \int_{x_0}^x \alpha$, where $\int_{x_0}^x \alpha$ is just the path integral for any path beginning at a fixed x_0 and ending at x. With this definition one can show that $df = \alpha$ and so α would be exact. In our example, the form ϑ is not exact, and so there must be a failure of path independence.

Exercise 10.2. A smooth fixed endpoint homotopy between a path $c_0 : [a, b] \to M$ and $c_1 : [a, b] \to M$ is a one parameter family of paths h_s such that $h_0 = c_0$ and $h_1 = c_1$ and such that the map $H(t, s) := h_s(t)$ is smooth on $[a, b] \times [0, 1]$. Show that if α is an exact 1-form, then $\frac{d}{ds} \int_{h_s} \alpha = 0$.

10. De Rham Cohomology

The de Rham complex and its cohomology can be viewed in terms of differential equations. For example, the task of finding a closed 1-form $f\,dx + g\,dy$ on an open set $U \subset \mathbb{R}^2$ amounts to finding a solution of the differential equation
$$\frac{\partial g}{\partial x} - \frac{\partial f}{\partial y} = 0.$$
The "trivial" solutions are the exact forms since they are automatically closed. Thus the de Rham cohomology $H^1(U)$ is a space of solutions modulo the "uninteresting" solutions. Similar statements apply to the higher cohomology groups.

Since we have found a closed 1-form on $\mathbb{R}^2 \setminus \{0\}$ that is not exact, we know that $H^1(\mathbb{R}^2 \setminus \{0\}) \neq 0$. We are not yet in a position to determine $H^1(\mathbb{R}^2 \setminus \{0\})$ completely. We will start out with even simpler spaces and eventually develop the machinery to bootstrap our way up to more complicated situations.

First, let $M = \{p\}$. That is, M consists of a single point and is hence a 0-dimensional manifold. In this case,
$$\Omega^k(\{p\}) = Z^k(\{p\}) = \begin{cases} \mathbb{R} & \text{for } k = 0, \\ 0 & \text{for } k > 0. \end{cases}$$
Furthermore, $B^k(\{p\}) = 0$ and so
$$H^k(\{p\}) = \begin{cases} \mathbb{R} & \text{for } k = 0, \\ 0 & \text{for } k > 0. \end{cases}$$

Next we consider the case $M = \mathbb{R}$. Here, $Z^0(\mathbb{R})$ is clearly just the constant functions and so is (isomorphic to) $\mathbb{R}$. On the other hand, $B^0(\mathbb{R}) = 0$ and so
$$H^0(\mathbb{R}) = \mathbb{R}.$$
Now since $d : \Omega^1(\mathbb{R}) \to \Omega^2(\mathbb{R}) = 0$, we see that $Z^1(\mathbb{R}) = \Omega^1(\mathbb{R})$. If $g(x)dx \in \Omega^1(\mathbb{R})$, then letting
$$f(x) := \int_0^x g(x)\,dx$$
we get $df = g(x)dx$. Thus, every $\Omega^1(\mathbb{R})$ is exact; $B^1(\mathbb{R}) = \Omega^1(\mathbb{R})$. We are led to
$$H^1(\mathbb{R}) = 0.$$
From this modest beginning we will be able to compute the de Rham cohomology for a large class of manifolds. Our first goal is to compute $H^k(\mathbb{R}^n)$ for all k. In order to accomplish this, we will need some preparation. The methods are largely algebraic, and so we will need to introduce a bit of "homological algebra".

Definition 10.3. Let R be a commutative ring. A **differential R-complex** is a direct sum of modules $C = \bigoplus_{k \in \mathbb{Z}} C^k$ together with a linear map $d : C \to C$ such that $d \circ d = 0$ and such that $d(C^k) \subset C^{k+1}$. Thus we have a sequence of linear maps

$$\cdots \to C^{k-1} \xrightarrow{d} C^k \xrightarrow{d} C^{k+1} \to \cdots$$

where we have denoted the restrictions $d_k = d|C^k$ all simply by the single letter d.

Let $A = \bigoplus_{k \in \mathbb{Z}} A^k$ and $B = \bigoplus_{k \in \mathbb{Z}} B^k$ be differential complexes. A map $f : A \to B$ is called a **chain map** if f is a (degree 0) graded map such that $d \circ f = f \circ d$. In other words, if we let $f|A^k := f_k$, then we require that $f_k(A^k) \subset B^k$ and that the following diagram commutes for all k:

$$\begin{array}{ccccccccc} \xrightarrow{d} & A^{k-1} & \xrightarrow{d} & A^k & \xrightarrow{d} & A^{k+1} & \xrightarrow{d} & \\ & \downarrow f_{k-1} & & \downarrow f_k & & \downarrow f_{k+1} & & \\ \xrightarrow{d} & B^{k-1} & \xrightarrow{d} & B^k & \xrightarrow{d} & B^{k+1} & \xrightarrow{d} & \end{array}$$

Notice that if $f : A \to B$ is a chain map, then $\mathrm{Ker}(f)$ and $\mathrm{Im}(f)$ are complexes with $\mathrm{Ker}(f) = \bigoplus_{k \in \mathbb{Z}} \mathrm{Ker}(f_k)$ and $\mathrm{Im}(f) = \bigoplus_{k \in \mathbb{Z}} \mathrm{Im}(f_k)$. Thus the notion of **exact sequence of chain maps** may be defined in the obvious way.

Definition 10.4. The k-th cohomology of the complex $C = \bigoplus_{k \in \mathbb{Z}} C^k$ is

$$H^k(C) := \frac{\mathrm{Ker}(d|C^k)}{\mathrm{Im}(d|C^{k-1})}.$$

The elements of $\mathrm{Ker}(d|C^k)$ (also denoted $Z^k(C)$) are called k-**cocycles**, while the elements of $\mathrm{Im}(d|C^{k-1})$ (also denoted $B^k(C)$) are called k-**coboundaries**.

If $f : A \to B$ is a chain map, then it is easy to see that there is a natural (degree 0) graded map $f_* : H(A) \to H(B)$ defined by $f_*([x]) := [f(x)]$ for $x \in A^k$.

Definition 10.5. An exact sequence of chain maps of the form

$$0 \to A \xrightarrow{f} B \xrightarrow{g} C \to 0$$

is called a **short exact sequence**.

10. De Rham Cohomology

Associated to every short exact sequence of chain maps is a **long exact sequence** of cohomology groups:

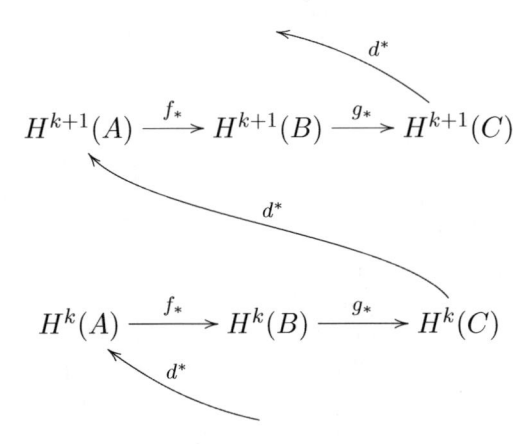

$$H^{k+1}(A) \xrightarrow{f_*} H^{k+1}(B) \xrightarrow{g_*} H^{k+1}(C)$$

$$H^k(A) \xrightarrow{f_*} H^k(B) \xrightarrow{g_*} H^k(C)$$

with connecting maps d^* as indicated.

The maps f_* and g_* are the maps induced by f and g, and the "**coboundary map**" or **connecting homomorphism** $d^* : H^k(C) \to H^{k+1}(A)$ is defined as follows: Let $c \in Z^k(C) \subset C^k$ represent the class $[c] \in H^k(C)$ so that $dc = 0$. Starting with this we hope to end up with a well-defined element of $H^{k+1}(A)$, but we must refer to the following diagram with exact rows to explain how we arrive at a choice of representative of our class $d^*([c])$:

$$\begin{array}{ccccccccc}
0 & \longrightarrow & A^{k+1} & \xrightarrow{f} & B^{k+1} & \xrightarrow{g} & C^{k+1} & \longrightarrow & 0 \\
& & \uparrow d & & \uparrow d & & \uparrow d & & \\
0 & \longrightarrow & A^k & \xrightarrow{f} & B^k & \xrightarrow{g} & C^k & \longrightarrow & 0
\end{array}$$

By the surjectivity of g there is a $b \in B^k$ with $g(b) = c$. Also, since $g(db) = d(g(b)) = dc = 0$, it must be that $db = f(a)$ for some $a \in A^{k+1}$. The scheme of the process is

$$c \dashrightarrow b \dashrightarrow a.$$

Certainly $f(da) = d(f(a)) = ddb = 0$, and so since f is 1-1, we must have $da = 0$, which means that $a \in Z^{k+1}(A)$. We would like to define $d^*([c])$ to be $[a]$, but we must show that this is well-defined. Suppose that we repeat this process starting with $c' = c + d\alpha$ for some $\alpha \in C^{k-1}$. In our first step, we find $b' \in B^k$ with $g(b') = c'$ and then a' with $f(a') = db'$. We wish to show that $[a] = [a']$. We have $g(b - b') = c - c' = d\alpha$. But there must be an element $\beta \in B^{k-1}$ such that $g(\beta) = \alpha$. Now we have

$$\begin{aligned}
g(b - b' - d\beta) &= g(b) - g(b') - g(d\beta) \\
&= g(b) - g(b') - dg(\beta) \\
&= c - c' - d\alpha = 0.
\end{aligned}$$

By exactness at B^k, there must be a $\gamma \in A^k$ such that $f(\gamma) = b - b' - d\beta$. Now we have

$$f(a - a' - d\gamma) = f(a) - f(a') - df(\gamma)$$
$$= db - db' - d(b - b' - d\beta) = 0,$$

and since f is injective, we have $a - a' - d\gamma = 0$, which means that $[a] = [a']$. Thus our definition $d^*([c]) := [a]$ is independent of the choices. We leave it to the reader to check (if there is any doubt) that d^*, so defined, is linear.

Let us review the situation we have with de Rham cohomology noticing how it fits the abstract homological algebra above. We let $\Omega^k(M) := 0$ for $k < 0$ and we have a differential complex $d : \Omega(M) \to \Omega(M)$, where d is the exterior derivative and $\Omega(M)$ is the direct sum of the $\Omega^k(M)$. In this case, $H^k(\Omega(M)) = H^k(M)$ by definition. If $f : M \to N$ is a C^∞ map, then we have $f^* : \Omega(N) \to \Omega(M)$. Since pull-back commutes with exterior differentiation d and preserves the degree of differential forms, f^* is a chain map. Thus we have the induced map on cohomology denoted by f^* so that

$$f^* : H^*(N) \to H^*(M),$$
$$f^* : [\alpha] \mapsto [f^*\alpha],$$

where we have used $H^*(M)$ to denote the direct sum $\bigoplus_i H^i(M)$. Notice that $f \mapsto f^*$ together with $M \mapsto H^*(M)$ is a contravariant functor, which means that for $f : M \to N$ and $g : N \to P$ we have

$$(g \circ f)^* = f^* \circ g^*.$$

(This is why we have now put the stars in as superscripts.) In particular, if $\iota_U : U \to M$ is the inclusion of an open set U in M, then $\iota_U^* \alpha$ is the same as the restriction of the form α to U. If $[\alpha] \in H^*(M)$, then $\iota_U^*([\alpha]) \in H^*(U)$;

$$\iota_U^* : H^*(M) \to H^*(U).$$

Remark 10.6. If $\alpha \in Z^k(M)$ and $\beta \in Z^l(M)$, then for any $\alpha' \in \Omega^{k-1}(M)$ and any $\beta' \in \Omega^{l-1}(M)$ we have

$$(\alpha + d\alpha') \wedge \beta = \alpha \wedge \beta + d\alpha' \wedge \beta$$
$$= \alpha \wedge \beta + d(\alpha' \wedge \beta) - (-1)^{k-1}\alpha' \wedge d\beta$$
$$= \alpha \wedge \beta + d(\alpha' \wedge \beta)$$

and similarly $\alpha \wedge (\beta + d\beta') = \alpha \wedge \beta + d(\alpha \wedge \beta')$. Thus we may define a product $H^k(M) \times H^l(M) \to H^{k+l}(M)$ by $[\alpha] \wedge [\beta] := [\alpha \wedge \beta]$. This gives $H^*(M)$ a graded ring structure.

10.1. The Mayer-Vietoris Sequence

Suppose that $M = U \cup V$ for open sets U and V. We have the following commutative diagram of inclusions:

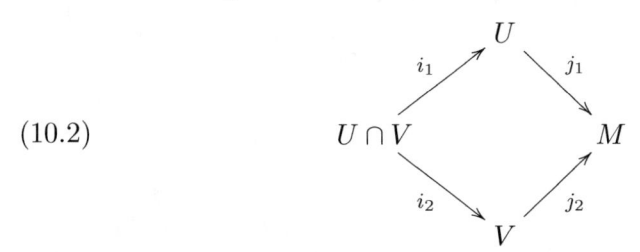

(10.2)
$$\begin{array}{ccc} & U & \\ {}^{i_1}\nearrow & & \searrow{}^{j_1} \\ U \cap V & & M \\ {}^{i_2}\searrow & & \nearrow{}^{j_2} \\ & V & \end{array}$$

This gives rise to the **Mayer-Vietoris short exact sequence**

$$0 \to \Omega(M) \xrightarrow{j^*} \Omega(U) \oplus \Omega(V) \xrightarrow{\partial i^*} \Omega(U \cap V) \to 0,$$

where
$$j^*(\omega) := (j_1^*\omega, j_2^*\omega)$$

and
$$\partial i^*(\alpha, \beta) := i_2^*(\beta) - i_1^*(\alpha) = \beta|_{U \cap V} - \alpha|_{U \cap V}.$$

Note that $j_1^*\omega \in \Omega(U)$ while $j_2^*\omega \in \Omega(V)$. Note also that $i_1^*(\alpha) = \alpha|_{U \cap V}$ and $i_1^*(\beta) = \beta|_{U \cap V}$ live in $\Omega(U \cap V)$. Let us write j^* suggestively as (j_1^*, j_2^*).

Let us show that this sequence is exact. First if $(j_1^*, j_2^*)(\omega) := (j_1^*\omega, j_2^*\omega) = (0, 0)$, then $\omega|_U = \omega|_V = 0$ and so $\omega = 0$ on $M = U \cup V$. Thus (j_1^*, j_2^*) is 1-1 and the exactness at $\Omega(M)$ is demonstrated. Next, if $\eta \in \Omega(U \cap V)$, then we take a smooth partition of unity $\{\rho_U, \rho_V\}$ subordinate to the cover $\{U, V\}$ and let $\xi := (-(\rho_V\eta)^U, (\rho_U\eta)^V)$, where $(\rho_V\eta)^U$ is the extension of the form $\rho_V|_{U \cap V}\eta$ by zero to U, and $(\rho_U\eta)^V$ likewise extends $\rho_U|_{U \cap V}\eta$ to V. This may look backwards at first, so note carefully that we use ρ_V, which has support in V, to get a function on U, and similarly ρ_U is used to define $(\rho_U\eta)^V$ on V! Figure 10.1 shows the circle as the union of two open sets U and V and serves as a schematic of what we have in mind. In the figure $\eta \in \Omega^0$ is 1 on one connected component of $U \cap V$ and 0 on the other component. Now we have

$$\partial i^*(-(\rho_V\eta)^U, (\rho_U\eta)^V)$$
$$= (\rho_U\eta)^V|_{U \cap V} + (\rho_V\eta)^U|_{U \cap V}$$
$$= \rho_U\eta|_{U \cap V} + \rho_V\eta|_{U \cap V}$$
$$= (\rho_U + \rho_V)\eta = \eta.$$

Perhaps the notation is too pedantic. If we let the restrictions and extensions by zero take care of themselves, so to speak, then the idea is expressed by saying that ∂i^* maps the element $(-\rho_V\eta, \rho_U\eta) \in \Omega(U) \oplus \Omega(V)$ to $\rho_U\eta - (-\rho_V\eta) = \eta \in \Omega(U \cap V)$. Thus we see that ∂i^* is surjective.

It is easy to see that $\partial i^* \circ (j_1^*, j_2^*) = 0$ so that $\mathrm{Im}(j_1^*, j_2^*) \subset \mathrm{Ker}(\partial i^*)$. Now let $(\alpha, \beta) \in \Omega(U) \oplus \Omega(V)$ and suppose that $\partial i^*(\alpha, \beta) = 0$. This translates to $\alpha|_{U \cap V} = \beta|_{U \cap V}$, which means that there is a form $\omega \in \Omega(U \cup V) = \Omega(M)$ such that ω coincides with α on U and with β on V. Thus

$$(j_1^*, j_2^*)\omega = (j_1^*\omega, j_2^*\omega) = (\alpha, \beta),$$

so that $\mathrm{Ker}(\partial i^*) \subset \mathrm{Im}(j_1^*, j_2^*)$, which, together with the reverse inclusion, gives $\mathrm{Ker}(\partial i^*) = \mathrm{Im}(j_1^*, j_2^*)$.

Following the general algebraic pattern, the Mayer-Vietoris short exact sequence gives rise to the **Mayer-Vietoris long exact sequence**:

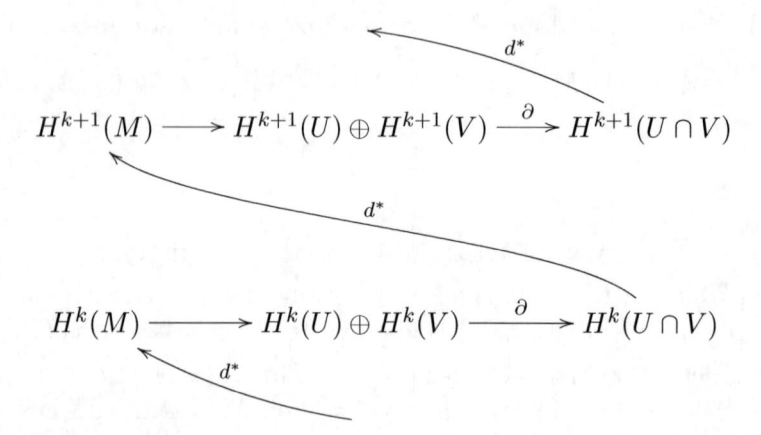

$$H^{k+1}(M) \longrightarrow H^{k+1}(U) \oplus H^{k+1}(V) \xrightarrow{\partial} H^{k+1}(U \cap V)$$

$$H^k(M) \longrightarrow H^k(U) \oplus H^k(V) \xrightarrow{\partial} H^k(U \cap V)$$

with connecting maps d^*.

Since our description of the coboundary map in the algebraic case was rather abstract, we will do well to take a closer look at the connecting homomorphism d^* in the present context. Referring to the diagram below, $\eta \in \Omega^k(U \cap V)$ represents a cohomology class $[\eta] \in H^k(U \cap V)$ so that in particular $d\eta = 0$:

$$\begin{array}{ccccccccc} 0 & \longrightarrow & \Omega^{k+1}(M) & \xrightarrow{j^*} & \Omega^{k+1}(U) \oplus \Omega^{k+1}(V) & \xrightarrow{\partial i^*} & \Omega^{k+1}(U \cap V) & \longrightarrow & 0 \\ & & \uparrow d & & \uparrow d \oplus d & & \uparrow d & & \\ 0 & \longrightarrow & \Omega^k(M) & \xrightarrow{j^*} & \Omega^k(U) \oplus \Omega^k(V) & \xrightarrow{\partial i^*} & \Omega^k(U \cap V) & \longrightarrow & 0 \end{array}$$

We will abbreviate the map $d \oplus d$ to just d so that $d(\eta_1, \eta_2) = (d\eta_1, d\eta_2)$. By the exactness of the rows we can find a form $\xi \in \Omega^k(U) \oplus \Omega^k(V)$ which maps to η. In fact, we may take $\xi = (-\rho_V \eta, \rho_U \eta)$ as before. Since $d\eta = 0$ and the diagram commutes, we see that $d\xi$ must map to 0 in $\Omega^{k+1}(U \cap V)$. This just tells us that $-d(\rho_V \eta)$ and $d(\rho_U \eta)$ agree on the intersection $U \cap V$. (Refer again to Figure 10.1.) Thus there is a well-defined form in $\Omega^{k+1}(M)$

10.2. Homotopy Invariance

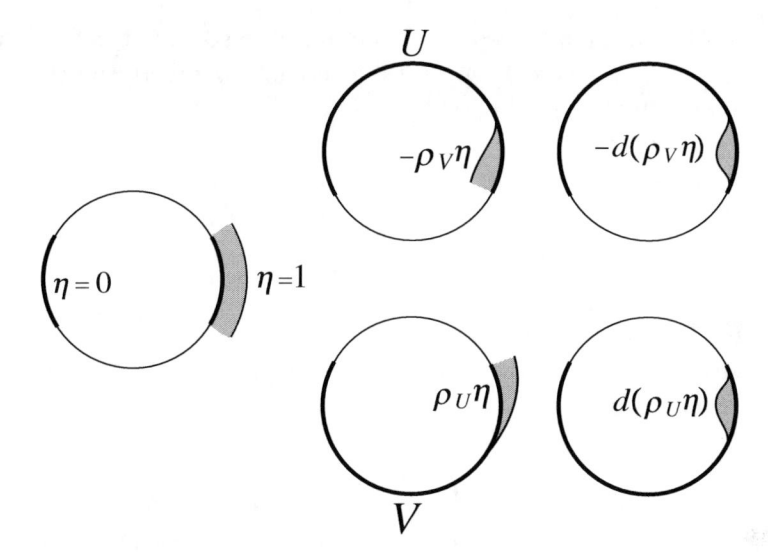

Figure 10.1. Scheme for d^* on the circle

which maps to $d\xi$. This global form γ is given by

$$\gamma = \begin{cases} -d(\rho_V \eta) & \text{on } U, \\ d(\rho_U \eta) & \text{on } V, \end{cases}$$

and then by definition $d^*[\eta] = [\gamma] \in H^{k+1}(M)$.

Exercise 10.7. Let the circle S^1 be parametrized by the angle θ in the usual way. Let U be the part of a circle with $-\pi/6 < \theta < 7\pi/6$ and let V be given by $5\pi/6 < \theta < 13\pi/6$.

(a) Show that $H^0(U) \cong H^0(V) \cong \mathbb{R}$.

(b) Show that the "difference map" $H^0(U) \oplus H^0(V) \xrightarrow{d^*} H^0(U \cap V)$ has 1-dimensional image.

(c) What is the cohomology of S^1?

10.2. Homotopy Invariance

Now we come to a result about the relation between $H^*(M)$ and $H^*(M \times \mathbb{R})$ which provides the leverage needed to compute the cohomology of some higher-dimensional manifolds based on that of lower-dimensional manifolds. One of our main goals in this section is to prove the homotopy invariance of de Rham cohomology and also the Poincaré lemma. We start with a theorem which is of interest in its own right. Let M be an n-manifold and also consider the product manifold $M \times \mathbb{R}$. We then have the projection

$\pi : M \times \mathbb{R} \to M$, and for a fixed $a \in \mathbb{R}$ we have the section $s_a : M \to M \times \mathbb{R}$ given by $x \mapsto (x, a)$. We can apply the cohomology functor to this pair of maps to obtain the maps π^* and s_a^*:

$$
\begin{array}{cc}
M \times \mathbb{R} & H^*(M \times \mathbb{R}) \\
\pi \updownarrow s_a & \pi^* \updownarrow s_a^* \\
M & H^*(M)
\end{array}
$$

Theorem 10.8. *Given M and the maps defined above, we have that $\pi^* : H^*(M) \to H^*(M \times \mathbb{R})$ and $s_a^* : H^*(M \times \mathbb{R}) \to H^*(M)$ are mutual inverses for each a. In particular,*

$$H^*(M \times \mathbb{R}) \cong H^*(M).$$

Proof. In what follows we let id denote the identity map on $\Omega(M \times \mathbb{R})$, $H^*(M)$ and $H^*(M \times \mathbb{R})$ as determined by the context. We already know that $\mathrm{id} = s_a^* \circ \pi^*$. The main idea of the proof is the use of a so-called homotopy operator, which in the present case is a degree -1 map $K : \Omega(M \times \mathbb{R}) \to \Omega(M \times \mathbb{R})$ with the property that

$$(10.3) \qquad \mathrm{id} - \pi^* \circ s_a^* = \pm(d \circ K - K \circ d).$$

The point is that $d \circ K - K \circ d$ must send closed forms to exact forms. Thus on the level of cohomology $\pm(d \circ K - K \circ d) = 0$, and hence $\mathrm{id} - \pi^* \circ s_a^*$ must be the zero map. Thus we also have $\mathrm{id} = \pi^* \circ s_a^*$ on $H^*(M)$, which gives the result.

So let us then prove equation (10.3) above. Let ∂_t denote the obvious vector field that is related to the standard coordinate field on $\mathbb{R}$ under the projection $M \times \mathbb{R} \to \mathbb{R}$, and let i_{∂_t} denote interior product with respect to ∂_t. The map K is defined on $\Omega^k(M \times \mathbb{R})$ by

$$K(\omega) = (-1)^{k-1} \int_a^t \pi^* s_\tau^* (i_{\partial_t} \omega) \, d\tau = (-1)^{k-1} \int_a^t (s_\tau \circ \pi)^* (i_{\partial_t} \omega) \, d\tau.$$

More explicitly, for $v_1, ..., v_{k-1} \in T_{(q,t)} M \times \mathbb{R}$,

$$K(\omega)|_{(q,t)} (v_1, \ldots, v_{k-1}) = \int_a^t \omega(Ts_\tau \circ T\pi\, (v_1), \ldots, Ts_\tau \circ T\pi\, (v_{k-1}), \partial_t|_\tau)\, d\tau.$$

This operator and d are both local and linear over $\mathbb{R}$, and so if $\mathrm{id} - \pi^* \circ s_a^* = \pm(d \circ K - K \circ d)$ is true on charts, then it is true in general. Thus we may as well assume that M is an open set U in $\mathbb{R}^n$. For any $\omega \in \Omega^k(U \times \mathbb{R})$, we can find a pair of functions $f_1(x,t)$ and $f_2(x,t)$ such that

$$\omega = f_1(x,t) \pi^* \alpha + f_2(x,t) \pi^* \beta \wedge dt$$

for some forms $\alpha \in \Omega^k(U)$ and $\beta \in \Omega^{k-1}(U)$. This decomposition is unique in the sense that if $f_1(x,t)\pi^*\alpha + f_2(x,t)\pi^*\beta \wedge dt = 0$, then $f_1(x,t)\pi^*\alpha = 0$

10.2. Homotopy Invariance

and $f_2(x,t)\pi^*\beta \wedge dt = 0$. In what follows we will abuse notation a bit. The standard coordinates on U will be denoted by $x^1, \ldots, x^n$, and we write $\pi^* x^i$ simply as x^i so that with t the standard coordinate of $\mathbb{R}^1$, we have coordinates $(x^1, \ldots, x^n, t)$ on $U \times \mathbb{R}$. Using the decomposition above, one can see that $K(f_1(x,t)\pi^*\alpha)$ is zero and in general

$$K : \omega \mapsto \left(\int_a^t f_2(x,\tau)\, d\tau \right) \times \pi^*\beta.$$

This map is our proposed homotopy operator.

Let us now check that K has the required properties. It is clear from what we have said that we may check the action of K separately on forms of the types $f_1(x,t)\pi^*\alpha$ and $f_2(x,t)\pi^*\beta \wedge dt$.

Case I (type $f(x,t)\pi^*\alpha$). If $\omega = f(x,t)\pi^*\alpha$, then $K\omega = 0$ and so $(d \circ K - K \circ d)\omega = -K(d\omega)$. Then

$$\begin{aligned}
K(d\omega) &= K(d(f(x,t)\pi^*\alpha)) \\
&= K(df(x,t) \wedge \pi^*\alpha + (-1)^k f(x,t)\pi^* d\alpha) \\
&= K\left(\sum \frac{\partial f_1}{\partial x^i} dx^i \wedge \pi^*\alpha + \frac{\partial f}{\partial t} dt \wedge \pi^*\alpha + (-1)^k f(x,t)\pi^* d\alpha \right) \\
&= (-1)^k K\left(\frac{\partial f}{\partial t}(x,t)\pi^*\alpha \wedge dt \right) \\
&= (-1)^k \int_a^t \frac{\partial f}{\partial t}(x,\tau) d\tau \times \pi^*\alpha = (-1)^k (f(x,t) - f(x,a))\pi^*\alpha.
\end{aligned}$$

On the other hand, since

$$\begin{aligned}
\pi^* s_a^* f(x,t)\,\pi^*\alpha &= \pi^* [f(x,a)\,(s_a^* \circ \pi^*)\,\alpha] \\
&= \pi^*[(f \circ s_a)\alpha] = (f \circ s_a \circ \pi^*)\,\pi^*\alpha = f(x,a)\pi^*\alpha,
\end{aligned}$$

we have

$$\begin{aligned}
(\mathrm{id} - \pi^* \circ s_a^*)\omega &= (\mathrm{id} - \pi^* \circ s_a^*) f(x,t)\pi^*\alpha \\
&= f(x,t)\pi^*\alpha - f(x,a)\pi^*\alpha \\
&= (f(x,t) - f(x,a))\,\pi^*\alpha
\end{aligned}$$

as above. So in this case we get $(d \circ K - K \circ d)\omega = \pm(\mathrm{id} - \pi^* \circ s_a^*)\omega$.

Case II (type $\omega = f(x,t)\pi^*\beta \wedge dt$). We have

$$d \circ K(\omega) = d \circ K(f(x,t)\pi^*\beta \wedge dt) = d\left(\pi^*\beta \int_a^t f(x,\tau)\,d\tau\right)$$

$$= \pi^* d\beta \left(\int_a^t f(x,\tau)\,d\tau\right) + (-1)^{k-1}\pi^*\beta \wedge f(x,t)\,dt$$

$$+ (-1)^{k-1}\pi^*\beta \wedge \sum \left(\int_a^t \frac{\partial f}{\partial x^i}(x,\tau)\,d\tau\right) dx^i$$

and

$$K \circ d\omega = Kd(f\pi^*\beta \wedge dt) = Kd(\pi^*\beta \wedge f dt)$$

$$= K\left(\pi^* d\beta \wedge f dt + (-1)^{k-1}\pi^*\beta \wedge df \wedge dt\right)$$

$$= K\left(\pi^* d\beta \wedge f dt + (-1)^{k-1}\pi^*\beta \wedge \sum \frac{\partial f}{\partial x^i} dx^i \wedge dt\right)$$

$$= \left(\int_a^t f(x,\tau)\,d\tau\right)\pi^* d\beta + (-1)^{k-1}\pi^*\beta \wedge \sum \left(\int_a^t \frac{\partial f}{\partial x^i}(x,\tau)d\tau\right)dx^i.$$

Thus $(d \circ K - K \circ d)\omega = (-1)^{k-1}\pi^*\beta \wedge f(x,t)\,dt = (-1)^{k-1}\omega$. On the other hand, we also have $(\mathrm{id} - \pi^* \circ s_a^*)\omega = \omega$ since $s_a^* dt = 0$ and so $(d \circ K - K \circ d) = \pm(\mathrm{id} - \pi^* \circ s_a^*)$, which is equation (10.3). As explained at the beginning of the proof, this implies that π^* and s_a^* are inverses and so $H^*(M \times \mathbb{R}) \cong H^*(M)$. $\square$

Corollary 10.9 (Poincaré lemma).

$$H^k(\mathbb{R}^n) = H^k(\mathrm{point}) = \begin{cases} \mathbb{R} & \text{if } k = 0, \\ 0 & \text{otherwise.} \end{cases}$$

Proof. One first verifies that the statement is true for $H^k(\mathrm{point})$. Then the remainder of the proof is a simple induction:

$$H^k(\mathrm{point}) \cong H^k(\mathrm{point} \times \mathbb{R}) = H^k(\mathbb{R})$$
$$\cong H^k(\mathbb{R} \times \mathbb{R}) = H^k(\mathbb{R}^2)$$
$$\cong \ldots$$
$$\cong H^k(\mathbb{R}^{n-1} \times \mathbb{R}) = H^k(\mathbb{R}^n). \quad \square$$

Corollary 10.10 (Homotopy axiom). *If $f : M \to N$ and $g : M \to N$ are homotopic, then the induced maps $f^* : H^*(N) \to H^*(M)$ and $g^* : H^*(N) \to H^*(M)$ are equal.*

10.2. Homotopy Invariance

Proof. By extending the homotopy as in Exercise 1.77 we may assume that we have a map $F : M \times \mathbb{R} \to N$ such that
$$F(x,t) = f(x) \text{ for } t \geq 1,$$
$$F(x,t) = g(x) \text{ for } t \leq 0.$$
If $s_1(x) := (x,1)$ and $s_0(x) := (x,0)$, then $f = F \circ s_1$ and $g = F \circ s_0$, and so
$$f^* = s_1^* \circ F^*,$$
$$g^* = s_0^* \circ F^*.$$
It is easy to check that s_1^* and s_0^* are one-sided inverses of π^*, where $M \times \mathbb{R} \to M$ is the projection as before. But we have shown that π^* is an isomorphism. It follows that $s_1^* = s_0^*$ in cohomology, and then from the above we have $f^* = g^*$. □

Homotopy plays a central role in algebraic topology, and so the last corollary is very important. Recall that if there exist maps $f : M \to N$ and $g : N \to M$ such that both $f \circ g$ and $g \circ f$ are defined and homotopic to id_N and id_M respectively, then f (or g) is called a homotopy equivalence, and M and N are said to have the same homotopy type. In particular, if a topological space has the same homotopy type as a single point, then we say that the space is **contractible**. If we are dealing with smooth manifolds, we may take the maps to be smooth. In fact, any continuous map between two smooth manifolds is continuously homotopic to a smooth map. We shall use this fact often without comment. The following corollaries follow easily.

Corollary 10.11 (Homotopy invariance). *If M and N are smooth manifolds which are of the same homotopy type, then $H^*(M) \cong H^*(N)$.*

Corollary 10.12. *If M is a contractible n-manifold, then*
$$H^0(M) \cong \mathbb{R},$$
$$H^k(M) = 0 \text{ for } 0 < k \leq n.$$

Next consider the situation where A is a subset of M and $i : A \hookrightarrow M$ is the inclusion map. If there exists a map $r : M \to A$ such that $r \circ i = \mathrm{id}_A$, then we say that r is a retraction of M onto A. If A is a regular submanifold of a smooth manifold M, then in case there is a retraction r of M onto A, we may assume that r is smooth. If we can find a smooth retraction r such that $i \circ r$ is smoothly homotopic to the identity id_M, then we say that r is a (smooth) **deformation retraction**, and this homotopy itself is also called a **deformation retraction**. In this case, A is said to be a (smooth) **deformation retract** of M.

Corollary 10.13. *If A is a smooth deformation retract of M, then A and M have isomorphic cohomologies.*

Exercise 10.14. Let U_+ and U_- be open subsets of the sphere $S^n \subset \mathbb{R}^{n+1}$ given by

$$U_+ := \{(x^i) \in S^n : -\varepsilon < x^{n+1} \le 1\},$$
$$U_- := \{(x^i) \in S^n : -1 \le x^{n+1} < \varepsilon\},$$

where $0 < \varepsilon < 1/2$. Show that there is a deformation retraction of $U_+ \cap U_-$ onto the equator $x^{n+1} = 0$ in S^n. Notice that the equator is a two point set in case $n = 0$. Show that S^n is not contractible. (See also Problem 6.)

Theorem 10.15 (Hairy sphere theorem). *A nowhere vanishing smooth vector field exists on the sphere S^n if and only if n is odd.*

Proof. If X is a nowhere vanishing vector field on S^n, then define $\nu := X/\|X\|$. We introduce a map $[0,1] \times S^n \to S^n$ by

$$F(x,t) := (\cos \pi t)\, x + (\sin \pi t)\, \nu(x).$$

This is a smooth homotopy between the identity map and the antipodal map $a : x \mapsto -x$. Now if vol_{S^n} is the volume form on S^n, then $[\mathrm{vol}_{S^n}]$ is not zero in $H^n(S^n)$. Indeed, suppose that $\mathrm{vol}_{S^n} = d\omega$. Then $0 \ne \mathrm{vol}(S^n) = \int_{S^n} \mathrm{vol} = \int_\emptyset \omega = 0$, which is a contradiction. Recall the special n-form on $\mathbb{R}^{n+1}$ given in Cartesian coordinates by

$$\sigma := \sum_{i=1}^{n+1} (-1)^{n-1} x^i dx^1 \wedge \cdots \wedge \widehat{dx^i} \wedge \cdots \wedge dx^{n+1}.$$

It was shown previously that vol_{S^n} is the restriction of σ to S^n. But if n is even, then $n+1$ is odd, and it is clear from the above formula that $a^* \sigma = -\sigma$. Since vol_{S^n} is obtained from σ taking restriction, we have $a^* \mathrm{vol}_{S^n} = -\mathrm{vol}_{S^n}$, and so also on the level of cohomology:

$$a^* [\mathrm{vol}_{S^n}] = -[\mathrm{vol}_{S^n}].$$

But this is impossible since if the homotopy exists, we must have $a^* = \mathrm{id}^*$ on $H^n(S^n)$. We conclude that a nonvanishing vector field does not exist on S^n if n is even.

If n is odd, then we can easily construct a nowhere vanishing vector field on S^n. For example, if $n = 3$, then let

$$X = x_2 \frac{\partial}{\partial x_1} - x_1 \frac{\partial}{\partial x_2} + x_4 \frac{\partial}{\partial x_3} - x_3 \frac{\partial}{\partial x_4},$$

where x_1, x_2, x_3, x_4 are the standard coordinates on $\mathbb{R}^4$, and restrict X to S^3. Notice that X is indeed tangent to S^3. The generalization to higher odd dimensions should be clear. $\square$

10.2. Homotopy Invariance

A trivial consequence of the Poincaré lemma is that the cohomology spaces of a Euclidean space are finite-dimensional. Below we use an induction that shows that this is true for a large class of manifolds which include all compact manifolds. For this and later purposes, we introduce a technical condition. An open cover $\{U_\alpha\}_{\alpha \in A}$ of an n-manifold M is called a **good cover** if for every choice $\alpha_0, \ldots, \alpha_k$ the set $U_{\alpha_0} \cap \cdots \cap U_{\alpha_k}$ is diffeomorphic to $\mathbb{R}^n$ (or empty).

Proposition 10.16. *If $\{O_\beta\}_{\beta \in B}$ is any open cover of an n-manifold M, then there exists a good cover $\{U_\alpha\}_{\alpha \in A}$ such that each U_α is contained in some O_β.*

Proof. The proof requires a result from Riemannian geometry (see Problem 2 of Chapter 13). We know that every manifold can be given a Riemannian metric. Each point of a Riemannian manifold has a neighborhood system made up of small geodesically convex sets. What matters to us now is that the intersection of any finite number of such sets is geodesically convex and hence diffeomorphic to $\mathbb{R}^n$. Actually, the assertion that an open geodesically convex set in a Riemannian manifold is diffeomorphic to $\mathbb{R}^n$ is common in the literature, but it is a more subtle issue than it may seem, and references to a complete proof are hard to find (but see [**Grom**]). Granted this claim, the result follows. $\square$

Many of the results we develop below will be proved for orientable manifolds that possess a finite good cover. This includes all orientable compact manifolds.

Theorem 10.17. *If an n-manifold has a finite good cover, then its de Rham cohomology spaces are all finite-dimensional.*

Proof. The proof is an induction argument that uses the Mayer-Vietoris sequence. Our k-th induction hypothesis is the following:

P(k): *Every smooth manifold that has a good cover consisting of k open sets has finite-dimensional de Rham cohomologies.*

Since we know that an open set diffeomorphic to $\mathbb{R}^n$ has finite-dimensional de Rham cohomology spaces, the statement P(1) is true by Corollary 10.9. Suppose that P(k) is true. We wish to show that this implies P($k+1$). So suppose that M has a good cover $\{U_1, \ldots, U_{k+1}\}$ and let $M_k := U_1 \cup \cdots \cup U_k$.

Since we are assuming P(k), M_k has finite-dimensional de Rham cohomology spaces. Note that $M_k \cap U_{k+1}$ has a good cover, which is just $\{U_1 \cap U_{k+1}, \ldots, U_k \cap U_{k+1}\}$. From the long Mayer-Vietoris sequence we have for a given q

$$\longrightarrow H^q(M_k \cap U_{k+1}) \xrightarrow{d^*} H^q(M_{k+1}) \xrightarrow{\iota_0^* + \iota_1^*} H^{q+1}(M_k) \oplus H^{q+1}(U_{k+1}) \longrightarrow,$$

which gives the exact sequence

$$0 \longrightarrow \operatorname{Ker} d^* \xrightarrow{d^*} H^q(M_{k+1}) \xrightarrow{\iota_0^* + \iota_1^*} \operatorname{Im}(\iota_0^* + \iota_1^*) \longrightarrow 0.$$

Since $H^q(M_k \cap U_{k+1})$ and $H^{q+1}(M_k) \oplus H^{q+1}(U_{k+1})$ are finite-dimensional by hypothesis, the same is true of $\operatorname{Ker} d^*$ and $\operatorname{Im}(\iota_0^* + \iota_1^*)$. It follows that $H^q(M_{k+1})$ is finite-dimensional and the induction is complete. □

10.3. Compactly Supported Cohomology

Let $\Omega_c(M)$ denote the algebra of compactly supported differential forms on a manifold M. Obviously, if M is compact, then $\Omega_c(M) = \Omega(M)$, and so our main interest here is the case where M is not compact. We now have a slightly different complex

$$\cdots \xrightarrow{d} \Omega_c^k(M) \xrightarrow{d} \Omega_c^{k+1}(M) \xrightarrow{d} \cdots,$$

which has a corresponding cohomology $H_c^*(M)$ called the **de Rham cohomology with compact support**. By definition

$$H_c^k(M) = \frac{Z_c^k(M)}{B_c^k(M)},$$

where $Z_c^k(M)$ is the vector space of closed k-forms with compact support and $B_c^k(M)$ is the space of all k-forms $d\omega$ where ω has compact support. Note carefully that $B_c^k(M)$ is *not* the set of exact k-forms with compact support. To drive the point home, consider $f \in C^\infty(\mathbb{R}^n)$ with compact support and with $f \geq 0$ and $f > 0$ at some point. Then

$$\omega = f \, dx^1 \wedge \cdots \wedge dx^n$$

is exact since every closed form on $\mathbb{R}^n$ is exact. However, ω cannot be $d\alpha$ for an α with compact support, since then we would have

$$\int_{\mathbb{R}^n} \omega = \int_{\mathbb{R}^n} d\alpha = \int_{\partial \mathbb{R}^n} \alpha = 0,$$

which contradicts the assumption that f is nonnegative with $f > 0$ at some point. This already shows that $H_c^n(\mathbb{R}^n) \neq 0$. Using a bump function with support inside a chart, one can similarly show that $H_c^n(M) \neq 0$ for any orientable manifold M.

Exercise 10.18. Let M be an oriented n-manifold. If $\omega \in \Omega_c^n(M)$ and $\int_M \omega \neq 0$, then $[\omega] \neq 0$ in $H_c^n(M)$.

Exercise 10.19. Show that $H_c^1(\mathbb{R}) \cong \mathbb{R}$ and that $H_c^0(M) = 0$ whenever $\dim M > 0$ and M is connected.

10.3. Compactly Supported Cohomology

If we look at the behavior of differential forms under the operation of pull-back, we immediately realize that the pull-back of a differential form with compact support may not have compact support. In order to get desired functorial properties, we consider the class of smooth proper maps. Recall that a smooth map $f : P \to M$ is called a proper map if $f^{-1}(K)$ is compact whenever K is compact. It is easy to verify that the set of all smooth manifolds together with proper smooth maps is a category and the assignments $M \mapsto \Omega_c(M)$ and $f \mapsto \{\alpha \mapsto f^*\alpha\}$ give a contravariant functor. In plain language, this means that if $f : P \to M$ is a proper map, then $f^* : \Omega_c(M) \to \Omega_c(P)$ and for two such maps we have $(f \circ g)^* = g^* \circ f^*$ as before, but now the assumption that f and g are proper maps is essential.

We will use a *different* functorial behavior associated with forms of compact support. The first thing we need is a new category (which is fortunately easy to describe). The category we have in mind has as objects the set of all open subsets of a fixed manifold M. The morphisms are the inclusion maps $j_{V,U} : V \hookrightarrow U$, which are only defined in case $V \subset U$. For any such inclusion $j_{V,U}$, we define a map $(j_{V,U})_* : \Omega_c(V) \to \Omega_c(U)$ according to the following simple prescription: For $\alpha \in \Omega_c(V)$, let $(j_{V,U})_*\alpha$ be the form in $\Omega_c(U)$ which is equal to α at all points in V and equal to zero otherwise (extension by zero). Since the support of α is neatly inside the open set V, the extension $(j_{V,U})_*\alpha$ is smooth. In what follows, we take this category whenever we employ the functor Ω_c.

Let U and V be open subsets which together cover M. Recall the commutative diagram of inclusion maps (10.2). Now for each k, let us define a map $(-i_{1*}, i_{2*}) : \Omega_c^k(U \cap V) \to \Omega_c^k(U) \oplus \Omega_c^k(V)$ by $\alpha \mapsto (-i_{1*}\alpha, i_{2*}\alpha)$. Now we also have the map $j_{1*} + j_{2*} : \Omega_c^k(V) \oplus \Omega_c(U) \to \Omega_c^k(M)$ given by $(\alpha_1, \alpha_2) \mapsto j_{1*}\alpha_1 + j_{2*}\alpha_2$. Notice that if $\{\phi_U, \phi_V\}$ is a partition of unity subordinate to $\{U, V\}$, then for any $\omega \in \Omega_c^k(M)$ we can define $\omega_U := \phi_U\omega|_U$ and $\omega_V := \phi_V\omega|_V$ so that we have

$$(j_{1*} + j_{2*})(\omega_U, \omega_V) = j_{1*}\omega_U + j_{2*}\omega_V = \phi_U\omega + \phi_V\omega = \omega.$$

Notice that ω_U and ω_V have compact support. For example,

$$\mathrm{Supp}(\omega_U) = \mathrm{Supp}(\phi_U\omega) \subset \mathrm{Supp}(\phi_U) \cap \mathrm{Supp}(\omega).$$

Since $\mathrm{Supp}(\phi_U) \cap \mathrm{Supp}(\omega)$ is compact, $\mathrm{Supp}(\omega_U)$ is also compact. Thus $j_{1*} + j_{2*}$ is surjective. We can associate to the diagram (10.2), the new sequence

(10.4) $\quad 0 \to \Omega_c^k(V \cap U) \xrightarrow{(-i_{1*}, i_{2*})} \Omega_c^k(V) \oplus \Omega_c^k(U) \xrightarrow{j_{1*}+j_{2*}} \Omega_c^k(M) \to 0.$

This is the short Mayer-Vietoris sequence for differential forms with compact support.

Theorem 10.20. *The sequence* (10.4) *is exact.*

We have shown the surjectivity of $j_{1*} + j_{2*}$ above. The rest of the proof is also easy and left as Problem 4.

Corollary 10.21. *There is a long exact sequence*

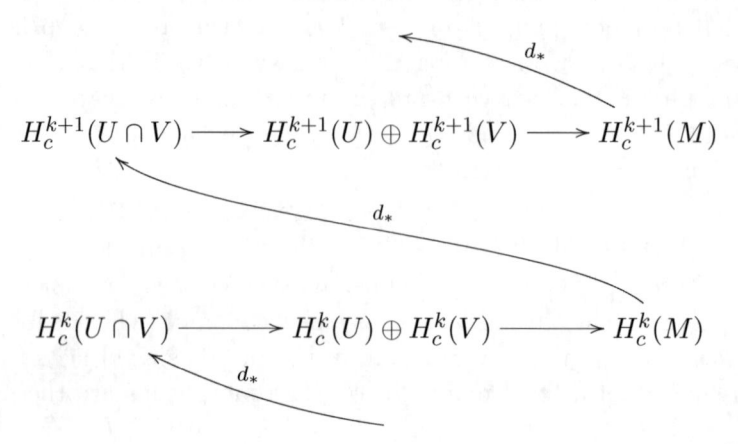

$$\begin{array}{ccccc}
H_c^{k+1}(U \cap V) & \longrightarrow & H_c^{k+1}(U) \oplus H_c^{k+1}(V) & \longrightarrow & H_c^{k+1}(M) \\
& & & & \\
H_c^k(U \cap V) & \longrightarrow & H_c^k(U) \oplus H_c^k(V) & \longrightarrow & H_c^k(M)
\end{array}$$

with connecting maps d_*,

which is called the (long) **Mayer-Vietoris sequence for cohomology with compact supports**.

Notice that we have denoted the connecting homomorphism by d_*. We will need to have a more explicit understanding of d_*: If $[\omega] \in H_c^k(M)$, then using a partition of unity $\{\rho_U, \rho_V\}$ as above we write $\omega = j_{1*}\omega_U + j_{2*}\omega_V$ and then $d\,j_{1*}\omega_U = -d\,j_{2*}\omega_V$ on $U \cap V$. Then

$$(10.5) \qquad d_*[\omega] = -\,d\,j_{1*}\omega_U|_{U \cap V} = d j_{2*}\omega_V|_{U \cap V}.$$

Next we prove a version of the Poincaré lemma for compactly supported cohomology. For a given n-manifold M, we consider the projection $\pi : M \times \mathbb{R} \to M$. We immediately notice that π^* does not map $\Omega_c^k(M)$ into $\Omega_c^k(M \times \mathbb{R})$. What we need is a map $\pi_* : \Omega_c^k(M \times \mathbb{R}) \to \Omega_c^{k-1}(M)$ called integration along the fiber. Before giving the definition of π_* we first note that every element of $\Omega_c^k(M \times \mathbb{R})$ is locally a sum of forms of the following types:

$$\text{Type I:} \quad f\,\pi^*\lambda,$$
$$\text{Type II:} \quad f\,\pi^*\varphi \wedge dt,$$

where $\lambda \in \Omega_c^k(M)$, $\varphi \in \Omega_c^{k-1}(M)$ and f is a smooth function with compact support on $M \times \mathbb{R}$. By definition π_* sends all forms of Type I to zero, and for Type II forms, we define

$$\pi_*\left(\pi^*\varphi \cdot f \wedge dt\right) = \varphi \int_{-\infty}^{\infty} f(\cdot, t)\,dt.$$

By linearity this defines π_* on all forms. In Problem 2 we ask the reader to show that $\pi_* \circ d = d \circ \pi_*$ so that π_* is a chain map. Thus we get a map on

cohomology:
$$\pi_* : H^k_c(M \times \mathbb{R}) \to H^{k-1}_c(M).$$

Next choose $e \in \Omega^1_c(\mathbb{R})$ with $\int e = 1$ and introduce the map $e_* : \Omega^k_c(M) \to \Omega^{k+1}_c(M \times \mathbb{R})$ by

$$e_* : \omega \mapsto \pi^*\omega \wedge \pi_2^* e,$$

where $\pi_2 : M \times \mathbb{R} \to \mathbb{R}$ is the projection on the second factor. It is easy to check that e_* commutes with d, so once again we get a map on the level of cohomology:

$$e_* : H^k_c(M) \to H^{k+1}_c(M \times \mathbb{R}) \text{ for all } k.$$

Our immediate goal is to show that $e_* \circ \pi_*$ and $\pi_* \circ e_*$ are both identity operators on the level of cohomology. In fact, it is not hard to see that $\pi_* \circ e_* = \mathrm{id}$ already on $\Omega^k_c(M)$. We need to construct a homotopy operator K between $e_* \circ \pi_*$ and id. The map $K : \Omega^k_c(M \times \mathbb{R}) \to \Omega^{k-1}_c(M \times \mathbb{R})$ is given by requiring that K is linear, maps Type I forms to zero, and if $\omega = \pi^*\varphi \cdot f \wedge dt$ is Type II, then

$$K(\omega) = K(\pi^*\varphi \cdot f \wedge dt) := \pi^*\varphi \left(\int_{-\infty}^t f(x, u)\, du - T(t) \int_{-\infty}^\infty f(x, u)\, du \right),$$

where $T(t) = \int_{-\infty}^t e$. In Problem 3, we ask the reader to show that

(10.6) $\qquad \mathrm{id} - e_* \circ \pi_* = (-1)^{k-1}(dK - Kd)$

on $\Omega^k_c(M \times \mathbb{R})$. It follows that $\mathrm{id} = e_* \circ \pi_*$ on $H^k_c(M \times \mathbb{R})$, and so finally we have the following result:

Theorem 10.22. *With notation as above the following maps are isomorphisms and mutual inverses:*

$$\pi_* : H^k_c(M \times \mathbb{R}) \to H^{k-1}_c(M),$$
$$e_* : H^{k-1}_c(M) \to H^k_c(M \times \mathbb{R}).$$

Corollary 10.23 (Poincaré lemma for compactly supported cohomology). *We have*

$$H^n_c(\mathbb{R}^n) = \mathbb{R},$$
$$H^k_c(\mathbb{R}^n) = 0 \text{ if } k \neq n.$$

Proof. By Exercise 10.19 we have $H^1_c(\mathbb{R}) \cong \mathbb{R}$. But from the previous theorem $H^n_c(\mathbb{R}^n) \cong H^{n-1}_c(\mathbb{R}^{n-1}) \cong \cdots \cong H^1_c(\mathbb{R})$. Also, trivially, $H^k_c(\mathbb{R}^n) = 0$ for $k > n$, and if $k < n$, then

$$H^k_c(\mathbb{R}^n) \cong H^{k-1}_c(\mathbb{R}^{n-1}) \cong \cdots \cong H^0_c(\mathbb{R}^{n-k}) = 0$$

by Exercise 10.19. $\square$

The isomorphism $H_c^n(\mathbb{R}^n) = \mathbb{R}$ is given by repeated application of π_*, which is just integration over the last coordinate of $\mathbb{R}^n$.

Remark: Notice that $\mathbb{R}^n$ is homotopy equivalent to $\mathbb{R}^{n-1}$ and yet, $H_c^n(\mathbb{R}^n) \neq H_c^n(\mathbb{R}^{n-1})$. This contrasts the compactly supported cohomology with the ordinary cohomology, since the latter is a homotopy invariant.

10.4. Poincaré Duality

Looking back at our results for $\mathbb{R}^n$ we notice that $H^k(\mathbb{R}^n) = H_c^{n-k}(\mathbb{R}^n)^*$. This is a special case of Poincaré duality which is proved in this section. Recall that we have already met Poincaré duality for compact manifolds in Section 9.6. The version we develop here will be valid for noncompact manifolds also. So let M be an oriented n-manifold which is not necessarily compact. For each k, we have a bilinear pairing

$$\Omega^k(M) \times \Omega_c^{n-k}(M) \to \mathbb{R},$$

$$(\omega_1, \omega_2) \mapsto \int_M \omega_1 \wedge \omega_2,$$

and as in Section 9.6 this defines a pairing on cohomology:

$$H^k(M) \times H_c^{n-k}(M) \to \mathbb{R},$$

$$([\omega_1], [\omega_2]) \mapsto \int_M \omega_1 \wedge \omega_2.$$

In turn, this provides a linear map $PD_k : H^k(M) \to \left(H_c^{n-k}(M)\right)^*$ defined as

$$PD_k([\omega_1])([\omega_2]) := \int_M \omega_1 \wedge \omega_2.$$

Our goal is to show that this map is an isomorphism. We prove this for orientable manifolds with a *finite* cover, but the result remains valid more generally (see [**Madsen**]). Notice also that unless the cohomologies are finite-dimensional, we may have $H^k(\mathbb{R}^n)^* \neq H_c^{n-k}(\mathbb{R}^n)$. The reason is that for an infinite-dimensional vector spaces, $V^* \cong W$ does not imply $V \cong W^*$.

In what follows, we will denote all the duality maps PD_k by the single designation PD. We use a theorem from algebra called the **five lemma**: Consider the following diagram of vector spaces (or abelian groups) and homomorphisms:

$$\begin{array}{ccccccccc} A & \xrightarrow{h_1} & B & \xrightarrow{h_2} & C & \xrightarrow{h_3} & D & \xrightarrow{h_4} & E \\ \downarrow{\alpha} & & \downarrow{\beta} & & \downarrow{\gamma} & & \downarrow{\delta} & & \downarrow{\varepsilon} \\ A' & \xrightarrow{h_1'} & B' & \xrightarrow{h_2'} & C' & \xrightarrow{h_3'} & D' & \xrightarrow{h_4'} & E' \end{array}$$

10.4. Poincaré Duality

Lemma 10.24 (Five lemma). *Suppose that the above diagram commutes up to sign (so for example $h'_1 \circ \alpha = \pm \beta \circ h_1$). Suppose also that the rows are exact and that $\alpha, \beta, \delta,$ and ε are isomorphisms. Then the middle map γ is also an isomorphism.*

The five lemma is usually stated for the case when the diagram commutes, but a simple modification of the usual proof (see [**L2**]) gives the version above.

If $\{U, V\}$ is a cover of the oriented n-manifold M, then consider the following diagram:

$$\begin{array}{ccccccccc}
\to & H^{k-1}(U) \oplus H^{k-1}(V) & \to & H^{k-1}(U \cap V) & \xrightarrow{d^*} & H^k(M) & \to & H^k(U) \oplus H^k(V) & \to \\
& \downarrow & & \downarrow & & \downarrow & & \downarrow & \\
\to & H_c^{n-k+1}(U)^* \oplus H_c^{n-k+1}(V)^* & \to & H_c^{n-k+1}(U \cap V)^* & \xrightarrow[(d_*)^*]{} & H_c^{n-k}(M)^* & \to & H_c^{n-k}(U)^* \oplus H_c^{n-k}(V)^* & \to
\end{array}$$

Here the bottom row is the sequence of dual maps from the Mayer-Vietoris long exact sequence and is exact itself (easy check). The vertical maps are the duality maps PD (or obvious direct sums of such).

Lemma 10.25. *The above diagram commutes up to sign.*

Proof. First let us consider the diagram

$$\begin{array}{ccc}
H^k(M) & \xrightarrow{j_1^* + j_2^*} & H^k(U) \oplus H^k(V) \\
\downarrow & & \downarrow \\
H_c^{n-k}(M)^* & \xrightarrow[(j_{1*})^* + (j_{2*})^*]{} & H_c^{n-k}(U)^* \oplus H_c^{n-k}(V)^*
\end{array}$$

In order to show that $(PD \oplus PD) \circ (j_1^* + j_2^*) = ((j_{1*})^* + (j_{2*})^*) \circ PD$, it is enough to show that $PD \circ j_1^* = (j_{1*})^* \circ PD$ and $PD \circ j_2^* = (j_{2*})^* \circ PD$. For a given $[\omega] \in H^k(M)$, the linear form $PD \circ j_1^*([\omega])$ takes an element $[\theta] \in H^k(U)$ to

$$\int_U j_1^* \omega \wedge \theta.$$

On the other hand, $((j_{1*})^* \circ PD)([\omega])$ maps $[\theta]$ to

$$\int_M \omega \wedge j_{1*} \theta.$$

But $\omega \wedge j_{1*}\theta$ and $j_1^*\omega \wedge \theta$ both have support in U and agree on this set. Therefore the above two integrals are equal. Similarly, $PD \circ j_2^* = (j_{2*})^* \circ PD$.

Next we show that the following square commutes up to sign:

$$\begin{array}{ccc} H^{k-1}(U \cap V) & \xrightarrow{d^*} & H^k(M) \\ \downarrow & & \downarrow \\ H_c^{n-k+1}(U \cap V)^* & \xrightarrow[(d_*)^*]{} & H_c^{n-k}(M)^* \end{array}$$

Let $\{\rho_U, \rho_V\}$ be a partition of unity subordinate to $\{U, V\}$ as before. If $[\omega] \in H^{k-1}(U \cap V)$, then $d^*[\omega]$ is represented by a form (which we denote by $d^*\omega$) that has the properties

$$d^*\omega|_U = -d(\rho_V \omega) \text{ on } U,$$
$$d^*\omega|_V = d(\rho_U \omega) \text{ on } V.$$

On the other hand, if $[\tau] \in H_c^{n-k}(M)$, then $d_*[\tau]$ is represented by a form $d_*\tau$ which has the properties

$$-i_{1*}d_*\tau = d(\rho_U \omega) \text{ on } U,$$
$$i_{2*}d_*\tau = d(\rho_V \omega) \text{ on } V.$$

Using the fact that ω is closed, we have

$$PD \circ d^*([\omega])([\tau]) = \int_M d^*\omega \wedge \tau = \int_{U \cap V} d^*\omega \wedge \tau$$
$$= \int_{U \cap V} d(\rho_U \omega) \wedge \tau = \int_{U \cap V} d\rho_U \wedge \omega \wedge \tau,$$

where have used the fact that $d^*\omega \wedge \tau$ has support in $U \cap V$. Meanwhile,

$$(d_*)^* \circ PD([\omega])([\tau]) = \int_{U \cap V} \omega \wedge d_*\tau = -\int_{U \cap V} \omega \wedge d(\rho_U \tau)$$
$$= -\int_{U \cap V} \omega \wedge d\rho_U \wedge \tau,$$

so the two integrals are equal up to sign. We leave the sign commutativity of the remaining square to the reader. □

Theorem 10.26 (Poincaré duality). *Let M be an oriented n-manifold with a finite good cover. Then for each k the map $PD_k : H^k(M) \to \left(H_c^{n-k}(M)\right)^*$ is an isomorphism.*

Proof. We will prove that PD_k is an isomorphism by induction. The inductive statement $P(N)$ is that PD_k is an isomorphism for all orientable manifolds which have a good cover consisting of at most N open sets. By the two Poincaré lemmas we already know that if U is diffeomorphic to $\mathbb{R}^n$, then $H^k(U) \cong \left(H_c^{n-k}(U)\right)^*$, where both sides are either zero or isomorphic to $\mathbb{R}$. It is then an easy exercise to show that in fact $PD_k : H^k(U) \to \left(H_c^{n-k}(U)\right)^*$ is an isomorphism. This verifies the statement $P(1)$. Now assume that

10.4. Poincaré Duality

$P(N)$ is true and suppose we are given an oriented manifold M with a good cover $\{U_1, \ldots, U_{N+1}\}$. Let $M_N := U_1 \cup \cdots \cup U_N$ and use Lemma 10.25 with $U := M_N$, $V := U_{N+1}$ and $U \cup V = M$. Then by assumption PD is an isomorphism for U, V and $U \cap V$, so the hypotheses of Lemma 10.24 are satisfied. But then the middle homomorphism $H^k(M) \to H_c^{n-k}(M)^*$ is an isomorphism (k was arbitrary). $\square$

Corollary 10.27. *If M is a connected oriented n-manifold with finite good cover, then $H_c^n(M) \cong \mathbb{R}$. This isomorphism is given by integration over M.*

This last corollary allows us to define the important concept of degree of a proper map. Let M and N be *connected* oriented n-manifolds and suppose that $f : M \to N$ is a proper map. Then a pull-back of a compactly supported form is compactly supported so that we obtain a map $f^* : H_c^n(N) \to H_c^n(M)$ and the following commutative diagram:

$$\begin{array}{ccc} H_c^n(N) & \xrightarrow{f^*} & H_c^n(M) \\ \downarrow{\scriptstyle \int_N} & & \downarrow{\scriptstyle \int_M} \\ \mathbb{R} & \xrightarrow{\mathrm{Deg}_f} & \mathbb{R} \end{array}$$

The induced map $\mathrm{Deg}_f : \mathbb{R} \to \mathbb{R}$ must be multiplication by a real number. This number is called the **degree** of f and is denoted $\deg(f)$. If $[\omega] \in H_c^n(N)$ is chosen so that $\int_N \omega = 1$, then

$$\int_M f^*\omega = \deg(f).$$

Such a form is called a generator and can be chosen to have support in an arbitrary open subset V of N. To see this, just choose a chart $(U, \mathbf{x})$ in N with $U \subset \overline{U} \subset V$ and let ϕ be a cut-off function with $\mathrm{supp}\,\phi \subset U$. Then let $\omega := \phi\,dx^1 \wedge \cdots \wedge dx^n$ and then scale ϕ so that $\int_N \omega = 1$. Of course, ω is closed, but also $[\omega] \neq 0$. Indeed, if we had $[\omega] = 0$, then there would be an $(n-1)$-form η with *compact support* and with $\omega = d\eta$. But then Stokes' theorem applies, and so $\int_N \omega = \int_N d\eta = \int_{\partial N} \eta = 0$ since $\partial N = \emptyset$. This would contradict $\int_N \omega = 1$.

Theorem 10.28. *Let M and N be connected n-manifolds oriented by volume forms μ_M and μ_N respectively. If $f : M \to N$ is a proper map and $y \in N$ is a regular value, then*

$$\deg(f) = \sum_{x \in f^{-1}(y)} \mathrm{sign}_x f,$$

where $\mathrm{sign}_x f = 1$ if $(f^*\mu_N)(x)$ is a positive multiple of $\mu_M(x)$ and -1 otherwise. In particular, $\deg(f)$ is an integer.

Proof. Using Sard's Theorem 2.34 and the fact that f is proper, we may choose a neighborhood V of y such that $f^{-1}(V) = \bigcup_{i=1}^{N} U_i$, where $U_i \cap U_j = \emptyset$ for $i \neq j$ and such that $f|_{U_i}$ is a diffeomorphism onto V for each i. Choose $[\omega] \in H_c^n(N)$ with $\int_N \omega = 1$, and such that ω has support in V. We may arrange that V and all U_i are connected so that $\text{sign}_x f$ is constant on each U_i. In this case, $f|_{U_i}$ is orientation preserving or reversing according to whether $\text{sign}_x f$ is 1 or -1. Let $x_i \in U_i$ be the inverse image of y in U_i. We have

$$\deg(f) = \int_M f^*\omega = \sum_{i=1}^{N} \int_{U_i} f^*\omega = \sum_{i=1}^{N} (\text{sign}_{x_i} f) \int_V \omega$$
$$= \sum_{x \in f^{-1}(y)} \text{sign}_x f. \qquad \square$$

We close this chapter with a quick explanation of the Poincaré dual of a submanifold. Let Y be an oriented regular submanifold of an oriented n-manifold. Let the dimension of Y be $n-k$. Then we obtain a linear map

$$\int_Y : \Omega_c^{n-k}(M) \longrightarrow \mathbb{R},$$
$$\omega \mapsto \int_Y \iota^*\omega.$$

where $\iota : Y \hookrightarrow M$ is the inclusion map. Since this map is zero on exact forms (by Stokes' theorem), it passes to the quotient giving a linear functional

$$\int_Y : H_c^{n-k}(M) \longrightarrow \mathbb{R}.$$

By Poincaré duality $PD : H^k(M) \cong \left(H_c^{n-k}(M)\right)^*$, there must be a unique class $[\theta_Y] \in H^k(M)$ such that this linear form is given by

$$\int_Y [\omega] = \int_Y \theta_Y \wedge \omega \text{ for all } [\omega] \in H_c^{n-k}(M).$$

This class (or a given representative θ_Y) is called the **Poincaré dual of the submanifold** Y.

For more on this and other important topics such as the Thom isomorphism, the Leray-Hirsch theorem, and Čech cohomology see [**Bo-Tu**].

Problems

(1) Show that formula (10.1) defines a smooth function θ on the open set $U := \mathbb{R}^2\setminus\{x \leq 0, y = 0\}$ and show that
$$d\theta = \frac{xdy - ydx}{x^2 + y^2} \text{ on } U.$$

(2) Show that $\pi_* : \Omega_c^*(M \times \mathbb{R}) \to \Omega_c^{*-1}(M)$ as defined in the discussion leading to Theorem 10.22 is a chain map.

(3) Prove the homotopy formula (10.6) (or see [**Bo-Tu**], pages 38–39).

(4) Prove exactness of (10.4).

(5) Prove Corollary 10.11.

(6) Use Exercise 10.14 and the long Mayer-Vietoris sequence to show that
$$H^k(S^n) = \begin{cases} \mathbb{R} & \text{if } k = 0 \text{ or } n, \\ 0 & \text{otherwise.} \end{cases}$$

(7) Show that $\omega \in \Omega^n(S^n)$ is exact if and only if $\int_{S^n} \omega = 0$.

(8) Determine the cohomology spaces for the punctured Euclidean spaces $\mathbb{R}^n\setminus\{0\}$. Show that if B is a ball, then $\mathbb{R}^n\setminus\overline{B}$ has the same cohomology.

(9) Let B_x be a ball centered at $x \in \mathbb{R}^n$. Show that a closed $(n-1)$-form on $\mathbb{R}^n\setminus\overline{B_x}$ is exact if and only if $\int_S \iota^*\omega = 0$ for some sphere centered at x. Here $\iota : S \hookrightarrow \mathbb{R}^n$ is the inclusion map. Show that this is true for all spheres centered at x.

(10) Let $n \geq 2$ and suppose that ω is a compactly supported n-form on $\mathbb{R}^n$ such that $\int_{\mathbb{R}^n} \omega = 0$.
   (a) Let B_1 and B_2 be open balls centered at the origin of $\mathbb{R}^n$ with $B_1 \subset \overline{B_1} \subset B_2$ and $\operatorname{supp}\omega \subset B_1$. By the (first) Poincaré lemma there is a form α such that $d\alpha = \omega$. Show that $\int_{\partial B_2} \alpha = 0$.
   (b) Continuing from (a), show that α is exact on $\mathbb{R}^n\setminus\overline{B}$ (use Problem 9 above).
   (c) Let $\rho \in \Omega^{n-2}(\mathbb{R}^n\setminus\overline{B})$ be such that $d\rho = \alpha$ where α is as above. Show that there is a function g such that if $\beta := \alpha - d(g\rho)$, then β is smooth, compactly supported and $d\beta = \omega$. Deduce that $H_c^n(\mathbb{R}^n) = 0$.

(11) Find $H^*(M)$ where M is $\mathbb{R}^n\setminus\{p, q\}$ for some distinct points $p, q \in \mathbb{R}^n$.

Chapter 11

Distributions and Frobenius' Theorem

The theory of distributions is a geometric formulation of the classical theory of certain systems of partial differential equations. The solutions are immersed submanifolds called integral manifolds. The Frobenius theorem gives necessary and sufficient conditions for the existence of such integral manifolds. One of the most important applications is to show that a subalgebra of the Lie algebra of a Lie group corresponds to a Lie subgroup. We give this application near the end of this chapter. We also develop a classic version of the Frobenius theorem and use it to prove the basic existence theorem for surfaces.

One can view the theory studied in this chapter as a higher-dimensional analogue of the study of vector fields and integral curves. If X is a smooth vector field on an n-manifold M, then we know that integral curves exist through each point, and if X never vanishes, then the integral curves are immersions. Nonvanishing vector fields do not even exist on many manifolds, and so we consider a somewhat different question. A **one-dimensional distribution** assigns to each $p \in M$ a one-dimensional subspace E_p of the tangent space T_pM. We say that the assignment is smooth if for each $p \in M$ there is a smooth vector field X defined in an open set U containing p such that $X(p)$ spans E_p for each $p \in U$. It is easy to see that a one-dimensional distribution is essentially the same thing as a rank one subbundle of the tangent bundle.

A curve $c : (a,b) \to M$ with $\dot{c} \neq 0$ is called an integral curve of the one-dimensional distribution if $\dot{c}(t) = T_t c \cdot \frac{\partial}{\partial t}$ is contained in $E_{c(t)}$ for each

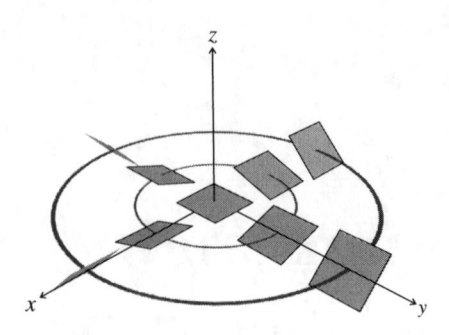

Figure 11.1. No integral manifold through the origin

$t \in (a,b)$. The curve is an immersion, and its restrictions to small enough subintervals are integral curves of vector fields that locally span the distribution. Since integral curves cannot cross themselves or each other, the curve is an injective immersion. The images of such curves are called integral manifolds (they are immersed submanifolds). Of course, every nonvanishing global vector field defines a one-dimensional distribution, but some one-dimensional distributions do not arise in this way. For example, the tangent spaces to the fibers of the Möbius band bundle MB $\to S^1$ form a distribution not spanned by any globally defined vector field (see Example 6.9). The integral manifolds are the fibers themselves.

11.1. Definitions

We wish to generalize the above idea to higher-dimensional distributions.

Definition 11.1. A smooth rank k **distribution** on an n-manifold M is a (smooth) rank k vector subbundle $E \to M$ of the tangent bundle.

Sometimes we refer to a distribution as a *tangent* distribution to distinguish it from the notion of a distributional function on a manifold. Recalling the definition of a vector subbundle (Definition 6.26) and the criteria given in Exercise 6.27, we see that a smooth rank k **distribution** on an n-manifold M gives a k-dimensional subspace $E_p \subset T_pM$ for each $p \in M$ such that for each fixed $p \in M$ there is a family of smooth vector fields $X_1, \ldots, X_k$ defined on some neighborhood U_p of p and such that $X_1(q), \ldots, X_k(q)$ are linearly independent and span E_q for each $q \in U_p$. We say that such a set of vector fields **locally spans** the distribution. In other words, $X_1, \ldots, X_k$ can be viewed as a local frame field for the subbundle E.

Consider the punctured 3-space $M = \mathbb{R}^3 \backslash \{0\}$. The level sets of the function $\varepsilon : (x,y,x) \mapsto x^2 + y^2 + x^2$ are spheres whose union is all of $\mathbb{R}^3 \backslash \{0\}$. Now define a distribution on M by the rule that E_p is the tangent space at

11.1. Definitions

p to the sphere containing p. Dually, we can define this distribution to be the one given by the rule

$$E_p = \{v \in T_pM : d\varepsilon(v) = 0\}.$$

Thus we see that a distribution of rank $n-1$ can arise as the family of tangent spaces to the level sets of nondegenerate smooth functions. More generally, rank k distributions can arise from the level sets of sufficiently nondegenerate $\mathbb{R}^{n-k}$-valued functions. On the other hand, not all distributions arise in this way, even locally.

Definition 11.2. Let $E \to M$ be a distribution. An immersed submanifold S of M is called an **integral manifold** (or integral submanifold) for the distribution if for each $p \in S$ we have $T\iota(T_pS) = E_p$, where $\iota : S \hookrightarrow M$ is the inclusion map. The distribution is called **integrable** if each point of M is contained in an integral manifold of the distribution.[1]

If $\iota : S \hookrightarrow M$ is the inclusion map, then we usually identify $T\iota(T_pS)$ with T_pS so the condition which makes S an integral manifold is stated simply as $E_p = T_pS$ for all $p \in S$.

Consider the distribution in $\mathbb{R}^3$ suggested by Figure 11.1. For each p, E_p is the subspace of $T_p\mathbb{R}^3$ orthogonal to

$$y \left.\frac{\partial}{\partial x}\right|_p - x \left.\frac{\partial}{\partial y}\right|_p + \left.\frac{\partial}{\partial z}\right|_p.$$

If there were an integral manifold though the origin, then it would clearly be tangent to the x, y plane at the origin. But if one tries to imagine following a closed path on such an integral manifold around the z axis, a problem appears. Namely, the curve would have to be tangent to the distribution, but this would clearly force the curve to spiral upward and never return to the same point.

A distribution as defined above is sometimes called a **regular distribution** to distinguish the concept from that of a singular distribution which is defined in a similar manner, but allows for the dimension of the subspaces to vary from point to point. Singular distributions are studied in the online supplement [**Lee, Jeff**].

Definition 11.3. Let $E \to M$ be a distribution on an n-manifold M and let X be a vector field defined on an open subset $U \subset M$. We say that X **lies in the distribution** if $X(p) \in E_p$ for each p in the domain of X; that is, if X takes values in E. In this case, we sometimes write $X \in E$ (a slight abuse of notation).

---

[1] In other contexts, integral manifold might be defined by the weaker condition $T\iota(T_pS) \subset E_p$. We will always require the stronger condition $T\iota(T_pS) = E_p$.

Definition 11.4. If for every pair of locally defined vector fields X and Y with common domain that lie in a distribution $E \to M$, the bracket $[X, Y]$ also lies in the distribution, then we say that the distribution is **involutive**.

It is easy to see that a vector field X lies in a distribution if and only if for any spanning local frame field $X_1, \ldots, X_k$ we have $X = \sum_{i=1}^{k} f^i X_i$ for smooth functions f^i defined on the common domain of $X, X_1, \ldots, X_k$.

Locally, finding integral curves of vector fields is the same as solving a system of ordinary differential equations. The local model for finding integral manifolds of a distribution is that of solving a system of partial differential equations. Consider the following system of partial differential equations defined on some neighborhood U of the origin in $\mathbb{R}^2$:

$$(11.1) \quad \begin{aligned} z_x &= F(x, y, z), \\ z_y &= G(x, y, z). \end{aligned}$$

Here the functions F and G are assumed to be smooth and defined on $U \times \mathbb{R}$. We might attempt to find a solution defined near the origin by integrating twice. Suppose we want a solution with $z(0,0) = z_0$. First, solve $f'(x) = F(x, 0, f(x))$ with $f(0) = z_0$. Then for each fixed x in the domain of f solve $\partial_y z(x, y) = G(x, y, z(x, y))$ with $z(x, 0) = f(x)$. This function $z(x, y)$ may not be a C^2 solution of our system of partial differential equations since, if it were, we would have $z_{xy} = z_{yx}$. But

$$z_{xy} = (z_x)_y = \frac{\partial}{\partial y} F(x, y, z) = F_y + G F_y,$$

and similarly, $z_{yx} = G_x + F G_x$. Thus the equality of mixed partials gives a necessary integrability condition;

$$(11.2) \quad F_y + G F_y = G_x + F G_x.$$

We shall see that this condition is also sufficient for the local solvability of our system.

Exercise 11.5.

(a) Show that solving the differential equations above is equivalent to finding the integral curves of the following pair of vector fields.

$$X = \frac{\partial}{\partial x} - F(x, y, z) \frac{\partial}{\partial z} \text{ and } Y = \frac{\partial}{\partial x} - G(x, y, z) \frac{\partial}{\partial z}.$$

(b) Show that the graph of a solution to the system (11.1) is an integral curve of the distribution spanned by X and Y.

(c) Show that $[X, Y] = 0$ is the same condition as (11.2).

There are no integrability conditions for finding integral curves of a vector field. But now we see that we should expect integrability conditions for the existence of integral manifolds of higher rank distributions.

11.2. The Local Frobenius Theorem

We will relate the notion of involutivity to the existence of integral manifolds.

Lemma 11.6. *Let $E \to M$ be a distribution (tangent subbundle) on an n-manifold M. Then $E \to M$ is involutive if and only if for every local frame field $X_1, \ldots, X_k$ for the subbundle $E \to M$ we have*

$$[X_i, X_j] = \sum_{s=1}^{k} c_{ij}^s X_s$$

for some smooth functions c_{ij}^s ($1 \le i, j, s \le k$).

Proof. It is clear that such functions exist if E is involutive. Now, suppose that such functions exist. If X and Y lie in the distribution, then

$$X = \sum_{i=1}^{k} f^i X_i,$$

$$Y = \sum_{j=1}^{k} g^j X_j$$

for smooth functions f^i and g^j. We wish to show that $[X, Y]$ lies in the distribution. This follows by direct calculation:

$$[X, Y] = \left[\sum_i f^i X_i, \sum_j g^j X_j \right]$$
$$= \sum_{i,j} f^i g^j [X_i, X_j] + \sum_{i,j} f^i \left(X_i g^j \right) X_j - \sum_{i,j} g^j \left(X_j f^i \right) X_i. \quad \square$$

We will show that a distribution is involutive if and only if it is integrable. In fact, we shall see that the integral manifolds fit together nicely. In the following discussion we identify $\mathbb{R}^k \times \mathbb{R}^{n-k}$ with $\mathbb{R}^n$.

Definition 11.7. A rank k distribution $E \to M$ on an n-manifold M is called **completely integrable** if for each $p \in M$ there is a chart $(U, \mathbf{x})$ centered at p such that $\mathbf{x}(U) = V \times W \subset \mathbb{R}^k \times \mathbb{R}^{n-k}$ and such that each set of the form $\mathbf{x}^{-1}(V \times \{y\})$ for each $y \in W$ is an integral manifold of the distribution.

If $\mathbf{x}(p) = (a^1, \ldots, a^n)$, then

$$\mathbf{x}^{-1}(V \times \{y\}) = \{q \in U : x^{k+1}(q) = a^{k+1}, \ldots, x^n(q) = a^n\},$$

where $y = (a^{k+1}, \ldots, a^n)$. For a chart as in the last definition we have

$$T\left(\mathbf{x}^{-1}(V \times \{y\})\right) = E|_{\mathbf{x}^{-1}(V \times \{y\})}$$

for each $y \in W$. We can always arrange that V and W in the above definition are connected (say open cubes or open balls). In this case, we refer to such a chart as a **distinguished chart** for the distribution. The integral manifolds of the form $\mathbf{x}^{-1}(V \times \{y\})$ for a distinguished chart are called **plaques**. Notice that since V and W are assumed path connected for a distinguished chart, the plaques are path connected. The reader was asked to prove the following result in Problem 4 of Chapter 3, but we will state and prove the result here for the sake of completeness.

Lemma 11.8. *Suppose that $f : N \to M$ is a one-one immersion and that Y is a vector field on M such that $Y(f(p)) \in T_p f(T_p N)$ for every $p \in N$. Then there is a unique vector field X on N that is f-related to Y.*

Proof. It is clear that we can define a field X such that $X(p)$ is the unique element of $T_p N$ with $Y(f(p)) = T_p f \cdot X(p)$. The task is to show that X so defined is smooth. For $p \in N$, let $(U, \mathbf{x})$ be a chart centered at p and $(V, \mathbf{y})$ a chart centered around $f(p)$ such that $\mathbf{y} \circ f \circ \mathbf{x}^{-1}$ has the form
$$(a^1, \ldots, a^n) \mapsto (a^1, \ldots, a^n, 0, \ldots, 0).$$
Then it is easy to check that
$$T_p f \cdot \left.\frac{\partial}{\partial x^i}\right|_p = \left.\frac{\partial}{\partial y^i}\right|_{f(p)}.$$
So if
$$Y = \sum Y^i \frac{\partial}{\partial y^i}$$
for smooth functions Y^i, then
$$X = \sum X^i \frac{\partial}{\partial x^i},$$
where we must have $X^i = Y^i \circ f$. This shows that the X^i are smooth and hence the field X is smooth. $\square$

We apply the previous lemma to the situation where S is an immersed submanifold of M and $\iota : S \hookrightarrow M$ is the inclusion. In this case, if Y is a vector field on M which is tangent to S, then there is a unique smooth vector field on S, denoted $Y|_S$, such that $Y|_S(q) = Y(q)$ for all $q \in S$. The vector field $Y|_S$ is called the restriction of Y to S and it is ι-related to Y.

Theorem 11.9 (Local Frobenius theorem). *A distribution is involutive if and only if it is completely integrable and if and only if it is integrable.*

Proof. First, completely integrable clearly implies integrable. Now assume that $E \to M$ is a rank k distribution which is integrable and let $\dim(M) = n$. Let X and Y be smooth vector fields defined on U which both lie in the distribution. Let $p \in U$ and let S be an integral manifold that contains p. Then $X_q, Y_q \in T_q S$ for all $q \in U \cap S$. Thus the vector fields are tangent to $U \cap S$ and hence if $\iota : U \cap S \hookrightarrow U$ is the inclusion map, then X and Y are ι-related to their restrictions to $U \cap S$ as in the previous lemma. This means that $[X|_{U \cap S}, Y|_{U \cap S}]$ is ι-related to $[X, Y]$ and hence $[X, Y]_q \in T_q S = E_q$ for all $q \in U \cap S$ and in particular $[X, Y]_p \in E_p$. Since p was an arbitrary element of U, we conclude that the distribution is involutive.

Now suppose that $E \to M$ is involutive. Pick a point $p_0 \in M$ and let $(U, \mathbf{x})$ be a chart centered at p_0. By shrinking U if necessary we can find fields $X_1, \ldots, X_k$ defined on U that span the distribution. Notice that if we have a chart of the type produced by Theorem 2.113, then we can easily arrange that the image of the chart is of the form $V \times W \subset \mathbb{R}^k \times \mathbb{R}^{n-k}$, and it is then easy to see that we have exactly the kind of chart needed (see the comments immediately following the proof of Theorem 2.113). Thus the strategy is to replace the frame field $X_1, \ldots, X_k$ by a commuting frame field.

Construct a chart $(U, \mathbf{x})$ centered at p_0 such that $X_j(p_0) = \left.\frac{\partial}{\partial x^j}\right|_{p_0}$ for $1 \leq j \leq k$. Let $\pi : \mathbb{R}^n \to \mathbb{R}^k$ be the map $(a^1, \ldots, a^n) \mapsto (a^1, \ldots, a^k)$ and let $f := \pi \circ \mathbf{x}$. We have

$$(T_{p_0} f)(X_j(p_0)) = T_{p_0} f \left.\frac{\partial}{\partial x^j}\right|_{p_0} = \left.\frac{\partial}{\partial t^j}\right|_0,$$

where $(t^1, \ldots, t^k)$ are standard coordinates on $\mathbb{R}^k$. Then $T_{p_0} f : T_{p_0} M \to T_0 \mathbb{R}^k$ maps E_{p_0} isomorphically onto $T_0 \mathbb{R}^k$. It follows that $T_p f$ maps E_p isomorphically onto $T_{f(p)} \mathbb{R}^k$ for all p in some neighborhood of p_0. Hence, for each such p, there are unique vectors $Y_1(p), \ldots, Y_k(p)$ in E_p with $T_p f(Y_j(p)) = \left.\frac{\partial}{\partial t^j}\right|_{f(p)}$. Thus we get vector fields $Y_1, \ldots, Y_k$ whose smoothness we leave to the reader to check. By shrinking U if necessary, we may assume that these are defined on U. It is clear that $Y_1, \ldots, Y_k$ form a local frame for E and Y_j is f-related to $\frac{\partial}{\partial t^j}$. Thus $[Y_i, Y_j]$ is f-related to $\left[\frac{\partial}{\partial t^i}, \frac{\partial}{\partial t^j}\right]$ and in particular

$$T_p f([Y_i, Y_j]_p) = \left[\frac{\partial}{\partial t^i}, \frac{\partial}{\partial t^j}\right]_0 = 0.$$

Since the distribution E is involutive, we have $[Y_i, Y_j]_p \in E_p$ for each p. Then since $T_p f$ is injective on E_p, we see that $[Y_i, Y_j]_p = 0$ for all p. Thus we arrive at a local frame field $Y_1, \ldots, Y_k$ for E with $[Y_i, Y_j] = 0$.

By Theorem 2.113 and the comments above, we are done. $\square$

11.3. Differential Forms and Integrability

One can use differential forms to effectively discuss distributions and involutivity. What is interesting is the way the exterior derivative comes into play.

The first thing to notice is that a smooth distribution is locally defined by 1-forms. We need a bit of creative terminology:

Definition 11.10. If $\alpha^1, \ldots, \alpha^j$ are 1-forms on an n-manifold M and S is a subset of M, then we say that $\alpha^1, \ldots, \alpha^j$ (defined at least on some open set containing S) are **independent on** S if $\{\alpha^1(p), \ldots, \alpha^j(p)\}$ is an independent set in $T_p^* M$ for each $p \in S$.

Proposition 11.11. *Let $E \to M$ be a rank k distribution on an n-manifold M. Every $p \in M$ has an open neighborhood U and 1-forms $\alpha^1, \ldots, \alpha^{n-k}$ independent on U such that*
$$E_q = \operatorname{Ker} \alpha^1(q) \cap \cdots \cap \operatorname{Ker} \alpha^{n-k}(q) \text{ for all } q \in U.$$

Proof. Let $p \in M$ and let $X_1, \ldots, X_k$ span E on a neighborhood O of p. We may extend to a frame field $X_1, \ldots, X_k, \ldots, X_n$ defined on some possibly smaller neighborhood U of p. Indeed, choose any other frame field $Y_1, \ldots, Y_n$ and then rearrange indices so that $X_1(p), \ldots, X_k(p), Y_{k+1}(p), \ldots, Y_n(p)$ are linearly independent. The vectors $X_1(q), \ldots, X_k(q), Y_{k+1}(q), \ldots, Y_n(q)$ must be independent for all q in some neighborhood of p, which we may as well take to be O. Indeed, $q \mapsto X_1(q) \wedge \cdots \wedge X_k(q) \wedge Y_{k+1}(q) \wedge \cdots \wedge Y_n(q)$ is nonzero at p and so remains nonzero in a neighborhood. In any case, let $X_1, \ldots, X_k, \ldots, X_n$ be our extended local frame and let $\theta^1, \ldots, \theta^n$ be the dual frame. Then it is easy to check that $v \in T_q M$ is in E_q if and only if $\theta_q^{k+1}(v) = \cdots = \theta_q^n(v) = 0$. Thus we just let $(\alpha^1, \ldots, \alpha^{n-k}) := (\theta^{k+1}, \ldots, \theta^n)$. $\square$

The forms described in the previous proposition are sometimes called **local defining forms** for the distribution.

The set of all differential forms vanishing on a distribution has a convenient algebraic structure. Recall that $\Omega(M)$ is a ring under the wedge product. If $\mathcal{J}$ is the ideal in $\Omega(M)$ and $U \subset M$ is open, then $\mathcal{J}|_U$ is the set of restrictions to U of elements of $\mathcal{J}$ and is an ideal in $\Omega(U)$ since the wedge product is natural with respect to restrictions (that is, with respect to pull-back by inclusion maps).

Definition 11.12. Let $\mathcal{J}$ be an ideal in $\Omega(M)$; we say that $\mathcal{J}$ is **locally generated** by $n-k$ smooth 1-forms if there is an open cover of M such that for each open U from the cover, there are 1-forms $\alpha^1, \ldots, \alpha^{n-k}$ independent on U such that $\mathcal{J}|_U$ is an ideal of $\Omega(U)$ generated by $\alpha^1, \ldots, \alpha^{n-k}$.

Exercise 11.13. Show that if $\mathcal{J}$ is locally generated by $n-k$ smooth 1-forms as above, then it defines a unique smooth rank k distribution E on M such that for each U in the cover, and corresponding $\alpha^1, \ldots, \alpha^{n-k}$, we have
$$E_q = \operatorname{Ker} \alpha^1(q) \cap \cdots \cap \operatorname{Ker} \alpha^{n-k}(q) \text{ for all } q \in U.$$

11.3. Differential Forms and Integrability

Definition 11.14. Let $E \to M$ be a rank k distribution on an n-manifold M. Let U be open in M. A j-form $\omega \in \Omega^j(U)$ is said to **annihilate** the distribution on U if for each $p \in U$

$$\omega_p(v_1, \ldots, v_j) = 0 \text{ whenever } v_1, \ldots, v_j \in E_p.$$

An element $\omega \in \Omega(U)$ is said to **annihilate** the distribution if each homogeneous component of ω annihilates the distribution. The subset of all elements of $\Omega(U)$ that annihilate the distribution $E \to M$ is denoted $\mathcal{I}(U)$.

It is easy to check that $\mathcal{I}(U)$ is an ideal in $\Omega(U)$. Clearly, local defining forms for a distribution annihilate the distribution.

Lemma 11.15. *The following conditions on a distribution $E \to M$ are equivalent:*

(i) *$E \to M$ is an involutive distribution.*

(ii) *For every open $U \subset M$, and 1-form $\omega \in \Omega^1(U)$ that annihilates the distribution, the form $d\omega$ also annihilates the distribution on U.*

Proof. The proof that these two conditions are equivalent follows easily from the formula

(11.3) $\qquad d\omega(X, Y) = X(\omega(Y)) - Y\omega(X) - \omega([X, Y]).$

Suppose that (i) holds. If ω annihilates E and X, Y lie in E, then the above formula becomes

$$d\omega(X, Y) = -\omega([X, Y]),$$

which shows that $d\omega$ vanishes on E since $[X, Y] \in E$ by condition (i). Conversely, suppose that (ii) holds and that $X, Y \in E$. Let $\alpha^1, \ldots, \alpha^{n-k}$ be local defining 1-forms. Then each form $d\alpha^i$ annihilates the distribution and also $\alpha^j(X) = \alpha^j(Y) = 0$ for all j. Using equation (11.3) again, we have $\alpha^i([X, Y]) = -d\alpha^i(X, Y) = 0$ for all i so that $[X, Y] \in E$. $\square$

Lemma 11.16. *Let $E \to M$ be a rank k distribution on the n-manifold M. A j-form ω annihilates E if and only if, whenever $\alpha^1, \ldots, \alpha^{n-k}$ are local defining 1-forms on some open U, we have*

(11.4) $\qquad \omega|_U = \sum_{i=1}^{n-k} \alpha^i \wedge \beta^i$

for some $\beta^1, \ldots, \beta^{n-k} \in \Omega^{j-1}(U)$.

Proof. Recalling the definition of exterior product makes it clear that if ω satisfies (11.4), then it annihilates E on U. If ω satisfies such an equation near each point for some generating 1-forms, then ω annihilates E on all of M.

Suppose that $\alpha^1, \ldots, \alpha^{n-k}$ are local defining forms for the distribution defined on an open set U and let $p \in U$. Then on some neighborhood V of p we can extend the coframe field $\alpha^1, \ldots, \alpha^{n-k}$ to $\alpha^1, \ldots, \alpha^n$ in such a way that $\alpha^1, \ldots, \alpha^n$ is independent on V. Define $X_1, \ldots, X_n$ as dual to the frame $\alpha^1|_V, \ldots, \alpha^n|_V$ so that $X_{n-k+1}, \ldots, X_n$ span E on V. Now for any $\omega \in \Omega^j(U)$, we have

$$\omega|_V = \sum_{i_1 < \cdots < i_j} \omega_{i_1 \cdots i_j} \, \alpha^{i_1}|_V \wedge \cdots \wedge \alpha^{i_j}|_V,$$

where $\omega_{i_1 \cdots i_j} = \omega(X_{i_1}, \ldots, X_{i_j})$ on V. Thus ω annihilates E on V if and only if $\omega(X_{i_1}, \ldots, X_{i_j}) = 0$ on V whenever $n - k + 1 \leq i_1 < \cdots < i_j \leq n$. So if ω annihilates E, then in the expansion of ω we need only include those $\omega_{i_1 \cdots i_j}$ such that $i_1 < \cdots < i_j$ and such that at least one index is less than or equal to $n - k$. But in the latter case we would have $i_1 \leq n - k$. Thus we can write

$$\omega|_V = \sum_{\vec{I}: i_1 \leq n-k} \omega_{i_1 \cdots i_j} \, \alpha^{i_1}|_V \wedge \cdots \wedge \alpha^{i_j}|_V$$

$$= \sum_{i=1}^{n-k} \alpha^i|_V \wedge \left(\sum_{i_2 < \cdots < i_j} \omega_{ii_2 \cdots i_j} \, \alpha^{i_2}|_V \wedge \cdots \wedge \alpha^{i_j}|_V \right)$$

$$:= \sum_{i=1}^{n-k} \alpha^i|_V \wedge \beta^i_V.$$

This expression holds on $V \subset U$, but there is a cover of U by such V on which similar expressions hold. Denote this cover by $\mathcal{V}$. By using a partition of unity $\{\phi_V : V \in \mathcal{V}\}$ subordinate to this cover we can patch these expressions together. Notice that for any $V \in \mathcal{V}$ we have $\phi_V(\alpha^i|_V) = \phi_V \alpha^i$ for $1 \leq i \leq n - k$, so

$$\omega|_U = \sum_{V \in \mathcal{V}} \phi_V \omega|_V = \sum_{V \in \mathcal{V}} \phi_V \sum_{i=1}^{n-k} \alpha^i|_V \wedge \beta^i_V$$

$$= \sum_{V \in \mathcal{V}} \sum_{i=1}^{n-k} \left(\phi_V \, \alpha^i|_V \right) \wedge \beta^i_V$$

$$= \sum_{V \in \mathcal{V}} \sum_{i=1}^{n-k} \phi_V \alpha^i \wedge \beta^i_V = \sum_{i=1}^{n-k} \alpha^i \wedge \sum_{V \in \mathcal{V}} \phi_V \beta^i_V$$

$$= \sum_{i=1}^{n-k} \alpha^i \wedge \beta^i$$

for $\beta^i := \sum_{V \in \mathcal{V}} \phi_V \beta^i_V$. $\square$

11.3. Differential Forms and Integrability

Corollary 11.17. *A distribution $E \to M$ is involutive if and only if for every local defining set of 1-forms $\alpha^1, \ldots, \alpha^{n-k}$ we have*

$$d\alpha^i = \sum_{j=1}^{n-k} \alpha^j \wedge \gamma^i_j$$

for some 1-forms γ^i_j with $1 \leq i, j \leq n-k$.

Proof. Just use the above theorem together with Lemma 11.15. □

Theorem 11.18. *A distribution $E \to M$ on M is involutive (and hence completely integrable) if and only if $d\mathcal{I}(E) \subset \mathcal{I}(E)$.*

The condition $d\mathcal{I}(E) \subset \mathcal{I}(E)$ is expressed by saying that $\mathcal{I}(E)$ is a **differential ideal**.

Proof. Suppose that $d\mathcal{I}(E) \subset \mathcal{I}(E)$. If ω is any 1-form that annihilates E on an open set U, then pick an arbitrary point $p \in U$ and a smooth cut-off function ϕ with support in U and equal to one on a neighborhood of p. Then $d(\phi\omega) \in \mathcal{I}(E)$. But $d(\phi\omega)(p) = d\omega(p)$, and since p was arbitrary, we see that $d\omega$ annihilates E on U. The open set U was also arbitrary so we can apply Lemma 11.15.

Now suppose that $E \to M$ is involutive and choose $\eta \in \mathcal{I}(E)$. Without loss of generality we may assume that η is homogeneous of degree j. Notice that the case $j = 0$ is trivial, while the case $j = 1$ is Lemma 11.15. For $j \geq 2$, consider a local defining set of 1-forms $\alpha^1, \ldots, \alpha^{n-k}$ on an open set U. Since by assumption η annihilates E, we can use Lemma 11.16 and Corollary 11.17 to write

$$\begin{aligned}
d\eta|_U &= d\left(\sum_{i=1}^{n-k} \alpha^i \wedge \beta^i\right) \\
&= \sum_{i=1}^{n-k} d\alpha^i \wedge \beta^i - \sum_{i=1}^{n-k} \alpha^i \wedge d\beta^i \\
&= \sum_{i=1}^{n-k}\sum_j \alpha^j \wedge \gamma^i_j \wedge \beta^i - \sum_{i=1}^{n-k} \alpha^i \wedge d\beta^i.
\end{aligned}$$

Now use Lemma 11.16 again to conclude that $d\eta$ annihilates E and so is in $\mathcal{I}(E)$. □

We close this section with a convenient application of vector-valued forms.

Proposition 11.19. *Let ω be a 1-form on an n-manifold with values in a finite-dimensional vector space V. Then $E_x = \operatorname{Ker} \omega_x$ defines a distribution*

E if $\dim E_x = k$ is constant for $x \in M$. This distribution is involutive and hence integrable if and only if $d\omega(X,Y) = 0$ whenever X, Y lie in E.

Proof. The first statement is clear once we choose a basis $\mathbf{e}_1, \ldots, \mathbf{e}_n$ for V for then we can write $\omega = \sum \mathbf{e}_i \omega^i$ for some linearly independent smooth 1-forms $\omega^1, \ldots, \omega^k$ which clearly define the distribution.

Choose local spanning fields $X_1, \ldots, X_k$ defined on an open set U. Then E is integrable on U if and only if
$$[X_i, X_j] \in \mathrm{span}\{X_1, \ldots, X_k\} \text{ for } 1 \leq i, j \leq k,$$
which is true if and only if $\omega([X_i, X_j]) = 0$ for $1 \leq i, j \leq k$. But as before,
$$d\omega(X_i, X_j) = X_i(\omega(X_j)) - X_j(\omega(X_i)) - \omega([X_i, X_j])$$
$$= -\omega([X_i, X_j]).$$
Thus E is integrable on U if and only if $d\omega(X_i, X_j) = 0$ for $1 \leq i, j \leq k$. This implies the result. $\square$

11.4. Global Frobenius Theorem

First notice that if $(U, \mathbf{x})$ is a distinguished chart for a distribution on an n-manifold with $\mathbf{x}(U) = V \times W \subset \mathbb{R}^k \times \mathbb{R}^{n-k}$, then the plaques $\mathbf{x}^{-1}(V \times \{y\})$ for each $y \in W$ are embedded integral manifolds. If $O \subset V$ and $y \in W$, then $\mathbf{x}^{-1}(O \times \{y\})$ is an open subset of the plaque $\mathbf{x}^{-1}(V \times \{y\})$ (in its submanifold topology). Clearly, all open subsets of a plaque are of this form.

Proposition 11.20. *Let $E \to M$ be an integrable distribution of rank k on the n-manifold M. Let $(U, \mathbf{x})$ be a fixed distinguished chart for E where $\mathbf{x}(U) = V \times W \subset \mathbb{R}^k \times \mathbb{R}^{n-k}$. If S is a connected integral manifold for E, then $S \cap U$ is a countable disjoint union of connected open subsets of the plaques of $(U, \mathbf{x})$. These open connected subsets are open in S and embedded in M.*

Proof. Since the inclusion $\iota : S \hookrightarrow M$ is an immersion, the sets $S \cap U = \iota^{-1}(U)$ are open in S. The set $S \cap U$ is a disjoint union of connected components each of which is open in S. Since S is second countable, this union must be a countable union. Let C be a connected component of $S \cap U$. We show first that $C \subset \mathbf{x}^{-1}(V \times \{y\})$ for some $y \in W$. Since $(U, \mathbf{x})$ is a distinguished chart, the 1-forms $dx^{k+1}, \ldots, dx^n$ are local defining forms for the distribution and so $dx^{k+1} = \cdots = dx^n = 0$ on $S \cap U$ and hence on C. But C is connected, so $x^{k+1}, \ldots, x^n$ are all constant on C, which puts it in a single plaque, say $\mathbf{x}^{-1}(V \times \{y\})$.

We know that $C \hookrightarrow M$ is smooth. Since $\mathbf{x}^{-1}(V \times \{y\})$ is embedded in M, the inclusion $C \hookrightarrow \mathbf{x}^{-1}(V \times \{y\})$ is also smooth by Corollary 3.15 and is thus an injective immersion. Since C and $\mathbf{x}^{-1}(V \times \{y\})$ are manifolds of the

11.4. Global Frobenius Theorem

same dimension, the inclusion $C \hookrightarrow \mathbf{x}^{-1}(V \times \{y\})$ is a local diffeomorphism and hence an open map. It follows that $C \hookrightarrow \mathbf{x}^{-1}(V \times \{y\})$ is an embedding. Since the inclusion $C \hookrightarrow M$ is the composition of embeddings $C \hookrightarrow \mathbf{x}^{-1}(V \times \{y\}) \hookrightarrow M$, it is itself an embedding. □

If $E \to M$ is an integrable distribution of rank k on the n-manifold M and $p \in M$, then we would like to consider the union

$$N = \bigcup_{S \in \mathcal{F}(p)} S,$$

where $\mathcal{F}(p)$ is the family of all *connected* integral manifolds containing p. We would like to say that N is the maximal integral manifold that contains p, but we must show that N is indeed an integral manifold. Let $\{(U_\alpha, \mathbf{x}_\alpha)\}_{\alpha \in A}$ be an atlas of *distinguished* charts for the distribution where $\mathbf{x}_\alpha(U_\alpha) = V_\alpha \times W_\alpha \subset \mathbb{R}^k \times \mathbb{R}^{n-k}$. On each plaque $\mathbf{x}^{-1}(V_\alpha \times \{y\})$ of $(U_\alpha, \mathbf{x}_\alpha)$ which is in N, we define

$$\mathrm{pr}_1 \circ \mathbf{x}_\alpha : \mathbf{x}^{-1}(V_\alpha \times \{y\}) \to V_\alpha \subset \mathbb{R}^k,$$

and this map is taken as a chart on N whose domain is a plaque. Charts obtained in this way will be called **plaque charts** and the family of all such charts provides N with an atlas. The question of the smoothness of overlaps is routine and is left to the reader. Thus we have a smooth atlas on N which induces a topology such that each plaque, and indeed, each member of $\mathcal{F}(p)$, is open. The topology induced by this smooth structure is often called the **leaf topology**. It is also evident that N is connected since it is the union of connected sets containing a fixed point. We can also see directly that N is path connected. Indeed, if q is in N, then it must be in some $S_1 \in \mathcal{F}(p)$. But also $p \in S_1$ by definition, and so since S_1 is connected (and hence smoothly path connected), there is a smooth curve connecting p and q. The questions that remain are whether the topology is Hausdorff and whether it is paracompact. Once again the proof that N is Hausdorff is easy and is left to the reader. The harder question is paracompactness. But this follows from the fact that N clearly satisfies property W(k) of Definition 3.18, and so N is a weakly embedded submanifold by Proposition 3.20. On the other hand, the part of the proof of that proposition that established paracompactness appealed to induced Riemannian metrics and material about topologies induced by metrics that we still have not covered. For this reason we offer a direct and traditional proof. First, since N is connected, the goal is to prove that it is second countable. The inclusion $N \hookrightarrow M$ is continuous, so N is contained in a connected component of M. Thus we may assume without loss of generality that M is second countable, and so also that the atlas $\{(U_\alpha, \mathbf{x}_\alpha)\}_{\alpha \in A}$ of distinguished charts is countable. This latter fact is one of the key points. Notice that each plaque is a regular

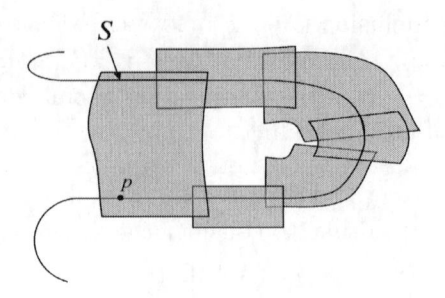

Figure 11.2. Slice accessible from point p

submanifold, so each such plaque is itself second countable. Observe that by construction, if a set $N \cap U_\alpha$ intersects a plaque in U_α, then that plaque is actually contained in $N \cap U_\alpha$. Indeed, if a plaque of U_α intersects N, then there is an element $S \in \mathcal{F}(p)$ such that S meets the plaque in an open set in the topology of the plaque (Proposition 11.20). But then the union of this plaque with S is another element of $\mathcal{F}(p)$ and so occurs in the union that defines N. This reduces the problem to that of showing that N contains only a countable number of such plaques.

We consider all sequences of the form $(U_{\alpha(1)}, P_1), \ldots, (U_{\alpha(m)}, P_m)$, where P_i is a plaque for a chart $(U_{\alpha(i)}, \mathbf{x}_{\alpha(i)})$ taken from our countable atlas of distinguished charts, such that $P_m = P$ is a plaque contained in $U_{\alpha(m)} \cap N$,

$$p \in P_1, \ P_m = P, \ P_i \subset N \text{ and } P_i \cap P_{i+1} \neq \emptyset \text{ for } i = 1, \ldots, m-1.$$

Notice that m is not fixed, but must be finite. Let us call such a sequence a *finite connection* from p to the plaque P. We first show that for each plaque P contained in some $U_\alpha \cap N$, there is at least one such finite connection from p to P. To see this, consider a point $q \in U_\alpha \cap N$ and suppose that $q \in P$. Let $c : [0,1] \to N$ be a smooth curve with $c(0) = p$ and $c(1) = q$. Now $c([0,1])$ is a compact set, so there is a sequence $t_1 < t_2 < \cdots < t_m$ such that for each i, $c([t_i, t_{i+1}])$ is contained in a distinguished chart, say $(U_{\alpha(i)}, \mathbf{x}_{\alpha(i)})$. Of course, each $c([t_i, t_{i+1}])$ is also connected and so must be contained in a connected component of $U_{\alpha(i)} \cap N$ and thus in some plaque which we call P_i. Clearly we have created a finite connection from p to P.

Now consider the family $\mathcal{C}$ of *all* finite connections of p to the plaques contained in the various $U_\alpha \cap N$. Each finite connection can be mapped to the plaque on which it terminates, and we have just seen that this map is surjective. Thus the set of plaques of the various $U_\beta \cap N$ that are finitely connectable to p must have cardinality less than or equal to that of $\mathcal{C}$. It may already seem that since the atlas is countable, $\mathcal{C}$ must also be countable. However, while the set of all finite sequences $U_{\alpha(1)}, \ldots, U_{\alpha(m)}$ is certainly countable, there is still the question of how many ways there are to choose the P_i inside each $U_{\alpha(i)}$ to build a connection of p to a plaque. However, since each plaque P_i is certainly an integral manifold, Proposition 11.20 tells us that it can only intersect $U_{\alpha(i+1)}$ in a countable number of plaques. Thus for a given sequence $U_{\alpha(1)}, \ldots, U_{\alpha(m)}$ there is only a countable number of ways to choose the corresponding P_i. It follows that N contains only a countable number of plaque charts and so is a second countable connected integral manifold for the distribution. It is clearly maximal by construction.

One final thing to notice is that unless the rank of the distribution is equal to the dimension of the manifold, M has an uncountable number of maximal integral manifolds, and they partition M. We have now proved the following:

Theorem 11.21. *Let $E \to M$ be an integrable distribution of rank k on the n-manifold M. There is a set of maximal integral manifolds for E which partition N. These are weakly embedded submanifolds.*

There is another slightly different way of looking at the above theorem. Namely, the set of all plaque charts puts a new smooth structure on the set M. We have seen that the overlaps are only nonempty if the plaques are on the same integral manifold. This smooth structure and topology gives us a new manifold which we might denote by $M(E)$. It has the same underlying set as M, but has a finer topology. The connected components of $M(E)$ are exactly the maximal integral manifolds.

Proposition 11.22. *If $\mathcal{L}_1$ and $\mathcal{L}_2$ are connected integral manifolds of an integrable distribution, then either $\mathcal{L}_1 \cap \mathcal{L}_2$ is empty or it is an open subset of the maximal integral manifold containing any point of $\mathcal{L}_1 \cap \mathcal{L}_2$. The set $\mathcal{L}_1 \cap \mathcal{L}_2$ is also open in the integral manifolds $\mathcal{L}_1$ and $\mathcal{L}_2$.*

Proof. Assume that $\mathcal{L}_1 \cap \mathcal{L}_2$ is not empty and let $p \in \mathcal{L}_1 \cap \mathcal{L}_2$. Then there is a distinguished chart $(U, \mathbf{x})$ for the distribution containing p such that both $\mathcal{L}_1 \cap U$ and $\mathcal{L}_2 \cap U$ are plaques containing p. Clearly, the plaques must be the same. Since plaques are open by definition in the maximal integral manifold, and since p was an arbitrary point in $\mathcal{L}_1 \cap \mathcal{L}_2$, it follows that $\mathcal{L}_1 \cap \mathcal{L}_2$ is open in this maximal integral manifold. The last statement is obvious. $\square$

Corollary 11.23. *Suppose that M and N are smooth manifolds and that there is an integrable distribution E on $M \times N$. Suppose that for some connected open subset U of M there are smooth functions $f_1 : U \to N$ and $f_2 : U \to N$ such that the graphs of f_1 and f_2 are integral manifolds. Then these graphs are either disjoint or equal. In the latter case, $f_1 = f_2$.*

Proof. Let Γ_{f_1} and Γ_{f_2} the graphs and suppose that $\Gamma_{f_1} \cap \Gamma_{f_2}$ is not empty. Let $(p_0, f_1(p_0)) = (p_0, f_2(p_0)) \in \Gamma_{f_1} \cap \Gamma_{f_2}$ and consider the set
$$S = \{p \in U : f_1(p) = f_2(p)\}.$$
Then S is clearly closed. We show that it is also open. Pick $p \in S$. Then Γ_{f_1} and Γ_{f_2} intersect at the point $(p, f_1(p)) = (p, f_2(p))$. But since Γ_{f_1} and Γ_{f_2} are integral manifolds of an integrable distribution on $M \times N$, the previous proposition tells us that $\Gamma_{f_1} \cap \Gamma_{f_2}$ is open in Γ_{f_1}, and so, since $\mathrm{pr}_1 : U \times N \to U$ restricts to a diffeomorphism on Γ_{f_1} (as for all graphs of smooth functions), we see that $S = \mathrm{pr}_1(\Gamma_{f_1} \cap \Gamma_{f_2})$ is an open set in U. Since U is connected, we have $S = U$. Thus $f_1 = f_2$ and the graphs are equal. $\square$

If we consider what we have discovered about integrable distributions, we arrive naturally at the following definition (and new terminology).

Definition 11.24. Let M be an n-manifold. A k-dimensional **foliation** $\mathcal{F}_M$ of M (or on M) is a partition of M into a family of disjoint connected subsets $\{\mathcal{L}_\alpha\}_{\alpha \in A}$ such that for every $p \in M$, there is a chart $(U, \mathbf{x})$ centered at p of the form $\mathbf{x} : U \to V \times W \subset \mathbb{R}^k \times \mathbb{R}^{n-k}$ for connected V and W with the property that for each $\mathcal{L}_\alpha$ the connected components $(U \cap \mathcal{L}_\alpha)_\beta$ of $U \cap \mathcal{L}_\alpha$ are given by
$$\varphi((U \cap \mathcal{L}_\alpha)_\beta) = V \times \{c_{\alpha,\beta}\},$$
where $c_{\alpha,\beta} \in W \subset F$ are constants. These charts are called **distinguished charts** for the foliation or **foliation charts**. The connected sets $\mathcal{L}_\alpha$ are called the **leaves** of the foliation while for a given chart as above, the connected components $(U \cap \mathcal{L}_\alpha)_\beta$ are called **plaques**.

It is easy to see that if $\mathcal{F}_M$ is a foliation as above, then there is a unique integrable distribution $E \to M$ on M such that if p is in the domain of a foliation chart $(U, \mathbf{x})$, then
$$E_p = \mathrm{Ker}\, dx|_p^{k+1} \cap \cdots \cap \mathrm{Ker}\, dx|_p^n.$$
Furthermore the leaves of the foliation are exactly the maximal integral manifolds of the distribution. We call this the distribution generated by the foliation. Combining this observation with Theorem 11.21 we obtain

Theorem 11.25. *Every integrable rank k distribution gives rise to a unique k-dimensional foliation whose leaves are the maximal integral manifolds. Conversely, a foliation generates an integrable distribution whose maximal integral manifolds are exactly the leaves of the original foliation.*

Thus the theory of foliations is essentially the study of integrable distributions. Now we see that the leaves are weakly embedded submanifolds. The connected components $(U \cap \mathcal{L}_\alpha)_\beta$ of $U \cap \mathcal{L}_\alpha$ are of the form $C_x(U \cap \mathcal{L}_\alpha)$

for some $x \in \mathcal{L}_\alpha$ (recall Definition 3.17). An important point anticipated by our concern above is that a fixed leaf $\mathcal{L}_\alpha$ may intersect a given chart domain U in many, even a countably infinite number of disjoint connected pieces no matter how small U is taken to be. In fact, it may be that $U \cap \mathcal{L}_\alpha$ is dense in U. On the other hand, each $\mathcal{L}_\alpha$ is connected by definition. The usual first example of this behavior is given by the irrationally foliated torus. Here we take $M = T^2 := S^1 \times S^1$ and let the leaves be given as the image of the immersions $\iota_a : t \mapsto (e^{ia\pi t}, e^{i\pi t})$, where a is a real number. If a is irrational, then the image $\iota_a(\mathbb{R})$ is a (connected) dense subset of $S^1 \times S^1$. On the other hand, even in this case there are infinitely many distinct leaves.

It may seem that a foliated manifold is just a special manifold, but from one point of view, a foliation is a generalization of a manifold. For instance, we can think of a manifold M as a foliation where the points are the leaves. This is called the **discrete foliation** on M. At the other extreme a manifold may be thought of as a foliation with a single leaf $\mathcal{L} = M$ (the **trivial foliation**). We also have the following special cases:

Example 11.26. On a product manifold, say $M \times N$, we have two complementary foliations:
$$\{\{p\} \times N\}_{p \in M}$$
and
$$\{M \times \{q\}\}_{q \in N}.$$

Example 11.27. Given any submersion $f : M \to N$, the level sets $\{f^{-1}(q)\}_{q \in N}$ form the leaves of a foliation.

Example 11.28. The fibers of any vector bundle foliate the total space.

Example 11.29 (Reeb foliation). Consider the strip in the plane given by $\{(x, y) : |x| \leq 1\}$. For $a \in \mathbb{R} \cup \{\pm\infty\}$, we form leaves $\mathcal{L}_a$ as follows:
$$\mathcal{L}_a := \{(x, a + f(x)) : |x| \leq 1\} \text{ for } a \in \mathbb{R},$$
$$\mathcal{L}_{\pm\infty} := \{(\pm 1, y) : |y| \leq 1\},$$
where $f(x) := \exp(\frac{x^2}{1-x^2}) - 1$. By rotating this symmetric foliation about the y axis we obtain a foliation of the solid cylinder. This foliation is such that translation of the solid cylinder C in the y direction maps leaves diffeomorphically onto leaves, and so we may let $\mathbb{Z}$ act on C by $(x, y, z) \mapsto (x, y+n, z)$ and then $C/\mathbb{Z}$ is a solid torus with a foliation called the Reeb foliation.

Example 11.30. The one point compactification of $\mathbb{R}^3$ is homeomorphic to $S^3 \subset \mathbb{R}^4$. Thus $S^3 - \{p\} \cong \mathbb{R}^3$ and so there is a copy of the Reeb foliated solid torus inside S^3. The complement of a solid torus in S^3 is another solid torus. It is possible to put another Reeb foliation on this complement and thus foliate all of S^3. The only compact leaf is the torus that is the common boundary of the two complementary solid tori.

Theorem 11.31. *Let $\mathcal{F}_M$ be a foliation on M and $\mathcal{L}$ a leaf of the foliation. Then for any smooth map $f : N \to M$ with $f(N) \subset \mathcal{L}$ the map $f : N \to \mathcal{L}$ is smooth.*

Proof. Just use Theorem 3.14. □

11.5. Applications to Lie Groups

First we fulfill the promise of proving Theorem 5.66, which we repeat here for convenience.

Theorem 11.32. *Let G be a Lie group with Lie algebra $\mathfrak{g}$. If $\mathfrak{h}$ is a Lie subalgebra of $\mathfrak{g}$, then there is a unique connected Lie subgroup H of G whose Lie algebra is $\mathfrak{h}$.*

Proof. For $a \in G$, we let Δ_a denote the subspace of $T_a G$ that is the set of all $X(a)$ for left invariant vector fields X with $X(e) \in \mathfrak{h}$. Thus $v_a \in \Delta_a$ if and only if $v_a = T_e L_a \cdot v$ for some $v \in \mathfrak{h}$. If $X_1, \ldots, X_k$ are left invariant fields such that $X_1(e), \ldots, X_k(e)$ is a basis for $\mathfrak{h}$, then $X_1, \ldots, X_k$ span the distribution which is therefore smooth. Since $\mathfrak{h}$ is a subalgebra, it follows that $\Delta : a \mapsto \Delta_a$ is an integrable distribution. Let H be the maximal (connected) integral manifold containing e. Note that for any $b \in G$, we have $T_e L_b(\Delta_a) = \Delta_{ba}$ so that $TL_b : TG \to TG$ leaves the distribution invariant. Thus L_b induces a permutation of the set of maximal integral manifolds and takes the maximal integral manifold through a diffeomorphically to that through ba. Thus if $h \in H$, then $L_{h^{-1}}$ takes H to the maximal integral manifold containing e, which is just H. In other words, $L_{h^{-1}}(H) = H$. This shows that H is a subgroup. It remains to show that the multiplication map $\mu : H \times H \to H$ is smooth. But the multiplication map *into* G, $H \times H \to G$, is smooth and so by Theorem 11.31 the map $H \times H \to H$ is smooth. □

Theorem 11.33. *Let G and H be Lie groups with respective Lie algebras $\mathfrak{g}$ and $\mathfrak{h}$. If $h : \mathfrak{g} \to \mathfrak{h}$ is a Lie algebra homomorphism, then there is a neighborhood U of the identity $e \in G$ and a smooth map $f : U \to H$ such that*

$$f(xy) = f(x)f(y)$$

whenever $x, y \in U$ and $xy \in U$, and such that

$$T_e f \cdot v = h(v)$$

for every $v \in \mathfrak{g}$.

Proof. Let $\mathfrak{k} \subset \mathfrak{g} \times \mathfrak{h}$ be defined by

$$\mathfrak{k} := \{(v, h(v)) : v \in \mathfrak{g}\}.$$

11.5. Applications to Lie Groups

The fact that h is a homomorphism implies that $\mathfrak{k}$ is a Lie subalgebra of $\mathfrak{g} \times \mathfrak{h}$. Thus by Theorem 11.32, there is a connected Lie subgroup K of $G \times H$ with Lie algebra $\mathfrak{k}$. Now let $\iota : K \hookrightarrow G \times H$ be inclusion and define a homomorphism $\rho : K \to G$ by

$$\rho := \mathrm{pr}_1 \circ \iota.$$

If $v \in \mathfrak{g}$, then

$$T\rho \cdot (v, h(v)) = v,$$

and this means that $T\rho : T_{(e,e)}K \to T_eG$ is a linear isomorphism. Thus by the inverse mapping theorem, there is a neighborhood V of $(e, e) \in K$ such that $\rho|_V$ is a diffeomorphism onto an open neighborhood U of $e \in G$. Define the homomorphism $\psi : K \to H$ by

$$\psi := \mathrm{pr}_2 \circ \iota,$$

where $\mathrm{pr}_2 : G \times H \to H$ is the second factor projection. Notice that $T_e\psi \cdot (v, h(v)) = h(v)$. Now let

$$f := \psi \circ \rho|_V^{-1}.$$

A straightforward diagram chase argument shows that $f(xy) = f(x)f(y)$ if $x, y \in U$ and $xy \in U$.

If $v \in \mathfrak{g}$, then $T\rho(v, h(v)) = v$ implies that $T(\rho|_V^{-1}) \cdot v = (v, h(v))$ so

$$Tf \cdot v = T\psi \circ T\left(\rho|_V^{-1}\right) \cdot v = T\psi \cdot (v, h(v)) = h(v). \qquad \square$$

Theorem 11.34. *If $f_1 : G \to H$ and $f_2 : G \to H$ are Lie group homomorphisms with $df_1 = df_2 : \mathfrak{g} \to \mathfrak{h}$ and G is connected, then $f_1 = f_2$.*

Proof. Let $h := df_1 = df_2$ and define $\mathfrak{k} := \{(v, h(v)) : v \in \mathfrak{g}\}$ and $K \subset G \times H$ as in the proof of the previous theorem. Now define $\theta : G \to G \times H$ by $\theta(x) := (x, f_1(x))$. The image of θ is a subgroup $K_1 \subset G \times H$. For $v \in \mathfrak{g}$, we have $T_e\theta \cdot v = (v, h(v))$ so the Lie algebra of K_1 must be $\mathfrak{k}$. Since G is connected, we must have $K = K_1$ which implies that $f_1 = f$ on U, where f and U are constructed from h as in the last theorem. But equally, $f_2 = h$ on U, and so by Proposition 5.40 we have $f_1 = f_2$. $\qquad \square$

We say that two Lie groups G and H are **locally isomorphic** if there is a diffeomorphism f from a neighborhood U of the identity of G onto a neighborhood V of the identity of H such that $f(xy) = f(x)f(y)$ whenever x, y and xy are contained in U.

Corollary 11.35. *The following assertions hold:*

(i) *Two Lie groups with isomorphic Lie algebras are locally isomorphic.*

(ii) *A connected Lie group with abelian Lie algebra is abelian.*

Proof. (i) Let $h : \mathfrak{g} \to \mathfrak{h}$ be a Lie algebra isomorphism. Then if f is the map constructed in Theorem 11.33, then f is a diffeomorphism on some possibly smaller neighborhood of the identity since $T_e f = h$ is an isomorphism.

(ii) By (i) a connected Lie group G of dimension n must be locally isomorphic to the (additive) abelian Lie group $\mathbb{R}^n$. But a neighborhood of the identity generates the whole group, and so G is abelian. $\square$

11.6. Fundamental Theorem of Surface Theory

In this section we state and outline the proof of a fundamental theorem concerning the existence of surfaces with prescribed first and second fundamental form. Our proof of the main theorem follows [**Pa2**]. To begin with, we need a few results about certain systems of partial differential equations. The first is equivalent to the local Frobenius theorem. For an open set $U \subset \mathbb{R}^k \times \mathbb{R}^m$, denote standard coordinates by $(x, z) = (x^1, \ldots, x^k, z^1, \ldots, z^m)$.

Proposition 11.36. *Let U be an open set in $\mathbb{R}^k \times \mathbb{R}^m$ and let (A^i_j) be an $m \times k$ matrix of smooth functions on U. Then the following assertions are equivalent:*

(i) *For every $(x_0, z_0) \in U$, there is a neighborhood V of x_0 in $\mathbb{R}^k$ and a unique smooth map $f : V \to \mathbb{R}^m$ with $f(x_0) = z_0$ such that*

$$(11.5) \qquad \frac{\partial f^i}{\partial x^j}(x) = A^i_j(x, f(x)) \text{ for all } i, j.$$

(ii) *The functions A^i_j satisfy the following system of equations on U:*

$$(11.6) \qquad \frac{\partial A^i_j}{\partial x^k} + \sum_l A^l_k \frac{\partial A^i_j}{\partial z^l} = \frac{\partial A^i_k}{\partial x^j} + \sum_l A^l_j \frac{\partial A^i_k}{\partial z^l} \text{ for all } i, j, k.$$

Proof. If (i) is true, then we obtain (ii) by equality of mixed partials of the f^i and the chain rule (see the comments following the proof).

Conversely, consider the vector fields on U defined by

$$X_j := \left.\frac{\partial}{\partial x^j}\right|_p + \sum_r A^r_j(p) \left.\frac{\partial}{\partial z^r}\right|_p.$$

A bit of linear algebra tells us that these are everywhere independent. Let $E \to U$ be the distribution spanned by these fields. A straightforward check using (11.6) shows that

$$[X_i, X_j] = 0,$$

so there is an integral manifold through each p. Let N be the integral manifold through (x_0, z_0). Using the last m coordinate functions of some distinguished chart, we obtain a map $\Phi : U' \to \mathbb{R}^m$ for some connected open $U' \subset U$ so that the level sets of Φ are integral manifolds. The tangent map

11.6. Fundamental Theorem of Surface Theory

$T_p\Phi$ has kernel E_p at each $p \in N$, and since $\frac{\partial}{\partial z^j}\big|_p$ never lies in E, we see that
$$\frac{\partial \Phi}{\partial z^j} \neq 0 \text{ on } N \cap U' \text{ for all } j.$$
In particular, this holds at (x_0, z_0), and the implicit mapping theorem tells us that a neighborhood of (x_0, z_0) in a plaque of N is the graph of a function $f : V \to \mathbb{R}^m$ with $f(x_0) = z_0$. Define a function $F : V \to \mathbb{R}^k \times \mathbb{R}^m$ by $F(x) := (x, f(x))$. Writing $p = F(x)$, we see that for each i the vector
$$TF \cdot \frac{\partial}{\partial x^i}\bigg|_x = \frac{\partial}{\partial x^i}\bigg|_p + \sum_r \frac{\partial f^r}{\partial x^i}(x) \frac{\partial}{\partial z^r}\bigg|_p$$
is a linear combination of vectors X_j defined above:
$$\frac{\partial}{\partial x^i}\bigg|_{f(x)} + \sum_r \frac{\partial f^r}{\partial x^i}(x) \frac{\partial}{\partial z^r}\bigg|_{f(x)}$$
$$= \sum_{s=1}^k c_i^s \left(\frac{\partial}{\partial x^s}\bigg|_{(x,f(x))} + \sum_r A_s^r(p) \frac{\partial}{\partial z^r}\bigg|_{(x,f(x))} \right).$$
Collecting terms and comparing we see that $c_i^s = \delta_i^s$ and
$$\frac{\partial f^r}{\partial x^j}(x) = A_j^r(x, f(x)).$$
It follows from Corollary 11.23 that f is uniquely determined on a sufficiently small connected neighborhood of (x,y). □

It is often convenient to be able to come up with the integrability conditions for a given application without trying to match indexing and notation with the above theorem. The basis of the procedure is to set mixed partials equal to each other. We demonstrate this using the notation of the theorem. We start with
$$\frac{\partial}{\partial x^k} A_j^i(x, f(x)) = \frac{\partial}{\partial x^j} A_k^i(x, f(x)).$$
Apply the chain rule:
$$\frac{\partial A_j^i}{\partial x^k}(x, f(x)) + \sum_r \frac{\partial}{\partial z^r} A_j^i(x, f(x)) \frac{\partial f^r}{\partial x^k}(x)$$
$$= \frac{\partial A_k^i}{\partial x^j}(x, f(x)) + \sum_s \frac{\partial}{\partial z^s} A_k^i(x, f(x)) \frac{\partial f^s}{\partial x^j}(x).$$

Finally, substitute back using the original equations (11.5) and replace all occurrences of $f(x)$ in the arguments with the independent variable z. We arrive at the integrability conditions:
$$\frac{\partial A_j^i}{\partial x^k}(x, z) + \sum_l A_k^l(x, z) \frac{\partial A_j^i}{\partial z^l}(x, z) = \frac{\partial A_k^i}{\partial x^j}(x, z) + \sum_l A_j^l(x, z) \frac{\partial A_k^i}{\partial z^l}(x, z).$$

The convenience of this may not be clear yet, but we shall shortly demonstrate the usefulness of this method.

Proposition 11.37. *Let U be open in $\mathbb{R}^k \times \mathbb{R}^m$ and (A^i_j) an $m \times k$ matrix of smooth functions on U. Let $(x_0, z_0) \in U$ and suppose that for some connected open set V, both $f_1 : V \to U \subset \mathbb{R}^m$ and $f_2 : V \to U \subset \mathbb{R}^m$ are solutions of*
$$\frac{\partial f^i}{\partial x^j}(x) = A^i_j(x, f(x)) \text{ for all } i, j,$$
$$f(x_0) = z_0.$$
Then $f_1 = f_2$.

Proof. This follows from Corollary 11.23 and the considerations in the proof of the previous proposition. □

Lemma 11.38. *Let V be an open set in $\mathbb{R}^k$ and let (A^i_j) be an $m \times k$ matrix of smooth functions on $V \times \mathbb{R}^m$ that are linear in the second argument and satisfy the integrability conditions (11.6) on $V \times \mathbb{R}^m$. Then for any $x_0 \in V$ there is a ball $B_{x_0} \subset V$ such that for any $(a, b) \in B_{x_0} \times \mathbb{R}^m$ there is a solution defined on B_{x_0} with $f(a) = b$.*

Proof. Let f_i be the solution with $f_i(x_0) = \mathbf{e}_i$, where $\mathbf{e}_i$ is the i-th standard basis vector of $\mathbb{R}^m$. Then $f_1, \ldots, f_m$ are defined and linearly independent on some ball B_{x_0} containing x_0 and contained in the intersection of the domains of the f_i. Choose $(a, b) \in B_{x_0} \times \mathbb{R}^m$ and note that $b = \sum b^r f_r(a)$ for some uniquely defined numbers b^i. Now define $f = \sum b^i f_i$ on B_{x_0}. Then writing $f = (f^1, \ldots, f^m)$, we have for any $x \in V$,
$$A^i_j(x, f(x)) = A^i_j\left(x, \sum b^r f_r(x)\right) = \sum b^r A^i_j(x, f_r(x))$$
$$= \sum b^r \frac{\partial f^i_r}{\partial x^j}(x) = \frac{\partial f^i}{\partial x^j}(x)$$
and
$$f(a) = \sum b^r f_r(a) = b. \qquad \square$$

Corollary 11.39. *Let V be a simply connected open set in $\mathbb{R}^k$ and let (A^i_j) be an $m \times k$ matrix of smooth functions on $U = V \times \mathbb{R}^m$. Suppose that each A^i_j is linear in its second argument. If*
$$\frac{\partial A^i_j}{\partial x^k} + \sum_l A^l_k \frac{\partial A^i_j}{\partial z^l} = \frac{\partial A^i_k}{\partial x^j} + \sum_l A^l_j \frac{\partial A^i_k}{\partial z^l} \qquad \text{for all } i, j, k$$

11.6. Fundamental Theorem of Surface Theory

on $V \times \mathbb{R}^m$, then given any $(x_0, z_0) \in V \times \mathbb{R}^m$, there exists a unique smooth map $f : V \to \mathbb{R}^m$ such that

$$\frac{\partial f^i}{\partial x^j}(x) = A^i_j(x, f(x)) \quad \text{for all } i, j,$$
$$f(x_0) = z_0.$$

Proof. Let $X_j := \frac{\partial}{\partial x^j}\big|_p + \sum_r A^r_j(p) \frac{\partial}{\partial z^r}\big|_p$ be the fields that span an integrable distribution on $V \times \mathbb{R}^m$ as in Proposition 11.36. Let $L_{(x_0,z_0)}$ be the maximal integral manifold through the point (x_0, z_0). Let $\wp$ denote the restriction of the projection $\mathrm{pr}_1 : V \times \mathbb{R}^m \to V$ to $L_{(x_0,z_0)}$. Let $(a_1, b_1) \in L_{(x_0,z_0)}$ and consider the set

$$F_{a_1} = \wp^{-1}(a_1).$$

By Lemma 11.38 above, there is a fixed open set U containing a_1 such that for every $(a_1, b) \in F_{a_1}$ there is a solution $f_b : U \to \mathbb{R}^m$ with $f(a_1) = b$. By Corollary 11.23, the graphs of these solutions are all disjoint open sets in $L_{(x_0,z_0)}$ and $\wp$ restricts to a diffeomorphism on each such graph. Thus $\wp : L_{(x_0,z_0)} \to V$ is a smooth covering map. The local solutions guaranteed to exist by Proposition 11.36 are local sections of this covering. Thus since V is simply connected, we know from Theorem 1.95 that there is a smooth lift $\widetilde{\wp}$ of $\mathrm{id}_V : V \to V$ such that $\widetilde{\wp}(x_0) = (x_0, z_0)$, which in this case means that we have a global section: $\wp \circ \widetilde{\wp} = \mathrm{id}_V$. Now let $f := \mathrm{pr}_2 \circ \widetilde{\wp} : V \to \mathbb{R}^m$. Then $\widetilde{\wp}(x) = (x, f(x))$ and $f(x)$ must be smooth and for every $a \in V$ the function f must agree with the unique local solution which takes the value $f(a)$ at a. $\square$

We return to the situation studied in Chapter 4. Consider an immersion $\mathbf{x} : V \to \mathbb{R}^3$, where V is an open set in $\mathbb{R}^2$ whose standard coordinates will be denoted by u^1, u^2. Let (f_1, f_2, f_3) be the frame fields along $\mathbf{x}$ defined by

$$f_1 := \mathbf{x}_{u^1}, \quad f_2 := \mathbf{x}_{u^2},$$
$$f_3 := N = \mathbf{x}_{u^1} \times \mathbf{x}_{u^2} / \|\mathbf{x}_{u^1} \times \mathbf{x}_{u^2}\|,$$

where $\mathbf{x}_{u^1} = \partial \mathbf{x}/\partial u^1$, etc. The first fundamental form is given by the matrix entries

$$g_{ij} = \langle \mathbf{x}_{u^i}, \mathbf{x}_{u^j} \rangle \text{ for } 1 \leq i, j \leq 2,$$

while the second fundamental form is given by the matrix entries

$$\ell_{ij} = -\langle N_{u^i}, \mathbf{x}_{u^j} \rangle = \langle N, \mathbf{x}_{u^i u^j} \rangle = \langle f_3, \mathbf{x}_{u^i u^j} \rangle.$$

Let us consider $\mathbf{f} = (f_1, f_2, f_3)$ as a matrix function of (u^1, u^2) that takes values in $\mathrm{GL}(3)$. We have

$$(f_i)_{u^1} = \sum_{r=1}^{3} P^r_i f_r \quad \text{and} \quad (f_i)_{u^2} = \sum_{r=1}^{3} Q^r_i f_r$$

for some matrix functions P and Q. In matrix notation, we have
$$\mathbf{f}_{u^1} = \mathbf{f}P,$$
(11.7)
$$\mathbf{f}_{u^2} = \mathbf{f}Q,$$

and these are called the **frame equations**. For convenience, we define a 3×3 matrix function G by $G_{ij} := \langle f_i, f_j \rangle$ for $1 \leq i, j \leq 3$ so that

(11.8)
$$G = \begin{pmatrix} g_{11} & g_{12} & 0 \\ g_{21} & g_{22} & 0 \\ 0 & 0 & 1 \end{pmatrix}.$$

For any given $x = \sum_i^3 x^i f_i$, we have
$$\xi_i = \langle x, f_i \rangle = \left\langle \sum x^k f_k, f_i \right\rangle = x^k \sum \langle f_k, f_i \rangle = G_{ki} x^k,$$
and so
$$\xi_i = \sum (G^t)_{ik} x^k = \sum G_{ik} x^k.$$

Now let $x = (f_i)_{u_1} = \sum f_k P_i^k$. Then $x^k = P_i^k$, so if we define $B_{ij} := \langle f_i, (f_j)_{u^1} \rangle$, then we have

(11.9)
$$\langle f_i, (f_1)_{u^1} \rangle = B_{i1} = \sum G_{ik} P_1^k,$$
$$\langle f_i, (f_2)_{u^1} \rangle = B_{i2} = \sum G_{ik} P_2^k,$$
$$\langle f_i, (f_3)_{u^1} \rangle = B_{i3} = \sum G_{ik} P_3^k,$$

or $B = GP$. Similarly, if $C = (C_{ij}) = \langle f_i, (f_j)_{u^2} \rangle$, then

(11.10)
$$\langle f_i, (f_1)_{u^2} \rangle = C_{i1} = \sum G_{ik} Q_1^k,$$
$$\langle f_i, (f_2)_{u^2} \rangle = C_{i2} = \sum G_{ik} Q_2^k,$$
$$\langle f_i, (f_3)_{u^2} \rangle = C_{i3} = \sum G_{ik} Q_3^k,$$

or $C = GQ$. We arrive at
$$P = G^{-1} B,$$
$$Q = G^{-1} C.$$

We denote the entries of G^{-1} by g^{ij} so that

(11.11)
$$G^{-1} := \begin{pmatrix} g^{11} & g^{12} & 0 \\ g^{21} & g^{22} & 0 \\ 0 & 0 & 1 \end{pmatrix}.$$

Proposition 11.40. *We have*

(11.12)
$$B = \begin{pmatrix} \frac{1}{2}(g_{11})_{u^1} & \frac{1}{2}(g_{11})_{u^2} & -\ell_{11} \\ (g_{12})_{u^1} - \frac{1}{2}(g_{11})_{u^2} & \frac{1}{2}(g_{22})_{u^1} & -\ell_{12} \\ \ell_{11} & \ell_{12} & 0 \end{pmatrix}$$

11.6. Fundamental Theorem of Surface Theory

and

$$(11.13) \quad C = \begin{pmatrix} \frac{1}{2}(g_{11})_{u^2} & \frac{1}{2}(g_{12})_{u^2} - \frac{1}{2}(g_{22})_{u^1} & -\ell_{12} \\ (g_{22})_{u^1} - \frac{1}{2}(g_{11})_{u^2} & \frac{1}{2}(g_{22})_{u^2} & -\ell_{22} \\ \ell_{12} & \ell_{22} & 0 \end{pmatrix}.$$

In particular, P and Q can be written in terms of the matrix entries of the first and second fundamental forms.

Proof. The proof is just a calculation, and we only do part. For example, if i is either 1 or 2, then

$$B_{ii} = \langle f_i, (f_i)_{u^1} \rangle = \langle \mathbf{x}_{u^i}, \mathbf{x}_{u^i u^1} \rangle = \frac{1}{2} \langle \mathbf{x}_{u^i}, \mathbf{x}_{u^i} \rangle_{u^1} = \frac{1}{2}(g_{ii})_{u^1}.$$

Similarly, for $i = 1$ or 2 we have

$$C_{ii} = \langle f_i, (f_i)_{u^2} \rangle = \langle \mathbf{x}_{u^i}, \mathbf{x}_{u^i u^2} \rangle = \frac{1}{2}(g_{ii})_{u^2}.$$

Now, $\frac{1}{2}(g_{12})_{u^1} = \langle \mathbf{x}_{u^1}, \mathbf{x}_{u^2} \rangle_{u^1} = \langle \mathbf{x}_{u^1 u^1}, \mathbf{x}_{u^2} \rangle + \frac{1}{2}(g_{11})_{u^2}$ from above, and so

$$B_{21} = \langle f_2, (f_1)_{u^1} \rangle = \langle \mathbf{x}_{u^1 u^1}, \mathbf{x}_{u^2} \rangle = \frac{1}{2}(g_{12})_{u^1} - \frac{1}{2}(g_{11})_{u^2}.$$

The entries B_{12}, C_{12}, C_{21} are calculated similarly. Next consider B_{i3} for $i = 1$ or 2. We have

$$B_{i3} = \langle f_i, (f_3)_{u^1} \rangle = \langle \mathbf{x}_{u^i}, N_{u^1} \rangle = -\ell_{1i}$$

and

$$0 = \langle f_3, f_i \rangle_{u^1} = \langle (f_3)_{u^1}, f_i \rangle + \langle f_3, (f_i)_{u^1} \rangle,$$

so

$$B_{3i} = -B_{i3}.$$

The entries C_{3i} and C_{i3} are obtained in the same way. Lastly, since $\langle f_3, f_3 \rangle = 1$,

$$0 = \frac{1}{2}\langle f_3, f_3 \rangle_{u^k} = \langle (f_3)_{u^k}, f_3 \rangle = \begin{cases} B_{33} & \text{if } k = 1, \\ C_{33} & \text{if } k = 2. \end{cases} \qquad \square$$

We record an observation to be used later:

$$(11.14) \quad \begin{aligned} B + B^t &= G_{u^1}, \\ C + C^t &= G_{u^2}. \end{aligned}$$

The frame equations (11.7) are a system to which Proposition 11.36 applies. Rather than trying to rewrite the equations in a form that matches that proposition we obtain the integrability conditions by setting

$$(\mathbf{f}_{u^1})_{u^2} = (\mathbf{f}_{u^2})_{u^1}$$

and then

$$\mathbf{f}_{u^2} P + \mathbf{f} P_{u^2} = \mathbf{f}_{u^1} Q + \mathbf{f} Q_{u^1}.$$

Substituting from the frame equations we obtain

$$\mathbf{f}\left(P_{u^2} - Q_{u^1} - (PQ - QP)\right) = 0.$$

Now $\mathbf{f}$ is a nonsingular matrix, so we have the equivalent integrability equation

(11.15) $$P_{u^2} - Q_{u^1} - (PQ - QP) = 0.$$

At this point we pause to appreciate an important fact. Namely, direct calculation reveals that these equations are equivalent to the combination of the Codazzi-Mainardi equation and the Gauss curvature equation, which we now see are integrability conditions (see Problem 7). We thus refer to the above integrability equations (11.15) as the **Gauss-Codazzi equations** with apologies to Gaspare Mainardi (1800–1879).

We now turn things around. Rather than assuming that we have a surface, we take the (g_{ij}) and (ℓ_{ij}) as some given symmetric smooth matrix-valued functions defined on a connected open $V \subset \mathbb{R}^2$ with the assumption that (g_{ij}) is positive definite. Furthermore, we now assume that G, P, and Q are actually defined in terms of these by the formulas above, which we found to be true in the case where we started with a surface. We will show that we can obtain a surface with these as first and second fundamental form.

Theorem 11.41 (Fundamental existence theorem for surfaces). *The following assertions hold:*

(i) *Let V be an open set in $\mathbb{R}^2$ diffeomorphic to an open disk and let $\mathbf{x} : V \to \mathbb{R}^3$ be an immersion with the corresponding first and second fundamental forms given in matrix form as (g_{ij}) and (ℓ_{ij}). Let $\mathbf{y} : V \to \mathbb{R}^3$ be another immersion with the corresponding forms $(\widetilde{g}_{ij})$ and $(\widetilde{\ell}_{ij})$. If*

$$\mathbf{y} = f \circ \mathbf{x}$$

for some proper Euclidean motion $f : \mathbb{R}^3 \to \mathbb{R}^3$, then

$$\widetilde{g}_{ij} = g_{ij},$$
$$\widetilde{\ell}_{ij} = \ell_{ij}.$$

Conversely, if the last equations hold, then $\mathbf{y} = f \circ \mathbf{x}$ for some Euclidean motion f.

(ii) *Suppose that (g_{ij}) and (ℓ_{ij}) are given symmetric matrix-valued functions defined on V with (g_{ij}) positive definite and suppose that G, B and C are defined in terms of the entries of (g_{ij}) and (ℓ_{ij}) as in*

11.6. Fundamental Theorem of Surface Theory

formulas (11.8), (11.11), (11.12) *and* (11.13). *Then if*
$$P = G^{-1}B,$$
$$Q = G^{-1}C,$$
and if $P_{u^2} - Q_{u^1} - (PQ - QP) = 0$, *then there exists an embedding* $\mathbf{x} : V \to \mathbb{R}^3$ *such that* (g_{ij}) *and* (ℓ_{ij}) *are the corresponding first and second fundamental forms.*

Proof. We leave the proof of the first part of (i) to the reader, but note that it can be proved using direct calculation or it can be derived from Theorem 4.22.

For the rest of (i), note that by composing with a translation we may assume that both $\mathbf{x}$ and $\mathbf{y}$ map some fixed point $u \in V$ to the origin in $\mathbb{R}^3$. Let (f_1, f_2, f_3) be the natural frame for $\mathbf{x}$ as above and let $(\widetilde{f}_1, \widetilde{f}_2, \widetilde{f}_3)$ be that of $\mathbf{y}$. By making a rotation we may assume that these two frames agree at p. But since we are assuming that $\widetilde{g}_{ij} = g_{ij}$ and $\widetilde{\ell}_{ij} = \ell_{ij}$, it follows that both frames satisfy the same frame equations and so by Proposition 11.37 they must agree on the connected set V. In particular, $\mathbf{x}_{u^i} = \mathbf{y}_{u^i}$ for $i = 1, 2$. Thus $\mathbf{x}$ and $\mathbf{y}$ only differ by a constant, which must be zero since $\mathbf{x}(u) = \mathbf{y}(u)$.

Next we consider (ii) where (g_{ij}) and (ℓ_{ij}) are given. We want to construct a surface, but first we construct the frame for the desired surface. Since it is assumed that $g = (g_{ij})$ is positive definite, g and the extended matrix G are both invertible and positive definite. Thus P and Q are well-defined. Since we assume that the integrability equations $P_{u^2} - Q_{u^1} - (PQ - QP) = 0$ hold, Theorem 11.36 tells us that we can solve the frame equations locally, near any point $u \in V$ and with any initial conditions $\mathbf{f}(u) = \mathbf{f}_0$ holding as desired. But the system is linear and our domain is diffeomorphic to a disk so we can use Corollary 11.39 to obtain a solution on all of V. Since G is positive definite, we may choose these initial conditions so that
$$\langle f_i(u), f_j(u) \rangle = G_{ij}(u) \quad (ij\text{-th entry of } G \text{ at } u).$$
Having obtained the f_i near u, we now wish to obtain a surface. This means solving the system

(11.16)
$$\mathbf{x}_{u^1} = f_1,$$
$$\mathbf{x}_{u^2} = f_2,$$

and this time the integrability conditions are derived from
$$(f_1)_{u^2} = (f_2)_{u^1}.$$
Using the frame equations, we obtain integrability conditions
$$\sum Q_1^j f_j = \sum P_2^j f_j.$$

This just says that the second column of P is equal to the first column of Q, which is true. Thus we can find $\mathbf{x} : V \to \mathbb{R}^3$ with $\mathbf{x}(u) = 0$ so that (11.16) holds.

Next we show that $\langle f_i, f_j \rangle = G_{ij}$ on all of V. We compute as follows:
$$\begin{aligned}\langle f_i, f_j \rangle_{u^1} &= \langle (f_i)_{u^1}, f_j \rangle + \langle f_i, (f_j)_{u^1} \rangle \\ &= \sum_r Q_i^r \langle f_r, f_j \rangle + \sum_s Q_j^s \langle f_i, f_s \rangle = (QG + (QG)^t)_{ij} \\ &= (GQ + (GQ)^t)_{ij} = (B + B^t)_{ij} = (G_{u^1})_{ij} = (G_{ij})_{u^1}\end{aligned}$$

by equations (11.14). Similarly for u^2. Thus $\langle f_i, f_j \rangle - G_{ij}$ is a constant, which must be zero since it is zero at u. From $\langle f_i, f_j \rangle = G_{ij}$ it follows that $\langle f_3, f_3 \rangle = 1$ and that f_1, f_2 are independent and orthogonal to f_3. It remains to show that $\langle (f_3)_{u^i}, f_j \rangle = -\ell_{ij}$. We compute as follows:
$$\begin{aligned}-\langle (f_3)_{u^1}, f_j \rangle &= (g^{11}\ell_{11} + g^{12}\ell_{12})\langle f_1, f_j \rangle + (g^{12}\ell_{11} + g^{22}\ell_{12})\langle f_2, f_j \rangle \\ &= (g^{11}\ell_{11} + g^{12}\ell_{12})g_{1j} + (g^{12}\ell_{11} + g^{22}\ell_{12})g_{2j} \\ &= \ell_{11}(g^{11}g_{1j} + g^{12}g_{2j}) + \ell_{12}(g^{21}g_{1j} + g^{22}g_{2j}) \\ &= \ell_{11}\delta_j^1 + \ell_{12}\delta_j^2.\end{aligned}$$

This shows that $\langle (f_3)_{u^1}, f_j \rangle = -\ell_{1j}$ for $j = 1, 2$. The computation of $-\langle (f_3)_{u^2}, f_j \rangle$ is similar and left for the reader. $\square$

11.7. Local Fundamental Theorem of Calculus

Recall the structure equations (8.15) satisfied by the Maurer-Cartan form ω_G for a Lie group G:
$$d\omega_G = -\frac{1}{2}[\omega_G, \omega_G]^\wedge.$$

If $v_1 = X_1(e), \ldots, v_n = X_n(e)$ is a basis for the Lie algebra $\mathfrak{g}$ which extends to left invariant vector fields $X_1, \ldots, X_n$, then the above equation is equivalent to
$$d\omega^k = -\sum_{i<j} c_{ij}^k \omega^i \wedge \omega^j = -\frac{1}{2}\sum_{i,j} c_{ij}^k \omega^i \wedge \omega^j,$$

where the c_{ij}^k are the structure constants associated to $\omega^1, \ldots, \omega^n$, which is the left invariant frame field dual to $X_1, \ldots, X_n$. If M is some m-manifold and $f : M \to G$ is a smooth map, then $\omega_f = f^*\omega_G$ is a $\mathfrak{g}$-valued 1-form on M. By naturality we have
$$d\omega_f = -\frac{1}{2}[\omega_f, \omega_f]^\wedge$$

or equivalently
$$d\omega_f^k = -\frac{1}{2}\sum_{i,j} c_{ij}^k \omega_f^i \wedge \omega_f^j,$$

11.7. Local Fundamental Theorem of Calculus

where $\omega_f^i = f^*\omega^i$ for $i = 1, \ldots, n$. The $\mathfrak{g}$-valued 1-form ω_f is sometimes called the (left) Darboux derivative of f. The right Darboux derivative is defined similarly using the right Maurer-Cartan form.

If we think of a $\mathfrak{g}$-valued 1-form on a manifold M as a map $TM \to \mathfrak{g}$, then $\omega_f = f^*\omega_G = \omega_G \circ Tf$. From this point of view we can understand why ω_f is a kind of derivative of f by considering the special case where G is a vector space V with its abelian (additive) Lie group structure. In this case, the Lie algebra is V itself and the Maurer-Cartan form is just the canonical map $\mathrm{pr}_2 : T\mathrm{V} = \mathrm{V} \times \mathrm{V} \to \mathrm{V}$, and so for a smooth map $f : M \to \mathrm{V}$, the Darboux derivative is the differential $df = \mathrm{pr}_2 \circ Tf$. Just as for the differential, the Darboux derivative embodies less information than the tangent map since the values that the map takes are "forgotten" and only tangential information is retained. Indeed, notice that if $L_g : G \to G$ is a left translation and $F := L_g \circ f$, then

$$\omega_F = F^*\omega_G = f^*L_g^*\omega_G = f^*\omega_G = \omega_f$$

since ω_G is left invariant. Hence two smooth maps into G that differ by left translation have the same (left) Darboux derivative. This generalizes the fact that two functions that differ by an additive constant have the same differential.

For a smooth 1-form $\vartheta = g\,dt$ on $\mathbb{R}$, we can always find a smooth function f with $df = \vartheta$ since by the Fundamental theorem of calculus one need only choose $f(t) = \int_0^t g(\tau)d\tau$. More generally, if M is simply connected and G is a vector space V, then the fact that $H^1(M) = 0$ means that every V-valued 1-form is the differential of some smooth $f : M \to \mathrm{V}$. For a general G, if ϑ is a $\mathfrak{g}$-valued 1-form on M, then we may ask for an f such that $\vartheta = \omega_f$. But there is no reason to expect a general ϑ to satisfy the above structure equation that ω_f satisfies. Now if we choose a basis $\{v_i\}$ for $\mathfrak{g}$, then there must be 1-forms $\vartheta^1, \ldots, \vartheta^n \in \Omega^1(M)$ such that

$$\vartheta = \sum_{i=1}^n v_i \vartheta^i.$$

Then $d\vartheta = -\frac{1}{2}[\vartheta, \vartheta]^\wedge$ is equivalent to

$$d\vartheta^k = -\frac{1}{2}\sum_{i,j} c_{ij}^k \vartheta^i \wedge \vartheta^j,$$

where c_{ij}^k are the structure constants. As we said, this may or may not hold. These equations are the integrability conditions for the existence of an f such that $\vartheta = \omega_f$. More precisely, we have the following theorem.

Theorem 11.42. *Let M be an m-manifold and G an n-dimensional Lie group. If ϑ is a $\mathfrak{g}$-valued 1-form on M that satisfies the structural equation $d\vartheta = -\frac{1}{2}[\vartheta,\vartheta]^\wedge$, then for every $p_0 \in M$ there is a neighborhood U_{p_0} of p_0 such that given any $(a,b) \in U_{p_0} \times G$ there is a smooth function $f: U_{x_o} \to G$ with $f(a) = b$ and $\vartheta = \omega_f$.*

Proof. Let $\mathrm{pr}_1 : M \times G \to M$ and $\mathrm{pr}_2 : M \times G \to G$ be the canonical projections and define a $\mathfrak{g}$-valued 1-form on $M \times G$ by

$$\Omega := \mathrm{pr}_1^* \vartheta - \mathrm{pr}_2^* \omega_G.$$

For each $(p,g) \in M \times G$, let $E_{(p,g)} = \mathrm{Ker}\,\Omega_{(p,g)}$. Now define a vector bundle homomorphism $T(M \times G) \to (M \times G) \times \mathfrak{g}$ by $v_{(p,g)} \mapsto ((p,g), \Omega_{(p,g)}(v_{(p,g)}))$. By Proposition 6.28, if this homomorphism has constant rank, then the kernel is a subbundle which clearly has fiber $E_{(p,g)}$ at (p,g). By linear algebra, this is equivalent to showing that the dimension of $E_{(p,g)}$ is independent of (p,g). This will follow if we show that $T\mathrm{pr}_1|_{E_{(p,g)}} : E_{(p,g)} \to T_pM$ is an isomorphism for all (p,g). If we identify $T_{(p,g)}(M \times G)$ with $T_pM \times T_gG$, then $T\mathrm{pr}_1$ is just the projection $(v,w) \mapsto v$ and similarly for $T\mathrm{pr}_2$. Now if $(v,w) \in E_{(p,g)}$ and $T\mathrm{pr}_1 \cdot (v,w) = 0$ then $v = 0$. But, since $\vartheta(v) = \omega_G(w)$, we have $w = 0$ also. Thus $T\mathrm{pr}_1|_{E_{(p,g)}}$ is injective. It is also surjective since for any $v \in T_pM$, we clearly have $(v, T_eL_g(\vartheta(v))) \in E_{(p,g)}$ and this has v as its image.

Now we use Proposition 11.19 to show that E is integrable:

$$d\Omega = \mathrm{pr}_1^* d\vartheta - \mathrm{pr}_2^* d\omega_G = \mathrm{pr}_1^*\left(-\frac{1}{2}[\vartheta,\vartheta]^\wedge\right) - \mathrm{pr}_2^*\left(-\frac{1}{2}[\omega_G,\omega_G]^\wedge\right)$$

$$= -\frac{1}{2}[\mathrm{pr}_1^*\vartheta, \mathrm{pr}_1^*\vartheta]^\wedge + \frac{1}{2}[\mathrm{pr}_2^*\omega_G, \mathrm{pr}_2^*\omega_G]^\wedge.$$

But $\mathrm{pr}_1^*\vartheta = \Omega + \mathrm{pr}_2^*\omega_G$, so

$$d\Omega = -\frac{1}{2}[(\Omega + \mathrm{pr}_2^*\omega_G), (\Omega + \mathrm{pr}_2^*\omega_G)]^\wedge - \frac{1}{2}[\mathrm{pr}_2^*\omega_G, \mathrm{pr}_2^*\omega_G]^\wedge$$

$$= -\frac{1}{2}[\Omega,\Omega]^\wedge - \frac{1}{2}[\Omega, \mathrm{pr}_2^*\omega_G]^\wedge - \frac{1}{2}[\mathrm{pr}_2^*\omega_G, \Omega]^\wedge,$$

which makes it clear that $d\Omega(X,Y) = 0$ whenever $\Omega(X) = 0$ and $\Omega(Y) = 0$.

Now we use the leaves of the distribution to construct the solution. Let $x_0 \in M$ and fix $g_0 \in G$. Then let $L_{(x_0,g_0)}$ be the maximal integral manifold through (x_0, g_0). The map $T\mathrm{pr}_1|\,E_{(p_0,g_0)} : E_{(p_0,g_0)} \to T_pM$ is an isomorphism so the inverse mapping theorem tells us that $\mathrm{pr}_1|\,L_{(p_0,g_0)}$ restricts to a diffeomorphism on some neighborhood O of (p_0, g_0) in $L_{(p_0,g_0)}$. Let $\Phi : U \to O \subset L_{(p_0,g_0)}$ denote the inverse of this diffeomorphism. Since $\mathrm{pr}_1 \circ \Phi = \mathrm{id}_U$, there must be a smooth function f such that $\Phi(p) = (p, f(p))$

for all $p \in U$. Observe that $\Phi^*(\Omega) = 0$ since the image of $T\Phi$ lies in the distribution by construction. Thus we have

$$0 = \Phi^*(\Omega) = \Phi^*(\mathrm{pr}_1^*\vartheta - \mathrm{pr}_2^*\omega_G)$$
$$= \Phi^*\mathrm{pr}_1^*\vartheta - \Phi^*\mathrm{pr}_2^*\omega_G) = \vartheta - f^*\omega_G,$$

or $\vartheta|_U = f^*\omega_G = \omega_f$. Let $(a,b) \in U \times G$. Now we argue that we may modify f so that we still have $\vartheta|_U = \omega_f$ while now $f(a) = b$. In fact, if $f(a) = b_1$, then let $g = b_1^{-1}b$ and replace f by $L_g \circ f$. □

Proposition 11.43. *Let M be an m-manifold and G an n-dimensional Lie group. Let ϑ be a $\mathfrak{g}$-valued 1-form on M and suppose that $f_1 : U \to G$ and $f_2 : U \to G$ are smooth maps such that $\vartheta|_U = f_i^*\omega_G$ for $i = 1, 2$. Then if U is connected and $f_1(p_0) = f_2(p_0)$ for some $p_0 \in U$, then $f_1 = f_2$.*

Proof. Let $\Phi_i := (\mathrm{id}_U, f_i)$. Then

$$\Phi_i^*(\Omega) = \Phi_i^*(\mathrm{pr}_1^*\vartheta - \mathrm{pr}_2^*\omega_G)$$
$$= \Phi_i^*\mathrm{pr}_1^*\vartheta - \Phi_i^*\mathrm{pr}_2^*\omega_G) = \vartheta - f^*\omega_G = 0,$$

so $\Phi_i : U \to M \times G$ is an embedding for $i = 1, 2$ whose image is the graph of f_i and an integral manifold of the distribution generated by Ω that contains $(p_0, f(p_0))$. Corollary 11.23 applies to show that $\Phi_1(U) = \Phi_2(U)$ and $f_1 = f_2$. □

Corollary 11.44. *Let M, G, ϑ and Ω be as above and suppose that $d\vartheta = -\frac{1}{2}[\vartheta, \vartheta]^\wedge$. Then the restriction of the map $\mathrm{pr}_1 : M \times G \to M$ to any leaf of the distribution given by Ω is a covering map.*

Proof. Let L be a leaf and choose $p_0 \in M$. Choose $(p_0, g_0) \in L$ and let $\Phi : U \to O \subset L_{(p_0,g_0)} = L$ be the diffeomorphism constructed as in the proof of the previous theorem where U is connected. Now $\Phi = (\mathrm{id}_U, f)$ where $f(p_0) = g_0$. If (p_0, g_1) is any other point in the leaf, then for $g = g_1 g_0^{-1}$ the map $\Phi_1 := (\mathrm{id}_U, L_g \circ f)$ is a diffeomorphism $U \to O_1$ where $(p_0, g_1) \in O_1$ and $\mathrm{pr}_1(O_1) = U$. In fact, it is easy to see that $\Phi_1 \circ \Phi^{-1} : O \to O_1$ is a diffeomorphism. If (p_0, g_1) and (p_0, g_2) are distinct points in the leaf, then we construct diffeomorphisms $\Phi_1 : O \to O_1$ and $\Phi_2 : O \to O_2$ as above, and Corollary 11.23 applies to show that O_1 and O_2 are disjoint and each map diffeomorphically onto O under pr_1. It is now clear that M is evenly covered by $\mathrm{pr}_1|L$. □

Corollary 11.45. *Let M be a simply connected m-manifold and G an n-dimensional Lie group. If ϑ is a $\mathfrak{g}$-valued 1-form on M that satisfies the structural equation $d\vartheta = -\frac{1}{2}[\vartheta, \vartheta]^\wedge$, then for every $(p_0, g_0) \in M \times G$ there is a smooth function $f : M \to G$ with $f(p_0) = g_0$ and $\vartheta = \omega_f$.*

Proof. Let L be the leaf of the distribution determined by Ω that contains (p_0, g_0). Since M is simply connected, we can lift $\mathrm{id}_M : M \to M$ to a section $\Phi : M \to L$ of $\mathrm{pr}_1|L$ such that $\Phi(p_0) = g_0$. Once again $\Phi = (\mathrm{id}_M, f)$ for some smooth function f with $f(p_0) = g_0$ and we argue as before to conclude that $\vartheta = \omega_f$ (this time globally). $\square$

Problems

(1) Show that the following vector fields define a rank 2 distribution on $\mathbb{R}^3$ which is not involutive (and hence not integrable):

$$X = \frac{\partial}{\partial x} + y\frac{\partial}{\partial z},$$
$$Y = \frac{\partial}{\partial y}.$$

Draw a picture of the portion of this distribution that sits at points in the x, y-plane and try to see geometrically why the distribution is not integrable.

(2) Show that the distribution on $\mathbb{R}^4$ given by $X = \frac{\partial}{\partial y} + x\frac{\partial}{\partial z}$ and $Y = \frac{\partial}{\partial x} + y\frac{\partial}{\partial w}$, where (x, y, z, w) are standard coordinates, has no integral manifolds.

(3) Let θ be a 1-form. Show that a 2-form η is in the ideal generated by θ if and only if $\eta \wedge \theta = 0$.

(4) Consider again the system of partial differential equations:

$$z_x = F(x, y, z),$$
$$z_y = G(x, y, z).$$

Show that the graphs of solutions of these equations are integral manifolds of the distribution defined by the 1-form $\theta = dz - Fdx - Gdy$. Use Theorem 11.18 to deduce the integrability conditions for this system. [Hint: Use Problem 3.]

(5) Let H be the Heisenberg group consisting of all matrices of the form

$$A = \begin{bmatrix} 1 & x_{12} & x_{13} \\ 0 & 1 & x_{23} \\ 0 & 0 & 1 \end{bmatrix}.$$

The x_{ij} give global coordinates and a diffeomorphism with $\mathbb{R}^3$. Let V_{12}, V_{13}, V_{23} be the left invariant vector fields on H that have values at the identity with components with respect to the coordinate fields given by $(1, 0, 0)$, $(0, 1, 0)$, and $(0, 0, 1)$, respectively. Let $\Delta_{\{V_{12}, V_{13}\}}$ and

$\Delta_{\{V_{12},V_{23}\}}$ be the 2-dimensional distributions generated by the indicated pairs of vector fields. Show that $\Delta_{\{V_{12},V_{13}\}}$ is integrable and $\Delta_{\{V_{12},V_{23}\}}$ is not.

(6) Prove (i) of Theorem 11.41.

(7) Show that for a given surface, equations (11.15) are equivalent to a combination of the Codazzi-Mainardi equations and the Gauss curvature equations defined in Chapter 4.

Chapter 12

Connections and Covariant Derivatives

The terms "covariant derivative" and "connection" are sometimes treated as synonymous. In fact, a covariant derivative is sometimes called a Koszul connection. From one point of view, the central idea is that of measuring the rate of change of sections of bundles in the direction of a vector or vector field on the base manifold. Here the derivative viewpoint is prominent. From another related point of view, a connection provides an extra structure that gives a principled way of lifting curves from the base to the total space. The lifts are parallel sections along the curve. In this chapter, we will always take the typical fiber of an $\mathbb{F}$-vector bundle of rank k to be $\mathbb{F}^k$. We let $(\mathbf{e}_1, \ldots, \mathbf{e}_k)$ be the standard basis of $\mathbb{F}^k$. Thus every vector bundle chart (U, ϕ) is associated with a frame field $(e_1, \ldots, e_k)$ where $e_i(x) := \phi^{-1}(x, \mathbf{e}_i)$.

12.1. Definitions

Let $\pi : E \to M$ be a smooth $\mathbb{F}$-vector bundle of rank k over a manifold M. A covariant derivative can either be defined as a map $\nabla : \mathfrak{X}(M) \times \Gamma(M, E) \to \Gamma(M, E)$ with certain properties from which one derives a well-defined map $\nabla : TM \times \Gamma(M, E) \to \Gamma(M, E)$ with nice properties or the other way around. We rather arbitrarily start with the first of these definitions.

Definition 12.1. A **covariant derivative** or **Koszul connection** on a smooth $\mathbb{F}$-vector bundle $E \to M$ is a map $\nabla : \mathfrak{X}(M) \times \Gamma(M, E) \to \Gamma(M, E)$ (where $\nabla(X, s)$ is written as $\nabla_X s$) satisfying the following four properties:

(i) $\nabla_{fX}(s) = f \nabla_X s$ for all $f \in C^\infty(M)$, $X \in \mathfrak{X}(M)$ and $s \in \Gamma(M, E)$;

(ii) $\nabla_{X_1+X_2} s = \nabla_{X_1} s + \nabla_{X_2} s$ for all $X_1, X_2 \in \mathfrak{X}(M)$ and $s \in \Gamma(M, E)$;

(iii) $\nabla_X(s_1 + s_2) = \nabla_X s_1 + \nabla_X s_2$ for all $X \in \mathfrak{X}(M)$ and $s_1, s_2 \in \Gamma(M, E)$;

(iv) $\nabla_X(fs) = (Xf)s + f\nabla_X s$ for all $f \in C^\infty(M; \mathbb{F})$, $X \in \mathfrak{X}(M)$ and $s \in \Gamma(M, E)$.

For a fixed $X \in \mathfrak{X}(M)$, the map $\nabla_X : \Gamma(M, E) \to \Gamma(M, E)$ is called the **covariant derivative with respect to** X.

As we will see below, this definition is enough to imply that ∇ induces maps $\nabla^U : \mathfrak{X}(U) \times \Gamma(U, E) \to \Gamma(U, E)$, one for each open $U \subset M$, that are naturally related in a sense we make precise below (this is not necessarily true for infinite-dimensional manifolds). Furthermore, we also prove that for a fixed $p \in M$, the value $(\nabla_X s)(p)$ depends only on the value of X at p and only on the values of s along any smooth curve c representing X_p. Thus we get a well-defined map $\nabla : TM \times \Gamma(M, E) \to \Gamma(M, E)$ such that $\nabla_v s = (\nabla_X s)(p)$ for any extension of $v \in T_p M$ to a vector field X with $X_p = v$. The resulting properties are

(i') $\nabla_{av}(s) = a\nabla_v s$ for all $a \in \mathbb{R}$, $v \in TM$ and $s \in \Gamma(M, E)$;

(ii') for all $p \in M$ we have $\nabla_{v_1+v_2} s = \nabla_{v_1} s + \nabla_{v_2} s$ for all $v_1, v_2 \in T_p M$, and $s \in \Gamma(M, E)$;

(iii') $\nabla_v(s_1 + s_2) = \nabla_v s_1 + \nabla_v s_2$ for all $v \in TM$ and $s_1, s_2 \in \Gamma(M, E)$;

(iv') for all $p \in M$ we have $\nabla_v(fs) = (vf)s(p) + f(p)\nabla_v s$ for all $v \in T_p M$, $s \in \Gamma(M, E)$ and $f \in C^\infty(M; \mathbb{F})$;

(v') $p \mapsto \nabla_{X_p} s$ is smooth for all smooth vector fields X.

A map satisfying these properties is *also* called a **covariant derivative** (or **Koszul connection**). Note that it is easy to obtain a Koszul connection in the first sense since we just let $(\nabla_X s)(p) := \nabla_{X_p} s$.

Definition 12.2. A covariant derivative on the tangent bundle TM of an n-manifold M is usually referred to as a covariant derivative on M.

In Chapter 4 have already met the Levi-Civita connection $\overline{\nabla}$ on $\mathbb{R}^n$, which is, from the current point of view, a Koszul connection on the tangent bundle of $\mathbb{R}^n$. The definition of this connection takes advantage of the natural identification of tangent spaces which makes taking the difference quotient possible:

$$\overline{\nabla}_v X = \lim_{t \to 0} \frac{X(p + tv) - X(p)}{t}.$$

In that same chapter we obtained, by a projection, a covariant derivative on (the tangent bundle of) any hypersurface in $\mathbb{R}^n$. A covariant derivative on

12.1. Definitions

a submanifold of arbitrary codimension can be obtained in the same way. Let M be a submanifold of $\mathbb{R}^n$ and let $X \in \mathfrak{X}(M)$ and $v \in T_pM$. We have

$$\nabla_v X := \left(\left.\frac{d}{dt}\right|_0 X \circ c \right)^\top \in T_pM.$$

One may easily verify that ∇, so defined, is a covariant derivative (in the second sense above).

Returning to the case of a general vector bundle, let us consider how covariant differentiation behaves with respect to restriction to open subsets of our manifold. Recall the restriction map $r_V^U : \Gamma(U, E) \to \Gamma(V, E)$ given by $r_V^U : \sigma \mapsto \sigma|_V$ and where $V \subset U$.

Definition 12.3. A **natural covariant derivative** ∇ on a smooth $\mathbb{F}$-vector bundle $E \to M$ is an assignment to each open set $U \subset M$ of a map $\nabla^U : \mathfrak{X}(U) \times \Gamma(U, E) \to \Gamma(U, E)$ written as $\nabla^U : (X, \sigma) \mapsto \nabla^U_X \sigma$ such that the following assertions hold:

(i) For every open $U \subset M$, the map ∇^U is a Koszul connection on the restricted bundle $E|_U \to U$.

(ii) For nested open sets $V \subset U$, we have $r_V^U(\nabla^U_X \sigma) = \nabla^V_{r_V^U X} r_V^U \sigma$ (naturality with respect to restrictions).

(iii) For $X \in \mathfrak{X}(U)$ and $\sigma \in \Gamma(U, E)$ the value $(\nabla^U_X \sigma)(p)$ only depends on the value of X at $p \in U$.

Here $\nabla^U_X \sigma$ is called the covariant derivative of σ with respect to X. We will denote all of the maps ∇^U by the single symbol ∇ when there is no chance of confusion. We have explicitly worked the naturality conditions (ii) and (iii) into the definition of a natural covariant derivative, so this definition is also appropriate for infinite-dimensional manifolds. The definition of Koszul connection did not specifically include these naturality features and was only defined for global sections. We shall now see that, in the case of finite-dimensional manifolds, a Koszul connection gives a natural covariant derivative for free.

Lemma 12.4. *Suppose* $\nabla : \mathfrak{X}(M) \times \Gamma(E) \to \Gamma(E)$ *is a Koszul connection for the vector bundle* $E \to M$. *Then if for some open* U *either* $X|_U = 0$ *or* $\sigma|_U = 0$, *then*

$$(\nabla_X \sigma)(p) = 0 \text{ for all } p \in U.$$

Proof. We prove the case of $\sigma|_U = 0$ and leave the case of $X|_U = 0$ to the reader.

Let $q \in U$. Then there is some relatively compact open set V with $q \in \overline{V} \subset U$ and a smooth function f that is identically one on V and zero outside of U. Thus $f\sigma \equiv 0$ on M and so since ∇ is linear, we have $\nabla(f\sigma) \equiv 0$ on M. Thus since (iv) of Definition 12.1 holds for global fields, we have

$$\nabla_X(f\sigma)(q) = f(p)(\nabla_X\sigma)(q) + (X_q f)\sigma(q)$$
$$= (\nabla_X\sigma)(q) = 0.$$

Since $q \in U$ was arbitrary, we have the result. □

We now define a natural covariant derivative derived from a given Koszul connection ∇. Given any open set $U \subset M$, we define $\nabla^U : \mathfrak{X}(U) \times \Gamma(E|_U) \to \Gamma(E|_U)$ by

(12.1) $$\left(\nabla^U_X \sigma\right)(p) := \left(\nabla_{\widetilde{X}} \widetilde{\sigma}\right)(p), \quad p \in U,$$

for $\widetilde{X} \in \mathfrak{X}(M)$ and $\widetilde{\sigma} \in \Gamma(E)$ chosen to be any sections which agree with X and σ on some open V with $p \in V \subset \bar{V} \subset U$. By the above lemma this definition does not depend on the choices of $\widetilde{X}$ and $\widetilde{\sigma}$.

Proposition 12.5. *Let $E \to M$ be a rank k vector bundle and suppose that $\nabla : \mathfrak{X}(M) \times \Gamma(E) \to \Gamma(E)$ is a Koszul connection. If for each open U we define ∇^U as in (12.1) above, then the assignment $U \mapsto \nabla^U$ is a natural covariant derivative as in Definition 12.3.*

Proof. We must show that (i), (ii) and (iii) of Definition 12.3 hold. It is easily checked that (i) holds, that is, that ∇^U is a Koszul connection for each U. The demonstration that (ii) holds is also easy and we leave it for the reader to check. Now since $X \to \nabla_X \sigma$ is linear over $C^\infty(U)$, (iii) follows by familiar arguments ($\nabla_X \sigma$ is linear over functions in the argument X). □

Because of this last lemma, we may define $\nabla_{v_p} \sigma$ for $v_p \in T_p M$ by

$$\nabla_{v_p} \sigma := (\nabla_X \sigma)(p),$$

where X is any vector field with $X(p) = v_p \in T_p M$. We say that $\nabla_X \sigma$ is "tensorial" in the variable X. The result can be seen as a special case of Proposition 6.55. We now see that it is safe to write expressions not directly justified by the definition of Koszul connection. For example, if $X \in \mathfrak{X}(U)$ and $\sigma \in \Gamma(V, E)$, where $U \cap V \neq \emptyset$, then $\nabla_X \sigma$ is taken to be an element of $\Gamma(U \cap V, E)$ defined by

$$(\nabla_X \sigma)(p) = \nabla_{X_p} \sigma := \nabla^U_{X_p} \sigma \text{ for all } p \in U \cap V.$$

This is a particularly useful convention when U is the domain of a chart $(U, \mathbf{x})$ and $X = \frac{\partial}{\partial x^i}$ and also when σ is a member of a frame field of the vector bundle defined on some open set.

In the same way that one extends a derivation on vector fields to a tensor derivation, one may show that a covariant derivative on a vector bundle induces naturally related covariant derivatives on all the multilinear bundles.

12.1. Definitions

In particular, if $\pi^* : E^* \to M$ denotes the dual bundle to $\pi : E \to M$ we may define connections on $\pi^* : E^* \to M$ and on $\pi \otimes \pi^* : E \otimes E^* \to M$. We do this in such a way that for $s \in \Gamma(M, E)$ and $s^* \in \Gamma(M, E^*)$ we have

$$\nabla_X^{E \otimes E^*}(s \otimes s^*) = \nabla_X s \otimes s^* + s \otimes \nabla_X^{E^*} s^*,$$

and

$$(\nabla_X^{E^*} s^*)(s) = X(s^*(s)) - s^*(\nabla_X s).$$

Of course, the second formula follows from the requirement that covariant differentiation commutes with contraction:

$$X(s^*(s)) = (\nabla_X C(s \otimes s^*)) = C(\nabla_X^{E \otimes E^*}(s \otimes s^*))$$
$$= C\left(\nabla_X s \otimes s^* + s \otimes \nabla_X^{E^*} s^*\right) = s^*(\nabla_X s) + (\nabla_X^{E^*} s^*)(s),$$

where C denotes the contraction given by $s \otimes \alpha \mapsto \alpha(s)$. All this works like the tensor derivation extension procedure discussed previously, and we often write all of these covariant derivatives with the single symbol ∇.

The bundle $E \otimes E^* \to M$ is naturally isomorphic to $\text{End}(E)$, and by this isomorphism we get a connection on $\text{End}(E)$. If we identify elements of $\Gamma(\text{End}(E))$ with $\text{End}(\Gamma E)$ (see Proposition 6.55), then we may use the following formula for the definition of the connection on $\text{End}(E)$:

$$(\nabla_X A)(s) := \nabla_X(A(s)) - A(\nabla_X s).$$

Indeed, since $C : s \otimes A \mapsto A(s)$ is a contraction, we must have

$$\nabla_X(A(s)) = C\left(\nabla_X s \otimes A + s \otimes \nabla_X A\right)$$
$$= A(\nabla_X s) + (\nabla_X A)(s).$$

Notice that if we fix $s \in \Gamma(E)$, then for each $p \in M$, we have an element $(\nabla s)(p)$ of $L(T_p M, E_p) \cong E \otimes T^* M$ given by

$$(\nabla s)(p) : v_p \mapsto \nabla_{v_p} s \text{ for all } v_p \in T_p M.$$

Thus we obtain a section ∇s of $E \otimes T^* M$ given by $p \mapsto (\nabla s)(p)$, which is easily shown to be smooth. In this way, we can also think of a connection as giving a map

$$\nabla : \Gamma(E) \to \Gamma(E \otimes T^* M)$$

with the property that

$$\nabla f s = f \nabla s + s \otimes df$$

for all $s \in \Gamma(E)$ and $f \in C^\infty(M)$. Since, by definition, $\Gamma(E) = \Omega^0(E)$ and $\Gamma(E \otimes T^* M) = \Omega^1(E)$, we really have a map $\Omega^0(E) \to \Omega^1(E)$. Later we will extend to a map $\Omega^k(E) \to \Omega^{k+1}(E)$ for all integral $k \geq 0$. Now if X is a smooth vector field, then $X \mapsto \nabla_X s \in \Gamma(E)$, so we may also view ∇s as an element of $\text{Hom}(\mathfrak{X}(M), \Gamma(E))$.

12.2. Connection Forms

Let $\pi : E \to M$ be a rank r vector bundle with a connection ∇. Recall that a choice of a local frame field over an open set $U \subset M$ is equivalent to a trivialization of the restriction $\pi_U : E|_U \to U$. Namely, if $\phi = (\pi, \Phi)$ is such a trivialization over U, then defining $e_i(x) = \phi^{-1}(x, \mathbf{e}_i)$, where $(\mathbf{e}_i)$ is the standard basis of $\mathbb{F}^n$, we have a frame field $(e_1, \ldots, e_k)$. We now examine the expression for a given covariant derivative from the viewpoint of such a local frame. It is not hard to see that for every such frame field there must be a matrix of 1-forms $\omega = (\omega_j^i)_{1 \leq i,j \leq r}$ such that for $X \in \Gamma(U, E)$ we may write

$$\nabla_X e_j = \sum_{i=1}^k \omega_j^i(X) e_i.$$

The forms ω_j^i are called **connection forms**.

Proposition 12.6. *If $s = \sum_i s^i e_i$ is the local expression of a section $s \in \Gamma(E)$ in terms of a local frame field $(e_1, \ldots, e_k)$, then the following local expression holds:*

$$\nabla_X s = \sum_i \left(X s^i + \sum_r \omega_r^i(X) s^r \right) e_i.$$

Proof. We simply compute:

$$\nabla_X s = \nabla_X \left(\sum_i s^i e_i \right)$$
$$= \sum_i (X s^i) e_i + \sum_i s^i \nabla_X e_i$$
$$= \sum_i (X s^i) e_i + \sum_{i,j} s^i \omega_i^j(X) e_j$$
$$= \sum_i (X s^i) e_i + \sum_{i,r} s^r \omega_r^i(X) e_i$$
$$= \sum_i \left(X s^i + \sum_r \omega_r^i(X) s^r \right) e_i. \qquad \square$$

So the i-th component of $\nabla_X s$ is

(12.2) $$(\nabla_X s)^i = X s^i + \sum_r \omega_r^i(X) s^r.$$

We may surely choose U small enough that it is also the domain of a coordinate frame $\{\partial_\mu\}$ for M. Thus we have

(12.3) $$\nabla_{\partial_\mu} e_j = \sum_k \omega_{\mu j}^k e_k,$$

12.3. Differentiation Along a Map

where $\omega_{\mu j}^k = \omega_i^j(\partial_\mu)$. We now have the local formula

$$(12.4) \qquad \nabla_X s = \sum_{i=1}^k \left(\sum_{\mu=1}^n X^\mu \partial_\mu s^i + \sum_{\mu=1}^n \sum_{r=1}^k X^\mu \omega_{\mu r}^i s^r \right) e_i.$$

Or, using the summation convention,

$$\nabla_X s = \left(X^\mu \partial_\mu s^i + X^\mu \omega_{\mu r}^i s^r \right) e_i.$$

Now suppose that we have two moving frames whose domains overlap, say $u = (e_1, \ldots, e_k)$ and $u' = (e_1', \ldots, e_k')$. Let us examine how the matrix of 1-forms $\omega = (\omega_i^j)$ is related to the corresponding $\omega' = (\omega_i'^j)$ defined in terms of the frame u'. The change of frame is

$$e_i' = \sum_j g_i^j e_j,$$

which in matrix notation is

$$u' = ug$$

for some smooth $g : U \cap U' \to \mathrm{GL}(n)$. (We treat u as a row vector of fields.) For a given moving frame, let $\nabla u := [\nabla e_1, \ldots, \nabla e_k]$. Differentiating both sides of $u' = ug$ and using matrix notation we have

$$u' = ug,$$
$$\nabla u' = \nabla(ug),$$
$$u'\omega' = (\nabla u)g + u\, dg,$$
$$u'\omega' = ugg^{-1}\omega g + ugg^{-1}dg,$$
$$u'\omega' = u'g^{-1}\omega g + u'g^{-1}dg,$$

and so we obtain the **transformation law for connection forms**:

$$\omega' = g^{-1}\omega g + g^{-1}dg.$$

12.3. Differentiation Along a Map

Once again let $\pi : E \to M$ be a vector bundle with a Koszul connection ∇. Consider a smooth map $f : N \to M$ and a section $\sigma : N \to E$ along f. Let $e_1, \ldots, e_k$ be a frame field defined over $U \subset M$. Since f is continuous, $O = f^{-1}(U)$ is open and $e_1 \circ f, \ldots, e_k \circ f$ are fields along f defined on O. We may write $\sigma = \sum_{a=1}^k \sigma^a\, e_a \circ f$ for some functions $\sigma^a : O \subset N \to \mathbb{F}$.

For any $p \in O$ and $v \in T_p N$, we define

$$(12.5) \qquad \nabla_v^f \sigma := \sum_{a=1}^k \left((d\sigma^a \cdot v) + \sum_{r=1}^k \omega_r^a (Tf(v))\, \sigma^r(p) \right) e_a(f(p)).$$

A direct albeit tedious calculation shows that this result is independent of the choice of frame field. Thus we obtain a global map

$$\nabla^f : TN \times \Gamma_f(E) \to \Gamma_f(E).$$

The map $\nabla^f : TN \times \Gamma_f(E) \to \Gamma_f(E)$ satisfies properties that qualify it as a **covariant derivative along** f. Namely, we have

Definition 12.7. Let $\pi : E \to M$ and $f : N \to M$ be as above. A **covariant derivative along** f is a map $\nabla^f : TN \times \Gamma_f(E) \to \Gamma_f(E)$ that satisfies

(i) $\nabla^f : TN \times \Gamma_f(E) \to \Gamma_f(E)$ is fiberwise linear in the first argument:

$$\nabla^f_{au+bv}\sigma = \nabla^f_{au}\sigma + \nabla^f_{bv}\sigma$$

for all $\sigma \in \Gamma_f(E)$, scalars a, b, and $u, v \in T_pN$ (p is arbitrary).

(ii) $\nabla^f_u \sigma = \nabla^f_u \sigma_1 + \nabla^f_u \sigma_2$ for any $u \in TN$ and any $\sigma_1, \sigma_2 \in \Gamma_f(E)$.

(iii) For $v \in T_pN$, $h \in C^\infty(N; \mathbb{F})$, and $\sigma \in \Gamma_f(E)$, we have

$$\nabla^f_v(h\sigma) = h(p)\nabla^f_v\sigma + v(h)\sigma(p).$$

(iv) If $U \in \mathfrak{X}(N)$, then $p \mapsto \nabla^f_{U(p)}\sigma$ is smooth for all $\sigma \in \Gamma_f(E)$.

(v) If $g : P \to N$ and $f : N \to M$ are smooth, then $\nabla^{f \circ g}$ is related to ∇^f by the following **chain rule**:

$$\nabla^{f \circ g}_u (\sigma \circ g) = (\nabla^f_{Tg \cdot u}\sigma)$$

for $u \in TP$; see the following diagram:

$$\begin{array}{ccc}
 & & E \\
 & \nearrow \sigma \circ g \quad \nearrow \sigma & \downarrow \pi \\
P & \xrightarrow{g} N \xrightarrow{f} & M
\end{array}$$

In the next section we give a more geometric view of covariant differentiation. Among other things, we obtain a more geometric way of producing ∇^f that does not appeal to a frame field. Also, we usually omit the superscript on ∇^f since it will be clear from context.

As a special case, consider a section σ of E along a smooth curve $c : (a, b) \to M$. Then we can form the operator $\nabla_{\frac{\partial}{\partial t}}\sigma$, which is also denoted by $\frac{D}{\partial t}$ or ∇_{∂_t}. We have the formula

$$\nabla_{\frac{\partial}{\partial t}}\sigma = \sum_i \left(\frac{d\sigma^i}{dt} + \sum_r (\omega^i_r \circ \dot{c})\sigma^r \right) e_i = \sum_i \left(\frac{d\sigma^i}{dt} + \sum_r c^*\omega^i_r\left(\frac{\partial}{\partial t}\right)\sigma^r \right) e_i,$$

where $e_i \circ c$ is abbreviated to e_i.

12.4. Ehresmann Connections

One way to get a natural covariant derivative on a vector bundle is through the notion of connection in another sense. What we will define in this section is a special case of what is called an Ehresmann connection. An Ehresmann connection is a structure that can be defined in the category of general smooth fiber bundles, but the two most common instances are defined for vector bundles and for principal bundles, and in each case extra hypotheses are added to the definition. We work with vector bundles here and give an explanation of the principal bundle approach in the online supplement [Lee, Jeff].

We first describe the construction in words since the notation tends to obscure what is really a simple geometric idea. First, notice that an ordinary function of one variable is constant on an interval if its graph is horizontal over that interval. For sections of a vector bundle, there is no a priori notion of "horizontal", so we have no principled way to decide what sections should be considered constant. Consider a moving frame field $e_1, \ldots, e_k$ on some vector bundle. If we had a reason to declare these frame fields to be constant, then we could differentiate a general section $s = \sum s^i e_i$ in a direction v by the rule $D_v s = \sum (v s^i) e_i$. So it seems clear that having a geometrically motivated way of picking out which sections should be constant will lead to a way of differentiating sections. The first step is to notice that we have a natural notion of vertical for a vector bundle $\pi : E \to M$. The tangent spaces of the fibers of E are to be considered vertical. Thus the vertical subspace of $T_y E$ is $T_y E_p \subset T_y E$, where $\pi(y) = p$.

Now if we have a vector in $T_y E$ for some $y \in E$, we would like to project it onto the vertical direction. However, this entails having a complementary horizontal space along which we project. The idea is then to assume that there is a distribution on E, that is, a subbundle of TE, that is everywhere complementary to the vertical directions. Thus we obtain a subspace complementary to the vertical, which allows projection onto the vertical. Once we have this we can say that a section $s \in \Gamma(E)$ is horizontal (i.e. constant) along a curve $c : I \to M$ if $Ts \cdot \dot{c}(t)$ has a vertical projection of zero for all t. Now we can define a covariant derivative as follows. If $v \in T_p M$ and s is a smooth section, then we can project $T_p s \cdot v$ onto the vertical space $T_{s(p)} E_p$ obtaining an element $\nabla_v s$. But E_p is a vector space, and so we may identify $T_{s(p)} E_p$ with E_p and take $\nabla_v s$ to be an element of E_p. The resulting map $v \mapsto \nabla_v s$ turns out to define a covariant derivative. We will also consider sections along general smooth maps $f : N \to M$ and obtain covariant differentiation for sections along f.

There is a canonical "parallelism" on $\mathbb{R}^n$. Recall the canonical map $j_p : \mathbb{R}^n \to T_p \mathbb{R}^n$. A vector $v_1 \in T_p \mathbb{R}^n$ is parallel to a vector $v_2 \in T_q \mathbb{R}^n$ exactly

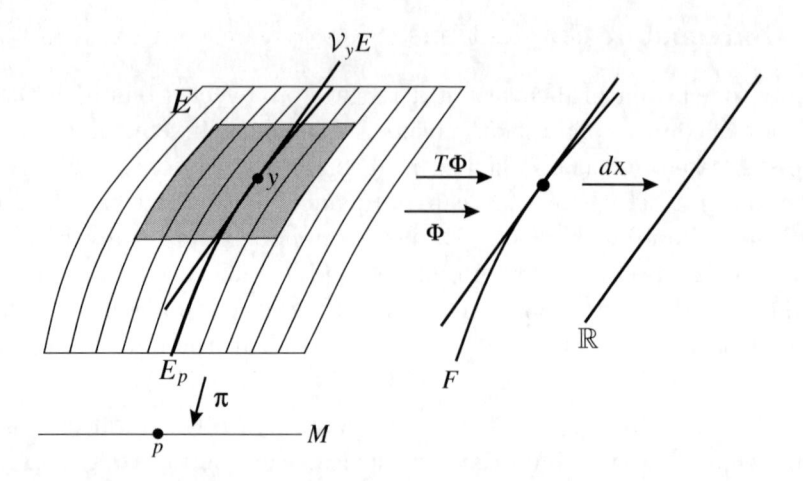

Figure 12.1. Vertical space

when $\left(\jmath_q \circ \jmath_p^{-1}\right)(v_1) = v_2$. The map $\jmath_q \circ \jmath_p^{-1}$ is a special example of what is called "parallel translation" and in this case establishes what is sometimes called "distant parallelism". In the presence of a connection, we obtain a map between fibers of a vector bundle which is called parallel translation. But this map may depend on a choice of smooth curve connecting the base points. Locally, this path dependence of parallel translation is due to the curvature of the connection.

We now proceed more formally. We give the definition of vertical bundle not just for a vector bundle, but also for a general fiber bundle. First a lemma:

Lemma 12.8. *Let $E \xrightarrow{\pi} M$ be a fiber bundle with typical fiber a k-manifold F. Fix $p \in M$ and let $\imath : E_p \hookrightarrow E$ be the inclusion. For all $y \in E_p$, we have*

$$T_y\imath\left(T_y E_p\right) = \mathrm{Ker}[T_y\pi : T_y E \to T_p M] = (T_y\pi)^{-1}(0_p) \subset T_y E,$$

where $0_p \in T_p M$ is the zero vector. If $\varphi : E_p \to F$ is a diffeomorphism and $(V, \mathbf{x})$ a chart on F, then for all $y \in \varphi^{-1}(V)$, $d\mathbf{x} \circ T_y\varphi$ maps $T_y E_p$ isomorphically onto $\mathbb{R}^k$.

Proof. $\pi \circ \imath \circ \gamma$ is constant for each smooth curve γ in N, so $T\pi \cdot (T\imath \cdot \dot{\gamma}(0)) = 0_p$. Thus $T_y\imath(T_y E_p) \subset (T_y\pi)^{-1}(0_p)$. On the other hand,

$$\dim((T_y\pi)^{-1}(0_p)) = \dim E - \dim M = \dim F = \dim T_y E_p,$$

so $(T_y\imath)(E_p) = (T_y\pi)^{-1}(0_p)$. The rest is clear since $d\mathbf{x}$ is just $T\mathbf{x}$ followed by the projection $T\mathbf{x}(U) = \mathbf{x}(U) \times \mathbb{R}^k \to \mathbb{R}^k$. $\square$

Note: As usual we identify $T_y\imath(T_y E_p)$ with $T_y E_p$.

12.4. Ehresmann Connections

Definition 12.9. Let $\pi : E \to M$ be a fiber bundle with typical fiber F and $\dim F = k$. Let $\mathcal{V}_y E := (T_y \pi)^{-1}(0_p)$ where $\pi(y) = p$. The **vertical bundle on** $\pi : E \to M$ is the real vector bundle $\pi_\mathcal{V} : \mathcal{V}E \to E$ with total space defined by the disjoint union

$$\mathcal{V}E := \bigcup_{y \in E} \mathcal{V}_y E \subset TE.$$

The projection map is defined by the restriction $\pi_\mathcal{V} := \pi_{TE}|_{\mathcal{V}E}$. A vector bundle atlas on $\mathcal{V}E$ is given by vector bundle charts of the form

$$(\pi_\mathcal{V}, d\mathbf{x} \circ T\Phi) : \pi_\mathcal{V}^{-1}(\pi^{-1}(U) \cap \Phi^{-1}(V)) \to (\pi^{-1}(U) \cap \Phi^{-1}(V)) \times \mathbb{R}^k,$$

where $\phi = (\pi, \Phi)$ is a bundle chart on E over U and $(V, \mathbf{x})$ a chart in F.

For the following, refer to the diagram:

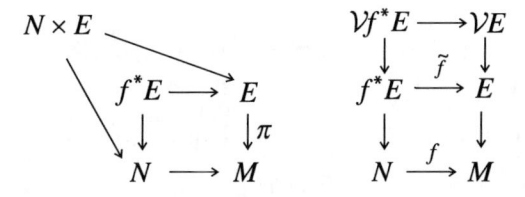

$$\begin{array}{ccc} N \times E & & \mathcal{V}f^*E \longrightarrow \mathcal{V}E \\ \searrow & & \downarrow \quad \tilde{f} \quad \downarrow \\ f^*E \longrightarrow E & & f^*E \longrightarrow E \\ \downarrow \quad \downarrow \pi & & \downarrow \quad \downarrow \\ N \longrightarrow M & & N \xrightarrow{f} M \end{array}$$

Exercise 12.10. Prove: Let $f : N \to M$ be a smooth map and $\pi : E \to M$ a fiber bundle with typical fiber F. Then $\mathcal{V}f^*E \to f^*E$ is bundle isomorphic to $\tilde{f}^*\mathcal{V}E \to f^*E$ where $\tilde{f} := \mathrm{pr}_2|_{f^*E} : f^*E \to E$ and $\mathrm{pr}_2 : M \times E \to E$.

Now consider the pull-back bundle f^*E where f is as above. Since f^*E is the submanifold of $N \times E$ defined by the condition that (q, y) is in f^*E if and only if $f(q) = \pi(y)$, a curve in f^*E must be of the form (c_1, c_2), where c_1 is a curve in N and c_2 is a curve in E, and it must also be the case that $f \circ c_1 = \pi \circ c_2$. Now if pr_1 and pr_2 are the first and second factor projections from $N \times E$, then $(T\mathrm{pr}_1, T\mathrm{pr}_2)$ gives a vector bundle isomorphism of the bundle $T(N \times E) \to N \times E$ with the bundle $TN \times TE \to N \times E$, and so we expect that under this isomorphism $T(f^*E)$ corresponds to a subbundle of $TN \times TE \to N \times E$.

Exercise 12.11. Show that under the bundle isomorphism $(T\mathrm{pr}_1, T\mathrm{pr}_2) : T(N \times E) \cong TN \times TE$, the tangent bundle $T(f^*E)$ corresponds to $\{(v, w) \in TN \times TE : Tf \cdot v = T\pi \cdot w\}$. Under this isomorphism $(\mathcal{V}f^*E)_{(q,y)}$ corresponds to $\{0_q\} \times \mathcal{V}_y E$.

We need to make an observation. If V is a complex vector bundle and $x \in$ V, then the tangent space T_xV has a natural complex structure. Indeed, the map $\jmath_x :$ V $\to T_x$V is used to transfer the complex structure of V to that of T_xV. In particular, if $\pi : E \to M$ is a complex vector bundle, then we may view $T_y E_p = (T_y \pi)^{-1}(0_p)$ as a complex vector space. Thus $\mathcal{V}E$ is a complex vector bundle.

The vertical vector bundle $\mathcal{V}E$ is isomorphic to the vector bundle π^*E over E (we say that $\mathcal{V}E$ is isomorphic to E along π). To see this, note that if $(v,w) \in \pi^*E$, then $\pi(v+tw)$ is constant in t. From this we see that the map from π^*E to TE given by $(v,w) \mapsto d/dt|_0(v+tw)$, maps into $\mathcal{V}E$. It is easy to see that this map is a vector bundle isomorphism:

$$\jmath : \pi^*E \cong \mathcal{V}E,$$

$$\jmath : (y,w) \mapsto \jmath_y w := \left.\frac{d}{dt}\right|_0 (y+tw) = w_y.$$

The meaning of the symbol $\jmath$ depends on the bundle we have in mind, but it is consistent with our previous use in the sense that if E is the tangent bundle of an open set in a vector space, then $\jmath_y$ is the canonical isomorphism as before.

Definition 12.12. A (linear Ehresmann) **connection** on a vector bundle $\pi : E \to M$ is a smooth distribution $\mathcal{H}$ on the total space E such that

(i) $\mathcal{H}$ is complementary to the vertical bundle:

$$TE = \mathcal{H} \oplus \mathcal{V}E.$$

(ii) $\mathcal{H}$ is homogeneous: $T_y\mu_r(\mathcal{H}_y) = \mathcal{H}_{ry}$ for all $y \in E$, $r \in \mathbb{R}$, where $\mu_r : E \to E$ is the multiplication map given by $\mu_r : y \mapsto ry$.

The subbundle $\mathcal{H}$ is called the **horizontal distribution** (or **horizontal subbundle**).

The statement $TE = \mathcal{H} \oplus \mathcal{V}E$ means that for every $y \in E$ we have the internal direct sum decomposition $T_yE = \mathcal{H}_y \oplus \mathcal{V}_yE$. Any $v \in TE$ has a corresponding decomposition $v = \mathrm{p}_\mathrm{v}v + \mathrm{p}_\mathrm{h}v$. Here, $\mathrm{p}_\mathrm{v} : v \mapsto \mathrm{p}_\mathrm{v}v$ and $\mathrm{p}_\mathrm{h} : v \mapsto \mathrm{p}_\mathrm{h}v$ are the obvious projections onto $\mathcal{V}_yE$ and $\mathcal{H}_y$ referred to respectively as the horizontal and vertical projections. Note well that without a choice of horizontal distribution $\mathcal{H}$ there is no vertical projection $\mathrm{p}_\mathrm{v} = 1 - \mathrm{p}_\mathrm{h}$. For $y \in E$, an individual element $w \in T_yE$ is **horizontal** if $w \in \mathcal{H}_y$ and **vertical** if $w \in \mathcal{V}_yE$. A vector field $\widetilde{X} \in \mathfrak{X}(E)$ is said to be a **horizontal vector field** (resp. **vertical vector field**) if $\widetilde{X}(y) \in \mathcal{H}_y$ (resp. $\widetilde{X}(y) \in \mathcal{V}_yE$) for all $y \in E$. On the right hand side of Figure 12.2 we see a schematic representation of the field of horizontal spaces. Notice that these spaces are tangent to the zero section (which we identify with M). This must be the case and is a consequence of the homogeneity condition (ii) from the definition. On the left hand side we see a particular tangent space to E together with a vector and its vertical projection.

Exercise 12.13. Show that $\mathcal{H} \cong \pi^*TM$.

Theorem 12.14. *Every vector bundle admits a connection.*

12.4. Ehresmann Connections

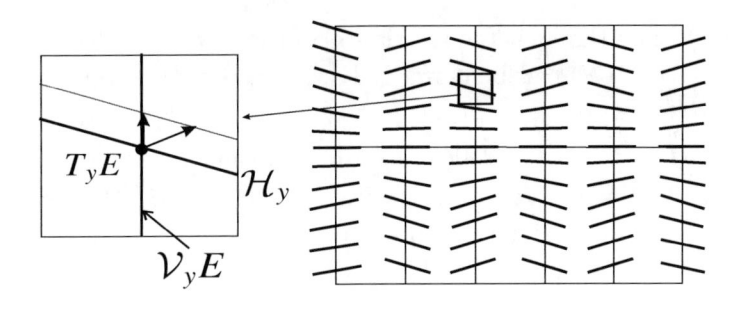

Figure 12.2. Horizontal distribution

Proof. First notice that we may easily define a connection on a trivial bundle $\text{pr}_1 : M \times V \to M$. Given a fixed $v \in V$, let $i_v : M \to M \times V$ be defined by $i_v(p) := (p, v)$. Next define $\mathcal{H}_{(p,v)} := Ti_v(T_pM)$ for each p. We then have $T\text{pr}_1(\mathcal{H}_{(p,v)}) = T_pM$. Since for any scalar a we have $\mu_a \circ i_v = i_{av}$, we also have $T\mu_a \circ Ti_v = Ti_{av}$ so that

$$T\mu_a(\mathcal{H}_{(p,v)}) = T\mu_a\left(Ti_v(T_pM)\right) = Ti_{av}(T_pM) = \mathcal{H}_{(p,av)} = \mathcal{H}_{a(p,v)}.$$

For a general vector bundle $\pi : E \to M$, let $\{U_\alpha\}$ be a locally finite cover of M such that the bundle is trivial over each U_α. Then we may choose a connection $\mathcal{H}^\alpha$ on each $\pi^{-1}(U_\alpha)$. Let $\{\rho_\alpha\}$ be a partition of unity subordinate to $\{U_\alpha\}$. For each $y \in E$, define $L_y : T_{\pi(y)}M \to T_yE$ by

$$L_y(v) := \sum_{\{\alpha : \pi(y) \in U_\alpha\}} \rho_\alpha(\pi(y))w_\alpha,$$

where for each α, the vector w_α is the unique vector in $\mathcal{H}^\alpha$ such that $T\pi \cdot w_\alpha = v$. It is easy to check that L_y is linear and satisfies $T_y\pi \circ L_y = \text{id}_{T_pM}$. The distribution we seek is then defined by $L_y\left(T_{\pi(y)}M\right)$ for each y. We leave it to the reader to check that this distribution is smooth and satisfies the required conditions. $\square$

Theorem 12.15. *If $\mathcal{H}$ is a connection on $\pi : E \to M$ and $f : N \to M$ a smooth map, then the distribution $f^*\mathcal{H} = (T\widetilde{f})^{-1}(\mathcal{H})$, where $\widetilde{f} := \text{pr}_2|_{f^*E}$, defines a connection on the pull-back bundle $f^*E \to N$. This is referred to as the **pull-back connection**:*

$$\begin{array}{ccc} f^*\mathcal{H} \subset Tf^*E & \xrightarrow{T\widetilde{f}} & TE \\ \downarrow & & \downarrow \\ TN & \xrightarrow{Tf} & TM \end{array}$$

Proof. First note that $\widetilde{f}$ is the restriction of the projection $N \times E \to E$ and so $f^*\mathcal{H} := (T\widetilde{f})^{-1}(\mathcal{H})$ can be defined: For $(q,y) \in f^*E$, we let $(f^*\mathcal{H})_{(q,y)} = (T_{(q,y)}\widetilde{f})^{-1}\mathcal{H}_y$. By Exercise 12.11 we can identify $T(f^*E)$ with $\{(v,w) \in TN \times TE : Tf \cdot v = T\pi \cdot w\}$. Under this isomorphism, $(\mathcal{V}f^*E)_{(q,y)}$ corresponds to $\{0_q\} \times \mathcal{V}_y E$ while $(f^*\mathcal{H})_{(q,y)}$ corresponds to $\{(v,w) \in TN \times \mathcal{H} : Tf \cdot u = T\pi \cdot v\}$. Thus we have the decomposition

$$(v, \mathrm{p_h} w) + (0, \mathrm{p_v} w),$$

which is easily seen to be unique. Hence the distribution $f^*\mathcal{H}$ is complementary to $\mathcal{V}f^*E$. Under the same identification, multiplication m_a on f^*E is $m_a(q,y) = (q, \mu_a y)$ and $Tm_a(v,w) = (v, T\mu_a w)$, which makes it clear that the second defining condition for a connection holds for $f^*\mathcal{H}$ since it holds for $\mathcal{H}$. □

Given a vector $v \in T_p M$ and a choice of $y \in E_p$, there is a unique vector $v_y \in \mathcal{H}_y \subset T_y E$ such that $T_y \pi \cdot v_y = v$. This vector is called the **horizontal lift** of v to $T_y E$. The idea works for fields too. Given a vector field $X \in \mathfrak{X}(M)$, there is a unique vector field $\widetilde{X} \in \mathfrak{X}(E)$ such that $\widetilde{X}(y)$ is horizontal for all $y \in M$ and $T_y \pi \cdot \widetilde{X}(y) = X(\pi(y))$. Thus $\widetilde{X} \in \Gamma(\mathcal{H}) \subset \mathfrak{X}(E)$. This horizontal vector field on the total space E is called the **horizontal lift** of X. Clearly $\widetilde{X}$ is π-related to X.

Proposition 12.16. *Let $\pi : E \to M$ be a vector bundle with connection $\mathcal{H}$. If $X, Y \in \mathfrak{X}(M)$ and $f \in C^\infty(M)$, then*

(i) $\widetilde{aX + bY} = a\widetilde{X} + b\widetilde{Y}$ *for all* $a, b \in \mathbb{R}$;
(ii) $\widetilde{fX} = (f \circ \pi)\widetilde{X}$;
(iii) $\widetilde{[X,Y]} = \mathrm{p_h}[\widetilde{X}, \widetilde{Y}]$.

Proof. (i) and (ii) are obvious. (iii) follows from the easy to check equalities $\pi_*[\widetilde{X}, \widetilde{Y}] = [X, Y] = \pi_* \widetilde{[X, Y]}$ and the uniqueness of horizontal lifts. □

Definition 12.17. Let $\sigma : N \to E$ be a section of E along a map $f : N \to M$. We say that σ is a **parallel section** if $T\sigma \cdot v$ is horizontal for all $v \in TN$. If s is a section of E and $c : I \to E$ is a curve, then we say that s is **parallel along** c provided $s \circ c$ is parallel.

Exercise 12.18. Recall that a section of f^*E must have the form $s : q \mapsto (q, \sigma_s(q))$, where σ is a section along f. Show that if s is parallel with respect to the pull-back connection on f^*E, then σ_s is parallel.

Exercise 12.19. Let $[0,b]$ be an interval and let t denote the standard coordinate on $[0,b]$. Suppose that $\pi : E \to [0,b]$ is a vector bundle over $[0,b]$ with connection. Let $\widetilde{\partial}$ denote the horizontal lift of $\partial/\partial t$.

12.4. Ehresmann Connections

(a) Show that if $c : [0, a] \to E$ is an integral curve of $\widetilde{\partial}$, then $c(a) \in E_a$. [Hint: Show that $\pi \circ c$ is an integral curve of $\partial/\partial t$.]

(b) Let $0 \le t_0 < b$. Show that there is a fixed $\epsilon > 0$ depending only on t_0 such that all maximal integral curves of $\widetilde{\partial}$ originating in the fixed fiber E_{t_0} are defined at least on $[t_0, \epsilon)$. [Hint: Endow E with a bundle metric and consider all integral curves originating in the unit sphere in E_{t_0} and then use property (ii) of Definition 12.12.]

(c) Using (a) and (b), show that integral curves of $\widetilde{\partial}$ all have domain equal to $[0, b]$.

Finding a horizontal lift of a vector field in $\mathfrak{X}(M)$ is trivial and automatic. On the other hand, finding a parallel section along a given map with prescribed value $\sigma(q)$ for some q is generally nontrivial and may not exist. However, we have the following

Theorem 12.20. *Let $\pi : E \to M$ be a (smooth) vector bundle with a connection $\mathcal{H}$. Suppose that $c : [a, b] \to M$ is a smooth curve. For each $u \in E_{c(a)}$, there is a unique parallel section $\sigma_{c,u}$ along c such that $\sigma_{c,u}(a) = u$. If $P_c : E_{c(a)} \to E_{c(b)}$ denotes the map which takes $u \in E_{c(a)}$ to $\sigma_{c,u}(b)$, then P_c is linear.*

Proof. Without loss of generality we may take $a = 0$. Let $\partial = \partial/\partial t$ denote the standard coordinate vector field on the interval $[0, b]$ and let $\widetilde{\partial}$ denote its horizontal lift in the pull-back bundle c^*E with respect to the pull-back connection $c^*\mathcal{H}$. Let c_u denote the maximal integral curve of $\widetilde{\partial}$ in c^*E with $c_u(0) = (0, u) \in c^*E$. We have

$$\frac{d}{dt}(\mathrm{pr}_1 \circ c_u) = T\mathrm{pr}_1 \circ \dot{c}_u = T\mathrm{pr}_1 \circ \widetilde{\partial} \circ c_u = \partial \circ \mathrm{pr}_1 \circ c_u.$$

Thus $\mathrm{pr}_1 \circ c_u$ is an integral curve of $\partial = \partial/\partial t$ and so $\mathrm{pr}_1 \circ c_u(t) = t$. From this we see that $c_u(t) = (t, \mathrm{pr}_2 \circ c_u)$. By Exercise 12.19 we know that c_u is defined on $[0, b]$ and that $c_u(b) \in (c^*E)_b$. Let $\sigma_{c,u} := \mathrm{pr}_2 \circ c_u$ on $[0, b]$. Then $\sigma_{c,u}$ is a section of $E \to M$ along c which is parallel since $\dot{c}_u$ is horizontal (see Exercise 12.18). We define $P_c u := \sigma_{c,u}(b)$ for $u \in E_{c(0)}$. The uniqueness follows from the uniqueness of integral curves and we leave this to the reader. For any $r \in \mathbb{R}$, the field $r\sigma_{c,u}$ is parallel. Indeed, $(r\sigma_{c,u})^{\cdot} = T\mu_r \circ \dot{\sigma}_{c,u}$ is horizontal since $T\mu_r$ preserves $\mathcal{H}$. But then $P_c(ru) = rP_c(u)$ so P_c is homogeneous. We aim to show that $P_c = \jmath_0^{-1} \circ TP_c \circ \jmath_0$, which is a composition of linear maps. Let $v_0 \in T_0 E_{c(0)}$ so that $v_0 = \dot{\gamma}(0)$, where γ is defined by $\gamma(t) = tv$ for an appropriate $v \in E_{c(0)}$. Then $\jmath_0^{-1} v_0 = v$. We have

$$T_0 P_c v_0 = \frac{d}{dt}(P_c \circ \gamma)(0).$$

But also $P_c \circ \gamma(t) = P_c(tv) = tP_c(v)$ so that
$$T_0 P_c v_0 = \jmath_0 (P_c(v)) = \jmath_0 \circ P_c \circ \jmath_0^{-1} v_0.$$
Thus $\jmath_0 \circ P_c \circ \jmath_0^{-1} = T_0 P_c$ or $P_c = \jmath_0^{-1} \circ T_0 P_c \circ \jmath_0$, and we conclude that P_c is linear.

Since P_c has inverse $P_{c^\leftarrow}$ where $c^\leftarrow(t) := c(b-t)$, we see that it is a linear isomorphism. $\square$

Definition 12.21. Let $c : [a, b] \to M$ be as in the theorem above. The map P_c is called **parallel translation** along c from $c(a)$ to $c(b)$. Let c be any smooth curve in M. For t_1 and t_2 in the domain of c, let $P(c)_{t_1}^{t_2} : E_{c(t_1)} \to E_{c(t_2)}$ be defined as $P(c)_{t_1}^{t_2} := P_{c|[t_1,t_2]}$ if $t_2 \geq t_1$ and $P(c)_{t_1}^{t_2} := P_{c|[t_2,t_1]}^{-1}$ if $t_1 \geq t_2$.

We also say that the curve $\sigma_{c,u}$ of Theorem 12.20 is a **parallel lift** or **horizontal lift** of the curve c. The map P_c is also sometimes called **parallel transport**. Suppose that $c : [a, b] \to M$ is a (continuous) piecewise smooth curve (Definition 2.121). Thus we may find a monotonic sequence $t_0, t_1, \ldots, t_j = t$ such that $c_i := c|_{[t_{i-1}, t_i]}$ (or $c|_{[t_i, t_{i-1}]}$) is smooth.[1] In this case we define
$$P(c)_{t_0}^{t} := P(c)_{t_{j-1}}^{t} \circ \cdots \circ P(c)_{t_0}^{t_1}.$$

Exercise 12.22. The map $P(c)_{t_0}^{t} : E_{c(t_0)} \to E_{c(t)}$ is a linear isomorphism for all t with inverse $P(c)_{t}^{t_0}$.

Parallel transport behaves nicely with respect to reparametrization. This is the content of the next theorem, which we ask the reader to prove in Problem 1.

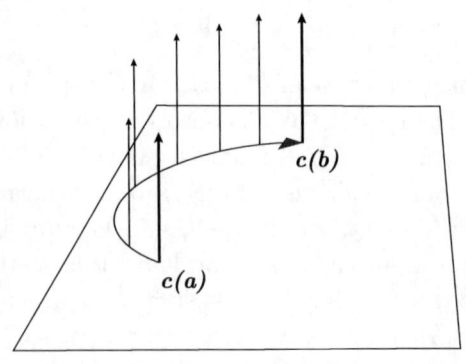

Parallel transport

Theorem 12.23. Let $c : [a, b] \to M$ be a smooth curve. If $r : [a', b'] \to [a, b]$ is a smooth map with $dr/dt > 0$, then for $\gamma := c \circ r$ we have $P_c = P_\gamma$.

---
[1] It may be that $t < t_0$.

12.4. Ehresmann Connections

It is often convenient to use special vector bundle charts constructed using parallel translation. Let $(U, \mathbf{x})$ be a chart on M centered at $p \in M$ and such that $\mathbf{x}(U)$ is a ball $B_R(0)$ of radius R. Consider the family of curves $c_\mathbf{u} : [0, R] \to M$ given for each $\mathbf{u} \in S^{n-1}(R) = \partial B_R(0)$ by

$$c_\mathbf{u}(t) := \mathbf{x}^{-1}(t\mathbf{u}).$$

Define a frame field $(e_1, \ldots, e_k)$ for E over U as follows: Let $(e_1(0), \ldots, e_k(0))$ be an ordered basis of E_p. If $q \in U$, then let $e_i(q)$ be defined as

$$e_i(q) := P(c_\mathbf{u})_0^t(e_i(0)),$$

where $(\mathbf{u}, t)$ is the unique element of $S^{n-1}(R) \times [0,1]$ such that $c_\mathbf{u}(t) = q$. We say that $(e_1, \ldots, e_k)$ is **radially parallel** with respect to the spherical chart $(U, \mathbf{x})$. Notice that by composing with a dilation of $\mathbb{R}^n$ if necessary, we may choose R to be any positive number. Also, if we use a radially parallel frame field to define a vector bundle chart $\phi = (\pi, \Phi)$ in the usual way, then Φ will be constant along each curve $c_\mathbf{u}$.

We still have more to learn about the geometric meaning of a connection. We will eventually be led to the notion of curvature. At this point let us just consider what it means when a connection is integrable as a distribution.

Definition 12.24. A connection on a vector bundle is called **flat** if it is integrable as a distribution.

Let us agree to call a connection on a vector bundle E a **trivial connection** if given any $u \in E$ there is a parallel section s such that $s(\pi(u)) = u$. Finding such sections is not always possible (even locally). In fact, we have the following characterization.

Theorem 12.25. *Let $\pi : E \to M$ be a vector bundle with connection $\mathcal{H}$. The following assertions are equivalent:*

(i) *For any simply connected open set $U \subset M$, the restriction of $\mathcal{H}$ to $\pi^{-1}(U)$ is a trivial connection on the restricted vector bundle $\pi^{-1}(U) \to U$.*

(ii) *$\mathcal{H}$ is flat.*

Proof. If (i) holds, then given any $u \in E$ there is a parallel section s defined on U. Then it is easy to see that $s(U)$ is an integral manifold of the distribution $\mathcal{H}$. Since we can find such an integral manifold through any u, we see that $\mathcal{H}$ is integrable (i.e. flat).

If $\mathcal{H}$ is flat, then certainly $\mathcal{H}|_U$ is flat for any open U. Suppose that U is simply connected. By the Frobenius theorem there is a maximal integral manifold L_u of $\mathcal{H}|_U$ through any $u \in \pi^{-1}(U)$. By Theorem 12.20, we see that any smooth path in U can be lifted uniquely to L_u. This is enough to imply that $\pi|_{L_u} : L_u \to U$ is a covering space (see [**Span**], 2.4.10). Since U is simply connected, the lifting theorem for covering spaces (Theorem 1.95) implies that $\pi|_{L_u}$ has an inverse. The desired parallel section is then $s := \left(\pi|_{L_u}\right)^{-1}$. $\square$

We now come to the task of relating covariant derivatives with Ehresmann connections on vector bundles. Denote the vector bundle isomorphism from $\mathcal{V}E$ to E along π by p:

$$\mathrm{p} : \mathcal{V}E \to E,$$
$$\mathrm{p} : w_y \mapsto w.$$

For each y, the map p just gives the canonical identification of $T_y E_p$ with E_p, and on each fiber, it is the inverse of $\jmath$. If we have a connection on $\pi : E \to M$, then we have an associated **connector** (or connection map), which is the map $\kappa : TE \to E$ defined by

$$\kappa(v) := \mathrm{p}(\mathrm{p}_{\mathcal{V}} v) = \jmath_y^{-1}(\mathrm{p}_{\mathcal{V}} v)$$

for $v \in T_y E$. The connector is a vector bundle homomorphism along the map $\pi : E \to M$:

$$\begin{array}{ccc} TE & \xrightarrow{\kappa} & E \\ \downarrow & & \downarrow \\ E & \xrightarrow{\pi} & M \end{array}$$

An interesting fact is that given the appropriate definition of vector space structure on the fibers, TE is also a vector bundle over TM via the map $T\pi$ (Problem 11 from Chapter 6). Recall that the addition and scalar multiplication on a fiber $T\pi^{-1}(x)$ of this bundle are defined by

$$u \boxplus v := T\alpha \cdot (u, v) \text{ for } u, v \in TE \text{ with } T\pi \cdot u = T\pi \cdot v = x,$$
$$c \odot v := T\mu_c \cdot v \text{ for } v \in TE \text{ and } c \in \mathbb{F},$$

where $\alpha(y_1, y_2) := y_1 + y_2$ for $(y_1, y_2) \in E \oplus E$ and $\mu_c y := cy$ for $y \in E$ and $c \in \mathbb{F}$.

Lemma 12.26. *Suppose that $f : \mathbb{R}^K \to \mathbb{R}^k$ is a smooth map such that $f(av) = af(v)$ for all $v \in \mathbb{R}^K$ and $a \in \mathbb{R}$. Then f is linear. Similarly, if $f : \mathbb{C}^K \to \mathbb{C}^k$ is a smooth map such that $f(av) = af(v)$ for all $v \in \mathbb{C}^K$ and $a \in \mathbb{C}$, then f is linear.*

Proof. Let $f : \mathbb{R}^K \to \mathbb{R}^k$ as in the statement. Then $Df(0)v = \frac{d}{dt}\big|_{t=0} f(tv) = \frac{d}{dt}\big|_{t=0} tf(v) = f(v)$. Thus $f = Df(0)$ and so f is linear. If $f : \mathbb{C}^K \to \mathbb{C}^k$ is a smooth map such that $f(av) = af(v)$ for all $v \in \mathbb{C}^K$ and $a \in \mathbb{C}$ then the first part shows that $f : \mathbb{C}^K \to \mathbb{C}^k$ is linear over $\mathbb{R}$. But by hypothesis $f(iv) = if(v)$, so f is actually complex linear. $\square$

12.4. Ehresmann Connections

Corollary 12.27. *Let $\pi_1 : E_1 \to M_1$ and $\pi_2 : E_2 \to M_2$ be $\mathbb{F}$-vector bundles. Let $\widehat{f} : E_1 \to E_2$ be a fiber bundle morphism over $f : M_1 \to M_2$. If $\widehat{f}$ is homogeneous on each fiber, so that $\widehat{f}(av) = a\widehat{f}(v)$ for all $v \in E_1$ and $a \in \mathbb{F}$, then $\widehat{f}$ is linear on fibers, and so it is a vector bundle morphism.*

Proof. Use vector bundle charts and Lemma 12.26. $\square$

Lemma 12.28. *Let $\mu_r : E \to E$ be multiplication by r. Then for any $p \in M$ and $y, w \in E_p$ we have*
$$T\mu_r(\jmath_y w) = \jmath_{ry}(rw) = r\jmath_{ry}w.$$

Proof. We have
$$T\mu_r(\jmath_y w) = \left.\frac{d}{dt}\right|_{t=0} \mu_r(y + tw) = \left.\frac{d}{dt}\right|_{t=0}(ry + trw)$$
$$= \jmath_{ry}(rw) = r\jmath_{ry}w. \quad \square$$

The connector κ gives a vector bundle homomorphism along $\pi_{TM} : TM \to M$. More precisely, we have

Theorem 12.29. *If κ is the connector for a connection on a vector bundle $\pi : E \to M$, then κ gives a vector bundle homomorphism from the bundle $T\pi : TE \to TM$ to the bundle $\pi : E \to M$ along the map $\pi_{TM} : TM \to M$,*

$$\begin{array}{ccc} TE & \xrightarrow{\kappa} & E \\ {\scriptstyle T\pi}\downarrow & & \downarrow \\ TM & \longrightarrow & M \end{array}$$

Proof. It is easy to check that the above diagram commutes. Thus we must show that κ is linear on fibers. Let $Y_y \in (T\pi)^{-1}(X_p)$ where $\pi(y) = p$. We may write $Y_y = H_y + V_y$, where H_y and V_y are horizontal and vertical respectively. Since $X_p = T\pi(Y_y) = T\pi(H_y)$, we see that H_y is the horizontal lift $\widetilde{X}_y$ of X_p to $T_y E$. Also, $V_y = \jmath_y w$ for a unique $w \in E_p$. So the decomposition may be written as $Y_y = \widetilde{X}_y + \jmath_y w$. By definition we have
$$\kappa(Y_y) = w.$$
Using the homogeneity property of the horizontal distribution together with Lemma 12.28 above we have
$$T\mu_r(Y_y) = T\mu_r\left(\widetilde{X}_y\right) + T\mu_r(\jmath_y w) = \widetilde{X}_{ry} + \jmath_{ry}rw.$$
Thus $\kappa(T\mu_r(Y_y)) = rw$. Therefore, we have $\kappa(T\mu_r(Y_y)) = r\kappa(Y_y)$. If we denote the scaling operation for the vector bundle structure on $TE \to TM$ using $\odot$ as before, then this last statement is just $\kappa(r \odot Y_y) = r\kappa(Y_y)$. The result now follows from the Corollary 12.27. $\square$

Once we have a connection on our bundle then the addition in the vector bundle $TE \to TM$ can be described in a convenient form. Any element of TE that lies over the same element $X_p \in TM$ can be written in the form $\widetilde{X}_y + \jmath_y w$ for some $w \in E_p$, where $\widetilde{X}_y$ is the horizontal lift of X_p to the point y. Let $\widetilde{X}_{y_1} + \jmath_{y_1} w_1$ and $\widetilde{X}_{y_2} + \jmath_{y_2} w_2$ be two such expressions. Then the sum of these two elements using the addition in the vector bundle $TE \to TM$ is given by $\widetilde{X}_{y_1+y_2} + \jmath_{y_1+y_2}(w_1 + w_2)$, where $\widetilde{X}_{y_1+y_2}$ is the horizontal lift of X_p to the point $y_1 + y_2$.

Exercise 12.30. Prove the last statement.

Exercise 12.31. Deduce from the previous theorem that $(\pi_{TE}, \kappa) : TE \to E \oplus E$ is a vector bundle isomorphism along the tangent bundle projection $\pi_{TM} : TM \to M$.

Using this notion of connection with associated connector κ we can get a covariant derivative. If $\mathcal{H}$ is a connection on a vector bundle $E \to M$ with connector κ, then for a section $\sigma \in \Gamma_f(E)$ along a smooth map $f : N \to M$ we make the definition

$$\nabla_v^f \sigma := \kappa(T_p \sigma \cdot v) \qquad \text{for } v \in T_p N.$$

If V is a vector field on N, then $(\nabla_V^f \sigma)(p) := \nabla_{V(p)}^f \sigma$.

Theorem 12.32. *Let $\pi : E \to M$ be a vector bundle. Suppose that the vector bundle is endowed with a connection $\mathcal{H}$ and associated connector κ. Then for each smooth map $f : N \to M$, the map ∇^f defined above is a covariant derivative along f (Definition 12.7). If $f = \mathrm{id}_M$, then we obtain a Koszul connection.*

Conversely, if ∇ is a Koszul connection on $\pi : E \to M$, then we may define a connection by

$$\mathcal{H}_y := \{Ts \cdot u - \jmath_y \nabla_u s : \; s \in \Gamma(E), \; s(\pi(y)) = y, \; u \in T_{\pi(y)} M\}.$$

The resulting connection, in turn, gives back the Koszul connection according to $\nabla_v s := \kappa(T_p s \cdot v)$ for $v \in T_p M$.

Proof. We write ∇ for ∇^f when no confusion is likely. We start by simply noting that (i) and (iv) of Definition 12.7 are easily seen to be true and leave these as an exercise. Notice that if $g : P \to N$ and $f : N \to M$ are smooth and $u \in TP$, then for each $\sigma \in \Gamma_f(E)$, we have $\nabla_u(s \circ g) = \nabla_{Tg \cdot u} s$. Indeed,

$$\nabla_u(\sigma \circ g) = (\kappa \circ T(\sigma \circ g)) u = \kappa(T\sigma(Tg \cdot u)) = \nabla_{Tg \cdot u} \sigma.$$

This gives (v) of Definition 12.7. Next we claim that if $\sigma_1, \sigma_2 \in \Gamma_f(E)$ and $u \in T_p N$, then

$$\nabla_u(\sigma_1 + \sigma_2) := \nabla_u \sigma_1 + \nabla_u \sigma_2,$$

which implies (ii) of Definition 12.7.

12.4. Ehresmann Connections

Proof of the claim: Recall the definition of addition in the vector bundle $T\pi : TE \to TM$ in terms of the tangent lift of $\alpha : (u, v) \mapsto u+v$. If $u = \dot{\gamma}(0)$ for a smooth curve γ in N with $\gamma(0) = p$, then

$$T\sigma_1 \cdot u \boxplus T\sigma_2 \cdot u := T\alpha(T\sigma_1 \cdot u, T\sigma_2 \cdot u) = \frac{d}{dt}\bigg|_0 (\sigma_1 \circ \gamma + \sigma_2 \circ \gamma)$$

$$= \frac{d}{dt}\bigg|_0 (\sigma_1 + \sigma_2) \circ \gamma = T(\sigma_1 + \sigma_2) \cdot u.$$

Now we use the above together with the fact that κ is a bundle homomorphism along π_{TM}. We have

$$\nabla_u (\sigma_1 + \sigma_2) = \kappa \left(T(\sigma_1 + \sigma_2)u \right) = \kappa(T\sigma_1 u \boxplus T\sigma_2 u) = \nabla_u \sigma_1 + \nabla_u \sigma_2.$$

Claim: If $h \in C^\infty(N, \mathbb{F})$ and $\sigma \in \Gamma_f(E)$, then $\nabla_u h\sigma = u(h)\sigma(p) + h(p)\nabla_u \sigma$, which is (iii) of Definition 12.7.

Proof: Let $u \in T_p N$ and let $\sigma : N \to E$ be a section along a smooth map $f : N \to M$. We begin by calculating $T\mu : T\mathbb{R} \times TE \to TE$, where $\mu : \mathbb{R} \times E \to E$ is scalar multiplication in the vector bundle $E \to M$. So let $(a, y) \in \mathbb{R} \times E$ and $(b \frac{d}{dt}|_a, v_y) \in T_a\mathbb{R} \times T_y E$. To find $T\mu \cdot (b \frac{d}{dt}|_a, v_y)$ we calculate $T\mu \cdot (0_a, v_y)$ and $T\mu \cdot (b \frac{d}{dt}|_a, 0_y)$ separately. Let c be a smooth curve with $c(0) = y$ and $\dot{c}(0) = v_y$. Then

$$T\mu \cdot (0_a, v_y) = \frac{d}{dt}\bigg|_0 \mu(a, c(t)) = \frac{d}{dt}\bigg|_0 \mu_a(c(t))$$

$$= T\mu_a \cdot v_y =: a \odot v_y,$$

where $\odot$ refers to multiplication in the vector bundle structure of $TE \to TM$. Next, let c be the curve in the manifold $\mathbb{R}$ given by $c(t) := a + tb$ so that $\dot{c}(0) = b \frac{d}{dt}|_a$. Then

$$T\mu \cdot \left(b \frac{d}{dt}\bigg|_a, 0_y \right) = \frac{d}{dt}\bigg|_0 \mu(c(t), y) = \frac{d}{dt}\bigg|_0 ((a + bt)y)$$

$$= \frac{d}{dt}\bigg|_0 (ay + tby) = \mathrm{J}_{ay}(by).$$

We now have

$$T\mu \cdot \left(b \frac{d}{dt}\bigg|_a, v_y \right) = a \odot v_y + \mathrm{J}_{ay}(by).$$

Next we let c be a curve in N with $c(0) = p$ and $\dot{c}(0) = u \in T_p N$. Then

$$T_p h \cdot u = \frac{d}{dt}\bigg|_0 h(c(t)) = (h \circ c)'(t) \frac{d}{dt}\bigg|_{h(c(0))}$$

$$= u[h] \frac{d}{dt}\bigg|_{h(p)}.$$

Now let $h \times \sigma : N \to \mathbb{R} \times E$ denote the map defined by $(h \times \sigma)(x) = (h(x), \sigma(x))$ and let $\widetilde{\sigma} := h\sigma$. Then using the linearity of κ and what we developed above we have

$$\nabla_u h\sigma = \kappa(T\widetilde{\sigma} \cdot u) = \kappa\left[T\mu \circ T(h \times \sigma)(u)\right]$$
$$= \kappa T\mu\left(u[h]\left.\frac{d}{dt}\right|_{h(p)}, T_p\sigma \cdot u\right)$$
$$= \kappa\left\{h(p) \odot (T\sigma \cdot u) + \jmath_{h(p)\sigma(p)}(u[h]\sigma(p))\right\}$$
$$= h(p)\kappa(T\sigma \cdot u) + u[h]\sigma(p) = h(p)\nabla_u\sigma + u[h]\sigma(p).$$

The proof that

$$\mathcal{H}_y := \{\jmath_y \nabla_u s - Ts \cdot u \mid s \in \Gamma(E), u \in T_{\pi(y)}M\}$$

defines a connection that gives back the original covariant derivative is left to the reader as Problem 9. $\square$

Example 12.33. In Chapter 4 we saw that each hypersurface in $\mathbb{R}^{n+1}$ has a natural covariant derivative (the Levi-Civita covariant derivative). This means that there is a corresponding horizontal distribution. Let us consider the canonical connection on $S^n \subset \mathbb{R}^{n+1}$. We may identify TS^n with the subset of $S^n \times \mathbb{R}^{n+1}$ given by $\{(p,u) \in S^n \times \mathbb{R}^{n+1} : \langle p, u\rangle = 0\}$. Under this identification, the velocity of a curve c has the form $\dot{c} = (c, c')$. Now let us find a representation of $T(TS^n)$ as a subset of $TS^n \times \mathbb{R}^{2n+2}$. A section of TS^n along a curve c has the form $\sigma : t \mapsto (c(t), x(t))$ and so we take $\dot{\sigma} = (c, x, c', x') \in TS^n \times \mathbb{R}^{2n+2}$. Note that since $\langle c, x\rangle = 0$, we have $\langle c, x'\rangle = -\langle c', x\rangle$. Also, since $\langle c, c\rangle = 1$, we have $\langle c, c'\rangle = 0$. It follows that we may make the identification

$$T(TS^n) = \{((p,u),(v,w)) \in TS^n \times \mathbb{R}^{2n+2} : \langle p, v\rangle = 0, \ \langle p, w\rangle + \langle u, v\rangle = 0\}.$$

The vertical space $\mathcal{V}_{(p,u)}$ in $T_{(p,u)}(TS^n)$ is

$$\mathcal{V}_{(p,u)} = \{((p,u),(0,w)) \in T_{(p,u)}(TS^n)\}.$$

Recall that in Chapter 4 we obtained the Levi-Civita covariant derivative by projection of the ambient derivative back into the tangent space. In the present context, this means that $((p,u),(v,w))$ is horizontal if w is a multiple of p. But since we also have $\langle p, w\rangle + \langle u, v\rangle = 0$, the criterion for being horizontal becomes $w = -\langle u, v\rangle p$. A little thought reveals that we should take $\mathcal{H}$ to be defined by

$$\mathcal{H}_{(p,u)} := \{((p,u),(v, -\langle u, v\rangle p)) \in T_{(p,u)}TS^n\}.$$

Exercise 12.34. Let $p, q \in S^2 \subset \mathbb{R}^3$ be such that $\langle p, q\rangle = 0$. The great circle containing p and q can be parametrized as

$$c(t) := (\cos t)\, p + (\sin t)\, q$$

12.4. Ehresmann Connections

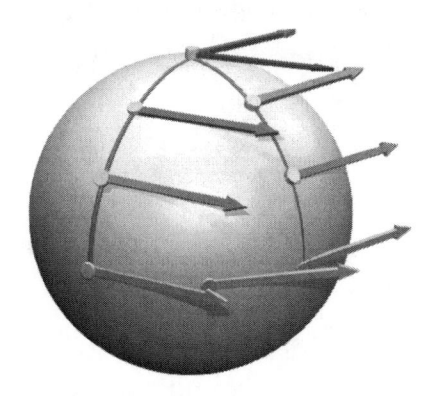

Figure 12.3. Parallel transport around path shows holonomy

for $0 \leq t \leq 2\pi$. Using the notation of the previous example, show that if $\sigma = (c, x)$ is a parallel section along c, then both $\langle x, x \rangle$ and $\langle c', x \rangle$ are constant along c.

Figure 12.3 shows parallel translation of a vector at the north pole around a loop comprised of segments of great circles. Notice that parallel translation around this loop results in a map which is not the identity map. In general, if $\pi : E \to M$ is a vector bundle with connection and $c : [a, b] \to M$ is a piecewise smooth closed curve with $p = c(a) = c(b)$, then we obtain an isomorphism $P_c \in \mathrm{GL}(E_p)$. This map is called the **holonomy** of the curve.

Definition 12.35. The subset $G_p \subset \mathrm{GL}(E_p)$ consisting of the maps P_c as c ranges over all piecewise smooth closed curves c with $p = c(a) = c(b)$ is called the **holonomy group** at p for the connection.

It is easy to see that any piecewise smooth curve $c : [a, b] \to M$ induces a group isomorphism between $G_{c(a)}$ and $G_{c(b)}$ given by

$$g \mapsto P_c \circ g \circ P_c^{-1}.$$

Thus for a connected manifold we may speak of the holonomy group of the manifold, and this is well-defined up to group isomorphism.

One may use parallel translation to recover the covariant derivative and this is a very natural viewpoint:

Theorem 12.36. *Let $\pi : E \to M$ be a vector bundle with connection $\mathcal{H}$. Let $f : N \to M$ be a smooth map. If $u \in T_pN$ and σ is a section of E along*

f, then for any smooth curve $c : (-\varepsilon, \varepsilon) \to N$ with $\dot{c}(0) = u$ we have

$$\nabla^f_u \sigma = \lim_{t \to 0} \frac{\left(P(f \circ c)_0^t\right)^{-1} \sigma(c(t)) - \sigma(c(0))}{t},$$

where $P(f \circ c)_0^t$ is the parallel transport along $f \circ c$ from $(f \circ c)(0)$ to $(f \circ c)(t)$.

Proof. Use parallel transport to obtain parallel frame fields $e_1, \ldots, e_k$ along $f \circ c$ with $e_1(0), \ldots, e_k(0)$ a basis of $E_{(f \circ c)(0)}$. Then we may write

$$\sigma \circ c = \sum \sigma^i e_i$$

for unique smooth functions σ^i defined on $(-\varepsilon, \varepsilon)$. Then

$$\left(P(f \circ c)_0^t\right)^{-1} \sigma \circ c(t) = \left(P(f \circ c)_0^t\right)^{-1} \sum \sigma^i(t) e_i(t)$$
$$= \sum \sigma^i(t) e_i(0).$$

Let D_0 denote $\frac{\partial}{\partial t}\big|_0$. Then using the chain rule for connections and the fact that $\nabla_{\partial_t} e_i(t) = 0$ we have

$$\nabla_u \sigma = \nabla_{\dot{c}(0)} \sigma = \nabla_{Tc \cdot D_0} \sigma$$
$$= \nabla_{D_0}(\sigma \circ c) = \nabla_{D_0} \sum \sigma^i(t) e_i(t)$$
$$= \sum \frac{d\sigma^i}{dt}\bigg|_0 e_i(0) = \frac{d}{dt}\bigg|_0 \left[\left(P(f \circ c)_0^t\right)^{-1} \sigma \circ c(t)\right]$$
$$= \lim_{t \to 0} \frac{\left(P(f \circ c)_0^t\right)^{-1} \sigma(c(t)) - \sigma(c(0))}{t}. \qquad \square$$

The most important special cases are the case of $f = \mathrm{id}_M$, where we recover a basic Koszul connection, and also the case where f itself is a curve $c : I \to M$. However, recall that we often omit the superscripts indicating maps on the various covariant derivative operators.

Corollary 12.37. *Let $\pi : E \to M$ be a vector bundle with connection $\mathcal{H}$. If $v \in T_p M$ and s is a section of E, then for any smooth curve $c : (-\varepsilon, \varepsilon) \to M$ with $\dot{c}(0) = v$ we have*

$$\nabla_v s = \lim_{t \to 0} \frac{\left(P(c)_0^t\right)^{-1} s(c(t)) - s(c(0))}{t},$$

where $P(c)_0^t$ is the parallel transport along c from $c(0)$ to $c(t)$.

Now recall that if $c : I \to M$ is a smooth curve and $\sigma : I \to E$ is a section along c, we define $\nabla_{\frac{\partial}{\partial t}} \sigma$ to be a section along c given by

$$\left(\nabla_{\frac{\partial}{\partial t}} \sigma\right)(t) := \nabla_{\frac{\partial}{\partial t}\big|_t} \sigma.$$

12.5. Curvature

We have

Corollary 12.38. *Let $\pi : E \to M$ be a vector bundle with connection $\mathcal{H}$. Let $c : I \to M$ be a smooth curve and suppose that $\sigma : I \to E$ is a section along c. Then we have*

$$\left(\nabla_{\frac{\partial}{\partial t}} \sigma\right)(t) = \lim_{\epsilon \to 0} \frac{\left(P(c)_t^{t+\epsilon}\right)^{-1} \sigma(t+\epsilon) - \sigma(t)}{\epsilon}.$$

Exercise 12.39. Derive the above two corollaries from Theorem 12.36.

Corollary 12.40. *Let $\pi : E \to M$ be a vector bundle with connection $\mathcal{H}$. Let $f : N \to M$ be a smooth map. If σ is a section of E along f, then σ is parallel if and only if $\nabla_u^f \sigma = 0$ for all $u \in TN$. In particular, a section $s \in \Gamma(E)$ is parallel along a curve c if and only if $\nabla_{\partial/\partial t} s \circ c \equiv 0$.*

If one approaches parallelism solely via covariant derivatives, then one could take $\nabla_u^f \sigma \equiv 0$ as the definition of parallel. This is a common equivalent approach.

Warning: There is a subtle point to be made here. If s is a section of E and $c : (-\varepsilon, \varepsilon) \to M$ is a curve, then $\nabla_{\dot{c}(t)} s$ is zero whenever $\dot{c}(t) = 0$, but for a section $\sigma : I \to E$ along c it is possible that $\left(\nabla_{\partial/\partial t} \sigma\right)(t)$ is nonzero even when $\dot{c}(t) = 0$. For example, if $c_p : I \to M$ is a constant curve taking the fixed value $p \in M$, and $\sigma : I \to T_p M$ is a smooth map, then σ can be considered a section along c_p and then $\nabla_{\partial/\partial t} \sigma = \sigma'$ (the ordinary derivative of a curve in the vector space $T_p M$).

Exercise 12.41. Show that $\sigma \in \Gamma(M, E)$ is a parallel section if and only if $\sigma \circ c$ is parallel along c for every curve $c : I \to M$.

Exercise 12.42. Show that if $t \mapsto \sigma(t)$ is a curve in E_p, then we can consider σ as a section along the constant map $c_p : t \mapsto p$ and then $\nabla_{\partial_t} \sigma(t) = \sigma'(t) \in E_p$.

Exercise 12.43. Let ∇ be a connection on $E \to M$ and let $\alpha : [a, b] \to M$ and $\beta : [a, b] \to E_{\alpha(t_0)}$. If $X(t) := P(\alpha)_{t_0}^t(\beta(t))$, then

$$\left(\nabla_{\partial_t} X\right)(t) = P(\alpha)_{t_0}^t(\beta'(t)).$$

Note: $\beta'(t) \in E_{\alpha(t_0)}$ for all t. [Hint: Use a parallel frame field along the curve.]

12.5. Curvature

An important fact about covariant derivatives is that they do not need to commute. If $\sigma : M \to E$ is a section and $X \in \mathfrak{X}(M)$, then $\nabla_X \sigma$ is a section also, and so we may take its covariant derivative $\nabla_Y \nabla_X \sigma$ with respect to some $Y \in \mathfrak{X}(M)$. In general, $\nabla_Y \nabla_X \sigma \neq \nabla_X \nabla_Y \sigma$. A measure of this lack of commutativity is the **curvature operator** that is defined for a pair $X, Y \in \mathfrak{X}(M)$ to be the map $F(X,Y) : \Gamma(E) \to \Gamma(E)$ given by

$$F(X,Y)\sigma := \nabla_X \nabla_Y \sigma - \nabla_Y \nabla_X \sigma - \nabla_{[X,Y]}\sigma,$$

i.e. $F(X,Y) := [\nabla_X, \nabla_Y] - \nabla_{[X,Y]}$.

Theorem 12.44. *For fixed σ, the map $(X,Y) \mapsto F(X,Y)\sigma$ is $C^\infty(M)$ bilinear and antisymmetric. Also, $F(X,Y) : \Gamma(E) \to \Gamma(E)$ is a $C^\infty(M)$ module homomorphism; that is, it is linear over the smooth functions,*

$$F(X,Y)(f\sigma) = fF(X,Y)(\sigma).$$

Proof. We leave the proof of the first part as an easy exercise. For the second part, we calculate:

$$\begin{aligned} F(X,Y)(f\sigma) &= \nabla_X \nabla_Y f\sigma - \nabla_Y \nabla_X f\sigma - \nabla_{[X,Y]} f\sigma \\ &= \nabla_X(f\nabla_Y \sigma + (Yf)\sigma) - \nabla_Y(f\nabla_X \sigma + (Xf)\sigma) \\ &\quad - f\nabla_{[X,Y]}\sigma - ([X,Y]f)\sigma \\ &= f\nabla_X \nabla_Y \sigma + (Xf)\nabla_Y \sigma + (Yf)\nabla_X \sigma + X(Yf) \\ &\quad - f\nabla_Y \nabla_X \sigma - (Yf)\nabla_X \sigma - (Xf)\nabla_Y \sigma - Y(Xf) \\ &\quad - f\nabla_{[X,Y]}\sigma - ([X,Y]f)\sigma \\ &= f[\nabla_X, \nabla_Y] - f\nabla_{[X,Y]}\sigma = fF(X,Y)\sigma. \qquad \square \end{aligned}$$

Thus we also have F as a map $F : \mathfrak{X}(M) \times \mathfrak{X}(M) \to \Gamma(M, \mathrm{End}(E))$. But F is $C^\infty(M)$ bilinear in the first two slots also.

Exercise 12.45. Show that F is $C^\infty(M)$ bilinear in the first two slots. Let $v_p, w_p \in T_pM$ and $\sigma_p \in E_p$ be given. Let $X, Y \in \mathfrak{X}(M)$ and $\sigma \in \Gamma(E)$ such that $X(p) = v_p, Y(p) = w_p$ and $\sigma(p) = \sigma_p$. Argue that $F(X_p, Y_p)\sigma_p$ is well-defined and that the map $(v_p, w_p, \sigma_p) \mapsto F(v_p, w_p)\sigma_p$ is multilinear.

In light of this exercise we see that for each $v_p, w_p \in T_pM$, $F(v_p, w_p)$ is a linear map $E_p \to E_p$. In other words, $F(X_p, Y_p) \in \mathrm{End}(E_p)$ (a.k.a. $L(E_p, E_p)$). But $F(X_p, Y_p)$ is antisymmetric, and so we obtain, for each p, an element of $\mathrm{End}(E_p) \otimes \wedge^2 T_p^*M$. Finally, we see that F may be considered as a section of $\mathrm{End}(E) \otimes \wedge^2 T^*M$. The space of such sections is denoted $\Omega^2(M, \mathrm{End}(E))$.

> Curvature is an $\mathrm{End}(E)$-valued 2-form.

We will have many occasions to consider differentiation along maps, and so we should also look at the curvature operator for sections along maps.

12.5. Curvature

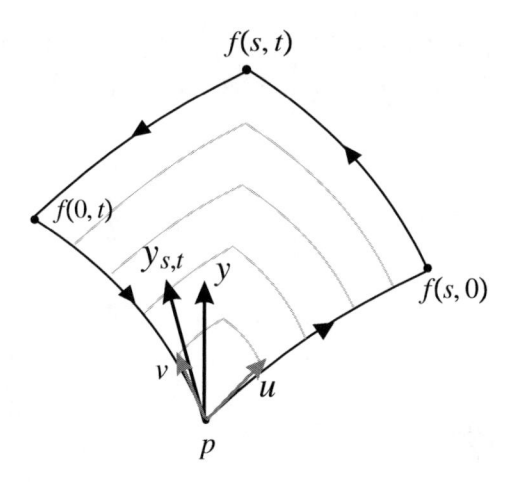

Figure 12.4. Parallel translation around loop

Let $f: N \to M$ be a smooth map and σ a section of $E \to M$ along f. For $U, V \in \mathfrak{X}(N)$, we define a map $F^f(U,V): \Gamma_f(E) \to \Gamma_f(E)$ by

$$F^f(U,V)\sigma := \nabla_U \nabla_V \sigma - \nabla_V \nabla_U \sigma - \nabla_{[U,V]} \sigma$$

for all $\sigma \in \Gamma_f(E)$.

Note that if $s \in \Gamma(E)$, then $s \circ f \in \Gamma_f(E)$, and if $U \in \mathfrak{X}(N)$, then $Tf \circ U \in \Gamma_f(TM) =: \mathfrak{X}_f(M)$ is a vector field along f. As a matter of notation, we let $F(Tf \circ U, Tf \circ V)s$ denote the map $p \mapsto F(Tf \cdot U_p, Tf \cdot V_p)s(p)$, which makes sense because of Theorem 12.44. Thus $F(Tf \circ U, Tf \circ V)s \in \Gamma_f(E)$. Then we have the following useful fact:

Proposition 12.46. *Let $X \in \Gamma_f(E)$ and $U, V \in \mathfrak{X}(N)$. Then*

$$F^f(U,V)X = F(Tf \circ U, Tf \circ V)X.$$

Proof. Exercise! □

The next theorem is an important step in understanding the geometric meaning of curvature in terms of parallel transport. Refer to Figure 12.4.

Theorem 12.47. *Let $\pi: E \to M$ be a vector bundle with connection. Let $u, v \in T_p M$ and $y \in E_p$. Let U be a neighborhood of the origin in $\mathbb{R}^2$ and denote standard coordinate frame fields on $\mathbb{R}^2$ by ∂_1, ∂_2 and their values at the origin by $\partial_1(0)$ and $\partial_2(0)$. Suppose that we have a smooth map $f: U \to M$ with $f(0) = p$, $T_p f \cdot \partial_1(0) = u$ and $T_p f \cdot \partial_2(0) = v$. For small s, t, let $y_{s,t}$ denote the element of E_p obtained by parallel translation of y around the piecewise smooth loop obtained by successively tracing the following four curves:*

(1) $c_1^{s,t}: \sigma \mapsto f(\sigma, 0)$, $0 \leq \sigma \leq s$;

(2) $c_2^{s,t} : \tau \mapsto f(s,\tau)$, $0 \leq \tau \leq t$;

(3) $c_3^{s,t} : \sigma \mapsto f(s-\sigma, t)$, $0 \leq \sigma \leq s$;

(4) $c_4^{s,t} : \tau \mapsto f(0, t-\tau)$, $0 \leq \tau \leq t$.

Then,
$$F(u,v)y = - \lim_{s,t \to 0} \frac{y_{s,t} - y}{st}.$$

Proof. Let Y be a section along f defined as follows: If we choose $\epsilon > 0$ sufficiently small, then for each $(s,t) \in [0,\epsilon] \times [0,\epsilon]$ the four curves described above will be defined. For each such (s,t), let $Y(s,t)$ be the result of parallel translation of y along the first two curves $c_1^{s,t}$ and $c_2^{s,t}$. We leave it to the reader to argue that Y is smooth. Notice that $y = Y(0,0)$. By Proposition 12.46 we have
$$F(u,v)y = F^f(\partial_1(0), \partial_2(0))y = \nabla_{\partial_1(0)} \nabla_{\partial_2} Y - \nabla_{\partial_2(0)} \nabla_{\partial_1} Y$$
since $[\partial_1, \partial_2] = 0$. Since Y is parallel along the curves $c_2^{s,t}$, we have $\nabla_{\partial_2} Y = 0$ and so
$$F(u,v)y = -\nabla_{\partial_2(0)} \nabla_{\partial_1} Y.$$
If we let P_t denote parallel translation along $\tau \mapsto f(0,\tau)$ from p to $f(0,t)$, then
$$\nabla_{\partial_2(0)} \nabla_{\partial_1} Y = \lim_{t \to 0} \frac{P_t^{-1} (\nabla_{\partial_1} Y)(0,t) - (\nabla_{\partial_1} Y)(0,0)}{t} = \lim_{t \to 0} \frac{P_t^{-1} (\nabla_{\partial_1(0,t)} Y)}{t}.$$
On the other hand, if $P_{s,t}$ denotes parallel translation along $\sigma \mapsto f(\sigma,t)$ from $f(0,t)$ to $f(s,t)$, then
$$\nabla_{\partial_1(0,t)} Y = \lim_{s \to 0} \frac{P_{s,t}^{-1} Y(s,t) - Y(0,t)}{s}.$$
Putting things together we have
$$F(u,v)y = - \lim_{s,t \to 0} \frac{P_t^{-1} P_{s,t}^{-1} Y(s,t) - P_t^{-1} Y(0,t)}{st} = - \lim_{s,t \to 0} \frac{y_{s,t} - y}{st}. \qquad \square$$

Next we will try to understand curvature in terms of the horizontal distribution. Let us define an object that in some sense measures how far the horizontal distribution is from being integrable. First note that the horizontal lifts of local frame fields on the base manifold M give vector fields on the total space that span the distribution. If the bracket of these lifted fields were always horizontal, then the distribution would be integrable—the connection would be flat. This suggests the following: For every pair of horizontal fields Z_1, Z_2 on E, let
$$C(Z_1, Z_2) := \mathrm{p_v}([Z_1, Z_2]).$$

12.5. Curvature

Lemma 12.48. *Let $\Gamma(\mathcal{H})$ denote the $C^\infty(E)$-module of horizontal fields and let $\Gamma(\mathcal{V})$ denote the $C^\infty(E)$-module of vertical fields. The map $C: \Gamma(\mathcal{H}) \times \Gamma(\mathcal{H}) \to \Gamma(\mathcal{V})$ given by $(Z_1, Z_2) \mapsto C(Z_1, Z_2)$ is a module homomorphism.*

Proof. Let $f \in C^\infty(E)$. Then we have
$$C(fZ_1, Z_2) = \mathrm{p_v}([fZ_1, Z_2]) = \mathrm{p_v}(f[Z_1, Z_2] - (Z_2 f)\, Z_2)$$
$$= f\,\mathrm{p_v}([Z_1, Z_2]) - \mathrm{p_v}((Z_2 f)\, Z_2) = f\,\mathrm{p_v}([Z_1, Z_2])$$
$$= fC(Z_1, Z_2).$$

The analogous result for the second factor follows because C is clearly skew-symmetric. Additivity is obvious. $\square$

Corollary 12.49. *For $y \in E$, the value $C(Z_1, Z_2)(y)$ depends only on the values $Z_1(y)$ and $Z_2(y)$. Also, C is well-defined on locally defined horizontal fields.*

Proof. The corollary says that C is "tensorial". The proof is just like the proof of Theorem 7.32 and can also be derived from Proposition 6.55. $\square$

Because of this corollary, we may think of C as a map $\mathcal{H} \times \mathcal{H} \to \mathcal{V}$.

Theorem 12.50. *Let $\pi: E \to M$ be a vector bundle with connection. Then for $u, v \in T_p M$ and $y \in E_p$ we have*
$$F(u,v)y = -\jmath_y^{-1}(C(u^y, v^y)),$$
where u^y, v^y denote the horizontal lifts of u and v to the point y. Here $\jmath_y: T_y E_p \to E_p$ is just the canonical map as usual.

Proof. Choose $U, V \in \mathfrak{X}(M)$ with $U(p) = u$ and $V(p) = v$. We may assume that $[U, V] = 0$ near p. Let φ_s and ψ_t be local flows of U and V respectively. Also, let $\widetilde{\varphi}_s$ and $\widetilde{\psi}_t$ be the flows of the horizontal lifts $\widetilde{U}$ and $\widetilde{V}$. Observe that since $\varphi_s \circ \pi = \pi \circ \widetilde{\varphi}_s$, we have that $\widetilde{\varphi}_s(y)$ is the parallel translation of y along the curve $\sigma \mapsto \varphi_\sigma \circ \pi(y)$, $0 \leq \sigma \leq s$. Similarly, $\widetilde{\psi}_t(y)$ is the parallel translation of y along the curve $\tau \mapsto \psi_\tau \circ \pi(y)$, $0 \leq \tau \leq t$.

Now consider the curve c defined for small t by
$$c(t) := \widetilde{\psi}_{-\sqrt{t}} \circ \widetilde{\varphi}_{-\sqrt{t}} \circ \widetilde{\psi}_{\sqrt{t}} \circ \widetilde{\varphi}_{\sqrt{t}}.$$

Then, by Theorem 12.47, we have
$$F(u,v)y = -\lim_{s,t \to 0} \frac{y_{s,t} - y}{st} = -\lim_{t \to 0^+} \frac{y_{\sqrt{t}, \sqrt{t}} - y}{t} = -\lim_{t \to 0^+} \frac{c(t) - c(0)}{t}.$$

But by Theorem 2.112,
$$-\lim_{t \to 0^+} \frac{c(t) - c(0)}{t} = -\jmath_y^{-1} \dot{c}(0) = -\jmath_y^{-1}[\widetilde{U}, \widetilde{V}](y).$$

Finally, notice that $[U,V](p) = 0$ so $[\widetilde{U},\widetilde{V}](y)$ is vertical and is equal to $\mathrm{p_v}([\widetilde{U},\widetilde{V}])(y) = C(\widetilde{U},\widetilde{V})(y) = C(u^y, v^y)$. □

Unwinding the definitions, the result of the previous theorem can be written as
$$(F(U,V)s)(p) = -\kappa([\widetilde{U},\widetilde{V}]_{s(p)}),$$
where $s \in \Gamma(E)$ and κ is the connector associated to the connection.

Theorem 12.51. *Let $\pi : E \to M$ be a vector bundle with connection. If the curvature F vanishes, then the connection is flat. In particular, given any $y \in E$, there is a locally defined parallel section s with $s(p) = y$ where $\pi(y) = p$.*

Proof. Let Z_1, Z_2 be locally defined horizontal vector fields. Then for any y in the common domain of Z_1, Z_2, let u, v be the images of $Z_1(y), Z_2(y)$ under $T\pi$. Thus $Z_1(y)$ and $Z_2(y)$ are the horizontal lifts of u and v. Then we have
$$-j_y^{-1}(\mathrm{p_v}([Z_1, Z_2]_y)) = F(u,v)y = 0.$$
But j_y^{-1} is an isomorphism, and so $\mathrm{p_v}([Z_1, Z_2]_y) = 0$, which means that $[Z_1, Z_2]_y$ is horizontal. Since y was arbitrary, $[Z_1, Z_2] = 0$. We conclude that $\mathcal{H}$ is integrable by the Frobenius theorem. Now apply Theorem 12.25. □

12.6. Connections on Tangent Bundles

A connection on the tangent bundle TM of a smooth manifold M is called a linear connection on M. It is traditional to use special notation for the components of the connection form when using coordinate frames $\{\frac{\partial}{\partial x^i}\}$. In this case, formula (12.3) becomes

$$(12.6) \qquad \nabla_{\frac{\partial}{\partial x^i}} \frac{\partial}{\partial x^j} = \sum_k \Gamma_{ij}^k \frac{\partial}{\partial x^k}.$$

The Γ_{ij}^k are a special case of the components of the connection forms. If $X = X^j \frac{\partial}{\partial x^j}$ and $Y = Y^j \frac{\partial}{\partial x^j}$, then

$$(12.7) \quad \nabla_X Y = \left(\frac{\partial Y^k}{\partial x^j} X^j + \Gamma_{ij}^k X^i Y^j\right) \frac{\partial}{\partial x^k} \qquad \text{(summation convention)},$$

which is formula (12.4) in the current context. The functions Γ_{ij}^k are called **Christoffel symbols**.

It is a consequence of Proposition 7.37 that for each $X \in \mathfrak{X}(M)$ there is a unique tensor derivation ∇_X on $\mathcal{T}_s^r(M)$ such that ∇_X commutes with contraction and coincides with the given covariant derivative on $\mathfrak{X}(M)$ (also denoted ∇_X) and with $\mathcal{L}_X$ on $C^\infty(M)$.

12.6. Connections on Tangent Bundles

To describe the covariant derivative on tensors more explicitly, consider $\omega \in \mathcal{T}_1^0(M)$. Since we have the contraction $Y \otimes \omega \mapsto C(Y \otimes \omega) = \omega(Y)$, we should have

$$\begin{aligned}
\nabla_X \omega(Y) &= \nabla_X C(Y \otimes \omega) \\
&= C(\nabla_X(Y \otimes \omega)) \\
&= C(\nabla_X Y \otimes \omega + Y \otimes \nabla_X \omega) \\
&= \omega(\nabla_X Y) + (\nabla_X \omega)(Y).
\end{aligned}$$

So we *define* $(\nabla_X \omega)(Y) := \nabla_X(\omega(Y)) - \omega(\nabla_X Y)$. This implies that for a local frame field $E_1, \ldots, E_n$ with dual frame field $\theta^1, \ldots, \theta^n$ we have

$$\nabla_X \theta^i = -\sum_k \omega_j^i(X) \theta^j,$$

where ω_j^i are the connection forms satisfying $\nabla_X E_j = \sum_k \omega_j^i(X) E_i$. More generally, if $\Upsilon \in \mathcal{T}_s^r$, then

(12.8)

$$(\nabla_X \Upsilon)(\omega_1, \ldots, \omega_r, Y_1, \ldots, Y_s)$$
$$= X(\Upsilon(\omega_1, \ldots, \omega_r, Y_1, \ldots, Y_s)) - \sum_{j=1}^r \Upsilon(\omega_1, . \nabla_X \omega_j, \ldots, \omega_r, Y_1, \ldots)$$
$$- \sum_{i=1}^s \Upsilon(\omega_1, \ldots, \omega_r, Y_1, \ldots, \nabla_X Y_i, \ldots).$$

Definition 12.52. The covariant differential of a tensor field $\Upsilon \in \mathcal{T}_l^k$ is denoted by $\nabla \Upsilon$ and is defined to be the element of $\mathcal{T}_{l+1}^k$ given by

$$\nabla \Upsilon(\omega^1, \ldots, \omega^1, X, Y_1, \ldots, Y_s) := \nabla_X \Upsilon(\omega^1, \ldots, \omega^1, Y_1, \ldots, Y_s).$$

For any fixed frame field $E_1, \ldots, E_n$, we denote the components of $\nabla \Upsilon$ by $\nabla_i \Upsilon_{j_1 \ldots j_s}^{i_1 \ldots i_r}$.

Remark 12.53. We have placed the new variable at the beginning as suggested by our notation $\nabla_i \Upsilon_{j_1 \ldots j_s}^{i_1}$ for the components of $\nabla \Upsilon$ but in opposition to the equally common notation $\Upsilon_{j_1 \ldots j_s;i}^{i_1}$. This has the advantage of meshing well with exterior differentiation, making the statement of Theorem 12.56 as simple as possible.

The reader is asked in Problem 2 to show that $T(X, Y) := \nabla_X Y - \nabla_Y X - [X, Y]$ defines a tensor and so $T(X_p, Y_p)$ is well-defined for $X_p, Y_p \in T_p M$. This tensor is called the **torsion tensor**. If T vanishes identically, then we say that the connection is **torsion free**. We have already noted that the

Levi-Civita connection defined in Chapter 4 for a hypersurface is torsion free.

We shall have more to say about torsion further on, but for now, notice that if the connection is torsion free, then

$$0 = \nabla_{\partial_i}\partial_j - \nabla_{\partial_j}\partial_i = \sum_k \left(\Gamma_{ij}^k - \Gamma_{ji}^k\right)\partial_k,$$

and so the Christoffel symbols are symmetric with respect to the lower two indices;

$$\Gamma_{ij}^k = \Gamma_{ji}^k.$$

However, even for torsion free connections, this is generally not true for $\omega_{\mu j}^k$ defined by $\nabla_{\partial_\mu}e_j = \sum \omega_{\mu j}^k e_k$ unless the elements $e_1, \ldots, e_n$ of the frame field have pairwise vanishing Lie brackets. But this amounts to saying that they are locally coordinate frame fields for some chart.

12.7. Comparing the Differential Operators

On a smooth manifold we have the Lie derivative $\mathcal{L}_X : \mathcal{T}_s^r(M) \to \mathcal{T}_s^r(M)$ and the exterior derivative $d : \Omega^k(M) \to \Omega^{k+1}(M)$, and in case we have a torsion free covariant derivative ∇, that makes three differential operators which we would like to compare. To this end, we restrict attention to purely covariant tensor fields $\mathcal{T}_s^0(M)$.

The extended covariant derivative on tensor fields $\nabla_X : \mathcal{T}_s^0(M) \to \mathcal{T}_s^0(M)$ respects the subspace consisting of alternating tensors, and so for each k we have a map

$$\nabla_X : L_{\text{alt}}^k(M) \to L_{\text{alt}}^k(M)$$

and these combine to give a degree preserving map

$$\nabla_X : L_{\text{alt}}(M) \to L_{\text{alt}}(M).$$

In other notation,

$$\nabla_X : \Omega(M) \to \Omega(M).$$

It is also easily seen that not only do we have $\nabla_X(\alpha \otimes \beta) = \nabla_X\alpha \otimes \beta + \alpha \otimes \nabla_X\beta$, but also

$$\nabla_X(\alpha \wedge \beta) = \nabla_X\alpha \wedge \beta + \alpha \wedge \nabla_X\beta.$$

Recall (12.8) and the similar formula (7.10) for the Lie derivative. If ∇ is torsion free so that $\mathcal{L}_X Y_i = [X, Y_i] = \nabla_X Y_i - \nabla_{Y_i} X$, then we obtain the following modification of formula (7.10) which incorporates ∇.

12.7. Comparing the Differential Operators

Proposition 12.54. *For a torsion free connection, we have the following equality for any* $S \in \mathcal{T}_s^0(M)$:

(12.9)
$$(\mathcal{L}_X S)(Y_1, \ldots, Y_s) = (\nabla_X S)(Y_1, \ldots, Y_s)$$
$$+ \sum_{i=1}^s S(Y_1, \ldots, Y_{i-1}, \nabla_{Y_i} X, Y_{i+1}, \ldots, Y_s).$$

Proof. See Problem 10. □

Corollary 12.55. $\mathcal{L}_X S - \nabla_X S$ *is a tensor.*

For $\omega \in \Omega^k(M)$, we have that $\nabla \omega$ is a covariant tensor field but now not necessarily alternating.[2] We search for a way to fix this. By antisymmetrizing, we get a map $\Omega^k(M) \to \Omega^{k+1}(M)$ which turns out to be none other than our old friend the exterior derivative as will be shown below.

Theorem 12.56. *If* ∇ *is a torsion free covariant derivative on* M, *then*

$$d = (k+1)\mathrm{Alt} \circ \nabla$$

or in other words, if $\omega \in \Omega^k(M)$, *then*

$$d\omega(X_0, X_1, \ldots, X_k) = \sum_{i=0}^k (-1)^i (\nabla_{X_i} \omega)(X_0, \ldots, \widehat{X_i}, \ldots, X_k).$$

Proof. First we show that

$$(\mathrm{Alt} \circ \nabla \omega)(X_0, X_1, \ldots, X_k) = \frac{1}{k+1} \sum_{i=0}^k (-1)^i (\nabla_{X_i} \omega)(X_0, \ldots, \widehat{X_i}, \ldots, X_k).$$

Consider the subgroup H consisting of the permutations of $\{0, 1, \ldots, k\}$ that fix 0. The cosets of H are $H = H_0, H_1, \ldots, H_k$ where H_j is the set of permutations that send 0 to j. If we decompose the group of permutations into these cosets, then

$$(\mathrm{Alt} \circ \nabla \omega)(X_0, X_1, \ldots, X_k) = \frac{1}{(k+1)!} \sum_\sigma \mathrm{sgn}(\sigma)(\nabla_{X_{\sigma_0}} \omega)(X_{\sigma_1}, \ldots, X_{\sigma_k})$$

$$= \frac{1}{(k+1)!} \sum_{i=0}^k \sum_{\sigma \in H_i} \mathrm{sgn}(\sigma)(\nabla_{X_{\sigma_0}} \omega)(X_{\sigma_1}, \ldots, X_{\sigma_k})$$

$$= \frac{k!}{(k+1)!} \sum_{i=0}^k (-1)^i (\nabla_{X_i} \omega)(X_0, \ldots, \widehat{X_i}, \ldots, X_k).$$

---

[2] However, $\nabla \omega$ can be viewed as being in $\Omega^k(M) \otimes_{C^\infty} \Omega^1(M)$.

To finish we calculate as follows:

$$d\omega(X_0, X_1, \ldots, X_k) = \sum_{i=0}^{k} (-1)^i X_i(\omega(X_0, \ldots, \widehat{X_i}, \ldots, X_k))$$
$$+ \sum_{1 \le r < s \le k} (-1)^{r+s} \omega([X_r, X_s], X_0, \ldots, \widehat{X_r}, \ldots, \widehat{X_s}, \ldots, X_k)$$
$$= \sum_{i=0}^{k} (-1)^i X_i(\omega(X_0, \ldots, \widehat{X_i}, \ldots, X_k))$$
$$+ \sum_{1 \le r < s \le k} (-1)^{r+s} \omega(\nabla_{X_r} X_s - \nabla_{X_s} X_r, X_0, \ldots, \widehat{X_r}, \ldots, \widehat{X_s}, \ldots, X_k),$$

which is equal to

$$\sum_{i=0}^{k} (-1)^i X_i(\omega(X_0, \ldots, \widehat{X_i}, \ldots, X_k))$$
$$+ \sum_{1 \le r < s \le k} (-1)^{r+1} \omega(X_0, \ldots, \widehat{X_r}, \ldots, \nabla_{X_r} X_s, \ldots, X_k)$$
$$- \sum_{1 \le r < s \le k} (-1)^s \omega(X_0, \ldots, \nabla_{X_s} X_r, \ldots, \widehat{X_s}, \ldots, X_k)$$
$$= \sum_{i=0}^{k} (-1)^i \nabla_{X_i} \omega(X_0, \ldots, \widehat{X_i}, \ldots, X_k) \quad \text{(by using 12.8)}. \quad \square$$

12.8. Higher Covariant Derivatives

Now let us suppose that we have a connection ∇^{E_i} on every vector bundle $E_i \to M$ in some family $\{E_i\}_{i \in I}$. We then also have the connections $\nabla^{E_i^*}$ induced on the duals $E_i^* \to M$. By demanding a product formula be satisfied as usual we can form a related family of connections on all bundles formed from tensor products of the bundles in the family $\{E_i, E_i^*\}_{i \in I}$. In this situation, it might be convenient to denote any and all of these connections by a single symbol as long as the context makes confusion unlikely. In particular, we have the following common situation: By the definition of a connection we have that $X \mapsto \nabla^E_X \sigma$ is $C^\infty(M)$ linear and so $\nabla^E \sigma$ is a section of the bundle $T^*M \otimes E$. We can use the Levi-Civita connection or any torsion free connection ∇ on M together with ∇^E to define a connection on $E \otimes T^*M$. To get a clear picture of this connection, we first notice that a section ξ of the bundle $E \otimes T^*M$ can be written locally in terms of a local frame field $\{\theta^i\}$ on T^*M and a local frame field $\{e_i\}$ on E. Namely, we may write $\xi = \sum \xi_j^i e_i \otimes \theta^j$. Then the connection $\nabla^{E \otimes T^*M}$ on $E \otimes T^*M$ is defined so that a product rule holds. Let γ_ν^μ and ω_j^i be the connection forms for ∇

12.8. Higher Covariant Derivatives

and ∇^E respectively, so that locally, using the summation convention, we have

$$\begin{aligned}
\nabla_X^{E\otimes T^*M}\xi &= \nabla_X^E\left(\xi_\mu^i e_i\right)\otimes\theta^\mu + \xi_\mu^i e_i\otimes\nabla_X\theta^\mu \\
&= \left(X\xi_\mu^i e_i + \xi_\mu^i \nabla_X^E e_i\right)\otimes\theta^\mu + \xi_\mu^i e_i\otimes\nabla_X\theta^\mu \\
&= \left(X\xi_\nu^r e_r + e_r\omega_i^r(X)\xi_\nu^i\right)\otimes\theta^\nu - \gamma_\nu^\mu(X)\xi_\mu^r e_r\otimes\theta^\nu \\
&= \left(X\xi_\mu^r + \omega_i^r(X)\xi_\nu^i - \gamma_\nu^\mu(X)\xi_\mu^r\right)e_r\otimes\theta^\nu.
\end{aligned}$$

Now let $\xi = \nabla^E\sigma$ for a given $\sigma\in\Gamma(E)$. The map $X\mapsto\nabla_X^{E\otimes T^*M}(\nabla^E\sigma)$ is $C^\infty(M)$ linear, and $\nabla^{E\otimes T^*M}(\nabla^E\sigma)$ is an element of $\Gamma(E\otimes T^*M\otimes T^*M)$, which can again be given the obvious connection. The process continues, and denoting all the connections by the same symbol we may consider the k-th covariant derivative $\nabla^k\sigma\in\Gamma(E\otimes T^*M^{\otimes k})$ for each $\sigma\in\Gamma(E)$.

It is sometimes convenient to change viewpoints just slightly and define the **covariant derivative operators** $\nabla_{X_1,X_2,\ldots,X_k}:\Gamma(E)\to\Gamma(E)$. The definition is given inductively as

$$\begin{aligned}
\nabla_{X_1,X_2}\sigma &:= \nabla_{X_1}\nabla_{X_2}\sigma - \nabla_{\nabla_{X_1}X_2}\sigma, \\
\nabla_{X_1,\ldots,X_k}\sigma &:= \nabla_{X_1}\left(\nabla_{X_2,X_3,\ldots,X_k}\sigma\right) \\
&\quad - \nabla_{\nabla_{X_1}X_2,X_3,\ldots,X_k}\sigma - \cdots - \nabla_{X_2,X_3,\ldots,\nabla_{X_1}X_k}\sigma.
\end{aligned}$$

Then we have the following convenient formula, which is true by definition:

$$\nabla^{(k)}\sigma(X_1,\ldots,X_k) = \nabla_{X_1,\ldots,X_k}\sigma.$$

Warning: $\nabla_{\partial_i}\nabla_{\partial_j}\tau$ is a section of E, but it is not the same section as $\nabla_{\partial_i\partial_j}\tau$ since in general

$$\nabla_{\partial_i}\nabla_{\partial_j}\tau \neq \nabla_{\partial_i}\nabla_{\partial_j}\sigma - \nabla_{\nabla_{\partial_i}\partial_j}\sigma.$$

Now we have that

$$\begin{aligned}
\nabla_{X_1,X_2}\sigma - \nabla_{X_2,X_1}\sigma &= \nabla_{X_1}\nabla_{X_2}\sigma - \nabla_{\nabla_{X_1}X_2}\sigma - \left(\nabla_{X_2}\nabla_{X_1}\sigma - \nabla_{\nabla_{X_2}X_1}\sigma\right) \\
&= \nabla_{X_1}\nabla_{X_2}\sigma - \nabla_{X_2}\nabla_{X_1}\sigma - \nabla_{(\nabla_{X_1}X_2-\nabla_{X_2}X_1)}\sigma \\
&= F(X_1,X_2)\sigma - \nabla_{T(X_1,X_2)}\sigma.
\end{aligned}$$

So, if ∇ (the connection on the base M) is torsion free, then we recover the curvature

$$\nabla_{X_1,X_2}\sigma - \nabla_{X_2,X_1}\sigma = F(X_1,X_2)\sigma.$$

One thing that is quite important to realize, is that F depends only on the connection on E, while the operators ∇_{X_1,X_2} involve a torsion free connection on the tangent bundle TM.

12.9. Exterior Covariant Derivative

The exterior covariant derivative essentially antisymmetrizes the higher covariant derivatives just defined in such a way that the dependence on the auxiliary torsion free linear connection on the base cancels out. Of course this means that there must be a definition that does not involve this connection on the base at all. We give the definitions below, but first we point out a little more algebraic structure. We can give the space of E-valued forms $\Omega(M, E)$ the structure of a $(\Omega(M), \Omega(M))$-"bi-module". This means that we define a product $\wedge : \Omega(M) \times \Omega(M, E) \to \Omega(M, E)$ and another product $\wedge : \Omega(M, E) \times \Omega(M) \to \Omega(M, E)$ which are compatible in the sense that

$$(\alpha \wedge \omega) \wedge \beta = \alpha \wedge (\omega \wedge \beta)$$

for $\omega \in \Omega(M, E)$ and $\alpha \in \Omega(M)$. These products are defined by extending linearly the rules

$$\alpha \wedge (\sigma \otimes \omega) := \sigma \otimes \alpha \wedge \omega \text{ for } \sigma \in \Gamma(E) \text{ and } \alpha, \omega \in \Omega(M),$$
$$(\sigma \otimes \omega) \wedge \alpha := \sigma \otimes \omega \wedge \alpha \text{ for } \sigma \in \Gamma(E) \text{ and } \alpha, \omega \in \Omega(M),$$
$$\alpha \wedge \sigma = \sigma \wedge \alpha = \sigma \otimes \alpha \text{ for } \alpha \in \Omega(M) \text{ and } \sigma \in \Gamma(E).$$

More precisely, we extend by using the universal properties of multilinear products. It follows that

$$\alpha \wedge \omega = (-1)^{kl} \omega \wedge \alpha \text{ for } \alpha \in \Omega^k(M) \text{ and } \omega \in \Omega^l(M, E).$$

In the situation where $\sigma \in \Gamma(E) := \Omega^0(M, E)$ and $\omega \in \Omega(M)$, we have all three of the following conventional equalities:

$$\sigma \omega = \sigma \wedge \omega = \sigma \otimes \omega \text{ (special case)}.$$

The proof of the following theorem is analogous to the proof of the existence of the exterior derivative.

Theorem 12.57. *Given a connection ∇ on a vector bundle $\pi : E \to M$, there exists a unique operator $d^\nabla : \Omega(M, E) \to \Omega(M, E)$ such that*

(i) $d^\nabla(\Omega^k(M, E)) \subset \Omega^{k+1}(M, E)$;

(ii) *For $\alpha \in \Omega^k(M)$ and $\omega \in \Omega^\ell(M, E)$ we have*

$$d^\nabla(\alpha \wedge \omega) = d\alpha \wedge \omega + (-1)^k \alpha \wedge d^\nabla \omega,$$
$$d^\nabla(\omega \wedge \alpha) = d^\nabla \omega \wedge \alpha + (-1)^\ell \omega \wedge d\alpha;$$

(iii) $d^\nabla \sigma = \nabla \sigma$ *for $\sigma \in \Gamma(E)$.*

In particular, if $\alpha \in \Omega^k(M)$ and $\sigma \in \Omega^0(M, E) = \Gamma(E)$, we have

$$d^\nabla(\sigma \otimes \alpha) = d^\nabla \sigma \wedge \alpha + \sigma \otimes d\alpha.$$

12.9. Exterior Covariant Derivative

It can be shown that if we use $\Omega^k(M;E) \cong L^k_{\text{alt}}(\mathfrak{X}(M), \Gamma(E))$, then we have the following formula:

$$d^\nabla \omega(X_0, \ldots, X_k) = \sum_{0 \leq i \leq k} (-1)^i \nabla_{X_i}(\omega(X_0, \ldots, \widehat{X_i}, \ldots, X_k))$$
$$+ \sum_{0 \leq i < j \leq k} (-1)^{i+j} \omega([X_i, X_j], X_0, \ldots, \widehat{X_i}, \ldots, \widehat{X_j}, \ldots, X_k).$$

Definition 12.58. The operator d^∇ whose existence is given by the previous theorem is called the **exterior covariant derivative**.

Exercise 12.59. We shall sometimes use the notation d^E rather than d^∇. This is especially useful when two possibly unrelated connections on possibly different bundles are involved in the discussion. For example, we may deal with a connection ∇^{TM} on TM and at the same time deal with a connection ∇^E on $E \to M$. Then it is conceivable that we may have to work with both d^E and d^{TM}. *Failure to notice these notational possibilities can result in serious confusion.*

Note that we have the following special case for $\mu \in \Omega^1(M, E)$:

$$\begin{aligned}(12.10) \quad d^\nabla \mu(X, Y) &= \nabla_X (\mu(Y)) - \nabla_Y (\mu(X)) - \mu([X,Y]) \\ &= d^\nabla (\mu(Y))(X) - d^\nabla(\mu(X))(Y) - \mu([X,Y]).\end{aligned}$$

Example 12.60. Let us consider the case where $E = TM$ with connection ∇^{TM}. Bundle maps $TM \to TM$ may be regarded as elements of $\Omega^1(M; TM)$ (think about this). In particular, the identity map $\text{id}_{TM}(v) = v$ can be viewed as a TM-valued 1-form. If we take the exterior covariant differential of $\theta := \text{id}_{TM}$, we obtain an element $d^{TM}\theta$ of $\Omega^2(M; TM)$. For $X, Y \in \mathfrak{X}(M)$, we compute

$$\begin{aligned} d^{TM}\theta(X, Y) &= d^{TM}(\theta(Y))(X) - d^{TM}(\theta(X))(Y) - \theta([X,Y]) \\ &= d^{TM}(Y)(X) - d^{TM}(X)(Y) - [X,Y] \\ &= \nabla_X Y - \nabla_Y X - [X,Y] = T(X,Y).\end{aligned}$$

So we see that $d^{TM}\theta$ gives the torsion of the connection ∇^{TM}.

If ∇^{TM} is a torsion free covariant derivative on M and ∇^E is a connection on the vector bundle $E \to M$, then as before we get a covariant derivative ∇ on all the bundles $E \otimes \bigwedge^k T^*M$ and as for the ordinary exterior derivative we have the formula

$$d^\nabla = d^E = (k+1) \text{Alt} \circ \nabla,$$

or in other words, if $\omega \in \Omega^k(M, E)$, then

$$d^\nabla \omega(X_0, X_1, \ldots, X_k) = \sum_{i=0}^{k}(-1)^i (\nabla_{X_i}\omega)(X_0, \ldots, \widehat{X_i}, \ldots, X_k).$$

Let us look at the local expressions with respect to a moving frame. Let $\{e_i\}$ be a frame field for E on $U \subset M$. Then locally, for $\omega \in \Omega^k(M, E)$, we may write

$$\omega = \sum e_i \otimes \omega^i,$$

where $\omega^i \in \Omega^k(M)$. Using the summation convention, we have

$$d^\nabla \omega = d^\nabla \left(e_i \otimes \omega^i \right) = e_i \otimes d\omega^i + d^\nabla e_i \wedge \omega^i$$
$$= e_j \otimes d\omega^j + \omega_i^j e_j \wedge \omega^i$$
$$= e_j \otimes d\omega^j + e_j \otimes \omega_i^j \wedge \omega^i$$
$$= e_j \otimes \left(d\omega^j + \omega_i^j \wedge \omega^i \right).$$

Thus the "$(k+1)$-form coefficients" of $d^\nabla \omega$ with respect to the frame $\{e_j\}$ are given by $d\omega^j + \omega_i^j \wedge \omega^i$.

We can extend our wedge operation to a bilinear operation between $\Omega(M, \text{End}(E))$ and $\Omega(M, E)$ in such a way that

$$(A \otimes \alpha) \wedge (\sigma \otimes \beta) = A(\sigma) \otimes \alpha \wedge \beta.$$

To understand what is going on a little better, let us consider how the action of $\Omega(M, \text{End}(E))$ on $\Omega(M, E)$ comes about from a slightly different point of view. We can identify $\Omega(M, \text{End}(E))$ with $\Omega(M, E \otimes E^*)$, and then the action of $\Omega(M, E \otimes E^*)$ on $\Omega(M, E)$ is given by tensoring and then contracting: For $\alpha, \beta \in \Omega(M)$, $s, \sigma \in \Gamma(E)$, and $s^* \in \Gamma(E)$, we have

$$(s \otimes s^* \otimes \alpha) \wedge (\sigma \otimes \beta) := C(s \otimes s^* \otimes \sigma \otimes (\alpha \wedge \beta))$$
$$= s^*(\sigma) s \otimes (\alpha \wedge \beta).$$

From ∇^E we get related connections on E^*, $E \otimes E^*$ and $E \otimes E^* \otimes E$. The connection on $E \otimes E^*$ is also a connection on $\text{End}(E)$. Of course, we have

$$\nabla_X^{\text{End}(E) \otimes E} (L \otimes \sigma) = (\nabla_X^{\text{End}(E)} L) \otimes \sigma + L \otimes \nabla_X^E \sigma,$$

and after contraction,

$$\nabla_X^E (L(\sigma)) = \nabla_X^E C(L \otimes \sigma) = C(\nabla_X^{\text{End}(E) \otimes E} L \otimes \sigma + L \otimes \nabla_X^E \sigma)$$
$$= (\nabla_X^{\text{End}(E)} L)(\sigma) + L(\nabla_X^E \sigma) = (\nabla_X^{\text{End}(E)} L)(\sigma) + L(\nabla_X^E \sigma).$$

But $X \mapsto L(\nabla_X^E \sigma)$ is just $L \wedge \nabla^E \sigma$. So we have

$$\nabla^E (L(\sigma)) = (\nabla^{\text{End}(E)} L)(\sigma) + L \wedge \nabla^E \sigma.$$

12.9. Exterior Covariant Derivative

The connections on $\mathrm{End}(E)$ and $\mathrm{End}(E) \otimes E$ give the corresponding exterior covariant derivative operators

$$d^{\mathrm{End}(E)} : \Omega^k(M, \mathrm{End}(E)) \to \Omega^{k+1}(M, \mathrm{End}(E))$$

and

$$d^{\mathrm{End}(E) \otimes E} : \Omega^k(M, \mathrm{End}(E) \otimes E) \to \Omega^{k+1}(M, \mathrm{End}(E) \otimes E).$$

It must be expected that many readers will find this formalism a bit daunting but it is not really as bad as it seems. Local calculations turn out to be quite natural as we shall see below. We encourage the reader to seek as much exposure as possible. To this end, we recommend [**BaMu**], [**Poor**], and [**Dar**]. Also, part of what might be intimidating is really just the bulkiness of the notations. The following notational convention makes things more palatable:

Notation 12.61. Let us now agree that, whenever convenient, we may write simply ∇ for $\nabla^{E^*}, \nabla^{\mathrm{End}(E)}, \nabla^{\mathrm{End}(E) \otimes E}$, etc. and d^∇ for what we have called d^{E^*}, $d^{\mathrm{End}(E)}$ etc.

Proposition 12.62. For $\Phi \in \Omega^k(M, \mathrm{End}(E))$ and $\omega \in \Omega^k(M, E)$, we have

$$d^\nabla(\Phi \wedge \omega) = d^\nabla \Phi \wedge \omega + (-1)^k \Phi \wedge d^\nabla \omega.$$

Proof. We have

$$d^E((A \otimes \alpha) \wedge (\sigma \otimes \beta)) := d^E(A(\sigma) \otimes (\alpha \wedge \beta))$$
$$= \nabla^E(A(\sigma)) \wedge (\alpha \wedge \beta) + A(\sigma) \otimes d(\alpha \wedge \beta)$$
$$= (-1)^k \left(\alpha \wedge \nabla^E(A(\sigma)) \wedge \beta \right) + A(\sigma) \otimes d(\alpha \wedge \beta)$$
$$= (-1)^k \left(\alpha \wedge \{\nabla^{\mathrm{End}(E)} A \wedge \sigma + A \wedge \nabla^E \sigma\} \wedge \beta \right)$$
$$\quad + A(\sigma) \otimes d\alpha \wedge \beta + (-1)^k A(\sigma) \otimes \alpha \wedge d\beta$$
$$= (\nabla^{\mathrm{End}(E)} A) \wedge \alpha \wedge \sigma \wedge \beta + (-1)^k A \wedge \alpha \wedge (\nabla^E \sigma) \wedge \beta$$
$$\quad + A(\sigma) \otimes d\alpha \wedge \beta + (-1)^k A(\sigma) \otimes \alpha \wedge d\beta$$
$$= (\nabla^{\mathrm{End}(E)} A \wedge \alpha + A \otimes d\alpha) \wedge (\sigma \otimes \beta)$$
$$\quad + (-1)^k (A \otimes \alpha) \wedge (\nabla^E \sigma \wedge \beta + \sigma \otimes d\beta)$$
$$= d^{\mathrm{End}(E)}(A \otimes \alpha) \wedge (\sigma \otimes \beta) + (-1)^k (A \otimes \alpha) \wedge d^E(\sigma \otimes \beta).$$

By linearity we conclude that for $\Phi \in \Omega^k(M, \mathrm{End}(E))$ and $\omega \in \Omega(M, E)$ we have

$$d^E(\Phi \wedge \omega) = d^{\mathrm{End}(E)} \Phi \wedge \omega + (-1)^k \Phi \wedge d^E \omega.$$

So in light of our notational conventions we are done. □

Remark 12.63. In the literature, it seems that the different natures of $(d^\nabla)^k$ and $(\nabla)^k$ are not always appreciated. For example, the higher derivatives given by $(d^\nabla)^k$ are not appropriate for defining k-th order Sobolev spaces since $(d^\nabla)^2$ is zero for any flat connection.

12.10. Curvature Again

The space $\Omega(M, \text{End}(E))$ is an algebra over $C^\infty(M)$ where the multiplication is according to $(L_1 \otimes \omega_1) \wedge (L_2 \otimes \omega_2) = (L_1 \circ L_2) \otimes \omega_1 \wedge \omega_2$. Now $\Omega(M, E)$ is a module over this algebra because we can multiply (using the symbol $\wedge$) as follows:

$$(L \otimes \alpha) \wedge (\sigma \otimes \beta) = L\sigma \otimes (\alpha \wedge \beta).$$

As usual, this definition on simple elements is sufficient. If $X, Y \in \mathfrak{X}(M)$ and $\Psi \in \Omega^2(M, \text{End}(E))$, then using $\Omega^2(M; E) \cong L^2_{\text{alt}}(\mathfrak{X}(M), \Gamma(E))$ we have

$$(\Psi \wedge \sigma)(X, Y) = \Psi(X, Y)\sigma$$

for any $\sigma \in \Omega(M, E)$.

Proposition 12.64. *The map $d^\nabla \circ d^\nabla : \Omega^k(M, E) \to \Omega^{k+2}(M, E)$ is given by the action of F, the curvature 2-form of ∇^E,*

$$d^\nabla \circ d^\nabla \mu = F \wedge \mu \text{ for } \mu \in \Omega^k(M, E).$$

Proof. Let us check the case where $k = 0$ first. From formula (12.10) above we have for $\sigma \in \Omega^0(M, E)$,

$$\begin{aligned}
(d^\nabla \circ d^\nabla \sigma)(X, Y) &= \nabla_X (d^\nabla \sigma(Y)) - \nabla_Y (d^\nabla \sigma(X)) - \sigma([X, Y]) \\
&= \nabla_X \nabla_Y \sigma - \nabla_X \nabla_Y \sigma - \sigma([X, Y]) \\
&= F(X, Y)\sigma = (F \wedge \sigma)(X, Y).
\end{aligned}$$

More generally, we just check $d^\nabla \circ d^\nabla$ on elements of the form $\mu = \sigma \otimes \theta$:

$$\begin{aligned}
d^\nabla \circ d^\nabla \mu &= d^\nabla \circ d^\nabla (\sigma \otimes \theta) \\
&= d^\nabla (d^\nabla \sigma \wedge \theta + \sigma \otimes d\theta) \\
&= (d^\nabla d^\nabla \sigma) \wedge \theta - d^\nabla \sigma \wedge d\theta + d^\nabla \sigma \wedge d\theta + 0 \\
&= (F \wedge \sigma) \wedge \theta = F \wedge (\sigma \wedge \theta) = F \wedge (\sigma \otimes \theta) \\
&= F \wedge \mu.
\end{aligned}$$
$\square$

Let us take a look at how curvature appears in a local frame field. As before restrict to an open set U on which $e_U = (e_1, \ldots, e_r)$ is a given local frame field and then write a typical element $s \in \Omega^k(M, E)$ as $s = e\eta$, where

12.11. The Bianchi Identity

$\eta = (\eta^1, \ldots, \eta^r)^t$ is a column vector of smooth k-forms. With $\omega_U = (\omega_j^i)$ the connection forms we have

$$\begin{aligned}
d^\nabla d^\nabla s &= d^\nabla d^\nabla \left(e_U \eta_U\right) = d^\nabla \left(e_U d\eta_U + d^\nabla e_U \wedge \eta_U\right) \\
&= d^\nabla \left(e_U \left(d\eta_U + \omega_U \wedge \eta_U\right)\right) \\
&= d^\nabla e_U \wedge \left(d\eta_U + \omega_U \wedge \eta_U\right) + e_U \wedge d^\nabla \left(d\eta_U + \omega_U \wedge \eta_U\right) \\
&= e_U \omega_U \wedge \left(d\eta_U + \omega_U \wedge \eta_U\right) + e_U \wedge d^\nabla \left(d\eta_U + \omega_U \wedge \eta_U\right) \\
&= e_U \omega_U \wedge \left(d\eta_U + \omega_U \wedge \eta_U\right) + e_U \wedge d\omega_U \wedge \eta_U - e_U \wedge \omega_U \wedge d\eta_U \\
&= e_U d\omega_U \wedge \eta_U + e_U \omega_U \wedge \omega_U \wedge \eta_U \\
&= e_U \left(d\omega_U + \omega_U \wedge \omega_U\right) \wedge \eta_U.
\end{aligned}$$

The matrix $d\omega_U + \omega_U \wedge \omega_U$ represents a section of $\text{End}(E)_U \otimes \bigwedge^2 T^*U$. In fact, we will now check that these local sections paste together to give a global section of $\text{End}(E) \otimes \bigwedge^2 T^*M$, or in other words, an element of $\Omega^2(M, \text{End}(E))$, which is clearly the curvature form: $F : (X, Y) \mapsto F(X, Y) \in \Gamma(\text{End}(E))$. Let $F_U = d\omega_U + \omega_U \wedge \omega_U$ and let $F_V = d\omega_V + \omega_V \wedge \omega_V$ be the corresponding form for a different moving frame $e_U := e_V g$, where $g : U \cap V \to \text{GL}(\mathbb{F}^r)$, and r is the rank of E. What we need to verify is the transformation law

$$F_V = g^{-1} F_U g,$$

which we met earlier in equation (6.2). Recall that $\omega_V = g^{-1}\omega_U g + g^{-1}dg$. Using $d\left(g^{-1}\right) = -g^{-1}dg g^{-1}$, we have

$$\begin{aligned}
F_V &= d\omega_V + \omega_V \wedge \omega_V \\
&= d\left(g^{-1}\omega_U g + g^{-1}dg\right) \\
&\quad + \left(g^{-1}\omega_U g + g^{-1}dg\right) \wedge \left(g^{-1}\omega_U g + g^{-1}dg\right) \\
&= d\left(g^{-1}\right) \wedge \omega_U g + g^{-1}d\omega_U g - g^{-1}\omega_U \wedge dg + d(g^{-1}) \wedge dg \\
&\quad + g^{-1}\omega_U \wedge \omega_U g + g^{-1}dg g^{-1} \wedge \omega_U g \\
&\quad + g^{-1}\omega_U \wedge dg + g^{-1}dg \wedge g^{-1}dg \\
&= g^{-1}d\omega_U g + g^{-1}\omega_U \wedge \omega_U g = g^{-1}F_U g,
\end{aligned}$$

where we have used that $g^{-1}dg \wedge g^{-1}dg = d(g^{-1}g) = 0$.

12.11. The Bianchi Identity

In this section we give several versions of the so-called Bianchi identity for a connection ∇ on a vector bundle $E \to M$. Perhaps the simplest version to understand is the following: If $U, V, W \in \mathfrak{X}(M)$, then

$$[\nabla_U, [\nabla_V, \nabla_W]] + [\nabla_V, [\nabla_W, \nabla_U]] + [\nabla_W, [\nabla_U, \nabla_V]] = 0 \text{ (Bianchi identity)},$$

where $[\nabla_U, \nabla_V] := \nabla_U \circ \nabla_V - \nabla_V \circ \nabla_U$. This identity follows trivially once we observe that the set of linear operators on any vector space is a Lie algebra under the commutator bracket operation $[A, B] := A \circ B - B \circ A$. So, in this form, the Bianchi identity is just an instance of the Jacobi identity. Let $(U, \mathbf{x})$ be a chart on M, and let $F_{\mu\nu} : \Gamma(E)|_U \to \Gamma(E)|_U$ be the local curvature operator defined by

$$F_{\mu\nu} = [\nabla_{\partial_\mu}, \nabla_{\partial_\nu}] = F(\partial_\mu, \partial_\nu),$$

where $\partial_\mu = \partial/\partial x^\mu$, etc. Then we have the following version of the Bianchi identity:

$$[\nabla_{\partial_\mu}, F_{\nu\lambda}] + [\nabla_{\partial_\nu}, F_{\lambda\mu}] + [\nabla_{\partial_\lambda}, F_{\mu\nu}] = 0 \qquad \text{(Bianchi identity)}.$$

Another revealing form of the Bianchi identity depends on our discussions in the last section and in particular on Proposition 12.62. We have, for any E-valued form η,

$$\left(d^\nabla\right)^3 \eta = d^\nabla \left(\left(d^\nabla\right)^2 \eta\right) = d^\nabla(F \wedge \eta) = d^\nabla F \wedge \eta + F \wedge d^\nabla \eta.$$

But equally,

$$\left(d^\nabla\right)^3 \eta = \left(d^\nabla\right)^2 \left(d^\nabla \eta\right) = F \wedge d^\nabla \eta$$

so it must be that $d^\nabla F \wedge \eta = 0$ for any η and so we obtain

$$d^\nabla F = 0 \qquad \text{(Bianchi identity)}.$$

Exercise 12.65. Use a calculation in local coordinates and with respect to a local frame $e_1, \ldots, e_k$ to show that the above versions of the Bianchi identity are equivalent.

12.12. G-Connections

If the vector bundle $\pi : E \to M$ has a G-bundle structure, then there should be certain connections that respect this structure. These are the G-connections. It is time to confess that we have come face to face with a weakness of our approach to connections. It is much easier and more natural to define the notion of G-connection if one first defines connections in terms of horizontal distributions on principal bundles. We treat this in the online supplement [**Lee, Jeff**]. However, we can still say what a G-connection should be from the current point of view without too much trouble.

Recall that if $\pi : E \to M$ is a rank k vector bundle with typical fiber V that has a G-bundle structure where G acts on V by the standard action as a subgroup of $\mathrm{GL}(\mathrm{V})$, then we have the bundle of G-frames

$$F_G(E) := \bigcup_{p \in M} F_G(E_p),$$

where $F_G(E_p) := \{u \in \mathrm{GL}(\mathrm{V}, E_p) : u = \phi^{-1}(p, \cdot) \text{ for some } (U, \phi) \in \mathcal{A}_G\}$ and $\mathcal{A}_G$ is the maximal G-atlas defining the structure. The elements of

12.12. G-Connections

$F_G(E_p)$ are called *G*-frames. Having fixed a basis $(\mathbf{e}_1, \ldots, \mathbf{e}_k)$ for V, each element of $F_G(E_p)$ can be identified as a basis $(u_1, \ldots, u_k)$ for E_p according to $u : x \mapsto \sum x^i u_i$. If $F(E_p)$ is the set of all frames at p, then $F_G(E_p) \subset F(E_p)$ and $F_G(E)$ is a subbundle of $F(E)$. Now let $c : [a,b] \to M$ be a smooth curve. If $(u_1, \ldots, u_k)$ is a frame, then $(P_c u_1, \ldots, P_c u_k)$ is also a frame. Thus we get a map $P_c : F(E_{c(a)}) \to F\left(E_{c(a)}\right)$. The following is then a workable definition of *G*-connection:

Definition 12.66. Let $\pi : E \to M$ be a vector bundle with a *G*-bundle structure as above. A connection on the bundle is called a *G*-**connection** if parallel transport P_c takes *G*-frames to *G*-frames for all piecewise smooth curves c.

A *G*-**frame field** is a frame field which is a *G*-frame at each point of its domain.

Exercise 12.67. Show that if ∇ is the covariant derivative associated to a *G*-connection on E, then for each *G*-frame field the associated connection forms take values in the Lie algebra of *G* thought of as a matrix subgroup of $\mathrm{GL}(n, \mathbb{F})$.

Let $\pi : E \to M$ be a vector bundle with metric $h = \langle \cdot, \cdot \rangle$ and standard fiber $\mathbb{R}^k$. The metric h gives a reduction of the structure group to $O(n)$. In this case, an $O(n)$-frame is an orthonormal frame. A connection on E is an $O(n)$-connection if and only if the associated covariant derivative ∇ satisfies

$$(12.11) \qquad v \langle s_1, s_2 \rangle = \langle \nabla_v s_1, s_2 \rangle + \langle s_1, \nabla_v s_2 \rangle$$

for all $s_1, s_2 \in \Gamma(E)$ and all $v \in TM$.

Exercise 12.68. Prove this last assertion.

One may also consider metrics which are nondegenerate but not necessarily positive definite. In this case, the structure group reduces to one of the semiorthogonal groups $O(k, n-k)$.

Definition 12.69. A covariant derivative satisfying (12.11) above for some metric h on a vector bundle $E \to M$ is called a **metric covariant derivative** (or **metric connection**).

A simple partition of unity argument shows that if h is a given metric, then there exists a (nonunique) metric connection for h (Problem 11). For a metric connection, parallel transport is an isometry (Problem 12). Furthermore if h is the metric and ∇ is metric with respect to h, then it is easy to check that $\nabla h = 0$.

Proposition 12.70. *Suppose that $h = \langle \cdot, \cdot \rangle$ is a metric on a vector bundle $E \to M$. If ∇ is a metric covariant derivative, then the corresponding curvature satisfies*

$$\langle F(X,Y)\sigma_1, \sigma_2 \rangle + \langle \sigma_1, F(X,Y)\sigma_2 \rangle = 0$$

for all $X, Y \in \mathfrak{X}(M)$ and all $\sigma_1, \sigma_2 \in \Gamma(E)$.

Proof. Let X and Y be fixed vector fields. Without loss of generality we may assume that $[X,Y] = 0$ (recall Exercise 7.34). It is also enough to show that $\langle F(X,Y)\sigma, \sigma \rangle = 0$ for all σ. We have

$$\begin{aligned}
\langle F(X,Y)\sigma, \sigma \rangle &= \langle \nabla_X \nabla_Y \sigma, \sigma \rangle - \langle \sigma, \nabla_X \nabla_Y \sigma \rangle \\
&= X \langle \nabla_Y \sigma, \sigma \rangle - \langle \nabla_Y \sigma, \nabla_X \sigma \rangle \\
&\quad - Y \langle \nabla_X \sigma, \sigma \rangle - \langle \nabla_X \sigma, \nabla_Y \sigma \rangle \\
&= \frac{1}{2} \left(XY \langle \sigma, \sigma \rangle - YX \langle \sigma, \sigma \rangle \right) = 0
\end{aligned}$$

since $[X,Y] = 0$. $\square$

We shall study metric connections on tangent bundles in the next chapter.

Problems

(1) Prove Theorem 12.23.

(2) Let M have a linear connection ∇ and let $T(X,Y) := \nabla_X Y - \nabla_Y X$. Show that T is $C^\infty(M)$-bilinear (tensorial). The resulting tensor is called the torsion tensor for the connection.

(3) Show that a holonomy group is indeed a group. Show that the holonomy at any point of a sphere is isomorphic to the special orthogonal group SO(2).

Second order differential equations and sprays

(4) A second order differential equation on a smooth manifold M is a vector field on TM, that is, a section X of the bundle TTM (second tangent bundle) such that every integral curve α of X is the velocity curve of its projection on M. In other words, $\alpha = \dot{\gamma}$ where $\gamma := \pi_{TM} \circ \alpha$. A solution curve $\gamma : I \to M$ for a second order differential equation X is, by definition, a curve with $\ddot{\alpha}(t) = X(\dot{\alpha}(t))$ for all $\tau \in I$.

In the case $M = \mathbb{R}^n$, show that this concept corresponds to the usual system of equations of the form

$$y' = v,$$
$$v' = f(y, v),$$

which is the reduction to a first order system of the second order equation $y'' = f(y, y')$. What is the vector field on $T\mathbb{R}^n = \mathbb{R}^n \times \mathbb{R}^n$ which corresponds to this system?

Notation: For a second order differential equation X, the maximal integral curve through $v \in TM$ will be denoted by α_v and its projection will be denoted by $\gamma_v := \pi_{TM} \circ \alpha$.

(5) A **spray** on M is a second order differential equation, that is, a section X of TTM as in the previous problem, such that for $v \in TM$ and $s \in \mathbb{R}$, a number $t \in \mathbb{R}$ belongs to the domain of γ_{sv} if and only if st belongs to the domain of γ_v, and in this case

$$\gamma_{sv}(t) = \gamma_v(st).$$

Show that there are infinitely many sprays on any smooth manifold M.

[Hint: (i) Show that a vector field $X \in \mathfrak{X}(TM)$ is a spray if and only if $T\pi \circ X = \text{id}_{TM}$, where $\pi = \pi_{TM}$:

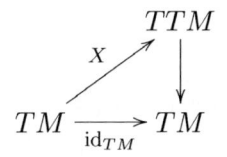

$$\begin{array}{ccc} & & TTM \\ & \nearrow X & \downarrow \\ TM & \xrightarrow{\text{id}_{TM}} & TM \end{array}$$

(ii) Show that $X \in \mathfrak{X}(TM)$ is a spray if and only if for any $s \in \mathbb{R}$ and any $v \in TM$,

$$X_{sv} = T\mu_s(sX_v),$$

where $\mu_s : v \mapsto sv$ is the multiplication map.

(iii) Show the existence of a spray on an open ball in a Euclidean space.

(iv) Show that if X_1 and X_2 both satisfy one of the two characterizations of a spray above, then so does any convex combination of X_1 and X_2.]

(6) Show that if one has a linear connection on M, then there is a spray whose solutions are the geodesics of the connection.

(7) Show that given a spray on a manifold there is a (not unique) linear connection ∇ on the manifold such that $\gamma : I \to M$ is a solution curve of the spray if and only if $\nabla_{\partial_t}\dot{\gamma} = 0$ for all $t \in I$. Note: γ is called a geodesic for the linear connection or the spray. Does the stipulation that the connection be torsion free force uniqueness?

(8) Let X be a spray on M. Equivalently, we may start with a connection on M (i.e. a connection on TM) which induces a spray. Show that the set $O_X := \{v \in TM : \gamma_v(1) \text{ is defined}\}$ is an open neighborhood of the zero section of TM.

(9) Finish the proof of Theorem 12.32.

(10) Prove formula (12.9).

(11) Let h be a metric on a vector bundle $E \to M$. Show that there exists a metric connection for h. [Hint: Use orthonormal frames defined on each open set of a locally finite cover.]

(12) Show that parallel transport along curves with respect to a metric connection in a vector bundle with metric is an isometry of scalar product spaces.

Chapter 13

Riemannian and Semi-Riemannian Geometry

> "The most beautiful thing we can experience is the mysterious.
> It is the source of all true art and science."
>
> – Albert Einstein

In this chapter we take up the subject of semi-Riemannian geometry, which includes Riemannian geometry and Lorentz geometry as important special cases. The exposition is inspired by [**ON1**], which we follow quite closely in some places (also see [**L1**]). Recall that by definition, a semi-Riemannian manifold (M, g) has a well-defined index denoted $\text{ind}(M)$ or $\text{ind}(g)$. In the case of an indefinite metric ($\text{ind}(M) > 0$), we will need a classification:

Definition 13.1. Let $(V, \langle \cdot, \cdot \rangle)$ be a scalar product space. A nonzero vector $v \in V$ is called

(1) **spacelike** if $\langle v, v \rangle > 0$;

(2) **lightlike** or **null** if $\langle v, v \rangle = 0$;

(3) **timelike** if $\langle v, v \rangle < 0$;

(4) **nonnull** if v is either timelike or spacelike.

The terms *spacelike*, *null (lightlike)*, and *timelike* indicate the **causal character** of a vector. The word causal comes from relativity theory and is most apropos in the context of Lorentz manifolds defined below. If (M, g) is a

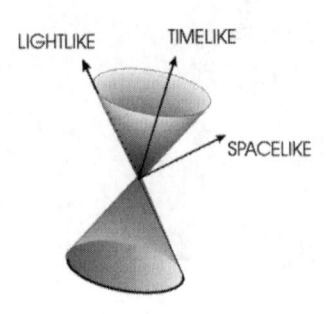

Figure 13.1. Lightcone

semi-Riemannian manifold, then each tangent space is a scalar product space and the above definition applies. Recall that we define $\|v\| = |\langle v,v \rangle|^{1/2}$, which we call the length of v.

Note: We have so far left the causal character of the zero vector undefined. It may seem reasonable that it should be considered null. A second possibility is that the zero vector should have all three causal characters. Actually, we shall see that if the index of the scalar product is one, then it is convenient to consider the zero vector as being spacelike.

Definition 13.2. The set of all null vectors in a scalar product space is called the **nullcone** or **lightcone**. If (M,g) is a semi-Riemannian manifold, then the nullcone in T_pM is called the **nullcone** at p.

Definition 13.3. Let $I \subset \mathbb{R}$ be some interval. A curve $c : I \to (M,g)$ is called **spacelike**, **null**, **timelike**, or **nonnull**, according as $\dot{c}(t) \in T_{c(t)}M$ is spacelike, null, timelike, or nonnull, respectively, for all $t \in I$.

While every smooth manifold supports Riemannian metrics by Proposition 6.45, the existence of an indefinite metric on a given smooth manifold has an obstruction:

Theorem 13.4. *A compact smooth manifold admits a continuous C^0 indefinite metric of index k if and only if its tangent bundle has a C^0 rank k subbundle.*

This result is Theorem 40.11 of [**St**].

Definition 13.5. Let (M,g) be semi-Riemannian. If $c : [a,b] \to M$ is a piecewise smooth curve, then

$$L_{c(a),c(b)}(c) = \int_a^b |\langle \dot{c}(t), \dot{c}(t) \rangle|^{1/2}\, dt$$

is called the **arc length** or simply **length** of the curve.

13. Riemannian and Semi-Riemannian Geometry

The word *length* could cause confusion since the length of a null curve is zero. Thus for indefinite metrics, arc length can have some properties that are decidedly not like our ordinary notion of length. In particular, a curve may connect two different points and the arc length might still be zero! The word length is therefore sometimes reserved for timelike or spacelike curves.

Definition 13.6. A **positive reparametrization** of a smooth curve $c : [a,b] \to M$ is a curve defined by composition $c \circ h : [a',b'] \to M$, where $h : [a',b'] \to [a,b]$ is a smooth monotonically increasing bijection. Similarly, a **negative reparametrization** is given by composition with a smooth monotonically decreasing bijection $h : [a',b'] \to [a,b]$. By a **reparametrization** we shall mean either a positive or negative reparametrization.

The above definition can be extended to piecewise smooth curves. Suppose $c : [a,b] \to M$ is a continuous curve such that, for some partition $a = t_0 < t_1 < \cdots < t_k = b$, we have that c is smooth on each $[t_{i-1}, t_i]$. A positive reparametrization of c is a curve $c \circ h : [a',b'] \to M$, where $h : [a',b'] \to [a,b]$ is a monotonically increasing continuous bijection that is smooth on each interval $h^{-1}([t_{i-1}, t_i])$. Negative reparametrization is defined similarly.

Remark 13.7 (Important fact). The integrals above are well-defined since $\dot{c}(t)$ is defined and continuous except for a finite number of points in $[a,b]$. Also, it is important to notice that by standard change of variable arguments, a reparametrization $\gamma = c \circ h$ does not change the arc length of the curve:

$$\int_a^b |\langle \dot{c}(t), \dot{c}(t) \rangle|^{1/2} \, dt = \int_{h^{-1}(a)}^{h^{-1}(b)} |\langle \dot{\gamma}(u), \dot{\gamma}(u) \rangle|^{1/2} du.$$

Thus the arc length of a piecewise smooth curve is a geometric property of the curve; i.e. a semi-Riemannian invariant.

Definition 13.8. Let (M,g) be semi-Riemannian. Let I be an interval (possibly infinite). If $c : I \to M$ is a smooth curve with $\|\dot{c}\| = 1$, then we say that c is a **unit speed curve**.

If $c : I \to M$ is a curve such that $\|\dot{c}\|$ is never zero, then choosing a reference $t_0 \in I$, we may define an **arc length function** $\ell : I \to I' \subset \mathbb{R}$ by

$$(13.1) \qquad \ell(t) := \int_{t_0}^{t} |\langle \dot{c}(t), \dot{c}(t) \rangle|^{1/2} \, dt.$$

For a finite interval of definition $[a,b]$, the reference t_0 is most often taken to be the left endpoint a. Since $d\ell/dt = |\langle \dot{c}(t), \dot{c}(t) \rangle|^{1/2} > 0$, we may invert to find ℓ^{-1} and then reparametrize:

$$\gamma(s) := c(\ell^{-1}(s)).$$

It is easy to see from the chain rule that the resulting curve is a unit speed curve. Conversely, if γ is a unit speed curve, then the arc length from $\gamma(s_1)$ to $\gamma(s_2)$ is $s_2 - s_1$. Often one abuses notation by writing $s = \ell(t)$ and then ds/dt instead of $d\ell/dt$. The use of the letter s for the parameter of a unit speed curve is traditional, and we say that the curve is **parametrized by arc length**. In the case of timelike curves, people sometimes use the letter τ instead of s and refer to it as a **proper time** parameter.

13.1. Levi-Civita Connection

In this chapter we will use the term "connection" to be synonymous with covariant derivative. Let (M, g) be a semi-Riemannian manifold and ∇ a metric connection for M (see Definition 12.11). By definition, we have

$$X\langle Y, Z\rangle = \langle \nabla_X Y, Z\rangle + \langle Y, \nabla_X Z\rangle$$

for all $X, Y, Z \in \mathfrak{X}(M)$. It is easy to show that the same formula holds for locally defined fields. Recall that the operator $T : \mathfrak{X}(M) \times \mathfrak{X}(M) \to \mathfrak{X}(M)$ defined by $T(X, Y) = \nabla_X Y - \nabla_Y X - [X, Y]$ is a tensor called the **torsion tensor** of ∇. From the previous chapter we know that $(X, Y) \mapsto T(X, Y)$ defines a $C^\infty(M)$-bilinear map $\mathfrak{X}(M) \times \mathfrak{X}(M) \to \mathfrak{X}(M)$. The isomorphism (7.6) implies that T gives a section of $T_2^0(TM; TM)$. That is, T can be thought of as defining a T_pM-valued 2-tensor field at each p, so if $X_p, Y_p \in T_pM$, then $T(X_p, Y_p)$ is a well-defined element of T_pM. Recall from our study of tensor fields that $T(X_p, Y_p)$ is defined to be $T(X, Y)(p)$ for any fields X, Y such that $X(p) = X_p$ and $Y(p) = Y_p$.

Requiring that a connection be both metric and torsion free, pins down the metric completely.

Theorem 13.9. *For a given semi-Riemannian manifold (M, g), there is a unique metric connection ∇ such that its torsion is zero, $T \equiv 0$. This unique connection is called the **Levi-Civita connection** for (M, g).*

Proof. We will derive a formula that must be satisfied by ∇ and that can be used to actually define ∇. Let X, Y, Z, W be arbitrary vector fields on M. If ∇ exists as stated, then we must have

$$X\langle Y, Z\rangle = \langle \nabla_X Y, Z\rangle + \langle Y, \nabla_X Z\rangle,$$
$$Y\langle Z, X\rangle = \langle \nabla_Y Z, X\rangle + \langle Z, \nabla_Y X\rangle,$$
$$Z\langle X, Y\rangle = \langle \nabla_Z X, Y\rangle + \langle X, \nabla_Z Y\rangle.$$

13.1. Levi-Civita Connection

Now add the first two equations and subtract the third to get

$$X\langle Y, Z\rangle + Y\langle Z, X\rangle - Z\langle X, Y\rangle$$
$$= \langle \nabla_X Y, Z\rangle + \langle Y, \nabla_X Z\rangle + \langle \nabla_Y Z, X\rangle + \langle Z, \nabla_Y X\rangle$$
$$- \langle \nabla_Z X, Y\rangle - \langle X, \nabla_Z Y\rangle.$$

If we assume the torsion zero hypothesis, then this reduces to

$$X\langle Y, Z\rangle + Y\langle Z, X\rangle - Z\langle X, Y\rangle$$
$$= \langle Y, [X, Z]\rangle + \langle X, [Y, Z]\rangle$$
$$- \langle Z, [X, Y]\rangle + 2\langle \nabla_X Y, Z\rangle.$$

Solving, we see that $\nabla_X Y$ must satisfy

$$\text{(13.2)} \quad \begin{aligned} 2\langle \nabla_X Y, Z\rangle &= X\langle Y, Z\rangle + Y\langle Z, X\rangle - Z\langle X, Y\rangle \\ &+ \langle Z, [X, Y]\rangle - \langle Y, [X, Z]\rangle - \langle X, [Y, Z]\rangle. \end{aligned}$$

Since knowing $\langle \nabla_X Y, Z\rangle$ for all Z is tantamount to knowing $\nabla_X Y$, we conclude that if ∇ exists, then it is unique. On the other hand, the patient reader can check that if we actually define $\langle \nabla_X Y, Z\rangle$ and hence $\nabla_X Y$ by this equation, then all of the defining properties of a covariant derivative are satisfied and furthermore T will be zero. □

Formula (13.2) above, which serves to determine the Levi-Civita connection, is called the **Koszul formula**. It is easy to see that the restriction of a Levi-Civita connection to any open submanifold is just the Levi-Civita connection on that open submanifold with the induced metric. It is a straightforward matter to show that the Christoffel symbols for the Levi-Civita connection in some chart are given by

$$\Gamma_{ij}^k = \frac{1}{2} g^{kl} \left(\frac{\partial g_{jl}}{\partial x^i} + \frac{\partial g_{li}}{\partial x^j} - \frac{\partial g_{ij}}{\partial x^l} \right),$$

where $g_{jk} g^{ki} = \delta_j^i$. (Recall that $g_{ij} = \langle \partial_i, \partial_j \rangle$ and Γ_{ij}^k are given by formula (12.6).)

We know from the study in the last chapter that we may take the covariant derivative of vector fields along maps. The most important cases for this chapter are fields along curves and fields along maps of the form $h : (a, b) \times (c, d) \to M$.

Exercise 13.10. Show that if $\alpha : I \to M$ is a smooth curve and X, Y are vector fields along α, then $\frac{d}{dt}\langle X, Y\rangle = \langle \nabla_{\partial/\partial t} X, Y\rangle + \langle X, \nabla_{\partial/\partial t} Y\rangle$.

Exercise 13.11. If $h : (a, b) \times (c, d) \to M$ is smooth, then $\partial h/\partial t$ and $\partial h/\partial s$ are vector fields along h. Show that $\nabla_{\partial/\partial t} \partial h/\partial s = \nabla_{\partial/\partial s} \partial h/\partial t$. [Hint: Use local coordinates and the fact that ∇ is torsion free.]

13.1.1. Covariant differentiation of tensor fields. Let ∇ be any natural covariant derivative on M. It is a consequence of Proposition 7.37 that for each $X \in \mathfrak{X}(U)$ there is a unique tensor derivation ∇_X on $\mathcal{T}^r_s(U)$ such that ∇_X commutes with contraction and coincides with the given covariant derivative on $\mathfrak{X}(U)$ (also denoted ∇_X) and with $\mathcal{L}_X f$ on $C^\infty(U)$.

Recall that if $\Upsilon \in \mathcal{T}^1_s$, then

$$(\nabla_X \Upsilon)(Y_1, \ldots, Y_s) = \nabla_X(\Upsilon(Y_1, \ldots, Y_s)) - \sum_{i=1}^s \Upsilon(\ldots, \nabla_X Y_i, \ldots).$$

If $Z \in \mathcal{T}^1_0$, we apply this to $\nabla Z \in \mathcal{T}^1_1$ and get

$$(\nabla_X \nabla Z)(Y) = X(\nabla Z(Y)) - \nabla Z(\nabla_X Y) = \nabla_X(\nabla_Y Z) - \nabla_{\nabla_X Y} Z,$$

from which we get the following definition:

Definition 13.12. The **second covariant derivative** of a vector field $Z \in \mathcal{T}^1_0$ is

$$\nabla^2 Z : (X, Y) \mapsto \nabla^2_{X,Y}(Z) = \nabla_X(\nabla_Y Z) - \nabla_{\nabla_X Y} Z.$$

Definition 13.13. A tensor field Υ is said to be **parallel** if $\nabla_\xi \Upsilon = 0$ for all $\xi \in TM$. Similarly, if $\sigma : I \to T^r_s(M)$ is a tensor field along a curve $c : I \to M$ that satisfies $\nabla_{\partial_t} \sigma = 0$ on I, then we say that σ is parallel along c. Just as in the case of a general connection on a vector bundle we then have a **parallel transport map** $P(c)^t_{t_0} : T^r_s(M)_{c(t_0)} \to T^r_s(M)_{c(t)}$.

From the previous chapter we know that

$$\nabla_{\partial_t} \sigma(t) = \lim_{\epsilon \to 0} \frac{P(c)^t_{t+\epsilon} \sigma(t+\epsilon) - \sigma(t)}{\epsilon}.$$

It is also true that if $\Upsilon \in \mathcal{T}^r_s$, and if c^X is the curve $t \mapsto \varphi^X_t(p)$, then

$$\nabla_X \Upsilon(p) = \lim_{\epsilon \to 0} \frac{P(c^X)^t_{t+\epsilon}(\Upsilon \circ \varphi^X_t(p)) - Y \circ \varphi^X_t(p)}{\epsilon}.$$

It is easy to see that the space of parallel tensor fields of type (r, s) is a vector space over $\mathbb{R}$.

Exercise 13.14. Show that if Υ is parallel, then for any smooth curve $c : [a, b] \to M$ such that $c(a) = p$ and $c(b) = q$ we have $P(c)^b_a \Upsilon_p = \Upsilon_q$. Deduce that if M is connected, then the dimension of the space of parallel tensor fields of type (r, s) has dimension less than or equal to $\dim T^r_s(M)_p$ for any fixed p.

The map $\nabla_X : \mathcal{T}^r_s M \to \mathcal{T}^r_s M$ just defined commutes with contraction by construction. Furthermore, if the connection we are extending is the Levi-Civita connection for a semi-Riemannian manifold (M, g), then

$$\nabla_\xi g = 0 \text{ for all } \xi \in TM.$$

To see this, recall that
$$\nabla_\xi(g \otimes Y \otimes W) = \nabla_\xi g \otimes X \otimes Y + g \otimes \nabla_\xi X \otimes Y + g \otimes X \otimes \nabla_\xi Y,$$
which upon contraction yields
$$\nabla_\xi(g(X,Y)) = (\nabla_\xi g)(X,Y) + g(\nabla_\xi X, Y) + g(X, \nabla_\xi Y),$$
$$\xi\langle X, Y\rangle = (\nabla_\xi g)(X,Y) + \langle \nabla_\xi X, Y\rangle + \langle X, \nabla_\xi Y\rangle.$$
We see that $\nabla_\xi g \equiv 0$ for all ξ if and only if $\langle X, Y\rangle = \langle \nabla_\xi X, Y\rangle + \langle X, \nabla_\xi Y\rangle$ for all ξ, X, Y. In other words, the statement that the metric tensor is parallel (constant) with respect to ∇ is the same as saying that the connection is a metric connection.

Exercise 13.15. Let ∇ be the Levi-Civita connection for a semi-Riemannian manifold (M, g). Prove the formula

(13.3) $\qquad (\mathcal{L}_X g)(Y, Z) = g(\nabla_X Y, Z) + g(Y, \nabla_X Z)$

for vector fields $X, Y, Z \in \mathfrak{X}(M)$.

13.2. Riemann Curvature Tensor

For (M, g) a Riemannian manifold with associated Levi-Civita connection ∇, the associated curvature tensor field is called the **Riemann curvature tensor**: For $X, Y \in \mathfrak{X}(M)$ we have the map $R(X, Y) : \mathfrak{X}(M) \to \mathfrak{X}(M)$ defined by

$$R(X,Y)Z := R_{X,Y} Z := \nabla_X \nabla_Y Z - \nabla_Y \nabla_X Z - \nabla_{[X,Y]} Z.$$

Exercise 13.16. Show that $\nabla^2_{X,Y}(Z) - \nabla^2_{Y,X}(Z) = R(X,Y)Z$ (recall Definition 13.12).

By direct calculation, or by appealing to Theorem 12.44 and Exercise 12.45 from the previous chapter, we find that $(X, Y, Z) \mapsto R(X, Y)Z$ is $C^\infty(M)$-multilinear (tensorial). Appealing to the isomorphism (7.7), we conclude that R gives a section of $T^0_3(TM; TM)$. That is, R can be thought of as defining a T_pM-valued tensor field at each p. In other words, if $X_p, Y_p, Z_p \in T_pM$, then $R(X_p, Y_p) Z_p$ is a well-defined element of T_pM and $(X_p, Y_p, Z_p) \mapsto R(X_p, Y_p) Z_p$ gives a multilinear map. Here, $R(X_p, Y_p) Z_p$ is defined to be $(R(X, Y)Z)(p)$ for any fields X, Y, Z such that $X(p) = X_p, Y(p) = Y_p$, and $Z(p) = Z_p$. Many interpretations of R arise. From the previous chapter we know that R is also to be thought of as a TM-valued 2-form. From this point on we will freely interpret elements of $\mathcal{T}^r_s(M)$ as elements of $T^r{}_s(\mathfrak{X}(M))$ when convenient.

Notice that we will use both the notation $R(X,Y)$ as well as $R_{X,Y}$.

Definition 13.17. A semi-Riemannian manifold (M, g) is called **flat** if the curvature tensor is identically zero.

Recall that if $f : (M, g) \to (N, h)$ is a local diffeomorphism between semi-Riemannian manifolds such that $f^*h = g$, then f is called a local isometry and we say that the manifolds are locally isometric.

Theorem 13.18. *Let (M, g) be a semi-Riemannian manifold of dimension n and index ν. If (M, g) is flat, that is, if the curvature tensor is identically zero, then (M, g) is locally isometric to the semi-Euclidean space $\mathbb{R}^n_\nu$.*

Proof. Let $p \in M$ given. If the curvature tensor vanishes, then by Theorem 12.51 we can find local parallel vector fields defined in a neighborhood of p with prescribed values at p. So we may find parallel fields $X_1, \ldots, X_n$ such that $X_1(p), \ldots, X_n(p)$ is an orthonormal basis. But since parallel translation preserves the various scalar products $\langle X_i, X_j \rangle$, we see that we actually have an orthonormal frame field in a neighborhood of p. Next we use the fact that the Levi-Civita connection is symmetric (torsion zero). We have $\nabla_{X_i} X_j - \nabla_{X_j} X_i - [X_i, X_j] = 0$ for all i, j. But since the X_i are parallel, this means that $[X_i, X_j] \equiv 0$. Therefore there exist coordinates $x^1, \ldots, x^n$ on a possibly smaller open set such that

$$\frac{\partial}{\partial x^i} = X_i \text{ for all } i.$$

The result is that these coordinates give a chart which is an isometry of a neighborhood of p with an open subset of the semi-Riemannian space $\mathbb{R}^n_\nu$. □

For another proof of the previous theorem see the online supplement [Lee, Jeff]. The next theorem exhibits the symmetries of the Riemann curvature tensor:

Theorem 13.19. *The map $X, Y, Z, W \mapsto \langle R_{X,Y} Z, W \rangle$ is tensorial in all variables. Furthermore, the following identities hold for all $X, Y, Z, W \in \mathfrak{X}(M)$:*

(i) $R_{X,Y} = -R_{Y,X}$.
(ii) $\langle R_{X,Y} Z, W \rangle = -\langle R_{X,Y} W, Z \rangle$.
(iii) $R_{X,Y} Z + R_{Y,Z} X + R_{Z,X} Y = 0$ *(First Bianchi identity)*.
(iv) $\langle R_{X,Y} Z, W \rangle = \langle R_{Z,W} X, Y \rangle$.

Proof. Tensorality is immediate from our previous observations. Also, (i) is immediate from the definition of R and (ii) is just a special case of Proposition 12.70 from the previous chapter. For (iii) we calculate:

$R_{X,Y} Z + R_{Y,Z} X + R_{Z,X} Y$
$= \nabla_X \nabla_Y Z - \nabla_Y \nabla_X Z + \nabla_Y \nabla_Z X - \nabla_Z \nabla_Y X + \nabla_Z \nabla_X Y - \nabla_X \nabla_Z Y = 0.$

13.2. Riemann Curvature Tensor

The proof of (iv) is rather unenlightening and is just some combinatorics. Since R is a tensor, we may assume without loss of generality that $[X, Y] = 0$. For any X, Y, Z, let $(\mathcal{C}R)_{X,Y} Z$ be defined by

$$(\mathcal{C}R)_{X,Y} Z := R_{X,Y} Z + R_{Y,Z} X + R_{Z,X} Y.$$

By (iii) we have $\langle (\mathcal{C}R)_{Y,Z} X, W \rangle = 0$ for any W. Summing over all cyclic permutations of Y, Z, X, W, we obtain

$$0 = \langle (\mathcal{C}R)_{Y,Z} X, W \rangle + \langle (\mathcal{C}R)_{W,Y} Z, X \rangle + \langle (\mathcal{C}R)_{X,W} Y, Z \rangle + \langle (\mathcal{C}R)_{Z,X} W, Y \rangle.$$

Expand this expression using the definition of $\mathcal{C}R$, and we have twelve terms. Four pairs of terms cancel due to (i) and (ii) resulting in

$$2 \langle R_{X,Y} Z, W \rangle + 2 \langle R_{W,Z} X, Y \rangle = 0.$$

Using (i) we obtain the result. $\square$

Theorem 13.20 (Second Bianchi identity). *For $X, Y, Z \in \mathfrak{X}(M)$, we have*

$$(\nabla_Z R)(X, Y) + \nabla_X R(Y, Z) + \nabla_Y R(Z, X) = 0.$$

Proof. This is the Bianchi identity for the Levi-Civita connection and in this context is also called the second Bianchi identity. We give an independent proof here. Since this is a tensor equation, we only need to prove it under the assumption that all brackets among the X, Y, Z are zero (recall Exercise 7.34). First we have

$$\begin{aligned}(\nabla_Z R)(X,Y)W &= \nabla_Z(R_{X,Y}W) - R(\nabla_Z X, Y)W \\ &\quad - R(X, \nabla_Z Y)W - R_{X,Y}\nabla_Z W \\ &= [\nabla_Z, R_{X,Y}]W - R(\nabla_Z X, Y)W - R(X, \nabla_Z Y)W.\end{aligned}$$

Using this, we calculate as follows:

$$\begin{aligned}&(\nabla_Z R)(X,Y)W + (\nabla_X R)(Y,Z)W + (\nabla_Y R)(Z,X)W \\ &= [\nabla_Z, R_{X,Y}]W + [\nabla_X, R_{Y,Z}]W + [\nabla_Y, R_{Z,X}]W \\ &\quad - R(\nabla_Z X, Y)W - R(X, \nabla_Z Y)W \\ &\quad - R(\nabla_X Y, Z)W - R(Y, \nabla_X Z)W \\ &\quad - R(\nabla_Y Z, X)W - R(Z, \nabla_Y X)W \\ &= [\nabla_Z, R_{X,Y}]W + [\nabla_X, R_{Y,Z}]W + [\nabla_Y, R_{Z,X}]W \\ &\quad + R([X,Z], Y)W + R([Z,Y], X)W + R([Y,X], Z)W \\ &= [\nabla_Z, [\nabla_X, \nabla_Y]] + [\nabla_X, [\nabla_Y, \nabla_Z]] + [\nabla_Y, [\nabla_Z, \nabla_X]] = 0.\end{aligned}$$

The last identity is the Jacobi identity for commutators and is true for purely algebraic reasons (see the next exercise). $\square$

Note: Given a semi-Riemannian manifold (M,g), the tensor R defined by $R(X,Y,Z,W) := \langle R_{X,Y}Z, W\rangle$ for all X,Y,Z,W, is also called the **Riemann curvature tensor**. Often this tensor is defined with a different ordering of the slots and one should always check which conventions are in use. One traditional ordering is $R(W,Z,X,Y) := \langle R_{X,Y}Z, W\rangle$.

Exercise 13.21. Show that if $L_i, i = 1,2,3$, are linear operators, and the commutator is defined as usual ($[A,B] = AB - BA$), then we always have the Jacobi identity $[L_1,[L_2,L_3]] + [L_2,[L_3,L_1]] + [L_3,[L_1,L_2]] = 0$.

We now introduce several objects which hold all or part of the information in the curvature tensor in different forms. First we mention that the reader should keep an eye out for expressions of the form $\langle R(v,w)v,w\rangle$ or $\langle R(v,w)w,v\rangle$ for v,w in the tangent space of a point on the semi-Riemannian manifold (M,g) under study.

It will be convenient to introduce a little linear algebra at this point. Recall that if $(V, \langle \cdot, \cdot \rangle)$ is a finite-dimensional scalar product space, then there is an associated natural scalar product on $\bigwedge^2 V$ defined so that for an orthonormal basis $\{e_i\}$, the basis $\{e_i \wedge e_j\}_{i,j}$ for $\bigwedge^2 V$ is also orthonormal. On simple elements we have

$$g(v_1 \wedge v_2, v_3 \wedge v_4) = \det \begin{pmatrix} \langle v_1, v_3\rangle & \langle v_1, v_4\rangle \\ \langle v_2, v_3\rangle & \langle v_2, v_4\rangle \end{pmatrix}.$$

We will use the angle brackets $\langle \cdot, \cdot \rangle$ for this scalar product also. The quantity $\langle v \wedge w, v \wedge w\rangle$ is important and needs special attention in the case of an indefinite scalar product. If $\langle \cdot, \cdot \rangle$ on V is indefinite, then the induced scalar product is also indefinite and $\langle v \wedge w, v \wedge w\rangle$ may be zero, even when v and w are linearly independent. For the next lemma, recall that a subspace W of a scalar product space $(V, \langle \cdot, \cdot \rangle)$ is called **nondegenerate** if $\langle \cdot, \cdot \rangle$ restricted to W is nondegenerate.

Lemma 13.22. *Let $(V, \langle \cdot, \cdot \rangle)$ be a finite-dimensional scalar product space. Let P be a plane spanned by v,w. Then,*

(i) *P is nondegenerate if and only if $\langle v \wedge w, v \wedge w\rangle \neq 0$.*

(ii) *$\langle v \wedge w, v \wedge w\rangle > 0$ if and only if $\langle \cdot, \cdot \rangle$ restricted to P is definite.*

(iii) *$\langle v \wedge w, v \wedge w\rangle < 0$ if and only if $\langle \cdot, \cdot \rangle$ restricted to P is indefinite.*

Proof. Exercise. □

If v and w span a nondegenerate plane, then $|\langle v \wedge w, v \wedge w\rangle|$ is the squared area of the parallelogram spanned by v and w.

Lemma 13.23. *Let $(V, \langle \cdot, \cdot \rangle)$ be a finite-dimensional scalar product space. If $v, w \in V$ are any two vectors, then there exist vectors $v', w' \in V$ arbitrarily close to v and w respectively such that v', w' span a nondegenerate plane.*

13.2. Riemann Curvature Tensor

Proof. Assume that $\langle v \wedge w, v \wedge w \rangle = 0$, or there is nothing to prove. Any pair of vectors is close to a pair of linearly independent vectors, so we may assume that v and w are linearly independent. There exists a vector x such that $\langle v \wedge x, v \wedge x \rangle < 0$. Indeed, if v is null, then we can pick x so that $\langle v, x \rangle \neq 0$, which means that $\langle v \wedge x, v \wedge x \rangle < 0$. If v is not null, then pick x of opposite causal character. That is, pick x to be spacelike if v is timelike and vice versa. Now if $w_\epsilon := w + \epsilon x$ for small $\epsilon > 0$, then $\langle v \wedge w_\epsilon, v \wedge w_\epsilon \rangle = 2\epsilon b + \epsilon^2 \langle v \wedge x, v \wedge x \rangle$ for some number b independent of ϵ. If $b = 0$, then $\langle v \wedge w_\epsilon, v \wedge w_\epsilon \rangle < 0$, and we are done by Lemma 13.22. If $b \neq 0$, then $\langle v \wedge w_\epsilon, v \wedge w_\epsilon \rangle$ is nonzero in case ϵ is sufficiently small, and then v, w_ϵ span a nondegenerate plane by Lemma 13.22 again. □

Note that in the previous lemma, closeness is measured in the standard topology of a finite-dimensional vectors space. One does not try to use an indefinite scalar product to define the topology!

The symmetry properties for the Riemann curvature tensor allow that we have a well-defined map

$$(13.4) \qquad \mathfrak{R} : \bigwedge^2(TM) \to \bigwedge^2(TM),$$

which is symmetric with respect to the natural extension of g to $\bigwedge^2(TM)$. The map $\mathfrak{R}$ is defined implicitly as follows:

$$g(\mathfrak{R}(v_1 \wedge v_2), v_3 \wedge v_4) := \langle R(v_1, v_2)v_4, v_3 \rangle.$$

Notice the switch in the indices 3 and 4. This hides a sign and must be remembered to avoid confusion later.

Another commonly used quantity is the **sectional curvature** K. If v and w span a nondegenerate plane in T_pM, then define

$$K(v \wedge w) := \frac{\langle R(v,w)w, v \rangle}{\langle v,v \rangle \langle w,w \rangle - \langle v,w \rangle^2}$$
$$= \frac{\langle \mathfrak{R}(v \wedge w), v \wedge w \rangle}{\langle v \wedge w, v \wedge w \rangle}.$$

The value $K(v \wedge w)$ only depends on the oriented plane spanned by the vectors v and w; therefore if $P = \text{span}\{v, w\}$ is such a nondegenerate plane, we also write $K(P)$ instead of $K(v \wedge w)$. The set of all planes in T_pM is denoted $Gr_p(2)$. We remark that if M is 2-dimensional, then K is a scalar function on M. It turns out that this function is exactly the Gauss curvature introduced in Chapter 4. There we showed that the Gauss curvature is intrinsic and we found an expression for it in terms of the metric, which is still valid in this situation.

In the following definition, V is an R-module. The two cases we have in mind are (1) V is $\mathfrak{X}(M)$, R $= C^\infty(M)$, and (2) V is T_pM, R $= \mathbb{R}$.

Definition 13.24. A multilinear function $F : V \times V \times V \times V \to \mathbb{R}$ is said to be **curvature-like** if it satisfies the symmetries proved for the curvature R above; namely, if for all $x, y, z, w \in V$ we have

  (i) $F(x, y, z, w) = -F(y, x, z, w)$;
  (ii) $F(x, y, z, w) = -F(x, y, w, z)$;
  (iii) $F(x, y, z, w) + F(y, z, x, w) + F(z, x, y, w) = 0$;
  (iv) $F(x, y, z, w) = F(w, z, x, y)$.

Exercise 13.25. Define the tensor C_g by
$$C_g(X, Y, Z, W) := g(Y, Z)g(X, W) - g(X, Z)g(Y, W).$$
Show that C_g is curvature-like.

Proposition 13.26. *If F is curvature-like and $F(v, w, v, w) = 0$ for all $v, w \in V$, then $F \equiv 0$.*

Proof. From (iv) it follows that for each v, the bilinear map $(w, z) \mapsto F(v, w, v, z)$ is symmetric, and so if $F(v, w, v, w) = 0$ for all $v, w \in V$, then $F(v, w, v, z) = 0$ for all $v, w, z \in V$. Now it is a simple matter to show that (i) and (ii) imply that $F \equiv 0$. □

Proposition 13.27. *If $\langle R_{v,w} v, w \rangle$ is known for all $v, w \in T_p M$, then R itself is determined at p. If $K(P)$ is known for all nondegenerate planes in $T_p M$, then R itself is determined at p.*

Proof. Let $R_2(v, w) := \langle R_{v,w} v, w \rangle$ for $v, w \in T_p M$. Using an orthonormal basis for $T_p M$, we see that K and R_2 contain the same information, so we will just show that R_2 determines R:

$$\left. \frac{\partial^2}{\partial s \partial t} \right|_{0,0} (R_2(v + tz, w + su) - R_2(v + tu, w + sz))$$
$$= \left. \frac{\partial^2}{\partial s \partial t} \right|_{0,0} \{g(R(v + tz, w + su)v + tz, w + su)$$
$$\quad - g(R(v + tu, w + sz)v + tu, w + sz)\}$$
$$= 6R(v, w, z, u).$$

The second part follows by continuity and the fact that $\langle R(v, w)w, v \rangle = \langle v \wedge w, v \wedge w \rangle K(v \wedge w)$ for v, w spanning a nondegenerate plane P. □

For each $v \in TM$, the **tidal operator** $R_v : T_p M \to T_p M$ is defined by
$$R_v(w) := R_{v,w} v.$$

We are now in a position to prove the following important theorem.

13.2. Riemann Curvature Tensor

Theorem 13.28. *The following assertions are all equivalent (κ is a constant):*

(i) $K(P) = \kappa$ for all $P \in Gr_p(2)$, P nondegenerate.

(ii) $\langle R_{v_1,v_2} v_3, v_4 \rangle = \kappa C_g(v_1, v_2, v_3, v_4)$ for all $v_1, v_2, v_3, v_4 \in T_p M$.

(iii) $-R_v(w) = \kappa(w - \langle w, v \rangle v)$ for all $w, v \in T_p M$ with $\|v\| = 1$.

(iv) $\mathfrak{R}(\xi) = \kappa \xi$ for all $\xi \in \bigwedge^2 T_p M$.

Proof. Let $p \in M$. The proof that (ii)$\Longrightarrow$(iii) and that (iii)$\Longrightarrow$(i) is left as an easy exercise. We prove that (i)$\Longrightarrow$(ii)$\Longrightarrow$(iv)$\Longrightarrow$(i).

(i)$\Longrightarrow$(ii): Let R be defined by $R(v_1, v_2, v_3, v_4) := \langle R_{v_1,v_2} v_3, v_4 \rangle$ and let $T_g := R - \kappa C_g$. Then T_g is curvature-like and $T_g(v, w, v, w) = 0$ for all $v, w \in T_p M$ by assumption. It follows from Proposition 13.26 that $T_g \equiv 0$.

(ii)$\Longrightarrow$(iv): Let $\{e_1, \ldots, e_n\}$ be an orthonormal basis for $T_p M$. Then $\{e_i \wedge e_j\}_{i<j}$ is an orthonormal basis for $\bigwedge^2 T_p M$. Using (ii), we see that

$$\langle \mathfrak{R}(e_i \wedge e_j), e_k \wedge e_l \rangle = \langle R_{e_i,e_j} e_k, e_l \rangle$$
$$= \langle R(e_i, e_j) e_k, e_l \rangle$$
$$= \kappa C_g(v_1, v_2, v_3, v_4)$$
$$= \kappa \langle e_i \wedge e_j, e_k \wedge e_l \rangle \text{ for all } k, l.$$

This implies that $\mathfrak{R}(e_i \wedge e_j) = \kappa e_i \wedge e_j$.

(iv)$\Longrightarrow$(i): This follows because if v, w are orthonormal, then we have $\kappa = \langle \mathfrak{R}(v \wedge w), v \wedge w \rangle = K(v \wedge w)$. $\square$

Definition 13.29. Let (M, g) be a semi-Riemannian manifold. The **Ricci curvature** is the $(1, 1)$-tensor Ric defined by

$$\mathrm{Ric}(v, w) := \sum_{i=1}^{n} \epsilon_i \langle R_{v,e_i} e_i, w \rangle,$$

where $(e_1, \ldots, e_n)$ is any orthonormal basis of $T_p M$ and $\epsilon_i := \langle e_i, e_i \rangle$.

We say that the Ricci curvature Ric is **bounded from below by** κ and write $\mathrm{Ric} \geq k$ if $\mathrm{Ric}(v, w) \geq k \langle v, w \rangle$ for all $v, w \in TM$. Similar and obvious definitions can be given for $\mathrm{Ric} \leq k$ and the strict bounds $\mathrm{Ric} > k$ and $\mathrm{Ric} < k$. Actually, it is often the case that the bound on Ricci curvature is given in the form $\mathrm{Ric} \geq \kappa(n-1)$, where $n = \dim(M)$.

In passing, let us mention that there is a very important and interesting class of manifolds called Einstein manifolds. A semi-Riemannian manifold (M, g) is called an **Einstein manifold** with Einstein constant k if and only if $\mathrm{Ric}(v, w) = k \langle v, w \rangle$ for all $v, w \in TM$. We write this as $\mathrm{Ric} = kg$ or even $\mathrm{Ric} = k$. For example, if (M, g) has constant sectional curvature κ, then it is an Einstein manifold with Einstein constant $k = \kappa(n-1)$. The effect of this condition depends on the signature of the metric. Particularly interesting is the case where the index is 0 (Riemannian) and also the case where the index is 1 (Lorentz manifold). Perhaps the first question one should ask is whether there exist any Einstein manifolds that do not have constant sectional curvature. It turns out that there are many interesting Einstein manifolds that do not have constant sectional curvature. For manifolds of dimension > 2 the Einstein manifold condition is natural and fruitful. Unfortunately, we do not have space to explore this fascinating topic (but see [**Be**]).

Exercise 13.30. Show that if M is connected and $\dim(M) > 2$, and $\mathrm{Ric}\,(\cdot,\cdot) = f\langle\cdot,\cdot\rangle$, where $f \in C^\infty(M)$, then $f \equiv k$ for some $k \in \mathbb{R}$ (so (M,g) is Einstein).

13.3. Semi-Riemannian Submanifolds

Let M be a d-dimensional submanifold of a semi-Riemannian manifold $\overline{M}$ of dimension n, where $d < n$. The metric $g(\cdot,\cdot) = \langle\cdot,\cdot\rangle$ on $\overline{M}$ restricts to a tensor on M, which we denote by h. Since h is a restriction of g, we shall also use the notation $\langle\cdot,\cdot\rangle$ for h. If the restriction h is nondegenerate on each space T_pM and has the same index for all p, then h is a metric tensor on M and we say that M is a **semi-Riemannian submanifold** of $\overline{M}$. If $\overline{M}$ is Riemannian, then this nondegeneracy condition is automatic and the metric h is automatically Riemannian.

More generally, if $\phi: N \to (\overline{M}, g)$ is an immersion, we can consider the pull-back tensor $\phi^* g$ defined by

$$\phi^* g(X, Y) = g(T\phi \cdot X, T\phi \cdot Y).$$

If $\phi^* g$ is nondegenerate on each tangent space, then it is a metric on N called the pull-back metric and we call ϕ a semi-Riemannian immersion. If N is already endowed with a metric g_N, and if $\phi^* g = g_N$, then we say that $\phi : (N, g_N) \to (\overline{M}, g)$ is an isometric immersion. Of course, if $\phi^* g$ is a metric at all, as it always is if $(\overline{M}, g)$ is Riemannian, then the map $\phi : (N, \phi^* g) \to (\overline{M}, g)$ is automatically an isometric immersion. Every immersion restricts locally to an embedding, and for the questions we study here there is not much loss in focusing on the case of a submanifold $M \subset \overline{M}$.

There is an obvious bundle on M which is the restriction of $T\overline{M}$ to M. This is the bundle $T\overline{M}\big|_M = \bigcup_{p\in M} T_p\overline{M}$. Recalling Lemma 7.47, we see that each tangent space $T_p\overline{M}$ decomposes as

$$T_p\overline{M} = T_pM \oplus (T_pM)^\perp,$$

where $(T_pM)^\perp = \{v \in T_p\overline{M} : \langle v, w\rangle = 0 \text{ for all } w \in T_pM\}$. Then $TM^\perp = \bigcup_{p\in M} (T_pM)^\perp$, with its natural structure as a smooth vector bundle, is called

13.3. Semi-Riemannian Submanifolds

the **normal bundle** to M in $\overline{M}$. The smooth sections of the normal bundle will be denoted by $\Gamma\left(TM^\perp\right)$ or $\mathfrak{X}(M)^\perp$. The orthogonal decomposition above is globalized as

$$T\overline{M}\big|_M = TM \oplus TM^\perp.$$

A vector field on M is always the restriction of some (not unique) vector field on a neighborhood of $\overline{M}$. The same is true of any, not necessarily tangent, vector field along M. The set of all vector fields along M will be denoted by $\mathfrak{X}(\overline{M})\big|_M$. Since any function on M is also the restriction of some function on $\overline{M}$, we may consider $\mathfrak{X}(M)$ as a submodule of $\mathfrak{X}(\overline{M})\big|_M$. If $\overline{X} \in \mathfrak{X}(\overline{M})$, then we denote its restriction to M by $\overline{X}\big|_M$ or sometimes just X. Notice that $\mathfrak{X}(M)^\perp$ is a submodule of $\mathfrak{X}(\overline{M})\big|_M$. We have two projection maps, nor: $T_p\overline{M} \to T_pM^\perp$ and tan: $T_p\overline{M} \to T_pM$ which in turn give module projections nor: $\mathfrak{X}(\overline{M})\big|_M \to \mathfrak{X}(M)^\perp$ and tan: $\mathfrak{X}(\overline{M})\big|_M \to \mathfrak{X}(M)$. We also have the pair of naturally related restrictions $C^\infty(\overline{M}) \overset{\text{restriction}}{\longrightarrow} C^\infty(M)$ and $\mathfrak{X}(\overline{M}) \overset{\text{restriction}}{\longrightarrow} \mathfrak{X}(\overline{M})\big|_M$. Note that $\mathfrak{X}(\overline{M})$ is a $C^\infty(\overline{M})$-module, while $\mathfrak{X}(\overline{M})\big|_M$ is a $C^\infty(M)$-module. We have an exact sequence of modules:

$$0 \to \mathfrak{X}(M)^\perp \to \mathfrak{X}(\overline{M})\big|_M \overset{\tan}{\to} \mathfrak{X}(M) \to 0.$$

Now we shall obtain a sort of splitting of the Levi-Civita connection of $\overline{M}$ along the submanifold M. First we notice that the Levi-Civita connection $\overline{\nabla}$ on $\overline{M}$ restricts nicely to a connection on the bundle $T\overline{M}\big|_M \to M$. The reader should be sure to realize that the space of sections of this bundle is exactly $\mathfrak{X}(\overline{M})\big|_M$. We wish to obtain a restricted covariant derivative

$$\overline{\nabla}\big|_M : \mathfrak{X}(M) \times \mathfrak{X}(\overline{M})\big|_M \to \mathfrak{X}(\overline{M})\big|_M.$$

If $X \in \mathfrak{X}(M)$ and $W \in \mathfrak{X}(\overline{M})\big|_M$, then $\overline{\nabla}_X W$ does not seem to be defined since X and W are not elements of $\mathfrak{X}(\overline{M})$. But we may extend X and W to elements of $\mathfrak{X}(\overline{M})$, use $\overline{\nabla}$, and then restrict again to get an element of $\mathfrak{X}(\overline{M})\big|_M$. Then recalling the local properties of a connection we see that the result does not depend on the extension.

Exercise 13.31. Use local coordinates to prove that $\overline{\nabla}_X W$ does not depend on the extensions used.

It is also important to observe that the restricted covariant derivative is exactly the covariant derivative obtained by the methods of Section 12.3 of the previous chapter. Namely, it is the covariant derivative along the inclusion map $M \hookrightarrow \overline{M}$. Thus, it is defined even without an appeal to the process of extending fields described above.

We shall write simply $\overline{\nabla}$ in place of $\overline{\nabla}\big|_M$ since the context will make it clear when the latter is meant. Thus for all $p \in M$, we have $\overline{\nabla}_X W(p) = \overline{\nabla}_{\overline{X}} \overline{W}(p)$, where $\overline{X}$ and $\overline{W}$ are any extensions of X and W respectively.

Clearly we have $\overline{\nabla}_X \langle Y_1, Y_2 \rangle = \langle \overline{\nabla}_X Y_1, Y_2 \rangle + \langle Y_1, \overline{\nabla}_X Y_2 \rangle$ and so $\overline{\nabla}$ is a metric connection on $T\overline{M}\big|_M$. For fixed $X, Y \in \mathfrak{X}(M)$, we have the decomposition of $\overline{\nabla}_X Y$ into tangent and normal parts. Similarly, for $V \in \mathfrak{X}(M)^\perp$, we can consider the decomposition of $\overline{\nabla}_X V$ into tangent and normal parts. Thus we have

$$\overline{\nabla}_X Y = (\overline{\nabla}_X Y)^{\tan} + (\overline{\nabla}_X Y)^\perp,$$
$$\overline{\nabla}_X V = (\overline{\nabla}_X V)^{\tan} + (\overline{\nabla}_X V)^\perp.$$

Proposition 13.32. *For a semi-Riemannian submanifold $M \subset \overline{M}$, we have*

$$\nabla_X Y = (\overline{\nabla}_X Y)^{\tan} \text{ for all } X, Y \in \mathfrak{X}(M),$$

where ∇ is the Levi-Civita covariant derivative on M with its induced metric.

Proof. It is straightforward to show that if we extend fields X, Y, Z to $\overline{X}, \overline{Y}, \overline{Z}$, then the Koszul formula (13.2) for $\overline{\nabla}_{\overline{X}} \overline{Y}$ implies that

(13.5)
$$2\langle (\overline{\nabla}_X Y)^{\tan}, Z \rangle = X\langle Y, Z \rangle + Y\langle Z, X \rangle - Z\langle X, Y \rangle$$
$$+ \langle Z, [X, Y] \rangle - \langle Y, [X, Z] \rangle - \langle X, [Y, Z] \rangle$$

for all Z. But the Koszul formula which determines $\nabla_X Y$ shows that $(\overline{\nabla}_X Y)^{\tan} = \nabla_X Y$. $\square$

Definition 13.33. Define maps $II: \mathfrak{X}(M) \times \mathfrak{X}(M) \to \mathfrak{X}(M)^\perp$, $\widetilde{II}: \mathfrak{X}(M) \times \mathfrak{X}(M)^\perp \to \mathfrak{X}(M)$ and $\nabla^\perp : \mathfrak{X}(M) \times \mathfrak{X}(M)^\perp \to \mathfrak{X}(M)^\perp$ according to

$$II(X, Y) := (\overline{\nabla}_X Y)^\perp \text{ for all } X, Y \in \mathfrak{X}(M),$$
$$\widetilde{II}(X, V) := (\overline{\nabla}_X V)^{\tan} \text{ for all } X \in \mathfrak{X}(M), V \in \mathfrak{X}(M)^\perp,$$
$$\nabla^\perp_X V := (\overline{\nabla}_X V)^\perp \text{ for all } X \in \mathfrak{X}(M), V \in \mathfrak{X}(M)^\perp.$$

It is easy to show that $\nabla^\perp$ defines a metric covariant derivative on the normal bundle $TM^\perp$. The map $(X, Y) \mapsto II(X, Y)$ is clearly $C^\infty(M)$-linear in the first slot. If $X, Y \in \mathfrak{X}(M)$ and $V \in \mathfrak{X}(M)^\perp$, then $0 = \langle Y, V \rangle$ and we have

$$0 = \overline{\nabla}_X \langle Y, V \rangle = \langle \overline{\nabla}_X Y, V \rangle + \langle Y, \overline{\nabla}_X V \rangle$$
$$= \langle (\overline{\nabla}_X Y)^\perp, V \rangle + \langle Y, (\overline{\nabla}_X V)^{\tan} \rangle$$
$$= \langle II(X, Y), V \rangle + \langle Y, \widetilde{II}(X, V) \rangle.$$

It follows that

(13.6) $$\langle II(X, Y), V \rangle = -\langle Y, \widetilde{II}(X, V) \rangle.$$

13.3. Semi-Riemannian Submanifolds

From this we see that $II(X,Y)$ is not only $C^\infty(M)$-linear in X, but also in Y. This means that II is tensorial, and so $II(X_p, Y_p)$ is a well-defined element of $T_pM^\perp$ for each $X_p, Y_p \in T_pM$. Thus II is a $TM^\perp$-valued tensor field and for each p we have an $\mathbb{R}$-bilinear map $II_p : T_pM \times T_pM \to T_pM^\perp$. (We often suppress the subscript p.)

Proposition 13.34. *II is symmetric.*

Proof. For any $X, Y \in \mathfrak{X}(M)$ we have
$$II(X,Y) - II(Y, X_1) = \left(\overline{\nabla}_X Y - \overline{\nabla}_Y X\right)^\perp$$
$$= ([X,Y])^\perp = 0. \qquad \square$$

We can also easily deduce that $\widetilde{II}$ is a symmetric $C^\infty(M)$-bilinear form with values in $\mathfrak{X}(M)$ and is similarly tensorial. So $\widetilde{II}(X_p, V_p)$ is a well-defined element of T_pM for each fixed $X_p \in T_pM$ and $V_p \in T_pM^\perp$. We thus obtain a bilinear form
$$\widetilde{II}_p : T_pM \times T_pM^\perp \to T_pM.$$

In summary, II and $\widetilde{II}$ are tensorial, $\nabla^\perp$ is a metric covariant derivative and we have the following formulas:
$$\overline{\nabla}_X Y = \nabla_X Y + II(X,Y),$$
$$\overline{\nabla}_X V = \nabla_X^\perp V + \widetilde{II}(X,V)$$
for $X, Y \in \mathfrak{X}(M)$ and $V \in \mathfrak{X}(M)^\perp$.

Recall that if $(V, \langle \cdot, \cdot \rangle_1)$ and $(W, \langle \cdot, \cdot \rangle_2)$ are scalar product spaces, then a linear map $A : V \to W$ has a **metric transpose** $A^t : W \to V$ uniquely defined by the requirement that
$$\langle Av, w \rangle_2 = \langle v, A^t w \rangle_1$$
for all $v \in V$ and $w \in W$.

Definition 13.35. For $v \in T_pM$, we define the linear map $B_v(\cdot) := II(v, \cdot)$.

Formula (13.6) shows that the map $\widetilde{II}(v, \cdot) : T_pM^\perp \to T_pM$ is equal to $-B_v^t : T_pM \to T_pM^\perp$. Writing any $Y \in \mathfrak{X}(\overline{M})\big|_M$ as a column vector $(Y^{\tan}, Y^\perp)^t$, we can write the map $\overline{\nabla}_X : \mathfrak{X}(\overline{M})\big|_M \to \mathfrak{X}(\overline{M})\big|_M$ as a matrix of operators:
$$\begin{bmatrix} \nabla_X & -B_X^t \\ B_X & \nabla_X^\perp \end{bmatrix}.$$

Next we define the **shape operator**, also called the **Weingarten map**. We have already met a special case of the shape operator in Chapter 4. The shape operator is sometimes defined with the opposite sign.

Definition 13.36. Let $p \in M$. For each unit vector u normal to M at p, we have a map called the **shape operator** S_u associated to u defined by
$$S_u(v) := -\left(\overline{\nabla}_v U\right)^{\text{tan}},$$
where U is any unit normal field defined near p such that $U(p) = u$.

Exercise 13.37. Show that the definition is independent of the choice of normal field U that extends u.

The family of shape operators $\{S_u : u \text{ a unit normal}\}$ contains essentially the same information as the second fundamental tensor II or the associated map B. This is because for any $X, Y \in \mathfrak{X}(M)$ and $U \in \mathfrak{X}(M)^\perp$, we have
$$\langle S_U X, Y \rangle = \langle \left(-\overline{\nabla}_X U\right)^{\text{tan}}, Y \rangle = \langle U, -\overline{\nabla}_X Y \rangle$$
$$= \langle U, \left(-\overline{\nabla}_X Y\right)^\perp \rangle = \langle U, -II(X,Y) \rangle.$$

In the case of a hypersurface, we have (locally) only two choices of unit normal. Once we have chosen a unit normal u, the shape operator is denoted simply by S rather than S_u.

Theorem 13.38. *Let M be a semi-Riemannian submanifold of $\overline{M}$. For any $V, W, X, Y \in \mathfrak{X}(M)$, we have*
$$\langle R_{VW} X, Y \rangle = \langle \overline{R}_{VW} X, Y \rangle$$
$$- \langle II(V,X), II(W,Y) \rangle + \langle II(V,Y), II(W,X) \rangle.$$

*This equation is called the **Gauss equation** or **Gauss curvature equation**.*

Proof. Since this is clearly a tensor equation, we may assume that $[V, W] = 0$ (see Exercise 7.34). We have $\langle \overline{R}_{VW} X, Y \rangle = \langle \overline{\nabla}_V \overline{\nabla}_W X, Y \rangle - \langle \overline{\nabla}_W \overline{\nabla}_V X, Y \rangle$. We calculate:
$$\langle \overline{\nabla}_V \overline{\nabla}_W X, Y \rangle = \langle \overline{\nabla}_V \nabla_W X, Y \rangle + \langle \overline{\nabla}_V (II(W,X)), Y \rangle$$
$$= \langle \nabla_V \nabla_W X, Y \rangle + \langle \overline{\nabla}_V (II(W,X)), Y \rangle$$
$$= \langle \nabla_V \nabla_W X, Y \rangle + V \langle II(W,X), Y \rangle - \langle II(W,X), \overline{\nabla}_V Y \rangle$$
$$= \langle \nabla_V \nabla_W X, Y \rangle - \langle II(W,X), \overline{\nabla}_V Y \rangle.$$

Since
$$\langle II(W,X), \overline{\nabla}_V Y \rangle = \langle II(W,X), \left(\overline{\nabla}_V Y\right)^\perp \rangle$$
$$= \langle II(W,X), II(V,Y) \rangle,$$

we have $\langle \overline{\nabla}_V \overline{\nabla}_W X, Y \rangle = \langle \nabla_V \nabla_W X, Y \rangle - \langle II(W,X), II(V,Y) \rangle$. Interchanging the roles of V and W and subtracting we get the desired conclusion. $\square$

13.3. Semi-Riemannian Submanifolds

The second fundamental form contains information about how the semi-Riemannian submanifold M bends in $\overline{M}$.

Definition 13.39. Let M be a semi-Riemannian submanifold of $\overline{M}$ and N a semi-Riemannian submanifold of $\overline{N}$. A **pair isometry** $\Phi : (\overline{M}, M) \to (\overline{N}, N)$ consists of an isometry $\Phi : \overline{M} \to \overline{N}$ such that $\Phi(M) = N$ and such that $\Phi|_M : M \to N$ is an isometry.

Proposition 13.40. *A pair isometry $\Phi : (\overline{M}, M) \to (\overline{N}, N)$ preserves the second fundamental tensor:*

$$T_p\Phi \cdot II(v,w) = II(T_p\Phi \cdot v, T_p\Phi \cdot w)$$

for all $v, w \in T_pM$ and all $p \in M$.

Proof. Let $p \in M$ and extend $v, w \in T_pM$ to smooth vector fields V and W. Since isometries respect Levi-Civita connections, we have $\Phi_*\overline{\nabla}_V W = \overline{\nabla}_{\Phi_*V}\Phi_*W$. Since Φ is a pair isometry, we have $T_p\Phi(T_pM) \subset T_{\Phi(p)}N$ and $T_p\Phi(T_pM^\perp) \subset (T_{\Phi(p)}N)^\perp$. This means that $\Phi_* : \mathfrak{X}(\overline{M})\big|_M \to \mathfrak{X}(\overline{N})\big|_N$ preserves normal and tangential components $\Phi_*(\mathfrak{X}(M)) \subset \mathfrak{X}(N)$ and $\Phi_*(\mathfrak{X}(M)^\perp) \subset \mathfrak{X}(N)^\perp$. We have

$$\begin{aligned}
T_p\Phi \cdot II(v,w) &= \Phi_* II(V,W)(\Phi(p)) = \Phi_* \left(\overline{\nabla}_V W\right)^\perp (\Phi(p)) \\
&= \left(\Phi_* \overline{\nabla}_V W\right)^\perp (\Phi(p)) = \left(\overline{\nabla}_{\Phi_*V} \Phi_* W\right)^\perp (\Phi(p)) \\
&= II(\Phi_*V, \Phi_*W)(\Phi(p)) = II(\Phi_*V, \Phi_*W)(\Phi(p)) \\
&= II(T_p\Phi \cdot v, T_p\Phi \cdot w).
\end{aligned}$$
$\square$

The following exercise gives a simple but conceptually important example.

Exercise 13.41. Let M be the 2-dimensional strip $\{(x,y,0) : -\pi < x < \pi\}$ considered as a submanifold of $\mathbb{R}^3$. Let N be the subset of $\mathbb{R}^3$ given by $\{(x,y,\sqrt{1-x^2}) : -1 < x < 1\}$. Show that M is isometric to N. Show that there is no pair isometry $(\mathbb{R}^3, M) \to (\mathbb{R}^3, N)$.

Definition 13.42. Let M be a semi-Riemannian submanifold of a semi-Riemannian manifold $\overline{M}$. Then for any $V, W, Z \in \mathfrak{X}(M)$ define $(\nabla_V II)$ by

$$(\nabla_V II)(W,Z) := \nabla_V^\perp (II(W,Z)) - II(\nabla_V X, Y) - II(X, \nabla_V Y).$$

Theorem 13.43. *With M, $\overline{M}$, and $V, W, Z \in \mathfrak{X}(M)$ as in the previous definition we have the following identity:*

$$\left(\overline{R}_{VW}Z\right)^\perp = (\nabla_V II)(W,Z) - (\nabla_W II)(V,Z) \qquad \text{(Codazzi equation)}.$$

Proof. Since both sides are tensorial, we may assume that $[V,W] = 0$. Then $\left(\bar{R}_{VW}Z\right)^\perp = \left(\overline{\nabla}_V\overline{\nabla}_W Z\right)^\perp - \left(\overline{\nabla}_W\overline{\nabla}_V Z\right)^\perp$. We have

$$\left(\overline{\nabla}_V\overline{\nabla}_W Z\right)^\perp = \left(\overline{\nabla}_V\left(\nabla_W Z\right)\right)^\perp - \left(\overline{\nabla}_V\left(II(W,Z)\right)\right)^\perp$$
$$= II(V, \nabla_W Z) - \left(\overline{\nabla}_V\left(II(W,Z)\right)\right)^\perp.$$

Now recall the definition of $\left(\nabla_V II\right)(W,Z)$ and find that

$$\left(\overline{\nabla}_V\overline{\nabla}_W Z\right)^\perp = II(V, \nabla_W Z) + \left(\nabla_V II\right)(W,Z) + II(\nabla_V W, Z) + II(W, \nabla_V Z).$$

Now compute $\left(\overline{\nabla}_V\overline{\nabla}_W Z\right)^\perp - \left(\overline{\nabla}_W\overline{\nabla}_V Z\right)^\perp$ and use the fact that $\nabla_V W - \nabla_W V = [V,W] = 0$. $\square$

The Gauss equation and the Codazzi equation belong together. If we have an isometric embedding $f : N \to \overline{M}$, then the Gauss and Codazzi equations on $f(N) \subset \overline{M}$ pull back to equations on N and the resulting equations are still called the Gauss and Codazzi equations. Obviously, these two equations simplify if the ambient manifold $\overline{M}$ is a Euclidean space. For a hypersurface existence theorem featuring these equations as integrability conditions see [**Pe**].

13.3.1. Semi-Riemannian hypersurfaces. A semi-Riemannian submanifold of codimension one is called a **semi-Riemannian hypersurface**. Let M be a semi-Riemannian hypersurface in $\overline{M}$. By definition, each tangent space $T_p M$ is a nondegenerate subspace of $T_p \overline{M}$. The complementary spaces $(T_p M)^\perp$ are easily seen to be nondegenerate, and $\mathrm{ind}\left(T_p M^\perp\right)$ is constant on M since we assume that $\mathrm{ind}\left(T_p M\right)$ is constant on M. The number $\mathrm{ind}\left(T_p M^\perp\right)$ called the **co-index** of M.

Exercise 13.44. Show that the co-index of a semi-Riemannian hypersurface must be either 0 or 1.

Definition 13.45. The **sign** ε of a hypersurface M is defined to be $+1$ if the co-index of M is 0 and is defined to be -1 if the co-index is 1. We denote it by $\mathrm{sgn}\, M$.

Notice that if $\varepsilon = 1$, then $\mathrm{ind}(M) = \mathrm{ind}(\overline{M})$, while if $\varepsilon = -1$, then $\mathrm{ind}(M) = \mathrm{ind}(\overline{M}) - 1$.

Proposition 13.46. *Let $f \in C^\infty(\overline{M})$ and $M := f^{-1}(c)$ for some $c \in \mathbb{R}$. Suppose that $M \neq \emptyset$ and that $\mathrm{grad}\, f \neq 0$ on M. Then M is a semi-Riemannian submanifold if and only if either $\langle \mathrm{grad}\, f, \mathrm{grad}\, f\rangle > 0$ on M, or $\langle \mathrm{grad}\, f, \mathrm{grad}\, f\rangle < 0$ on M. The sign of M is the sign of $\langle \mathrm{grad}\, f, \mathrm{grad}\, f\rangle$, and $\mathrm{grad}\, f / \|\mathrm{grad}\, f\|$ restricts to a unit normal field along M.*

13.4. Geodesics

Proof. The relation $\langle \operatorname{grad} f, \operatorname{grad} f \rangle \neq 0$ on M ensures that $df \neq 0$ on M, and it follows that M is a regular submanifold of codimension one. Now if $v \in TM$, then

$$\langle \operatorname{grad} f, v \rangle = df(v) = v(f)$$
$$= v(f|_M) = 0,$$

so $\operatorname{grad} f$ is normal to M. Thus for any $p \in M$ the space $(T_pM)^\perp$ is nondegenerate, and so the orthogonal complement T_pM is also nondegenerate. The rest is clear. $\square$

We now consider certain exemplary hypersurfaces in $\mathbb{R}^{n+1}_\nu$. Let $q : \mathbb{R}^{n+1}_\nu \to \mathbb{R}$ be the quadratic form defined by

$$q(x) := \langle x, x \rangle = -\sum_{i=1}^{\nu}(x^i)^2 + \sum_{i=\nu+1}^{n+1}(x^i)^2$$
$$= \sum_{i=1}^{n} \varepsilon_i (x^i)^2,$$

where the reader will recall that $\varepsilon_i = -1$ or $\varepsilon_i = 1$ as $1 \leq i \leq \nu$ or $1 + \nu \leq i \leq n + 1$. Hypersurfaces in $\mathbb{R}^{n+1}_\nu$ defined by $Q(n, r, \varepsilon) := \{x \in \mathbb{R}^{n+1}_\nu : q(x) = \varepsilon r^2\}$, where $\varepsilon = -1$ or $\varepsilon = 1$, are called **hyperquadrics**.

Exercise 13.47. Let $Q(n, r, \varepsilon)$ be a hyperquadric as defined above. Let $P := \sum x^i \partial_i$ be the position vector field in $\mathbb{R}^{n+1}_\nu$. Show that the restriction of P/r to $Q(n, r, \varepsilon)$ is a unit normal field along $Q(n, r, \varepsilon)$.

Exercise 13.48. Show that a hyperquadric $Q(n, r, \varepsilon)$ as defined above is a semi-Riemannian hypersurface with sign ε.

13.4. Geodesics

In this section, I will denote a nonempty interval assumed to be open unless otherwise indicated by the context. Usually, it would be enough to assume that I has nonempty interior. We also allow I to be infinite or "half-infinite". Let (M, g) be a semi-Riemannian manifold. Suppose that $\gamma : I \to M$ is a smooth curve that is **self-parallel** in the sense that

$$\nabla_{\partial_t} \dot\gamma = 0$$

along γ. We call γ a **geodesic**. To be precise, one should distinguish various cases as follows: If $\gamma : [a, b] \to M$ is a curve which is the restriction of a geodesic defined on an open interval containing $[a, b]$, then we call γ a (parametrized) **closed geodesic segment** or just a geodesic for short. If $\gamma : [a, \infty) \to M$ (resp. $\gamma : (-\infty, a] \to M$) is the restriction of a geodesic then we call γ a positive (resp. negative) **geodesic ray**.

If the domain of a geodesic is $\mathbb{R}$, then we call γ a **complete geodesic**. If M is an n-manifold and the image of a geodesic γ is contained in the domain of some chart with coordinate functions $x^1, \ldots, x^n$, then the condition for γ to be a geodesic is

$$\text{(13.7)} \quad \frac{d^2 x^i \circ \gamma}{dt^2}(t) + \sum \Gamma^i_{jk}(\gamma(t)) \frac{dx^j \circ \gamma}{dt}(t) \frac{dx^k \circ \gamma}{dt}(t) = 0$$

for all $t \in I$ and $1 \leq i \leq n$. This follows from formula (12.7) and this is a system of n second order equations often abbreviated to $\frac{d^2 x^i}{dt^2} + \sum \Gamma^i_{jk} \frac{dx^j}{dt} \frac{dx^k}{dt} = 0$, $1 \leq i \leq n$. These are the **local geodesic equations**. Now consider a smooth curve γ whose image is not necessarily contained in the domain of a chart. For every $t_0 \in I$, there is an $\epsilon > 0$ such that $\gamma|_{(t_0 - \epsilon, t_0 + \epsilon)}$ is contained in the domain of a chart, and thus it is not hard to see that γ is a geodesic if and only if each such restriction satisfies the corresponding local geodesic equations for each chart which meets the image of γ. We can convert the local geodesic equations (13.7) into a system of $2n$ *first order* equations by the usual reduction of order trick. We let v denote a new dependent variable and then we get

$$\frac{dx^i}{dt} = v^i, \, 1 \leq i \leq n,$$

$$\frac{dv^i}{dt} + \sum_{i,j} \Gamma^i_{jk} v^j v^k = 0, \, 1 \leq i \leq n.$$

We can think of x^i and v^i as coordinates on TM. Once we do this, we recognize that the first order system above is the local expression of the equations for the integral curves of a vector field on TM.

Exercise 13.49. Show that there is a vector field $G \in \mathfrak{X}(TM)$ such that α is an integral curve of G if and only if $\gamma := \pi_{TM} \circ \alpha$ is a geodesic. Show that the local expression for G is

$$\sum_i v^i \frac{\partial}{\partial x^i} + \sum_i \sum_{j,k} \Gamma^i_{jk} v^j v^k \frac{\partial}{\partial v^i}.$$

The vector field G from this exercise is an example of a spray (see Problems 4–8 from Chapter 12). The flow of G in the manifold TM is called the **geodesic flow**.

Lemma 13.50. *For each $v \in T_p M$, there is an open interval I containing 0 and a unique geodesic $\gamma : I \to M$, such that $\dot{\gamma}(0) = v$ (and hence $\gamma(0) = p$).*

Proof. This follows from standard existence and uniqueness results for differential equations. One may also deduce this result from the facts about flows since, as the exercise above shows, geodesics are projections of integral curves of the vector field G. The reader who did not do the problems

13.4. Geodesics

on sprays in Chapter 12 would do well to look at those problems at this time. □

Lemma 13.51. *Let γ_1 and γ_2 be geodesics $I \to M$. If $\dot{\gamma}_1(t_0) = \dot{\gamma}_2(t_0)$ for some $t_0 \in I$, then $\gamma_1 = \gamma_2$.*

Proof. If not, there must be $t' \in I$ such that $\gamma_1(t') \neq \gamma_2(t')$. Let us assume that $t' > t_0$ since the proof of the other case is similar. The set $A = \{t \in I : t > t_0 \text{ and } \gamma_1(t) \neq \gamma_2(t)\}$ has an infimum $b = \inf A$. Note that $b \geq t_0$.

Claim: $\dot{\gamma}_1(b) = \dot{\gamma}_2(b)$. Indeed, if $b = t_0$, there is nothing to prove. If $b > t_0$, then $\dot{\gamma}_1(t) = \dot{\gamma}_2(t)$ on the interval (t_0, b). By continuity $\dot{\gamma}_1(b) = \dot{\gamma}_2(b)$.

Now $t \mapsto \gamma_1(b+t)$ and $t \mapsto \gamma_2(b+t)$ are clearly geodesics with initial velocity $\dot{\gamma}_1(b) = \dot{\gamma}_2(b)$. Thus by Lemma 13.50, $\gamma_1 = \gamma_2$ for some open interval containing b. But this contradicts the definition of b as the infimum of A. □

A geodesic $\gamma : I \to M$ is called **maximal** if there is no other geodesic with open interval domain J strictly containing I that agrees with γ on I.

Theorem 13.52. *For any $v \in TM$, there is a unique **maximal geodesic** γ_v with $\dot{\gamma}_v(0) = v$.*

Proof. Take the class $\mathcal{G}_v$ of all geodesics with initial velocity v. This is not empty by Lemma 13.50. If $\alpha, \beta \in \mathcal{G}_v$ and the respective domains I_α and I_β have nonempty intersection, then α and β agree on this intersection by Lemma 13.51. From this we see that the geodesics in $\mathcal{G}_v$ fit together to form a manifestly maximal geodesic with domain $I = \bigcup_{\gamma \in \mathcal{G}_v} I_\gamma$. Obviously this geodesic has initial velocity v. □

Definition 13.53. If the domain of every maximal geodesic emanating from a point $p \in T_pM$ is all of $\mathbb{R}$, then we say that M is **geodesically complete at** p. A semi-Riemannian manifold is said to be **geodesically complete** if and only if it is geodesically complete at each of its points.

Exercise 13.54. Let $\mathbb{R}^n_\nu$ be the semi-Euclidean space of index ν. Show that all geodesics are of the form $t \mapsto x_0 + tw$ for $w \in \mathbb{R}^n_\nu$.

Definition 13.55. A continuous curve $\gamma : [a, b] \to M$ is called a **broken geodesic segment** if it is a piecewise smooth curve whose smooth segments are geodesic segments. If t_* is a point of $[a, b]$ where γ is not smooth, we call $\gamma(t_*)$ a **break point**. (A smooth geodesic segment is considered a special case.)

Exercise 13.56. Prove that a semi-Riemannian manifold is connected if and only if every pair of its points can be joined by a broken geodesic $\gamma : [a, b] \to M$.

Exercise 13.57. Show that if γ is a geodesic, then a reparametrization $c := \gamma \circ f$ is a geodesic if and only if $f(t) := at + b$ for some $a, b \in \mathbb{R}$ and $a \neq 0$. Show that if $\dot{\gamma}$ is never null, then we may choose a, b so that the geodesic is unit speed and hence parametrized by arc length.

The existence of geodesics passing through a point $p \in M$ at parameter value zero with any specified velocity allows us to define a very important map. Let $\widetilde{\mathcal{D}}_p$ denote the set of all $v \in T_p M$ such that the geodesic γ_v is defined at least on the interval $[0, 1]$. The **exponential map**, $\exp_p : \widetilde{\mathcal{D}}_p \to M$, is defined by

$$\exp_p v := \gamma_v(1).$$

Lemma 13.58. *If γ_v is the maximal geodesic with $\dot{\gamma}_v(0) = v \in T_p M$, then for any $c, t \in \mathbb{R}$, we have that $\gamma_{cv}(t)$ is defined if and only if $\gamma_v(ct)$ is defined. When either side is defined, we have*

$$\gamma_{cv}(t) = \gamma_v(ct).$$

Proof. Let $J_{v,c}$ be the maximal interval for which $\gamma_v(ct)$ is defined for all $t \in J_{v,c}$. Certainly $0 \in J$. Use the chain rule for covariant derivatives or calculate locally to see that $t \mapsto \gamma_v(ct)$ is a geodesic with initial velocity cv. But then by uniqueness and the maximality of γ_{cv}, the interval $J_{v,c}$ must be contained in the domain of γ_{cv} and for $t \in J_{v,c}$ we must have

$$\gamma_{cv}(t) = \gamma_v(ct).$$

In other words, if the right hand side is defined, then so is the left and we have equality. Now let $u = cv$, $s = ct$ and $b = 1/c$. Then we just as well have that

$$\gamma_{bu}(s) = \gamma_u(bs),$$

where if the right hand side is defined, then so is the left. But this is just

$$\gamma_v(ct) = \gamma_{cv}(t).$$

So left and right have reversed and we conclude that if either side is defined, then so is the other. $\square$

Corollary 13.59. *If γ_v is the maximal geodesic with $\dot{\gamma}_v(0) = v \in T_p M$, then*

(i) *t is in the domain of γ_v if and only if tv is in the domain of $\exp_p$;*

(ii) *$\gamma_v(t) = \exp_p(tv)$ for all t in the domain of γ_v.*

Proof. Suppose that tv is in the domain of $\exp_p$. Then $\gamma_{tv}(1)$ is defined. But $\gamma_{tv}(1) = \gamma_v(t)$ and t is in the domain of γ_v by the previous lemma. The converse is proved similarly and so we obtain $\exp_p(tv) = \gamma_{tv}(1) = \gamma_v(t)$. $\square$

13.4. Geodesics

Now we have a very convenient situation. The maximal geodesic through p with initial velocity v can always be written in the form $t \mapsto \exp_p tv$. Straight lines through $0_p \in T_pM$ are mapped by $\exp_p$ onto geodesics which we sometimes refer to as **radial geodesics** through p. Similarly, we have radial geodesic segments and radial geodesic rays emanating from p. The result of the following exercise is a fundamental observation.

Exercise 13.60. Show that $\langle \dot\gamma_v(t), \dot\gamma_v(t)\rangle = \langle v, v \rangle$ for all t in the domain of γ_v.

The exponential map has many uses. For example, it is used in comparing semi-Riemannian manifolds with each other. Also, it provides special coordinate charts. The basic theorem is the following:

Theorem 13.61. *Let (M, g) be a semi-Riemannian manifold and $p \in M$. There exists an open neighborhood $\widetilde{U}_p \subset \widetilde{\mathcal{D}}_p$ containing 0_p such that $\exp_p|_{\widetilde{U}_p}$ is a diffeomorphism onto its image U_p.*

Proof. The tangent space T_pM is a vector space, which is isomorphic to $\mathbb{R}^n$ and so has a standard differentiable structure. Using the results about smooth dependence on initial conditions for differential equations, we can easily see that $\exp_p$ is well-defined and smooth in some neighborhood of $0_p \in T_pM$. The main point is that the tangent map $T\exp_p : T_{0_p}(T_pM) \to T_pM$ is an isomorphism and so the inverse mapping theorem gives the result. To see that $T\exp_p$ is an isomorphism, let $v_{0_p} \in T_{0_p}(T_pM)$ be the velocity of the curve $t \mapsto tv$ in T_pM. Then, unraveling definitions, we have $T\exp_p v_{0_p} = \frac{d}{dt}\big|_0 \exp_p tv = v$. Thus $T\exp_p$ is just the canonical map $v_{0_p} \mapsto v$. $\square$

Definition 13.62. A subset C of a vector space V that contains 0 is called **star-shaped** about 0 if whenever $v \in C$, $tv \in C$ for all $t \in [0, 1]$.

Definition 13.63. If $\widetilde{U} \subset \widetilde{\mathcal{D}}_p$ is a star-shaped open set about 0_p in T_pM such that $\exp_p|_{\widetilde{U}}$ is a diffeomorphism as in the theorem above, then the image $\exp_p(\widetilde{U}) = U$ is called a **normal neighborhood of** p. In this case, U is also referred to as **star-shaped**.

Theorem 13.64. *If $U \subset M$ is a normal neighborhood about p with corresponding preimage $\widetilde{U} \subset T_pM$, then for every point $q \in U$ there is a unique geodesic $\gamma : [0, 1] \to U \subset M$ such that $\gamma(0) = p$, $\gamma(1) = q$, $\dot\gamma(0) \in \widetilde{U}$ and $\exp_p \dot\gamma(0) = q$. (Note that uniqueness here means unique among geodesics with image in U.)*

Proof. The preimage $\widetilde{U}$ corresponds diffeomorphically to U under $\exp_p$. Let $v = \exp_p|_{\widetilde{U}}^{-1}(q)$ so that $v \in \widetilde{U}$. By assumption $\widetilde{U}$ is star-shaped and so the map $\rho : [0, 1] \to T_pM$ given by $t \mapsto tv$ has image in $\widetilde{U}$. But then, the geodesic segment $\gamma : t \mapsto \exp_p tv$, $t \in [0,1]$ has its image inside U. Clearly, $\gamma(0) = p$ and $\gamma(1) = q$. Since $\dot\rho = v$, we get

$$\dot\gamma(0) = T\exp_p \dot\rho(0) = T\exp_p v = v$$

under the usual identifications in T_pM.

Now assume that $\gamma_1 : [0,1] \to U \subset M$ is some geodesic with $\gamma_1(0) = p$ and $\gamma_1(1) = q$. If $\dot\gamma_1(0) = w$, then $\gamma_1(t) = \exp_p tw$.

Claim: The ray $\rho_1 : t \mapsto tw$ ($t \in [0,1]$) stays inside $\widetilde{U}$. If not, then the set $A = \{t : tw \notin \widetilde{U}\}$ is nonempty. Let $t_* = \inf A$ and consider the set $\widetilde{C} := \{tw : t \in (0, t_*)\}$. Then $\widetilde{C} \subset \widetilde{U}$ and $\widetilde{U} \setminus \widetilde{C}$ is contractible (check this). But its image $\exp_p|_{\widetilde{U}}(\widetilde{U} \setminus \widetilde{C})$ is $U \setminus C$, where C is the image of $(0, t_*)$ under γ_1. Since $U \setminus C$ is certainly not contractible, we have contradiction. Thus the claim is true, and in particular $w = \rho_1(1) \in \widetilde{U}$. Therefore, both w and v are in $\widetilde{U}$. On the other hand,

$$\exp_p w = \gamma_1(1) = q = \exp_p v.$$

Thus, since $\exp_p|_{\widetilde{U}}$ is a diffeomorphism and hence injective, we conclude that $w = v$. By the basic uniqueness theorem for geodesics, the segments γ and γ_1 are equal and both given by $t \mapsto \exp_p tv$. □

Let (M, g) be a semi-Riemannian manifold of dimension n. Let $p_0 \in M$ and pick any orthonormal basis $(e_1, \ldots, e_n)$ for the semi-Euclidean scalar product space $(T_{p_0}M, \langle \cdot, \cdot \rangle_{p_0})$. This basis induces an isometry $I : \mathbb{R}^n_v \to T_{p_0}M$ by $(x^i) \mapsto \sum x^i e_i$. If U is a normal neighborhood centered at $p_0 \in M$, then $\mathrm{x}_{\mathrm{norm}} := I \circ \exp_{p_0}|_{\widetilde{U}}^{-1} : U \to \mathbb{R}^n_v = \mathbb{R}^n$ is a coordinate chart with domain U. These coordinates are referred to as **normal coordinates** centered at p_0. Normal coordinates have some very nice properties:

Theorem 13.65. *If* $\mathrm{x}_{\mathrm{norm}} = (x^1, \ldots, x^n)$ *are normal coordinates defined on U and centered at p_0, then*

$$g_{ij}(p_0) = \left\langle \frac{\partial}{\partial x^i}, \frac{\partial}{\partial x^j} \right\rangle_{p_0} = \epsilon_i \delta_{ij} \text{ for all } i,j,$$

$$\Gamma^i_{jk}(p_0) = 0 \text{ for all } i,j,k.$$

(When using normal coordinates, it should not be forgotten that the Γ^i_{jk} are only guaranteed to vanish at p_0.)

Proof. Let $v \in T_{p_0}M$ and let $\{e^i\}$ be the basis of $T^*_{p_0}M$ dual to $\{e_i\}$. We write $v = \sum a^i e_i$. We have that $e^i \circ \exp_{p_0}|_{\widetilde{U}}^{-1} = x^i$. Now $\gamma_v(t) = \exp_{p_0} tv$ and so

$$x^i(\gamma_v(t)) = e^i(tv) = te^i(v) = ta^i.$$

13.4. Geodesics

So we see that $v = \dot{\gamma}_v(0) = \sum a^i \frac{\partial}{\partial x^i}\big|_{p_0}$. In particular, if $a^i = \delta^i_j$, then $e_i = \frac{\partial}{\partial x^i}\big|_{p_0}$ and $\langle \frac{\partial}{\partial x^i}, \frac{\partial}{\partial x^j}\rangle_{p_0} = \epsilon_i \delta_{ij}$. Since γ_v is a geodesic and $x^i(\gamma_v(t)) = ta^i$, the coordinate expression for the geodesic equations reduces to

$$\sum \Gamma^i_{jk}(\gamma_v(t)) a^j a^k = 0$$

for all i. In particular, this holds at $p_0 = \gamma_v(0)$. But v is arbitrary, and hence the n-tuple (a^i) is arbitrary. Thus the quadratic form defined on $\mathbb{R}^n$ by $Q^i(u) = \sum \Gamma^i_{jk}(p_0) u^j u^k$ is identically zero, and by polarization, the bilinear form $Q^i : (u,v) \mapsto \sum \Gamma^i_{jk}(p_0) u^j v^k$ is identically zero. Of course this means that $\Gamma^i_{jk}(p_0) = 0$ for all j,k and arbitrary i. □

Notation 13.66. From now on, whenever we write $\exp_p^{-1}$, we must have in mind an open set $\widetilde{U}$ and an open set U such that $\exp_p|_{\widetilde{U}} : \widetilde{U} \to U$ is a diffeomorphism. Thus, $\exp_p^{-1}$ is an abbreviation for $\exp_p|_{\widetilde{U}}^{-1}$. Usually, $\widetilde{U}$ will be star-shaped and thereby U is a normal neighborhood of the point p.

Definition 13.67. Let U be a normal neighborhood of a point p_0 in a semi-Riemannian manifold M. The **radius function** $r : U \to \mathbb{R}$ is defined by

$$r_{p_0}(p) := \left\|\exp_{p_0}^{-1}(p)\right\|$$

for $p \in U$. We often write simply r if the central point p_0 is understood.

If $(x^1, \ldots, x^n)$ are normal coordinates defined on U and centered at p_0, then the radius function is given by

$$r = r_{p_0} = \left| -\sum_{i=1}^{\nu} (x^i)^2 + \sum_{i=\nu+1}^{n} (x^i)^2 \right|^{1/2}.$$

The radius function is smooth except on the set where it is zero. This zero set is called the **local nullcone** and is the image of the intersection of the nullcone in $T_{p_0}M$ with $\widetilde{U} = \exp_{p_0}^{-1}(U)$. In the Riemannian case, where the metric is definite, the radius function r is smooth except at the center point p_0. Note that in this case, r^2 is smooth even at p_0.

Now suppose that $\gamma : [0,1] \to U \subset M$ is the geodesic with $\gamma(0) = p_0$, $\gamma(1) = p$ and $\dot{\gamma}(0) = v$. Then we have the useful and often used fact that

$$L(\gamma) = r(p).$$

To see this, first note that $v = \exp_{p_0}^{-1}(p)$. Then, since $\|\dot{\gamma}\|$ is constant (by Exercise 13.60), we have

$$L(\gamma) = \int_0^1 \|\dot{\gamma}\| \, dt = \int_0^1 \|v\| \, dt = \|v\| = r(p).$$

More generally, if $r > 0$ is such that $\gamma_v : t \mapsto \exp_p tv$ is defined for $0 \le t \le r$, then

$$\int_0^r |\langle \gamma_v(t), \gamma_v(t) \rangle|^{1/2} \, dt = r \, |\langle v, v \rangle|^{1/2} = r \, \|v\|.$$

In particular, if v is a unit vector, then the length of the geodesic $\gamma_v|_{[0,r]}$ is equal to r.

Let $\widetilde{\mathcal{D}} := \bigcup_p \widetilde{\mathcal{D}}_p$. We can gather the maps $\exp_p : \widetilde{\mathcal{D}}_p \subset T_pM \to M$ together to get a map $\exp : \widetilde{\mathcal{D}} \to M$ defined by $\exp(v) := \exp_{\pi(v)}(v)$. The set $\widetilde{\mathcal{D}}$ is the set of $v \in TM$ such that the geodesic γ_v is defined at least on $[0,1]$.

Proposition 13.68. *$\widetilde{\mathcal{D}}$ is open and for each $p \in M$, $\widetilde{\mathcal{D}}_p$ is open and starshaped. Thus $\mathcal{D}_p := \exp_p(\widetilde{\mathcal{D}}_p)$ is a (maximal) normal neighborhood of p.*

Proof. Let $W \subset \mathbb{R} \times TM$ be the domain of the geodesic flow $(s, v) \mapsto \dot{\gamma}_v(s)$. This is the flow of a vector field on TM, and so W is open. W is also the domain of the map $(s, v) \mapsto \pi \circ \dot{\gamma}_v(s) = \gamma_v(s)$. The map $(1, v) \mapsto v$ is a diffeomorphism $\{1\} \times TM \longrightarrow TM$. Under this diffeomorphism, $\widetilde{\mathcal{D}}$ corresponds to the set $W \cap (\{1\} \times TM)$, and so it must be open in TM. It also follows that $\widetilde{\mathcal{D}}_p = \widetilde{\mathcal{D}} \cap T_pM$ is open in T_pM and $\mathcal{D}_p$ is open in M.

To see that $\widetilde{\mathcal{D}}_p$ is star-shaped, let $v \in \widetilde{\mathcal{D}}_p$. Then γ_v is defined for all $t \in [0,1]$. On the other hand, $\gamma_{tv}(1)$ is defined and equal to $\gamma_v(t)$ for all $t \in [0,1]$ by Lemma 13.58. Thus, by definition, $tv \in \widetilde{\mathcal{D}}_p$ for all $t \in [0,1]$. $\square$

Let $\Delta = \{(p,p) : p \in M\}$ be the diagonal subset of $M \times M$. Let $\mathrm{EXP} : \widetilde{\mathcal{D}} \subset TM \to M \times M$ be defined by

$$\mathrm{EXP} : v \mapsto (\pi(v), \exp_p v),$$

where $\pi : TM \to M$ is the tangent bundle projection.

Theorem 13.69. *If 0_p is the zero element of T_pM, then there is a neighborhood W of 0_p in TM such that $\mathrm{EXP}|_W$ is a diffeomorphism onto a neighborhood of $(p,p) \in \Delta \subset M \times M$.*

Proof. We first show that if $T_x \exp_p$ is nonsingular for some $x \in \widetilde{\mathcal{D}}_p \subset T_pM$, then $T_x \mathrm{EXP}$ is also nonsingular at x. So assume that $T_x \exp_p$ is nonsingular and suppose that $T_x \mathrm{EXP}(v_x) = 0$. We have $\pi = \mathrm{pr}_1 \circ \mathrm{EXP}$ and so $T\pi(v_x) = T\mathrm{pr}_1(T_x \mathrm{EXP}(v_x)) = 0$. This means that v_x is tangent to $\widetilde{\mathcal{D}}_p \subset T_pM$. But

13.4. Geodesics

the restricted map $\text{EXP}|_{\widetilde{\mathcal{D}}_p}$ is related to $\exp_p$ by trivial diffeomorphisms:

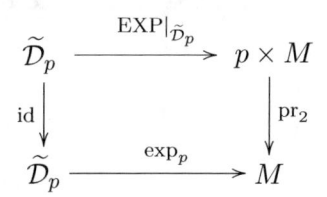

$$\begin{array}{ccc} \widetilde{\mathcal{D}}_p & \xrightarrow{\text{EXP}|_{\widetilde{\mathcal{D}}_p}} & p \times M \\ \text{id} \downarrow & & \downarrow \text{pr}_2 \\ \widetilde{\mathcal{D}}_p & \xrightarrow{\exp_p} & M \end{array}$$

Thus $T_x \exp_p(v_x) = 0$ and hence $v_x = 0$.

Since $T_{0_p} \exp_p$ is nonsingular at each point 0_p of the zero section, we see that the same is true for $T_{0_p} \text{EXP}$, and the result follows from the inverse mapping theorem. $\square$

Definition 13.70. An open subset U of a semi-Riemannian manifold will be said to be **totally star-shaped** if it is a normal neighborhood of each of its points.

Notice that U being totally star-shaped according to the above definition implies that for any two points $p, q \in U$, there is a geodesic segment $\gamma : [0,1] \to U$ such that $\gamma(0) = p$ and $\gamma(1) = q$, and this is the *unique* such geodesic with image in U. (One may always make an affine change of parameter, but then we have a different interval as the domain.) Thinking about the sphere makes it clear that even if U is totally star-shaped, there may be geodesic segments connecting p and q whose images do not lie in U.

Theorem 13.71. *Every $p \in M$ has a totally star-shaped neighborhood.*

Proof. Let $p \in M$ and choose a neighborhood W of 0_p in TM such that $\text{EXP}|_W$ is a diffeomorphism onto a neighborhood of $(p,p) \in M \times M$. By a simple continuity argument we may assume that $\text{EXP}|_W(W)$ is of the form $U(\delta) \times U(\delta)$ for $U(\delta) := \{q : \sum_{i=1}^n (x^i(q))^2 < \delta\}$ and $\mathbf{x} = (x^1, \ldots, x^n)$ is a normal coordinate system. Now consider the tensor b on $U(\delta)$ whose components with respect to $\mathbf{x}$ are $b_{ij} = \delta_{ij} - \sum_k \Gamma_{ij}^k x^k$. This is clearly symmetric and positive definite at p, and so by choosing δ smaller if necessary we may assume that this tensor is positive definite on $U(\delta)$. Let us show that $U(\delta)$ is a normal neighborhood of each of its points q. Let $W_q := W \cap T_q M$. We know that $\text{EXP}|_{W_q}$ is a diffeomorphism onto $\{q\} \times U(\delta)$, and it is easy to see that this means that $\exp_q|_{W_q}$ is a diffeomorphism onto $U(\delta)$. We now show that W_q is star-shaped about 0_q. Let $q' \in U(\delta)$, $q' \neq q$, and $v = \text{EXP}|_{W_q}^{-1}(q, q')$. This means that $\gamma_v : [0,1] \to M$ is a geodesic from q to q'. If $\gamma_v([0,1]) \subset U(\delta)$, then $tv \in W_q$ for all $t \in [0,1]$ and so we could conclude that W_q is star-shaped. Let us assume that $\gamma_v([0,1])$ is not contained in $U(\delta)$ and work for a contradiction.

If in fact γ_v leaves $U(\delta)$, then the function $f: t \mapsto \sum_{i=1}^{n}(x^i(\gamma_v(t)))^2$ has a maximum at some $t_0 \in (0, 1)$. Thus the second derivative of f cannot be positive at t_0. We have

$$\frac{d^2}{dt^2} f = 2 \sum_{i=1}^{n} \left(\frac{d\left(x^i \circ \gamma_v\right)}{dt} + x^i \circ \gamma_v \frac{d^2\left(x^i \circ \gamma_v\right)}{dt^2} \right).$$

But γ_v is a geodesic, and so using the geodesic equations we get

$$\frac{d^2}{dt^2} f = 2 \sum_{i,j} \left(\delta_{ij} - \sum_k \Gamma_{ij}^k x^k \right) \frac{d\left(x^i \circ \gamma_v\right)}{dt} \frac{d\left(x^j \circ \gamma_v\right)}{dt}.$$

Plugging in t_0 we get

$$\frac{d^2}{dt^2} f(t_0) = 2b(\dot{\gamma}_v(t_0), \dot{\gamma}_v(t_0)) > 0,$$

which contradicts f having a maximum at t_0. $\square$

Note: It follows from the proof that given $p \in M$, there is a $\delta > 0$ such that $\exp(\{v_p \in T_p M : \|v_p\| < \varepsilon\})$ is totally star-shaped for all $\varepsilon < \delta$.

Warning: Clearly a totally star-shaped open set is, in some sense, "convex". Indeed, some authors define convexity in this manner. However, notice that on the circle, open intervals are totally star-shaped while the intersection of two such intervals need not even be connected. For proper Riemannian manifolds, there is another definition of convexity that does not suffer from this defect (see Problem 2). Actually, there are several notions of convexity on manifolds and the terminology does not seem to be quite standardized.

Theorem 13.72 (Gauss lemma). *Let $p \in M$, $x \in T_p M$, with $x \neq 0_p$ in the domain of $\exp_p$. Choose $v_x, w_x \in T_x(T_p M)$, where v_x, w_x correspond to $v, w \in T_p M$ under the canonical isomorphism between $T_x(T_p M)$ and $T_p M$. If v_x is radial, i.e. if v is a scalar multiple of x, then*

$$\langle T_x \exp_p v_x, T_x \exp_p w_x \rangle = \langle v_x, w_x \rangle.$$

Proof. Clearly we may suppose that $x = v$. For small $\epsilon > 0$, we define $\widetilde{h} : [0, 1 + \epsilon) \times (-\epsilon, \epsilon) \to T_p M$ by

$$\widetilde{h}(t, s) = t(v + sw).$$

If we take ϵ sufficiently small, then on the same domain we may define h by $h(t, s) := \exp_p(t(v + sw))$. Then

$$\frac{\partial h}{\partial t}(t, s) := T_{(t,s)} h \cdot \frac{\partial}{\partial t},$$

$$\frac{\partial h}{\partial s}(t, s) := T_{(t,s)} h \cdot \frac{\partial}{\partial s}.$$

13.4. Geodesics

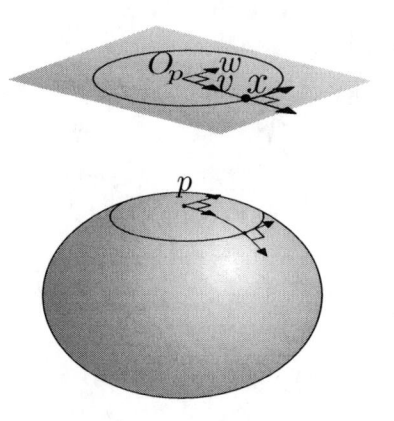

Figure 13.2. Gauss lemma

We have that $\frac{\partial \tilde{h}}{\partial t}(1,0) = v_v$ and $\frac{\partial \tilde{h}}{\partial s}(1,0) = w_v$, so that

$$\frac{\partial h}{\partial t}(1,0) = T_v \exp_p v_v,$$

$$\frac{\partial h}{\partial s}(1,0) = T_v \exp_p w_v.$$

We wish to show that $\langle \frac{\partial h}{\partial t}, \frac{\partial h}{\partial s} \rangle (1,0) = \langle v_v, w_v \rangle = \langle v, w \rangle$. Since the curve $t \mapsto \exp_p(t(v+sw))$ is a geodesic with initial velocity $v+sw$, we have

$$\left\langle \frac{\partial h}{\partial t}, \frac{\partial h}{\partial t} \right\rangle (t,s) = \left\langle \frac{\partial h}{\partial t}, \frac{\partial h}{\partial t} \right\rangle (0,s) = \langle v+sw, v+sw \rangle.$$

We have

$$\frac{\partial}{\partial t} \left\langle \frac{\partial h}{\partial t}, \frac{\partial h}{\partial s} \right\rangle = \left\langle \nabla_{\frac{\partial}{\partial t}} \frac{\partial h}{\partial t}, \frac{\partial h}{\partial s} \right\rangle + \left\langle \frac{\partial h}{\partial t}, \nabla_{\frac{\partial}{\partial t}} \frac{\partial h}{\partial s} \right\rangle$$

$$= \left\langle \frac{\partial h}{\partial t}, \nabla_{\frac{\partial}{\partial t}} \frac{\partial h}{\partial s} \right\rangle \text{ (since } t \mapsto h(t,s) \text{ is a geodesic)}$$

$$= \left\langle \frac{\partial h}{\partial t}, \nabla_{\frac{\partial}{\partial s}} \frac{\partial h}{\partial t} \right\rangle \text{ (by Exercise 13.11)}$$

$$= \frac{1}{2} \frac{\partial}{\partial s} \left\langle \frac{\partial h}{\partial t}, \frac{\partial h}{\partial t} \right\rangle = \langle v, w \rangle.$$

Since $h(0,s) = p$ for all s, we have $\langle \frac{\partial h}{\partial t}, \frac{\partial h}{\partial s} \rangle (0,0) = 0$ and so $\langle \frac{\partial h}{\partial t}, \frac{\partial h}{\partial s} \rangle (t,0) = t\langle v, w \rangle$. The result follows by letting $t = 1$. $\square$

If v_x is not assumed to be radial, then the above equality does not always hold. We need to have the dimension of the manifold greater than 3 in order to see what can go wrong. Figure 13.3 shows a unit sphere in the tangent space of a Riemannian manifold and a pair of orthogonal vectors tangent to this sphere. Under the exponential map these vectors map to vectors which need not be orthogonal.

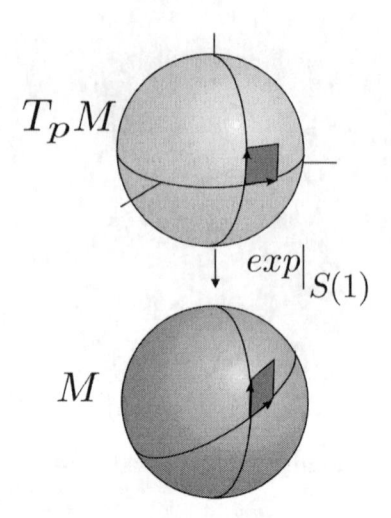

Figure 13.3. Distortions under the exponential map

We now introduce the position vector fields associated with a normal neighborhood of a point p. First, let us consider a vector space V with scalar product $\langle \cdot, \cdot \rangle$. Then V is a semi-Riemannian manifold with metric defined by $\langle v_x, w_x \rangle := \langle v, w \rangle$. Let $\overline{\nabla}$ denote the associated Levi-Civita connection. Because of the canonical isomorphisms $T_v V \cong V$, every vector field Y on V can be identified with a map $Y : V \to V$. Under this identification, $\overline{\nabla}_X Y$ is just the directional derivative of Y in the X direction. The **position vector field** on V is defined by $P : v \mapsto v_v$, and it is easy to check that $\overline{\nabla}_X P = X$ for any vector field X. Now consider the quadratic form q defined by $\mathrm{q}(v) = \langle v, v \rangle$. Unraveling the definitions we see that $\mathrm{q} = \langle P, P \rangle$. We have for any X

$$\langle \operatorname{grad} \mathrm{q}, X \rangle = X \mathrm{q} = X \langle P, P \rangle = 2 \langle \overline{\nabla}_X P, P \rangle = \langle X, 2P \rangle,$$

and we conclude that

$$\operatorname{grad} \mathrm{q} = 2P.$$

It follows that P is normal to every hyperquadric $\mathrm{q}^{-1}(c)$, $c \in \mathbb{R}$, $c \neq 0$.

Now we want a similar result in a normal neighborhood U of a point p on a semi-Riemannian manifold M. We consider $T_p M$ as a semi-Riemannian manifold in its own right. We have the position vector field

$$\widetilde{P} : v \mapsto v_v,$$

and we have the quadratic form $\widetilde{\mathrm{q}}$ defined by $\widetilde{\mathrm{q}}(v) = \langle v, v \rangle$. Let $\widetilde{U} = \exp_p^{-1}(U)$ and for each $c \neq 0$ let $\widetilde{Q}_c := \widetilde{\mathrm{q}}^{-1}(c)$ and $Q_c := \exp_p(\widetilde{U} \cap \widetilde{Q}_c)$. Corresponding to $\widetilde{\mathrm{q}}$ we have a function q defined on U by

$$\mathrm{q} := \widetilde{\mathrm{q}} \circ \exp_p^{-1},$$

so $Q_c = \mathrm{q}^{-1}(c)$.

13.4. Geodesics

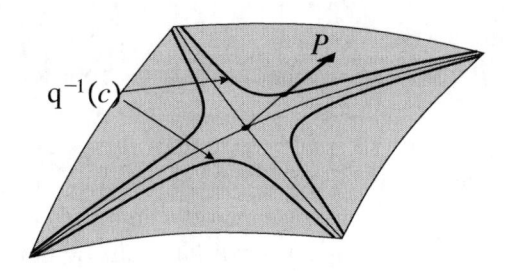

Definition 13.73. For $c \neq 0$, the sets $Q_c = \mathrm{q}^{-1}(c)$ are called **local hyperquadrics** associated to the normal neighborhood centered at p. The set $\Lambda = \mathrm{q}^{-1}(0) = \widetilde{\mathrm{q}} \circ \exp_p^{-1}(0)$ is called the **local nullcone** at p.

On the normal neighborhood U there is a unique vector field P that is $\exp_p$-related to $\widetilde{P}$. We refer to P as the **local position vector field** for the normal neighborhood at p,

$$P = T\exp_p \circ \widetilde{P} \circ \exp_p^{-1}.$$

Proposition 13.74. *Let U be a normal neighborhood of p and let q, $\widetilde{Q}_c$, Q_c, $\widetilde{P}$, and P be as above. Then*

  (i) *P is normal to each Q_c;*
  (ii) *$\langle P, P \rangle \circ \exp_p = \langle \widetilde{P}, \widetilde{P} \rangle$;*
  (iii) *$\mathrm{grad}\, \mathrm{q} = 2P$.*

Proof. (i) follows from the Gauss lemma (Lemma 13.72) and the corresponding fact that $\widetilde{P}$ is normal to each $\widetilde{Q}_c$ in T_pM. The Gauss lemma also immediately gives (ii).

(iii) For any v, let $\widetilde{v}$ be such that $T\exp_p \widetilde{v} = v$. Then

$$\begin{aligned}\langle \mathrm{grad}\, \mathrm{q}, v \rangle &= v(\mathrm{q}) = (T\exp_p \widetilde{v})\,\mathrm{q} \\ &= \widetilde{v}\,(\mathrm{q} \circ \exp_p) = \widetilde{v}(\widetilde{\mathrm{q}}) \\ &= \langle \mathrm{grad}\, \widetilde{\mathrm{q}}, \widetilde{v} \rangle = 2\langle \widetilde{P}, \widetilde{v} \rangle = 2\langle P, v \rangle,\end{aligned}$$

where we used the Gauss lemma in the last step. Since v was arbitrary, we obtain (iii). $\square$

Consider the unit sphere S^{n-1} and the map $\mathbb{R}^n \to (0, \infty) \times S^{n-1}$ given by $x \mapsto (\|x\|, x/\|x\|)$. Now put coordinates on the sphere, say $\theta^1, \ldots, \theta^{n-1}$. Composing, we obtain coordinates $(r, \theta^1, \ldots, \theta^{n-1})$ on an open subset of $\mathbb{R}^n \setminus \{0\}$, where r gives the distance to the origin and the θ directions are normal to the r direction. A standard method of choosing the angle functions $\theta^1, \ldots, \theta^{n-1}$ leads to what is sometimes called hyperspherical coordinates. If (M, g) is Riemannian, and if $(r, \theta^1, \ldots, \theta^{n-1})$ are "spherical" coordinates on $\mathbb{R}^n$ as above, then we can compose with normal coordinates centered at p to obtain coordinate functions on our normal neighborhood, which we again denote by $(r, \theta^1, \ldots, \theta^{n-1})$. These coordinates are called **geodesic spherical coordinates** or **geodesic polar coordinates**. As usual, the function r is extended to be zero at the center, and in the case of hyperspherical coordinates, the angle functions are extended to be multivalued. The resulting "coordinates" are not really proper coordinates on the normal neighborhood since they suffer from the usual defects. For example, r is not smooth at the center point where it is zero, and the angles $\theta^1, \ldots, \theta^{n-1}$ become ambiguous when $r = 0$. Thus a little care is need when using spherical coordinates. No matter how we choose the $\theta^1, \ldots, \theta^{n-1}$, the function r is the radial function introduced earlier. Whether or not angle functions are introduced, one often uses the notation $\frac{\partial}{\partial r}$ to denote the unit vector field defined as follows: If v is a unit vector in T_pM, then

$$\left.\frac{\partial}{\partial r}\right|_q := \left.\frac{d}{dt}\right|_{t=t_0} \exp_p(tv),$$

where $q = \exp_p(t_0 v)$ (and p is the center point of the normal coordinates). In fact, it is not hard to see that $\frac{\partial}{\partial r} = P/\|P\|$. One might use this last equation to define $\frac{\partial}{\partial r}$ in the case of an indefinite metric but note that $P/\|P\|$ is undefined when $\|P\| = 0$ and so is undefined on the local nullcone. We refer to $\frac{\partial}{\partial r}$ as the **unit radial vector field**. If $(r, \theta^1, \ldots, \theta^{n-1})$ are geodesic spherical coordinates centered at some point p_0 of a Riemannian manifold, then by the Gauss lemma

$$\left\langle \frac{\partial}{\partial r}, \frac{\partial}{\partial \theta^i} \right\rangle = 0 \text{ for } i = 1, 2, \ldots, n-1.$$

Exercise 13.75. Show that if a geodesic $\gamma : [a, b) \to M$ is extendable to a continuous map $\overline{\gamma} : [a, b] \to M$, then there is an $\varepsilon > 0$ such that $\gamma : [a, b) \to M$ is extendable further to a geodesic $\widetilde{\gamma} : [a, b + \varepsilon) \to M$ with $\widetilde{\gamma}|_{[a,b]} = \overline{\gamma}$.

Under certain conditions, geodesics can help us draw conclusions about maps. The following result is an example and a main ingredient in the proof of the Hadamard theorem to be given later.

Theorem 13.76. *Let $f : (M, g) \to (N, h)$ be a local isometry of semi-Riemannian manifolds with N connected. Suppose that f has the property that given any geodesic $\gamma : [0, 1] \to N$ and $p \in M$ with $f(p) = \gamma(0)$, there is a curve $\widetilde{\gamma} : [0, 1] \to M$ such that $p = \widetilde{\gamma}(0)$ and $\gamma = f \circ \widetilde{\gamma}$. Then ϕ is a semi-Riemannian covering.*

Proof. Since any two points of N can be joined by a broken geodesic, it is easy to see that the hypotheses imply that f is onto.

13.4. Geodesics

Let U be a normal neighborhood of an arbitrary point $q \in N$ and let $\widetilde{U} \subset T_q N$ be the open set such that $\exp_q(\widetilde{U}) = U$. We will show that U is evenly covered by f. Choose $p \in f^{-1}(q)$. Observe that $T_p f : T_p M \to T_q N$ is a linear isometry (the metrics on $T_p M$ and $T_q N$ are given by the scalar products $g(p)$ and $h(q)$). Thus $\widetilde{V}_p := T_p f^{-1}(\widetilde{U})$ is star-shaped about $0_p \in T_p M$. Now if $v \in \widetilde{V}_p$, then by hypothesis, the geodesic $\gamma(t) := \exp_q(t\,(T_p f\,(v)))$ has a lift to a curve $\widetilde{\gamma} : [0, 1] \to M$ with $\widetilde{\gamma}(0) = p$. But since f is a local isometry, this curve must be a geodesic. It is also easy to see that $T_p f(\widetilde{\gamma}'(0)) = \gamma'(0) = T_p f\,(v)$. It follows that $v = \widetilde{\gamma}'(0)$ and then $\exp_p(v) = \widetilde{\gamma}(1)$. Thus $\exp_p$ is defined on all of $\widetilde{V}$. In fact, it is clear that $f(\exp_p v) = \exp_{f(p)}(Tf\,(v))$ and so we see that f maps $V_p := \exp_p(\widetilde{V}_p)$ onto the set $\exp_q(\widetilde{U}) = U$. We show that V_p is a normal neighborhood of p. From $f \circ \exp_p = \exp_{f(p)} \circ Tf$ we see that $f \circ \exp_p$ is a diffeomorphism on $\widetilde{V}$. But then $\exp_p : \widetilde{V}_p \to V_p$ is bijective. Combining this with the fact that $Tf \circ T\exp_p$ is a linear isomorphism at each $v \in \widetilde{V}_p$ and the fact that Tf is a linear isomorphism, it follows that $T_v \exp_p$ is a linear isomorphism. It follows that V_p is open and $\exp_p : \widetilde{V}_p \longrightarrow V_p$ is a diffeomorphism. Composing, we obtain $f|_{V_p} = \exp_{f(p)}|_{\widetilde{U}} \circ Tf \circ \exp_p|_{V_p}^{-1}$, which is a diffeomorphism taking V_p onto U.

Now we show that if $p_i, p_j \in f^{-1}(q)$ and $p_i \neq p_j$, then the sets V_{p_i} and V_{p_j} (obtained for these points as we did for a generic p above) are disjoint. Suppose to the contrary that $m \in V_{p_i} \cap V_{p_j}$ and let $\gamma_{p_i m}$ and $\gamma_{p_j m}$ be the reverse radial geodesics from m to p_i and p_j respectively. Then $f \circ \gamma_{p_i m}$ and $f \circ \gamma_{p_j m}$ are both reversed radial geodesics from $f(x)$ to q, and so they must be equal. But then $\gamma_{p_i m}$ and $\gamma_{p_j m}$ are equal since they are both lifts of the same curve and start at the same point. It follows that $p_i = p_j$ after all. It remains to prove that $f^{-1}(U) \subset \bigcup_{p \in f^{-1}(q)} V_p$ since the reverse inclusion is obvious. Let $x \in f^{-1}(U)$ and let $\alpha : [0, 1] \to U$ be the reverse radial geodesic from $f(x)$ to the center point q. Now let γ be the lift of α starting at x and let $p = \gamma(1)$. Then $f(p) = \alpha(1) = q$, which means that $p \in f^{-1}(q)$. On the other hand, the image of γ must lie in V_p and so $x \in V_p$. $\square$

13.4.1. Geodesics on submanifolds. Let M be a semi-Riemannian submanifold of $\overline{M}$. For a smooth curve $\gamma : I \to M$, it is easy to show using Proposition 13.32 that $\nabla_{\partial_t} Y = (\overline{\nabla}_{\partial_t} Y)^{\mathrm{tan}}$ and that we have

$$\overline{\nabla}_{\partial_t} Y = \nabla_{\partial_t} Y + II(\dot{\gamma}, Y)$$

for any vector field Y along γ. If Y is a vector field in $\mathfrak{X}(\overline{M})|_M$ or in $\mathfrak{X}(\overline{M})$, then $Y \circ \gamma$ is a vector field along γ. In this case we shall still write $\overline{\nabla}_{\partial_t} Y = \nabla_{\partial_t} Y + II(\dot{\gamma}, Y)$ rather than $\overline{\nabla}_{\partial_t}(Y \circ \gamma) = \nabla_{\partial_t}(Y \circ \gamma) + II(\dot{\gamma}, Y \circ \gamma)$.

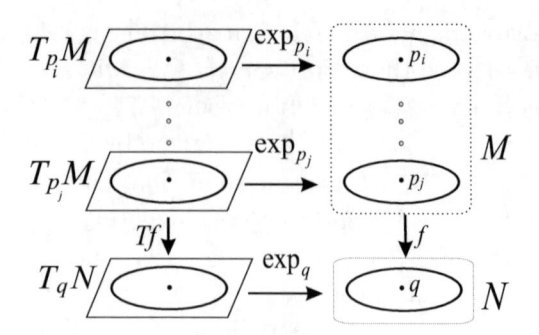

Figure 13.4. Semi-Riemannian covering

Recall that $\dot{\gamma}$ is a vector field along γ. We also have $\overline{\nabla}_{\partial_t}\dot{\gamma}$, which in this context will be called the **extrinsic acceleration** (or acceleration in $\overline{M}$). The **intrinsic acceleration** (acceleration in M) is $\nabla_{\partial_t}\dot{\gamma}$. Thus we have

$$\overline{\nabla}_{\partial_t}\dot{\gamma} = \nabla_{\partial_t}\dot{\gamma} + II(\dot{\gamma},\dot{\gamma}).$$

Since $II(\dot{\gamma},\dot{\gamma})$ is the normal part of $\overline{\nabla}_{\partial_t}\dot{\gamma}$, we immediately obtain the following:

Proposition 13.77. *If $\gamma : I \to M$ is a smooth curve where M is a semi-Riemannian submanifold of $\overline{M}$, then γ is a geodesic in M if and only if $\overline{\nabla}_{\partial_t}\dot{\gamma}$ is normal to M for every $t \in I$.*

Exercise 13.78. A constant speed parametrization of a great circle in $S^n(r)$ is a geodesic. Every geodesic in $S^n(r)$ is of this form.

Definition 13.79. A semi-Riemannian submanifold $M \subset \overline{M}$ is called **totally geodesic** if every geodesic in M is a geodesic in $\overline{M}$.

Theorem 13.80. *For a semi-Riemannian submanifold $M \subset \overline{M}$, the following conditions are equivalent:*

(i) *M is totally geodesic;*

(ii) *$II \equiv 0$;*

(iii) *For all $v \in TM$, the geodesic γ_v in $\overline{M}$ with initial velocity v is such that $\gamma_v([0,\epsilon]) \subset M$ for $\epsilon > 0$ sufficiently small;*

(iv) *For any curve $\alpha : I \to M$, parallel translation along α induced by $\overline{\nabla}$ in $\overline{M}$ is equal to parallel translation along α induced by ∇ in M.*

Proof. (i)$\Longrightarrow$(iii) follows from the uniqueness of geodesics with a given initial velocity.

(iii)$\Longrightarrow$(ii): Let $v \in TM$. Applying Proposition 13.77 to γ_v we see that $II(v,v) = 0$. Since v was arbitrary, we conclude that $II \equiv 0$.

13.4. Geodesics

(ii)⟹(iv): Suppose $v \in T_pM$. If V is a parallel vector field with respect to ∇ that is defined near p such that $V(p) = v$, then $\overline{\nabla}_{\partial_t}V = \nabla_{\partial_t}V + II(\dot{\gamma}, V) = 0$ for any γ with $\gamma(0) = p$ so that V is a parallel vector field with respect to $\overline{\nabla}$.

(iv)⟹(i): Assume (iv). If γ is a geodesic in M, then $\dot{\gamma}$ is parallel along γ with respect to ∇. Then by assumption $\dot{\gamma}$ is parallel along γ with respect to $\overline{\nabla}$. Thus γ is also an $\overline{M}$ geodesic. □

From Proposition 13.46, it follows that if $M = f^{-1}(c)$ is a semi-Riemannian hypersurface, then $U = \nabla f / \|\nabla f\|$ is a unit normal for M and $\langle U, U \rangle = \varepsilon = \text{sgn } M$. Notice that this implies that $M = f^{-1}(c)$ is orientable if $\overline{M}$ is orientable. Thus not every semi-Riemannian hypersurface is of the form $f^{-1}(c)$. On the other hand every hypersurface is locally of this form.

We are already familiar with the sphere $S^n(r)$, which is $f^{-1}(r^2)$ where $f(x) = \langle x, x \rangle = \sum_{i=1}^n (x^i)^2$.

Definition 13.81. For $n > 1$ and $0 \leq \nu \leq n$, we define
$$S^n_\nu(r) = \{x \in \mathbb{R}^{n+1}_\nu : \langle x, x \rangle_\nu = r^2\}.$$

$S^n_\nu(r)$ is called the **pseudo-sphere** of radius r and index ν.

Definition 13.82. For $n > 1$ and $0 \leq \nu \leq n$, we define
$$H^n_\nu(r) = \{x \in \mathbb{R}^{n+1}_{\nu+1} : \langle x, x \rangle_\nu = -r^2\}.$$

$H^n_\nu(r)$ is called the **pseudo-hyperbolic space** of radius r and index ν.

If Π is a two-dimensional plane through the origin in $\mathbb{R}^{n+1}_\nu$, and if $C \subset \Pi$ is a conic section (ellipse, straight line, hyperbola, etc.), then we shall say that C is a conic section in $\mathbb{R}^{n+1}_\nu$. If $Q \subset \mathbb{R}^{n+1}_\nu$ is a hyperquadric, then it is easy to show that $\Pi \cap Q$ is a conic section in Π and hence in $\mathbb{R}^{n+1}_\nu$. Problems 4 and 5 show that geodesics in hyperquadrics can be understood once we understand the case of $S^n_\nu(1)$. With this in mind, we have

Proposition 13.83. *All geodesics in $S^n_\nu(r)$ are parametrizations of the connected components of sets of the form $\Pi \cap S^n_\nu(r)$, where Π is a plane.*

a) *If γ is a timelike geodesic in $S^n_\nu(r)$, then it is a parametrization of one branch of a hyperbola.*

b) *If γ is a null geodesic in $S^n_\nu(r)$, then it is a parametrization of a straight line.*

c) *If γ is a spacelike geodesic in $S^n_\nu(r)$, then it is a parametrization of an ellipse (and hence periodic).*

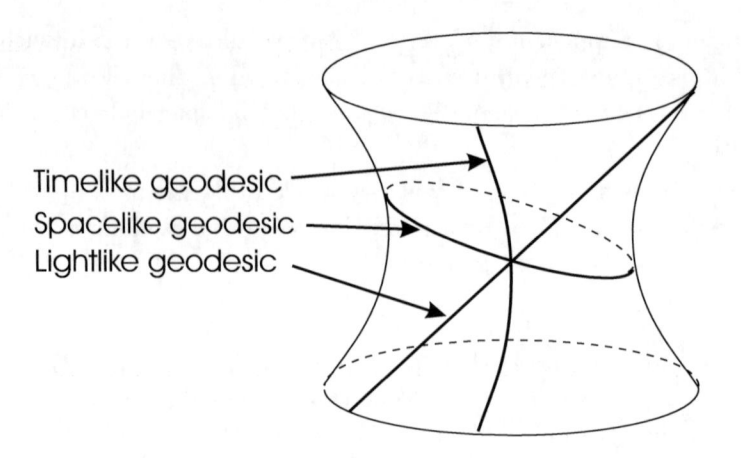

Figure 13.5. Geodesics in $S_1^2(1)$

Proof. We follow [**ON1**]. Let $p \in S_\nu^n(r)$ be given and let Π be a plane in $\mathbb{R}_\nu^{n+1}$ containing 0 and p. We will show that the conic section $\Pi \cap S_\nu^n(r)$ can be parametrized as a geodesic. We identify the type of conic section and we argue that these account for all geodesics on $S_\nu^n(r)$. We restrict the scalar product g on $\mathbb{R}_\nu^{n+1}$ to Π. Since p is spacelike from the definition of S_ν^n, we only have three possibilities for $g|_\Pi$. We handle these in turn:

(1) $g|_\Pi$ is positive definite. Choose an orthonormal basis e_1, e_2 for Π. Then a point $ae_1 + be_2$ of Π is also on S_ν^n only if $a^2 + b^2 = r^2$. Thus $S_\nu^n(r) \cap \Pi$ is a circle in Π and hence an ellipse in $\mathbb{R}_\nu^{n+1}$. Now $\gamma(t) := r \cos t\, e_1 + r \sin t\, e_2$ is a parametrization of $S_\nu^n(r) \cap \Pi$, and since $\langle \dot{\gamma}, \dot{\gamma} \rangle_\nu = r^2$, it is a constant speed spacelike curve. But also $\overline{\nabla}_{\partial_t} \dot{\gamma} = -P \circ \gamma$ so $\overline{\nabla}_{\partial_t} \dot{\gamma}$ is normal to S_ν^n and thus γ is a geodesic.

(2) $g|_\Pi$ is nondegenerate with index 1. Choose an orthonormal basis e_0, e_1 for Π such that $\langle e_0, e_0 \rangle = -1$ and $re_1 = p$ for some r. Observe that a point $ae_0 + be_1$ of Π is also on S_ν^n if and only if $-a^2 + b^2 = r^2$. Thus $S_\nu^n(r) \cap \Pi$ is both branches of a hyperbola. We can parametrize the branch through p as

$$\gamma(t) := (r \sinh t)\, e_0 + (r \cosh t)\, e_1.$$

This time $\langle \dot{\gamma}, \dot{\gamma} \rangle_\nu = -r^2$, so $\gamma(t)$ is timelike. Furthermore, $\overline{\nabla}_{\partial_t} \dot{\gamma} = P \circ \gamma$, so $\overline{\nabla}_{\partial_t} \dot{\gamma}$ is normal to S_ν^n and thus γ is a geodesic.

(3) $g|_\Pi$ is degenerate. In this case, the null space of $g|_\Pi$ must be of dimension 1. We choose a nonzero null vector v so that p, v is a basis for Π. Then a point $ap + bv$ of Π is also on S_ν^n only if $a = \pm 1$ which means that $S_\nu^n(r) \cap \Pi$ is a pair of lines. The line through p is parametrized as $t \mapsto p + tv$ and is a geodesic of $\mathbb{R}_\nu^{n+1}$ contained in S_ν^n and so is certainly a geodesic of S_ν^n.

13.5. Riemannian Manifolds and Distance

Finally, we argue that, up to reparametrization, this accounts for all geodesics in S_ν^n. Indeed, if γ is such a geodesic, then $v = \dot\gamma(0)$ is based at $p = \gamma(0)$ and there is a unique plane Π through the origin containing p and v. By uniqueness, γ must be a reparametrization of one of the geodesics already discovered above. $\square$

13.5. Riemannian Manifolds and Distance

In this section we consider only Riemannian manifolds (definite metrics). Then we have the notion of the length of a curve (Definition 13.5). Using this we can then define a distance function (a *metric* in the sense of "metric space") as follows: Let $p, q \in M$. Consider the set $\mathrm{Path}(p,q)$ consisting of all piecewise smooth[1] curves that begin at p and end at q. We define the **Riemannian distance from** p to q as

$$(13.8) \qquad \mathrm{dist}(p,q) = \inf\{L(c) : c \in \mathrm{Path}(p,q)\}.$$

On a general Riemannian manifold, $\mathrm{dist}(p,q) = r$ does **not necessarily** mean that there must be a curve connecting p to q having length r. To see this, just consider the points $(-1,0)$ and $(1,0)$ on the punctured plane $\mathbb{R}^2 \setminus \{0\}$.

Definition 13.84. If $p \in M$ is a point in a Riemannian manifold and $R > 0$, then the set $B_R(p)$ (also denoted $B(p,R)$) defined by $B_R(p) = \{q \in M : \mathrm{dist}(p,q) < R\}$ is called an open geodesic **ball** centered at p with radius R.

It is important to notice that unless R is small enough, $B_R(p)$ may not be homeomorphic to a ball in a Euclidean space. To see this just consider a ball of large radius on a circular cylinder of small diameter.

Proposition 13.85. *Let U be a normal neighborhood of a point p in a Riemannian manifold (M,g). If $q \in U$ and if $\gamma : [0,1] \to M$ is the radial geodesic such that $\gamma(0) = p$ and $\gamma(1) = q$, then γ is the unique shortest curve in U (up to reparametrization) connecting p to q.*

Proof. Let α be a curve connecting p to q (refer to Figure 13.6). Without loss of generality we may take the domain of α to be $[0,b]$. Let $\frac{\partial}{\partial r}$ be the radial unit vector field in U. Then if we define the vector field R along α by $t \mapsto \frac{\partial}{\partial r}\big|_{\alpha(t)}$, we may write $\dot\alpha = \langle R, \dot\alpha \rangle R + N$ for some field N normal to

---

[1] Recall that by our conventions, a piecewise smooth curve is assumed to be continuous.

R (but note that $N(0) = 0$). We now have

$$L(\alpha) = \int_0^b \langle \dot\alpha, \dot\alpha \rangle^{1/2} \, dt = \int_0^b \left[\langle R, \dot\alpha \rangle^2 + \langle N, N \rangle \right]^{1/2} dt$$
$$\geq \int_0^b |\langle R, \dot\alpha \rangle| \, dt \geq \int_0^b \langle R, \dot\alpha \rangle \, dt = \int_0^b \frac{d}{dt}(r \circ \alpha) \, dt$$
$$= r(\alpha(b)) = r(q).$$

On the other hand, if $v = \dot\gamma(0)$, then $r(q) = \int_0^1 \|v\| \, dt = \int_0^1 \langle \dot\gamma, \dot\gamma \rangle^{1/2} \, dt$ so $L(\alpha) \geq L(\gamma)$. Now we show that if $L(\alpha) = L(\gamma)$, then α is a reparametrization of γ. Indeed, if $L(\alpha) = L(\gamma)$, then all of the above inequalities must be equalities so that N must be identically zero and $\frac{d}{dt}(r \circ \alpha) = \langle R, \dot\alpha \rangle = |\langle R, \dot\alpha \rangle|$. It follows that $\dot\alpha = \langle R, \dot\alpha \rangle R = \left(\frac{d}{dt}(r \circ \alpha) \right) R$, and so α travels radially from p to q and must be a reparametrization of γ. □

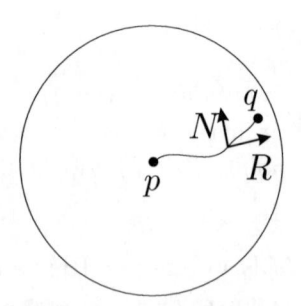

Figure 13.6. Normal neighborhood of p

It is important to notice that the uniqueness assertion of Theorem 13.85 only refers to curves with image in U. This is in contrast to the proposition below.

Proposition 13.86. *Let p_0 be a point in a Riemannian manifold M. There exists a number $\varepsilon_0(p) > 0$ such that for all ε, $0 < \varepsilon \leq \varepsilon_0(p)$ we have the following:*

  (i) *The open geodesic ball $B(p_0, \varepsilon)$ is normal and has the form*

  $$B(p_0, \varepsilon) = \exp_{p_0}\{v \in T_{p_0}M : |v| < \varepsilon\}.$$

  (ii) *For any $p \in B(p_0, \varepsilon)$, the radial geodesic segment connecting p_0 to p is the shortest curve in M, up to parametrization, from p_0 to p. (Note carefully that we now mean the shortest curve among curves into M rather than just the shortest among curves with image in $B(p_0, \varepsilon)$.)*

13.5. Riemannian Manifolds and Distance

Proof. Let $\widetilde{U} \subset T_{p_0}M$ be chosen so that $U = \exp_{p_0}\widetilde{U}$ is a normal neighborhood of p_0. Then for sufficiently small $\varepsilon > 0$ the ball

$$\widetilde{B}(0,\varepsilon) := \{v \in T_{p_0}M : \|v\| < \varepsilon\}$$

is a starshaped open set in $\widetilde{U}$, and so $A_{p_0,\varepsilon} = \exp_{p_0}(\widetilde{B}(0,\varepsilon))$ is a normal neighborhood of p_0. From Proposition 13.85 we know that the radial geodesic segment σ from p_0 to p is the shortest curve in $A_{p_0,\varepsilon}$ from p_0 to p. This curve has length less than ε. We claim that any curve from p_0 to p whose image leaves $A_{p_0,\varepsilon}$ must have length greater than ε. Once this claim is proved, it is easy to see that

$$A_{p_0,\varepsilon} = B(p_0,\varepsilon) = \{p \in M : \operatorname{dist}(p_0,p) < \varepsilon\}$$

and that (ii) holds. Now suppose that $\alpha : [a,b] \to M$ is a curve from p_0 to p which leaves $A_{p_0,\varepsilon}$. Then for any $r > 0$ with $r < \varepsilon$, the curve α must meet the set $S(r) := \exp_{p_0}(\{v \in T_{p_0}M : \|v\| = r\})$ at some first parameter value $t_1 \in [a,b]$. Then $\alpha_1 := \alpha|_{[a,t_1]}$ lies in $A_{p_0,\varepsilon}$, and Proposition 13.85 tells us that $L(\alpha) \geq L(\alpha|_{[a,t_1]}) \geq r$. Since this is true for all $r < \varepsilon$, we have $L(\alpha) \geq \varepsilon$, which is what was claimed. □

Theorem 13.87 (Distance topology). *Given a Riemannian manifold, define the distance function* dist *as before. Then* (M,dist) *is a metric space, and the metric topology coincides with the manifold topology on* M.

Proof. To show that dist is a true distance function (metric) we must prove that

(1) dist is symmetric; $\operatorname{dist}(p,q) = \operatorname{dist}(q,p)$;

(2) dist satisfies the triangle inequality $\operatorname{dist}(q,p) \leq \operatorname{dist}(p,x) + \operatorname{dist}(x,q)$;

(3) $\operatorname{dist}(p,q) \geq 0$; and

(4) $\operatorname{dist}(p,q) = 0$ if and only if $p = q$.

Now, (1) is obvious, and (2) and (3) are clear from the properties of the integral and the metric tensor. For (4) suppose that $p \neq q$. Then since M is Hausdorff, we can find a normal neighborhood U of p that does not contain q. In fact, by the previous proposition, we may take U to be of the form $B(p,\varepsilon)$. Since (by the proof of the previous proposition) every curve starting at p and leaving $B(p,\varepsilon)$ must have length at least $\varepsilon/2$, we see that $\operatorname{dist}(p,q) \geq \varepsilon/2$. □

By definition a curve segment in a Riemannian manifold, say $c : [a,b] \to M$, is a shortest curve if $L(c) = \operatorname{dist}(c(a),c(b))$. We say that such a curve is (absolutely) **length minimizing**. Such curves must be geodesics.

Proposition 13.88. *Let M be a Riemannian manifold. A length minimizing curve $c : [a,b] \to M$ must be an (unbroken) geodesic.*

Proof. There exist numbers t_i with $a = t_0 < t_1 < \cdots < t_k = b$ such that for each subinterval $[t_i, t_{i+1}]$, the restricted curve $c|_{[t_i,t_{i+1}]}$ has image in a totally star-shaped open set. Thus since $c|_{[t_i,t_{i+1}]}$ is minimizing, it must be a reparametrization of a unit speed geodesic (use the uniqueness part of Proposition 13.85). Thus there is a reparametrization of c that is a broken geodesic. But this new reparametrized curve is also length minimizing, and so by Problem 1 it is smooth. □

13.6. Lorentz Geometry

In this section we define and discuss a few aspects of Lorentz manifolds. Lorentz manifolds play a prominent role in physics and are often singled out for special study. We discuss the local length *maximizing* property of timelike geodesics in a Lorentz manifold and derive the Lorentzian analogue of Proposition 13.85.

Definition 13.89. A **Lorentz vector space** is a scalar product space with index equal to one and dimension greater than or equal to 2. A **Lorentz manifold** is a semi-Riemannian manifold such that each tangent space is a Lorentz space with the scalar product given by the metric tensor. Under our conventions, the signature of a Lorentz manifold is of the form $(-1, 1, , \ldots, 1, 1)$.[2]

Each tangent space of a Lorentz manifold is a Lorentz vector space, and so we first take a closer look at some of the distinctive features of Lorentz vector spaces. Let us now agree to classify the zero vector in a Lorentz space as *spacelike*. For Lorentz spaces, we may classify subspaces into three categories:

Definition 13.90. Let V be a Lorentz vector space such as a tangent space of a Lorentz manifold. A subspace $W \subset V$ is called

(1) spacelike if $g|_W$ is positive definite (or if W is the zero subspace);

(2) timelike if $g|_W$ nondegenerate with index 1;

(3) lightlike if $g|_W$ is degenerate.

Thus a subspace falls into one of the three types, which we refer to as its **causal character**.

If we take a timelike vector v in a Lorentz space V, then $\mathbb{R}v$, the space spanned by v, is nondegenerate and has index 1. By Lemma 7.47, $v^\perp$ is nondegenerate and $V = \mathbb{R}v \oplus v^\perp$. Since $1 = \operatorname{ind}(V) = \operatorname{ind}(\mathbb{R}v) + \operatorname{ind}(v^\perp)$, it follows that $\operatorname{ind}(v^\perp) = 0$, so that $v^\perp$ is spacelike. This little observation is useful enough to set out as a proposition.

---

[2]Some authors use $(1, -1, \ldots, -1, -1)$, but this does not really change the geometry.

13.6. Lorentz Geometry

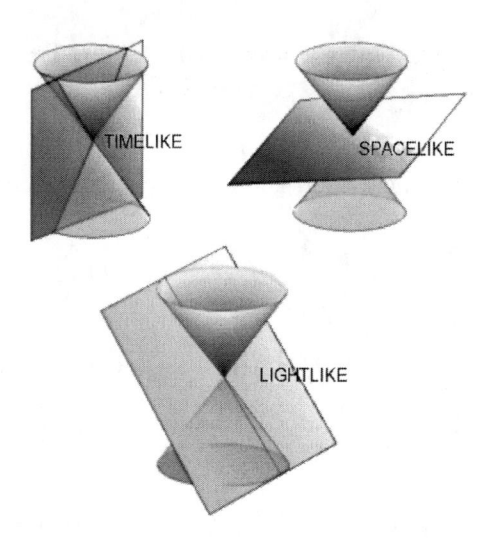

Figure 13.7. Causal character of a subspace

Proposition 13.91. *If* V *is a Lorentz vector space and* v *is a timelike element, then* $v^\perp$ *is spacelike, and we have the orthogonal direct sum* $V = \mathbb{R}v \oplus v^\perp$.

Exercise 13.92. Show that if W is a subspace of a Lorentz space, then W is timelike if and only if $W^\perp$ is spacelike.

Exercise 13.93. Suppose that v, w are linearly independent null vectors in a Lorentz space V. Show that $\langle v, w \rangle \neq 0$. [Hint: Use an orthonormal basis to orthogonally decompose; $V = \mathbb{R}e_0 \oplus P$, where e_0 is timelike and where $\langle \cdot, \cdot \rangle$ is positive definite on P. Suppose $\langle v, w \rangle = 0$; write $v = \alpha e_0 + p_1$ and $w = \beta e_0 + p_2$. Then show that $\langle p_1, p_2 \rangle = \alpha\beta$, $\langle p_i, p_i \rangle = \alpha^2 = \beta^2$ and $|\langle p_1, p_2 \rangle| = \|p_1\| \|p_2\|$.]

Lemma 13.94. *Let* W *be a subspace of a Lorentz space. Then the following conditions are equivalent:*

(i) W *is timelike and so a Lorentz space in its own right.*

(ii) *There exist null vectors* $v, w \in W$ *that are linearly independent.*

(iii) W *contains a timelike vector.*

Proof. Suppose (i) holds. Let $e_1, \ldots, e_m$ be an orthonormal basis for W with e_1 timelike. Then $e_1 + e_2$ and $e_1 - e_2$ are both null and, taken together, are a linearly independent pair so that (ii) holds. Now suppose that (ii) holds and let v, w be a linearly independent pair of null vectors. By Exercise 13.93 above, either $v+w$ or $v-w$ must be timelike so we have (iii). Finally, suppose (iii) holds and $v \in W$ is timelike. Since $v^\perp$ is spacelike and $W^\perp \subset v^\perp$, we see that $W^\perp$ is spacelike. But then W is timelike by Exercise 13.92 so that (i) holds. □

Exercise 13.95. Use the above lemma to prove that if W is a nontrivial subspace of a Lorentz space, then the following three conditions are equivalent:

(i) W contains a nonzero null vector but no timelike vector.

(ii) W is lightlike.

(iii) The intersection of W with the nullcone is one-dimensional.

Definition 13.96. The **timecone** determined by a timelike vector v is the set $C(v) := \{w \in V : \langle v, w \rangle < 0\}$.

In Problem 6 we ask the reader to show that timelike vectors v and w in a Lorentz space V are in the same timecone if and only if $\langle v, w \rangle < 0$.

Exercise 13.97. Show that there are exactly two timecones in a Lorentz vector space whose union is the set of all nonzero timelike vectors. Describe the relation of the nullcone to the timecones.

Now we come to an aspect of Lorentz spaces that underlies the twins paradox of special relativity.

Proposition 13.98. *If v, w are timelike elements of a Lorentz vector space then we have the **backward Schwartz inequality***

$$|\langle v, w \rangle| \geq \|v\| \|w\|,$$

with equality only if v is a scalar multiple of w. Also, if v and w are in the same timecone, then there is a uniquely determined number $\alpha \geq 0$, called the hyperbolic angle between v and w, such that

$$\langle v, w \rangle = - \|v\| \|w\| \cosh \alpha.$$

Note: The minus sign appears because of our convention that $\langle v, v \rangle = -1$ for timelike vectors.

Proof. We may write $w = av + z$ where $z \in v^\perp$. We have

$$a^2 \langle v, v \rangle + \langle z, z \rangle = \langle w, w \rangle < 0.$$

Using this and recalling that $\langle v, v \rangle < 0$, we have

$$\langle v, w \rangle^2 = a^2 \langle v, v \rangle = (\langle w, w \rangle - \langle z, z \rangle) \langle v, v \rangle$$
$$\geq \langle v, v \rangle \langle w, w \rangle = \|v\| \|w\|.$$

13.6. Lorentz Geometry

Equality holds exactly when $\langle z, z \rangle = 0$, but since $z \in v^\perp$, this implies that $z = 0$ so $w = av$. Using Problem 6, we see that since v and w are in the same timecone, we have $\langle v, w \rangle < 0$, and hence

$$-\frac{\langle v, w \rangle}{\|v\| \|w\|} \geq 1.$$

The properties of the function cosh now give a unique number $\alpha \geq 0$ such that $\langle v, w \rangle = -\|v\| \|w\| \cosh \alpha$ as required. $\square$

Corollary 13.99. *If v, w are timelike elements of a Lorentz vector space which are in the same timecone, then we have the backward triangle inequality:*

$$\|v\| + \|w\| \leq \|v + w\|.$$

Equality holds only if v is a scalar multiple of w.

Proof. Since $\langle v, w \rangle < 0$ by hypothesis, we have $-\langle v, w \rangle \geq \|v\| \|w\|$ by the proposition. Then

$$(\|v\| + \|w\|)^2 = \|v\|^2 + 2\|w\| \|v\| + \|w\|^2$$
$$\leq \|v\|^2 - 2\langle v, w \rangle + \|w\|^2 = \|v + w\|^2.$$

Equality happens only if $-\langle v, w \rangle = \|v\| \|w\|$, which, by the previous proposition, means that v is a scalar multiple of w. $\square$

In relativity theory, spacetime (the set of all "idealized" possible events) is modeled as a 4-dimensional Lorentz manifold and the paths of massive bodies are to be timelike curves. But we have yet to talk about what distinguishes the past from the future! In each tangent space we have two timecones and we could arbitrarily choose one of them to be the future timecone. But what we really want is a smooth way of choosing a future timecone in each tangent space. This leads to the notion of time orientability. First we say that a vector field X on a Lorentz manifold M is **timelike** if $X(p)$ is timelike for each $p \in M$.

Definition 13.100. A Lorentz manifold M is said to be **time-orientable** if and only if there exists a timelike vector field $X \in \mathfrak{X}(M)$. A **time orientation** of M is a choice of timecone $C(p) \in T_pM$ for each p such that there exists a timelike $X \in \mathfrak{X}(M)$ with $X_p \in C(p)$ for each p. In the latter case, $C(p)$ is referred to as the **positive** or **future timecone** at p. The other timecone at T_pM is called the negative timecone. Timelike vectors in the positive timecone are said to be **future pointing** (and those in the negative timecone are **past pointing**).

Definition 13.101. A lightlike vector in a tangent space T_pM of a time-oriented Lorentz manifold M is said to be future pointing if it is the limit of a sequence of future pointing timelike vectors in T_pM. Thus the lightcone (nullcone) in T_pM is partitioned into future lightcone and past lightcone.

Based on these definitions we can speak of timelike or lightlike curves as being either future pointing or "past pointing". Time orientability is certainly a global condition since we need a choice of timecone in every tangent space and this choice must be made smoothly. However, the following exercise shows that the smoothness condition can be described in terms of local vector fields:

Exercise 13.102. Suppose that it is possible to choose a timecone $C(p)$ in every tangent space T_pM of a Lorentz manifold M in such a way that in a neighborhood of each point there is a *local* smooth vector field with values in these timecones. Show that this implies that M is orientable. [Hint: Use a partition of unity argument.]

Exercise 13.103. Show that $S_1^n(r) := \{p \in \mathbb{R}_1^n : \langle p,p \rangle = r^2\}$ is a time-orientable Lorentz manifold. [Hint: consider the restriction to $S_1^n(r)$ of the first coordinate vector field from $\mathbb{R}_1^n$.]

Now let us consider timelike curves in a Lorentz manifold M. Since we want to include piecewise smooth curves, we have to decide what timelike should mean. If $\gamma : I \to M$ is a piecewise smooth curve and $t_i \in I$ is a parameter value at which γ is not smooth, then what condition is appropriate if we are to refer to the curve as a timelike curve? We consider the one-sided limits

$$\dot{\gamma}(t_i^+) := \lim_{\varepsilon \downarrow 0} \dot{\gamma}(t_i + \varepsilon) \quad \text{and} \quad \dot{\gamma}(t_i^-) := \lim_{\varepsilon \downarrow 0} \dot{\gamma}(t_i - \varepsilon).$$

For many purposes, the following definition is appropriate:

Definition 13.104. Let M be a Lorentz manifold. A piecewise smooth curve $\gamma : I \to M$ is called **timelike** if

(i) $\dot{\gamma}$ is timelike where it is smooth, and

(ii) for every t_i where γ is not smooth, we have that $\dot{\gamma}(t_i^+)$ and $\dot{\gamma}(t_i^-)$ are timelike and the following further condition holds at each such t_i:

$$\langle \dot{\gamma}(t_i^+), \dot{\gamma}(t_i^-) \rangle < 0.$$

Thus, for timelike curves, $\dot{\gamma}(t_i^+)$ are $\dot{\gamma}(t_i^-)$ are in the same timecone. Following [**ON1**], we next prove a useful technical lemma.

Lemma 13.105. *Let p be a point in a Lorentz manifold M and U a normal neighborhood of p. Let $\widetilde{U}$ be the corresponding starshaped open set in T_pM with $U = \exp_p(\widetilde{U})$. Let $\gamma : [0,b] \to \widetilde{U} \subset T_pM$ be a piecewise smooth curve such that $\alpha := \exp_p \circ \gamma : [0,b] \to M$ is timelike (in the sense of Definition*

13.6. Lorentz Geometry

13.104). Then the image of γ is contained in a single timecone of T_pM and $\langle \dot{\alpha}, P \rangle < 0$.

Proof. Let us first handle the case where γ is smooth. Then since $\dot{\gamma}(0)$ is timelike, γ is initially in one of the timecones which we denote by $C(p)$. Here and below, "initially" is taken to mean "for all sufficiently small positive parameter values t". Let $\widetilde{P}$ be the position vector field on T_pM and let P be the local position vector field which is $\exp_p$-related to $\widetilde{P}$ and defined on the normal neighborhood U. Note that $\widetilde{P}$ is timelike and outward radial at each point of $\widetilde{U} \cap C(p)$. Thus $\langle \dot{\gamma}, \widetilde{P} \rangle$ is initially negative. Letting $\widetilde{q}(x) := \langle x, x \rangle$ in T_pM and considering T_pM as a Lorentz manifold itself we have $\operatorname{grad} \widetilde{q} = 2\widetilde{P}$ and hence $\frac{d}{dt}\widetilde{q} \circ \gamma = 2\langle \dot{\gamma}, \widetilde{P} \rangle$. The Gauss lemma (Lemma 13.72) gives

$$\langle \dot{\gamma}, \widetilde{P} \rangle = \langle \dot{\alpha}, P \rangle,$$

which implies that $\langle \dot{\alpha}, P \rangle$ and hence $\frac{d}{dt}\widetilde{q} \circ \gamma$ are initially negative. For any $t > 0$ such that $\gamma(t)$ is in $C(p)$, the vector $\widetilde{P}(\gamma(t))$, and hence $P(\alpha(t))$, must be timelike. For such t, $\langle \dot{\alpha}, P \rangle < 0$ which implies that $\langle \dot{\gamma}, \widetilde{P} \rangle < 0$ and hence $\frac{d}{dt}\widetilde{q} \circ \gamma < 0$. So $\widetilde{q} \circ \gamma$ starts out negative and goes down hill as long as $\gamma(t)$ is in $C(p)$. Since γ can only exit $C(p)$ by reaching the nullcone (or 0) where $\widetilde{q}$ vanishes, we see that γ must remain inside $C(p)$.

Now we consider what happens if γ is timelike but merely piecewise smooth. The first segment remains in $C(p)$, and at the first parameter value t_1 where γ fails to be smooth, we must have $\langle \dot{\gamma}(t_1^-), \widetilde{P} \rangle < 0$. But then by the Gauss lemma again $\langle \dot{\alpha}(t_1^-), P \rangle < 0$. The technical restriction of Definition 13.104 forces $\langle \dot{\alpha}(t_1^+), P \rangle < 0$ so that $\dot{\alpha}(t_1^+) \in C(p)$. Applying the Gauss lemma gives $\langle \dot{\gamma}(t_1^+), \widetilde{P} \rangle < 0$ and so $\frac{d}{dt}\widetilde{q} \circ \gamma$ cannot change sign at t_1. We are now set up to repeat the argument for the next segment. The result follows inductively. $\square$

The following proposition for Lorentz manifolds should be compared to Proposition 13.85 proved for Riemannian manifolds. In this proposition we find that the geodesics are locally *longer* than nearby curves.

Proposition 13.106. *Let U be a normal neighborhood of a point p in a Lorentz manifold. If the radial geodesic γ connecting p to $q \in U$ is timelike, then it is the unique longest geodesic segment in U that connects p to q. Once again, uniqueness is up to reparametrization.*

Proof. Let $\widetilde{U}$ be related to U as usual. Take any timelike curve $\alpha : [0, b] \to U$ segment in U that connects p to q. By the previous lemma, $\beta := \exp_p^{-1} \circ \alpha$ stays inside a single timecone $C(p)$ and so also inside $C(p) \cap \widetilde{U}$. Thus α stays inside $\exp_p(C(p) \cap \widetilde{U})$ where it is timelike and where the field $R = (P/r) \circ \alpha$ is a unit timelike field along α. We now seek to imitate the proof of Proposition 13.85. We may decompose $\dot{\alpha}$ as

$$\dot{\alpha} = -\langle R, \dot{\alpha} \rangle R + N,$$

where N is a spacelike field along α that is orthogonal to R. We have

$$\|\dot{\alpha}\| = \sqrt{-\langle \dot{\alpha}, \dot{\alpha} \rangle} = \sqrt{\langle R, \dot{\alpha} \rangle^2 - \langle N, N \rangle} \leq |\langle R, \dot{\alpha} \rangle|.$$

Recall that $\widetilde{\mathsf{q}}(\cdot) = \langle \cdot, \cdot \rangle$ and $\mathsf{q} := \widetilde{\mathsf{q}} \circ \exp_p^{-1}$. Since $r = \sqrt{-\mathsf{q}}$ and so $\operatorname{grad} r = -P/r$, we have $(\operatorname{grad} r) \circ \alpha = -R$. By the previous lemma $\langle \dot{\alpha}, R \rangle$ is negative. Then

$$|\langle \dot{\alpha}, R \rangle| = -\langle R, \dot{\alpha} \rangle = \frac{d(r \circ \alpha)}{dt}.$$

Thus we have

$$L(\alpha) = \int_0^b \|\dot{\alpha}(t)\| \, dt \leq r(q) = L(\gamma).$$

If $L(\alpha) = L(\gamma)$, then $N = 0$ and we argue as in the Riemannian case to conclude that α is the same as γ up to reparametrization. $\square$

Recall that the arc length of a timelike curve is often called the curve's **proper time** and is thought of as the intrinsic duration of the curve. We may reparametrize a timelike geodesic to have unit speed. The parameter is then an **arc length parameter**, which is often referred to as a **proper time parameter**. We may restate the previous theorem to say that the unit speed geodesic connecting p to q in U is the unique curve of maximum proper time among curves connecting p to q in U.

13.7. Jacobi Fields

Once again we consider a semi-Riemannian manifold (M, g) of arbitrary index. We shall be dealing with smooth two-parameter maps $h : (-\epsilon, \epsilon) \times [a, b] \to M$. The partial maps $t \mapsto h_s(t) = h(s, t)$ are called the longitudinal curves, and the curves $s \mapsto h(s, t)$ are called the transverse curves. Let α be the center longitudinal curve $t \mapsto h_0(t)$. The vector field along α defined by $V(t) = \frac{d}{ds}\big|_{s=0} h_s(t)$ is called the **variation vector field** along α. We will use the following important result more than once:

Lemma 13.107. *Let Y be a vector field along the smooth map $h : (-\epsilon, \epsilon) \times [a, b] \to M$. Then*

$$\nabla_{\partial_s} \nabla_{\partial_t} Y - \nabla_{\partial_t} \nabla_{\partial_s} Y = R(\partial_s h, \partial_t h) Y.$$

Proof. If one computes in a local chart, the result falls out after a mildly tedious computation, which we leave to the curious reader. $\square$

13.7. Jacobi Fields

Suppose we have the special situation that, for each s, the partial maps $t \mapsto h_s(t)$ are geodesics. In this case, let us denote the center geodesic $t \mapsto h_0(t)$ by γ. We call h a variation of γ through geodesics. Let h be such a special variation and V the variation vector field. Using Lemma 13.107 and the result of Exercise 13.11 we compute

$$\nabla_{\partial_t}\nabla_{\partial_t}V = \nabla_{\partial_t}\nabla_{\partial_t}\partial_s h = \nabla_{\partial_t}\nabla_{\partial_s}\partial_t h$$
$$= \nabla_{\partial_s}\nabla_{\partial_t}\partial_t h + R(\partial_t h, \partial_s h)\partial_t h$$
$$= R(\partial_t h, \partial_s h)\partial_t h$$

and evaluating at $s = 0$ we get $\nabla_{\partial_t}\nabla_{\partial_t}V(t) = R(\dot{\gamma}(t), V(t))\dot{\gamma}(t)$. This equation is important and shows that V is a Jacobi field as per the folowing definition:

Definition 13.108. Let $\gamma : [a,b] \to M$ be a geodesic and let $J \in \mathfrak{X}_\gamma(M)$ be a vector field along γ. The field J is called a **Jacobi field** if

$$\nabla_{\partial_t}\nabla_{\partial_t}J = R(\dot{\gamma}(t), J(t))\dot{\gamma}(t)$$

for all $t \in [a,b]$.

In local coordinates, we recognize the above as a second order system of linear differential equations and we easily arrive at the following

Theorem 13.109. *Let (M,g) and the geodesic $\gamma : [a,b] \to M$ be as above. Given $w_1, w_2 \in T_{\gamma(a)}M$, there is a unique Jacobi field $J^{w_1,w_2} \in \mathfrak{X}_\gamma(M)$ such that $J(a) = w_1$ and $\nabla_{\partial_t}J(a) = w_2$. The set $\mathrm{Jac}\,(\gamma)$ of all Jacobi fields along γ is a vector space isomorphic to $T_{\gamma(a)}M \times T_{\gamma(a)}M$.*

We now examine the more general case of a Jacobi field J^{w_1,w_2} along a geodesic $\gamma : [a,b] \to M$. First notice that for any curve $\alpha : [a,b] \to M$ with $|\langle \dot{\alpha}(t), \dot{\alpha}(t)\rangle| > 0$ for all $t \in [a,b]$, any vector field Y along α decomposes into an orthogonal sum $Y^\top + Y^\perp$. This means that $Y^\top$ is a multiple of $\dot{\alpha}$ and that $Y^\perp$ is normal to $\dot{\alpha}$. If $\gamma : [a,b] \to M$ is a geodesic, then $\nabla_{\partial_t}Y^\perp$ is also normal to $\dot{\gamma}$ since $0 = \frac{d}{dt}\langle Y^\perp, \dot{\gamma}\rangle = \langle \nabla_{\partial_t}Y^\perp, \dot{\gamma}\rangle + \langle Y^\perp, \nabla_{\partial_t}\dot{\gamma}\rangle = \langle \nabla_{\partial_t}Y^\perp, \dot{\gamma}\rangle$. Similarly, $\nabla_{\partial_t}Y^\top$ is parallel to $\dot{\gamma}$ all along γ.

Theorem 13.110. *Let $\gamma : [a,b] \to M$ be a geodesic segment.*

(i) *If $Y \in \mathfrak{X}_\gamma(M)$ is tangent to γ, then Y is a Jacobi field if and only if $\nabla^2_{\partial_t}Y = 0$ along γ. In this case, $Y(t) = (at+b)\dot{\gamma}(t)$.*

(ii) *If J is a Jacobi field along γ and there are some distinct $t_1, t_2 \in [a,b]$ with $J(t_1)\perp\dot{\gamma}(t_1)$ and $J(t_2)\perp\dot{\gamma}(t_2)$, then $J(t)\perp\dot{\gamma}(t)$ for all $t \in [a,b]$.*

(iii) *If J is a Jacobi field along γ and there is some $t_0 \in [a,b]$ with $J(t_0)\perp\dot{\gamma}(t_0)$ and $\nabla_{\partial_t}J(t_0)\perp\dot{\gamma}(t_0)$, then $J(t)\perp\dot{\gamma}(t)$ for all $t \in [a,b]$.*

(iv) *If γ is not a null geodesic, then Y is a Jacobi field if and only if both $Y^\top$ and $Y^\perp$ are Jacobi fields.*

Proof. (i) Let $Y = f\dot\gamma$. Then the Jacobi equation reads
$$\nabla^2_{\partial_t} f\dot\gamma(t) = R(\dot\gamma(t), f\dot\gamma(t))\dot\gamma(t) = 0.$$
Since γ is a geodesic, this implies that $f'' = 0$ and (i) follows.

(ii) and (iii) We have $\frac{d^2}{dt^2}\langle J, \dot\gamma\rangle = \langle R(\dot\gamma(t), J(t))\dot\gamma(t), \dot\gamma(t)\rangle = 0$ (from the symmetries of the curvature tensor). Thus $\langle J(t), \dot\gamma(t)\rangle = at + b$ for some $a, b \in \mathbb{R}$. The reader can now easily deduce both (ii) and (iii).

(iv) The operator $\nabla^2_{\partial_t}$ preserves the normal and tangential parts of Y. We now show that the same is true of the map $Y \mapsto R(\dot\gamma(t), Y)\dot\gamma(t)$. Since we assume that γ is not null, we have $Y^\top = f\dot\gamma$ for some $\dot\gamma$. Thus $R(\dot\gamma(t), Y^\top)\dot\gamma(t) = R(\dot\gamma(t), f\dot\gamma(t))\dot\gamma(t) = 0$, which is trivially tangent to $\dot\gamma(t)$. On the other hand, $\langle R(\dot\gamma(t), Y^\perp(t))\dot\gamma(t), \dot\gamma(t)\rangle = 0$ by symmetries of the curvature tensor. We have
$$\left(\nabla^2_{\partial_t}Y\right)^\top + \left(\nabla^2_{\partial_t}Y\right)^\perp = \nabla^2_{\partial_t}Y = R(\dot\gamma(t), Y(t))\dot\gamma(t)$$
$$= R(\dot\gamma(t), Y^\top(t))\dot\gamma(t) + R(\dot\gamma(t), Y^\perp(t))\dot\gamma(t)$$
$$= 0 + R(\dot\gamma(t), Y^\perp(t))\dot\gamma(t).$$
So the Jacobi equation $\nabla^2_{\partial_t}Y(t) = R(\dot\gamma(t), Y(t))\dot\gamma(t)$ splits into two equations
$$\nabla^2_{\partial_t}Y^\top(t) = 0,$$
$$\nabla^2_{\partial_t}Y^\perp(t) = R_{\dot\gamma(t), Y^\perp(t)}\dot\gamma(t),$$
and the result follows from this. □

Corollary 13.111. *Let $\gamma = \gamma_v$ and $J^{0,w}_\gamma$ be as above. Then $J^{0,rv}_\gamma(t) = rt\dot\gamma_v(t)$. If $w \perp v$, then $\langle J^{0,w}(t), \dot\gamma_v(t)\rangle = 0$ for all $t \in [0, b]$. Furthermore, every Jacobi field $J^{0,w}$ along $\exp_v tv$ with $J^{0,w}(0) = 0$ has the form $J^{0,w} := rt\dot\gamma_v + J^{0,w_1}$, where $w = \nabla_{\partial_t}J(0) = rv + w_1$, $w_1 \perp v$ and $J^{0,w_1}(t) \perp \dot\gamma_v(t)$ for all $t \in [0, b]$.*

The proof of the last result shows that a Jacobi field decomposes into a parallel vector field along γ, which is just a multiple of the velocity $\dot\gamma$, and a "normal Jacobi field" $J^\perp$, which is normal to γ at each of its points. Of course, the important part is the normal part since the tangential part is merely the infinitesimal model for a variation through geodesics which are merely reparametrizations of the $s = 0$ geodesic. Thus we focus attention on the Jacobi fields that are normal to the geodesics along which they are defined. Thus we consider the Jacobi equation $\nabla^2_{\partial_t}J(t) = R(\dot\gamma(t), J(t))\dot\gamma(t)$ with initial conditions such as in (ii) or (iii) of Theorem 13.110.

Exercise 13.112. For $v \in T_pM$, let $v^\perp := \{w \in T_pM : \langle w, v\rangle = 0\}$. Prove that the tidal operator $R_v : w \mapsto R_{v,w}v$ maps $v^\perp$ to itself.

In light of this exercise, we make the following definition.

13.7. Jacobi Fields

Definition 13.113. For $v \in T_pM$, the (restricted) **tidal force operator** $F_v : v^\perp \to v^\perp$ is the restriction of R_v to $v^\perp \subset T_pM$.

Notice that in terms of the tidal force operator the Jacobi equation for normal Jacobi fields is

$$\nabla^2_{\partial_t} J(t) = F_{\dot\gamma(t)}(J(t)) \text{ for all } t.$$

If J is the variation vector field of a geodesic variation, then it is an infinitesimal model of the separation of nearby geodesics. In general relativity, one thinks of a one-parameter family of freely falling particles. Then $\nabla_{\partial_t} J$ is the *relative* velocity field and $\nabla^2_{\partial_t} J$ is the *relative* acceleration. Thus the Jacobi equation can be thought of as a version of Newton's second law with the curvature term playing the role of a force.

Proposition 13.114. *For $v \in T_pM$, the tidal force operator $F_v : v^\perp \to v^\perp$ is self-adjoint and* $\operatorname{Trace}(F_v) = -\operatorname{Ric}(v,v)$.

Proof. First, $\langle F_v w_1, w_2 \rangle = \langle R_{v,w_1} v, w_2 \rangle = \langle R_{v,w_2} v, w_1 \rangle = \langle F_v w_2, w_1 \rangle$ by (iv) of Theorem 13.19. The proof that $\operatorname{Trace} F_v = -\operatorname{Ric}(v,v)$ is easy for definite metrics but for indefinite metrics the possibility that v may be a null vector involves a little extra work. If v is not null, then letting $e_2, \ldots, e_n$ be an orthonormal basis for $v^\perp$ we have

$$\operatorname{Ric}(v,v) = -\sum \varepsilon_i \langle R_{v,e_i} v, e_i \rangle = -\sum \varepsilon_i \langle F_v e_i, e_i \rangle = -\operatorname{Trace} F_v.$$

If v is null, then we can find a vector w such that $\langle w, v \rangle = -1$ and w, v span a Lorentz plane L in T_pM. Define $e_1 := (v+w)/\sqrt{2}$ and $e_2 := (v-w)/\sqrt{2}$. One checks that e_1 is timelike while e_2 is spacelike. Now choose an orthonormal basis $e_3, \ldots, e_n$ for $L^\perp \subset v^\perp$ so that $e_1, \ldots, e_n$ is an orthonormal basis for T_pM. Then we have

$$\operatorname{Ric}(v,v) = \langle R_{v,e_1} v, e_1 \rangle - \langle R_{v,e_2} v, e_2 \rangle - \sum_{i>2} \varepsilon_i \langle R_{v,e_i} v, e_i \rangle.$$

But $\langle R_{v,e_1} v, e_1 \rangle = \frac{1}{2} \langle R_{v,w} v, w \rangle = \langle R_{v,e_2} v, e_2 \rangle$ and so we are left with

$$\operatorname{Ric}(v,v) = -\sum_{i>2} \varepsilon_i \langle R_{v,e_i} v, e_i \rangle = -\sum_{i>2} \varepsilon_i \langle F_v e_i, e_i \rangle.$$

Since $v, e_3, \ldots, e_n$ is an orthonormal basis for $v^\perp$ and $F_v v = 0$, we have

$$\operatorname{Ric}(v,v) = -\sum_{i>2} \varepsilon_i \langle F_v e_i, e_i \rangle$$
$$= -\langle F_v v, v \rangle - \sum_{i>2} \varepsilon_i \langle F_v e_i, e_i \rangle = -\operatorname{Trace} F_v. \quad \square$$

Definition 13.115. Let $\gamma : [a,b] \to M$ be a geodesic. Let $\mathcal{J}_0(\gamma, a, b)$ denote the set of all Jacobi fields J such that $J(a) = J(b) = 0$.

Definition 13.116. Let $\gamma : [a, b] \to M$ be a geodesic. If there exists a nonzero Jacobi field $J \in \mathcal{J}_0(\gamma, a, b)$, then we say that $\gamma(a)$ is **conjugate** to $\gamma(b)$ along γ.

From standard considerations in the theory of linear differential equations it follows that the set $\mathcal{J}_0(\gamma, a, b)$ is a vector space. The dimension of the vector space $\mathcal{J}_0(\gamma, a, b)$ is the **order** of the conjugacy. Since the Jacobi fields in $\mathcal{J}_0(\gamma, a, b)$ vanish twice, and, as we have seen, this means that such fields are normal to $\dot{\gamma}$ all along γ, it follows that the dimension of $\mathcal{J}_0(\gamma, a, b)$ is at most $n - 1$, where $n = \dim M$. We have seen that a variation through geodesics is a Jacobi field; so if we can find a nontrivial variation h of a geodesic γ such that all of the longitudinal curves $t \mapsto h_s(t)$ begin and end at the same points $\gamma(a)$ and $\gamma(b)$, then the variation vector field will be a nonzero element of $\mathcal{J}_0(\gamma, a, b)$. Thus we conclude that $\gamma(a)$ is conjugate to $\gamma(b)$. We will see that we may obtain a Jacobi field by more general variations, where the endpoints of the curves meet at time b only to first order.

Let us bring the exponential map into play. Let $\gamma : [0, b] \to M$ be a geodesic as above. Let $v = \dot{\gamma}(0) \in T_pM$. Then $\gamma : t \mapsto \exp_p tv$ is exactly our geodesic γ which begins at p and ends at q at $t = b$. Now we create a variation of γ by
$$h(s, t) = \exp_p t(v + sw),$$
where $w \in T_pM$ and s ranges in $(-\epsilon, \epsilon)$ for some sufficiently small ϵ. We know that $J(t) = \frac{\partial}{\partial s}\big|_{s=0} h(s, t)$ is a Jacobi field, and it is clear that $J(0) := \frac{\partial}{\partial s}\big|_{s=0} h(s, 0) = 0$. If w_{bv} is the vector tangent in $T_{bv}(T_pM)$ which canonically corresponds to w, in other words, if w_{bv} is the velocity vector at $s = 0$ for the curve $s \mapsto b(v + sw)$ in T_pM, then
$$J(b) = \frac{\partial}{\partial s}\bigg|_{s=0} h(s, b) = \frac{\partial}{\partial s}\bigg|_{s=0} \exp_p b(v + sw) = T_{bv} \exp_p(w_{bv}).$$
(We have just calculated the tangent map of $\exp_p$ at $x = bv$!) Also,
$$\nabla_{\partial_t} J(0) = \nabla_{\partial_t} \frac{\partial}{\partial s} \exp_p t(v + sw) \bigg|_{s=0, t=0}$$
$$= \nabla_{\partial_s}\big|_{s=0} \frac{\partial}{\partial t}\bigg|_{t=0} \exp_p t(v + sw).$$
But $X(s) := \frac{\partial}{\partial t}\big|_{t=0} \exp_p t(v + sw) = v + sw$ is a vector field along the constant curve $t \mapsto p$, and so by Exercise 12.42 we have $\nabla_{\partial_s}\big|_{s=0} X(s) = X'(0) = w$. The equality $J(b) = T_{bv} \exp_p(v_{bv})$ is important because it shows that if $T_{bv} \exp_p : T_{bv}(T_pM) \to T_{\gamma(b)}M$ is not an isomorphism, then we can find a vector $w_{bv} \in T_{bv}(T_pM)$ such that $T_{bv} \exp_p(w_{bv}) = 0$. But then if w is the vector in T_pM which corresponds to w_{bv} as above, then for this choice of w,

the Jacobi field constructed above is such that $J(0) = J(b) = 0$ so that $\gamma(0)$ is conjugate to $\gamma(b)$ along γ. Also, if J is a Jacobi field with $J(0) = 0$ and $\nabla_{\partial_t} J(0) = w$, then this uniquely determines J and it must have the form $\frac{\partial}{\partial s}\big|_{s=0} \exp_p t(v+sw)$ as above.

Theorem 13.117. *Let $\gamma : [0,b] \to M$ be a geodesic. Then the following are equivalent:*

(i) *$\gamma(0)$ is conjugate to $\gamma(b)$ along γ.*

(ii) *There is a nontrivial variation h of γ through geodesics that all start at $p = \gamma(0)$ such that $J(b) := \frac{\partial h}{\partial s}(0, b) = 0$.*

(iii) *If $v = \dot\gamma(0)$, then $T_{bv} \exp_p$ is singular.*

Proof. (ii)$\Longrightarrow$(i): We have already seen that a variation through geodesics is a Jacobi field J and that if (ii) holds, then by assumption $J(0) = J(b) = 0$, and so we have (i).

(i)$\Longrightarrow$(iii): If (i) is true, then there is a nonzero Jacobi field J with $J(0) = J(b) = 0$. Now let $w = \nabla_{\partial_t} J(0)$ and $h(s,t) = \exp_p t(v+sw)$. Then $h(s,t)$ is a variation through geodesics and $0 = J(b) = \frac{\partial}{\partial s}\big|_0 \exp_p b(v+sw) = T_{bv} \exp_p(w_{bv})$ so that $T_{bv} \exp_p$ is singular.

(iii)$\Longrightarrow$(ii): Let $v = \dot\gamma(0)$. If $T_{bv} \exp_p$ is singular, then there is a w with $T_{bv} \exp_p w_{bv} = 0$. Thus the variation $h(s,t) = \exp_p t(v+sw)$ does the job. $\square$

13.8. First and Second Variation of Arc Length

Let us restrict attention to the case where α is either spacelike or timelike (not necessarily geodesic). This is just the condition that $|\langle \dot\alpha(t), \dot\alpha(t)\rangle| > 0$. Let $\varepsilon = +1$ if α is spacelike and $\varepsilon = -1$ if α is timelike. We call ε the **sign** of α and write $\varepsilon = \operatorname{sgn}\alpha$. Consider the arc length functional defined by

$$L(\alpha) = \int_a^b (\varepsilon \langle \dot\alpha(t), \dot\alpha(t)\rangle)^{1/2}\, dt = \int_a^b |\langle \dot\alpha(t), \dot\alpha(t)\rangle|^{1/2}\, dt.$$

If $h : (-\epsilon, \epsilon) \times [a,b] \to M$ is a variation of α as above with variation vector field V, then formally V is a tangent vector at α in the space of curves $[a,b] \to M$. By a simple continuity and compactness argument we may choose a real number $\epsilon > 0$ small enough that $\left|\langle \dot h_s(t), \dot h_s(t)\rangle\right| > 0$ for all $s \in (-\epsilon, \epsilon)$. Then we have the variation of the arc length functional defined by

$$\delta L\big|_\alpha (V) := \frac{d}{ds}\bigg|_{s=0} L(h_s) := \frac{d}{ds}\bigg|_{s=0} \int_a^b \left(\varepsilon \langle \dot h_s(t), \dot h_s(t)\rangle\right)^{1/2} dt.$$

Thus, we are interested in studying the critical points of $L(s) := L(h_s)$, and so we need to find $L'(0)$ and $L''(0)$. For the proof of the following proposition we use the result of Exercise 13.11 to the effect that $\nabla_{\partial_s} \partial_t h = \nabla_{\partial_t} \partial_s h$.

Proposition 13.118. *Let $h : (-\epsilon, \epsilon) \times [a, b] \to M$ be a variation of a curve $\alpha := h_0$ such that $|\langle \dot{h}_s(t), \dot{h}_s(t) \rangle| > 0$ for all $s \in (-\epsilon, \epsilon)$. Then*

$$L'(s) = \int_a^b \varepsilon \langle \nabla_{\partial_s} \partial_t h(s,t), \partial_t h(s,t) \rangle \left(\varepsilon \langle \partial_t h(s,t), \partial_t h(s,t) \rangle \right)^{-1/2} dt.$$

Proof. We have

$$L'(s) = \frac{d}{ds} \int_a^b \left\| \dot{h}_s(t) \right\| dt$$

$$= \int_a^b \frac{d}{ds} \left(\varepsilon \langle \dot{h}_s(t), \dot{h}_s(t) \rangle \right)^{1/2} dt$$

$$= \int_a^b 2\varepsilon \langle \nabla_{\partial_s} \dot{h}_s(t), \dot{h}_s(t) \rangle \frac{1}{2} \left(\varepsilon \langle \dot{h}_s(t), \dot{h}_s(t) \rangle \right)^{-1/2} dt$$

$$= \int_a^b \varepsilon \langle \nabla_{\partial_s} \partial_t h(s,t), \partial_t h(s,t) \rangle \left(\varepsilon \langle \partial_t h(s,t), \partial_t h(s,t) \rangle \right)^{-1/2} dt$$

$$= \varepsilon \int_a^b \langle \nabla_{\partial_t} \partial_s h(s,t), \partial_t h(s,t) \rangle \left(\varepsilon \langle \partial_t h(s,t), \partial_t h(s,t) \rangle \right)^{-1/2} dt. \quad \square$$

Corollary 13.119. *We have*

$$\delta L|_\alpha (V) = L'(0) = \varepsilon \int_a^b \langle \nabla_{\partial_t} V(t), \dot{\alpha}(t) \rangle \left(\varepsilon \langle \dot{\alpha}(t), \dot{\alpha}(t) \rangle \right)^{-1/2} dt.$$

Let us now consider a more general situation where $\alpha : [a, b] \to M$ is only piecewise smooth (but still continuous). Let us be specific by saying that there is a partition $a = t_0 < t_1 < \cdots < t_k < t_{k+1} = b$ so that α is smooth on each $[t_i, t_{i+1}]$. A variation appropriate to this situation is a continuous map $h : (-\epsilon, \epsilon) \times [a, b] \to M$ with $h(0, t) = \alpha(t)$ such that h is smooth on each set of the form $(-\epsilon, \epsilon) \times [t_i, t_{i+1}]$. This is what we mean by a piecewise smooth variation of a piecewise smooth curve. The velocity $\dot{\alpha}$ and the variation vector field $V(t) := \frac{\partial h(0,t)}{\partial s}$ are only piecewise smooth. At each "kink" point t_i we have the jump vector $\triangle \dot{\alpha}(t_i) := \dot{\alpha}(t_i+) - \dot{\alpha}(t_i-)$, which measures the discontinuity of $\dot{\alpha}$ at t_i. Using this notation, we have the following theorem which gives the **first variation formula**:

Theorem 13.120. *Let $h : (-\epsilon, \epsilon) \times [a, b] \to M$ be a piecewise smooth variation of a piecewise smooth curve $\alpha : [a, b] \to M$ with variation vector field V. If α has constant speed $c = (\varepsilon \langle \dot{\alpha}, \dot{\alpha} \rangle)^{1/2}$, then*

$$\delta L|_\alpha (V) = L'(0) = -\frac{\varepsilon}{c} \int_a^b \langle \nabla_{\partial_t} \dot{\alpha}, V \rangle \, dt - \frac{\varepsilon}{c} \sum_{i=1}^k \langle \triangle \dot{\alpha}(t_i), V(t_i) \rangle + \frac{\varepsilon}{c} \langle \dot{\alpha}, V \rangle \Big|_a^b$$

Proof. Since $c = (\varepsilon \langle \dot{\alpha}, \dot{\alpha} \rangle)^{1/2}$, Proposition 13.119 gives

$$L'(0) = \frac{\varepsilon}{c} \sum \int_{t_i}^{t_{i+1}} \langle \nabla_{\partial_t} V(t), \dot{\alpha}(t) \rangle \, dt.$$

Since we have $\langle \dot{\alpha}, \nabla_{\partial_t} V \rangle = \frac{d}{dt} \langle \dot{\alpha}, V \rangle - \langle \nabla_{\partial_t} \dot{\alpha}, V \rangle$, we can employ integration by parts: On each interval $[t_i, t_{i+1}]$ we have

$$\frac{\varepsilon}{c} \int_{t_i}^{t_{i+1}} \langle \nabla_{\partial_t} V, \dot{\alpha} \rangle \, dt = \frac{\varepsilon}{c} \langle \dot{\alpha}, V \rangle \Big|_{t_i}^{t_{i+1}} - \frac{\varepsilon}{c} \int_{t_i}^{t_{i+1}} \langle \nabla_{\partial_t} \dot{\alpha}, V \rangle \, dt.$$

We sum from $i = 0$ to $i = k$ to get

$$L'(0) = \frac{\varepsilon}{c} \langle \dot{\alpha}, V \rangle \Big|_a^b - \frac{\varepsilon}{c} \sum_{i=1}^k \langle \triangle \dot{\alpha}(t_i), V(t_i) \rangle - \frac{\varepsilon}{c} \int_a^b \langle \nabla_{\partial_t} \dot{\alpha}, V \rangle \, dt,$$

which is the required result. $\square$

A variation $h : (-\epsilon, \epsilon) \times [a, b] \to M$ of α is called a **fixed endpoint variation** if $h(s, a) = \alpha(a)$ and $h(s, b) = \alpha(b)$ for all $s \in (-\epsilon, \epsilon)$. In this situation, the variation vector field V is zero at a and b.

Corollary 13.121. *A piecewise smooth curve $\alpha : [a, b] \to M$ with constant speed $c > 0$ on each subinterval where α is smooth is a (nonnull) geodesic if and only if $\delta L|_\alpha (V) = 0$ for all fixed endpoint variations of α. In particular, if M is a Riemannian manifold and $\alpha : [a, b] \to M$ minimizes length among nearby curves, then α is an (unbroken) geodesic.*

Proof. If α is a geodesic, then it is smooth and so $\triangle \dot{\alpha}(t_i) = 0$ for all t_i (even though α is smooth, the variation still only needs to be piecewise smooth). It follows that $L'(0) = 0$.

Now if we suppose that α is a piecewise smooth curve and that $L'(0) = 0$ for any variation, then we can conclude that α is a geodesic by picking some clever variations. As a first step we show that $\alpha|_{[t_i, t_{i+1}]}$ is a geodesic for each segment $[t_i, t_{i+1}]$. Let $t \in (t_i, t_{i+1})$ be arbitrary and let v be any nonzero vector in $T_{\alpha(t)} M$. Let β be a cut-off function on $[a, b]$ with support in $(t - \delta, t + \delta)$ and δ chosen sufficiently small. Then let $V(t) := \beta(t) Y(t)$, where Y is the parallel translation of y along α. We can now easily produce a fixed endpoint variation with variation vector field V by the formula

$$h(s, t) := \exp_{\alpha(t)} sV(t).$$

With this variation the last theorem gives

$$L'(0) = -\frac{\varepsilon}{c} \int_a^b \langle \nabla_{\partial_t} \dot{\alpha}, V \rangle \, dt = -\frac{\varepsilon}{c} \int_{t-\delta}^{t+\delta} \langle \nabla_{\partial_t} \dot{\alpha}, \beta(t) Y(t) \rangle \, dt,$$

which must hold no matter what our choice of y and for any $\delta > 0$. From this it is straightforward to show that $\nabla_{\partial_t} \dot{\alpha}(t) = 0$, and since t was an arbitrary element of (t_i, t_{i+1}), we conclude that $\alpha|_{[t_i,t_{i+1}]}$ is a geodesic. All that is left is to show that there can be no discontinuities of $\dot\alpha$. Once again we choose a vector y, but this time $y \in T_{\alpha(t_i)}M$, where t_i is a potential kink point. Take another cut-off function β with $\operatorname{supp}\beta \subset [t_{i-1}, t_{i+1}] = [t_{i-1}, t_i] \cup [t_i, t_{i+1}]$, $\beta(t_i) = 1$, and i a fixed but arbitrary element of $\{1, 2, \ldots, k\}$. Extend y to a field Y as before and let $V = \beta Y$. Since we now have that α is a geodesic on each segment, and we are assuming that the variation is zero, the first variation formula for any variation with variation vector field V reduces to

$$0 = L'(0) = -\frac{\varepsilon}{c}\langle \triangle\dot\alpha(t_i), y\rangle$$

for all y. This means that $\triangle\dot\alpha(t_i) = 0$, and since i was arbitrary, we are done. $\square$

We now see that, for fixed endpoint variations, $L'(0) = 0$ implies that α is a geodesic. The geodesics are the critical "points" (or curves) of the arc length functional restricted to all curves with fixed endpoints. In order to classify the critical curves, we look at the *second* variation but we only need the formula for variations of geodesics. For a variation h of a geodesic γ, we have the variation vector field V as before, but now we also consider the transverse acceleration vector field $A(t) := \nabla_{\partial_s}\partial_s h(0, t)$. Recall that for a curve γ with $|\langle\dot\gamma, \dot\gamma\rangle| > 0$, a vector field Y along γ has an orthogonal decomposition $Y = Y^\top + Y^\perp$ (tangent and normal to γ). Also we have $(\nabla_{\partial_t} Y)^\perp = \nabla_{\partial_t} Y^\perp$, and so we can use $\nabla_{\partial_t} Y^\perp$ to denote either of these without ambiguity.

We now have the **second variation formula of Synge**:

Theorem 13.122. *Let $\gamma : [a,b] \to M$ be a (nonnull) geodesic of speed $c > 0$. Let ε be the sign of γ as before. If $h : (-\epsilon, \epsilon) \times [a, b]$ is a variation of γ with variation vector field V and acceleration vector field A, then the second variation of $L(s) := L(h_s(t))$ at $s = 0$ is*

$$L''(0) = \frac{\varepsilon}{c}\int_a^b \left(\langle\nabla_{\partial_t}V^\perp, \nabla_{\partial_t}Y^\perp\rangle + \langle R_{\dot\gamma, V}\dot\gamma, V\rangle\right) dt + \frac{\varepsilon}{c}\langle\dot\gamma, A\rangle\bigg|_a^b.$$

Proof. Let $H(s,t) := \left|\langle\frac{\partial h}{\partial s}(s,t), \frac{\partial h}{\partial s}(s,t)\rangle\right|^{1/2} = \left(\varepsilon\langle\frac{\partial h}{\partial s}(s,t), \frac{\partial h}{\partial s}(s,t)\rangle\right)^{1/2}$. We have $L'(s) = \int_a^b \frac{\partial}{\partial s} H(s,t)\, dt$. Computing as before, we see that

$$\frac{\partial H(s,t)}{\partial s} = \frac{\varepsilon}{H}\left\langle\frac{\partial h}{\partial s}(s,t), \nabla_{\partial_s}\frac{\partial h}{\partial t}(s,t)\right\rangle.$$

13.8. First and Second Variation of Arc Length

Taking another derivative, we have

$$\frac{\partial^2 H(s,t)}{\partial s^2} = \frac{\varepsilon}{H^2}\left(H\frac{\partial}{\partial s}\langle\frac{\partial h}{\partial t},\nabla_{\partial_s}\frac{\partial h}{\partial t}\rangle - \langle\frac{\partial h}{\partial t},\nabla_{\partial_s}\frac{\partial h}{\partial t}\rangle\frac{\partial H}{\partial s}\right)$$

$$= \frac{\varepsilon}{H}\left(\langle\nabla_{\partial_s}\frac{\partial h}{\partial t},\nabla_{\partial_s}\frac{\partial h}{\partial t}\rangle + \langle\frac{\partial h}{\partial t},\nabla^2_{\partial_s}\frac{\partial h}{\partial t}\rangle - \frac{1}{H}\langle\frac{\partial h}{\partial t},\nabla_{\partial_s}\frac{\partial h}{\partial t}\rangle\frac{\partial H}{\partial s}\right)$$

$$= \frac{\varepsilon}{H}\left(\langle\nabla_{\partial_s}\frac{\partial h}{\partial t},\nabla_{\partial_s}\frac{\partial h}{\partial t}\rangle + \langle\frac{\partial h}{\partial t},\nabla^2_{\partial_s}\frac{\partial h}{\partial t}\rangle - \frac{\varepsilon}{H^2}\langle\frac{\partial h}{\partial t},\nabla_{\partial_s}\frac{\partial h}{\partial t}\rangle^2\right).$$

Using $\nabla_{\partial_t}\partial_s h = \nabla_{\partial_s}\partial_t h$ and Lemma 13.107, we obtain

$$\nabla_{\partial_s}\nabla_{\partial_s}\frac{\partial h}{\partial t} = \nabla_{\partial_s}\nabla_{\partial_t}\frac{\partial h}{\partial s} = R\left(\frac{\partial h}{\partial s},\frac{\partial h}{\partial t}\right)\frac{\partial h}{\partial s} + \nabla_{\partial_t}\nabla_{\partial_s}\frac{\partial h}{\partial s},$$

and then

$$\frac{\partial^2 H}{\partial s^2} = \frac{\varepsilon}{H}\left\{\langle\nabla_{\partial_t}\frac{\partial h}{\partial s},\nabla_{\partial_t}\frac{\partial h}{\partial s}\rangle + \langle\frac{\partial h}{\partial t},R\left(\frac{\partial h}{\partial s},\frac{\partial h}{\partial t}\right)\frac{\partial h}{\partial s}\rangle\right.$$
$$\left. + \langle\frac{\partial h}{\partial t},\nabla_{\partial_t}\nabla_{\partial_s}\frac{\partial h}{\partial s}\rangle - \frac{\varepsilon}{H^2}\langle\frac{\partial h}{\partial t},\nabla_{\partial_t}\frac{\partial h}{\partial s}\rangle^2\right\}.$$

Now we let $s = 0$ and get

$$\frac{\partial^2 H}{\partial s^2}(0,t) = \frac{\varepsilon}{c}\left\{\langle\nabla_{\partial_t}V,\nabla_{\partial_t}V\rangle + \langle\dot{\gamma},R(V,\dot{\gamma})V\rangle \right.$$
$$\left. + \langle\dot{\gamma},\nabla_{\partial_t}A\rangle - \frac{\varepsilon}{c^2}\langle\dot{\gamma},\nabla_{\partial_t}V\rangle^2\right\}.$$

Before we integrate the above expression, we use the fact that $\langle\dot{\gamma},\nabla_{\partial_t}A\rangle = \frac{d}{dt}\langle\dot{\gamma},A\rangle$ (γ is a geodesic) and the fact that the orthogonal decomposition of $\nabla_{\partial_t}V$ is

$$\nabla_{\partial_t}V = \frac{\varepsilon}{c^2}\langle\dot{\gamma},\nabla_{\partial_t}V\rangle\dot{\gamma} + \nabla_{\partial_t}V^\perp,$$

so that $\langle\nabla_{\partial_t}V,\nabla_{\partial_t}V\rangle = \frac{\varepsilon}{c^2}\langle\dot{\gamma},\nabla_{\partial_t}V\rangle^2 + \langle\nabla_{\partial_t}V^\perp,\nabla_{\partial_t}V^\perp\rangle$. Plugging these identities in, observing the cancellation, and integrating, we get

$$L''(0) = \int_a^b \frac{\partial^2 H}{\partial s^2}(0,t)\,dt = \frac{\varepsilon}{c}\int_a^b\left(\langle\nabla_{\partial_t}V^\perp,\nabla_{\partial_t}V^\perp\rangle + \langle\dot{\gamma},R(V,\dot{\gamma})V\rangle\right)dt$$
$$+ \frac{\varepsilon}{c}\langle\dot{\gamma},A\rangle\Big|_a^b. \qquad \square$$

The right hand side of the main equation of the second variation formula just proved depends only on V except for the last term. But if the variation is a fixed endpoint variation, then this dependence drops out.

It is traditional to think of the set $\Omega_{a,b}(p,q)$ of all piecewise smooth curves $\alpha : [a,b] \to M$ from p to q as an infinite-dimensional manifold. Then a variation vector field V along a curve $\alpha \in \Omega(p,q)$ which is zero at the endpoints is the "tangent vector" at α to the curve in $\Omega_{a,b}(p,q)$ given by the corresponding fixed endpoint variation h. Thus the "tangent space"

$T_\alpha \Omega = T_\alpha(\Omega_{a,b}(p,q))$ at α is the set of all piecewise smooth vector fields V along α such that $V(a) = V(b) = 0$. We then think of L as being a function on $\Omega_{a,b}(p,q)$ whose constant speed and nonnull critical points we have discovered to be nonnull geodesics beginning at p and ending at q at times a and b respectively. Further thinking along these lines leads to the idea of the index form. Let us abbreviate $\Omega_{a,b}(p,q)$ to $\Omega_{a,b}$ or even to Ω. For our present purposes, we will not lose anything by assuming that $a = 0$ whenever convenient. On the other hand, it will pay to refrain from assuming that $b = 1$.

Definition 13.123. For a given nonnull geodesic $\gamma\colon [0,b] \to M$, the **index form** $I_\gamma \colon T_\gamma \Omega \times T_\gamma \Omega \to \mathbb{R}$ is defined by $I_\gamma(V,V) = L''_\gamma(0)$, where $L_\gamma(s) = \int_0^b |\langle \dot{h}_s(t), \dot{h}_s(t)\rangle|^{1/2} \, dt$ and $\nabla_{\partial_s} h(0,t) = V$.

Of course this definition makes sense because $L''_\gamma(0)$ only depends on V and not on h itself. Also, we have defined the quadratic form $I_\gamma(V,V)$, but not directly $I_\gamma(V,W)$. Of course, polarization gives $I_\gamma(V,W)$, but if $V, W \in T_\gamma\Omega$, then it is not hard to see from the second variation formula that

$$(13.9) \qquad I_\gamma(V,W) = \frac{\varepsilon}{c} \int_0^b \left\{ \langle \nabla_{\partial_t} V^\perp, \nabla_{\partial_t} W^\perp \rangle + \langle R(\dot\gamma, V)\dot\gamma, W\rangle \right\} dt.$$

It is important to notice that the right hand side of the above equation is in fact symmetric in V and W.

It is important to remember that the variations and variation vector fields we are dealing with are allowed to be only piecewise smooth even if the center curve is smooth. So let $0 = t_0 < t_1 < \cdots < t_k < t_{k+1} = b$ as before and let V and W be vector fields along a geodesic γ. We now derive another formula for $I_\gamma(V,W)$. Rewrite formula (13.9) as

$$I_\gamma(V,W) = \frac{\varepsilon}{c} \sum_{i=0}^{k} \int_{t_i}^{t_{i+1}} \left\{ \langle \nabla_{\partial_t} V^\perp, \nabla_{\partial_t} W^\perp \rangle + \langle R(\dot\gamma, V)\dot\gamma, W\rangle \right\} dt.$$

On each interval $[t_i, t_{i+1}]$ we have

$$\langle \nabla_{\partial_t} V^\perp, \nabla_{\partial_t} W^\perp \rangle = \nabla_{\partial_t} \langle \nabla_{\partial_t} V^\perp, W^\perp \rangle - \langle \nabla_{\partial_t}^2 V^\perp, W^\perp \rangle,$$

and substituting this into the above formula we obtain

$$I_\gamma(V,W) = \frac{\varepsilon}{c} \sum_{i=0}^{k} \int_{t_i}^{t_{i+1}} \Big\{ \nabla_{\partial_t} \langle \nabla_{\partial_t} V^\perp, W^\perp \rangle - \langle \nabla_{\partial_t}^2 V^\perp, W^\perp \rangle$$
$$+ \langle R(\dot\gamma, V)\dot\gamma, W\rangle \Big\} dt.$$

As for the last term, we use symmetries of the curvature tensor to see that

$$\langle R(\dot\gamma, V)\dot\gamma, W\rangle = \langle R(\dot\gamma, V^\perp)\dot\gamma, W^\perp\rangle.$$

13.8. First and Second Variation of Arc Length

Substituting we get

$$I_\gamma(V,W) = \frac{\varepsilon}{c} \sum_{i=0}^{k} \int_{t_i}^{t_{i+1}} \left\{ \nabla_{\partial_t}\langle \nabla_{\partial_t} V^\perp, W^\perp \rangle - \langle \nabla^2_{\partial_t} V^\perp, W^\perp \rangle \right.$$
$$\left. + \langle R(\dot\gamma, V^\perp)\dot\gamma, W^\perp \rangle \right\} dt.$$

Using the fundamental theorem of calculus on each interval $[t_i, t_{i+1}]$, and the fact that W vanishes at a and b, we obtain the following alternative formula:

Proposition 13.124 (Formula for index form)**.** *Let $\gamma : [0,b] \to M$ be a nonnull geodesic. Then for $V, W \in T_\gamma \Omega_{a,b}$,*

$$I_\gamma(V,W) = -\frac{\varepsilon}{c} \int_0^b \langle \nabla^2_{\partial_t} V^\perp + R(\dot\gamma, V^\perp)\dot\gamma, W^\perp \rangle \, dt$$
$$- \frac{\varepsilon}{c} \sum_{i=1}^{k} \langle \triangle \nabla_{\partial_t} V^\perp(t_i), W^\perp(t_i) \rangle,$$

where $\triangle \nabla_{\partial_t} V^\perp(t_i) = \nabla_{\partial_t} V^\perp(t_i+) - \nabla_{\partial_t} V^\perp(t_i-)$.

Letting $V = W$ we have

$$I_\gamma(V,V) = -\frac{\varepsilon}{c} \int_0^b \langle \nabla^2_{\partial_t} V^\perp + R(\dot\gamma, V^\perp)\dot\gamma, V^\perp \rangle \, dt - \frac{\varepsilon}{c} \sum_{i=1}^{k} \langle \triangle \nabla_{\partial_t} V^\perp(t_i), V^\perp(t_i) \rangle,$$

and the presence of the term $\langle R(\dot\gamma, V^\perp)\dot\gamma, V^\perp \rangle$ indicates a connection with Jacobi fields.

Definition 13.125. A geodesic segment $\gamma : [a,b] \to M$ is said to be **relatively length minimizing** (resp. **relatively length maximizing**) if for all piecewise smooth fixed endpoint variations h of γ the function $L(s) := \int_a^b |\langle \dot h_s(t), \dot h_s(t) \rangle|^{1/2} \, dt$ has a local minimum (resp. local maximum) at $s = 0$ (where $\gamma = h_0(t) := h(0,t)$).

If $\gamma : [a,b] \to M$ is a relatively length minimizing nonnull geodesic, then $L''(0) = 0$, which means that $I_\gamma(V,V) = 0$ for any $V \in T_\gamma \Omega_{a,b}$. The adverb "relatively" is included in the terminology because of the possibility that there may be curves in $\Omega_{a,b}$ which are "far away" from γ and which have smaller length than γ. A simple example of this is depicted in Figure 13.8, where γ_2 has greater length than γ even though γ_2 is relatively length minimizing. We assume that the metric on $(0,1) \times S^1$ is the usual definite metric $dx^2 + dy^2$ induced from that on $\mathbb{R} \times (0,1)$, where we identify $S^1 \times (0,1)$ with the quotient $\mathbb{R} \times (0,1)/((x,y) \sim (x+2\pi,y))$. On the other hand, one sometimes hears the statement that geodesics in a Riemannian manifold are **locally length minimizing**. This means that for any geodesic $\gamma :$

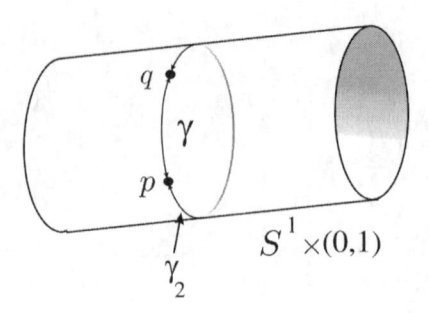

Figure 13.8. Geodesic segments on a cylinder

$[a, b] \to M$, the restrictions to small intervals are always relatively length minimizing. But note that this is only true for Riemannian manifolds. For a semi-Riemannian manifold with indefinite metric, a small geodesic segment can have nearby curves that *decrease* the length. For example, consider the metric $-dx^2 + dy^2$ on $\mathbb{R} \times (0,1)$ and the induced metric on the quotient $S^1 \times (0,1) = \mathbb{R} \times (0,1)/\sim$. In this case, the geodesic γ in the figure has length greater than all nearby geodesics; the index form I_γ is now negative semidefinite.

Exercise 13.126. Prove the above statement concerning I_γ for $S^1 \times (0,1)$ with the index 1 metric $-dx^2 + dy^2$.

It is not hard to see that if even one of V or W is tangent to $\dot{\gamma}$, then $I_\gamma(V, W) = 0$ and so $I_\gamma(V, W) = I_\gamma(V^\perp, W^\perp)$. Thus, we may as well restrict I_γ to
$$T_\gamma^\perp \Omega = \{V \in T_\gamma \Omega : V \perp \dot{\gamma}\}.$$

Notation 13.127. The restriction of I_γ to $T_\gamma^\perp \Omega$ will be called the **restricted index** and will be denoted by $I_\gamma^\perp$. The **nullspace** $\mathcal{N}(I_\gamma^\perp)$ is then defined by
$$\mathcal{N}(I_\gamma^\perp) := \left\{V \in T_\gamma^\perp \Omega : I_\gamma^\perp(V, W) = 0 \text{ for all } W \in T_\gamma^\perp \Omega\right\}.$$

Theorem 13.128. *Let $\gamma : [0, b] \to M$ be a nonnull geodesic. The nullspace $\mathcal{N}(I_\gamma^\perp)$ of $I_\gamma^\perp : T_\gamma^\perp \Omega \to R$ is exactly the space $\mathcal{J}_0(\gamma, 0, b)$ of Jacobi fields vanishing at $\gamma(0)$ and $\gamma(b)$.*

Proof. The formula of Proposition 13.124 makes it clear that $\mathcal{J}_0(\gamma, 0, b) \subset \mathcal{N}(I_\gamma^\perp)$.

Suppose that $V \in \mathcal{N}(I_\gamma^\perp)$. Let $t \in (t_i, t_{i+1})$, where the t_i determine a partition of $[0, b]$ such that V is potentially nonsmooth at the t_i as before. Pick an arbitrary nonzero element $y \in (\gamma(t))^\perp \subset T_{\gamma(t)}M$ and let Y be the unique parallel field along $\gamma|_{[t_i, t_{i+1}]}$ such that $Y(t) = y$. Picking a cut-off

13.8. First and Second Variation of Arc Length

function β with support in $[t + \delta, t - \delta] \subset (t_i, t_{i+1})$ as before we extend βY to a field W along γ with $W(t) = y$. Now V is normal to the geodesic and so $I_\gamma(V, W) = I_\gamma^\perp(V, W)$ and

$$I_\gamma(V, W) = -\frac{\varepsilon}{c} \int_{t-\delta}^{t+\delta} \langle \nabla_{\partial_t}^2 V + R(\dot\gamma, V)\dot\gamma, \beta Y \rangle \, dt.$$

For small δ, $\beta Y^\perp$ is approximately the arbitrary nonzero y and it follows that $\nabla_{\partial_t}^2 V + R(\dot\gamma, V)\dot\gamma$ is zero at t. Since t was arbitrary, $\nabla_{\partial_t}^2 V + R(\dot\gamma, V)\dot\gamma$ is identically zero on (t_i, t_{i+1}). Thus V is a Jacobi field on each interval (t_i, t_{i+1}), and since V is continuous on $[0, b]$, it follows from the standard theory of differential equations that V is a smooth Jacobi field along all of γ. Since $V \in T_\gamma \Omega$, we already have $V(0) = V(b) = 0$. We conclude that $V \in \mathcal{J}_0(\gamma, 0, b)$. □

Proposition 13.129. *Let (M, g) be a semi-Riemannian manifold of index $\mathrm{ind}(g)$ and $\gamma : [a, b] \to M$ a nonnull geodesic. If the index form I_γ is positive semidefinite, then $\mathrm{ind}(g) = 0$ or n (thus the metric is definite and so, up to the sign of g, the manifold is Riemannian). On the other hand, if I_γ is negative semidefinite, then $\mathrm{ind}(g) = 1$ or $n-1$ (so that up to the sign convention, M is a Lorentz manifold).*

Proof. For simplicity we assume that $a = 0$ so that $\gamma : [0, b] \to M$. Let I_γ be positive semi-definite and assume that $0 < \nu < n$ ($\nu = \mathrm{ind}(M)$). In this case, there must be a unit vector u in $T_{\gamma(0)}M$ which is normal to $\dot\gamma(0)$ and has the opposite causal character of $\dot\gamma(0)$. This means that if $\varepsilon = \langle \dot\gamma(0), \dot\gamma(0) \rangle / \|\dot\gamma(0)\|$, then $\varepsilon \langle u, u \rangle = -1$. Let U be the field along γ which is the parallel translation of u. By choosing $\delta > 0$ appropriately we can arrange that δ is as small as we like and simultaneously that $\sin(t/\delta)$ is zero at $t = 0$ and $t = b$. Let $V := \delta \sin(t/\delta) U$ and make the harmless assumption that $\|\dot\gamma\| = 1$. Notice that by construction $V \perp \dot\gamma$. We compute:

$$I_\gamma(V, V) = \varepsilon \int_0^b \{\langle \nabla_{\partial_t} V, \nabla_{\partial_t} V \rangle + \langle R(\dot\gamma, V)\dot\gamma, V \rangle\} \, dt$$

$$= \varepsilon \int_0^b \{\langle \nabla_{\partial_t} V, \nabla_{\partial_t} V \rangle - \langle R(\dot\gamma, V)V, \dot\gamma \rangle\} \, dt$$

$$= \varepsilon \int_0^b \{\langle \nabla_{\partial_t} V, \nabla_{\partial_t} V \rangle - K(V \wedge \dot\gamma)\langle V \wedge \dot\gamma, V \wedge \dot\gamma \rangle\} \, dt$$

$$= \varepsilon \int_0^b \{\langle \nabla_{\partial_t} V, \nabla_{\partial_t} V \rangle - K(V \wedge \dot\gamma)\langle V, V \rangle \varepsilon\} \, dt,$$

where

$$K(V \wedge \dot\gamma) := \frac{\langle \mathfrak{R}(V \wedge \dot\gamma), V \wedge \dot\gamma \rangle}{\langle V \wedge \dot\gamma, V \wedge \dot\gamma \rangle} = \frac{\langle \mathfrak{R}(V \wedge \dot\gamma), V \wedge \dot\gamma \rangle}{\varepsilon \langle V, V \rangle}$$

as defined earlier. Continuing the computation we have

$$I_\gamma(V,V) = \varepsilon \int_0^b \left\{ \langle u, u \rangle \cos^2(t/\delta) + K(V \wedge \dot{\gamma}) \delta^2 \sin^2(t/\delta) \right\} dt$$

$$= \int_0^b \left\{ -\cos^2(t/\delta) + \varepsilon K(V \wedge \dot{\gamma}) \delta^2 \sin^2(t/\delta) \right\} dt$$

$$= -b/2 + \delta^2 \int_0^b \varepsilon K(V \wedge \dot{\gamma}) \sin^2(t/\delta) \, dt.$$

Now as we said, we can choose δ as small as we like, and since $K(V(t) \wedge \dot{\gamma}(t))$ is bounded on the (compact) interval $[0, b]$, this clearly means that $I_\gamma(V, V) < 0$, which contradicts the fact that I_γ is positive semidefinite. Thus our assumption that $0 < \nu < n$ is impossible.

Now let I_γ be negative semidefinite. Suppose that we assume that contrary to what we wish to show, ν is not 1 or $n-1$. In this case, one can find a unit vector $u \in T_{\gamma(0)}M$ normal to $\dot{\gamma}(0)$ such that $\varepsilon \langle u, u \rangle = +1$. The same sort of calculation as we just did shows that I_γ cannot be semidefinite; again a contradiction. $\square$

By changing the sign of the metric the cases handled by this last theorem boil down to the two important cases: 1) where (M, g) is Riemannian, γ is arbitrary, and 2) where (M, g) is Lorentz and γ is timelike. We consolidate these two cases by a definition:

Definition 13.130. A geodesic $\gamma : [a, b] \to M$ is **cospacelike** if the subspace $\dot{\gamma}(s)^\perp \subset T_{\gamma(s)}M$ is spacelike for some (and consequently all) $s \in [a, b]$.

Exercise 13.131. Show that if $\gamma : [a, b] \to M$ is cospacelike, then γ is nonnull, $\dot{\gamma}(s)^\perp \subset T_{\gamma(s)}M$ is spacelike for all $s \in [a, b]$, and also show that (M, g) is either Riemannian or Lorentz.

A useful observation about Jacobi fields along a geodesic is the following:

Lemma 13.132. *If we have two Jacobi fields J_1 and J_2 along a geodesic γ, then $\langle \nabla_{\partial_t} J_1, J_2 \rangle - \langle J_1, \nabla_{\partial_t} J_2 \rangle$ is constant along γ.*

Proof. To see this, we note that

$$\nabla_{\partial_t} \langle \nabla_{\partial_t} J_1, J_2 \rangle = \langle \nabla_{\partial_t}^2 J_1, J_2 \rangle + \langle \nabla_{\partial_t} J_1, \nabla_{\partial_t} J_2 \rangle$$
$$= \langle R(\dot{\gamma}, J_1)\dot{\gamma}, J_2 \rangle + \langle \nabla_{\partial_t} J_1, \nabla_{\partial_t} J_2 \rangle$$
$$= \langle R(\dot{\gamma}, J_2)\dot{\gamma}, J_1 \rangle + \langle \nabla_{\partial_t} J_2, \nabla_{\partial_t} J_1 \rangle = \nabla_{\partial_t} \langle \nabla_{\partial_t} J_2, J_1 \rangle.$$

Similarly, we compute $\nabla_{\partial_t} \langle J_1, \nabla_{\partial_t} J_2 \rangle$ and subtract the result from the above to obtain the conclusion. $\square$

13.8. First and Second Variation of Arc Length

In particular, if $\langle \nabla_{\partial_t} J_1, J_2 \rangle = \langle J_1, \nabla_{\partial_t} J_2 \rangle$ at $t = 0$, then $\langle \nabla_{\partial_t} J_1, J_2 \rangle - \langle J_1, \nabla_{\partial_t} J_2 \rangle = 0$ for all t.

We need another simple technical lemma:

Lemma 13.133. *If $J_1, \ldots, J_k$ are Jacobi fields along a geodesic γ such that $\langle \nabla_{\partial_t} J_i, J_j \rangle = \langle J_i, \nabla_{\partial_t} J_j \rangle$ for all $i, j \in \{1, \ldots, k\}$, then any field Y which can be written as $Y = \sum \varphi^i J_i$ has the property that*

$$\langle \nabla_{\partial_t} Y, \nabla_{\partial_t} Y \rangle + \langle R(Y, \dot{\gamma})Y, \dot{\gamma} \rangle = \langle (\partial_t \varphi^i) J_i, (\partial_t \varphi^i) J_i \rangle + \partial_t \langle Y, \varphi^r (\nabla_{\partial_t} J_r) \rangle.$$

Proof. We have $\nabla_{\partial_t} Y = (\partial_t \varphi^i) J_i + \varphi^r (\nabla_{\partial_t} J_r)$ and so using the summation convention,

$$\begin{aligned}
\partial_t \langle Y, \varphi^r (\nabla_{\partial_t} J_r) \rangle &= \langle (\nabla_{\partial_t} Y), \varphi^r (\nabla_{\partial_t} J_r) \rangle + \langle Y, \nabla_{\partial_t} [\varphi^r (\nabla_{\partial_t} J_r)] \rangle \\
&= \langle (\partial_t \varphi^i) J_i, \varphi^r (\nabla_{\partial_t} J_r) \rangle + \langle \varphi^r (\nabla_{\partial_t} J_r), \varphi^k (\nabla_{\partial_t} J_k) \rangle \\
&\quad + \langle Y, \partial_t \varphi^r \nabla_{\partial_t} J_r \rangle + \langle Y, \varphi^r \nabla^2_{\partial_t} J_r \rangle.
\end{aligned}$$

The last term $\langle Y, \varphi^r \nabla^2_{\partial_t} J_r \rangle$ equals $\langle R(Y, \dot{\gamma})Y, \dot{\gamma} \rangle$ by the Jacobi equation. Using this and the fact that $\langle Y, \partial_t \varphi^r \nabla_{\partial_t} J_r \rangle = \langle (\partial_t \varphi^i) J_i, \varphi^r (\nabla_{\partial_t} J_r) \rangle$, which follows from a short calculation using the hypotheses on the J_i, we arrive at

$$\begin{aligned}
\partial_t \langle Y, \varphi^r (\nabla_{\partial_t} J_r) \rangle &= 2 \langle (\partial_t \varphi^i) J_i, \varphi^r (\nabla_{\partial_t} J_r) \rangle + \langle \varphi^r (\nabla_{\partial_t} J_r), \varphi^r (\nabla_{\partial_t} J_r) \rangle \\
&\quad + \langle R(Y, \dot{\gamma})Y, \dot{\gamma} \rangle.
\end{aligned}$$

Using the last equation together with $\nabla_{\partial_t} Y = (\partial_t \varphi^i) J_i + \varphi^r (\nabla_{\partial_t} J_r)$ gives the result (check it!). $\square$

Exercise 13.134. Work through the details of the proof of the lemma above.

Throughout the following discussion, $\gamma : [0, b] \to M$ will be a cospacelike geodesic with sign ε and speed c.

Suppose that there are no conjugate points of $p = \gamma(0)$ along γ. There exist Jacobi fields $J_1, \ldots, J_{n-1}$ along γ which vanish at $t = 0$ and are such that the vectors $\nabla_{\partial_t} J_1(0), \ldots, \nabla_{\partial_t} J_{n-1}(0) \in T_p M$ are a basis for the space $\dot{\gamma}(0)^\perp \subset T_{\gamma(0)} M$.

Claim: $J_1(t), \ldots, J_{n-1}(t)$ are linearly independent for each $t > 0$.

Indeed, suppose that $c_1 J_1(t) + \cdots + c_2 J_{n-1}(t) = 0$ for some t. Then, $Z := \sum_{i=1}^{n-1} c_i J_i$ is a normal Jacobi field with $Z(0) = Z(t) = 0$. But then, since there are no conjugate points, $Z = 0$ identically and so $0 = \nabla_{\partial_t} Z(0) := \sum_{i=1}^{n-1} c_i \nabla_{\partial_t} J_i(0)$. Since the $\nabla_{\partial_t} J_i(0)$ are linearly independent, we conclude that $c_i = 0$ for all i and the claim is proved.

It follows that at each t with $0 < t \leq b$ the vectors $J_1(t), \ldots, J_{n-1}(t)$ form a basis of $\dot{\gamma}(t)^\perp \subset T_{\gamma(t)} M$. Now let $Y \in T_\gamma(\Omega)$ be a piecewise smooth variation vector field along γ and write $Y = \sum \varphi^i J_i$ for some piecewise smooth functions φ^i on $(0, b]$, which can be shown to extend continuously to $[0, b]$ (see Problem 3). Since $\langle \nabla_{\partial_t} J_i, J_j \rangle = \langle J_i, \nabla_{\partial_t} J_j \rangle = 0$ at $t = 0$, we have $\langle \nabla_{\partial_t} J_i, J_j \rangle - \langle J_i, \nabla_{\partial_t} J_j \rangle = 0$ for all t by Lemma 13.132. This allows for the use of Lemma 13.133 to arrive at

$$\langle \nabla_{\partial_t} Y, \nabla_{\partial_t} Y \rangle + \langle R(Y, \dot\gamma)Y, \dot\gamma \rangle$$
$$= \left\langle \sum (\partial_t \varphi^i) J_i, \sum (\partial_t \varphi^i) J_i \right\rangle + \partial_t \left\langle Y, \sum \varphi^r (\nabla_{\partial_t} J_r) \right\rangle$$

and then
(13.10)
$$\varepsilon I_\gamma(Y, Y) = \frac{1}{c} \int_0^b \left\langle \sum (\partial_t \varphi^i) J_i, \sum (\partial_t \varphi^i) J_i \right\rangle dt + \frac{1}{c} \left. \langle Y, \varphi^r (\nabla_{\partial_t} J_r) \rangle \right|_0^b.$$

On the other hand, Y is zero at a and b and so the last term above vanishes. Now we notice that since γ is cospacelike and the J_i are normal to the geodesic, we must have that the integrand in equation (13.10) above is nonnegative. We conclude that $\varepsilon I_\gamma(Y, Y) \geq 0$. On the other hand, if $I_\gamma(Y, Y) = 0$ identically, then $\int_0^b \langle \sum (\partial_t \varphi^i) J_i, \sum (\partial_t \varphi^i) J_i \rangle dt = 0$ and $\langle \sum (\partial_t \varphi^i) J_i, \sum (\partial_t \varphi^i) J_i \rangle = 0$. In turn, this implies that $\sum (\partial_t \varphi^i) J_i \equiv 0$ and that each φ^i is constant, in fact zero, and finally that Y itself is identically zero along γ. All we have assumed about Y is that it is in the domain of the restricted index $I_\gamma^\perp$ and so we have proved the following:

Proposition 13.135. *If $\gamma \in \Omega$ is cospacelike and there is no conjugate points to $p = \gamma(0)$ along γ, then $\varepsilon I_\gamma^\perp(Y, Y) \geq 0$ and $Y = 0$ along γ if and only if $I_\gamma^\perp(Y, Y) = 0$.*

We may paraphrase the above result as follows: For a cospacelike geodesic γ without conjugate points, the restricted index $I_\gamma^\perp$ is definite; it is positive definite if $\varepsilon = +1$ and negative definite if $\varepsilon = -1$. The first case ($\varepsilon = +1$) is exactly the case where (M, g) is Riemannian (Exercise 13.131).

Next we consider the situation where the cospacelike geodesic $\gamma : [0, b] \to M$ is such that $\gamma(b)$ is the only point conjugate to $p = \gamma(0)$ along γ. In this case, Theorem 13.128 tells us that $I_\gamma^\perp$ has a nontrivial nullspace and so I_γ cannot be definite.

Claim: I_γ is semidefinite. To see this, let $Y \in T_\gamma \Omega$ and write Y in the form $(b - t)Z(t)$ for some (continuous) piecewise smooth Z. Let $b_i \to b$ and define Y_i to be $(b_i - t)Z(t)$ on $[0, b_i]$. Our last proposition applied to $\gamma_i := \gamma|_{[0,b_i]}$ shows that $\varepsilon I_{\gamma_i}(Y_i, Y_i) \geq 0$. Now $\varepsilon I_{\gamma_i}(Y_i, Y_i) \to \varepsilon I_\gamma(Y, Y)$ (some uninteresting details are omitted) and so the claim is true.

Now we consider the case where there is a conjugate point to p before $\gamma(b)$. Suppose that J is a nonzero Jacobi field along $\gamma|_{[0,r]}$ with $0 < r < b$

13.8. First and Second Variation of Arc Length

such that $J(0) = J(r) = 0$. We can extend J to a field J_{ext} on $[0, b]$ by defining it to be 0 on $[r, b]$. Notice that $\nabla_{\partial_t} J_{ext}(r-)$ is equal to $\nabla_{\partial_t} J(r)$, which is not 0 since otherwise J would be identically zero (over determination). On the other hand, $\nabla_{\partial_t} J_{ext}(r+) = 0$ and so the "kink" $\triangle J'_{ext}(r) := \nabla_{\partial_t} J_{ext}(r+) - \nabla_{\partial_t} J_{ext}(r-)$ is not zero. Notice that $\triangle J'_{ext}(r)$ is normal to γ (why?). We will now show that if $W \in T_\gamma(\Omega)$ is such that $W(r) = \triangle J'_{ext}(r)$ (and there are plenty of such fields), then $\varepsilon I_\gamma(J_{ext} + \delta W, J_{ext} + \delta W) < 0$ for small enough $\delta > 0$. This will allow us to conclude that I_γ cannot be definite since by Proposition 13.135 we can always find a Z with $\varepsilon I_\gamma(Z, Z) > 0$. We have

$$\varepsilon I_\gamma(J_{ext} + \delta W, J_{ext} + \delta W) = \varepsilon I_\gamma(J_{ext}, J_{ext}) + 2\delta\varepsilon I_\gamma(J_{ext}, W) + \varepsilon\delta^2 I_\gamma(W, W).$$

It is not hard to see from the formula of Theorem 13.124 that $I_\gamma(J_{ext}, J_{ext})$ is zero since it is piecewise Jacobi and is zero at the single kink point r. But using the formula again, $\varepsilon I_\gamma(J_{ext}(r), W(r))$ reduces to

$$-\frac{1}{c}\langle \triangle J'_{ext}(r), W(r)\rangle = -\frac{1}{c}\left|\triangle J'_{ext}(r)\right|^2 < 0,$$

and so taking δ small enough gives the desired conclusion.

Summarizing the conclusions of the above discussion (together with the result of Proposition 13.135) yields the following nice theorem:

Theorem 13.136. *If $\gamma : [0, b] \to M$ is a cospacelike geodesic of sign ε, then (M, g) is either Riemannian or Lorentz and we have the following three cases:*

(i) *If there are no points conjugate to $\gamma(0)$ along γ, then $\varepsilon I_\gamma^\perp$ is positive definite.*

(ii) *If $\gamma(b)$ is the only conjugate point to $\gamma(0)$ along γ, then I_γ is not definite, but must be semidefinite.*

(iii) *If there is a point $\gamma(r)$ conjugate to $\gamma(0)$ with $0 < r < b$, then I_γ is not semidefinite (or definite).*

As we mentioned the Jacobi equation can be written in terms of the tidal force operator $R_v : T_pM \to T_pM$ as

$$\nabla^2_{\partial_t} J(t) = R_{\dot\gamma(t)}(J(t)).$$

The meaning of the term force here is that $R_{\dot\gamma(t)}$ controls the way nearby families of geodesics attract or repel each other. Attraction tends to create conjugate points, while repulsion tends to prevent conjugate points. If γ is cospacelike, then we take any unit vector u normal to $\dot\gamma(t)$ and look at the component of $R_{\dot\gamma(t)}(u)$ in the u direction. Up to sign this is

$$\langle R_{\dot\gamma(t)}(u), u\rangle u = \langle R_{\dot\gamma(t),u}(\dot\gamma(t)), u\rangle u = -\langle \mathfrak{R}(\dot\gamma(t) \wedge u), \dot\gamma(t) \wedge u\rangle u.$$

In terms of sectional curvature,
$$\langle R_{\dot{\gamma}(t)}(u), u\rangle u = K(\dot{\gamma}(t) \wedge u) \langle \dot{\gamma}(t), \dot{\gamma}(t)\rangle.$$
It follows from the Jacobi equation that if $\langle R_{\dot{\gamma}(t)}(u), u\rangle \geq 0$, i.e., if $K(\dot{\gamma}(t) \wedge u)\langle \dot{\gamma}(t), \dot{\gamma}(t)\rangle \leq 0$, then we have repulsion, and if this always happens anywhere along γ, we expect that $\gamma(0)$ has no conjugate point along γ. This intuition is indeed correct.

Proposition 13.137. *Let $\gamma : [0, b] \to M$ be a cospacelike geodesic. If for every t and every vector $v \in \gamma(t)^\perp$ we have $\langle R_{\dot{\gamma}(t)}(v), v\rangle \geq 0$ (i.e. if $K(\dot{\gamma}(t) \wedge v)\langle \dot{\gamma}(t), \dot{\gamma}(t)\rangle \leq 0$), then $\gamma(0)$ has no conjugate point along γ.*

In particular, a Riemannian manifold with sectional curvature $K \leq 0$ has no conjugate pairs of points. Similarly, a Lorentz manifold with sectional curvature $K \geq 0$ has no conjugate pairs along any timelike geodesics.

Proof. Take J to be a Jacobi field along γ such that $J(0) = 0$ and $J \perp \dot{\gamma}$. We have $\frac{d}{dt}\langle J, J\rangle = 2\langle \nabla_{\partial_t} J, J\rangle$ and
$$\begin{aligned}\frac{d^2}{dt^2}\langle J, J\rangle &= 2\langle \nabla_{\partial_t} J, \nabla_{\partial_t} J\rangle + 2\langle \nabla^2_{\partial_t} J, J\rangle \\ &= 2\langle \nabla_{\partial_t} J, \nabla_{\partial_t} J\rangle + 2\langle R_{\dot{\gamma}(t), J}(\dot{\gamma}(t)), J\rangle \\ &= 2\langle \nabla_{\partial_t} J, \nabla_{\partial_t} J\rangle + 2\langle R_{\dot{\gamma}(t)}(J), J\rangle,\end{aligned}$$
and by the hypotheses $\frac{d^2}{dt^2}\langle J, J\rangle \geq 0$. On the other hand, $\langle J(0), J(0)\rangle = 0$ and $\frac{d}{dt}\big|_0 \langle J, J\rangle = 0$. It follows that since $\langle J, J\rangle$ is not identically zero we must have $\langle J, J\rangle > 0$ for all $t \in (0, b]$ and the result follows. $\square$

13.9. More Riemannian Geometry

Recall that a manifold is geodesically complete at p if and only if $\exp_p$ is defined on all of T_pM. The following lemma is the essential ingredient in the proof of the Hopf-Rinow theorem stated and proved below. Note that this is a theorem about Riemannian manifolds.

Lemma 13.138. *Let (M, g) be a connected Riemannian manifold. Suppose that $\exp_p$ is defined on the ball of radius $\rho > 0$ centered at $0 \in T_pM$. Let $B_\rho(p) := \{x : \mathrm{dist}(p, x) < \rho\}$. Then each point $q \in B_\rho(p)$ can be connected to p by an absolutely minimizing geodesic. In particular, if M is geodesically complete at $p \in M$, then each point $q \in M$ can be connected to p by an absolutely minimizing geodesic.*

Proof. Let $q \in B_\rho(p)$ with $p \neq q$ and let $R = \mathrm{dist}(p, q)$. Choose $\epsilon > 0$ small enough that $B_{2\epsilon}(p)$ is the domain of a normal coordinate system. (Refer to Figure 13.9.) By Lemma 13.85, we already know the theorem is true if $B_\rho(p) \subset B_\epsilon(p)$, so we will assume that $\epsilon < R < \rho$. Because $\partial B_\epsilon(p)$ is

13.9. More Riemannian Geometry

diffeomorphic to $S^{n-1} \subset \mathbb{R}^n$, it is compact and so there is a point $p_\epsilon \in \partial B_\epsilon(p)$ such that $x \mapsto \mathrm{dist}(x,q)$ achieves its minimum at p_ϵ. This means that

$$\mathrm{dist}(p,q) = \mathrm{dist}(p,p_\epsilon) + \mathrm{dist}(p_\epsilon,q) = \epsilon + \mathrm{dist}(p_\epsilon,q).$$

Let $\gamma : [0,\rho] \to M$ be the geodesic with $|\dot\gamma| = 1$, $\gamma(0) = p$, and $\gamma(\epsilon) = p_\epsilon$.

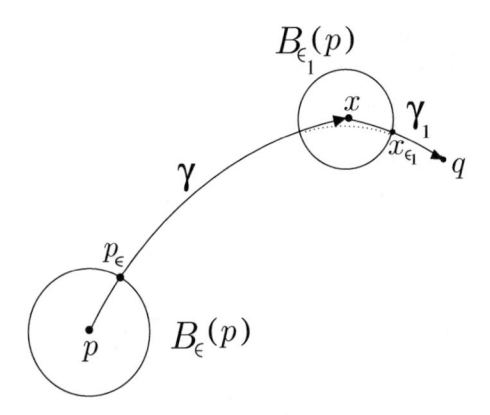

Figure 13.9

It is not difficult to see that the set

$$T = \{t \in [0,R] : \mathrm{dist}(p,\gamma(t)) + \mathrm{dist}(\gamma(t),q) = \mathrm{dist}(p,q)\}$$

is closed in $[0,R]$ and is nonempty since $\epsilon \in T$. Let $t_{\sup} = \sup T > 0$. We will show that $t_{\sup} = R$ from which it will follow that $\gamma|_{[0,R]}$ is a minimizing geodesic from p to q since then $\mathrm{dist}(\gamma(R),q) = 0$ and so $\gamma(R) = q$. With an eye toward a contradiction, assume that $t_{\sup} < R$. Let $x := \gamma(t_{\sup})$ and choose ϵ_1 with $0 < \epsilon_1 < R - t_{\sup}$ and small enough that $B_{2\epsilon_1}(x) \subset B_\rho(p)$ is the domain of normal coordinates about x. Arguing as before we see that there must be a point $x_{\epsilon_1} \in \partial B_{\epsilon_1}(x)$ such that

$$\mathrm{dist}(x,q) = \mathrm{dist}(x,x_{\epsilon_1}) + \mathrm{dist}(x_{\epsilon_1},q) = \epsilon_1 + \mathrm{dist}(x_{\epsilon_1},q).$$

Now let γ_1 be the unit speed geodesic such that $\gamma_1(0) = x$ and $\gamma_1(\epsilon_1) = x_{\epsilon_1}$. But since $t_{\sup} \in T$ and $x = \gamma(t_{\sup})$, we also have

$$\mathrm{dist}(p,x) + \mathrm{dist}(x,q) = \mathrm{dist}(p,q).$$

Combining, we now have

$$\mathrm{dist}(p,q) = \mathrm{dist}(p,x) + \mathrm{dist}(x,x_{\epsilon_1}) + \mathrm{dist}(x_{\epsilon_1},q).$$

By the triangle inequality, $\mathrm{dist}(p,q) \leq \mathrm{dist}(p,x_{\epsilon_1}) + \mathrm{dist}(x_{\epsilon_1},q)$ and so

$$\mathrm{dist}(p,x) + \mathrm{dist}(x,x_{\epsilon_1}) \leq \mathrm{dist}(p,x_{\epsilon_1}).$$

But also $\operatorname{dist}(p,x_{\epsilon_1}) \le \operatorname{dist}(p,x) + \operatorname{dist}(x,x_{\epsilon_1})$ and so
$$\operatorname{dist}(p,x_{\epsilon_1}) = \operatorname{dist}(p,x) + \operatorname{dist}(x,x_{\epsilon_1}).$$

Examining the implications of this last equality, we see that the concatenation of $\gamma|_{[0,t_{\sup}]}$ with γ_1 forms a curve from p to x_{ϵ_1} of length $\operatorname{dist}(p,x_{\epsilon_1})$, which must therefore be a minimizing curve. By Problem 1, this potentially broken geodesic must in fact be smooth and so must actually be the geodesic $\gamma|_{[0,t_{\sup}+\epsilon_1]}$. Then, $t_{\sup}+\epsilon_1 \in T$ which contradicts the definition of $t_{\sup}$. This contradiction forces us to conclude that $t_{\sup} = R$ and we are done. □

Theorem 13.139 (Hopf-Rinow). *If (M,g) is a connected Riemannian manifold, then the following statements are equivalent:*

(i) *The metric space (M, dist) is complete. That is, every Cauchy sequence is convergent.*

(ii) *There is a point $p \in M$ such that M is geodesically complete at p.*

(iii) *M is geodesically complete.*

(iv) *Every closed and bounded subset of M is compact.*

Proof. (iv)$\Longrightarrow$(i): The set of points of a Cauchy sequence is bounded and so has compact closure. Thus there is a subsequence converging to some point. Since the original sequence was Cauchy, it must converge to this point.

(i)$\Longrightarrow$(iii): Let p be arbitrary and let $\gamma_v(t)$ be the geodesic with $\dot{\gamma}_v(0) = v$ and J its maximal domain of definition. We can assume without loss of generality that $\langle v,v \rangle = 1$ so that $L(\gamma_v|_{[t_1,t_2]}) = t_2 - t_1$ for all relevant t_1, t_2. We want to show that there can be no upper bound for the set J. We argue by contradiction: Assume that $t_+ = \sup J$ is finite. Let $\{t_n\} \subset J$ be a Cauchy sequence such that $t_n \to t_+ < \infty$. Since $\operatorname{dist}(\gamma_v(t), \gamma_v(s)) \le |t-s|$, it follows that $\gamma_v(t_n)$ is a Cauchy sequence in M, which by assumption must converge. Let $q := \lim_{n\to\infty} \gamma_v(t_n)$ and choose a small ball $B_\epsilon(q)$ which is small enough to be a normal neighborhood. Take t_1 with $0 < t_+ - t_1 < \epsilon/2$ and let γ_1 be the (maximal) geodesic with initial velocity $\dot{\gamma}_v(t_1)$. Then in fact $\gamma_1(t) = \gamma_v(t_1 + t)$ and so γ_1 is defined for $t_1 + \epsilon/2 > t_+$ and this is a contradiction.

(iii)$\Longrightarrow$(ii) is a trivial implication.

(ii)$\Longrightarrow$(iv): Let K be a closed and bounded subset of M. For $x \in M$, Lemma 13.138 tells us that there is a minimizing geodesic $\alpha_x : [0,1] \to M$ connecting p to x. Then $\|\dot{\alpha}_x(0)\| = \operatorname{dist}(p,x)$ and $\exp(\dot{\alpha}_x(0)) = x$. Using the triangle inequality, one sees that

$$\sup_{x \in K}\{\|\dot{\alpha}_x(0)\|\} \le r$$

13.9. More Riemannian Geometry

for some $r < \infty$. From this we obtain $\{\dot{\alpha}_x(0) : x \in K\} \subset B_r := \{v \in T_pM : \|v\| \leq r\}$. The set B_r is compact. Now $\exp(B_r)$ is compact and contains the closed set K, so K is also compact.

(ii)$\Longrightarrow$(i): Suppose M is geodesically complete at p. Now let $\{x_n\}$ be any Cauchy sequence in M. For each x_n, there is (by assumption) a minimizing geodesic from p to x_n, which we denote by γ_{px_n}. We may assume that each γ_{px_n} is unit speed. It is easy to see that the sequence $\{l_n\}$, where $l_n := L(\gamma_{px_n}) = \mathrm{dist}(p, x_n)$, is a Cauchy sequence in $\mathbb{R}$ with some limit, say l. The key fact is that the vectors $\dot{\gamma}_{px_n}$ are all unit vectors in T_pM and so form a sequence in the (compact) unit sphere in T_pM. Replacing $\{\dot{\gamma}_{px_n}\}$ by a subsequence if necessary we have $\dot{\gamma}_{px_n} \to u \in T_pM$ for some unit vector u. Continuous dependence on initial velocities implies that $\{x_n\} = \{\gamma_{px_n}(l_n)\}$ has the limit $\gamma_u(l)$. $\square$

Let (M, g) be a complete connected Riemannian manifold with sectional curvature $K \leq 0$. By Proposition 13.137, for each point $p \in M$, the geodesics emanating from p have no conjugate points and so $T_{v_p}\exp_p : T_{v_p}T_pM \to M$ is nonsingular for each $v_p \in T_pM$. This means that $\exp_p$ is a local diffeomorphism. If we give T_pM the metric $\exp_p^*(g)$, then $\exp_p$ is a local isometry. It now follows from Theorem 13.76 that $\exp_p : T_pM \to M$ is a Riemannian covering. Thus we arrive at the Hadamard theorem.

Theorem 13.140 (Hadamard). *If (M, g) is a complete simply connected Riemannian manifold with sectional curvature $K \leq 0$, then $\exp_p : T_pM \to M$ is a diffeomorphism and each two points of M can be connected by a unique geodesic segment.*

Definition 13.141. If (M, g) is a Riemannian manifold, then the **diameter** of M is defined to be

$$\mathrm{diam}(M) := \sup\{\mathrm{dist}(p, q) : p, q \in M\}.$$

The injectivity radius at $p \in M$, denoted $\mathrm{inj}(p)$, is the supremum over all $\epsilon > 0$ such that $\exp_p : \widetilde{B}(0_p, \epsilon) \to B(p, \epsilon)$ is a diffeomorphism. The injectivity radius of M is $\mathrm{inj}(M) := \inf_{p \in M}\{\mathrm{inj}(p)\}$.

The Hadamard theorem above has as a hypothesis that the sectional curvature is nonpositive. A bound on the sectional curvature is stronger than a bound on Ricci curvature since the latter is a sort of average sectional curvature. In the sequel, statements like $\mathrm{Ric} \geq C$ should be interpreted to mean $\mathrm{Ric}(v, v) \geq C\langle v, v\rangle$ for all $v \in TM$.

Lemma 13.142. *Let (M, g) be an n-dimensional Riemannian manifold and let $\gamma : [0, L] \to M$ be a unit speed geodesic. Suppose that $\mathrm{Ric} \geq (n-1)\kappa > 0$ for some constant $\kappa > 0$ (at least along γ). If the length L of γ is greater than or equal to $\pi/\sqrt{\kappa}$, then there is a point conjugate to $\gamma(0)$ along γ.*

Proof. Suppose $0 < \pi/\sqrt{\kappa} \leq L$. If we can show that $I_\gamma^\perp$ is not positive definite, then Theorem 13.136 implies the result. To show that $I_\gamma^\perp$ is not positive definite, we find an appropriate vector field $V \neq 0$ along γ such that $I(V,V) \leq 0$. Choose orthonormal fields $E_2, \ldots, E_n$ so that $\dot{\gamma}, E_2, \ldots, E_n$ is an orthonormal frame along γ. For a function $f : [0, \pi/\sqrt{\kappa}] \to \mathbb{R}$ that vanishes at endpoints, we form the fields fE_i. Using (13.9), we have

$$I_\gamma(fE_j, fE_j) = \int_0^{\pi/\sqrt{\kappa}} \left\{ f'(s)^2 + f(s)^2 \langle R_{E_j, \dot{\gamma}}(E_j(s)), \dot{\gamma}(s) \rangle \right\} ds,$$

and then

$$\sum_{j=2}^n I_\gamma(fE_j, fE_j) = \int_0^{\pi/\sqrt{\kappa}} \left\{ (n-1)\left(f'\right)^2 - f^2 \operatorname{Ric}(\dot{\gamma}, \dot{\gamma}) \right\} ds$$

$$\leq (n-1) \int_0^{\pi/\sqrt{\kappa}} \left(\left(f'\right)^2 - \kappa f^2 \right) ds.$$

Letting $f(s) = \sin(\sqrt{\kappa}s)$, we get

$$\sum_{j=2}^n I(fE_j, fE_j) \leq (n-1) \int_0^{\pi/\sqrt{\kappa}} \kappa \left(\cos^2(\sqrt{\kappa}s) - \sin^2(\sqrt{\kappa}s) \right) ds = 0,$$

and so $I(fE_j, fE_j) \leq 0$ for some j. $\square$

The next theorem also assumes only a bound on the Ricci curvature and is one of the most celebrated theorems of Riemannian geometry. A weaker version involving sectional curvature was first proved by Ossian Bonnet (see [**Hicks**], page 165).

Theorem 13.143 (Myers). *Let (M, g) be a complete connected Riemannian manifold of dimension n. If $\operatorname{Ric} \geq (n-1)\kappa > 0$, then*

   (i) $\operatorname{diam}(M) \leq \pi/\sqrt{\kappa}$, M *is compact, and*
   (ii) $\pi_1(M)$ *is finite.*

Proof. Since M is complete, there is always a shortest geodesic γ_{pq} between any two given points p and q. We can assume that γ_{pq} is parametrized by arc length:

$$\gamma_{pq} : [0, \operatorname{dist}(p,q)] \to M.$$

It follows that $\gamma_{pq}|_{[0,a]}$ is arc length minimizing for all $a \in [0, \operatorname{dist}(p,q)]$. From Proposition 13.129 we see that the only possible conjugate to p along γ_{pq} is q. The preceding lemma shows that $\pi/\sqrt{\kappa} > \operatorname{dist}(p,q)$ is impossible.

Since the points p and q were arbitrary, we must have $\operatorname{diam}(M) \leq \pi/\sqrt{\kappa}$. It follows from the Hopf-Rinow theorem that M is compact.

For (ii) we consider the simply connected covering $\wp : \widetilde{M} \to M$ (which is a local isometry). Since $\wp$ is a local diffeomorphism, it follows that $\wp^{-1}(p)$

13.10. Cut Locus   617

has no accumulation points for any $p \in M$. But also, because $\widetilde{M}$ is complete and has the same Ricci curvature bound as M, it is compact. It follows that $\wp^{-1}(p)$ is finite for any $p \in M$, which implies (ii). □

The reader may check that if $S^n(R)$ is a sphere of radius R in $\mathbb{R}^{n+1}$, then $S^n(R)$ has constant sectional curvature $\kappa = 1/R^2$ and the distance from any point to its cut locus (defined below) is $\pi/\sqrt{\kappa}$. A result of S. Y. Cheng states that with the curvature bound of the theorem above, if $\mathrm{diam}(M) = \pi/\sqrt{\kappa}$, then M is a sphere of constant sectional curvature κ. See [**Cheng**].

13.10. Cut Locus

In this section we consider Riemannian manifolds. Related to the notion of conjugate point is the notion of a cut point. For a point $p \in M$ and a geodesic γ emanating from $p = \gamma(0)$, a **cut point** of p along γ is the first point $q = \gamma(t')$ along γ such that for any point $r = \gamma(t'')$ beyond p (i.e. $t'' > t'$) there is a geodesic shorter than $\gamma|_{[0,t'']}$ which connects p with r. To see the difference between this notion and that of a point conjugate to p, it suffices to consider the example of a cylinder $S^1 \times \mathbb{R}$ with the obvious flat metric. If $p = (1,0) \in S^1 \times \mathbb{R}$, then for any $x \in \mathbb{R}$, the point $(e^{i\pi}, x)$ is a cut point of p along the geodesic $\gamma(t) := (e^{it\pi}, tx)$. We know that beyond a conjugate point, a geodesic is not (locally) minimizing. In our cylinder example, for any $\epsilon > 0$, the point $q = \gamma(1+\epsilon)$ can be reached by the geodesic segment $\gamma_2 : [0, 1-\epsilon] \to S^1 \times \mathbb{R}$ given by $\gamma(t) := (e^{it\pi}, ax)$, where $a = (1+\epsilon)/(1-\epsilon)$. It can be checked that γ_2 is shorter than $\gamma|[0, 1+\epsilon]$. However, the last example shows that a cut point need not be a conjugate point. In fact, $S^1 \times \mathbb{R}$ has no conjugate points along any geodesic. Let us agree that all geodesics referred to in this section are parametrized by arc length unless otherwise indicated.

Definition 13.144. Let (M,g) be a complete Riemannian manifold and let $p \in M$. The set $C(p)$ of all cut points to p along geodesics emanating from p is called the **cut locus** of p.

For a point $p \in M$, the situation is summarized by the fact that if $q = \gamma(t')$ is a cut point of p along a geodesic γ, then for any $t'' > t'$ there is a geodesic connecting p with q which is shorter than $\gamma|_{[0,t'']}$, while if $t'' < t'$, then not only is there no geodesic connecting p and $\gamma(t'')$ with shorter length but there is no geodesic connecting p and $\gamma(t'')$ whose length is even equal to that of $\gamma|_{[0,t'']}$. (Why?)

Consider the following two conditions:

(C1): $\gamma(t_0)$ is the first conjugate point of $p = \gamma(0)$ along γ.

(C2): There is a unit speed geodesic α from $\gamma(0)$ to $\gamma(t_0)$ that is different from $\gamma|_{[0,t_0]}$ such that $L(\alpha) = L(\gamma|_{[0,t_0]})$.

Proposition 13.145. *Let M be a complete Riemannian manifold.*

(i) *If for a given unit speed geodesic γ, either condition (C1) or (C2) holds, then there is a $t_1 \in (0, t_0]$ such that $\gamma(t_1)$ is the cut point of p along γ.*

(ii) *If $\gamma(t_0)$ is the cut point of $p = \gamma(0)$ along the unit speed geodesic ray γ, then either condition (C1) or (C2) holds.*

Proof. (i) This is already clear from our discussion: For suppose (**C1**) holds, then $\gamma|_{[0,t']}$ cannot minimize for $t' > t_0$ and so the cut point must be $\gamma(t_1)$ for some $t_1 \in (0, t_0]$. Now if (**C2**) holds, then choose $\epsilon > 0$ small enough that $\alpha(t_0 - \epsilon)$ and $\gamma(t_0 + \epsilon)$ are both contained in a convex neighborhood of $\gamma(t_0)$. The concatenation of $\alpha|_{[0,t_0]}$ and $\gamma|_{[t_0, t_0+\epsilon]}$ is a curve, say c, that has a kink at $\gamma(t_0)$. But there is a unique minimizing geodesic τ joining $\alpha(t_0 - \epsilon)$ to $\gamma(t_0 + \epsilon)$, and we can concatenate the geodesic $\alpha|_{[0,t_0-\epsilon]}$ with τ to get a curve with arc length strictly less than $L(c) = t_0 + \epsilon$. It follows that the cut point to p along γ must occur at $\gamma(t')$ for some $t' \leq t_0 + \epsilon$. But ϵ can be taken arbitrarily small and so the result (i) follows.

(ii) Suppose that $\gamma(t_0)$ is the cut point of $p = \gamma(0)$ along a unit speed geodesic ray γ. We let $\epsilon_i \to 0$ and consider a sequence $\{\alpha_i\}$ of minimizing geodesics with α_i connecting p to $\gamma(t_0 + \epsilon_i)$. We have a corresponding sequence of initial velocities $u_i := \dot{\alpha}_i(0) \in S^1 \subset T_pM$. The unit sphere in T_pM is compact, so replacing u_i by a subsequence we may assume that $u_i \to u \in S^1 \subset T_pM$. Let α be the unit speed segment joining p to $\gamma(t_0 + \epsilon_i)$ with initial velocity u. Arguing from continuity, we see that α is also minimizing and $L(\alpha) = L(\gamma|_{[0,t_0]})$. If $\alpha \neq \gamma|_{[0,t_0]}$, then we are done. If $\alpha = \gamma|_{[0,t_0]}$, then since $\gamma|_{[0,t_0]}$ is minimizing, it will suffice to show that $T_{t_0\dot{\gamma}(0)}\exp_p$ is singular because that would imply that condition (**C1**) holds. The proof of this last statement is by contradiction: Suppose that $\alpha = \gamma|_{[0,t_0]}$ (so that $\dot{\gamma}(0) = u$) and that $T_{t_0\dot{\gamma}(0)}\exp_p$ is not singular. Take U to be an open neighborhood of $t_0\dot{\gamma}(0)$ in T_pM such that $\exp_p|_U$ is a diffeomorphism. Now $\alpha_i(t_0 + \epsilon'_i) = \gamma(t_0 + \epsilon_i)$ for $0 < \epsilon'_i \leq \epsilon_i$ since the α_i are minimizing. We now restrict attention to i such that ϵ_i is small enough that $(t_0 + \epsilon'_i)u_i$ and $(t_0 + \epsilon_i)u$ are in U. Then we have

$$\exp_p(t_0 + \epsilon_i)u = \gamma(t_0 + \epsilon_i)$$
$$= \alpha_i(t_0 + \epsilon'_i) = \exp_p(t_0 + \epsilon'_i)u_i,$$

and so $(t_0 + \epsilon_i)u = (t_0 + \epsilon'_i)u_i$, and then, since $\epsilon_i \to 0$ and both u and u_i are unit vectors, we have $\dot{\gamma}(0) = u = u_i$ for sufficiently large i. But then for such i, we have $\alpha_i = \gamma$ on $[0, t_0 + \epsilon_i]$, which contradicts the fact that $\gamma|_{[0,t_0+\epsilon_i]}$ is not minimizing. $\square$

Exercise 13.146. Show that if q is the cut point of p along γ, then p is the cut point of q along $\gamma^{\leftarrow}$ (where $\gamma^{\leftarrow}(t) := \gamma(L-t)$ and $L = L(\gamma)$).

It follows from the development so far that if $q \in M \setminus C(p)$, then there is a unique minimizing geodesic joining p to q, and that if $B(p, R)$ is the ball of radius R centered at p, then $\exp_p$ is a diffeomorphism on $B(p, R)$ provided $R \le d(p, C(p))$. In fact, an alternative definition of the injectivity radius at p is $d(p, C(p))$ and the injectivity radius of M is

$$\mathrm{inj}(M) = \inf_{p \in M} \{d(p, C(p))\}.$$

Intuitively, the complexities of the topology of M begin at the cut locus of a given point.

Let $T^1 M$ denote the **unit tangent bundle** of the Riemannian manifold:

$$T^1 M = \{u \in TM : \|u\| = 1\}.$$

Define a function $c_M : T^1 M \to (0, \infty]$ by

$$c_M(u) := \begin{cases} t_0 & \text{if } \gamma_u(t_0) \text{ is the cut point of } \pi_{TM}(u) \text{ along } \gamma_u, \\ \infty & \text{if there is no cut point in the direction } u. \end{cases}$$

Recall that the topology on $(0, \infty]$ is such that a sequence t_k converges to the point ∞ if $\lim_{k \to \infty} t_k = \infty$ in the usual sense. It can be shown that if (M, g) is a complete Riemannian manifold, then the function $c_M : T^1 M \to (0, \infty]$ is continuous (see [**Kob**]).

13.11. Rauch's Comparison Theorem

In this section we deal strictly with Riemannian manifolds.

Definition 13.147. Let $\gamma : [a, b] \to M$ be a smooth curve. For X, Y piecewise smooth vector fields along γ, define

$$\mathcal{I}_\gamma(X, Y) := \int_a^b \langle \nabla_{\partial_t} X, \nabla_{\partial_t} Y \rangle + \langle R_{\dot\gamma, X} \dot\gamma, Y \rangle \, dt.$$

The map $X, Y \mapsto \mathcal{I}_\gamma(X, Y)$ is symmetric and bilinear. In defining $\mathcal{I}_\gamma$ we have used a formula for the index I_γ valid for fields which vanish at the endpoints and with γ a nonnull geodesic. Thus when γ is a nonnull geodesic, the restriction of $\mathcal{I}_\gamma$ to variation vector fields which vanish at endpoint is the index I_γ. Thus $\mathcal{I}_\gamma$ is a sort of extended index form.

Corollary 13.148. Let $\gamma : [0, b] \to M$ be a cospacelike geodesic of sign ε with no points conjugate to $\gamma(0)$ along γ. Suppose that Y is a piecewise smooth vector field along γ and that J is a Jacobi field along γ such that

$$Y(0) = J(0),\ Y(b) = J(b),\ \text{and}\ (Y - J) \perp \dot{\gamma}.$$

Then $\varepsilon \mathcal{I}_\gamma(J, J) \le \varepsilon \mathcal{I}_\gamma(Y, Y)$.

Proof of the corollary. From Theorem 13.136, we have $0 \le \varepsilon I_\gamma^\perp(Y - J, Y - J) = \varepsilon \mathcal{I}_\gamma(Y - J, Y - J)$ and so

$$0 \le \varepsilon \mathcal{I}_\gamma(Y, Y) - 2\varepsilon \mathcal{I}_\gamma(J, Y) + \varepsilon \mathcal{I}_\gamma(J, J).$$

Integrating by parts, we have

$$\varepsilon \mathcal{I}_\gamma(J, Y) = \varepsilon \left\langle \nabla_{\partial_t} J, Y \right\rangle \big|_0^b - \int_0^b \left\langle \nabla_{\partial_t}^2 J, Y \right\rangle - \left\langle R_{\dot{\gamma}, J}\dot{\gamma}, Y \right\rangle dt$$

$$= \varepsilon \left\langle \nabla_{\partial_t} J, Y \right\rangle \big|_0^b = \varepsilon \left\langle \nabla_{\partial_t} J, J \right\rangle \big|_0^b$$

$$= \varepsilon \mathcal{I}_\gamma(J, J) \quad \text{(since J is a Jacobi field)}.$$

Thus $0 \le \varepsilon \mathcal{I}_\gamma(Y, Y) - 2\varepsilon \mathcal{I}_\gamma(J, Y) + \varepsilon \mathcal{I}_\gamma(J, J) = \varepsilon \mathcal{I}_\gamma(Y, Y) - \varepsilon \mathcal{I}_\gamma(J, J)$. $\square$

Recall that for a Riemannian manifold M, the sectional curvature $K^M(P)$ of a 2-plane $P \subset T_p M$ is

$$\left\langle \mathfrak{R}(e_1 \wedge e_2), e_1 \wedge e_2 \right\rangle$$

for any orthonormal pair e_1, e_2 that spans P.

Definition 13.149. Let M, g and N, h be Riemannian manifolds and let $\gamma^M : [a, b] \to M$ and $\gamma^N : [a, b] \to N$ be unit speed geodesics defined on the same interval $[a, b]$. We say that $K^M \ge K^N$ **along the pair** (γ^M, γ^N) if $K^M(Q_{\gamma^M(t)}) \ge K^N(P_{\gamma^N(t)})$ for all $t \in [a, b]$ and every pair of 2-planes $Q_t \in T_{\gamma^M(t)} M$, $P_t \in T_{\gamma^N(t)} N$.

We develop some notation to be used in the proof of Rauch's theorem. Let M be a given Riemannian manifold. If Y is a piecewise smooth vector field along a unit speed geodesic γ^M such that $Y(a) = 0$, then let

$$\mathcal{I}_s^M(Y, Y) := \int_a^s \left\langle \nabla_{\partial_t} Y(t), \nabla_{\partial_t} Y(t) \right\rangle + \left\langle R_{\dot{\gamma}^M, Y}\dot{\gamma}^M, Y \right\rangle (t) dt$$

$$= \int_a^s -\left\langle \nabla_{\partial_t}^2 Y(t), Y(t) \right\rangle + \left\langle R_{\dot{\gamma}^M, Y}\dot{\gamma}^M, Y \right\rangle (t) dt + \left\langle \nabla_{\partial_t} Y, Y \right\rangle (s).$$

If Y is an orthogonal Jacobi field, then

$$\mathcal{I}_s^M(Y, Y) = \left\langle \nabla_{\partial_t} Y, Y \right\rangle (s).$$

13.11. Rauch's Comparison Theorem

Theorem 13.150 (Rauch). *Let M, g and N, h be Riemannian manifolds of the same dimension and let $\gamma^M : [a,b] \to M$ and $\gamma^N : [a,b] \to N$ be unit speed geodesics defined on the same interval $[a,b]$. Let J^M and J^N be Jacobi fields along γ^M and γ^N respectively and orthogonal to their respective curves. Suppose that the following four conditions hold:*

(i) $J^M(a) = J^N(a) = 0$ and neither of $J^M(t)$ or $J^N(t)$ is zero for $t \in (a, b]$.

(ii) $\left\|\nabla_{\partial_t} J^M(a)\right\| = \left\|\nabla_{\partial_t} J^N(a)\right\|$.

(iii) $L(\gamma^M) = \text{dist}(\gamma^M(a), \gamma^M(b))$.

(iv) $K^M \geq K^N$ along the pair (γ^M, γ^N).

Then $\left\|J^M(t)\right\| \leq \left\|J^N(t)\right\|$ for all $t \in [a, b]$.

Proof. Let f_M be defined by $f_M(s) := \left\|J^M(s)\right\|^2$ and h_M by $h_M(s) := \mathcal{I}_s^M(J^M, J^M)/\left\|J^M(s)\right\|^2$ for $s \in (a, b]$. Define f_N and h_N analogously. We have
$$f_M'(s) = 2\mathcal{I}_s^M(J^M, J^M) \text{ and } f_M'/f_M = 2h_M$$
and the analogous equalities for f_N and h_N. If $c \in (a, b)$, then
$$\ln(\left\|J^M(s)\right\|^2) = \ln(\left\|J^M(c)\right\|^2) + 2\int_c^s h_M(s')ds'$$
with the analogous equation for N. Thus
$$\ln\left(\frac{\left\|J^M(s)\right\|^2}{\left\|J^N(s)\right\|^2}\right) = \ln\left(\frac{\left\|J^M(c)\right\|^2}{\left\|J^N(c)\right\|^2}\right) + 2\int_c^s [h_M(s') - h_N(s')]ds'.$$
From the assumptions (i) and (ii) and L'Hôpital's rule, we have
$$\lim_{c \to a+} \frac{\left\|J^M(c)\right\|^2}{\left\|J^N(c)\right\|^2} = 1,$$
and so
$$\ln\left(\frac{\left\|J^M(s)\right\|^2}{\left\|J^N(s)\right\|^2}\right) = 2\lim_{c \to a+} \int_c^s [h_M(s') - h_N(s')]ds'.$$
If we can show that $h_M(s) - h_N(s) \leq 0$ for $s \in (a, b]$, then the result will follow. So fix $r \in (a, b]$ and let $Z^M(s) := J^M(s)/\left\|J^M(r)\right\|$ and $Z^N(s) := J^N(s)/\left\|J^N(r)\right\|$. Notice the r in the denominator; $Z^N(s)$ is not necessarily of unit length for $s \neq r$. We now define a parametrized families of sub-tangent spaces along γ^M by $W_M(s) := \dot{\gamma}^M(s)^\perp \subset T_{\gamma^M(s)}M$ and similarly for $W_N(s)$. We can choose a linear isometry $L_r : W_N(r) \to W_M(r)$ such that $L_r(Z^N(r)) = Z^M(r)$. We now want to extend L_r to a family of linear isometries $L_s : W_N(s) \to W_M(s)$. We do this using parallel transport by
$$L_s := P(\gamma^M)_r^s \circ L_r \circ P(\gamma^N)_s^r.$$

Define a vector field Y along γ^M by $Y(s) := L_s(Z^N(s))$. Check that

$$Y(a) = Z^M(a) = 0,$$
$$Y(r) = Z^M(r),$$
$$\|Y\|^2 = \|Z^N\|^2,$$
$$\|\nabla_{\partial_t} Y\|^2 = \|\nabla_{\partial_t} Z^N\|^2.$$

The last equality is a result of Exercise 12.43 where in the notation of that exercise $\beta(t) := P(\gamma^M)_t^r \circ Y(t)$. Since (iii) holds, there can be no conjugates along γ^M up to r. Now $Y - Z^M$ is orthogonal to the geodesic γ^M and so by Corollary 13.148 we have $\mathcal{I}_r^M(Z^M, Z^M) \leq \mathcal{I}_r^M(Y, Y)$ and in fact, using (iv) and the list of equations above, we have

$$\mathcal{I}_r^M(Z^M, Z^M) \leq \mathcal{I}_r^M(Y, Y) = \int_a^r \|\nabla_{\partial_t} Y\|^2 + R^M(\dot{\gamma}^M, Y, \dot{\gamma}^M, Y)$$
$$= \int_a^r \|\nabla_{\partial_t} Y\|^2 - K(\dot{\gamma}^M, Y, \dot{\gamma}^M, Y) \|\dot{\gamma}^M \wedge Y\|^2$$
$$= \int_a^r \|\nabla_{\partial_t} Y\|^2 - K(\dot{\gamma}^M, Y, \dot{\gamma}^M, Y) \|\dot{\gamma}^M\|^2 \|Y\|^2$$
$$\leq \int_a^r \|\nabla_{\partial_t} Z^N\|^2 - K(\dot{\gamma}^M, Z^N, \dot{\gamma}^M, Z^N) \|\dot{\gamma}^M\|^2 \|Z^N\|^2 \quad \text{(by (iv))}$$
$$= \int_a^r \|\nabla_{\partial_t} Z^N\|^2 + R^N(\dot{\gamma}^N, Z^N, \dot{\gamma}^N, Z^N) = \mathcal{I}_r^N(Z^N, Z^N).$$

Recalling the definition of Z^M and Z^N we obtain

$$\mathcal{I}_r^M(J^M, J^M)/\|J^M(r)\|^2 \leq \mathcal{I}_r^N(J^N, J^N)/\|J^N(r)\|^2,$$

and so $h_M(r) - h_N(r) \leq 0$. But r was arbitrary, and so we are done. $\square$

Corollary 13.151. *Let M, g and N, h be Riemannian manifolds of the same dimension and let $\gamma^M : [a,b] \to M$ and $\gamma^N : [a,b] \to N$ be unit speed geodesics defined on the same interval $[a,b]$. Assume that $K^M \geq K^N$ along the pair (γ^M, γ^N). Then if $\gamma^M(a)$ has no conjugate point along γ^M, then $\gamma^N(a)$ has no conjugate point along γ^N.*

The above corollary is easily deduced from the Rauch comparison theorem above and we invite the reader to prove it. The following famous theorem is also proved using the Rauch comparison theorem but the proof is quite difficult and also uses Morse theory. The proof may be found in **[C-E]**.

13.12. Weitzenböck Formulas

Theorem 13.152 (The sphere theorem). *Let M, g be a complete simply connected Riemannian manifold with sectional curvature satisfying the condition*

$$\frac{1}{4}\frac{1}{R^2} < K \leq \frac{1}{R^2}$$

for some $R > 0$. Then M is homeomorphic to the sphere S^n.

13.12. Weitzenböck Formulas

The **divergence** of a vector field in terms of Levi-Civita connection is given by

$$\operatorname{div}(X) = \operatorname{trace}(\nabla X).$$

Thus if $(e_1, \ldots, e_n)$ and $(\theta^1, \ldots, \theta^n)$ are dual bases for $T_p M$ and $T_p^* M$, then

$$(\operatorname{div} X)_p = \sum_{j=1}^n \theta^j \left(\nabla_{e_j} X \right).$$

Exercise 13.153. Show that this definition is compatible with our previous definition by showing that, with the above definition, we have $L_X \operatorname{vol} = (\operatorname{div} X) \operatorname{vol}$, where vol is the metric volume element for M. Hint: Use $L_X = d\, i_X + i_X\, d$.

Definition 13.154. Let ∇ be a torsion free covariant derivative on M. The divergence of a (k, ℓ) tensor field A is a $(k-1, \ell)$ tensor field is defined by

$$(\operatorname{div} A)_p (\alpha_1 \ldots, \alpha_{k-1}, v_1, \ldots, v_\ell) = \sum_{j=1}^n \left(\nabla_{e_j} A \right) (\theta^j, \alpha_1 \ldots, \alpha_{k-1}, v_1, \ldots, v_\ell),$$

where $(e_1, \ldots, e_n)$ and $(\theta^1, \ldots, \theta^n)$ are dual as above.

Notice that the above definition depends on the choice of covariant derivative but if M is semi-Riemannian, then we will use the Levi-Civita connection. In index notation the definition is quite simple in appearance. For example, if A^{ij}_{kl} are the components of a $(2,2)$ tensor field, then we have

$$(\operatorname{div} A)^j_{kl} = \nabla_r A^{rj}_{kl},$$

where $\nabla_a A^{bj}_{kl}$ are the components of ∇A. Also, the definition given is really for the divergence with respect to the first contravariant slot, but we could use other slots (the last slot being popular). Thus $\nabla_r A^{jr}_{kl}$ is also a divergence. If we are to define a divergence with respect to a covariant slot (i.e., a lower index), then we must use a metric to raise the index. This leads to the following definition appropriate in the presence of a metric.

Definition 13.155. Let ∇ be the Levi-Civita covariant derivative on a manifold M. The (metric) divergence of a function is defined to be zero and the divergence of a $(0, \ell)$ tensor field A is a $(0, \ell - 1)$ tensor field defined at any $p \in M$ by

$$(\operatorname{div} A)_p (v_1, \ldots, v_{\ell-1}) = \sum_{j=1}^{n} \langle e_j, e_j \rangle (\nabla_{e_j} A)(e_j, v_1, \ldots, v_{\ell-1}),$$

where $(e_1, \ldots, e_n)$ is an orthonormal basis for T_pM. Note that the factors $\langle e_j, e_j \rangle$ are equal to 1 in the case of a definite metric.

On may check that if $A_{i_1 \ldots i_\ell}$ are the components of A in some chart, then the components of $\operatorname{div} A$ are given by

$$(\operatorname{div} A)_{i_1 \ldots i_{\ell-1}} = \nabla_r A^r_{i_1 \ldots i_{\ell-1}}.$$

Recall that the formal adjoint $\delta : \Omega(M) \to \Omega(M)$ of the exterior derivative on a Riemannian manifold M is given on $\Omega^k(M)$ by $\delta := (-1)^{n(k+1)+1} * d *$. In Problem 8 we ask the reader to show that the restriction of div to $\Omega^k(M)$ is $-\delta$:

(13.11) $\qquad \operatorname{div} = -\delta \quad \text{on } \Omega^k(M) \text{ for each } k.$

In the same problem the reader is asked to show that if $\mu \in \Omega^k(M)$ is parallel ($\nabla \mu = 0$), then it is harmonic (recall Definition 9.46).

In what follows we simplify calculations by the use of a special kind of orthonormal frame field. If (M, g) is a Riemannian manifold and $p \in M$, choose an orthonormal basis $(e_1, \ldots, e_n)$ in the tangent space T_pM and parallel translate each e_i along the radial geodesics $t \mapsto \exp_p(tv)$ for $v \in T_pM$. This results in an orthonormal frame field $(E_1, \ldots, E_n)$ on some normal neighborhood centered at p. Smoothness is easy to prove. The resulting fields are radially parallel and satisfy $E_i(p) = e_i$ and $(\nabla E_i)(p) = 0$ for every i. Furthermore we have

$$[E_i, E_j](p) = \nabla_{E_i} E_j(p) - \nabla_{E_j} E_i(p) = 0 \quad \text{for all } i, j.$$

We refer to an orthonormal frame field with these properties as an **adapted orthonormal frame field centered** at p.

Before proceeding we need an exercise to set things up.

Exercise 13.156. Describe how a connection ∇ (say the Levi-Civita connection) on M extends to a connection on the bundle $\bigwedge T^*M$ in such a way that

$$\nabla_X (\alpha^1 \wedge \cdots \wedge \alpha^k) = \sum_{i=1}^{k} \alpha^1 \wedge \cdots \wedge \nabla_X \alpha^i \wedge \cdots \wedge \alpha^k$$

13.12. Weitzenböck Formulas

for $\alpha^i \in \Omega^1(M)$. Show that the curvature of the extended connection is given by

$$R(X,Y)\mu = \nabla_X \nabla_Y \mu - \nabla_Y \nabla_X \mu - \nabla_{[X,Y]}\mu.$$

Relate this curvature operator to the curvature operator on $\mathfrak{X}(M)$ and show that the extended connection is flat if the original connection on M is flat.

Let M be Riemannian and $\mu \in \Omega^k(M)$. Define R_μ by

$$R_\mu(v_1, \ldots, v_k) = \sum_{i=1}^n \sum_{j=1}^k (R(e_i, v_j)\mu)(v_1, \ldots, v_{j-1}, e_i, v_{j+1}, \ldots, v_k).$$

Theorem 13.157 (Weitzenböck formulas). *Let M be Riemannian and $\mu \in \Omega^k(M)$. Then we have*

$$\langle \Delta \mu | \mu \rangle = \frac{1}{2} \Delta \|\mu\|^2 + \|\nabla \mu\|^2 + \langle R_\mu | \mu \rangle,$$
$$\Delta \mu = -\operatorname{div} \nabla \mu + R_\mu,$$

where $\|\nabla \mu\|^2(p) := \sum_i \langle \nabla_{e_i} \mu | \nabla_{e_i} \mu \rangle$ for any orthonormal basis $(e_1, \ldots, e_n)$ for $T_p M$.

Proof. Using the formula of Theorem 12.56, we see that for $U, V_1, \ldots, V_k \in \mathfrak{X}(M)$, we have

$$(\nabla \mu - d\mu)(U, V_1, \ldots, V_k) = \nabla \mu(U, V_1, \ldots, V_k) - d\mu(U, V_1, \ldots, V_k)$$
$$= \sum_{j=1}^k (\nabla_{V_j} \mu)(V_1, \ldots, V_{j-1}, U, V_{j+1}, \ldots, V_k).$$

With this in mind, we fix $p \in M$ and $v_1, \ldots, v_k \in T_p M$. We may choose $V_1, \ldots, V_k \in \mathfrak{X}(M)$ such that $V_i(p) = v_i$ and may assume that $(\nabla V_i)(p) = 0$. Now choose an adapted orthonormal frame field $E_1, \ldots, E_n$ centered at p so that $(\nabla E_i)(p) = 0$ for all i and $[E_i, E_j](p) = 0$ for all i, j. In the following calculation, several steps may appear at first to be wrong. However, if one begins to write out the missing terms, one sees that they vanish because of how the fields were chosen to behave at p. Using equation (13.11) we have

$$(\operatorname{div} \nabla \mu + \delta d\mu)(v_1, \ldots, v_k) = \operatorname{div}(\nabla \mu - d\mu)(v_1, \ldots, v_k)$$
$$= \sum_{i=1}^n \nabla_{e_i}(\nabla \mu - d\mu)(e_i, v_1, \ldots, v_k)$$
$$= \sum_{i=1}^n e_i \left[(\nabla \mu - d\mu)(E_i, V_1, \ldots, V_k)\right]$$

and using what we know about the fields at p, this is equal to

$$\sum_{i=1}^{n} e_i \left(\sum_{j=1}^{k} \left(\nabla_{V_j}\mu\right)(V_1,\ldots,V_{i-1},E_j,V_{i+1},\ldots,V_k) \right)$$

$$= \sum_{i=1}^{n}\sum_{j=1}^{k} e_i \left(\left(\nabla_{V_j}\mu\right)(V_1,\ldots,V_{i-1},E_j,V_{i+1},\ldots,V_k)\right)$$

$$= \sum_{i=1}^{n}\sum_{j=1}^{k} \left(\nabla_{E_i}\left(\nabla_{V_j}\mu\right)\right)(V_1,\ldots,V_{i-1},E_j,V_{i+1},\ldots,V_k)(p)$$

$$= \sum_{i=1}^{n}\sum_{j=1}^{k} \left(\nabla_{e_i}\nabla_{V_j}\mu\right)(v_1,\ldots,v_{i-1},e_j,v_{i+1},\ldots,v_k).$$

We also have

$$d\delta\mu(v_1,\ldots,v_k) = \sum_{j=1}^{k}(-1)^{j+1}\left(\nabla_{v_j}\delta\mu\right)(v_1,\ldots,\widehat{v_j},\ldots,v_k)$$

$$= \sum_{j=1}^{k}(-1)^{j} v_j \left(\sum_{i=1}^{n} \nabla_{E_i}\mu\left(E_i,V_1,\ldots,\widehat{V_j},\ldots,V_k\right) \right)$$

$$= -\sum_{j=1}^{k}\sum_{i=1}^{n} \nabla_{v_i}\nabla_{E_j}\mu(v_1,\ldots,v_{i-1},e_j,v_{i+1},\ldots,v_k).$$

Since $[E_i,V_j] = \nabla_{E_i}V_j - \nabla_{V_j}E_i = 0$, we can add the above results to obtain $\Delta\mu + \operatorname{div}\nabla\mu = R_\mu$.

For the second part we again take advantage of our arrangement $\nabla E_i = 0$, $\nabla V_j = 0$ at p; we have

$$(\operatorname{div}\nabla\mu)(v_1,\ldots,v_k) = \sum_{i=1}^{n}\left(\nabla_{e_i}\nabla\mu\right)(e_i,v_1,\ldots,v_k)$$

$$= \sum_{i=1}^{n} e_i\left(\left(\nabla_{E_i}\mu\right)(V_1,\ldots,V_k)\right)$$

$$= \sum_{i=1}^{n}\left(\nabla_{e_i}\nabla_{E_i}\mu\right)(v_1,\ldots,v_k).$$

13.13. Structure of General Relativity

From this we obtain

$$\begin{aligned}\langle -\operatorname{div}\nabla\mu \,|\, \mu\rangle(p) &= -\sum_{i=1}^{n}\langle\nabla_{e_i}\nabla_{E_i}\mu \,|\, \mu(p)\rangle \\ &= -\sum_{i=1}^{n}\left(e_i\langle\nabla_{E_i}\mu\,|\,\mu\rangle - \langle\nabla_{e_i}\mu\,|\,\nabla_{e_i}\mu\rangle\right) \\ &= -\frac{1}{2}\sum_{i=1}^{n}e_iE_i\,\|\mu\|^2 + \|\nabla\mu\|^2\,(p) \\ &= \left(\frac{1}{2}\Delta\,\|\mu\|^2 + \|\nabla\mu\|^2\right)(p). \quad\square\end{aligned}$$

We give only one application (but see [**Pe**] or [**Poor**]).

Proposition 13.158. *If (M, g) is a flat connected compact Riemannian manifold (without boundary), then a form $\mu \in \Omega^k(M)$ is parallel if and only if it is harmonic.*

Proof. We have observed the inclusion {parallel k-forms} $\subset$ {harmonic k-forms} (Problem 8). On the other hand, if (M, g) is flat, then $R_\mu = 0$ and so by Theorem 13.157 we have $\langle\Delta\mu|\mu\rangle = \frac{1}{2}\Delta\,\|\mu\|^2 + \|\nabla\mu\|^2$. So if $\Delta\mu = 0$, we have

$$\int_M \|\nabla\mu\|^2\, dV = -\frac{1}{2}\int_M \Delta\,\|\mu\|^2\, dV = 0$$

by Stokes' theorem. Thus $\nabla\mu = 0$. $\quad\square$

Corollary 13.159. *Let M be a connected compact n-manifold. If M admits a flat Riemannian metric g, then $\dim H^k(M) \leq \binom{n}{k}$.*

Proof. Pick any $p \in M$. From the Hodge theorem, and the previous proposition, we have

$$\dim H^k(M) = \dim\{\text{harmonic } k\text{-forms}\} = \dim\{\text{parallel } k\text{-forms}\}$$

$$\leq \dim\bigwedge\nolimits^k T_p^* M = \binom{n}{k}.$$

The inequality follows from Exercise 13.14. $\quad\square$

13.13. Structure of General Relativity

The reader is now in a position to appreciate the basic structure of Einstein's general theory of relativity. We can only say a few words about this wonderful part of physics. General Relativity is a theory of gravity based on the mathematics of semi-Riemannian geometry. In a nutshell, the theory models spacetime as a four dimensional Lorentz manifold M^4 that is usually assumed to be time oriented. The points of spacetime are idealized events.

The motion of a test particle subject only to gravity is along a geodesic in spacetime and the metric is subject to a nonlinear tensor differential equation that involves the curvature and a physical tensor T that describes the local flow of energy-momentum. We consider the following equations to be central:

(13.12) $$\text{Ric} - \frac{1}{2}Rg = 8\pi\kappa T \qquad \text{(Einstein's equation)},$$

(13.13) $$\text{div}\, T = 0 \qquad \text{(continuity)},$$

(13.14) $$\nabla_{\dot{\alpha}}\dot{\alpha} = 0 \qquad \text{(geodesic equation)},$$

(13.15) $$\nabla_{\dot{\alpha}}\nabla_{\dot{\alpha}}V + R(V,\dot{\alpha})\dot{\alpha} = 0 \qquad \text{(Jacobi equation)},$$

where in the geodesic equation and Jacobi equation, $\lambda \mapsto \alpha(\lambda)$ is a parametrized curve that represents the career of a test particle subject only to gravitation. In the Jacobi equation we are to imagine a smooth family of geodesic curves $t \mapsto h_s(t) = h(t,s)$ such that $h_0 = \alpha$. Then $V = \frac{\partial h}{\partial s}$ is the variation vector field. The first equation, Einstein's equation, is the centerpiece of Einstein's theory. On the right hand side of this equation we see a tensor T called the stress-energy-momentum tensor; it represents the matter and energy that generate the gravitational field. The constant κ is Newton's gravitational constant, which is also often denoted by the letter G. The left hand side is the Einstein curvature tensor and is built from the Riemann curvature tensor and so ultimately from the metric tensor. There, "Ric" is the Ricci curvature whose index form is written $R_{\mu\nu}$; R is the scalar curvature defined by contraction $R := g^{\mu\nu}R_{\mu\nu}$, and g is the metric tensor. Einstein's equation can be seen as an equation for the metric of spacetime; it shows how the distribution of matter and energy influences the metric and the resulting curvature of spacetime. We have already studied the last two equations, but we will say something below about their role in gravitational theory.

We will base our explanations on the following two-pronged incantation:

Matter tells spacetime how to curve.

Spacetime tells matter how to move.

Newtonian gravity. To fully appreciate Einstein's theory of gravity one must compare it to Newton's theory. In Newton's theory, the equations of motion of a test particle moving in (flat Euclidean) space and subject to a gravitational field **g** is described by

(13.16) $$m_I \frac{d^2\mathbf{x}(t)}{dt^2} = m_G \mathbf{g}(\mathbf{x}(t)).$$

Here $\mathbf{x}(t)$ is a vector-valued function of t that gives the location of the test particle relative to a fixed inertial frame (which entails a choice of origin and system of rectangular coordinates). The field **g** is a vector-valued function of position in space. Here m_I is the inertial mass of the particle, which is the m in Newton's $\mathbf{F} = m\mathbf{a}$. The constant m_G is the gravitational mass, which plays the role of gravitational charge. As demonstrated by Galileo and later by Eötvös, we actually have equality $m_I = m_G$. This equality is in fact one of the key influences on Einstein's thinking, and it led him to assert that, for sufficiently small regions of spacetime, gravitational forces and inertial forces (as perceived in an accelerating frame) are indistinguishable.

If ρ represents the mass density in a region of space, then the **gravitational potential** ϕ produced by this matter is given by

$$\nabla^2 \phi = 4\pi\kappa\rho.$$

The gravitational field is then

$$\mathbf{g} = -\operatorname{grad} \phi.$$

Thus we have the pair of equations

(13.17) $$\nabla^2 \phi = 4\pi\kappa\rho,$$

(13.18) $$\frac{d^2 \mathbf{x}}{dt^2} = -\operatorname{grad} \phi,$$

where $\nabla^2 \phi = \operatorname{div}(\operatorname{grad} \phi)$ and where the right hand side is evaluated at $\mathbf{x}(t)$. The first equation tells how matter creates the Newtonian gravity field, and the second describes how the field tells matter how to move. This is the Newtonian analogue of the two-pronged incantation above. In Newton's picture, gravity is unambiguously treated as a field created by mass that induces a force on test particles.

Free fall. Let α be a curve parametrized by proper time τ that represents the path of a test particle. In Einstein's theory, what corresponds to equation (13.18) is the geodesic equation $\ddot{\alpha} := \nabla_{\dot\alpha}\dot\alpha = 0$. According to Einstein, if the particle is subject only to gravity, then α is a geodesic. If we choose a coordinate system (x^0, x^1, x^2, x^3), then the geodesic equation gives four differential equations (the geodesic equations), which can be written as

(13.19) $$\frac{d^2 x^\mu}{d\tau^2} = -\frac{1}{2} g^{\mu\delta} \left(\frac{\partial g_{\beta\delta}}{\partial x^\alpha} + \frac{\partial g_{\delta\alpha}}{\partial x^\beta} - \frac{\partial g_{\alpha\beta}}{\partial x^\delta} \right) \frac{dx^\alpha}{d\tau} \frac{dx^\beta}{d\tau}, \quad \mu = 0, 1, 2, 3,$$

where we have written out the formula for $\Gamma^\mu_{\alpha\beta}$ explicitly. (We use the common convention that the Greek indices run over $0, 1, 2, 3$, while the Latin indices run over $1, 2, 3$.) The above equations have a similarity to (13.18) when the latter are written in the form

$$\frac{d^2 x^i}{dt^2} = -\frac{\partial \phi}{\partial x^i}, \quad i = 1, 2, 3.$$

From this point of view, (13.19) looks like a force law and the metric components $g_{\alpha\beta}$ play a role analogous to the potential ϕ in the Newtonian theory. From the point of view of the chosen coordinate system, these equations appear to tell the particle how to accelerate with respect to the coordinate system. However, if we choose normal coordinates at an event $p \in M^4$, then at p the left hand side of (13.19) is zero! In fact, from the intrinsic point of view, the law is simply that the career of the particle in spacetime is a geodesic and therefore represents a state of *zero* intrinsic acceleration. Rephrasing the second prong of our incantation, we say that spacetime tells free test particles how to *curve* or *accelerate*. Namely, not at all. This is a wonderfully simple geometric law of motion. Freely falling bodies are described by geodesics in spacetime.

Tidal forces. Imagine yourself in free fall in a uniform gravitational field. Imagine that you are surrounded by a spherical array of apples in free fall which are initially stationary with respect to you. All the apples appear motionless against the starry background of space. If the field was truly uniform, the spherical swarm would remain spherical. However, this situation corresponds to no spacetime curvature, and so from the intrinsic point of view, is no gravitational field at all. A realistic gravitational field such as that produced by the Earth is not uniform, and our sphere of apples would deform becoming elongated along a line passing through the center of mass of the gravitating body (the Earth, say) and passing through the center of the array. If you were in free fall at the center of the spherical array with your feet toward the earth, then apples that are roughly in a plane perpendicular to the axis of your body would be seen to accelerate towards you, while those below your feet and above your head would be seen to recede. Each apple follows a geodesic in spacetime (not in space), and so we have a family of geodesics. This situation, properly idealized, is described by the Jacobi equation (13.15). The vector field V can be thought of as describing the separation of nearby geodesics, and the Jacobi equation describes the *relative acceleration* of nearby geodesics. Curvature is the "force" behind this relative acceleration. It is this relative acceleration, positive in some directions and negative in others, that is responsible for the distortion of our initially spherical array of free falling apples. If we neglect the attraction that the apples have for each other, then the volume of the array remains constant. This is because we are in a region where the Einstein tensor (and hence the Ricci tensor) vanishes.

Energy-momentum tensor. Fields and particles carry 4-momentum. Let α be a unit speed timelike curve giving the career of a particle of (rest) mass m. The 4-momentum of the particle is $\mathbf{p} = m\dot\alpha$. In a flat spacetime we may choose Lorentz coordinates (x^0, x^1, x^2, x^3) so that the corresponding coordinate frame field (∂_μ) is oriented orthonormal with ∂_0

13.13. Structure of General Relativity

timelike. We choose units so that the speed of light is unity; $c = 1$ so that $x^0 = t$ is coordinate time. Let $\mathbf{v} := (\frac{dx^1}{dt}, \frac{dx^2}{dt}, \frac{dx^3}{dt})$ be the "ordinary" spatial velocity or "3-velocity" as viewed in this Lorentz frame and let $v := |\mathbf{v}| = [\sum_{i=1}^{3}(dx^i/dt)^2]^{1/2}$. Then the 4-momentum has components $(E, \gamma \mathbf{p})$, where E is the relativistic energy of the particle, $\gamma = (1-v^2)^{-1/2}$, and we shall refer to $\gamma \mathbf{p}$ as the relativistic 3-momentum. Notice that if we change to a different Lorentz coordinate, then we will have a new energy and 3-momentum, but the geometric 4-momentum vector $m\dot{\alpha}$ is an "invariant" notion defined without reference to a specific coordinate frame. Thus 4-momentum unifies the notions of momentum and energy. Furthermore, this relativistic energy is really mass-energy. Indeed, if we choose a frame in which the particle is (momentarily) at rest, then the 4-momentum has components $(m, \mathbf{0})$.

Now when we consider a region in spacetime filled with particles and fields, it is appropriate to go to the continuum approximation. The right hand side of Einstein's equation features the $(0,2)$-tensor T that keeps track of the flow of 4-momentum produced by all the matter and non-gravitational energy. It will pay to first think about charge and then consider the meaning of T in the setting of special relativity. In this setting of special relativity we are assuming that T does not produce curvature (contrary to fact). In standard Lorentz coordinates (x^0, x^1, x^2, x^3), the tensor T has 16 components $T_{\mu\nu}$. Let us do a type change, $T^\mu_\nu := g^{\mu\alpha}T_{\alpha\nu}$. Then $(T^0_0, T^0_1, T^0_2, T^0_3)$ represents the density of 4-momentum and $(T^i_0, T^i_1, T^i_2, T^i_3)$ represents the flux of 4-momentum in the spatial direction i (so $i = 1, 2,$ or 3). By this we mean that the result of a flux integral should be a 4-vector quantity, while the flux of a vector field in the usual calculus sense is a *scalar* (such as charge or mass per unit time).

In electrodynamics we describe the flow of a charge by a time dependent vector field $\mathbf{J}$, and if ρ is the charge density, then local conservation of charge is given by the continuity equation

$$\frac{\partial \rho}{\partial t} = -\operatorname{div} \mathbf{J}.$$

We can combine ρ and $\mathbf{J}$ into a unified notion of 4-current J that has components in a Lorentz frame given by $(J^0, \ldots, J^3) = (\rho, \mathbf{J})$ and the corresponding covector field $\mathcal{J}$ has components $(J_0, \ldots, J_3) = (-\rho, \mathbf{J})$. The continuity equation can then be written as

$$d * \mathcal{J} = 0,$$

and follows from Maxwell's equations. The corresponding integral version of conservation of charge is simply

$$\int_{\partial R} *\mathcal{J} = 0,$$

where ∂R is the boundary of a region of spacetime R. For a general volume Ω that does not necessarily bound an open region of spacetime, the integral $\int_\Omega *\mathcal{J}$ gives the total charge "crossing" Ω. It is the fact that charge is a scalar quantity that allows us to do a continuous sum over Ω.

In the Newtonian theory, mass is the analogue of charge and we have a similar continuity equation. However, in relativity, rest mass is not a conserved quantity and energy is not a scalar. Somehow charge is to be replaced by energy-momentum as we cross the bridge of analogy to the land of gravity. If spacetime is Minkowski space, then there is a quantity we can integrate to get a total energy-momentum. The integral implies a sum, and we can add tensors located at different points if we take advantage of distant parallelism (see [**M-T-W**]). If we express T in a Lorentz frame, then we can integrate to get a total energy-momentum $\mathbf{p}_{tot}$ crossing a "3-volume" V. The covariant components of $\mathbf{p}_{tot}$ in the Lorentz frame are given by

$$\mathbf{p}^\mu_{tot} = \int_V T^\mu_\nu d\Sigma^\nu,$$

where $d\Sigma^\nu = *dx^\nu = \frac{1}{3!}\varepsilon^\nu{}_{\alpha\beta\gamma}dx^\alpha \wedge dx^\beta \wedge dx^\gamma$. Then, the conservation law says that $\int_V T^\mu_\nu d\Sigma^\nu = 0$ if $V = \partial\Omega$ is the boundary of a spacetime 4-volume Ω. The differential form of the conservation law is $\operatorname{div} T = 0$, which in a Lorentz frame is just $\partial_\mu T^\mu_\nu = 0$.

Now, for general relativity we retain the local version of conservation by assuming that $\operatorname{div} T = 0$, which is now defined in terms of the covariant derivative. In general coordinates we have

$$\nabla_\mu T^\mu_\nu = 0.$$

However, we must give up the integral version, although it would still hold approximately for sufficiently small regions of spacetime since the latter would be approximately flat. We shall find that this continuity equation is automatically satisfied if Einstein's equation holds since it turns out that the divergence of the left hand side is zero for purely geometric reasons!

The Einstein tensor. The left hand side of Einstein's equation features the tensor $\operatorname{Ric} - \frac{1}{2}Rg$ given in index notation as $R_{\mu\nu} - \frac{1}{2}Rg_{\mu\nu}$. Here, $R_{\mu\nu} = R^\alpha{}_{\mu\alpha\nu}$ and $R = R^\mu{}_\mu$ (sum). This tensor is denoted by the letter G and is called the **Einstein tensor**. Let us show that $\operatorname{div} G = 0$. This fact shows that the conservation law is forced by the geometry (assuming that Einstein's equation holds). In this sense we may take this to be another manifestation of the second prong of our incantation in that the geometry

13.13. Structure of General Relativity

tells matter to behave in accordance with local conservation. We have not seen many examples of tensor calculations using index notation, so we take this opportunity to do a calculation. In index notation, what we wish to show is that $\nabla_\mu G^\mu_\nu = \nabla^\mu G_{\mu\nu} = 0$. We start with the Bianchi identity, make a switch in the first two indices of the first term and then raise indices and contract:

$$0 = \nabla_\mu R_{\alpha\beta\gamma\delta} + \nabla_\alpha R_{\beta\mu\gamma\delta} + \nabla_\beta R_{\mu\alpha\gamma\delta},$$
$$0 = -\nabla^\mu R^\beta{}_{\alpha\gamma\delta} + \nabla_\alpha R^{\beta\mu}{}_{\gamma\delta} + \nabla^\beta R^\mu{}_{\alpha\gamma\delta},$$
$$0 = -\nabla^\gamma R^\delta{}_{\alpha\gamma\delta} + \nabla_\alpha R^{\delta\gamma}{}_{\gamma\delta} + \nabla^\delta R^\gamma{}_{\alpha\gamma\delta},$$
$$0 = -\nabla^\gamma R^\delta{}_{\alpha\gamma\delta} - \nabla_\alpha R^{\gamma\delta}{}_{\gamma\delta} + \nabla^\delta R^\gamma{}_{\alpha\gamma\delta},$$
$$0 = \nabla^\gamma R_{\alpha\gamma} - \nabla_\alpha R + \nabla^\delta R_{\alpha\delta}.$$

This last equation gives $\nabla^\gamma R_{\alpha\gamma} - \frac{1}{2}\nabla_\alpha R = 0$. But since $\nabla^\mu g_{\mu\nu} = 0$, we have

$$\nabla^\mu G_{\mu\nu} = \nabla^\mu \left(R_{\mu\nu} - \frac{1}{2}Rg_{\mu\nu}\right) = \left(\nabla^\mu R_{\mu\nu} - \frac{1}{2}\nabla^\mu(Rg_{\mu\nu})\right)$$
$$= \left(\nabla^\mu R_{\mu\nu} - \frac{1}{2}g^{\mu\alpha}\nabla_\alpha(Rg_{\mu\nu})\right) = \left(\nabla^\mu R_{\mu\nu} - \frac{1}{2}g^{\mu\alpha}(g_{\mu\nu}\nabla_\alpha R)\right)$$
$$= \nabla^\gamma R_{\alpha\gamma} - \frac{1}{2}\delta^\alpha_\nu \nabla_\alpha R = \nabla^\gamma R_{\alpha\gamma} - \frac{1}{2}\nabla_\alpha R = 0.$$

The Schwarzschild metric. If we contract both sides of Einstein's equation, we obtain $R = -8\pi\kappa T^\mu_\mu$. Plugging this back into Einstein's equation and rearranging we obtain an equivalent form of Einstein's equation $R_{\mu\nu} = 8\pi\kappa\left(T_{\mu\nu} - \frac{1}{2}T^\alpha_\alpha g_{\mu\nu}\right)$. Thus in case the tensor T vanishes in the region of interest, we obtain the **vacuum field equation**

$$R_{\mu\nu} = 0.$$

There is a famous metric defined in terms of spherical coordinates (t, r, θ, ϕ), given by

$$ds^2 = -\left(1 - \frac{2\kappa M}{r}\right)dt^2 + \left(1 - \frac{2\kappa M}{r}\right)^{-1} dr^2 + r^2(d\theta^2 + \sin^2\theta \, d\phi^2),$$

where θ is the polar angle $0 \leq \theta \leq \pi$. Notice that this expression for our metric is undefined at both $r = 0$ and $r = 2\kappa M$. But this is only the coordinate expression for a metric that intrinsically may be quite nice at $r = 2\kappa M$ and/or $r = 0$. It can be shown that there is a metric perfectly well-defined on $r = 2\kappa M$ whose expression in (t, r, θ, ϕ) just happens to be the above away from $r = 2\kappa M$ and $r = 0$. The fact that the above expression blows up as we approach $r = 2\kappa M$ is a failure of the coordinates and not a feature of the intrinsic metric. On the other hand, if one calculated the scalar curvature, then it can be seen to blow up as $r \to 0$, which makes

$r = 0$ a true singularity. This metric is called the **Schwarzschild metric** and describes a spherically symmetric solution of Einstein's equation that is taken to be due to a spherical distribution of matter concentrated near $r = 0$ (such as a star). We should mention that this metric only satisfies Einstein's equation in the vacuum away from the star. In most cases the radius of the star is larger than $2\kappa M$, and then the Schwarzschild metric is not a solution inside the star anyway. If the radius of the star is less than $2\kappa M$, then we have a nonrotating black hole, and $r = 2\kappa M$ defines the famous event horizon.

Problems

(1) Show that in a Riemannian manifold, a length minimizing piecewise smooth curve must be a smooth geodesic. [Hint: Each potential kink point has a totally star-shaped neighborhood. Use Proposition 13.85.]

(2) A set U in a Riemannian manifold M is said to be **geodesically convex** if for each pair of points $p, q \in U$, there is a unique length minimizing geodesic segment connecting them, and this unique geodesic segment lies completely in U.
  (a) Show that the intersection of geodesically convex sets is geodesically convex.
  (b) Show that given $p \in M$ there is a $\delta > 0$ such that $\exp(\{v_p \in T_pM : \|v_p\| < \varepsilon\})$ is geodesically convex for all $\varepsilon < \delta$.

(3) Referring to the discussion leading up to Proposition 13.135, let $Y \in T_\gamma(\Omega)$ be a piecewise smooth variation vector field along γ and write $Y = \sum \varphi^i J_i$ for some piecewise smooth functions φ^i on $(0, b]$. Show that the φ^i can be extended continuously to $[0, b]$.

(4) A smooth map $f : (M, g) \to (N, h)$ of semi-Riemannian manifolds is called a **homothety** if $f^*h = cg$ for some constant $c \neq 0$. The case of $c = -1$ is called an **anti-isometry**. Show that an anti-isometry preserves covariant derivatives and geodesics.

(5) Show that the mapping $\sigma : \mathbb{R}^{n+1}_\nu \to \mathbb{R}^{n+1}_{n-\nu+1}$ given by
$$\sigma(a_1, \ldots, a_{n+1}) := (a_{\nu+1}, \ldots, a_{n+1}, a_1, \ldots, a_\nu)$$
is an anti-isometry (see above) and that its restriction to $S^n_\nu(r)$ is an anti-isometry from $S^n_\nu(r)$ onto $H^n_{n-\nu}(r)$.

(6) Show that timelike vectors v and w in a Lorentz space V are in the same timecone if and only if $\langle v, w \rangle < 0$.

(7) Construct examples sufficient to make the point that time orientability (of Lorentz manifolds) and orientability are unrelated.

(8) Show that the restriction of div to $\Omega^k(M)$ is $-\delta$. Show that if $\mu \in \Omega^k(M)$ is parallel ($\nabla_X \mu = 0$ for all $X \in \mathfrak{X}(M)$), then it is harmonic.

(9) Let $H := \{(u,v) \in \mathbb{R}^2 : v > 0\}$ be the upper half-plane endowed with the metric
$$g := \frac{1}{v}(du \otimes du + dv \otimes dv).$$
Show that H has constant curvature $K = -1$. Find which curves are geodesics.

(10) (Killing fields) On a semi-Riemannian manifold (M, g), a vector field X is called a **Killing field** if $\mathcal{L}_X g = 0$. Show that the local flows of a Killing field are isometries. Show that X is a Killing field if and only if $X \langle V, W \rangle = \langle \mathcal{L}_X V, W \rangle + \langle V, \mathcal{L}_X W \rangle$ for all $V, W \in \mathfrak{X}(M)$. Show that X is a Killing field if and only if $\langle \nabla_V X, W \rangle = - \langle \nabla_W X, V \rangle$ for all $V, W \in \mathfrak{X}(M)$.

(11) Show that if γ is a geodesic in M and X is a Killing field (see the previous problem), then $X \circ \gamma$ is a Jacobi field along γ and that $\langle X \circ \gamma, \dot{\gamma} \rangle$ is constant.

(12) Show that an $\mathbb{R}$-linear combination of Killing fields is a Killing field and that if X and Y are Killing fields, then $\mathcal{L}_{[X,Y]} = [\mathcal{L}_X, \mathcal{L}_Y]$. Deduce that the space of Killing fields is a real Lie algebra. What are the Killing fields of $\mathbb{R}^3$?

(13) Let (M, g) and (N, h) be semi-Riemannian manifolds. Let f be a positive smooth function on M. The warped product metric on $M \times N$ is defined by
$$(g \times_f h) := \mathrm{pr}_1^* g + (f \circ \mathrm{pr}_2) \mathrm{pr}_2^* g,$$
where $\mathrm{pr}_1 : M \times N \to M$ and $\mathrm{pr}_2 : M \times N \to N$ are the first and second factor projections.
(a) Show that this is indeed a metric.
(b) Show that for each $p \in M$, the map $\mathrm{pr}_2|_{p \times N}$ is a homothety.
(c) Show that each $M \times q$ is normal to each $p \times N$.
(d) Let R^M be the curvature on (M, g) and $R^{M \times N}$ be the curvature tensor on $(M \times N, g \times_f h)$. If $\widetilde{X}, \widetilde{Y}$, and $\widetilde{Z}$ are lifts of $X, Y, Z \in \mathfrak{X}(M)$ as described in Problem 30 of Chapter 2, then what is the relationship between $R^{M \times N}_{\widetilde{X}, \widetilde{Y}} \widetilde{Z}$ and $R^M_{X,Y} Z$?

Appendix A

The Language of Category Theory

Category theory provides a powerful means of organizing our thinking in mathematics. Some readers may be put off by the abstract nature of category theory. To such readers, I can only say that it is not really difficult to catch on to the spirit of category theory and the payoff in terms of organizing mathematical thinking is considerable. I encourage these readers to give it a chance. In any case, it is not strictly necessary for the reader to be completely at home with category theory before going further into the book. In particular, physics and engineering students may not be used to this kind of abstraction and should simply try to gradually become accustomed to the language. Feel free to defer reading this appendix on category theory until it seems necessary.

Roughly speaking, category theory is an attempt at clarifying structural similarities that tie together different parts of mathematics. A category has "objects" and "morphisms". The prototypical category is just the category **Set** which has for its objects ordinary *sets* and for its morphisms *maps* between sets. The most important category for differential geometry is what is sometimes called the "smooth category" consisting of smooth manifolds and smooth maps. (The definition of these terms is given in the text proper, but roughly speaking, smooth means differentiable.)

Now on to the formal definition of a category.

Definition A.1. A **category** $\mathfrak{C}$ is a collection of objects $\text{Ob}(\mathfrak{C}) = \{X, Y, Z, \ldots\}$ and for every pair of objects X, Y, a set $\text{Hom}_{\mathfrak{C}}(X, Y)$ called the set of **morphisms** from X to Y. The family of all morphisms in a category $\mathfrak{C}$ will be denoted Mor($\mathfrak{C}$). In addition, a category is required to have a **composition law** which is defined as a map $\circ : \mathrm{Hom}_\mathfrak{C}(X,Y) \times \mathrm{Hom}_\mathfrak{C}(Y,Z) \to \mathrm{Hom}_\mathfrak{C}(X,Z)$ such that for every three objects $X, Y, Z \in \mathrm{Ob}(\mathfrak{C})$ the following axioms hold:

(1) $\mathrm{Hom}_\mathfrak{C}(X,Y)$ and $\mathrm{Hom}_\mathfrak{C}(Z,W)$ are disjoint unless $X = Z$ and $Y = W$, in which case $\mathrm{Hom}_\mathfrak{C}(X,Y) = \mathrm{Hom}_\mathfrak{C}(Z,W)$.

(2) The composition law is associative: $f \circ (g \circ h) = (f \circ g) \circ h$.

(3) Each set of morphisms of the form $\mathrm{Hom}_\mathfrak{C}(X,X)$ must contain a necessarily unique element id_X, the identity element, such that $f \circ \mathrm{id}_X = f$ for any $f \in \mathrm{Hom}_\mathfrak{C}(X,Y)$ (and any Y), and $\mathrm{id}_X \circ f = f$ for any $f \in \mathrm{Hom}_\mathfrak{C}(Y,X)$.

Notation A.2. A morphism is sometimes written using an arrow. For example, if $f \in \mathrm{Hom}_\mathfrak{C}(X,Y)$ we would indicate this by writing $f : X \to Y$ or by $X \xrightarrow{f} Y$.

The notion of category is typified by the case where the objects are sets and the morphisms are maps between the sets. In fact, subject to putting extra structure on the sets and the maps, this will be almost the only type of category we shall need to talk about. On the other hand there are plenty of interesting categories of this type. Examples include the following.

(1) **Grp**: The objects are groups and the morphisms are group homomorphisms.

(2) **Rng** : The objects are rings and the morphisms are ring homomorphisms.

(3) **Lin**$_\mathbb{F}$: The objects are vector spaces over the field $\mathbb{F}$ and the morphisms are linear maps. This category is referred to as the linear category or the vector space category (over the field $\mathbb{F}$).

(4) **Top**: The objects are topological spaces and the morphisms are continuous maps.

(5) **Man**^{r}: The category of C^r differentiable manifolds and C^r maps: One of the main categories discussed in this book. This is also called the smooth or differentiable category, especially when $r = \infty$.

Notation A.3. If for some morphisms $f_i : X_i \to Y_i$, $(i = 1, 2)$, $g_X : X_1 \to X_2$ and $g_Y : Y_1 \to Y_2$ we have $g_Y \circ f_1 = f_2 \circ g_X$, then we express this by saying that the following diagram "commutes":

$$\begin{array}{ccc} X_1 & \xrightarrow{f_1} & Y_1 \\ {\scriptstyle g_X}\downarrow & & \downarrow{\scriptstyle g_Y} \\ X_2 & \xrightarrow{f_2} & Y_2 \end{array}$$

A. The Language of Category Theory

Similarly, if $h \circ f = g$, we say that the diagram

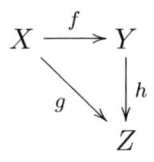

$$\begin{array}{ccc} X & \xrightarrow{f} & Y \\ & \searrow_{g} & \downarrow_{h} \\ & & Z \end{array}$$

commutes. More generally, tracing out a path of arrows in a diagram corresponds to composition of morphisms, and to say that such a diagram **commutes** is to say that the compositions arising from two paths of arrows that begin and end at the same objects are equal.

Definition A.4. Suppose that $f : X \to Y$ is a morphism from some category $\mathfrak{C}$. If f has the property that for any two (parallel) morphisms g_1, $g_2 : Z \to X$ we always have that $f \circ g_1 = f \circ g_2$ implies $g_1 = g_2$, i.e. if f is "left cancellable", then we call f a **monomorphism**. Similarly, if $f : X \to Y$ is "right cancellable", we call f an **epimorphism**. A morphism that is both a monomorphism and an epimorphism is called an **isomorphism**. If the category needs to be specified, then we talk about a $\mathfrak{C}$-**monomorphism**, $\mathfrak{C}$-**epimorphism** and so on).

In some cases we will use other terminology. For example, an isomorphism in the smooth category is called a **diffeomorphism**. In the linear category, we speak of linear maps and linear isomorphisms. Morphisms which comprise $\mathrm{Hom}_{\mathfrak{C}}(X, X)$ are also called endomorphisms and so we also write $\mathrm{End}_{\mathfrak{C}}(X) := \mathrm{Hom}_{\mathfrak{C}}(X, X)$. The set of all isomorphisms in $\mathrm{Hom}_{\mathfrak{C}}(X, X)$ is sometimes denoted by $\mathrm{Aut}_{\mathfrak{C}}(X)$, and these morphisms are called **automorphisms**.

We single out the following: In many categories such as the above, we can form a new category that uses the notion of pointed space and pointed map. For example, we have the "pointed topological category". A pointed topological space is a topological space X together with a distinguished point p. Thus a typical object in the pointed topological category would be written as (X, p). A morphism $f : (X, p) \to (W, q)$ is a continuous map such that $f(p) = q$.

A **functor** $\mathcal{F}$ is a pair of maps, both denoted by the same letter $\mathcal{F}$, that map objects and morphisms from one category to those of another,

$$\mathcal{F} : \mathrm{Ob}(\mathfrak{C}_1) \to \mathrm{Ob}(\mathfrak{C}_2),$$
$$\mathcal{F} : \mathrm{Mor}(\mathfrak{C}_1) \to \mathrm{Mor}(\mathfrak{C}_2),$$

so that composition and identity morphisms are respected. This means that for a morphism $f : X \to Y$, the morphism

$$\mathcal{F}(f) : \mathcal{F}(X) \to \mathcal{F}(Y)$$

is a morphism in the second category and we must have

(1) $\mathcal{F}(\mathrm{id}_{\mathfrak{C}_1}) = \mathrm{id}_{\mathfrak{C}_2}$.

(2) If $f : X \to Y$ and $g : Y \to Z$, then $\mathcal{F}(f) : \mathcal{F}(X) \to \mathcal{F}(Y)$, $\mathcal{F}(g) : \mathcal{F}(Y) \to \mathcal{F}(Z)$ and

$$\mathcal{F}(g \circ f) = \mathcal{F}(g) \circ \mathcal{F}(f).$$

Example A.5. Let **Lin**$_\mathbb{R}$ be the category whose objects are real vector spaces and whose morphisms are real linear maps. Similarly, let **Lin**$_\mathbb{C}$ be the category of complex vector spaces with complex linear maps. To each real vector space V, we can associate the complex vector space $\mathbb{C} \otimes_\mathbb{R} V$, called the complexification of V, and to each linear map of real vector spaces $\ell : V \to W$ we associate the complex extension $\ell_\mathbb{C} : \mathbb{C} \otimes_\mathbb{R} V \to \mathbb{C} \otimes_\mathbb{R} W$. Here, $\mathbb{C} \otimes_\mathbb{R} V$ is easily thought of as the vector space V where now complex scalars are allowed. Elements of $\mathbb{C} \otimes_\mathbb{R} V$ are generated by elements of the form $c \otimes v$, where $c \in \mathbb{C}$, $v \in V$ and we have $i(c \otimes v) = ic \otimes v$, where $i = \sqrt{-1}$. The map $\ell_\mathbb{C} : \mathbb{C} \otimes_\mathbb{R} V \to \mathbb{C} \otimes_\mathbb{R} W$ is defined by the requirement $\ell_\mathbb{C}(c \otimes v) = c \otimes \ell v$. Now the assignments

$$\ell \mapsto \ell_\mathbb{C},$$
$$V \mapsto \mathbb{C} \otimes_\mathbb{R} V$$

define a functor from **Lin**$_\mathbb{R}$ to **Lin**$_\mathbb{C}$. In practice, complexification amounts to simply allowing complex scalars. For instance, we might just write cv instead of $c \otimes v$.

Actually, what we have defined here is a **covariant functor**. A **contravariant functor** is defined similarly except that the order of composition is reversed so that instead of (2) above we would have $\mathcal{F}(g \circ f) = \mathcal{F}(f) \circ \mathcal{F}(g)$. An example of a contravariant functor is the dual vector space functor, which is a functor from the category of vector spaces **Lin**$_\mathbb{R}$ to itself that sends each space V to its dual V* and each linear map to its dual (or transpose). Under this functor a morphism $V \xrightarrow{L} W$ is sent to the morphism

$$V^* \xleftarrow{L^*} W^*.$$

Notice the arrow reversal.

One of the most important functors for our purposes is the **tangent functor** defined in Chapter 2. Roughly speaking this functor replaces differentiable maps and spaces by their linear parts.

Example A.6. Consider the category of real vector spaces and linear maps. To every vector space V, we can associate the dual of the dual V^{**}. This is

A. The Language of Category Theory

a *covariant* functor which is the composition of the dual functor with itself:

$$
\begin{array}{ccccc}
V & & W^* & & V^{**} \\
A \downarrow & \mapsto & A^* \downarrow & \mapsto & A^{**} \downarrow \\
W & & V^* & & W^{**}
\end{array}
$$

Now suppose we have two functors,

$$\mathcal{F}_1 : \mathrm{Ob}(\mathfrak{C}_1) \to \mathrm{Ob}(\mathfrak{C}_2),$$
$$\mathcal{F}_1 : \mathrm{Mor}(\mathfrak{C}_1) \to \mathrm{Mor}(\mathfrak{C}_2)$$

and

$$\mathcal{F}_2 : \mathrm{Ob}(\mathfrak{C}_1) \to \mathrm{Ob}(\mathfrak{C}_2),$$
$$\mathcal{F}_2 : \mathrm{Mor}(\mathfrak{C}_1) \to \mathrm{Mor}(\mathfrak{C}_2).$$

A **natural transformation** $\mathcal{T}$ from $\mathcal{F}_1$ to $\mathcal{F}_2$ is given by assigning to each object X of $\mathfrak{C}_1$, a morphism $\mathcal{T}(X) : \mathcal{F}_1(X) \to \mathcal{F}_2(X)$ such that for every morphism $f : X \to Y$ of $\mathfrak{C}_1$, the following diagram commutes:

$$
\begin{array}{ccc}
\mathcal{F}_1(X) & \xrightarrow{\mathcal{T}(X)} & \mathcal{F}_2(X) \\
\mathcal{F}_1(f) \downarrow & & \downarrow \mathcal{F}_2(f) \\
\mathcal{F}_1(Y) & \xrightarrow[\mathcal{T}(Y)]{} & \mathcal{F}_2(Y)
\end{array}
$$

A common first example is the natural transformation ι between the identity functor $I : \mathbf{Lin}_\mathbb{R} \to \mathbf{Lin}_\mathbb{R}$ and the double dual functor $** : \mathbf{Lin}_\mathbb{R} \to \mathbf{Lin}_\mathbb{R}$:

$$
\begin{array}{ccc}
V & \xrightarrow{\iota(V)} & V^{**} \\
f \downarrow & & \downarrow f^{**} \\
W & \xrightarrow[\iota(W)]{} & W^{**}
\end{array}
$$

The map $V \to V^{**}$ sends a vector to a linear function $\tilde{v} : V^* \to \mathbb{R}$ defined by $\tilde{v}(\alpha) := \alpha(v)$ (the hunted becomes the hunter, so to speak). If there is an inverse natural transformation $\mathcal{T}^{-1}$ in the obvious sense, then we say that $\mathcal{T}$ is a natural isomorphism, and for any object $X \in \mathfrak{C}_1$ we say that $\mathcal{F}_1(X)$ is naturally isomorphic to $\mathcal{F}_2(X)$. The natural transformation just defined is easily checked to have an inverse, so it is a natural isomorphism. The point here is not just that V is isomorphic to V^{**} in the category $\mathbf{Lin}_\mathbb{R}$, but that the isomorphism exhibited is natural. It works for all the spaces V in a uniform way that involves no special choices. This is to be contrasted with the fact that V is isomorphic to V^*, where the construction of such an isomorphism involves an arbitrary choice of a basis.

Appendix B

Topology

B.1. The Shrinking Lemma

We first state and prove a simple special case of the shrinking lemma since it makes clear the main idea at the root of the fancier versions.

Lemma B.1. *Let X be a normal topological space and $\{U_1, U_2\}$ an open cover of X. There exists an open set V with $\overline{V} \subset U_1$ such that $\{V, U_2\}$ is still a cover of X.*

Proof. Since $U_1 \cup U_2 = X$, we have $(X \backslash U_1) \cap (X \backslash U_2) = \emptyset$. Using normality, we find disjoint open sets O and V such that $X \backslash U_1 \subset O$ and $X \backslash U_2 \subset V$. Then it follows that $X \backslash O \subset U_1$ and $X \backslash V \subset U_2$ and so $X = U_2 \cup V$. But $O \cap V = \emptyset$ so $V \subset X \backslash O$. Thus $\overline{V} \subset X \backslash O \subset U_1$. □

Proposition B.2. *Let X be a normal topological space and $\{U_1, U_2, \ldots, U_n\}$ a finite open cover of X. Then there exists an open cover $\{V_1, V_2, \ldots, V_n\}$ such that $\overline{V}_i \subset U_i$ for $i = 1, 2, \ldots, n$.*

Proof. Simple induction using Lemma B.1. □

The proof we give of the shrinking lemma uses transfinite induction. A different proof may be found in the online supplement [**Lee, Jeff**]. It is a fact that any set A can be well ordered, which means that we may impose a partial order $\prec$ on the set so that each nonempty subset $S \subset A$ has a least element. Every well-ordered set is in an order preserving isomorphism with an ordinal ω (which, by definition, is itself the set of ordinals strictly less than ω). For purposes of transfinite induction, we may as well assume that the given indexing set is such an ordinal. Let A be the ordinal which is the indexing set. Each $\alpha \in A$ has a unique successor, which is written as $\alpha + 1$.

The successor of $\alpha+1$ is written as $\alpha+2$, and so on. If $\beta \in A$ has the form $\beta = \{\alpha, \alpha+1, \alpha+2, \ldots\}$, then we say that β is a limit ordinal and of course $a \prec \beta$ for all $a \in \beta$. (This may seem confusing if one is not familiar with ordinals.) Suppose we have a statement $P(\alpha)$ for all $\alpha \in A$. Let 0 denote the first element of A. The principle of transfinite induction on A says that if $P(0)$ is true and if the truth of $P(\alpha)$ for all $\alpha \prec \beta$ can be shown to imply the truth of $P(\beta)$ for arbitrary β, then $P(\alpha)$ is in fact true for all $\alpha \in A$. In most cases, a transfinite induction proof has three steps:

(1) **Zero case:** Prove that $P(0)$ is true.

(2) **Successor case:** Prove that for any successor ordinal $\alpha + 1$, the assumption that $P(\theta)$ is true for all $\theta \prec \alpha+1$ implies that $P(\alpha+1)$ is true.

(3) **Limit case:** Prove that for any limit ordinal ω, $P(\omega)$ follows from the assumption that $P(\alpha)$ is true for all $\alpha \prec \omega$.

Definition B.3. A cover $\{U_\alpha\}_{\alpha \in A}$ of a topological space X is called **point finite** if for every $p \in X$ the family $A(p) = \{\alpha : p \in U_\alpha\}$ is finite.

Clearly, a locally finite cover is point finite.

Theorem B.4. *Let X be a normal topological space and $\{U_\alpha\}_{\alpha \in A}$ a point finite open cover of X. Then there exists an open cover $\{V_\alpha\}_{\alpha \in A}$ of X such that $\overline{V}_\alpha \subset U_\alpha$ for all $\alpha \in A$.*

Proof. Assume that A is an ordinal. The goal is to construct a cover $\{V_\alpha\}_{\alpha \in A}$ of X such that $\overline{V}_\alpha \subset U_\alpha$ for all $\alpha \in A$. We do transfinite induction on A. Let $P(\alpha)$ be the statement

(*) For all $\theta \prec \alpha$ there exists V_θ with $\overline{V}_\theta \subset U_\theta$ such that $\{V_\theta\}_{\theta \prec \alpha} \cup \{U_\theta\}_{\alpha \preceq \theta}$ is a cover of X.

The goal is to prove that $P(\alpha)$ is true for all $\alpha \prec A+1$ since this entails that $P(A)$ is true.

Case 1: α has a successor $\alpha + 1$. Assume that $P(\alpha)$ holds. We show that $P(\alpha+1)$ is true, which entails constructing V_α. Let

$$F = X \setminus \left\{ \left(\bigcup_{\theta \prec \alpha} V_\theta\right) \cup \left(\bigcup_{\alpha+1 \preceq \theta} V_\theta\right) \right\}.$$

Clearly, F is closed and $F \subset U_\alpha$. By normality there is a set V_α with

$$F \subset V_\alpha \subset \overline{V}_\alpha \subset U_\alpha.$$

Then $\{V_\theta\}_{\theta \prec \alpha+1} \cup \{U_\theta\}_{\alpha+1 \preceq \theta}$ is a cover of X, which is the statement $P(\alpha+1)$.

Case 2: Limit ordinal case. Suppose that ω is a limit ordinal and that $P(\theta)$ is true for all $\theta \prec \omega$. We want to show that this implies that $\{V_\theta\}_{\theta \prec \omega} \cup \{U_\theta\}_{\omega \preceq \theta}$ is a cover of X. Suppose this is not the case. Then there is an element $x \in X$ which is not in the union of this family of open sets. We know that there exists a finite collection of sets $U_{\alpha_1}, \ldots, U_{\alpha_n}$, each containing x and such that no other U_α contains x. We have that $\alpha_i \prec \omega$ for each $i = 1, \ldots, n$, and since ω is a limit ordinal, there exists an ordinal δ such that $\alpha_i \prec \delta \prec \omega$ for all i. Then the point x is in the union $\{V_\theta\}_{\theta \prec \delta} \cup \{U_\theta\}_{\delta \preceq \theta}$, and since we know that x is not in any of the sets in $\{U_\theta\}_{\delta \preceq \theta}$, we must have $x \in V_\theta$ for some $\theta \prec \delta$ so that

$$x \in \left(\bigcup_{\theta \prec \omega} V_\theta\right) \cup \left(\bigcup_{\omega \preceq \theta} U_\theta\right),$$

which contradicts our assumption that $\{V_\theta\}_{\theta \prec \omega} \cup \{U_\theta\}_{\omega \preceq \theta}$ is not a cover of X. Thus $P(\alpha)$ is true for all $\alpha \prec A + 1$. □

B.2. Locally Euclidean Spaces

If every point of a topological space X has an open neighborhood that is homeomorphic to an open set in a Euclidean space, then we say that X is **locally Euclidean**. A locally Euclidean space need not be Hausdorff. For example, if we take the spaces $\mathbb{R} \times \{0\}$ and $\mathbb{R} \times \{1\}$ and give them the relative topologies as subsets of $\mathbb{R} \times \mathbb{R}$, then they are both homeomorphic to $\mathbb{R}$. Now on the (disjoint) union $(\mathbb{R} \times \{0\}) \cup (\mathbb{R} \times \{1\})$ define an equivalence relation by requiring $(x, 0) \sim (x, 1)$ except when $x = 0$. The quotient topological space thus obtained in locally Euclidean, but not Hausdorff. Indeed, the two points $[(0,0)]$ and $[(0,1)]$ are distinct but cannot be separated. It is as if they both occupy the origin.

A **refinement** of a cover $\{U_\beta\}_{\beta \in B}$ of a topological space X is another cover $\{V_i\}_{i \in I}$ such that every set from the second cover is contained in at least one set from the original cover. We say that a cover $\{V_i\}_{i \in I}$ of a topological space X is a **locally finite** cover if every point of X has a neighborhood that intersects only a finite number of sets from the cover. A topological space X is called **paracompact** if every *open* cover of X has a refinement which is a locally finite open cover.

Proposition B.5. *If X is a locally Euclidean Hausdorff space, then the are following properties equivalent:*

(1) *X is paracompact.*

(2) *X is metrizable.*

(3) *Each connected component of X is second countable.*

(4) *Each connected component of X is σ-compact.*

(5) *Each connected component of X is separable.*

For a proof see [**Spv**, volume I] and [**Dug**].

Appendix C

Some Calculus Theorems

For a review of multivariable calculus and the proofs of the theorems below, see the online supplement [**Lee, Jeff**].

Theorem C.1 (Inverse mapping theorem). *Let U be an open subset of $\mathbb{R}^n$ and let $f : U \to \mathbb{R}^n$ be a C^r mapping for $1 \leq r \leq \infty$. Suppose that $x_0 \in U$ and that $Df(x_0) : \mathbb{R}^n \to \mathbb{R}^n$ is a linear isomorphism. Then there exists an open set $V \subset U$ with $x_0 \in V$ such that $f(V)$ is open and $f : V \to f(V) \subset \mathbb{R}^n$ is a C^r diffeomorphism. Furthermore the derivative of f^{-1} at y is given by $Df^{-1}\big|_y = (Df|_{f^{-1}(y)})^{-1}$.*

Theorem C.2 (Implicit mapping theorem). *Let $O \subset \mathbb{R}^k \times \mathbb{R}^l$ be open. Let $f : O \to \mathbb{R}^m$ be a C^r mapping such that $f(x_0, y_0) = 0$. If $D_2 f(x_0, y_0) : \mathbb{R}^l \to \mathbb{R}^m$ is an isomorphism, then there exist open sets $U_1 \subset \mathbb{R}^k$ and $U_2 \subset \mathbb{R}^l$ such that $U_1 \times U_2 \subset O$ with $x_0 \in U_1$ and a C^r mapping $g : U_1 \to U_2$ with $g(x_0) = y_0$ such that for all $(x, y) \in U_1 \times U_2$ we have*

$$f(x, y) = 0 \text{ if and only if } y = g(x).$$

We may take U_1 to be connected. The function g in the theorem satisfies $f(x, g(x)) = 0$, which says that the graph of g is contained in $(U_1 \times U_2) \cap f^{-1}(0)$, but the conclusion of the theorem is stronger since it says that in fact the graph of g is exactly equal to $(U_1 \times U_2) \cap f^{-1}(0)$.

Corollary C.3. *If U is an open neighborhood of $0 \in \mathbb{R}^k$ and $f : U \subset \mathbb{R}^k \to \mathbb{R}^n$ is a smooth map with $f(0) = 0$ such that $Df(0)$ has rank k, then there is an open neighborhood V of $0 \in \mathbb{R}^n$, an open neighborhood W of $0 \in \mathbb{R}^n$,* and a diffeomorphism $g : V \to W$ such that $g \circ f : f^{-1}(V) \to W$ is of the form $(a^1, \ldots, a^k) \mapsto (a^1, \ldots, a^k, 0, \ldots, 0)$.

Corollary C.4. *If U is an open neighborhood of $0 \in \mathbb{R}^n$ and $f : U \subset \mathbb{R}^k \times \mathbb{R}^{n-k} \to \mathbb{R}^k$ is a smooth map with $f(0) = 0$ and if the partial derivative $D_1 f(0,0)$ is a linear isomorphism, then there exist a diffeomorphism $h : V \subset \mathbb{R}^n \to U_1$, where V is an open neighborhood of $0 \in \mathbb{R}^n$ and U_1 is an open neighborhood of $0 \in \mathbb{R}^k$ such that the composite map $f \circ h$ is of the form*

$$(a^1, \ldots, a^n) \mapsto (a^1, \ldots, a^k).$$

Theorem C.5 (The constant rank theorem). *Let $f : (\mathbb{R}^n, p) \to (\mathbb{R}^m, q)$ be a local map such that Df has constant rank r in an open set containing p. Then there are local diffeomorphisms $g_1 : (\mathbb{R}^n, p) \to (\mathbb{R}^n, q)$ and $g_2 : (\mathbb{R}^m, q) \to (\mathbb{R}^m, 0)$ such that $g_2 \circ f \circ g_1^{-1}$ has the form*

$$(x^1, \ldots, x^n) \mapsto (x^1, \ldots, x^r, 0, \ldots, 0)$$

on a sufficiently small neighborhood of 0.

Theorem C.6 (Mean value). *Let U be an open subset of $\mathbb{R}^n$ and let $f : U \to \mathbb{R}^m$ be of class C^1. Suppose that for $x, z \in U$ the line segment L given by $x + t(z - x)$ (for $0 \leq t \leq 1$) is contained in U. Then*

$$\|f(z) - f(x)\| \leq \|z - x\| \sup\{\|Df(y)\| : y \in L\},$$

where $\|Df(y)\| := \sup_{\|v\|=1} \{\|Df(y) \cdot v\|\}$.

Appendix D

Modules and Multilinearity

A module is an algebraic object that shows up quite a bit in differential geometry and analysis (at least implicitly). A module is a generalization of a vector space where the field $\mathbb{F}$ is replaced by a ring or an algebra over a field. For the definition of ring and field consult any book on abstract algebra. The definition of algebra is given below. The modules that occur in differential geometry are almost always finitely generated projective modules over the algebra of C^r functions, and these correspond to the spaces of C^r sections of vector bundles. We give the abstract definitions but we ask the reader to keep two cases in mind. The first is just the vector spaces which are the fibers of vector bundles. In this case, the ring in the definition below is the field $\mathbb{F}$ (the real numbers $\mathbb{R}$ or the complex numbers $\mathbb{C}$), and the module is just a vector space. The second case, already mentioned, is where the ring is the algebra $C^r(M)$ for a C^r manifold and the module is the set of C^r sections of a vector bundle over M.

As we have indicated, a module is similar to a vector space with the differences stemming from the use of elements of a ring R as the scalars rather than the field of complex $\mathbb{C}$ or real numbers $\mathbb{R}$. For an element v of a module V, one still has $0v = 0$, and if the ring is a ring with unity 1, then we usually require $1v = v$. Of course, every vector space is also a module since the latter is a generalization of the notion of vector space. We also have maps between modules, the module homomorphisms (see Definition D.5 below), which make the class of modules and module homomorphisms into a category.

Definition D.1. Let R be a ring. A left R-**module** (or a left **module** over R) is an abelian group $(V, +)$ together with an operation $R \times V \to V$ written as $(a, v) \mapsto av$ and such that

1) $(a + b)v = av + bv$ for all $a, b \in R$ and all $v \in V$;
2) $a(v_1 + v_2) = av_1 + av_2$ for all $a \in R$ and all $v_2, v_1 \in V$;
3) $(ab)v = a(bv)$ for all $a, b \in R$ and all $v \in V$.

A right R-**module** is defined similarly with the multiplication on the right so that

1) $v(a + b) = va + vb$ for all $a, b \in R$ and all $v \in V$;
2) $(v_1 + v_2)a = v_1 a + v_2 a$ for all $a \in R$ and all $v_2, v_1 \in V$
3) $v(ab) = (va)b$ for all $a, b \in R$ and all $v \in V$.

If R has an identity and $1v = v$ (or $v1 = v$) for all $v \in V$, then we say that V is a **unitary R-module**.

If R has an identity, then by "R-module" we shall always mean a **unitary R-module** unless otherwise indicated. Also, if the ring is commutative (the usual case for us), then we may write $av = va$ and consider any right module as a left module and vice versa. Even if the ring is not commutative, we will usually stick to left modules in this appendix and so we drop the reference to "left" and refer to such as R-modules. We do also use right modules in the text. For example, we consider right modules over the quaternions.

Remark D.2. We shall often refer to the elements of R as **scalars**.

Example D.3. An abelian group $(A, +)$ is a $\mathbb{Z}$-module, and a $\mathbb{Z}$-module is none other than an abelian group. Here we take the product of $n \in \mathbb{Z}$ with $x \in A$ to be $nx := x + \cdots + x$ if $n \geq 0$ and $nx := -(x + \cdots + x)$ if $n < 0$ (in either case we are adding $|n|$ terms).

Example D.4. The set of all $m \times n$ matrices with entries that are elements of a commutative ring R is an R-module with scalar multiplication.

Definition D.5. Let V_1 and V_2 be modules over a ring R. A map $L : V_1 \to V_2$ is called a **module homomorphism** or a **linear map** if

$$L(av_1 + bv_2) = aL(v_1) + bL(v_2).$$

By analogy with the case of vector spaces we often characterize a module homomorphism L by saying that L is **linear over** R.

Example D.6. The set of all module homomorphisms of a module V over a *commutative* ring R to another R-module M is also a (left) R-module in its own right and is denoted $\mathrm{Hom}_R(V, M)$ or $L_R(V, M)$ (we mainly use the latter). The scalar multiplication and addition in $L_R(V, M)$ are defined by

$$(f + g)(v) := f(v) + g(v) \text{ for } f, g \in L_R(V, M) \text{ and all } v \in V;$$
$$(af)(v) := af(v) \text{ for } a \in R.$$

Note that $((ab)f)(v) := (ab)f(v) = a(bf(v)) = a((bf)(v)) = (a(bf))(v)$. Also, in order to show that af is linear we argue as follows: $(af)(cv) = af(cv) = acf(v) = caf(v) = c(af)(v)$. This argument fails if R is not commutative! Indeed, if R is not commutative, then $L_R(V, M)$ is not a module but rather only an abelian group.

Example D.7. Let V be a vector space and $\ell : V \to V$ a linear operator. Using ℓ, we may consider V as a module over the ring of polynomials $\mathbb{R}[t]$ by defining the "scalar" multiplication by the rule

$$p(t)v := p(\ell)v$$

for $p \in \mathbb{R}[t]$, $v \in V$. Here, if $p(t) = \sum a_n t^n$, then $p(\ell)$ is the linear map $\sum a_n \ell^n$.

Since the ring is usually fixed, we often omit mentioning the ring. In particular, we often abbreviate $L_R(V, W)$ to $L(V, W)$. Similar omissions will be made without further mention.

Remark D.8. If the modules are infinite-dimensional topological vector spaces such as Banach space, then we must distinguish between the bounded linear maps and simply linear maps. If E and F are infinite-dimensional Banach spaces, then $L(E; F)$ would normally denote **bounded** linear maps.

A **submodule** is defined in the obvious way as a subset $S \subset V$ that is closed under the operations inherited from V so that S itself is a module. The intersection of all submodules containing a subset $A \subset V$ is called the **submodule generated by** A and is denoted $\langle A \rangle$. In this case, A is called a **generating set**. If $\langle A \rangle = V$ for a *finite* set A, then we say that V is **finitely generated**.

Let S be a submodule of V and consider the quotient abelian group V/S consisting of cosets, that is, sets of the form $[v] := v + S = \{v + x : x \in S\}$ with addition given by $[v] + [w] = [v + w]$. We define scalar multiplication by elements of the ring R by $a[v] := [av]$ for $a \in R$. In this way, V/S is a module called a **quotient module**.

Many of the operations that exist for vector spaces have analogues in the module category. For example, if V and W are R-modules, then the set $V \times W$ can be made into an R-module by defining

$$(v_1, w_1) + (v_2, w_2) := (v_1 + v_2, w_1 + w_2) \text{ for } (v_1, w_1) \text{ and } (v_2, w_2) \text{ in } V \times W$$

and
$$a(v,w) := (av, aw) \text{ for } a \in \mathsf{R} \text{ and } (v,w) \in \mathrm{V} \times \mathrm{W}.$$
This module is sometimes written as $\mathrm{V} \oplus \mathrm{W}$, especially when taken together with the injections $\mathrm{V} \to \mathrm{V} \times \mathrm{W}$ and $\mathrm{W} \to \mathrm{V} \times \mathrm{W}$ given by $v \mapsto (v,0)$ and $w \mapsto (0,w)$ respectively. Also, for any module homomorphism $L : \mathrm{V}_1 \to \mathrm{V}_2$ we have the usual notions of kernel and image:
$$\operatorname{Ker} L = \{v \in \mathrm{V}_1 : L(v) = 0\} \subset \mathrm{V}_1,$$
$$\operatorname{Im}(L) = L(\mathrm{V}_1) = \{w \in \mathrm{V}_2 : w = Lv \text{ for some } v \in \mathrm{V}_1\} \subset \mathrm{V}_2.$$
These are submodules of V_1 and V_2 respectively.

On the other hand, modules are generally not as simple to study as vector spaces. For example, there are several notions of dimension. The following notions for a vector space all lead to the same notion of dimension. For a completely general module, these are all potentially different notions:

(1) The length n of the longest chain of submodules
$$0 = \mathrm{V}_n \subsetneq \cdots \subsetneq \mathrm{V}_1 \subsetneq \mathrm{V}.$$

(2) The cardinality of the largest linearly independent set (see below).

(3) The cardinality of a basis (see below).

For simplicity, in our study of dimension, let us now assume that R is commutative.

Definition D.9. A set of elements $\{e_1, \ldots, e_k\}$ of a module are said to be **linearly dependent** if there exist ring elements $r_1, \ldots, r_k \in \mathsf{R}$ not all zero, such that $r_1 e_1 + \cdots + r_k e_k = 0$. Otherwise, they are said to be **linearly independent**. We also speak of the set $\{e_1, \ldots, e_k\}$ as being a **linearly independent set**.

So far so good, but it is important to realize that just because $e_1, \ldots, e_k$ are linearly dependent does not mean that we may write each of these e_i as a linear combination of the others. It may even be that some single element v forms a linearly dependent set since there may be a nonzero r such that $rv = 0$ (such a v is said to be a **torsion element**).

If a linearly independent set $\{e_1, \ldots, e_k\}$ is maximal in size, then we say that the module has **rank** k. Another strange possibility is that a maximal linearly independent set may not be a generating set for the module and hence may not be a basis in the sense to be defined below. The point is that although for an arbitrary $w \in \mathrm{V}$ we must have that $\{e_1, \ldots, e_k\} \cup \{w\}$ is linearly dependent and hence there must be a nontrivial expression $rw + r_1 e_1 + \cdots + r_k e_k = 0$, it does **not** follow that we may solve for w since r may not be an invertible element of the ring. In other words, it may not be a **unit**.

D. Modules and Multilinearity

Definition D.10. If B is a generating set for a module V such that every element of V has a *unique* expression as a *finite* R-linear combination of elements of B, then we say that B is a **basis** for V.

Definition D.11. If an R-module has a basis, then it is referred to as a **free module**. If this basis is finite we indicate this by referring to the module as a **finitely generated free module**.

It turns out that just as for vector spaces the cardinality of a basis for a finitely generated free module V is the same as that of every other basis for V. If a module over a (commutative) ring R has a basis, then the number of elements in the basis is called the **dimension** and must in this case be the same as the rank (the size of a maximal linearly independent set). Thus a finitely generated free module is also called a finite-dimensional free module.

Exercise D.12. Show that every finitely generated R-module is the homomorphic image of a finitely generated free module.

If R is a field, then every module is free and is a vector space by definition. In this case, the current definitions of dimension and basis coincide with the usual ones.

The ring R is itself a free R-module with standard basis given by $\{1\}$. Also, $\mathsf{R}^n := \mathsf{R} \oplus \cdots \oplus \mathsf{R}$ is a finitely generated free module with standard basis $\{\mathbf{e}_1, \ldots, \mathbf{e}_n\}$, where, as usual, $\mathbf{e}_i := (0, \ldots, 1, \ldots, 0)$; the only nonzero entry is in the i-th position. Up to isomorphism, these account for all finitely generated free modules: If a module V is free with basis $e_1, \ldots, e_n$, then we have an isomorphism $\mathsf{R}^n \cong \mathsf{V}$ given by

$$(r_1, \ldots, r_n) \mapsto r_1 e_1 + \cdots + r_n e_n.$$

Definition D.13. Let V_i, $i = 1, \ldots, k$, and W be modules over a ring R. A map $\mu : \mathsf{V}_1 \times \cdots \times \mathsf{V}_k \to \mathsf{W}$ is called **multilinear** (k-multilinear) if for each i, $1 \leq i \leq k$, and each fixed $(v_1, \ldots, \widehat{v_i}, \ldots, v_k) \in \mathsf{V}_1 \times \cdots \times \widehat{\mathsf{V}_i} \times \cdots \times \mathsf{V}_k$ we have that the map

$$v \mapsto \mu(v_1, \ldots, v_{i-1}, \underset{i\text{-th}}{v}, v_{i+1} \ldots, v_k),$$

obtained by fixing all but the i-th variable, is a module homomorphism. In other words, we require that μ be R-linear in each slot separately. The set of all multilinear maps $\mathsf{V}_1 \times \cdots \times \mathsf{V}_k \to \mathsf{W}$ is denoted $L_\mathsf{R}(\mathsf{V}_1, \ldots, \mathsf{V}_k; \mathsf{W})$. If $\mathsf{V}_1 = \cdots = \mathsf{V}_k = \mathsf{V}$, then we abbreviate this to $L_\mathsf{R}^k(\mathsf{V}; \mathsf{W})$.

If R is commutative, then the space of multilinear maps $L_\mathsf{R}(\mathsf{V}_1, \ldots, \mathsf{V}_k; \mathsf{W})$ is itself an R-module in a fairly obvious way: If $a, b \in \mathsf{R}$ and $\mu_1, \mu_2 \in L_\mathsf{R}(\mathsf{V}_1, \ldots, \mathsf{V}_k; \mathsf{W})$, then $a\mu_1 + b\mu_2$ is defined pointwise in the usual way.

Note: For the remainder of this chapter, all modules will be taken to be over a fixed *commutative* ring R.

Let us agree to use the following abbreviation: $V^k = V \times \cdots \times V$ (k-fold Cartesian product).

Definition D.14. The dual of an R-module V is the module $V^* := L_R(V, R)$ of all R-linear functionals on V.

Any element $w \in V$ can be thought of as an element of $V^{**} := L_R(V^*, R)$ according to $w(\alpha) := \alpha(w)$ where $\alpha \in V^*$. This provides a map $V \hookrightarrow V^{**}$, and if this map is an isomorphism, then we say that V is **reflexive**.

If V is reflexive, then we are free to identify V with V^{**}.

Exercise D.15. Show that if V is a finitely generated free module, then V is reflexive.

For completeness, we include the definition of a projective module but what is important for us is that the finitely generated projective modules over $C^\infty(M)$ correspond to spaces of sections of smooth vector bundles. These modules are not necessarily free but are reflexive and have many other good properties such as being "locally free".

Definition D.16. A module V is projective if, whenever V is a quotient of a module W, there exists a module U such that the direct sum $V \oplus U$ is isomorphic to W.

Given two modules V and W over some commutative ring R, consider the class $\mathcal{C}_{V \times W}$ consisting of all bilinear maps $V \times W \to X$ where X varies over all R-modules, but V and W are fixed. We take members of $\mathcal{C}_{V \times W}$ as the objects of a category (see Appendix A). A morphism from, say $\mu_1 : V \times W \to X_1$ to $\mu_2 : V \times W \to X_2$ is defined to be a homomorphism $\ell : X_1 \to X_2$ such that $\mu_2 = \ell \circ \mu_1$. There exists a vector space $T_{V,W}$ together with a bilinear map $\otimes : V \times W \to T_{V,W}$ that has the following **universal property**: For every bilinear map $\mu : V \times W \to X$, there is a unique linear map $\widetilde{\mu} : T_{V,W} \to X$ such that $\mu = \widetilde{\mu} \circ \otimes$. If a pair $(T_{V,W}, \otimes)$ with this property exists, then it is unique up to isomorphism in $\mathcal{C}_{V \times W}$. We refer to such a universal object as a **tensor product** of V and W. We will indicate the construction of a specific tensor product that we denote by $V \otimes W$ with the corresponding map $\otimes : V \times W \to V \otimes W$. The idea is simple: We let $V \otimes W$ be the set of all linear combinations of symbols of the form $v \otimes w$ for $v \in V$ and $w \in W$, subject to the relations

$$(v_1 + v_2) \otimes w = v_1 \otimes w + v_2 \otimes w,$$
$$v \otimes (w_1 + w_2) = v \otimes w_1 + v \otimes w_2,$$
$$r(v \otimes w) = rv \otimes w = v \otimes rw, \text{ for } r \in \mathbb{F}.$$

D. Modules and Multilinearity

The map $\otimes$ is then simply $\otimes : (v, w) \to v \otimes w$. Let us generalize this idea to tensor products of several vector spaces at a time, and also, let us be a bit more pedantic about the construction. We seek a universal object for the category of k-multilinear maps of the form $\mu : V_1 \times \cdots \times V_k \to W$ with $V_1, \ldots, V_k$ fixed.

Definition D.17. A module $T = T_{V_1,\ldots,V_k}$ together with a multilinear map $\otimes : V_1 \times \cdots \times V_k \to T$ is called **universal for k-multilinear maps on $V_1 \times \cdots \times V_k$** if for every multilinear map $\mu : V_1 \times \cdots \times V_k \to W$ there is a unique linear map $\widetilde{\mu} : T \to W$ such that the following diagram commutes:

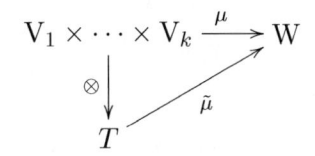

$$\begin{array}{ccc} V_1 \times \cdots \times V_k & \xrightarrow{\mu} & W \\ {\scriptstyle \otimes}\downarrow & \nearrow{\scriptstyle \widetilde{\mu}} & \\ T & & \end{array}$$

i.e. we must have $\mu = \widetilde{\mu} \circ \otimes$. If such a universal object exists, it will be called a **tensor product** of $V_1, \ldots, V_k$, and the module itself $T = T_{V_1,\ldots,V_k}$ is also referred to as a tensor product of the modules $V_1, \ldots, V_k$.

The tensor product is again unique up to isomorphism:

Proposition D.18. *If $(T_1, \otimes_1)$ and $(T_2, \otimes_2)$ are both universal for k-multilinear maps on $V_1 \times \cdots \times V_k$, then there is a unique isomorphism $\Phi : T_1 \to T_2$ such that $\Phi \circ \otimes_1 = \otimes_2$:*

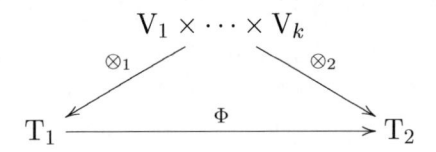

$$\begin{array}{ccc} & V_1 \times \cdots \times V_k & \\ {\scriptstyle \otimes_1}\swarrow & & \searrow{\scriptstyle \otimes_2} \\ T_1 & \xrightarrow{\Phi} & T_2 \end{array}$$

Proof. By the assumption of universality, there are maps $\otimes_1$ and $\otimes_2$ such that $\Phi \circ \otimes_1 = \otimes_2$ and $\bar{\Phi} \circ \otimes_2 = \otimes_1$. We thus have $\bar{\Phi} \circ \Phi \circ \otimes_1 = \otimes_1$, and by the uniqueness part of the definition of universality of $\otimes_1$ we must have $\bar{\Phi} \circ \Phi = \mathrm{id}$ or $\bar{\Phi} = \Phi^{-1}$. $\square$

The usual specific realization of the tensor product of modules $V_1, \ldots, V_k$ is, roughly, the set of all linear combinations of symbols of the form $v_1 \otimes \cdots \otimes v_k$ subject to the obvious multilinear relations:

$$v_1 \otimes \cdots \otimes av_i \otimes \cdots \otimes v_k = a(v_1 \otimes \cdots \otimes v_i \otimes \cdots \otimes v_k)$$

and

$$v_1 \otimes \cdots \otimes (v_i + v_i') \otimes \cdots \otimes v_k \\ = v_1 \otimes \cdots \otimes v_i \otimes \cdots \otimes v_k + v_1 \otimes \cdots \otimes v_i' \otimes \cdots \otimes v_k.$$

This space is denoted by $V_1 \otimes \cdots \otimes V_k$ or by $\bigotimes_{i=1}^k V_i$ and called the **tensor product** of $V_1, \ldots, V_k$. Also, we will use $V^{\otimes k}$ or $\bigotimes^k V$ to denote $V \otimes \cdots \otimes V$ (k-fold tensor product of V). The associated map $\otimes : V_1 \times \cdots \times V_k \to V_1 \otimes \cdots \otimes V_k$ is simply

$$\otimes : (v_1, \ldots, v_k) \mapsto v_1 \otimes \cdots \otimes v_k.$$

A more pedantic description is as follows. We take $V_1 \otimes \cdots \otimes V_k := F(V_1 \times \cdots \times V_k)/U_0$, where $F(V_1 \times \cdots \times V_k)$ is the free module on the *set* $V_1 \times \cdots \times V_k$ and U_0 is the submodule generated by the set of all elements of the form

$$(v_1, \ldots, av_i, \ldots, v_k) - a(v_1, \ldots, v_i, \ldots, v_k)$$

and

$$(v_1, \ldots, (v_i + v_i'), \ldots, v_k)$$
$$- (v_1, \ldots, v_i, \ldots, v_k) - (v_1, \ldots, v_i', \ldots, v_k),$$

where $v_i, v_i' \in V_i$, and $a \in \mathsf{R}$. Each element $(v_1, \ldots, v_k)$ of the set $V_1 \times \cdots \times V_k$ is naturally identified with a generator of the free module $F(V_1 \times \cdots \times V_k)$ and we have the obvious injection $V_1 \times \cdots \times V_k \hookrightarrow F(V_1 \times \cdots \times V_k)$. Its equivalence class is denoted $v_1 \otimes \cdots \otimes v_k$ and the map $\otimes$ is then the composition $V_1 \times \cdots \times V_k \hookrightarrow F(V_1 \times \cdots \times V_k) \to F(V_1 \times \cdots \times V_k)/U_0$.

Proposition D.19. $\otimes : V_1 \times \cdots \times V_k \to V_1 \otimes \cdots \otimes V_k$ *is universal for multilinear maps on* $V_1 \times \cdots \times V_k$.

Exercise D.20. Prove the above proposition.

Proposition D.21. *If* $f : V_1 \to W_1$ *and* $g : V_2 \to W_2$ *are module homomorphisms, then there is a unique homomorphism* $f \otimes g : V_1 \otimes V_2 \to W_1 \otimes W_2$, *the tensor product, which has the characterizing properties that* $f \otimes g$ *is linear and that* $(f \otimes g)(v_1 \otimes v_2) = (fv_1) \otimes (gv_2)$ *for all* $v_1 \in V_1, v_2 \in V_2$. *Similarly, if* $f_i : V_i \to W_i$, *we may obtain* $\otimes_i f_i : \bigotimes_{i=1}^k V_i \to \bigotimes_{i=1}^k W_i$.

Proof. Exercise. □

Definition D.22. Elements of $\bigotimes_{i=1}^k V_i$ that may be written as $v_1 \otimes \cdots \otimes v_k$ for some v_i are called **simple** or **decomposable**.

Remark D.23. It is clear from our specific realization of $\bigotimes_{i=1}^k V_i$ that elements in the image of $\otimes : V_1 \times \cdots \times V_k \to \bigotimes_{i=1}^k V_i$ span $\bigotimes_{i=1}^k V_i$. I.e., decomposable elements span the space.

Exercise D.24. Not all elements are decomposable but the decomposable elements generate $V_1 \otimes \cdots \otimes V_k$.

It may be that the V_i are modules over more than one ring. For example, any complex vector space is a module over both $\mathbb{R}$ and $\mathbb{C}$. Also, the module of

D. Modules and Multilinearity

smooth vector fields $\mathfrak{X}_M(U)$ is a module over $C^\infty(U)$ and a module (actually a vector space) over $\mathbb{R}$. Thus it is sometimes important to indicate the ring involved, and so we write the tensor product of two R-modules V and W as $V \otimes_R W$. For instance, there is a big difference between $\mathfrak{X}_M(U) \otimes_{C^\infty(U)} \mathfrak{X}_M(U)$ and $\mathfrak{X}_M(U) \otimes_\mathbb{R} \mathfrak{X}_M(U)$.

Lemma D.25. *There are the following natural isomorphisms:*

(1) $(V \otimes W) \otimes U \cong V \otimes (W \otimes U) \cong V \otimes W \otimes U$, *and under these isomorphisms,* $(v \otimes w) \otimes u \longleftrightarrow v \otimes (w \otimes u) \longleftrightarrow v \otimes w \otimes u$.

(2) $V \otimes W \cong W \otimes V$, *and under this isomorphism* $v \otimes w \longleftrightarrow w \otimes v$.

Proof. We prove (1) and leave (2) as an exercise.

Elements of the form $(v \otimes w) \otimes u$ generate $(V \otimes W) \otimes U$, so any map that sends $(v \otimes w) \otimes u$ to $v \otimes (w \otimes u)$ for all v, w, u must be unique. Now we have compositions

$$(V \times W) \times U \xrightarrow{\otimes \times \mathrm{id}_U} (V \otimes W) \times U \xrightarrow{\otimes} (V \otimes W) \otimes U$$

and

$$V \times (W \times U) \xrightarrow{\mathrm{id}_V \times \otimes} V \times (W \otimes U) \xrightarrow{\otimes} V \otimes (W \otimes U).$$

It is a simple matter to check that these composite maps have the same universal property as the map $V \times W \times U \xrightarrow{\otimes} V \otimes W \otimes U$. The result now follows from the existence and essential uniqueness (Propositions D.19 and D.18). □

We shall use the first isomorphism and the obvious generalizations to identify $V_1 \otimes \cdots \otimes V_k$ with all legal parenthetical constructions such as $(((V_1 \otimes V_2) \otimes \cdots \otimes V_j) \otimes \cdots) \otimes V_k$ and so forth. In short, we may construct $V_1 \otimes \cdots \otimes V_k$ by tensoring spaces two at a time. In particular, we assume the isomorphisms (as identifications)

$$(V_1 \otimes \cdots \otimes V_k) \otimes (W_1 \otimes \cdots \otimes W_k) \cong V_1 \otimes \cdots \otimes V_k \otimes W_1 \otimes \cdots \otimes W_k,$$

where $(v_1 \otimes \cdots \otimes v_k) \otimes (w_1 \otimes \cdots \otimes w_k)$ maps to $v_1 \otimes \cdots \otimes v_k \otimes w_1 \otimes \cdots \otimes w_k$.

Proposition D.26. *If* V *is an* R-*module, then we have natural isomorphisms*

$$V \otimes R \cong V \cong R \otimes V$$

given on decomposable elements as $v \otimes r \mapsto rv \mapsto r \otimes v$. *(Recall that we are assuming that* R *is commutative.)*

The proof is left to the reader. The following proposition gives a basic and often used isomorphism.

Proposition D.27. *For R-modules W, V, U, we have*
$$L_R(W \otimes V, U) \cong L(W, V; U).$$
More generally,
$$L_R(W_1 \otimes \cdots \otimes W_k, U) \cong L(W_1, \ldots, W_k; U).$$

Proof. This is more or less just a restatement of the universal property of $W \otimes V$. One should check that this association is indeed an isomorphism. □

Exercise D.28. Show that if W is free with basis $(f_1, \ldots, f_n)$, then W^* is also free and has a dual basis $(f^1, \ldots, f^n)$, that is, $f^i(f_j) = \delta^i_j$.

Theorem D.29. *If $V_1, \ldots, V_k$ are free R-modules and if $(e_1^j, \ldots, e_{n_j}^j)$ is a basis for V_j, then the set of all decomposable elements of the form $e_{i_1}^1 \otimes \cdots \otimes e_{i_k}^k$ is a basis for $V_1 \otimes \cdots \otimes V_k$.*

Proof. We prove this for the case of $k=2$. The general case is similar. We wish to show that if $(e_1, \ldots, e_{n_1})$ is a basis for V_1 and $(f_1, \ldots, f_{n_2})$ is a basis for V_2, then $\{e_i \otimes f_j\}$ is a basis for $V_1 \otimes V_2$. Define $\phi_{lk} : V_1 \times V_2 \to R$ by $\phi_{lk}(e_i, f_j) = \delta^l_i \delta^k_j 1$, where 1 is the identity in R and

$$\delta^l_i \delta^k_j 1 := \begin{cases} 1 & \text{if } (l,k) = (i,j), \\ 0 & \text{otherwise.} \end{cases}$$

Extend this definition bilinearly. These maps are linearly independent in $L(V_1, V_2; R)$ since if $\sum_{lk} a_{lk} \phi_{lk} = 0$ in R, then for any i, j we have

$$0 = \sum_{lk} a_{lk} \phi_{lk}(e_i, f_j) = \sum_{lk} a_{lk} \delta^l_i \delta^k_j 1$$
$$= a_{ij}.$$

Thus $\dim(V_1 \otimes V_2) = \dim((V_1 \otimes V_2)^*) = \dim L(V_1, V_2; R) \geq n_1 n_2$. On the other hand, $\{e_i \otimes f_j\}$ spans the set of all decomposable elements and hence the whole space $V_1 \otimes V_2$, so that $\dim(V_1 \otimes V_2) \leq n_1 n_2$ and it follows that $\{e_i \otimes f_j\}$ is a basis. □

Proposition D.30. *There is a unique R-module map $\iota : L(V_1, W_1) \otimes \cdots \otimes L(V_k, W_k) \to L(V_1 \otimes \cdots \otimes V_k, W_1 \otimes \cdots \otimes W_k)$ such that if $f_1 \otimes \cdots \otimes f_k$ is a (decomposable) element of $L(V_1, W_1) \otimes \cdots \otimes L(V_k, W_k)$ then*

$$\iota(f_1 \otimes \cdots \otimes f_k)(v_1 \otimes \cdots \otimes v_k) = f_1(v_1) \otimes \cdots \otimes f_k(v_k).$$

If the modules are all finitely generated and free, then this is an isomorphism.

D. Modules and Multilinearity

Proof. If such a map exists, it must be unique since the decomposable elements span $L(V_1, W_1) \otimes \cdots \otimes L(V_k, W_k)$. To show the existence, we define a multilinear map
$$\vartheta : L(V_1, W_1) \times \cdots \times L(V_k, W_k) \times V_1 \times \cdots \times V_k \to W_1 \otimes \cdots \otimes W_k$$
by the recipe
$$(f_1, \ldots, f_k, v_1, \ldots, v_k) \mapsto f_1(v_1) \otimes \cdots \otimes f_k(v_k).$$
By the universal property there must be a linear map
$$\widetilde{\vartheta} : V_1^* \otimes \cdots \otimes V_k^* \otimes V_1 \otimes \cdots \otimes V_k \to W_1 \otimes \cdots \otimes W_k$$
such that $\widetilde{\vartheta} \circ \otimes = \vartheta$, where $\otimes$ is the universal map. Now define
$$\iota(f_1 \otimes \cdots \otimes f_k)(v_1 \otimes \cdots \otimes v_k)$$
$$:= \widetilde{\vartheta}(f_1 \otimes \cdots \otimes f_k \otimes v_1 \otimes \cdots \otimes v_k).$$
The fact that ι is an isomorphism in case the V_i are all free follows easily from Exercise D.28 and Theorem D.29. $\square$

Since $R \otimes R = R$, we obtain

Corollary D.31. *There is a unique R-module map* $\iota : V_1^* \otimes \cdots \otimes V_k^* \to (V_1 \otimes \cdots \otimes V_k)^*$ *such that if* $\alpha_1 \otimes \cdots \otimes \alpha_k$ *is a (decomposable) element of* $V_1^* \otimes \cdots \otimes V_k^*$, *then*
$$\iota(\alpha_1 \otimes \cdots \otimes \alpha_k)(v_1 \otimes \cdots \otimes v_k) = \alpha_1(v_1) \cdots \alpha_k(v_k).$$
If the modules are all finitely generated and free, then this is an isomorphism.

Corollary D.32. *There is a unique module map* $\iota_0 : W \otimes V^* \to L(V, W)$ *such that if* $v \otimes \beta$ *is a (decomposable) element of* $W \otimes V^*$, *then*
$$\iota_0(w \otimes \beta)(v) = \beta(v)w.$$
If V and W are finitely generated free modules, then this is an isomorphism.

Proof. If we associate to every $w \in W$ the map $w^{\mathrm{map}} \in L(R, W)$ given by $w^{\mathrm{map}}(r) := rw$, then we obtain an isomorphism $W \cong L(R, W)$. Use this and then compose
$$W \otimes V^* \to L(R, W) \otimes L(V, R)$$
$$\to L(R \otimes V, W \otimes R) \cong L(V, W),$$

$$V^* \otimes W \to L(V, R) \otimes L(R, W)$$
$$\to L(V \otimes R, R \otimes W) \cong L(V, W). \quad \square$$

By combining Corollary D.31 with Proposition D.27 and taking $U = R$ we obtain the following assertion.

Corollary D.33. *There is a unique R-module map $\iota : V_1^* \otimes \cdots \otimes V_k^* \to L(V_1, \ldots, V_k; R)$ such that if $\alpha_1 \otimes \cdots \otimes \alpha_k$ is a (decomposable) element of $V_1^* \otimes \cdots \otimes V_k^*$, then*

$$\iota(\alpha_1 \otimes \cdots \otimes \alpha_k)(v_1, \ldots, v_k) = \alpha_1(v_1) \cdots \alpha_k(v_k).$$

If the modules are all finitely generated and free, then this is an isomorphism.

Theorem D.34. *If $\varphi_i : V_i \times W_i \to U_i$ are bilinear maps for $i = 1, \ldots, k$, then there is a unique bilinear map*

$$\varphi : \bigotimes_{i=1}^k V_i \times \bigotimes_{i=1}^k W_i \to \bigotimes_{i=1}^k U_i$$

such that for $v_i \in V_i$ and $w_i \in W_i$,

$$\varphi(v_1 \otimes \cdots \otimes v_k, w_1 \otimes \cdots \otimes w_k) = \varphi_1(v_1, w_1) \otimes \cdots \otimes \varphi_k(v_k, w_k).$$

Proof. We sketch the proof in the $k = 2$ case. If φ exists, it is unique since elements of the form $v_1 \otimes v_2$ span $V_1 \otimes V_2$ and similarly for $W_1 \otimes W_2$. Now by the universal property of tensor products, associated to φ_i for $i = 1, 2$, we have unique linear maps $f_i : V_i \otimes W_i \to U_i$ with $f_i \circ \otimes = \varphi_i$. Then we obtain the linear map $f_1 \otimes f_2 : (V_1 \otimes W_1) \otimes (V_2 \otimes W_2) \to U_1 \otimes U_2$. On the other hand we have the natural isomorphism $S : (V_1 \otimes V_2) \otimes (W_1 \otimes W_2) \to (V_1 \otimes W_1) \otimes (V_2 \otimes W_2)$ induced by the obvious switching of factors of simple elements. Now define $\varphi : (V_1 \otimes V_2) \times (W_1 \otimes W_2) \to U_1 \otimes U_2$ by $\varphi(x, y) = (f_1 \otimes f_2)(S(x \otimes y))$ for $x \in V_1 \otimes V_2$ and $y \in W_1 \otimes W_2$. Then we have

$$\begin{aligned}\varphi(v_1 \otimes v_2, w_1 \otimes w_2) &= (f_1 \otimes f_2)(v_1 \otimes w_1 \otimes v_2 \otimes w_2) \\ &= f_1(v_1 \otimes w_1) \otimes f_2(v_2 \otimes w_2) \\ &= \varphi_1(v_1, w_1) \otimes \varphi_2(v_2, w_2).\end{aligned}$$
$\square$

Corollary D.35. *Let V_i and W_i be R-modules for $i = 1, \ldots, k$. If $\varphi_i : V_i \times W_i \to R$ are bilinear maps, $i = 1, \ldots, k$, then there is a unique bilinear map*

$$\varphi : \bigotimes_{i=1}^k V_i \times \bigotimes_{i=1}^k W_i \to R$$

such that for $v_i \in V_i$ and $w_i \in W_i$,

$$\varphi(v_1 \otimes \cdots \otimes v_k, w_1 \otimes \cdots \otimes w_k) = \varphi_1(v_1, w_1) \cdots \varphi_k(v_k, w_k).$$

Proof. Imitate the proof of the previous theorem or just use the previous theorem together with the natural isomorphism $R \otimes \cdots \otimes R \cong R$. $\square$

D.1. R-Algebras

Definition D.36. Let R be a commutative ring. An (associative) **R-algebra** $\mathfrak{A}$ is a unitary R-module that is also a ring with identity $1_\mathfrak{A}$, where the ring addition and the module addition coincide and where $r(a_1 a_2) = (ra_1)a_2 = a_1(ra_2)$ for all $a_1, a_2 \in \mathfrak{A}$ and all $r \in R$.

D.1. R-Algebras

As defined above, an algebra is associative. However, one can also define nonassociative algebras, and a Lie algebra is an example of such.

Definition D.37. Let $\mathfrak{A}$ and $\mathfrak{B}$ be R-algebras. A module homomorphism $h : \mathfrak{A} \to \mathfrak{B}$ that is also a ring homomorphism is called an **R-algebra homomorphism**. Epimorphism, monomorphism, and isomorphism are defined in the obvious way.

If a submodule $\mathfrak{J}$ of an algebra $\mathfrak{A}$ is also a two-sided ideal with respect to the ring structure on $\mathfrak{A}$, then $\mathfrak{A}/\mathfrak{J}$ is also an algebra.

Example D.38. Let U be an open subset of a C^r manifold M. The set of all smooth functions $C^r(U)$ is an $\mathbb{R}$-algebra ($\mathbb{R}$ is the real numbers) with unity being the function constantly equal to 1.

Example D.39. The set of all complex $n \times n$ matrices is an algebra over $\mathbb{C}$ with the product being matrix multiplication.

Example D.40. The set of all complex $n \times n$ matrices with real polynomial entries is an algebra over the ring of polynomials $\mathbb{R}[x]$.

Definition D.41. The set of all endomorphisms of an R-module W is an R-algebra denoted $\text{End}_R(W)$ and called the **endomorphism algebra** of W. Here, the sum and scalar multiplication are defined as usual and the product is composition. Note that for $r \in R$

$$r(f \circ g) = (rf) \circ g = f \circ (rg),$$

where $f, g \in \text{End}_R(W)$.

Definition D.42. A set A together with a binary operation $* : A \times A \to A$ is called a **monoid** if the operation is associative and there exists an element e (the identity) such that $a * e = e * a = a$ for all $a \in A$.

With the operation of addition, $\mathbb{N}$, $\mathbb{Z}$ and $\mathbb{Z}_2$ are all commutative monoids.

Definition D.43. Let $(A, *)$ be a monoid and R a ring. An A-**graded R-algebra** is an R-algebra with a direct sum decomposition $\mathfrak{A} = \sum_{i \in A} \mathfrak{A}_i$ such that $\mathfrak{A}_i \mathfrak{A}_j \subset \mathfrak{A}_{i*j}$. An $\mathbb{N}$-graded algebra is sometimes simply referred to as a **graded algebra**. A **superalgebra** is a $\mathbb{Z}_2$-graded algebra.

Definition D.44. Let $\mathfrak{A} = \sum_{i \in \mathbb{Z}} \mathfrak{A}_i$ and $\mathfrak{B} = \sum_{i \in \mathbb{Z}} \mathfrak{B}_i$ be $\mathbb{Z}$-graded algebras. An R-algebra homomorphism $h : \mathfrak{A} \to \mathfrak{B}$ is called a $\mathbb{Z}$-graded homomorphism if $h(\mathfrak{A}_i) \subset \mathfrak{B}_i$ for each $i \in \mathbb{Z}$.

We now construct the tensor algebra on a fixed R-module W. This algebra is important because using it we may construct by quotients many important algebras. Consider the following situation: $\mathfrak{A}$ is an R-algebra, W an R-module, and $\phi : W \to \mathfrak{A}$ is a module homomorphism. If $h : \mathfrak{A} \to \mathfrak{B}$ is an algebra homomorphism, then of course $h \circ \phi : W \to \mathfrak{B}$ is an R-module homomorphism.

Definition D.45. Let W be an R-module. An R-algebra $\mathfrak{A}$ together with a map $\phi : W \to \mathfrak{A}$ is called universal with respect to W if for any R-module homomorphism $\psi : W \to \mathfrak{B}$ there is a unique algebra homomorphism $h : \mathfrak{A} \to \mathfrak{B}$ such that $h \circ \phi = \psi$.

Again if such a universal object exists, it is unique up to isomorphism. We now exhibit the construction of this type of universal algebra. First we define $\bigotimes^0 W := R$ and $\bigotimes^1 W := W$. Then we define $\bigotimes^k W := W^{\otimes k} = W \otimes \cdots \otimes W$. The next step is to form the direct sum $\bigotimes W := \sum_{i=0}^{\infty} \bigotimes^i W$. In order to make this a $\mathbb{Z}$-graded algebra, we define $\bigotimes^i W := 0$ for $i < 0$ and then define a product on $\bigotimes W := \sum_{i \in \mathbb{Z}} \bigotimes^i W$ as follows: We know that for $i, j > 0$ there is an isomorphism $W^{i\otimes} \otimes W^{\otimes j} \to W^{\otimes(i+j)}$ and so a bilinear map $W^{i\otimes} \times W^{\otimes j} \to W^{\otimes(i+j)}$ such that

$$(w_1 \otimes \cdots \otimes w_i) \times (w_1' \otimes \cdots \otimes w_j') \mapsto w_1 \otimes \cdots \otimes w_i \otimes w_1' \otimes \cdots \otimes w_j'.$$

Similarly, we define $\bigotimes^0 W \times W^{\otimes i} = R \times W^{\otimes i} \to W^{\otimes i}$ by scalar multiplication. Also, $W^{\otimes i} \times W^{\otimes j} \to 0$ if either i or j is negative. Now we may use the symbol $\otimes$ to denote these multiplications without contradiction and put them together to form a product on $\bigotimes W := \sum_{i \in \mathbb{Z}} \bigotimes^i W$. It is now clear that

$$\bigotimes^i W \times \bigotimes^j W \to \bigotimes^{i+j} W,$$

where we make the needed trivial definitions for the negative powers:

$$\bigotimes^i W = 0, \qquad i < 0.$$

Definition D.46. The algebra $\bigotimes W$ is a graded algebra called the **R-tensor algebra**.

Bibliography

[Arm]   M.A. Armstrong, *Basic Topology*, Springer-Verlag (1983).

[At]   M.F. Atiyah, *K-Theory*, W.A. Benjamin (1967).

[A-M-R]   R. Abraham, J.E. Marsden, and T. Ratiu, *Manifolds, Tensor Analysis, and Applications*, Addison Wesley, Reading, (1983).

[Arn]   V.I. Arnold, *Mathematical Methods of Classical Mechanics*, Graduate Texts in Math. **60**, Springer-Verlag, 2nd edition (1989).

[BaMu]   John Baez and Javier P. Muniain, *Gauge Fields, Knots and Gravity*, World Scientific (1994).

[Be]   Arthur L. Besse, *Einstein Manifolds*, Classics in Mathematics, Springer-Verlag (1987).

[BishCr] R. L. Bishop and R. J. Crittenden, *Geometry of Manifolds*, Academic Press (1964).

[Bo-Tu]   R. Bott and L. Tu, *Differential Forms in Algebraic Topology*, Springer-Verlag GTM 82 (1982).

[Bre]   G. Bredon, *Topology and Geometry*, Springer-Verlag GTM 139, (1993).

[BrCl]   F. Brickell and R. Clark, *Differentiable Manifolds: An Introduction*, Van Nostran Rienhold (1970).

[Bro-Jan] Th. Bröcker and K. Jänich, *Introduction to Differential Topology*, Cambridge University Press (1982).

[Brou]   L.E.J. Brouwer, *Über Abbildung von Mannigfaltigkeiten*, Mathematische Annalen 71 (1912), 97–115.

[Chavel] I. Chavel, *Eigenvalues in Riemannian Geometry,* Academic Press, (1984).

[C-E]   J. Cheeger and D. Ebin, *Comparison Theorems in Riemannian Geometry*, North-Holland (1975).

[Cheng]   S.Y. Cheng, *Eigenvalue comparison theorems and its geometric applications*, Math. Z. 143 (1975), 289–297.

[Clark]   C.J.S. Clark, *On the global isometric embedding of pseudo-Riemannian manifolds*, Proc. Roy. Soc. A314 (1970), 417–428.

[Dar] R.W.R. Darling, *Differential Forms and Connections*, Cambridge University Press (1994).

[Dieu] J. Dieudonné, *A History of Algebraic and Differential Topology 1900-1960*, Birkhäuser (1989).

[Dod-Pos] C.T.J. Dodson and T. Poston, *Tensor Geometry: The Geometric Viewpoint and Its Uses*, Springer-Verlag (2000).

[Dol] A. Dold, *Lectures on Algebraic Topology*, Springer-Verlag (1980).

[Donaldson] S. Donaldson, *An application of gauge theory to the topology of 4-manifolds*, J. Diff. Geo. 18 (1983), 269–316.

[Dug] J. Dugundji, *Topology*, Allyn & Bacon (1966).

[Eil-St] S. Eilenberg and N. Steenrod, *Foundations of Algebraic Topology*, Princeton Univ. Press (1952).

[Fen] R. Fenn, *Techniques of Geometric Topology*, Cambridge Univ. Press (1983).

[Freedman] M. Freedman, *The topology of four-dimensional manifolds*, J. Diff. Geo. 17 (1982), 357–454.

[Fr-Qu] M. Freedman and F. Quinn, *Topology of 4-Manifolds*, Princeton Univ. Press (1990).

[Fult] W. Fulton, *Algebraic Topology: A First Course*, Springer-Verlag (1995).

[G2] V. Guillemin, and S. Sternberg, *Symplectic Techniques in Physics*, Cambridge Univ. Press (1984).

[Gray] B. Gray, *Homotopy Theory*, Academic Press (1975).

[Gre-Hrp] M. Greenberg and J. Harper, *Algebraic Topology: A First Course*, Addison-Wesley (1981).

[Grom] M. Gromov, *Convex sets and Kähler manifolds*, in *Advances in Differential Geometry and Topology*, World Sci. Publ. (1990), pp. 1–38.

[Helg] S. Helgason, *Differential Geometry, Lie Groups and Symmetric Spaces*, Amer. Math. Soc. (2001).

[Hicks] Noel J. Hicks, *Notes on Differential Geometry*, D. Van Nostrand Company Inc (1965).

[Hilt1] P.J. Hilton, *An Introduction to Homotopy Theory*, Cambridge University Press (1953).

[Hilt2] P.J. Hilton and U. Stammbach, *A Course in Homological Algebra*, Springer-Verlag (1970).

[Hus] D. Husemoller, *Fibre Bundles*, McGraw-Hill, (1966) (later editions by Springer-Verlag).

[Kirb-Seib] R. Kirby and L. Siebenmann, *Foundational Essays on Topological Manifolds, Smoothings, and Triangulations*, Ann. of Math. Studies 88 (1977).

[Klein] Felix Klein, *Gesammelte Abhandlungen*, Vol. I, Springer, Berlin (1971).

[Kob] S. Kobayashi, *On conjugate and cut Loci*, in Global Differential Geometry, M.A.A. Studies in Math, Vol. 27, S.S. Chern, Editor, Prentice Hall (1989).

[K-N] S. Kobayashi and K. Nomizu, *Foundations of Differential Geometry*. I, II, J. Wiley Interscience (1963), (1969).

[L1] S. Lang, *Fundamentals of Differential Geometry*, Springer-Verlag GTM Vol. 191 (1999).

[L2] S. Lang, *Algebra,* Springer-Verlag GTM Vol. 211 (2002).

[L-M]	H. Lawson and M. Michelsohn, *Spin Geometry*, Princeton University Press, (1989).
[Lee, Jeff]	Jeffrey M. Lee, Online Supplement to the present text, Internet: http://webpages.acs.ttu.edu/jlee/Supp.pdf (see also http://www.ams.org/bookpages/gsm-107).
[Lee, John]	John Lee, *Introduction to Smooth Manifolds*, Springer-Verlag GTM Vol. 218 (2002).
[L-R]	David Lovelock and Hanno Rund, Tensors, *Differential Forms, and Variational Principles*, Dover Publications (1989).
[Madsen]	Id Madsen and Jørgen Tornehave, *From Calculus to Cohomology*, Cambridge University Press (1997).
[Matsu]	Y. Matsushima, *Differentiable Manifolds*, Marcel Dekker (1972).
[M-T-W]	C. Misner, J. Wheeler, and K. Thorne, *Gravitation*, Freeman (1974).
[MacL]	S. MacLane, *Categories for the Working Mathematician*, Springer-Verlag GTM Vol. 5 (1971).
[Mass]	W. Massey, *Algebraic Topology: An Introduction*, Harcourt, Brace & World (1967) (reprinted by Springer-Verlag).
[Mass2]	W. Massey, *A Basic Course in Algebraic Topology*, Springer-Verlag (1993).
[Maun]	C.R.F. Maunder, *Algebraic Topology*, Cambridge Univ. Press (1980) (reprinted by Dover Publications).
[Mich]	P. Michor, *Topics in Differential Geometry*, Graduate Studies in Mathematics, Vol. 93, Amer. Math. Soc. (2008)
[Mil]	J. Milnor, *Morse Theory*, Annals of Mathematics Studies **51**, Princeton University Press (1963).
[Miln1]	J. Milnor, *Topology from the Differentiable Viewpoint*, University Press of Virginia (1965).
[Mil-St]	J. Milnor and J. Stasheff, *Characteristic Classes*, Ann. of Math. Studies 76, Princeton University Press (1974).
[Molino]	P. Molino, *Riemannian foliations,* Progress in Mathematics, Birkhäuser Boston (1988).
[My-St]	S.B. Myers, N.E. Steenrod, *The group of isometries of a Riemannian manifold*, Annals of Mathematics 40 (1939), no. 2.
[Nash1]	John Nash, C^1-*isometric imbeddings*, Annals of Mathematics 60 (1954), pp 383-396.
[Nash2]	John Nash, *The imbedding problem for Riemannian manifolds*, Annals of Mathematics 63 (1956), 20–63.
[ON1]	B. O'Neill, *Semi-Riemannian Geometry*, Academic Press (1983).
[ON2]	B. O'Neill, *Elementary differential geometry*, Academic Press (1997).
[Pa]	Richard, Palais, *Natural Operations on Differential Forms*, Trans. Amer. Math. Soc. 92 (1959), 125-141.
[Pa2]	Richard Palais, *A Modern Course on Curves and Surfaces*, Online Book at http://www.math.uci.edu/~cterng/NotesByPalais.pdf.
[Pen]	Roger Penrose, *The Road to Reality*, Alfred A. Knopf (2005).
[Pe]	Peter Peterson, *Riemannian Geometry*, Springer-Verlag (1991).
[Poor]	Walter Poor, *Differential Geometric Structures*, McGraw Hill (1981).

[Roe]   J. Roe, *Elliptic Operators, Topology and Asymptotic methods*, Longman (1988).

[Ros]   W. Rossmann, *Lie Groups. An Introduction through Linear Groups*, Oxford University Press (2002).

[Shrp]   R. Sharpe, *Differential Geometry; Cartan's Generalization of Klein's Erlangen Program*, Springer-Verlag (1997).

[Span]   Edwin H. Spanier, *Algebraic Topology*, Springer-Verlag (1966).

[Spv]   M. Spivak, *A Comprehensive Introduction to Differential Geometry* (5 volumes), Publish or Perish Press (1979).

[St]   N. Steenrod, *Topology of Fiber Bundles*, Princeton University Press (1951).

[Stern]   Shlomo Sternberg, *Lectures on Differential Geometry*, Prentice Hall (1964).

[Tond]   Ph. Tondeur, *Foliations on Riemannian manifolds*, Springer-Verlag (1988).

[Tond2]   Ph. Tondeur, *Geometry of Foliations*, Monographs in Mathematics, Vol. 90, Birkhäuser (1997).

[Wal]   Gerard Walschap, *Metric Structures in Differential Geometry*, Springer-Verlag (2004).

[War]   Frank Warner, *Foundations of Differentiable Manifolds and Lie Groups*, Springer-Verlag (1983).

[Wh]   John A. Wheeler, *A Journey into Gravity and Spacetime*, Scientific American Library, A division of HPHLP, New York (1990).

[We1]   A. Weinstein, *Lectures on Symplectic Manifolds*, Regional Conference Series in Mathematics 29, Amer. Math. Soc. (1977).

Index

adapted chart, 46
adjoint map, 220
adjoint representation, 220, 221
admissible chart, 13
algebra bundle, 287
alternating, 345
associated bundle, 300
asymptotic curve, 159
asymptotic vector, 158
atlas, 11
   submanifold, 46
automorphism, 201

base space, 258
basis criterion, 6
bilinear, 3
boundary, 9
   manifold with, 48, 49
bundle atlas, 260, 263
bundle chart, 260
bundle morphism, 258
bundle-valued forms, 370

canonical parametrization, 180
Cartan's formula, 374
causal character, 547
chart, 11
   centered, 11
Christoffel symbols, 168
closed form, 366
closed Lie subgroup, 192
coboundary, 444
cocycle, 263, 444
   conditions, 263
Codazzi-Mainardi equation, 173
codimension, 46

coframe, 121
complete vector field, 97
complex, 444
   orthogonal group, 196
conjugation, 202
connected, 8
connection forms, 506
conservative, 118
   locally, 119
consolidated tensor product, 311
consolidation maps, 309
contraction, 317
coordinate frame, 87
cotangent bundle, 85
cotangent space, 65
covariant derivative, 501, 503
covector field, 110
cover, 6
covering, 33
   map, 33
   space, 33
critical point, 74
critical value, 74
curvature function, 146
curvature vector, 146
cut-off function, 28

Darboux frame, 160
de Rham cohomology, 367
deck transformation, 34
decomposable, 656
deformation retraction, 453
degree, 463
derivation, 61, 89
   of germs, 64
determinant, 353

diffeomorphism, 4, 25
differentiable, 3
  manifold, 14
  structure, 12
differential, 81, 111
  Lie, 211
differential complex, 444
differential form, 359
discrete group action, 42
distant parallelism, 56
distinguished Frenet frame, 148
distribution (tangent), 468
divergence, 398, 623

effective action, 41
embedding, 128
endomorphism algebra, 661
equivariant rank theorem, 230
exact 1-form, 112
exact form, 366
exponential map, 213
exterior derivative, 363, 366
exterior product, 347
  wedge product, 359

faithful representation, 247
fiber, 138, 258
  bundle, 257
flat point, 162
flatting, 334
flow, 95
  box, 96
  complete, 95
  local, 97
  maximal, 100
foliation, 482
  chart, 482
form, 1-form, 110
frame, 277
  bundle, 292
  field, 120, 277
free action, 41
free module, 653
Frenet frame, 146
functor, 640
fundamental group, 36

G-structure, 264
gauge transformation, 297
Gauss curvature, 157, 557
Gauss curvature equation, 172
Gauss formula, 166
Gauss' theorem, 399
Gauss-Bonnet, 186
general linear group, 193
geodesic, 567

geodesic curvature, 160
  vector, 161
geodesically complete, 569
geodesically convex, 634
germ, 28
global, 30
good cover, 455
graded algebra, 661
graded commutative, 359
Grassmann manifold, 21
group action, 40

half-space, 9
  chart, 49
harmonic form, 419
Hermitian metric, 280
Hessian, 76, 123
homogeneous component, 359
homogeneous coordinates, 19
homogeneous space, 241
homologous, 402
homomorphism presheaf, 289
homothety, 634
homotopy, 32
Hopf bundles, 295
Hopf map, 238, 239
hypersurface, 143, 153

identity component, 191
immersed submanifold, 130
immersion, 127
implicit napping theorem, 647
index, 332, 337
  of a critical point, 77
  raising and lowering, 335
induced orientaiton, 382
integral curve, 95
interior, 9
interior product, 374
intrinsic, 169
inversion, 190
isometry, 338
isotropy group, 228
isotropy representation, 248

Jacobi identity, 92
Jacobian, 3

Koszul connection, 501

Lagrange identity, 164
left invariant, 204
left translation, 191
length, 333
lens space, 52
Levi-Civita derivative, 550

Index

Lie algebra, 92
   automorphism, 206
   homomorphism, 206
Lie bracket, 91
Lie derivative, 89, 330
   of a vector field, 103
   on functions, 89
Lie differential, 211
Lie group, 189
   homomorphism, 201
Lie subgroup, 191
   closed, 192
lift, 37, 126
lightcone, 548
lightlike curve, 548
lightlike subspace, 588
line bundle, 281
line integral, 117
linear frame bundle, 292
linear Lie group, 193
local flow, 97
local section, 259
local trivialization, 260
locally connected, 8
locally finite, 6
long exact, 445
Lorentz manifold, 337

Möbius band, 261
manifold topology, 13
manifold with corners, 54
Maurer-Cartan equation, 387
Maurer-Cartan form, 224
maximal flow, 99
maximal integral curve, 98
Mayer-Vietoris, 447
mean curvature, 157
measure zero, 75
meridians, 160
metric, 279
   connection, 543
minimal hypersurface, 179
Minkowski space, 337
module, 649
morphism, 638
moving frame, 120
multilinear, 653
musical isomorphism, 334

n-manifold, 16
neighborhood, 2
nice chart, 383
nonnull curve, 548
nonnull vector, 547
normal coordinates, 572
normal curvature, 158
normal field, 153

normal section, 159
nowhere vanishing, 277
null vector, 547

one-parameter subgroup, 202
open manifold, 50
open submanifold, 16
orbit, 41
   map, 232
orientable manifold, 377
orientation cover, 380
orientation for a vector space, 353
orientation of a vector bundle, 375
oriented manifold, 377
orthogonal group, 195
orthonormal frame field, 165, 279
outward pointing, 381
overlap maps, 11

paracompact, 6, 645
parallel, 171
   translation, 516
parallelizable, 278
parallels, 160
partial tangent map, 72
partition of unity, 30
path component, 8
path connected, 8
Pauli matrices, 203
piecewise smooth, 117
plaque, 482
point derivation, 61
point finite cover, 644
presheaf, 289
principal bundle, 293
   atlas, 293
   morphism, 297
principal curve, 159
principal frame field, 165
principal normal, 146
principal part, 56
principal vector, 158
product group, 191
product manifold, 20
projective plane, 18
projective space, 18
proper action, 231
proper map, 33
proper time, 550
property W, 133
pull-back, 92, 114, 322, 323
   bundle, 268
   vector bundle, 276
push-forward, 92, 115, 323

R-algebra, 660
radial geodesic, 571

radially parallel, 517
rank, 78
 of a linear map, 127
real projective space, 18
refinement, 6
reflexive, 654
reflexive module, 309
regular point, 74
regular submanifold, 46
regular value, 74
related vector fields, 93
Riemannian manifold, 337
Riemannian metric, 279

Sard's theorem, 76
scalar product, 193, 331
 space, 193
second fundamental form, 156
section, 87, 259
 along a map, 269
sectional curvature, 557
self-avoiding, 42
semi-Euclidean motion., 339
semi-Riemannian, 337
semiorthogonal, 195
shape operator, 155
sharping, 334
sheaf, 289
short exact, 444
shuffle, 347
sign, 599
simple tensor, 656
simply connected, 37
single-slice chart, 46
singular homology, 402
singular point, 74
smooth functor, 283
smooth manifold, 15
smooth map, 22
smooth structure, 12
smoothly universal, 129
spacelike curve, 548
spacelike subspace, 588
spacelike vector, 547
sphere theorem, 622
spin-j, 250
spray, 544
stabilizer, 228
standard action, 250
standard transition maps, 274
stereographic projection, 17
Stiefel manifold, 51, 243
Stokes' theorem, 396
straightening, 102
structural equations, 386
structure constants, 205
subgroup (Lie), 191

submanifold, 46
 property, 46
submersion, 138
submodule, 651
summation convention, 5
support, 28, 101, 391
surface of revolution, 160
symplectic group, 196

tangent bundle, 81
tangent functor, 71, 84
tangent map, 67, 68, 81
tangent space, 58, 61, 65
tangent vector, 58, 60, 61
tautological bundle, 281
tensor (algebraic), 308
 bundle, 319
 derivation, 327
 field, 320
 map, 308
tensor product, 251, 319, 654, 655
 bundle, 282
 of tensor fields, 319
 of tensor maps, 311
 of tensors, 311
theorema egregium, 176
tidal operator, 558
time dependent vector field, 110
timelike curve, 548
timelike subspace, 588
timelike vector, 547
TM-valued tensor field, 320
top form, 378
topological manifold, 7
torsion (of curve), 150
total space, 258
totally geodesic, 582
totally umbilic, 162
transition maps, 261, 265
 standard, 274
transitive, 228
transitive action, 41
transversality, 80
trivialization, 84, 260
typical fiber, 258

umbilic, 162
unitary group, 196
universal, 655
 cover, 39
 property, bilinear, 251, 654

VB-chart, 270
vector bundle, 270
 morphism, 271
vector field, 87
 along, 89

Index

vector subbundle, 271
velocity, 69
volume form, 378

weak embedding, 129

weakly embedded, 132
wedge product, 347
Whitney sum bundle, 276

zero section, 276

Titles in This Series

108 **Enrique Outerelo and Jesús M. Ruiz,** Mapping degree theory, 2009
107 **Jeffrey M. Lee,** Manifolds and differential geometry, 2009
106 **Robert J. Daverman and Gerard A. Venema,** Embeddings in manifolds, 2009
105 **Giovanni Leoni,** A first course in Sobolev spaces, 2009
104 **Paolo Aluffi,** Algebra: Chapter 0, 2009
103 **Branko Grünbaum,** Configurations of points and lines, 2009
102 **Mark A. Pinsky,** Introduction to Fourier analysis and wavelets, 2009
101 **Ward Cheney and Will Light,** A course in approximation theory, 2009
100 **I. Martin Isaacs,** Algebra: A graduate course, 2009
99 **Gerald Teschl,** Mathematical methods in quantum mechanics: With applications to Schrödinger operators, 2009
98 **Alexander I. Bobenko and Yuri B. Suris,** Discrete differential geometry: Integrable structure, 2008
97 **David C. Ullrich,** Complex made simple, 2008
96 **N. V. Krylov,** Lectures on elliptic and parabolic equations in Sobolev spaces, 2008
95 **Leon A. Takhtajan,** Quantum mechanics for mathematicians, 2008
94 **James E. Humphreys,** Representations of semisimple Lie algebras in the BGG category $\mathcal{O}$, 2008
93 **Peter W. Michor,** Topics in differential geometry, 2008
92 **I. Martin Isaacs,** Finite group theory, 2008
91 **Louis Halle Rowen,** Graduate algebra: Noncommutative view, 2008
90 **Larry J. Gerstein,** Basic quadratic forms, 2008
89 **Anthony Bonato,** A course on the web graph, 2008
88 **Nathanial P. Brown and Narutaka Ozawa,** C^*-algebras and finite-dimensional approximations, 2008
87 **Srikanth B. Iyengar, Graham J. Leuschke, Anton Leykin, Claudia Miller, Ezra Miller, Anurag K. Singh, and Uli Walther,** Twenty-four hours of local cohomology, 2007
86 **Yulij Ilyashenko and Sergei Yakovenko,** Lectures on analytic differential equations, 2007
85 **John M. Alongi and Gail S. Nelson,** Recurrence and topology, 2007
84 **Charalambos D. Aliprantis and Rabee Tourky,** Cones and duality, 2007
83 **Wolfgang Ebeling,** Functions of several complex variables and their singularities (translated by Philip G. Spain), 2007
82 **Serge Alinhac and Patrick Gérard,** Pseudo-differential operators and the Nash–Moser theorem (translated by Stephen S. Wilson), 2007
81 **V. V. Prasolov,** Elements of homology theory, 2007
80 **Davar Khoshnevisan,** Probability, 2007
79 **William Stein,** Modular forms, a computational approach (with an appendix by Paul E. Gunnells), 2007
78 **Harry Dym,** Linear algebra in action, 2007
77 **Bennett Chow, Peng Lu, and Lei Ni,** Hamilton's Ricci flow, 2006
76 **Michael E. Taylor,** Measure theory and integration, 2006
75 **Peter D. Miller,** Applied asymptotic analysis, 2006
74 **V. V. Prasolov,** Elements of combinatorial and differential topology, 2006
73 **Louis Halle Rowen,** Graduate algebra: Commutative view, 2006
72 **R. J. Williams,** Introduction the the mathematics of finance, 2006
71 **S. P. Novikov and I. A. Taimanov,** Modern geometric structures and fields, 2006
70 **Seán Dineen,** Probability theory in finance, 2005
69 **Sebastián Montiel and Antonio Ros,** Curves and surfaces, 2005

TITLES IN THIS SERIES

68 **Luis Caffarelli and Sandro Salsa,** A geometric approach to free boundary problems, 2005
67 **T.Y. Lam,** Introduction to quadratic forms over fields, 2004
66 **Yuli Eidelman, Vitali Milman, and Antonis Tsolomitis,** Functional analysis, An introduction, 2004
65 **S. Ramanan,** Global calculus, 2004
64 **A. A. Kirillov,** Lectures on the orbit method, 2004
63 **Steven Dale Cutkosky,** Resolution of singularities, 2004
62 **T. W. Körner,** A companion to analysis: A second first and first second course in analysis, 2004
61 **Thomas A. Ivey and J. M. Landsberg,** Cartan for beginners: Differential geometry via moving frames and exterior differential systems, 2003
60 **Alberto Candel and Lawrence Conlon,** Foliations II, 2003
59 **Steven H. Weintraub,** Representation theory of finite groups: algebra and arithmetic, 2003
58 **Cédric Villani,** Topics in optimal transportation, 2003
57 **Robert Plato,** Concise numerical mathematics, 2003
56 **E. B. Vinberg,** A course in algebra, 2003
55 **C. Herbert Clemens,** A scrapbook of complex curve theory, second edition, 2003
54 **Alexander Barvinok,** A course in convexity, 2002
53 **Henryk Iwaniec,** Spectral methods of automorphic forms, 2002
52 **Ilka Agricola and Thomas Friedrich,** Global analysis: Differential forms in analysis, geometry and physics, 2002
51 **Y. A. Abramovich and C. D. Aliprantis,** Problems in operator theory, 2002
50 **Y. A. Abramovich and C. D. Aliprantis,** An invitation to operator theory, 2002
49 **John R. Harper,** Secondary cohomology operations, 2002
48 **Y. Eliashberg and N. Mishachev,** Introduction to the h-principle, 2002
47 **A. Yu. Kitaev, A. H. Shen, and M. N. Vyalyi,** Classical and quantum computation, 2002
46 **Joseph L. Taylor,** Several complex variables with connections to algebraic geometry and Lie groups, 2002
45 **Inder K. Rana,** An introduction to measure and integration, second edition, 2002
44 **Jim Agler and John E. McCarthy,** Pick interpolation and Hilbert function spaces, 2002
43 **N. V. Krylov,** Introduction to the theory of random processes, 2002
42 **Jin Hong and Seok-Jin Kang,** Introduction to quantum groups and crystal bases, 2002
41 **Georgi V. Smirnov,** Introduction to the theory of differential inclusions, 2002
40 **Robert E. Greene and Steven G. Krantz,** Function theory of one complex variable, third edition, 2006
39 **Larry C. Grove,** Classical groups and geometric algebra, 2002
38 **Elton P. Hsu,** Stochastic analysis on manifolds, 2002
37 **Hershel M. Farkas and Irwin Kra,** Theta constants, Riemann surfaces and the modular group, 2001
36 **Martin Schechter,** Principles of functional analysis, second edition, 2002
35 **James F. Davis and Paul Kirk,** Lecture notes in algebraic topology, 2001
34 **Sigurdur Helgason,** Differential geometry, Lie groups, and symmetric spaces, 2001
33 **Dmitri Burago, Yuri Burago, and Sergei Ivanov,** A course in metric geometry, 2001
32 **Robert G. Bartle,** A modern theory of integration, 2001

For a complete list of titles in this series, visit the
AMS Bookstore at **www.ams.org/bookstore/**.